MOLECULAR BIOLOGY AND TOXICOLOGY OF METALS

MOLECULAR BIOLOGY AND TOXICOLOGY OF METALS

Edited by
Rudolfs K. Zalups
Mercer University School of Medicine,
Macon, GA, USA

and

James Koropatnick
London Regional Cancer Centre,
London, Ontario, Canada

TAYLOR & FRANCIS
ALERE FLAMMAM
Founded 1798
London and New York

First published 2000
by Taylor & Francis
11 New Fetter Lane, London EC4P 4EE

Simultaneously published in the USA and Canada
by Taylor & Francis Inc,
29 West 35th Street, New York, NY 10001

Taylor & Francis is an imprint of the Taylor & Francis Group

Typeset in Sabon,
by Keystroke, Jacaranda Lodge, Wolverhampton
Printed and bound in Great Britain
by TJ International Ltd, Padstow, Cornwall

Every effort has been made to ensure that the advice and information in this book is true and accurate at the time of going to press. However, neither the publisher nor the authors can accept any legal responsibility or liability for any errors or omissions that may be made. In the case of drug administration, any medical procedure or the use of technical equipment mentioned within this book, you are strongly advised to consult the manufacturer's guidelines.

British Library Cataloguing in Publication Data
A catalogue record for this book is available from the British Library

Library of Congress Cataloguing in Publication Data
Molecular biology and toxicology of metals / edited by R. K. Zalups and D. J. Koropatnick.
p. cm.
Includes bibliographical references and index.
ISBN 0-7484-0798-7 (hbk. : alk. paper)
1. Metals—Toxicology. 2. Metal ions—Physiological transport. 3. Metal ions—Physiological effect. I. Zalups, Rudolfs K. II. Koropatnick, Donald James, 1952– .
RA1231.M52M65 2000
615.9'253—dc21 99-31652
CIP

ISBN 0-7484-0798-7

CONTENTS

CONTRIBUTORS

H. Vasken Aposhian Departments of Molecular and Cellular Biology and Pharmacology, The University of Arizona, Tucson, Arizona 85721-0106, USA

Michael Aschner Department of Physiology and Pharmacology, Wake Forest University School of Medicine, Winston-Salem, NC 27157-1083, USA

Nazzareno Ballatori Department of Environmental Medicine, University of Rochester School of Medicine, Rochester, NY 14642, USA

M. H. Bhattacharyya Argonne National Laboratory, Center for Mechanistic Biology, 9700 South Cass Avenue, Argonne, IL 60439-4833, USA

Gregory S. Buzard Intramural Research Support Program, SAIC-Frederick, Frederick Cancer Research and Development Center, Frederick, MD 21702, USA

Max Costa Nelson Institute of Environmental Medicine and Kaplan Comprehensive Cancer Center, New York University Medical Center, 550 First Avenue, New York, NY10016, USA

Gary L. Diamond Syracuse Research Corporation, 6225 Running Ridge Road, North Syracuse, NY 13212, USA

Michael DiDonato Structural Biology and Biochemistry, The Hospital for Sick Children, 555 University Avenue, Toronto, Ontario M5G 1X8, Canada

Beth A. Hart Department of Biochemistry, University of Vermont College of Medicine, Burlington, Vermont 05405-0068, USA

Sheila M. Healy Department of Molecular and Cellular Biology, The University of Arizona, Tucson, Arizona 85721-0106, USA

Laran T. Jensen University of Utah Health Sciences Center, Salt Lake City, UT 84132, USA

Margaret M. Jonah Dominican University, 7900 West Division Street, River Forest, Illinois 60305, USA

Kazimierz S. Kasprzak Department of Human Health, Frederick Cancer Research and Development Center, Frederick, Maryland, 21702-12-1, USA

Laura E. Kerper Dept of Pharmacology and Physiology, University of Rochester, School of Medicine and Dentistry, Rochester, NY, USA

James Koropatnick Cancer Research Laboratories, London Regional Cancer Centre, 790 Commissioners Road East, London, Ontario N6A 4L6, Canada

Lawrence H. Lash Department of Pharmacology, Wayne State University, School of Medicine, 540 East Canfield Avenue, Detroit, Michigan 48201, USA

Raül Pérez-Ollé Inorganic Carcinogenesis Section, Laboratory of Comparative Carcinogenesis, National Cancer Institute at NIEHS, Research Triangle Park, NC 27709-2233, USA

Anthony Presta Department of Chemistry, The University of Western Ontario, London, Ontario N6A 5B7, Canada

S. S. Rajan Argonne National Laboratory, Center for Mechanistic Biology, 9700 South Cass Avenue, Argonne, IL 60439-4833, USA

Christopher Rensing Department of Biochemistry and Molecular Biology, Wayne State University, School of Medicine, Detroit, MI 48201, USA

Barry P. Rosen Department of Biochemistry and Molecular Biology, Wayne State University, School of Medicine, Detroit, MI 48201, USA

Bibudhendra Sarkar Department of Biochemistry, University of Toronto, Toronto, Ontario M5S 1A8, Canada

Timothy J. Shafer Neurotoxicology Division (MD-74B), US Environmental Protection Agency, Research Triangle Park, NC 27711, USA

Martin J. Stillman Department of Chemistry, The University of Western Ontario, London, Ontario N6A 5B7, Canada

Douglas M. Templeton Department of Laboratory Medicine and Pathobiology, University of Toronto, 100 College Street, Toronto, Ontario M5G 1L5, Canada

Michael P. Waalkes Inorganic Carcinogenesis Section, Laboratory of Comparative Carcinogenesis, National Cancer Institute at NIEHS, Research Triangle Park, NC 27709-2233, USA

Eric Wildfang Department of Pharmacology and Toxicology, The University of Arizona, Tucson, Arizona 85721-0106, USA

A. K. Wilson Department of Biological Sciences, Benedictine University, 5700 College Road, Lisle, IL 60532-0900, USA

Dennis R. Winge University of Utah Health Sciences Center, Salt Lake City, UT 84132

Rudolfs K. Zalups Divisions of Basic Medical Sciences, Mercer University School of Medicine, Macon, Georgia 31207, USA

1

CHARACTERIZING METAL ION INTERACTIONS WITH BIOLOGICAL MOLECULES – THE SPECTROSCOPY OF METALLOTHIONEIN

Martin J. Stillman and Anthony Presta

1.1 General introduction – metals in biology

Metal ions regulate a vast array of essential physiological mechanisms with unprecedented specificity and selectivity. The amounts of each metal required to maintain normal function of the body range from the bulk nutrient level for calcium of over 1 kg, potassium of about 250 g and sodium of over 50 g to the ultratrace level for elements such as chromium (~3 mg), selenium (~2 mg) and vanadium (~2 mg). However, under conditions of metal overload, toxic effects are observed. Overload can be caused by external means or by genetic means. Release and transport of metals through the environment by natural (e.g. erosion, volcanic activity) and human (e.g. mining, agriculture, industrial) activities have exacerbated metal ion toxicological effects through the ages. The mechanism of action of both excess concentrations of beneficial metals, and the presence of toxic metals is complex and has been subject to much study. The genetic origins of the incorrect metabolism of copper and iron are beginning to be understood. These two topics form the core of this book. At each point in the discovery of these mechanisms, new reactivity is described that frequently involves complexation with small peptides, proteins, enzymes and/or nucleic acids for transport, storage or reaction. The characterization of these interactions is illustrated later in this chapter, where we discuss the metal-binding properties of the metalloprotein, metallothionein. The spectral data are complicated to interpret and extensive in information, yet carry the key to both the steady-state, metal-dependent structures formed and the temporal reaction kinetics that are dependent on the actual step-by-step mechanism of the metal-binding reactions.

Both the essential and toxic reactivities of metals are consequences of their specific chemical properties, including reduction/oxidation potential, acid/base chemistry, and structural or ligand coordination properties. For example, copper displays a range of biological reactivities that may be beneficial or toxic to the cell. Ceruloplasmin is a blue copper protein found in plasma and contains six copper atoms per molecule. Ligation of the copper atoms occurs by histidine nitrogen, cysteine sulfur or methionine sulfur. The cysteine S-to-Cu(II) charge transfer electronic transition of the distorted tetrahedral type I copper atoms produces the blue copper of this protein. This copper protein catalyzes the oxidation of a number of substrates, including iron(II), phenols and ascorbic acid, proceeding by cycling of the copper ion redox state. By comparison, superoxide dismutase (SOD) is a dimeric protein containing one zinc and one copper protein in each subunit. In this enzyme, the copper atoms are coordinated by four histidine imidazole groups and one axial water ligand. The copper atoms of this enzyme catalyze the dismutation of superoxide via the cycling of copper redox states:

$$\text{SOD-Cu(II)} + O_2^- \rightarrow \text{SOD-Cu(I)} + O_2 \quad (1.1)$$

$$\text{SOD-Cu(I)} + O_2^- + 2H^+ \rightarrow \text{SOD-Cu(II)} + H_2O_2 \quad (1.2)$$

$$\text{Net:} \quad 2O_2^- + 2H^+ \rightarrow O_2 + H_2O_2 \quad (1.3)$$

Copper bound to metallothionein exists solely in the +1 oxidation state, due to complete ligation of the metal ions by the reducing cysteine thiolate ligands. The copper atoms are bound as discrete Cu(I)-thiolate clusters, with the Cu(I) coordination geometry ranging from trigonal to digonal at higher Cu loadings (>12 Cu mole equivalents). Hence, this protein functions to sequester the metal and thus eliminate any deleterious reactions of the free metal ion. In the unbound state, copper may propagate a Fenton-type reaction by cycling redox states, thus catalyzing the production of the highly toxic hydroxyl radical:

$$Cu^{2+} + O_2^- \rightarrow O_2 + Cu^+ \quad (1.4)$$

$$Cu^+ + H_2O_2 \rightarrow Cu^{2+} + OH^- + OH^{\cdot} \quad (1.5)$$

$$\text{Net:} \quad O_2^- + H_2O_2 \rightarrow O_2 + OH^- + OH^{\cdot} \quad (1.6)$$

In general, the reactivities of metal ions are defined by the electronic and nuclear properties of these elements. Elements with similar electronic configurations, especially with respect to the outermost or valance electron shell, are grouped together in columns as part of the periodic table of the elements. These reactions are described in more detail in connection with the redox chemistry of a range of metals by Kasprzak and Buzard (Chapter 16).

1.2 Groups 1A and 2A

Groups 1A and 2A of the periodic table (Fig. 1.1) contain the alkali and alkaline earth metals. Of these, sodium, potassium, magnesium and calcium are the most abundant elements in the natural environment, including biological systems. The alkali metal ions, Na^+ and K^+, commonly function as electrolytes, regulating metabolic and signaling systems via concentration gradients across cell membranes. Their respective concentrations inside and outside mammalian cells accentuate the electrolyte gradients. The intracellular $[Na^+]$ is 10 mM whereas the extracellular $[Na^+]$ is 145 mM. K^+ displays an opposite cellular concentration gradient, with an intracellular $[K^+]$ of 140 mM and an extracellular $[K^+]$ of 5 mM (Cowan, 1997). This concentration gradient is a driving force for Na^+ entry into the cell. However, in order to maintain the much higher $[Na^+]$ outside of the cell, there must be an active mechanism of Na^+ efflux. This occurs by the pumping of K^+ into the cell as Na^+ is pumped out, a reaction that is driven by ATP hydrolysis.

1 **H** 1.008																	2 **He** 4.003
3 **Li** 6.94	4 **Be** 9.01											5 **B** 10.81	6 **C** 12.011	7 **N** 14.01	8 **O** 16.00	9 **F** 19.00	10 **Ne** 20.18
11 **Na** 22.99	12 **Mg** 24.31											13 **Al** 26.98	14 **Si** 28.09	15 **P** 30.97	16 **S** 32.06	17 **Cl** 35.45	18 **Ar** 39.95
19 **K** 39.10	20 **Ca** 40.08	21 **Sc** 44.96	22 **Ti** 47.90	23 **V** 50.94	24 **Cr** 52.00	25 **Mn** 54.94	26 **Fe** 55.85	27 **Co** 58.93	28 **Ni** 58.71	29 **Cu** 63.55	30 **Zn** 65.37	31 **Ga** 69.72	32 **Ge** 72.59	33 **As** 74.92	34 **Se** 78.96	35 **Br** 79.90	36 **Kr** 83.80
37 **Rb** 85.47	38 **Sr** 87.62	39 **Y** 88.91	40 **Zr** 91.22	41 **Nb** 92.91	42 **Mo** 95.94	43 **Tc** 98.91	44 **Ru** 101.07	45 **Rh** 102.91	46 **Pd** 106.4	47 **Ag** 107.87	48 **Cd** 112.40	49 **In** 114.82	50 **Sn** 118.69	51 **Sb** 121.75	52 **Te** 127.60	53 **I** 126.90	54 **Xe** 131.30
55 **Cs** 132.91	56 **Ba** 137.34	Lanthanides	72 **Hf** 178.49	73 **Ta** 180.95	74 **W** 183.85	75 **Re** 186.2	76 **Os** 190.2	77 **Ir** 192.2	78 **Pt** 195.09	79 **Au** 196.97	80 **Hg** 200.59	81 **Tl** 204.37	82 **Pb** 207.19	83 **Bi** 208.98	84 **Po** (209)	85 **At** (210)	86 **Rn** (222)
87 **Fr** (223)	88 **Ra** 226.03	Actinides															

Figure 1.1 The periodic table of elements.

Unlike sodium, which usually binds nonspecifically to biological molecules, potassium is an essential cofactor for a number of enzymes. Potassium has specific binding sites, usually through oxygen donor ligation, thus providing a structural activation of these enzymes. Two examples of enzymes that require potassium for their biological activity are the pyruvate and aspartate kinases. In studying the roles of potassium in these enzymes, thallium(I) has been used as it is similar to potassium in size and charge density, can act as a radiotracer of potassium mobility, and as a structural probe for potassium.

Although the alkaline earth metals Mg^{2+} and Ca^{2+} are also ionic electrolytes in biological solutions, they more commonly play structural roles in many biological molecules. Calcium is the most abundant metal in the body by virtue of its structural roles in the skeletal matrices of bones and teeth. This function is due to the low solubility of calcium minerals, especially the carbonates and phosphates. Additionally, Mg^{2+} and Ca^{2+} activate many enzymes by specific and high-affinity binding, with oxygen donor amino-acid residues as the main ligating species (e.g. the EF-hand Ca^{2+}-binding proteins calmodulin and troponin C). The transport and function of Ca^{2+} within cells is often studied with the use of calcium-specific fluorescent probes. In Chapter 7, Shafer describes in detail the characteristics of the Ca^{2+} channels and the ion selectivity exhibited, in particular, how metal ions like Cd^{2+} utilize Ca^{2+} channels for entry into cells.

1.3 Transition metals

Many of the transition metals play a role throughout biological systems. We describe here just a sample of their chemistry.

1.3.1 Vanadium

Of the groups 3B to 5B, vanadium is the only known biologically essential element, although only at the ultratrace level. As such, its essential functions are not entirely understood. The inorganic chemistry of vanadium is dominated by three stable oxidation states, +2, +4 and +5, which tend to exist as polyoxovanadates at physiological pH conditions. Vanadium compounds are not readily absorbed by the body, so this metal is not toxic to mammals. There is some evidence that a higher intake of vanadium can lower serum cholesterol levels by inhibiting its synthesis, and is also associated with lower incidences of heart disease. Vanadium deficiency in mammals is also manifest through impaired bone development and decreased reproductive performance.

1.3.2 Chromium

Of the group 6B elements, chromium and molybdenum are essential to mammals at the ultratrace level. The essential and toxicological effects of chromium are covered in Chapter 4 by Costa. In general, chromium ions exist in one of two oxidation states, +3 or +6. In the trivalent state, chromium usually forms 6-coordinate complexes. These complexes are sparingly soluble in water, are not absorbed by the body, and so pose little toxicity risk. The hexavalent chromium is usually found as the chromate or dichromate anions. Cr(VI) complexes are more readily absorbed, passing into red blood cells and binding to the protein part of hemoglobin. Under physiological conditions, Cr(VI) is rapidly reduced to Cr(III). Chromium or a chromium complex has been shown to have an essential function in carbohydrate metabolism and glucose tolerance. Costa discusses the toxicity of chromium and nickel in Chapter 4.

1.3.3 Molybdenum

Molybdenum is a common cofactor in various oxidoreductase enzymes, including nitrogenase, sulfite oxidase, nitrate reductase and xanthine oxidase, usually in conjunction with flavin nucleotides and iron–sulfur clusters. Complexes of molybdenum most often contain the metal in the +5 or +6 oxidation state. Moreover, the larger atomic radius of molybdenum in comparison to chromium favors higher coordinate (especially CN = 8) complexes. These complexes show a preference for oxygen ligation, although thiolate and amine groups are also common biological ligands. Mo(V) is d^1 and so electron paramagnetic resonance (EPR) spectroscopy is a common probe of metal structure and reactivity. However, Mo(V) complexes have a strong tendency to dimerize, making the complex diamagnetic and EPR silent. Molybdenum is readily absorbed by the body in order to fulfill its essential roles.

1.3.4 Manganese

Manganese is the only group 7B element commonly found in biological environments. Although the inorganic chemistry of manganese displays an amazing range of stable oxidation states, the biological chemistry of manganese is dominated by the +2 state. Mn(II) is very similar in size and charge density to Mg(II) and Zn(II) and also favors the tetrahedral and octahedral geometric structures. Mn(II) can replace Mg(II) in the enzyme pyruvate carboxylase and Zn(II) in superoxide dismutase, with negligible effects on enzyme activities. Mn(II) has a d^5 electronic configuration. The paramagnetic nature of Mn(II) has been exploited by using nuclear magnetic resonance (NMR) spectroscopy to probe the metal–protein–substrate interactions in these enzymes.

1.3.5 *Iron*

Iron is the first member of the group 8B elements and is essential to the proper utilization of oxygen by biological systems. Iron is a cofactor in a number of oxidoreductase enzymes owing to its two stable oxidation states, +2 and +3. Iron is incorporated in enzymes often as unique structural motifs, for example the iron protoporphyrin IX or heme functional group and as the iron–sulfur clusters. The heme group is involved in reactions of oxygen activation, such as those of the cytochromes P_{450} and the nitric oxide synthases, and metabolism of oxygen-containing molecules, including horseradish peroxidase and catalase. The heme cofactor is also essential for the oxygen carrying ability of hemoglobin and myoglobin. Iron–sulfur proteins range in complexity from the $Fe_1S_0cys_4$ cluster of rubredoxin, which carries out the ω-hydroxylation of fatty acids, to the $Fe_6S_6cys_6$ ferredoxin cluster involved in bacterial nitrogen fixation.

Iron occurs in many foods of most diets and also in cooking utensils, thus deficiency in this essential element is usually the result of a hereditary metabolic disorder. The metabolism and transport of iron are discussed in Chapter 6 by DiDonato and Sarkar in relationship to hereditary hemochromatosis and thalassemias. Iron is readily absorbed in the intestinal mucosal cells as the +2 form. The absorbed Fe(II) is then believed to be oxidized by apoferritin to the +3 state and bound by this protein. Ferritin is the major iron storage protein in mammals. Iron is released from ferritin by enzymatic or nonenzymatic reduction to the Fe(II) state.

1.3.6 *Copper*

Copper is a group 11 (or 1B) element and is the third most abundant transition metal found in living things, at a concentration of about 2 p.p.m. (Linder, 1991). Copper displays two stable oxidation states in solution, +1 (cuprous) and +2 (cupric). Consequently, its biological chemistry is dominated by participation in redox reactions. Copper is an essential trace element, playing vital roles in many enzymatic processes. However, at elevated concentrations copper is toxic to living cells, presumably by binding indiscriminately to thiol moieties, or by catalyzing a Fenton-type reaction, producing the reactive hydroxyl radical. Binding of copper by biological ligands such as peptides, proteins or enzymes is required to minimize the deleterious effects of copper. The metallothioneins are a ubiquitous class of proteins exquisitely suited to the role of copper sequestration (Kagi and Nordberg, 1979; Kagi and Kojima, 1987; Stillman *et al.*, 1992; Suzuki *et al.*, 1993; Stillman, 1995). The fine balance required for copper in humans is also evident in genetic metabolic disorders. Menkes' disease is characterized by severe copper deficiency due to an error in copper transport from the intestine. Wilson's hepatolenticular degeneration is the result of a failure to excrete copper from the liver to the bile, resulting in copper overload. DiDonato and Sarkar present a detailed description of these two genetic disorders in Chapter 6. The complex chemistry involving copper is also discussed by Jensen and Winge in Chapter 17. Here the roles of the copper transcription factors and metallothionein are described in detail, illustrating the complexity involved in copper homeostasis.

In general, the dietary intake of copper far exceeds the trace amounts required for copper-dependent processes. The small amounts of copper that are absorbed from the diet (~0.5 mg/day) first enter the cells of the intestinal mucosa, and then into the bloodstream. Copper travels via the bloodstream to the liver and kidney. In the liver, some of the copper is excreted into the bile, some is used in the production of liver proteins (metallothionein, cytochrome oxidase, superoxide dismutase, tryptophan oxygenase, phenylalanine hydroxylase), and most is transferred to ceruloplasmin for transport to other tissues. For example, copper is also delivered to the enzymes lysyl oxidase (connective tissue), dopamine-β-hydroxylase (brain), tyrosinase (eyes, skin, hair) and the various diamine oxidases.

The majority of copper-dependent proteins and enzymes contain copper predominantly as the divalent ion. There are three basic types of copper(II) in these protein complexes (Solomon *et al.*, 1992; Kaim and Schwederski, 1994):

- type 1 Cu(II)- EPR-detectable, reoxidizable, produces the blue color;
- type 2 Cu(II)- EPR-detectable, not reoxidizable, not blue; and
- type 3 Cu(II)- EPR-silent, spin couple pair of Cu(II).

These copper proteins and enzymes contain a mixture of oxygen-, nitrogen- and sulfur-donating ligands, with predominantly tetra- and penta-coordinate Cu(II) geometry. By contrast, copper binds to metallothionein as the Cu(I) ion, and is chelated exclusively by cysteinyl sulfhydryl groups with trigonal planar or digonal Cu(I) coordination.

1.3.7 Zinc

Zinc is the first member of the group 12 (or 2B) elements, and is the second most abundant transition metal found *in vivo*. Biological complexes of this element contain zinc only in the divalent oxidation state with a d^{10} electronic configuration. Hence, the coordination chemistry of Zn(II) complexes is dominated by the tetrahedral geometry for this ion. Because of the single stable oxidation state for zinc in solution, this metal does not play a redox active role in biological processes, but nevertheless participates in a vast range of enzymatic reactions as a Lewis acid or as a structural cofactor. An example of the biological role of zinc is the enzyme carbonic anhydrase. This enzyme catalyzes the reversible hydration of CO_2 and contains one zinc atom per molecule. The Zn(II) coordination sphere comprises three histidine N ligands and a fourth water ligand. It is believed that the high positive charge density of the Zn(II) group (Lewis acid) greatly decreases the pK_a of the nearby aquo ligand, leading to its deprotonation to a bound hydroxide. The zinc-bound hydroxide group can then attack a molecule of CO_2 forming the bicarbonate anion (Zheng and Merz, 1992). Other Zn(II)-containing enzymes that catalyze similar hydrolysis reactions are carboxypeptidase and alkaline phosphatase. Two zinc-containing enzymes, alcohol dehydrogenase and superoxide dismutase, involve catalytic oxidation of the substrate, however the Zn(II) moiety is not involved as a redox cycling cofactor.

Zn(II) also displays a structural role in biological systems, as exemplified by the zinc finger transcription factors. These molecules bind to DNA and regulate the transcription of genetic information. In these molecules, each Zn(II) is coordinated via four cysteine sulfur groups or two cysteine sulfurs and two histidine nitrogen ligands (Kraulis *et al.*, 1992; Pavletich and Pabo, 1993). The tetrahedral nature of Zn(II) biocomplexes is further illustrated in the metallothionein molecules, that are postulated to act as storage and/or transport proteins for Zn(II) (Stillman *et al.*, 1992; Fowle and Stillman, 1997). In the mammalian proteins, seven Zn(II) are bound by a total of 20 cysteinyl thiol ligands, with exclusive tetrahedral coordination for the Zn(II) ion.

The electronic configuration of Zn(II) is the closed shell, d^{10}. Thus Zn(II) complexes do not offer a great deal of available spectroscopic signals. Indeed, much of the structural information, and by extension mechanistic insight, of Zn(II)-containing proteins and enzymes has come from techniques that probe the coordination sphere of the zinc atom rather than its electronic interactions. These techniques include NMR, X-ray absorption spectroscopy (XAS) and X-ray crystallographic methods.

1.3.8 Cadmium and mercury

The other members of group 12 exhibit toxicity at all levels of concentration. In Chapter 9, Zalups brings together a complete description of the reactions of mercury in the kidney, while Ballatori discusses the reactions of mercury in the liver (Chapter 12). Cadmium has been the focus of many monographs and reviews. Hart in Chapter 8 discusses the respiratory effects of cadmium. Webb (1979) includes extensive discussion of the biochemistry of cadmium, and this review serves as an excellent starting point in studies of cadmium toxicity. Bhattacharyya, Wilson and Rajan bring the description up to date in Chapter 2, in a detailed review of the biochemical pathways involved in cadmium toxicity. In Chapter 13, Lash describes the chemistry of glutathione with cadmium, mercury, methylmercury and arsenic. The physiological role of thiol-based peptides and proteins in chelating the group 12 metals is clear, especially for cadmium and mercury. Neither cadmium nor mercury has any demonstrated beneficial effect, and the chapters that follow show that both metals are generally disruptive towards cellular chemistry, for example Templeton in Chapter 15 describes the effect of methylmercury on tubule polymerization.

1.4 Characterization of metal-binding proteins

We can classify metal-binding proteins broadly as those proteins that incorporate metals for functional use and those that bind metals for storage. The protein metallothionein appears to fall into this latter class. Although hundreds of different metalloproteins from procaryotic, lower eucaryotic, plant and animal sources have been identified, associated with a wide variety of metals, including iron, copper, zinc, cobalt, mercury and cadmium, in which binding is through sulfur, oxygen and other ligands, the metal-binding properties of the metallothioneins offer a superb model from which to illustrate the various techniques available to determine the characteristics of a metalloprotein. Metallothionein covers the range of metal binding when sulfur is the ligand. The wide variety of metals, stoichiometric ratios and clustered binding structures makes this system important as an example. Much of the bioinorganic chemistry of metals in physiological systems is also dominated by metabolism and transport phenomena that are present in the cell. These metal transport mechanisms are complicated and, in many cases, are not always metal specific. However, as is clear in a number of the chapters presented in this book, the chemistry involved is critical to our understanding of the processes involved in the essential regulation of metal flow and those that are overwhelmed by metal-based chemistry in toxic response. In Chapter 12 Ballatori discusses the range of metal transport mechanisms in the liver. Lash (Chapter 13) focuses on the significant role played by the tripeptide glutathione in reactions with cadmium, mercury, methylmercury and arsenic. Diamond, in Chapter 11, describes the transport properties of renal tubules and the small intestine, beginning with the transport of calcium and continuing with a discussion of the transport of lead, cadmium and mercury.

1.5 Bioinorganic characterization of metallothionein

The characterization of the metal-binding properties of the metalloprotein metallothionein provides an ideal model for studies of the metal–protein interactions discussed throughout the book. The essence of the characterization step is to define the metal–protein interaction. This is a very large, multifaceted problem that requires information from multidisciplinary studies. Questions that must be addressed before the metal-binding reactions are fully described include:

- identification of the metal oxidation state
- identification of the metal-binding ligands
- determination of the number of binding sites
- determination of the metal to protein stoichiometry
- determination of the metal coordination geometry.

These results are usually obtained from steady-state data; however, in many cases, the dynamic data reveal information that provides the clues to the function of the metal in the protein. In addition, then, the following must also be included:

- determination of the rates of the metal-dependent reactions
- determination of the molecularity of the reactions.

Even with dynamic data, the relationship between structure and function may not be apparent. It is also likely that the electronic coupling between the metal and its coordination ligands plays a critical role in determining the function of the metal in its binding site. Therefore add to the list:

- determination of the electronic interactions between the metal and its coordinating ligands.

The chemical, spectroscopic, and kinetic analyses of the metal-binding properties of the metallothioneins over the past 25 years have provided a rich description of its properties.

1.5.1 Introduction to metallothionein

The metallothioneins (MTs) are a unique class of metalloproteins, characterized by an approximately 60-amino-acid sequence that is high in cysteine content and also is devoid of aromatic residues (Kagi and Nordberg, 1979; Kagi and Kojima, 1987; Riordan and Vallee, 1991; Stillman *et al.*, 1992; Suzuki *et al.*, 1993; Stillman, 1995). The role of metallothioneins is discussed in detail by Waalkes and Pérez-Ollé in Chapter 14.

Cadmium was the initial focus of the research into metallothionein, and the biochemical properties of cadmium are discussed in Chapters 2 and 8. The absence of aromatic amino acids in metallothioneins is an important and characteristic property of these peptides. It is significant spectroscopically because optical spectroscopy is able to measure absorption, circular dichroism (CD) and emission spectra for the thiolate to metal charge transfer transitions that occur between 220 and 350 nm. This region would normally be completely masked by the presence of aromatic groups and, therefore, unavailable for study by optical methods. As we will see below, the accessibility of these charge-transfer bands provides detail of the metal-binding chemistry unavailable by any other technique. The known structural chemistry of metallothioneins comes from a variety of sources but is currently based heavily on the detailed results from both NMR (^{111,113}Cd and ^{1}H) and X-ray diffraction (Otvos and Armitage, 1980; Otvos *et al.*, 1982; Robbins *et al.*, 1991; Braun *et al.*, 1992; Robbins and Stout, 1992). X-ray absorption spectroscopy (XAS) experiments using the extended X-ray absorption fine structure (EXAFS) technique carried out on the mammalian protein have provided important metal–sulfur bond length information that allows structures to be proposed for metallothioneins that do not include NMR-active nuclei (Riordan and Vallee, 1991; Stillman, 1995).

In the mammalian protein, the N terminus of the peptide is named the *β* metal-binding domain, in which, for Zn(II) and Cd(II), the M_3S_9 cluster forms. The C terminus of the peptide is named the *α* domain, in which the M_4S_{11} cluster again forms with the tetrahedrally directing Zn(II) and Cd(II) (Kagi and Nordberg, 1979; Kagi and Kojima,

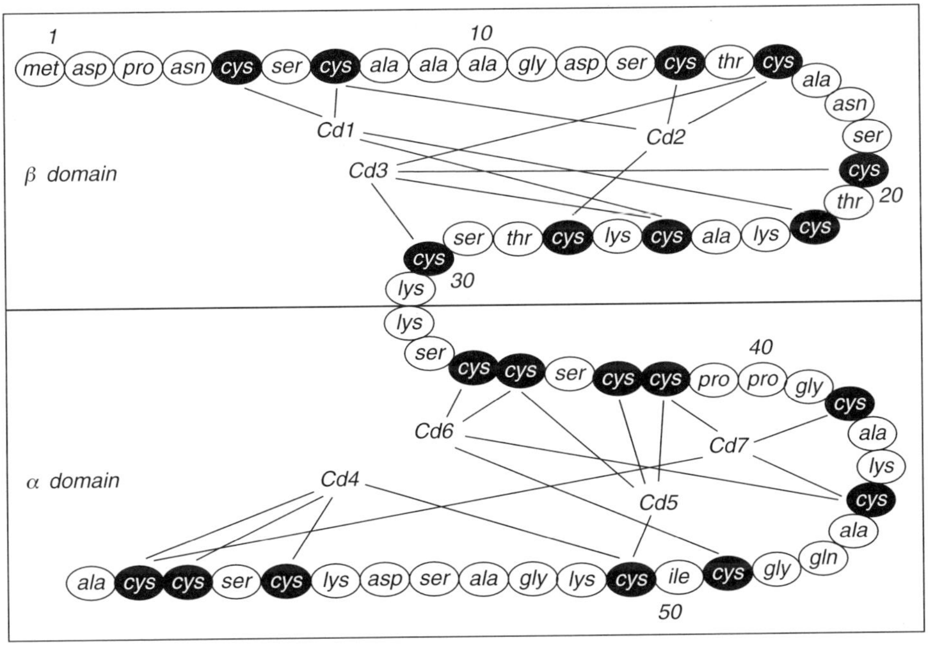

Figure 1.2 Amino-acid sequence of rabbit liver metallothionein isoform 2A. Cd(II) and Zn(II) bind in two domains, named *α* and *β*, that are associated with 31 residues of the N terminus and 31 residues of the C terminus, respectively. (The figure was constructed using the data of Kagi (1993).)

1987; Riordan and Vallee, 1991; Stillman *et al.*, 1992; Suzuki *et al.*, 1993; Stillman, 1995). Figure 1.2 shows the primary amino-acid sequence for isoform 2A of rabbit liver metallothionein, based on the data of Kagi (1993). Using the cluster units for Zn_4S_{11} and Zn_3S_9 shown in Fig. 1.3, a space-filling representation of the three-dimensional structure of rabbit liver Zn_7-MT 2A was calculated with the CAChe molecular modeling

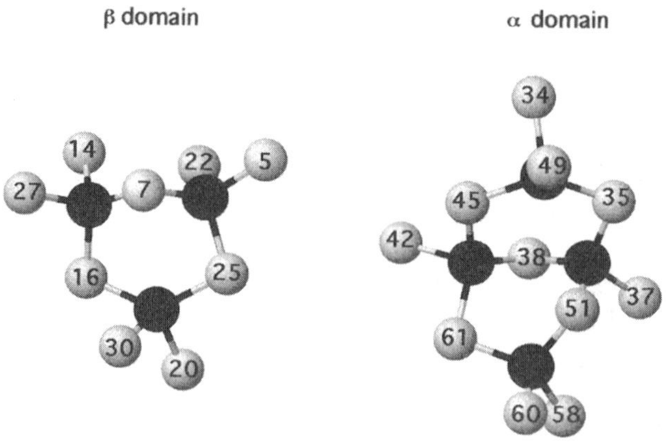

Figure 1.3 A three-dimensional representation of the Zn_3S_9 and Zn_4S_{11} metal-binding clusters. This image was calculated by molecular modeling techniques. (Reproduced with permission from Fowle and Stillman (1997).)

program, Fig. 1.4. The key metal–thiolate bond lengths used in the cluster structures shown in Fig. 1.3 were taken from EXAFS measurements of Jiang *et al.* (1994) and Gui *et al.* (1996). The metal–thiolate connectivities, especially the critically important differentiation between terminal thiolates (one sulfur binds to one metal) and bridging thiolates (one sulfur binds to two metals), were taken from NMR measurements of Braun *et al.* (1992). The overall structure of the two domains in the mammalian protein comes from the results of the X-ray diffraction measurements of Stout (Robbins *et al.*, 1991; Robbins and Stout, 1992). The incorporation of bridging sulfurs in the metal-binding site confers the 'cluster' chemistry, as now metals are connected through a single ligand – the prevalence of metal–S–metal units allows the metals to be connected electronically.

The formation of S_{cys}–M–S_{cys} bonds as part of digonal (2-), trigonal (3-) or tetrahedral (4-) coordination (Dance *et al.*, 1992) crosslinks the peptide, much like disulfide groups, resulting in formation of a three-dimensional metal-binding site that is wrapped by the hydrophilic envelope of the peptide chain (Fig. 1.4) (Fowle and Stillman, 1997).

1.5.2 Classes of metallothionein

Three classes of metallothionein have been identified from the very wide range of sources studied to date. Mammalian peptides, in which there are typically 60–68 amino acids, are in class I. The correspondence between cysteine locations, the Cys-X-Cys repeating unit (where X is a noncysteine residue) and the number of Cys–X–X–Cys units results in very similar properties being reported for the class I proteins. Class I proteins bind

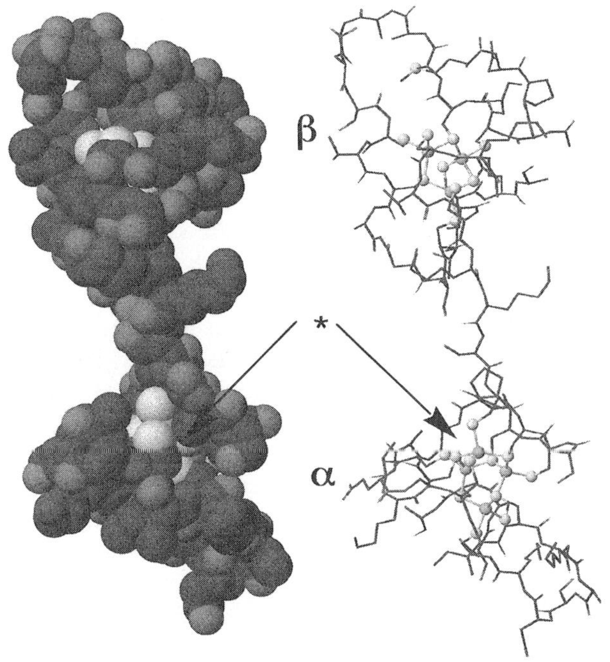

Figure 1.4 Space-filling and tube representation of the three-dimensional structure of rabbit liver Zn_7-MT 2A incorporating the Zn_3S_9 and Zn_4S_{11} metal-binding domains. The zinc–thiolate structures and the 62-amino-acid primary sequence were minimized using the MM2 method. The orientation shows the β domain at the top of the figure. *Denotes accessible sites of the zinc(II) cysteine thiolate. (Reproduced with permission from Fowle and Stillman (1997).)

metals in group 12. Class II metallothioneins occur in yeasts and fungi. These and the other class II metallothioneins contain roughly half the number of cysteines of class I peptides and comprise a single domain with a clustered metal–thiolate structure that has a M_8S_{12} core (George *et al.*, 1988; Suzuki *et al.*, 1993; Winge and Dameron, 1993; Stillman, 1995). The class III γ-glutamyl cysteinyl isopeptides (γ-EC peptides) are found in a range of sources, including yeasts (for example, *Schizosaccharomyces pombe*) and plants, and readily bind Zn(II), Cd(II) and Cu(I) (Hayashi and Winge, 1992).

The metal-binding properties of the class I mammalian metallothioneins, especially that isolated from rabbit liver, have been most widely studied. Rabbit liver MT contains 20 cysteines (as RSH) as part of a 61 (isoform 1) or 62 (isoform 2A) amino acid sequence (Kagi, 1993) Optical studies have identified a number of different metal–thiolate clusters that form within the metal-binding sites in the protein (Stillman, 1995). Well-characterized complexes include Zn_7-MT, Cd_7-MT, Cu_{12}-MT, Hg_7-MT and Ag_{12}-MT (Stillman *et al.*, 1987, 1988, 1989a, b, 1994; Cai and Stillman, 1988a, b; Stillman and Zelazowski, 1988, 1989; Zelazowski *et al.*, 1989; Lu *et al.*, 1990, 1993; Stillman and Gasyna, 1991; Stillman, 1992; Zelazowski and Stillman, 1992; Lu and Stillman, 1993; Green and Stillman, 1994, 1996; Green *et al.*, 1994; Jiang *et al.*, 1994, 1995; Presta and Stillman, 1994; Presta *et al.*, 1995; Gui *et al.*, 1996; Le Blanc, 1997).

Studies using X-ray absorption fine structure (XAFS) and NMR have provided descriptions of models for the metal–cysteine connectivities in both the copper- and silver-substituted yeast metallothionein (Stillman, 1995).

The goal of the following sections is to describe the very complicated metal-binding chemistry of mammalian metallothionein through a series of spectral data and derived calculated models of possible structures. Experiments carried out by the U.W.O. group use the change in optical properties as a function of metal loading (metal salt added that will bind in stoichiometric amounts, say, 2 metals per protein molecule). As many of the authors of the chapters in this book deal with the effects of metal binding, we will focus on how the metal-binding properties of a protein can be determined by analysis of a variety of spectroscopic techniques. The implementation of new molecular modeling techniques to visualize the structures based on the derived parameters provides new information on the structure–function relationships. The key is to obtain structural and chemical information that can eventually relate to the function of the metal-binding properties of the protein. This chapter will describe the characterization of the metal-binding site in mammalian metallothionein in terms of the metals bound, the metal to protein stoichiometry, and the possible geometry in the metal-binding sites. For examples, work from my laboratory will be referred to, but similar studies have been published by others and are cited in the recent series of monographs and reviews on the properties of the metallothioneins.

1.5.3 Occurrence, induction, isolation and purification of metallothionein

Naturally occurring metallothioneins are typically isolated as fully metallated species with Zn(II) (mammalian) or Cu(I) (yeast and fungus). In the mammalian protein, 20 cysteinyl sulfurs (S_{cys}) out of a total of between 60 and 62 amino acids in the peptide are available for coordination to metals (Shaw *et al.*, 1992; Kagi, 1993; Suzuki *et al.*, 1993). The key new tool in the characterization of metalloproteins is electrospray mass spectrometry. In this technique, the exact mass of the peptide can be obtained and with the appropriate knowledge of the sequence, the purity and metal-binding status can readily be determined (Yu *et al.*, 1993; Le Blanc *et al.*, 1994, 1997).

Electrospray (ES) mass spectrometry offers a soft ionization so that, usually, the entire peptide is detected. It offers a new method by which the molar mass of metallothionein, together with its complement of metals, can be determined with precision. Information on the distribution of isoforms and subisoforms is uniquely provided by the mass spectral patterns (Le Blanc *et al.*, 1997). Identification and subsequent characterization of the isoforms and subisoforms of metallothioneins from a variety of organisms has been an important task for bioanalytical chemistry in the study of these proteins. An ES mass spectrum comprises an envelope of peaks representing ions that differ by one proton, Fig. 1.5. ES-MS can uniquely identify the purity of samples that are to be used for structural and spectroscopic analysis, Table 1.1. ES-MS data for the chromatographically isolated Zn_7-MT 2A are shown in Fig. 1.5. Figure 1.5B shows that Zn_7-MT 2A is the sole protein present following separation under these conditions. A single mass of 6124 is calculated (Table 1.1). For the MT 2 fraction shown in Fig. 1.5, the protonated species distribute across the 3+ (2042), 4+ (1532), 5+ (1226), 6+ (1022), and 7+ (876) ions. These data are processed to yield a single mass peak that

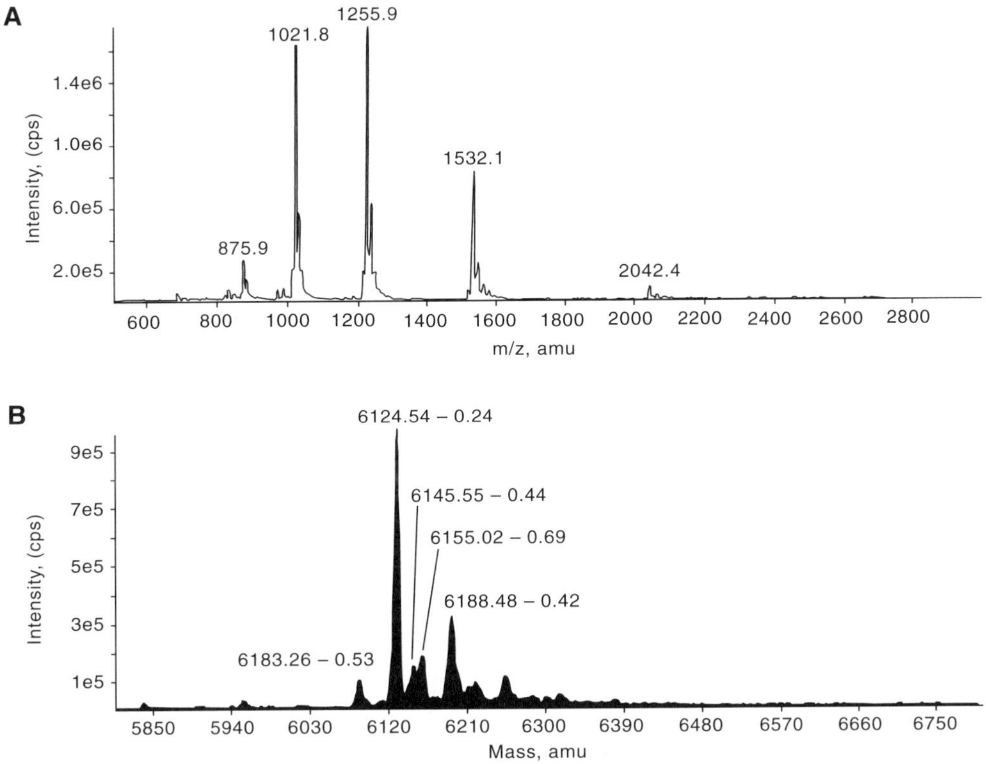

Figure 1.5 Electrospray mass spectrum (A) and BioSpec reconstructed mass spectrum (B) for Zn-MT 2A. The major ions (estimated charges) are at 875.9 (7), 1021.8 (6), 1225.9 (5), 1532.1 (4) and 2042.4 (3) mass-to-charge units. Based on the distribution of these ions, the BioSpec program reconstructed the mass of the parent molecule and other masses that were also identified in (A). The component at 6125 corresponds to 2A with the N terminus acetylated, whereas the components at 6083, 6146 and 6155 represent MT 2A (NH-terminal), 2B and 2C, respectively. The component at 6188 is assigned as Zn_1-MT 2A. Conditions: Sciex API 300; sample dissolved in distilled water with 0.1% formic acid; the average of five scans acquired in steps of 0.1 m/z is shown. (Reproduced with permission from Le Blanc *et al.* (1997).)

indicates the actual molecular mass of the protein, shown in Fig. 1.5B. Essentially complete isolation of Zn_7-MT 2A is achieved using ion exchange on a porous cartridge. The routine determination of molecular masses by the ES-MS technique provides the unambiguous data required to specify the peptide purity of the preparative samples of metallothionein. For spectroscopic analysis, isolation of homogeneous rabbit liver MT 2A is likely to provide the most homogeneous sample.

ES-MS data for metallothionein-containing metals other than Zn or Cd can be obtained under the optimum low pH conditions used to obtain the results shown in Fig 1.5 and Table 1.1. Figure 1.6 shows the ES-MS data of X. Guo, M. Siu, A. Presta and M.J. Stillman (unpublished) for silver-containing rabbit liver metallothionein. Under these conditions the silver is bound to the protein to form a number of different complexes. For the conditions used here, a number of species of Ag_n-MT (where n = 15, 16,

Table 1.1 Comparison of the molecular mass of each isoform reported to date for rabbit liver metallothionein (adapted from Le Blanc *et al.*, 1997)

Sample description based on the sequence derived from the molecular mass	*Theoretical mass from the sequence for the N-acetyl terminal (Da)*	*Number of amino acids in peptide chain*	*Reconstructed mass calculated from data shown in Le Blanc* et al. *(1997) (Da)*
MT 1A	6145.30	61	6145 ± 0.5
MT 2A	NH_2 terminal = 6083.2	62	(NH): 6083 ± 0.5
	N-acetyl terminal = 6125.2	62	(*N*-acetylated): 6125 ± 0.5
MT 2A + 1 Zn (–2H)		62	6188 ± 0.5
MT 2B	6146.30	61	6146 ± 0.5
MT 2C	6155.30	62	6155 ± 0.7
MT 2D	6215.40	61	6214 ± 1
MT 2E	6241.40	61	6244 ± 1

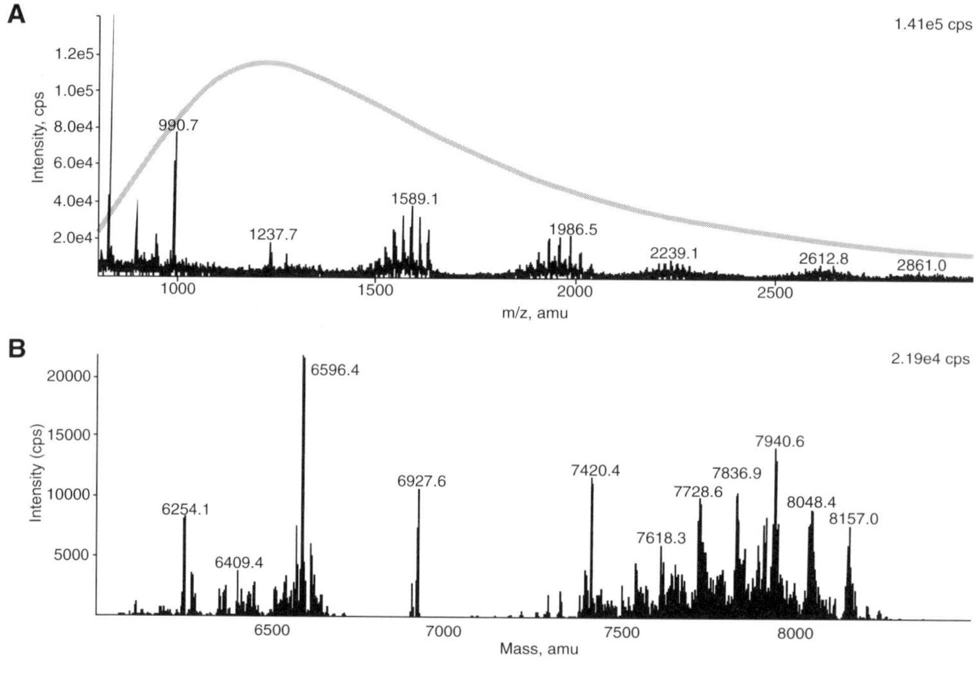

Figure 1.6 Electrospray mass spectrum (A) and BioSpec reconstructed mass spectrum (B) for Ag-containing rabbit liver MT 2A. The complicated patterns seen in (A) result from the co-existence of several metallated species, each with different numbers of silvers bound; (B) shows that species with 15, 16, 17, 18 and 19 silvers exist. Conditions: Sciex API 365; sample dissolved in distilled water with 0.1% formic acid. (Unpublished data of Presta, Xu, Siu and Stillman (1999).)

17, 18, 19) coexist, with masses between 7728 and 8157. The pattern observed in the mass spectrum is characteristic of the systematic binding of multiple metals to the protein.

1.5.4 Spectroscopic characterization of metal binding in vitro

A great many metals will bind to MT *in vitro*, including, but not restricted to, Ag(I), Au(I), Cd(II), Cu(I), Co(II), Fe(II), Hg(II), Ni(II) and Pt(II) (Stillman, 1995). The binding constants for metal binding to the metallothionein, cysteinyl thiolates follow the general order found for inorganic thiolates, Hg(II) > Ag(I) >> Cu(I) > Cd(II) > Zn(II) (Stillman, 1995).

Two strategies are used when studying the metal-binding reactions of metallothioneins *in vitro*:

1 Metals are mixed with metal-free or apo-protein in stoichiometric ratios that may or may not saturate all the thiolate groups in the peptide chain. In the reactions reported here, using mammalian metallothionein, the metal-free protein was prepared by gel filtration of the Zn_7-MT at low pH. Because apo-MT is so air sensitive precautions must be taken to ensure that oxidation of cysteine to cystine does not occur. The spectroscopic information obtained by titration of the metal-free strands will, of course, be determined largely by the spontaneous wrapping of the peptide around the metals – possibly in a metal-by-metal sequence. An example of this type of reaction is shown in Fig. 1.7 in which Hg(II) is added, metal by metal, in single mole equivalent steps to apo-MT 2A from rabbit liver.

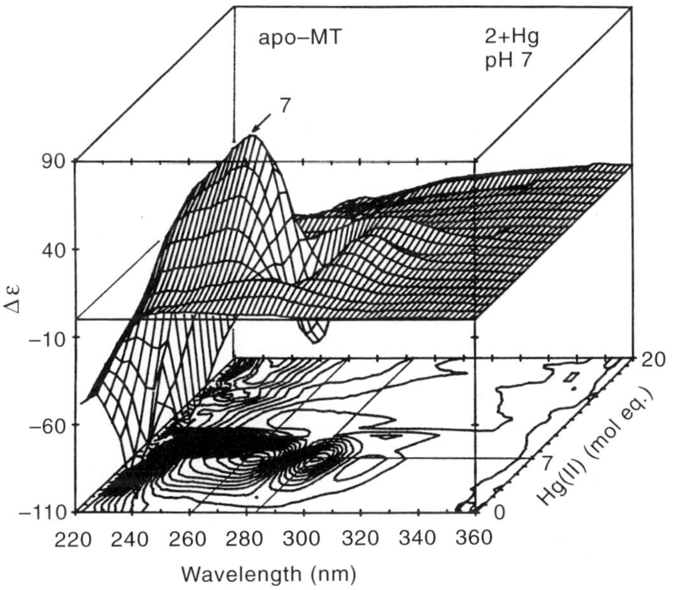

Figure 1.7 CD spectra recorded during a titration of rabbit liver apo-MT 2 with Hg(II) at room temperature and pH 7, showing formation of Hg_7-MT 2. Addition of Hg(II) > 7 mole equivalents unwinds the peptide chain; the Hg(II) binds to the protein but no chirality is detected for the binding site. (Reproduced with permission from Lu and Stillman (1993).)

2 Metals with greater binding constants than Zn(II) will displace the Zn(II) in Zn_7-MT at pH > 6. A large number of spectroscopic data have been reported that involve this type of metal exchange reaction. In this case there are complications because the Zn(II) in Zn_7-MT already holds the cysteinyl sulfurs in place in a three-dimensional structure that we will call the binding site. For Zn(II) and Cd(II), the four thiolate ligands are tetrahedrally arranged (Fig. 1.3). When an incoming metal that preferentially binds with a coordination geometry other than tetrahedral, considerable reorganization of the peptide chain must occur. Cu(I) and Ag(I) are metals that generally exhibit trigonal or digonal coordination geometries and, not surprisingly, the circular dichroism spectra recorded as Cu(I) is added to Zn_7-MT are complicated by changes in dichroic intensity, dichroic sign and band maxima wavelengths.

1.5.5 Circular dichriosm spectroscopy

The CD spectral data recorded when Cd(II) was added to a solution of Zn_7-MT 2A at pH 7 (data of Stillman *et al.* (1987)) show very clearly that the incoming Cd(II) displaces the bound Zn(II) because the Zn-S band near 235 nm is replaced by a new band at 250 nm. Notably, the new CD band envelope adopts the same overall shape as that for the Zn, in other words, the chirality of the metal-binding site does not change when Zn(II) is replaced by Cd(II). The red-shift observed is because Cd(II) has an atomic number greater than Zn(II) and ligand-to-metal charge transfer transitions will be red-shifted. In contrast to these data are the CD data obtained when Hg(II) was added to metal-free apo-MT 2 at pH 7 (Fig. 1.7). In these experiments of Lu *et al.* (1993) and Lu and Stillman (1993), the CD spectra show the changes induced by the incoming Hg(II) as it is added to the metal-free protein. Initially, a broad band appears in the CD spectrum, rising at about 270 nm; after two atoms of Hg(II) have been added, a new and much more intense envelope grows in intensity. The Hg:MT = 7 point is very well defined. However, the signal intensity collapses as more and more Hg(II) is added, until at the Hg:MT = 10 point the CD signal reverts to the baseline, indicating that there is no chiral structure. The situation at this point contrasts with that at the beginning of the titration, where the signal was also zero. Because no metal is bound to the apo-MT 2 at the start of the titration, there are no CD (or UV absorption) bands between 230 and 300 nm. As the titration proceeds, this region becomes occupied by bands arising from the charge transfer between the coordinating cysteinyl sulfurs and the metal, here Hg(II). The effect of the chain length is apparent in the titration of just the α fragment with Hg(II). Now the CD maximum appears at the 4 Hg:MT point, representing the formation of the Hg_4S_{11} species. However, as for the whole protein, addition of Hg(II) in excess of 4 mole equivalents leads to the cluster opening up and the CD intensity being quenched. Finally, for Hg(II) titrations, we see how the availability of different coordination geometries leads to formation of the Hg_{18}-MT 2 species when the titration is carried out at low pH (Fig. 1.8). Hg(II) is unusual in that there are three distinct maxima in the titrations with MT, at 7, 11 and 18 mole equivalents. Our current interpretation is that these three maxima represent saturation of the available binding sites for the Hg(II) with a defined coordination geometry, here 4, 3 and 2, respectively; that is, the tetrahedral coordination changes into a trigonal coordination and finally into a digonal coordination.

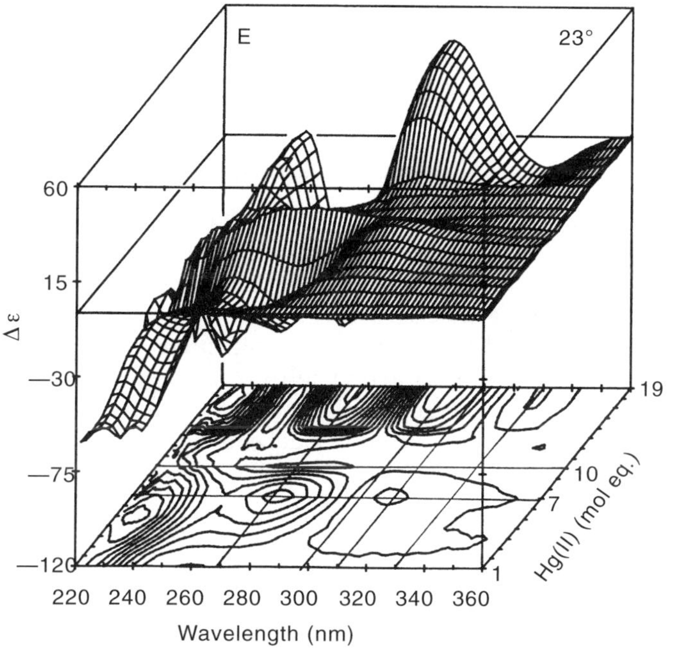

Figure 1.8 CD spectra recorded during a titration of rabbit liver apo-MT 2 with Hg(II) at room temperature and pH 2, showing formation first of Hg_7-MT 2, then Hg_{18}-MT 2. (Reproduced with permission from Lu *et al.* (1993).)

Analysis of CD spectral data for a number of different metals shows that metallothioneins containing Ag(I), Co(II), Cd(II), Cu(I), Fe(II) or Hg(II) exhibit spectroscopic signal maxima at metal to protein molar ratios of either 7 (for Hg(II), Cd(II), Co(II), and Fe(II)), 12 (for Cu(I) and Ag(I)), 17 (for Ag(I)) and 18 (for Hg(II)) (see the review by Stillman (1995) for a summary of these reports).

The data described above are for the equilibrated protein, the metals have been allowed sufficient time to locate the thermodynamically most stable site. Although the stability constants of metals bound in the metal–thiolate clusters of metallothionein are high, spectroscopic evidence suggests that these metals are quite labile. CD data indicate that at room temperature Cd(II) displaces Zn(II) from Zn_7-metallothionein in a random manner, with Cd(II) atoms binding evenly across the two domains of the protein (Stillman *et al.*, 1987; Stillman and Zelazowski, 1988). If the temperature is raised, the metallothionein-bound Cd(II) and Zn(II) rearrange from this distributed species into the domain-specific species with a filled Cd_4-α domain, as found naturally. The presence of a carrier such as ethylenediaminetetraacetic acid (EDTA) during the digestion of the Cd-substituted protein with subtilisin catalyzes the rearrangement of the metals from the kinetic product to the thermodynamically stable, domain-specific species (Stillman and Zelazowski, 1989). In experiments described by Green and Stillman (1994) the emission intensity of the copper-substituted metallothionein has been used to monitor the temporal properties of the formation of the copper–thiolate structures following addition of Cu(I) to Zn_7-MT 2. These results are described in more detail below.

The precise determination of the stoichiometric ratio between the bound metals and the number of accessible cysteinyl sulfurs (S_{cys}) plays a critical role in our understanding of the chemistry, biochemistry and physiological chemistry of this exceptional protein.

1.5.6 Emission spectroscopy

The most well-known emission of the metallothioneins is the orange luminescence observed at room temperature for copper-containing metallothioneins. First reported by Beltramini and Lerch in 1981, subsequent reports suggested that emission intensity in the 550–650 nm region is a general property of copper metallothioneins (Table 1.2) (Stillman, 1995). As described below, the emission intensity is entirely dependent on the molar ratio of Cu(I) to the cysteinyl sulfurs. In recent work by Green *et al.* (1994) it has been shown that the emission intensity for the two-domain rabbit liver metallothionein follows a much more complex dependence on the Cu(I):S_{cys} ratio.

Table 1.2 Examples of luminescent metallothioneins (adapted from Stillman, 1995)

Protein	*Spectral data measured*	*Reference*
Ag-MT; rabbit liver	Excitation and emission spectra recorded at 77 K during a titration of Zn_7-MT with Ag(I) for n = 1–24, wavelength and intensity dependence of luminescence on Ag:S_{cys} molar ratio; for $Ag_{10}S_{20}$ l_{ex} = 300 nm, l_{max} = 540 nm	Zelazowski *et al.* (1989)
Au-MT; rat liver	Emission and excitation spectra measured at 77 K; l_{ex} = 305 nm, l_{max} = 600 nm	Stillman *et al.* (1989b)
Cu-MT; rabbit liver	Emission and excitation spectra measured at 77 K for Cu_6-MT; l_{ex} = 300 nm, l_{max} = 530 nm	Stillman *et al.* (1989a)
	Study of emission properties during titrations of Zn_7-MT 2 with Cu(I) to form Cu_n-MT, n = 1–20, at room temperature; l_{ex} = 300 nm, l_{max} = 605 nm; temperature dependence of the emission intensity indicates site of initial Cu(I) binding to Zn-MT is in both domains; domain specificity observed only following equilibration	Green *et al.* (1994)
	Use of emission spectra to monitor the mobility of Cu(I) atoms between sites	Green and Stillman (1996)
Cu-MT: rat liver; Cu_nS_{20}	*In vivo* measurement of emission at 77 K directly from tissue following induction of Cu-MT with $CuCl_2$; l_{ex} = 310 nm, l_{max} = 605 nm; $t_{1/2}$ = 11, 98 ms. Lifetime of Cu-MT at 77 K, $t_{1/2}$ = 63, 132 ms.	Stillman *et al.* (1989a)
Pt-MT; rat liver	Emission and excitation spectra measured at 77 K for *cis*-Pt-MT, *trans*-Pt-MT and Pt-MT; l_{ex} = 305 nm, l_{max} = 600 nm (Pt-MT), 620 nm (*cis*-Pt-MT; *trans*-Pt-MT)	Stillman *et al.* (1989b)

Emission spectra in the 450–750 nm region have been reported for metallothioneins containing Ag(I), Au(I), Cu(I) and Pt(II), at both room temperature and cryogenic temperatures, but not for Cd-MT, Hg-MT or Zn-MT (Table 1.2) (Stillman *et al.*, 1989) d^{10} metals coordinated by thiolates are frequently luminescent at low temperatures as glasses or powders. Excitation in the 250–300 nm region results in emission intensity in the 500–700 nm region for copper(I), silver(I) and gold (I) metallothioneins. Table 1.2 illustrates the properties of the emission of silver-, gold-, copper- and platinum-containing metallothioneins.

Although many analogous inorganic compounds luminesce, at present a well-described assignment for the states involved is not available for complexes that do not contain aromatic ligands. We have suggested a scheme that involves excitation into both ligand to metal charge-transfer (LMCT) states and intrametal states in the 250–300 nm region, followed by intersystem crossing to a purely intrametal triplet state (Stillman and Gasyna, 1991). The unusually large Stoke's shift, the long-lived nature of the emission, and the observation that the excited state is readily quenched by oxygen, all support an assignment of the emissive state as a metal-based triplet in Cu-MT from rabbit liver. By extension, the origin of the orange emission from copper-containing metallothioneins from other sources, including the single domain yeast metallothioneins, may also be assigned to a triplet, intrametal state.

1.5.7 Metal replacement reactions – changes in the circular dichroism spectra

1.5.7.1 Cadmium and zinc metallothioneins – mammalian Cd_7-MT and Zn_7MT

The mixed-metal Cd,Zn-MT, and the single-metal Zn_7-MT and Cd_7-MT mammalian metallothioneins have been studied widely and continuously since the 1960s. The two metal-binding domains in mammalian Zn_7-MT and Cd_7-MT are based on metal–thiolate clusters with stoichiometries of M_3S_9 and M_4S_{11} (Fig. 1.3). The ^{113}Cd NMR data for Cd_6-MT isolated from crab and lobster have been interpreted in terms of two domains with metal to sulfur stoichiometry of M_3S_9. CD data for the Cd_6-MT from crab are quite unlike the data for the Cd_7-MT from rabbit livers (Stillman, 1992).

1.5.7.2 Mercury-containing metallothioneins – formation of Hg_7-MT, Hg_{11}-MT and Hg_{18}-MT

The extreme toxicity of mercury to most life forms has attracted wide interest and is discussed in Chapter 9. Mercury(II) binds strongly to thiolates forming complexes characterized by distorted tetrahedral, trigonal and digonal coordination of the Hg(II). In many cases, distant mercury–chloride bonds play an important role in the structures. Hg(II) also binds strongly to the cysteinyl thiolates in metallothionein both *in vivo* and *in vitro*. Structural information about mercury-containing metallothioneins is currently limited to optical spectroscopy, both absorption and CD, and X-ray absorption (mainly X-ray absorption near-edge structure (XANES) and EXAFS) studies. The absorption spectrum of metallothionein exhibits the red-shift expected for a sulfur to metal charge-transfer band when the metal changes from Zn(II) to Cd(II) to Hg(II). The CD spectra of mercury-containing metallothioneins are rich in detail and strongly dependent on the Hg to protein molar ratio and the coordination geometry of the mercury.

Hg(II) ion reacts with both metal-free and metallated forms of metallothionein in a complicated manner. In a series of spectroscopic studies, we have described CD spectral data that show the formation of Hg_n-MT species (n = 1–18) from rabbit liver metallothionein with maxima at Hg:MT molar ratios of 7, 11 and 18 (Lu and Stillman, 1993; Lu *et al.*, 1993). Hg(II) binds strongly to the apometallothionein with a molar ratio of 7 (Fig. 1.7). If the protein fragment is prepared, then a CD maximum at Hg:protein of 4 is evident. This supports a binding-site geometry similar to that shown in Fig. 1.3 for Zn_4S_{11} of the α domain. In Fig. 1.7, we see that the CD intensity collapses past the peak maximum, quite unlike the behavior observed when Cd(II) replaces the Zn in Zn_7-MT (Stillman *et al.*, 1987). However, the Hg(II) must still bind to the cysteinyl sulfurs at molar ratios greater than 7 Hg(II) per protein, the binding constant for Hg(II) and thiolate sulfur is simply too great for there to be free Hg(II). What is happening is that the binding site expands to accommodate the incoming Hg(II) so that the chiral structure is lost. The interpretation of these spectral effects is that the growth in signal intensity towards the 7 Hg:MT point depends on the formation of well-defined, three-dimensional structures that involve clustered Hg–S units, much like the structure shown in Fig. 1.3. At molar ratios greater than 7, the coordination geometry changes around the mercury as the tetrahedral arrangement of sulfurs gives way to first trigonal then digonal arrangements. If these arrangements do not naturally form a macroscopic three-dimensional structure, then there is no reason why a CD spectrum will be observed in the region of the sulfur to metal charge-transfer bands that lie in the 240–320 nm region of the optical spectrum. So at pH 7, random structures form. However, the CD spectra shown in Fig. 1.8 shows how under certain conditions (here low pH) a new and very much more extensive structure forms. Hg_{18}–S_{20} is the product of the addition of Hg(II) to metal-free metallothionein isoform 2A up to the saturation point at low pH.

1.5.8 Copper-binding to metallothioneins – emission and CD data

Excitation into one of the sulfur–copper charge-transfer bands results in the appearance of a strong broad-emission band centered near 600 nm at room temperature for all copper metallothioneins (Green *et al.*, 1994; Stillman, 1995). The emission is entirely dependent on the presence of Cu(I) bound to the cysteinyl thiolates of the metallothionein. Thus, emission spectral intensity reports on copper binding directly from the metal-binding site.

Reactions of Cu(I) with metallothionein are more complicated than these same reactions with divalent metals. Figures 1.9 and 1.10 illustrate the rich spectral properties that are exhibited by metallothionein during titrations with Cu(I). The emission data are described first.

Cu-MT frozen at 77 K glows an orange color when excited with light of wavelength near 300 nm. Since interactions with the solvent provide a radiationless path for degrading the energy of an excited state, room temperature emission of solvated complexes is unusual. Changes in the magnitude of the emission intensity in copper-substituted metallothioneins can be attributed largely to the efficiency with which a particular metal-binding structure shields the copper–thiolate cluster from the aqueous

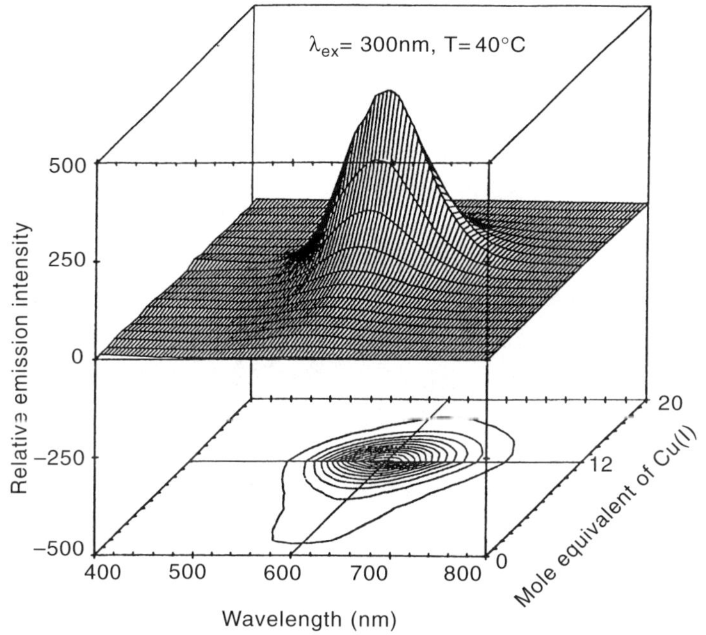

Figure 1.9 Emission spectra recorded during a titration of rabbit liver Zn_7-MT 2 with Cu(I) at 40 °C and pH 7. Aliquots of Cu(I) were added to a 10 mM solution of Zn_7-MT 2. The intensity in this figure has been normalized for the number of Cu(I) atoms added. At 40 °C the overall intensity is less than at 10 °C; however, significantly, there is negligible emission intensity between 1 and 6 Cu(I) added. (Reproduced with permission from Green *et al.* (1994).)

solvent. The more open the structure, or the more exposed the copper–thiolate cluster is, the lower the emission intensity expected.

Figures 1.9 and 1.10 (Green *et al.*, 1994) illustrate how the metal-binding reaction for Cu(I) depends on secondary reactions. That is, the initial binding of the metal is only the first of a number of steps required before the metal reaches a thermodynamically stable location within the binding site. In these titrations, Cu(I) was added to Zn_7-MT one metal (1 mole equivalent) at a time. The emission intensity shown in Fig. 1.9 at the elevated temperature of 40 °C increases up to the 12 Cu(I):MT point then decreases. There is a nonuniform dependence of the normalized intensities on the mole equivalent Cu(I) added, from 1 to 20. In addition, there is a significant effect of temperature. At low temperatures (below 10 °C) the quantum yield increases as a function of copper loading roughly linearly from Cu(I):S_{cys} = 1 up to 12, then the intensity collapses steeply towards the 20 Cu(I) point. If there were no increase in quantum yield as a function of Cu(I) loading, the intensity per Cu(I) would exhibit a flat line from Cu(I):S = 1 to 20. The low-temperature data support a model in which between 1 and 12 Cu(I) bind equally, and therefore statistically, in the two domains. Under these conditions the two domains emit light with an intensity that is approximately dependent on the number of domains occupied and their respective emissiveness. The systematic variation in band center confirms that at each temperature similar copper–thiolate clusters form with a saturation point occurring at the 12 Cu(I) point

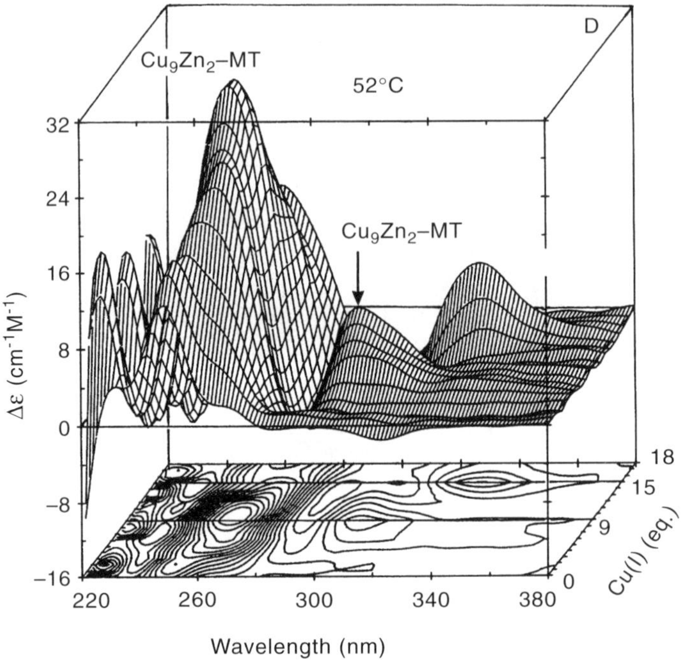

Figure 1.10 CD spectra recorded during a titration of rabbit liver Zn_7-MT 2 with Cu(I) at 52 °C and pH 7. At 52 °C the new Cu_9Zn_2-MT species forms with a unique and intense CD spectral pattern. The CD spectrum of Cu_{12}-MT is completely obscured by the strong bands associated with Cu_9Zn_2-MT. The elevated temperature allows formation of a species not accessible below room temperature. Cu_{15}-MT subsequently forms before the peptide chain unwinds under the influence of a Cu(I):MT molar ratio > 16. (Reproduced with permission from Presta *et al.* (1995).)

(Green 1995). The interpretation of these and other supporting data is that the copper–thiolate clusters in the β domain emit far less light than the copper–thiolate clusters in the α domain. At temperatures above 15 °C, the data support a model in which Cu(I) binds across both domains initially, then migrates to the β domain, which is associated with reduction in the emission intensity as a function of time (Green *et al.*, 1994). At low temperatures, the reaction rate is so slow that this migration essentially stops and only the kinetic product is found. The resulting copper–thiolate structures form in both domains, which results in a linear increase in emission intensity as there is no redistribution between highly emissive α and the not-so-emissive β binding sites. In addition to the complication of the initial and final binding sites, the metallated peptide folds in different ways at elevated temperatures. When the emission spectrum is recorded as a function of decreasing temperature starting at 50 °C, we observe that the intensity at the band maximum increases, as expected, linearly, until a sharp change occurs near 30 °C from which point the intensity increases more steeply as a function of temperature (Green and Stillman, unpublished data). These data support an interpretation that the water extrusion by the two metal-binding-site clusters depends on the folding of the protein backbone. At 30 °C there is a sharp contraction in the binding-site volumes and the emission intensity increases at a steeper rate.

The change in the emission spectral intensity at 600 nm also provides a continuous indication of the location of the Cu(I) added to metallothionein. When not bound to the cysteinyl sulfurs, the Cu(I) does not emit light. We have described the first direct evidence for the mobility of Cu(I) atoms within the metal-binding sites in mammalian metallothionein from the temporal changes observed in the emission spectrum of Cu_n-MT (where $n = 1–20$) in the 600 nm region (Green and Stillman, 1996). As we have described above, the emission intensities are dependent on the temperature and on the metal loading ratio. In these more detailed kinetic studies, it has been determined that this emission spectral intensity changes systematically with time in the period following the initial binding of Cu(I) to rabbit liver Zn_7-metallothionein. This effect can be interpreted in terms of the adoption by Cu(I) of different copper–thiolate cluster structures within the binding site over a period of 20 minutes or greater (at room temperature) following initial binding of the Cu(I) atoms.

Significant change in the CD spectral data recorded at temperatures between 5 °C and 50 °C also indicates that considerable thermally induced rearrangement takes place after the metals Cd(II), Cu(I) or Ag(I) initially bind to Zn_7-metallothionein (Stillman and Zelazowski, 1988; Presta *et al.*, 1995).

The CD spectra in Fig. 1.10 show how temperature influences the products that form when Cu(I) displaces Zn(II) in Zn_7-MT. In titrations of Zn_7-MT 2A with Cu(I) carried out at low temperatures (10 °C), Cu_{12}-MT and Cu_{15}-MT form as expected (Presta *et al.*, 1995). However, at elevated temperatures (in Fig. 1.10 data recorded at 52 °C are shown) a mixed-metal compound forms that exhibits far greater CD spectral intensity than the Cu_{12}-MT. Cu_9Zn_2-MT clearly adopts a well-defined, three-dimensional structure that is sufficiently different from that of Cu_{12}-MT that the Cu_{12}-MT cannot form. We interpret these data in terms of protein backbone folding. At low temperatures, the path of the peptide must remain close to that set by the seven Zn(II)s bound in the Zn_7-MT (Fig. 1.4). However, the Cu(I) binds trigonally rather than tetrahedrally, imposing considerable strain on the peptide as Cu(I) and Zn(II) must coexist in the same binding-site region. The models of the structure of Cu_{12}-MT (Presta *et al.*, 1997) show that the methionine group in the β fragment is located at the center of the molecule compared with the extremity shown in Fig. 1.4). At the elevated temperatures, there is sufficient thermal energy to allow the peptide to rearrange under the influence of the incoming Cu(I). This allows a new species, the mixed-metal CuZn-MT to form. The CD spectra show that the final product is still the Cu_{15}-MT, to which we assign a high fraction of digonal coordination.

Good models for the binding site in metallothionein for Cu(I) are the complexes formed with the tripeptide glutathione (Presta and Stillman, 1994). As with metallothionein, copper binds to glutathione as Cu(I), rather than Cu(II). Thus the major optical transitions observed in the visible–UV region for the Cu(I)–thiolate clusters formed between Cu(I) and the protonated (at pH 4) or deprotonated (at pH 9) metal-free glutathione (GSH), will be thiolate ligand to Cu(I) metal charge transfer in nature. All changes that are observed in the CD spectrum above 235 nm as Cu(I) binds to GSH are metal-dependent, as the changes in structure directly affect the asymmetry of the binding site and hence directly affect the sign and magnitude of the CD signal measured under the thiolate to Cu(I) charge-transfer bands. As is the case for metallothioneins, the CD spectrum reports on the geometry of the binding site through the asymmetry of the electric field that exists in the metal-binding site. These changes provide essential stoichiometric data since the band energies and the shape of the spectral envelope are dependent on the coordination environment around the metal ion in the binding site.

1.5.9 Silver-containing metallothioneins – emission and CD data

The emission spectra of silver-containing metallothionein recorded at 77 K as a function of silver loading are quite unlike the spectra recorded when Cu(I) is added to Zn_7-MT. When Cu(I) is added the intensity rises steeply as a function of the number of atoms of Cu(I) added up to the 12 Cu(I) point, then falls to almost zero as from 12 to 20 Cu(I) are added. The emission spectra recorded at 77 K from solutions of rabbit liver apo-MT 2A that had received increasing mole equivalents of Ag(I) show while Ag_{12}-MT is the first product, further Ag(I) forms Ag_{18}MT. The Ag_{18}-MT is the final product and this species is resistant to the addition of further Ag(I). The CD data in Figs 1.11 and 1.12 are significant in providing strong evidence that a species with a stoichiometry of 18 Ag(I) to the 20 cysteines is the final product. Figure 1.11 shows the CD spectral traces for a titration of Zn_7-MT 2A with Ag(I) carried out at 20 °C. Again, when compared with the CD spectra in Fig. 1.10 for Cu(I) binding, we find very little resolution. It was expected that the Ag_{12}-MT would be observed in these titrations, but as is evident, at 20 °C the 12 Ag(I) species is not observed. Figure 1.12 shows data

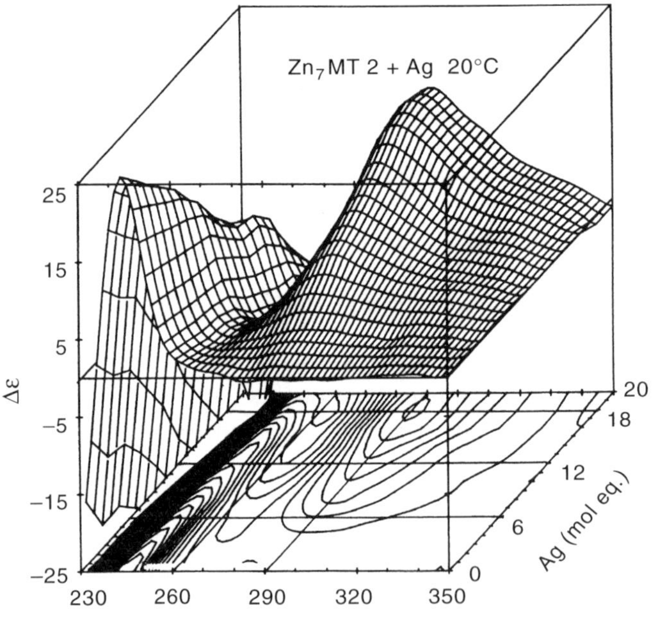

Figure 1.11 CD spectra recorded during a titration of rabbit liver Zn_7-MT 2A with Ag(I) at 20 °C and pH 7. At this temperature the spectral features of the Zn_7-MT are replaced very gradually by the spectral signature of Ag_{18}-MT. (Reproduced with permission from Zelazowski and Stillman, 1992.)

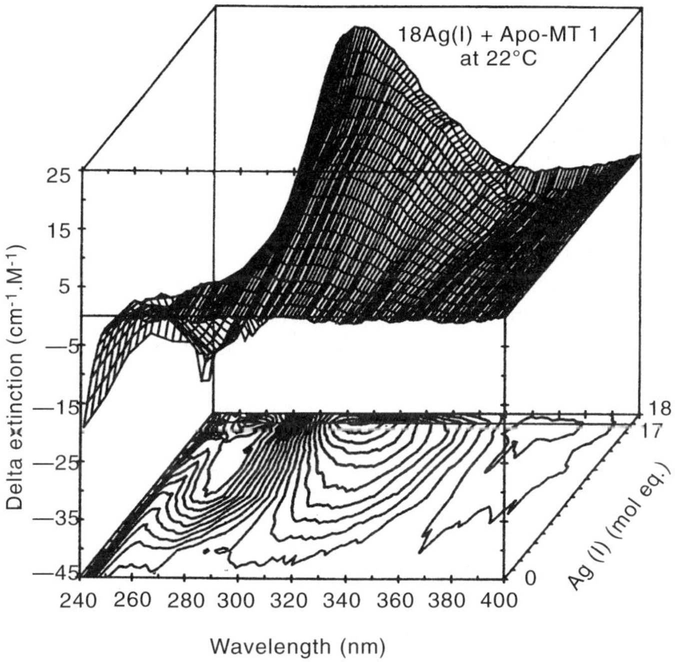

Figure 1.12 CD spectra recorded during a titration of metal-free, rabbit liver apo-MT 1 with Ag(I) at 22 °C and pH 2.7. Under these conditions, there is no CD spectral intensity to the red of 240 nm. The figure shows the growth of the spectral signature of Ag_{17}-MT. (Reproduced with permission from Gui *et al.* (1996).)

in which Ag(I) is added to the metal-free, apo-MT 1, again at 22 °C. Now there is an indication that an Ag_{12}-MT species forms. At elevated temperatures (55 °C) there is more evidence for the Ag_{12}MT species (Zelazowski *et al.*, 1989; Gui *et al.*, 1996).

What information do we obtain from a comparison between the CD data for two seemingly similar metals? First, these data tell us that Ag(I)- and Cu(I)-binding reactions to MT result in different products at all temperatures when the incoming metal is added to Zn_7-MT. We interpret the data for Ag(I) in terms of the energy requirements for a binding site that involves a much greater fraction of digonal Ag(I) coordination. In other words, the formation of S–Ag–S units in which the sulfur may be bridging to two or more Ag(I) involves dramatically greater rearrangement of the peptide than the trigonal coordination of the Cu(I). The CD data also provide strong evidence that a species forms in which 18 Ag(I) bind to the 20 cysteinyl sulfurs, much like the case with the 18 Hg(II).

1.5.10 Determination of metal–ligand bond information – use of the XAFS technique

XAFS (X-ray absorption fine structure) commonly referred to as EXAFS (extended XAFS) in the literature, has served as the only alternative direct source of this information for MT species containing metals other than Cd and Zn. The availability of XAFS data from both the coordinating thiolate sulfur and the bound metal provides information unavailable from other techniques; a comprehensive summary of published data is available (Stillman, 1995). An important caveat accompanying the use of XAFS data is that the bond lengths and coordination numbers are averages of all the properties of all metals bound, and unless large differences exist between metals bound in different sites, the presence of individual sites with unusual properties for a fraction of the metals can only be recognized from an unusually long or short average bond length. Thus coordination numbers obtained from XAFS data may represent the coordination of the metal by all ligands and may have wide confidence limits. Other spectroscopic techniques must be used to determine the 'coordination purity' of the clusters examined and inorganic models must be used to correlate calculated bond lengths with coordination geometry. The data obtained from XAFS experiments can be tested using molecular modeling calculations and two-dimensional NMR experiments.

1.5.11 Modeling the structures of metallothionein

The general structural properties that characterize all metallothioneins are:

1 an absence of disulfide bonds, which means that the three-dimensional structure of *metal-free* metallothionein is essentially that of a random chain (see Fig. 1.14);
2 formation of metal–thiolate clusters that involve terminal or bridging cysteinyl thiolate groups of the peptide chain; and
3 a tertiary structure of the metallated protein that is completely dominated by the manner in which the peptide chain wraps round these clusters.

Stout and co-workers have reported the only protein structure determined from a crystal (Robbins *et al.*, 1991; Robbins and Stout, 1992). The sample used here was the mixed-metal Cd_5Zn_2-MT prepared from rat liver. Significantly, the three-dimensional, space-filling model drawn for Cd_5Zn_2-MT showed clearly that deep crevices formed when the peptide chain was wrapped round the two metal–thiolate cores in mammalian metallothionein. Similar crevices are observed in the structure of Zn_7-MT 2A (Fig. 1.4), and also in the structure of Cu_{12}-MT constructed by molecular modeling (Fowle and Stillman, 1997; Presta *et al.*, 1997).

Minimizing the energy of the approximately 800 total atoms that comprise the primary sequence in mammalian metallothioneins, together with the 7, 12 or 18 group 11 and 12 metals, is a complicated and time-consuming task. We have carried out calculations using the rabbit liver MT 2A and 1 sequences for Zn_7-S_{20}, Cd_7-S_{20}, Hg_7-S_{20} and Cu_{12}-S_{20} assuming the presence of a two-domain structure that involves either M_3S_9 (in the β domain), M_4S_{11} (in the α domain) for Zn(II), Cd(II) and Hg(II), or M_6S_9 (in the β domain) and M_6S_{11} (in the α domain) for Cu(I) (Fowle and Stillman, 1997; Presta *et al.*, 1997). All calculations were performed using a CAChe 3-D Stereo Workstation (CAChe version 3.8; CAChe Scientific Inc. 1996) using Allinger's MM2 force field (Allinger, 1977) as augmented by CAChe.

Using the three-dimensional structure of Cd_5Zn_2-MT determined through ^{113}Cd NMR, two-dimensional ^1H NMR and X-ray analysis (Otvos and Armitage, 1980; Robbins *et al.*, 1991; Braun *et al.*, 1992) as guides, metal-free models of the α and β domains were individually constructed and approximate structures obtained for each. The minimized structure for α MT 2A is shown here as Fig. 1.14. Structures for

Zn_7-MT 2A (Fig. 1.4), Cd_7-MT 2A and Hg_7-MT 2A were constructed using the connectivities reported by Robbins *et al.* (1991) and Braun *et al.* (1992). The final molecule was subjected to a geometrical optimization using a combination of molecular mechanics and molecular dynamics. The conjugate gradient method was used in the minimization of the structure using the augmented (by CAChe) MM2 parameters and treating methylene and methyl groups explicitly. The structures were minimized in the gas phase, absent from solvent interaction at 0 K and refined until the final energy change was less than 0.001 kJ/mol. Figure 1.4 shows the structure of the Zn_7-MT 2A (Fowle and Stillman, 1997). Like the structure reported from the analysis of the X-ray diffraction data by Robbins *et al.* (1991), we observed two deep crevices that provide access to the Zn–thiolate clusters. The structures of the Cd_7-MT and the Hg_7-MT involve metals of increasing diameters and the cluster structures show signs of strain, especially for the Hg(II), that may be the origin of the unusual spectroscopic properties of Hg-MT discussed earlier.

The minimized structure for the binding sites in the two domains in rabbit liver Cu_{12}-MT 2A was calculated in a similar way using the CAChe implementation of the MM2 force fields (Presta *et al.*, 1997). Six cuprous ions of unit positive charge were added to each domain with each copper ion bonded to three cysteinyl residues in a trigonal coordination (following the information obtained from the XAFS experiments, for example, Gui *et al.*, 1996). The spectroscopic data obtained during the replacement of zinc by copper(I) (Presta *et al.*, 1995), indicate that little overall structural change takes place during metal replacement. The prevalence of Cys–Cys and Cys–X–Cys groups in the peptide chain, unique for metallothioneins among proteins, are of critical importance in metal–thiolate cluster formation. Thus, the final constraint was that these proximal cysteines were chosen to bind the same Cu(I) atom in each domain cluster.

As for the M_7-MT species, the final molecule was subjected to a geometrical optimization using molecular mechanics. The conjugate gradient method was used in the minimization of the structure using the augmented (by CAChe) MM2 parameters and treating methylene and methyl groups explicitly. The structures were minimized in the gas phase, absent from solvent interaction at 0 K and refined until the final energy change was less than 0.001 kJ/mol. The minimized structure for the binding sites in the two domains in rabbit liver Cu_{12}-MT 2A was calculated by the MM2 program (Presta *et al.*, 1997). In this structure a barrel-like feature for the Cu_6S_9 cluster in the β domain and a ladder-like

Figure 1.13 View of the Cu–thiolate cluster structures calculated for the Cu_6S_9 β domain that involves only bridging thiolates. (Unpublished results of Kowalchuk and Stillman (1999), based on the calculation of Presta *et al.* (1997).)

structure in the α domain have been calculated. The open architecture of the β domain means that water can enter this cluster, thus reducing the emission intensity compared with the more enclosed structure for the Cu_6S_{11} α domain. Figure 1.13 shows the view a water molecule would see of the β domain Cu_6S_9 cluster. Clearly, the coordinating sulfurs are very exposed. Indeed, there is access right through the center of the cluster that allows deactivation of the excited states, as observed in the emission data shown earlier. The presence of access to the clusters in this manner suggests a mechanism for metal replacement reactions in which Zn(II) or Cu(I) atoms are handed over to other biomolecules.

When the energies of all atoms in the 62-residue sequence of rabbit liver MT 2A were minimized, the calculated three-dimensional structure exhibited two features found by the X-ray diffraction studies of the Cd_5Zn_2-MT (and seen for Zn_7-MT, shown in Fig. 1.4), namely, two independent metal-binding domains and a deep crevice in each domain that exposes the copper–thiolate clusters. The metal–thiolate bond lengths calculated and reported here are a close match for trigonal coordination. Our XAFS data for Cu_{12}-MT provide an average Cu–S bond length of 225 pm (Gui *et al.*, 1996). By calculating the structure of the metal–thiolate core, using CH_3SH to model the cysteinyl thiols, we can determine the effect of thiolate binding when the thiolates are connected to a peptide chain through cysteines. The bond lengths obtained (Presta *et al.*, 1997) are very much more regular than those found with the protein, which can be interpreted in terms of the strain imposed on the structures by the presence of the peptide.

The final two structures shown are the metal-free apo-α fragment in Fig. 1.14 and the energy-minimized Zn_1-α fragment in Fig. 1.15. In these calculations, the CD data indicate that there is little three-dimensional structure in the metal-free peptide. The calculation shown in Fig. 1.15 supports the assumption that the metals provide all the structure. The metal-free α fragment shown in Fig. 1.14 simply exhibits no significant secondary structure. The key question of how MT binds metals is very hard to answer.

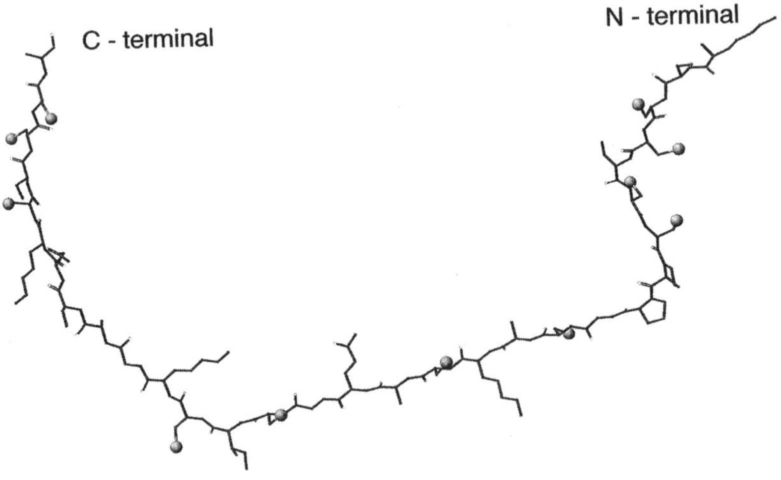

Figure 1.14 The energy-minimized structure of the α fragment of rabbit liver MT 2A. (Unpublished results of Kowalchuk and Stillman (1999).)

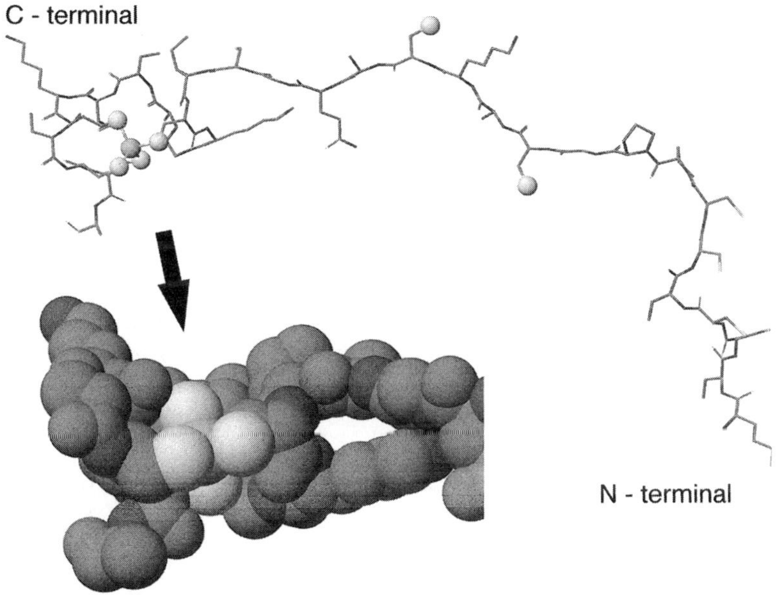

Figure 1.15 The energy-minimized structure of the α fragment of rabbit liver MT 2A bound to a single Zn(II). The site of binding represents the lowest energy of all permutations. The inset shows the structure drawn using space-filling atoms. (Unpublished results of Kowalchuk and Stillman (1999).)

This is because the spectral data show that the conditions greatly change the reaction. We have begun calculations that will determine the lowest thermodynamic product for metallation at each site in the protein. We show in Fig. 1.15 the lowest-energy site for the first zinc to bind. The C-terminus cysteines are predicted to bind a single zinc in competition with all other sites. Later additions of zinc show that the cluster that will eventually exhibit a stoichiometry of Zn_4S_{11} forms in a concerted manner, each additional zinc bringing the sulfurs closer together (Kowalchuk, Hunter and Stillman, unpublished). The coordination geometries of the well-characterized proteins are summarized in Table 1.3. Note that there is considerable uncertainty in the coordination numbers for the silver, gold and mercury species.

Acknowledgments

The research described here was funded by NSERC of Canada and the Academic Development Fund at the UWO. MJS is grateful to the energy and enthusiasm of his group who worked on these projects and whose work is described here. We would also like to thank Donald Thomas for his help in assembling the figures.

Table 1.3 Metal–thiolate coordination geometries reported for metals bound to mammalian and yeast metallothioneins (adapted from Stillman *et al.*, 1994); the data used to construct this table are described in Stillman (1995)

	Coordination geometry of the metals in the binding site of different metallothioneins – the stoichiometry in the binding site is shown			
	Tetrahedral	*Trigonal*	*Digonal*	*Mixed*
Cd(II) in mammalian liver protein	Cd_7-MT (Cd_3S_9 and Cd_4S_{11})			
Co(II) in mammalian liver protein	Co_7-MT (Co_3S_9 and Co_4S_{11})			
Zn(II) in mammalian liver protein	Zn_7-MT (Zn_3S_9 and Zn_4S_{11})			
Hg(II) in mammalian protein	Hg_7-MT (but see the XAFS results)	Hg_{11}-MT (Hg_5S_9 and Hg_6S_{11})	Hg_{18}-MT ($Hg_{18}S_{20}$)	
Ag(I) in mammalian liver protein			Ag_{17}-MT ($Ag_{17}S_{20}$)	Ag_{12}-MT (Ag_6S_9 and Ag_6S_{11})
Cu(I) in mammalian liver protein		Cu_{12}-MT (Cu_6S_9 and Cu_6S_{11})		Cu_{15}-MT (Cu_6S_9 and Cu_9S_{11})
Au(I) in horse kidney protein			Au,Zn,Cd-MT (Au-S = 229 pm) $(TmSAu)_x$MT (Au-S = 230 pm	
Cu(I) in yeast protein		Cu_8-MT		Cu_7-MT

References

Allinger, N.L. 1977. Conformational Analysis. 130. MM2. A hydrocarbon force field utilizing V_1 and V_2 torsional terms. *Journal of the American Chemical Society* **99**, 8127–8134.

Beltramini, M. and Lerch, K. 1981. Luminescence properties of Neurospora Cu(I)-MT. *FEBS Letters* **127**, 201–203.

Braun, W., Vasak, M., Robbins, A.H., Stout, C.D., Wagner, G., Kagi, J.H.R. and Wuthrich, K. 1992. Comparison of the NMR solution structure and the X-ray crystal structure of rat metallothionein-2. *Proceedings of the National Academy of Sciences (USA)* **89**, 10124–10128.

Cai, W. and Stillman, M.J. 1988a. Metal binding to MT: Competition for cadmium and zinc between Chelex-100 and metal binding sites in MT. *Inorganica Chimica Acta* **152**, 111–115.

Cai, W. and Stillman, M.J. 1988b. Hg_{18}-metallothionein. *Journal of the American Chemical Society* **110**, 7872–7873.

Cowan, J.A. 1997. *Inorganic biochemistry*, Wiley-VCH, NY, Chapter 6, p. 257.

Dance, I., Fisher, K. and Lee, G. 1992. Metal thiolate compounds, in *Metallothioneins*, (eds M.J. Stillman, C.F. Shaw III and K.T. Suzuki), VCH Publishers, New York, pp. 284–345.

Fowle, D.A. and Stillman, M.J. 1997. Comparison of the structures of the metal–thiolate binding sites in $Zn(II)_7$, $Cd(II)_7$, and $Hg(II)_7$-metallothioneins using molecular modeling techniques. *Journal of Biomolecular Structure and Dynamics* **14**, 393-406.

George, G.N., Byrd, J. and Winge D.R. 1988. X-ray absorption studies of yeast copper MT. *Journal of Biological Chemistry* **263**, 8199–8203.

Green, A.R. 1995. Spectroscopic properties of copper binding to metallothionein. Ph.D. Thesis, University of Western Ontario, London, Canada.

Green, A.R. and Stillman, M.J. 1994. Oxidative quenching of luminescence from copper metallothionein. *Inorganica Chimica Acta* **226**, 275–283.

Green, A.R. and Stillman, M.J. 1996. Mobility of copper in binding sites in rabbit liver metallothionein 2. *Inorganic Chemistry* **35**, 2799–2807.

Green, A.R., Presta, A., Gasyna, Z. and Stillman, M.J. 1994. Luminescent probe of copper–thiolate cluster formation within mammalian metallothionein. *Inorganic Chemistry* **33**, 4159–4168.

Green, A.R., Presta, P.A. and Stillman, M.J. 1999. Copper binding to metallothionein at low pH (manuscript submitted).

Gui, Z., Green, A.R., Kasrai, M., Bancrofi, G.M. and Stillman, M.J. 1996. Sulfur K-edge EXAFS studies of cadmium, zinc, copper, and silver rabbit liver metallothioneins. *Inorganic Chemistry* **35**, 6520–6529.

Hayashi, Y. and Winge, D.R. 1992 $(\gamma\text{-EC})_n$G peptides. Properties of cadmium: $(\gamma\text{-Glu-Cys})_n$Gly complexes in fission yeast, in *Metallothioneins*, (eds M.J. Stillman, C.F. Shaw III and K.T. Suzuki), VCH Publishers, New York, pp. 271–283.

Jiang, D.T., Heald, S.M., Sham, T.K. and Stillman, M.J. 1994. Structures of the cadmium, mercury, and zinc thiolate clusters in metallothionein: XAFS study of Zn_7-MT, Cd_7-MT, Hg_7-MT and Hg_{18}-MT formed from rabbit liver metallothionein. *Journal of the American Chemical Society* **116**, 11004–11013.

Jiang, D.T., Gui, Z.Q., Heald, S.M., Sham, T.K. and Stillman, M.J. 1995. XAFS of silver(I) metallothionein. *Physica B* **208/209**, 729–730.

Kagi, J.H.R. 1993. Evolution, structure and chemical reactivity of class I metallothioneins: An overview, in *Metallothionein III*, (eds K.T. Suzuki, N. Imura and M. Kimura), Birkhauser Verlag, Basel, pp. 29–55.

Kagi, J.H.R. and Kojima, Y. (eds) 1987. *Metallothionein II*, Birkhauser Verlag, Basel.

Kagi, J.H.R. and Nordberg, M. (eds) 1979. *Metallothionein*, Birkhauser Verlag, Basel.

Kaim, W. and Schwederski, B. 1994. Copper-containing proteins: An alternative to biological iron, in *Bioinorganic chemistry: Inorganic elements in the chemistry of life*, (eds W. Kaim and B. Schwederski), Wiley Publishing, New York.

Kraulis, P.J., Raine, A.R.C., Gadhavi, P.L. and Laue, E.D. 1992. Structure of the DNA-binding domain of zinc GAL4. *Nature* **356**, 448.

Le Blanc, J.C.Y., Siu, K.W.M. and Guevremont, R. 1994. Electrospray mass spectrometric study of protein–ketone equilibria in solution. *Analytical Chemistry* **66**, 3289–3296.

Le Blanc, Y.J.C., Presta, A., Veinot, J., Gibson, D., Siu, K.W.M. and Stillman, M.J. 1997. Identification of isoforms and subisoforms of rabbit liver metallothionein using electrospray mass spectrometry. *Protein Peptide Letters* **4**, 313–320.

Linder, M.C. (ed.) 1991. *Biochemistry of copper*, Plenum Press, New York.

Lu, W. and Stillman, M.J. 1993. Mercury–thiolate clusters in metallothioneins. Analysis of circular dichroism spectra of complexes formed between alpha-metallothionein, apometallothionein, zinc metallothionein, and cadmium metallothionein and Hg^{2+}. *Journal of the American Chemical Society* **115**, 3291–3299.

Lu, W., Kasrai, M., Bancroft, G.M., Stillman, M.J. and Tan, K.H. 1990. S L-edge XANES study of Zn, Cd, and Hg containing metallothionein and model compounds MT. *Inorganic Chemistry* **29**, 2561–2563.

Lu, W., Zelazowski, A.J. and Stillman, M.J. 1993. Mercury binding to metallothioneins: Formation of Hg_{18}-MT. *Inorganic Chemistry* **32**, 919–926.

Otvos J.D. and Armitage, I.M. 1980. Structure of the metal clusters in rabbit liver metallothionein MT, ^{113}Cd NMR. *Proceedings of the National Academy of Sciences (USA)* **77**, 7094–7098.

Otvos, J.D., Olafson, R.W. and Armitage, I.M. 1982. Structure of an invertebrate metallothionein from Scylla Serrata ^{113}Cd NMR. *Journal of Biological Chemistry* **257**, 2427–2431.

Pavletich, N.P. and Pabo, C.O. 1993. Crystal structure of a five-finger GLI-DNA complex: new perspectives on zinc fingers. *Science* **261**, 1701.

Presta, P.A. and Stillman, M.J. 1994. Chiral copper(I)-thiolate clusters in metallothionein and glutathione. *Chirality* **6**, 521–530.

Presta, P.A., Green, A.R., Zelazowski, A.J. and Stillman, M.J. 1995. Copper binding to rabbit liver metallothionein. Formation of a continuum of copper(I)–thiolate stoichiometric species. *European Journal of Biochemistry* **227**, 226–240.

Presta, P.A., Fowle, D.A. and Stillman, M.J. 1997. Structural model of mammalian copper metallothionein. *Dalton Transactions*, 977–984.

Riordan, J.F. and Vallee, B.L. (eds) 1991. Metallobiochemistry Part B. Metallothionein and related molecules. *Methods in Enzymology* **205**,

Robbins, A.H. and Stout, C.D. 1992. Crystal structure of metallothioneins, in *Metallothioneins*, (eds M.J. Stillman, C.F. Shaw III and K.T. Suzuki), VCH Publishers, New York, pp. 31–54.

Robbins, A.H., McRee, D.E., Williamson, M., Collett, S.A., Xuong, N.H., Furey, W.F., Wang, B.C. and Stout, C.D. 1991. Refined crystal structure of Cd,Zn-metallothionein at 2.0 Å resolution. *Journal of Molecular Biology* **221**, 1269–1293.

Shaw, C.F. III, Stillman, M.J. and Suzuki, K.J. 1992. Metallothioneins, in *Metallothioneins*, (eds M.J. Stillman, C.F. Shaw III and K.T. Suzuki), VCH Publishers, New York, pp. 1–13.

Solomon, E.I., Baldwin, M. J. and Lowery, M.D. 1992. Electronic structures of active sites in copper proteins: contributions to reactivity. *Chemical Reviews* **92**, 521.

Stillman, M.J. 1992. Optical spectroscopy of metallothioneins, in *Metallothioneins*, (eds M.J. Stillman, C.F. Shaw III and K.T. Suzuki), VCH Publishers, New York, pp. 55–127.

Stillman, M.J. 1995. Metallothioneins. *Coordination Chemistry Review* **144**, 461–571.

Stillman, M.J. and Gasyna, Z. 1991. Luminescence spectroscopy of metallothioneins. *Methods in Enzymology* **205**, 540–555.

Stillman, M.J. and Zelazowski, A.J. 1988. Domain specificity in metal binding to metallothionein. A CD and MCD study of Cd and Zn binding at temperature extremes. *Journal of Biological Chemistry* **263**, 6128–6133.

Stillman, M.J. and Zelazowski, A.J. 1989. Domain specificity of cadmium and zinc binding to rabbit liver metallothionein 2. Metal ion mobility in the formation of Cd_4-alpha MT. *Biochemical Journal* **262**, 181–188.

Stillman, M.J., Cai, W. and Zelazowski, A.J. 1987. Cadmium binding to metallothioneins domain specificity in reactions of a and b fragments, apometallothionein, and zinc metallothionein with Cd. *Journal of Biological Chemistry* **262**, 4538–4548.

Stillman, M.J., Zelazowski, A.J. and Gasyna, Z. 1988. Luminescent Ag-metallothionein: dependence of emission intensity on silver–thiolate cluster formation. *FEBS Letters* **240**, 159–162.

Stillman, M.J., Gasyna, Z., and Zelazowski, A.J. 1989a. A luminescence probe for MT in liver tissue: emission intensity measured directly from Cu(I)-MT induced in rat liver. *FEBS Letters* **257**, 283–286.

Stillman, M.J., Zelazowski, A.J., Szymanska, J., and Gasyna, Z. 1989b. Luminescent metallothioneins: Emission properties of copper, silver, gold and platinum complexes of MT. *Inorganica Chimica Acta* **161**, 275–279.

Stillman, M.J., Shaw, C.F. and Suzuki, K.T. (eds) 1992. *Metallothioneins*. V.C.H. Publishers, New York.

Stillman, M.J., Presta, A., Gui, Z. and Jiang, D.T. 1994. Spectroscopic studies of copper, silver and gold-metallothioneins. *Metal-Based Drugs* **1**, 375–393.

Suzuki, K.T., Imura, N. and Kimura, M. (eds) 1993. *Metallothionein III*, Birkhauser Verlag, Basel.

Webb, M. (ed.) 1979. *The chemistry, biochemistry and biology of cadmium*, Elsevier, Amsterdam.

Winge, D.H. and Dameron, C.T. 1993. The metallothionein structural motif involved in metalloregulation, in *Metallothionein III*, (eds K.T. Suzuki, N. Imura and M. Kimura), Birkhauser Verlag, Basel, pp. 381–397.

Yu, Z., Wojciechowski, M. and Fenselau, C. 1993. Assessment of metals in reconstituted metallothioneins by electrospray mass spectrometry. *Analytical Chemistry* **65**, 1355–1359.

Zelazowski, A.J. and Stillman, M.J. 1992. Ag(I) binding to metallothionein. CD study of silver–thiolate cluster formation formation with Zn-MT and Zn-alpha and Zn-beta MT. *Inorganic Chemistry* **31**, 3363–3370.

Zelazowski, A.J., Gasyna, Z. and Stillman, M.J. 1989. Ag binding to rabbit liver MT. CD and emission study of Ag–thiolate cluster formation with apo-MT and the alpha and beta fragments. *Journal of Biological Chemistry* **264**, 17091–17099.

Zheng, Y.J. and Merz, K.M. Jr 1992. Mechanism of the human carbonic anhydrase II catalyzed hydration of carbon dioxide. *Journal of the American Chemical Society* **114**, 10498.

2

BIOCHEMICAL PATHWAYS IN CADMIUM TOXICITY

M.H. Bhattacharyya, A.K. Wilson, S.S. Rajan and M. Jonah

2.1 Introduction

Cadmium is a relatively rare metal, present in the Earth's crust at an average concentration of 0.5 μg/g. In the metallic form, it is widely used as a protective coating on steel and in alloys such as solders and brazes. Cadmium compounds are used extensively as colored pigments (bright yellow, orange and red colors in paints and plastics), as organic stabilizers in certain plastics (e.g., cadmium stearate) and in rechargeable nickel–cadmium batteries (for portable power tools, laptop computers, etc).

The purpose of this chapter is to present an overview of cadmium toxicity in the context of:

1 the biochemical pathways for cellular uptake and transport; and
2 molecular events that lead to its toxicity.

The target organs considered are intestine, kidney, liver and bone. These organs are targets for both environmental and occupational exposure to cadmium. Not included are lungs, testes or prostate. Other chapters address respiratory and carcinogenic responses to cadmium and other metals (see Chapters 12 and 13).

Cadmium that enters the blood from either the lung or intestine clears from the circulation extremely rapidly, depositing mainly in the liver and kidney (Friberg *et al.*, 1986a). Cadmium resides in these organs with very long half-lives; excretion into urine or feces is extremely limited. As a result of this metabolic profile, cadmium accumulates in the body over time, increasing in concentration with age in a number of organs. The cadmium that we take in stays with us most of our lives. That makes it important to determine the sensitivities of organ systems to toxic responses and corresponding exposure levels that provide an adequate safety margin.

By 50 years of age, kidney cadmium concentrations reach approximately 12 μg/g in nonsmokers and 25μg/g in persons who smoke one pack of cigarettes or more per day, reflecting the fact that tobacco leaves concentrate cadmium from the soil (Friberg *et al.*, 1986a). The kidney cadmium concentration at which renal tubular dysfunction starts to appear is estimated to be about 200 μg/g for long-term exposures (Friberg *et al.*, 1986b). That means that in the US, there is currently a safety factor of 10 to 20 with respect to kidney cadmium concentrations achieved by environmental exposure routes. That fairly limited safety margin identifies the importance of determining and restricting those activities that might increase environmental pathways for cadmium exposure. In some countries, including Japan, environmental pathways for human exposure are already higher than in the US, with 50-year-olds showing average renal cadmium concentrations of 50 μg/g in nonsmokers and 65 μg/g in smokers (Friberg *et al.*, 1986a). It is in this context that this chapter is presented.

2.2 Pathways for cellular uptake and transport

Figure 2.1 presents a diagram of the cadmium pathways discussed in this section.

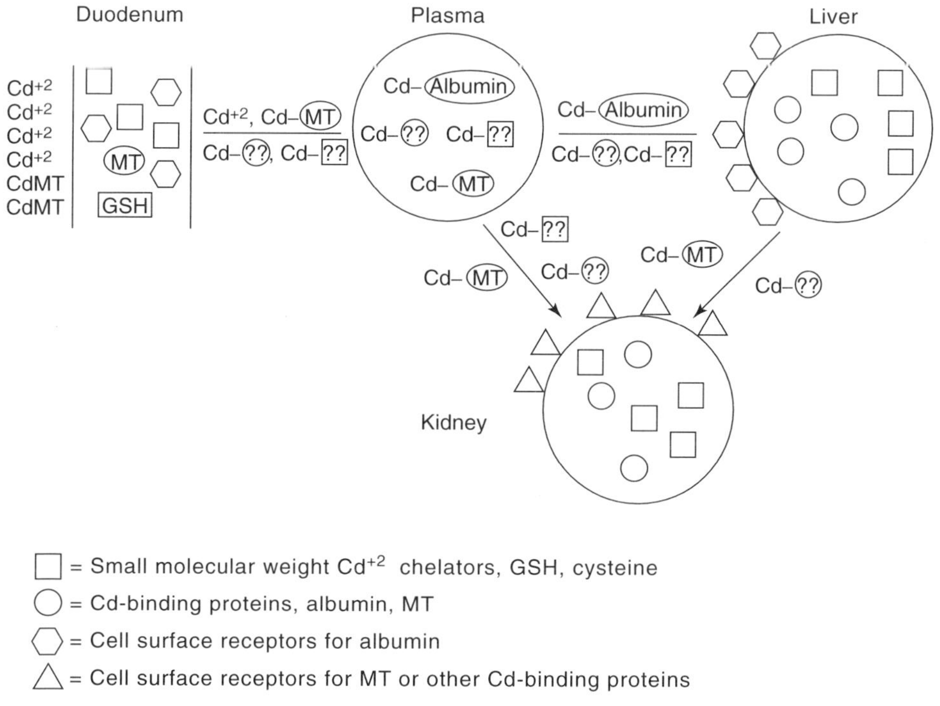

Figure 2.1 Possible pathways of cadmium uptake and distribution following oral exposure.

2.2.1 Intestinal uptake and transport

Cadmium from dietary sources can be presented to the intestinal lumen as free ions or as bound cadmium (Groten *et al.*, 1991, 1994). As a ligand, cadmium can bind to small molecules (such as amino acids or peptides such as glutathione (GSH)) or to proteins such as albumin or metallothionein (MT). MT is a cysteine-rich, metal-binding protein of low molecular weight which is synthesized by cells in response to various stimulants, including cadmium. In this section, we will discuss some of the pathways for the influx of oral cadmium into the gastrointestinal (GI) tract. These pathways involve several steps, including uptake at the mucosal surface of the intestinal cells, its binding to intracellular molecules, and its subsequent transfer across the basolateral membranes to the circulation (Andersen *et al.*, 1994).

2.2.1.1 *Step 1: Mucosal uptake*

The early uptake of cadmium ions by the cells of the small intestine is extensive and rapid even at very low lumenal concentrations (e.g. tracer doses) (Rajan *et al.*, 1999). Although this may indicate specificity, Foulkes has made a credible case for an early rapid binding of cadmium via nonspecific electrostatic interactions with the anionic sites on the lumenal surface (Figure 2.1) (Foulkes, 1985, 1989; Foulkes and McMullen, 1987). To illustrate duodenal uptake kinetics, nearly 15% of a 0.1 μg dose of cadmium is bound by the small intestine within 30 min of oral administration to young adult mice during their active feeding cycle (Fig. 2.2a). However, only a small fraction (less than 1%) of this administered dose is subsequently transferred to the circulation. Unlike zinc, cadmium ions deposit mainly in the duodenum, with much smaller amounts in jejunum and ileum (Fig. 2.2a) (Sorensen *et al.*, 1993; Lind and Wicklund-Glynn, 1997; Rajan *et al.*, 1999). Cadmium ion influx into the jejunum from the lumen has been shown to have a rapid, temperature-insensitive, saturable component and a second nonsaturable, temperature-dependent, slower component that probably relies on passive diffusion (Foulkes and Bergman, 1993).

The kinetics and magnitude of the rapid mucosal uptake of cadmium and the kinetics of its disappearance from the small intestine have been shown to vary little with intracellular MT status (Kello *et al.*, 1979; Ohta and Cherian, 1991; Rajan *et al.*, 1999). For example, these parameters are identical in normal mice and in mice deficient in *MT-1* and *MT-2* genes (Fig. 2.2a versus Fig. 2.2b) and are independent of the cadmium dosage (0.1–100 μg) (Rajan Wilson and Bhattacharyya, unpubl.). Furthermore, prior substantial induction of intestinal MT by zinc does not change the early, rapid intestinal uptake of a subsequent oral dose of cadmium (Ohta *et al.*, 1989; Min *et al.*, 1991; Ohta and Cherian, 1991). These results indicate that molecules other than MT are responsible for the early binding of cadmium ions in the duodenum.

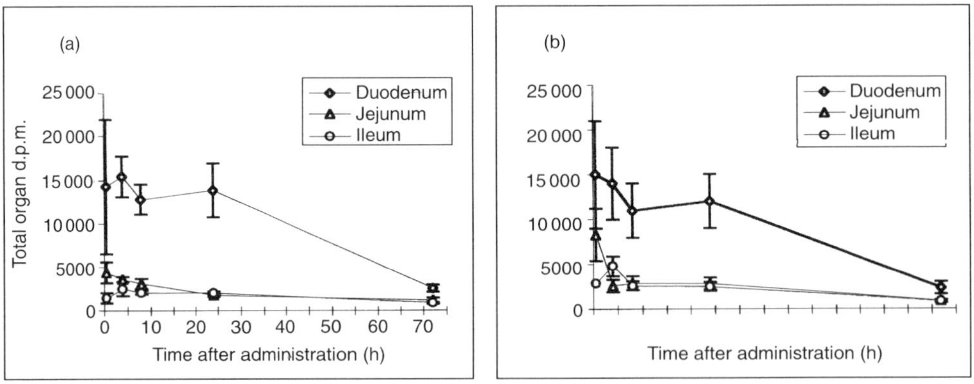

Figure 2.2 ^{109}CD uptake by the small intestine following a single gavage dose of 50 μCi (0.1μg) cadmium. A group of six mice was sacrificed at each time point, starting at 0.5 h after administration. Both MT-N (a) and MT-KO (b) mice show similar uptake patterns.

For cadmium administered orally in the form of Cd-MT, the initial uptake from lumen is much more limited, and there is clear evidence that the Cd-MT complex can be taken up intact by the intestinal cells (Fig. 2.1) (Ohta *et al.*, 1989; Ohta and Cherian, 1991; Sugawara and Sugawara, 1991).

Groten *et al.* (1992) demonstrated that, when compared to calcium and zinc, iron protected most effectively against intestinal absorption of cadmium from Cd-MT, decreasing cadmium uptake by 80%. Results indicate that intestinal absorption of cadmium from the Cd-MT complex utilizes pathways for the essential metal, iron.

2.2.1.2 Step 2: Intracellular complexation

Following its uptake into the intestinal cells, cadmium can bind to various chelators. One of the predominant cadmium-binding entities inside the duodenal cells is MT. Elsenhans *et al.* (1994) have shown that mucosal MT levels parallel those of orally administered cadmium. At low micromolar concentrations, cadmium in the cytosol binds to high molecular weight proteins and also to endogenous MT (Ohta *et al.*, 1989; Min *et al.*, 1991). At higher oral doses that induce intestinal MT, cytosolic cadmium is bound to high molecular weight proteins prior to MT induction. After MT is induced, most (about 80%) of the intestinal cadmium is found to be complexed with MT (Min *et al.*, 1991).

2.2.1.3 Step 3: Basolateral transfer to circulation

Unlike its rapid mucosal uptake, the transfer of cadmium across the basolateral side of the cell to blood takes place in a slow diffusion, being dependent on the intracellular ligands binding it in the intestine (Elsenhans *et al.*, 1997). Although initial cadmium uptake is not affected by the intracellular MT status, MT has been shown to be a major limiting factor in the rate and extent of release from the intestine and subsequent deposition in the tissues (Min *et al.*, 1991). Ohta and Cherian (1991) have shown that the very high-affinity binding of cadmium to intestinal MT can reduce the rate of release of cadmium from the intestinal cells and its subsequent transfer to liver.

Of the small fraction of cadmium that eventually passes into blood from intestine after a single oral dose, nearly half is absorbed early, within the first 30 minutes (Rajan *et al.*, 1999). This early fraction leaves the intestine in a form that goes almost entirely to liver (80–90%) (Jonah and Bhattacharyya, 1989; Rajan *et al.*, 1999). Subsequently, if the oral dose is low, cadmium leaving the intestine changes in chemical form. By 4–8 h, cadmium in the liver reaches a plateau, while cadmium in the kidneys rises steadily for several days. The form of cadmium that homes to the kidney from intestine has been postulated by many to be the cadmium–MT complex (Jonah and Bhattacharyya, 1989; Ohta *et al.*, 1989; Min *et al.*, 1991, 1992). A pathway has been defined in which intracellular cadmium binds to intestinal MT, exits across the basolateral membrane to blood, and subsequently deposits in kidney. Potentially, this pathway can explain the observation that renal tubular dysfunction, rather than hepatic dysfunction, predominates after continuous exposure to dietary cadmium (Dorian *et al.*, 1995).

The pathways described above are derived from experiments in which animals were administered cadmium in an inorganic form ($CdCl_2$). However, cadmium in the human diet may, in large part, be cadmium complexed to a plant or animal form of MT.

Therefore, several investigations have been undertaken to study intestinal absorption pathways of cadmium complexed to MT (Cherian *et al.*, 1978; Maitani *et al.*, 1984; Ohta *et al.*, 1989; Groten *et al.*, 1991, 1994; Ohta and Cherian, 1991; Min *et al.*, 1992). From these experiments, it is clear that:

1 the Cd-MT complex can be taken into intestinal cells and transferred to blood intact;
2 kidney deposition is favored after oral exposure to Cd-MT; and
3 the absorption of cadmium as Cd-MT is significantly less than that of $CdCl_2$.

2.2.1.4 Step 4: Excretion

Extensive, rapid lumenal uptake of cadmium by intestinal cells, combined with much slower transfer across the basolateral membranes to blood, creates a residing pool of unabsorbed cadmium in the small intestine. This pool serves two functions. It provides a continuous source of cadmium to the bloodstream over time (Jonah and Bhattacharyya, 1989). More importantly, it retains cadmium in a form that is very poorly absorbed into blood, resulting in cadmium being sloughed off and excreted into feces, along with the intestinal cells during their natural cycle of death and renewal (Min *et al.*, 1991).

2.2.2 Blood transport

After oral or inhalation exposure, cadmium must transit in blood before depositing in tissues beyond the intestine or lung. Because cadmium clears from the blood so quickly (Frazier, 1980, 1984; Friberg *et al.*, 1986a; Jonah and Bhattacharyya, 1989), few investigators have definitely identified the biochemical forms(s) of cadmium in blood. This is especially true for exposures to cadmium at levels encountered by humans via their diet – the main environmental source of cadmium exposure in nonsmoking members of the general population.

However, some hypotheses have developed from the data that have been obtained. It is clear, for example, that a sizeable fraction of cadmium ions that enter the blood transits through blood within the first minutes of exposure and deposits very rapidly in liver (liver : kidney amount ratio of approximately 8–10 : 1) (Frazier, 1980, 1984; Jonah and Bhattacharyya, 1989; Rajan *et al.*, 1999). The small fraction remaining in blood appears to be bound to albumin (Fig. 2.1). If exposure is low and via the oral route, cadmium leaving the intestine begins to change its biochemical form to one that prefers kidney to liver (Jonah and Bhattacharyya, 1989). With time, the relative levels of cadmium in liver and kidney change such that, after continuous oral exposure, the liver : kidney amount ratio falls, in some cases to nearly 1 : 1 (Bhattacharyya *et al.*, 1982). Many investigators hypothesize that the new form of cadmium coming from the intestine to the blood is cadmium bound to metallothionein (Jonah and Bhattacharyya, 1989; Ohta *et al.*, 1989; Min *et al.*, 1991, 1992).

A third reaction that occurs involving cadmium in blood is its transfer to red blood cells. This transfer has been shown to occur via a plasma membrane anion exchanger (Dawson and Ballatori, 1995). The reaction is not as fast as the extremely rapid uptake of cadmium by liver and kidney, but by 24 h after a low oral dose of cadmium (e.g.

4 μg/kg), 65% of the cadmium in blood is bound to the red blood cell compartment, further reducing the availability of diffusible cadmium to other tissues (Rajan *et al.*, 1999).

In summary, three major biochemical reactions involving cadmium in blood have been hypothesized (Fig. 2.1): liver deposition, possibly from circulating cadmium–albumin; kidney deposition from circulating Cd-MT; and cadmium transfer to red blood cells. Other unidentified reactions may be involved, because most of the cadmium that enters blood has already been deposited in tissues whenever blood is sampled.

2.2.3 *Kidney uptake and transport*

Biochemical reactions involved in cadmium uptake and retention by kidney have been studied primarily in three systems – whole animal; kidney cells in culture, including both primary cell cultures and immortalized cell lines; and brush-border membrane vesicles (BBMVs). Because the proximal convoluted tubules of the kidney cortex are a primary site of renal damage after cadmium exposure, proximal tubule cells (PTCs), either in primary culture or represented by the pig cell line LLC-PK1 have been utilized.

2.2.3.1 *Renal uptake of Cd-MT*

Based on the hypothesis that Cd-MT plays a critical role cadmium renal toxicity, investigations have been carried out to identify pathways for renal uptake and metabolism of the Cd-MT complex. Experiments where Cd-MT was administered intravenously to mice at a non-nephrotoxic dose (0.1 mg Cd/kg) demonstrate that 80% of injected Cd-MT is in the kidney in 15 min (Dorian *et al.*, 1992a, b; Dorian *et al.*, 1995). The primary site of uptake is the S1 and S2 segments of the proximal tubules. This rapid uptake of Cd-MT from blood appears to occur via a transport pathway for low molecular weight proteins. This hypothesis is based on at least three observations:

1. both Zn-MT and Cd-MT deposit MT in the kidney with the same kinetics (80% in kidney in 15 min), indicating that uptake is a function of the protein portion of the molecule (Dorian and Klaassen, 1995);
2. renal uptake of Cd-MT can be blocked by the low molecular weight protein, β_2μG (Nordberg *et al.*, 1992); and
3. when a high nephrotoxic dose of Cd-MT is injected into rats (30 or 60 μg Cd/kg), immediate uptake of Cd-MT by the kidney is accompanied by an immediate sharp peak of β_2-microglobulin (β_2μG) excretion into urine (Bernard *et al.*, 1987); not until 20 h is there a broad peak of urinary β_2μG indicative of renal tubular dysfunction. The early sharp peak indicates that Cd-MT and β_2μG may compete for reabsorption sites on the kidney tubules.

With ^{35}S-labeled Cd-MT, Dorian *et al.* (1992b) have demonstrated that the MT portion of Cd-MT is degraded with a half-life of approximately 2.5 h, releasing cadmium rapidly to the cells, such that, by 4 h, 70% of cadmium from Cd-MT is not bound to MT. However, this released cadmium is retained by the kidney with a long half-life, ultimately causing kidney cell damage and induction of endogenous renal MT.

Although several articles report preferential uptake or toxicity of cadmium from Cd-MT compared to $CdCl_2$ in a setting other than *in vivo* (Cherian, 1985; Abel *et al.*, 1987), almost all investigators have found that PTCs in culture, either as primary cell cultures or cell lines (for example, Liu *et al.*, 1994a), show much lower uptake of cadmium from Cd-MT than from $CdCl_2$. The conclusion is drawn that the pathway for specific Cd-MT uptake is not retained when the proximal tubule cells are prepared for culture or are immortalized into cell lines. The report of Abel *et al.* (1987) that primary PTCs obtained from rats pretreated with cadmium to induce MT show significant increases in Cd-MT uptake in primary cell culture is potentially significant (Jin *et al.*, 1986; Kimura *et al.*, 1997). Maybe the low molecular weight protein reabsorption pathway is also induced when MT is induced in the kidney, and more activity is retained *in vitro*.

2.2.3.2 Renal uptake of $CdCl_2$

An early burst of cadmium appears in blood after intestinal absorption, and this cadmium behaves much like free cadmium ions injected via a parenteral route. After parenteral injection, cadmium ions are rapidly, but not extensively, taken up by the kidney (80–90% goes to the liver), and there is no specific localization in the S1 and S2 segments of the proximal tubules (Dorian *et al.*, 1992a, 1995). Cadmium ions are taken up generally by all segments of the proximal tubules and are much less acutely nephrotoxic than is Cd-MT (Dorian and Klaassen, 1995; Dorian *et al.*, 1995).

Many insights into biochemical pathways for renal uptake of cadmium ions come from studies with cell culture and brush border membrane vesicle (BBMV) systems (Jin *et al.*, 1986; Endo *et al.*, 1995, 1997). Results of these studies demonstrate that cadmium ions can enter renal tubular cells via multiple pathways. These include:

1 saturable, energy-dependent, carrier-mediated transport in proximal tubule cells;
2 parathyroid-hormone-stimulated transport via dihydropyridine-sensitive calcium channels in distal tubule cells; and
3 passive diffusion at both basolateral and apical surfaces of both cell types. In general, cadmium ion entry has been shown to be about twofold higher at basolateral than apical surfaces in polarized cell culture systems.

K_m and V_{max} values for the saturable pathways in renal PTCs have been determined in primary cultures and in cultured LLC-PK1 cells. For example, Endo and Shaikh (1993) reported values of V_{max} = 0.13 nmol/min/mg protein and K_m = 7 μM for confluent cultures of rat PTCs, while Endo *et al.* (1995) last reported values of V_{max} = 0.16 nmol/min/mg protein and K_m = 24.6 μM for LLC-PK1 cells. A K_m value of 13.8 μM was reported for BBMVs from rat PTCs (Endo *et al.*, 1997). These saturable pathways have been shown to be sensitive to inhibition by low temperature, metabolic inhibitors, zinc ions, and replacement of sodium by potassium ions in the medium, with a K_i for zinc in range of the K_m for cadmium (Endo *et al.*, 1995; Shaikh *et al.*, 1995). In addition, Felley-Bosco and Diezi (1991) have shown that, at 5 μM $CdCl_2$, metallothionein is synthesized in 3–6 h in LLC-PK1 cells. In the BBMV system, the results of Endo *et al.* (1997) indicate that diffusion accounts for about 20% of total uptake at 1 μM cadmium and increases to 60–70% at 100 μM cadmium. Translated to the *in vivo* setting, these kinetic values characterize pathways active at blood cadmium concentrations in the micromolar range – ones that would clearly produce short-term toxic responses (1 μM Cd ≃ 100 ng/L). To identify renal uptake pathways active in handling environmental levels of cadmium, concentrations in the picomolar range would need to be studied. (A blood cadmium concentration of 5 μg/L (~50 nM) triggers medical surveillance in industry (USOSHA, 1992; 29CFR1910.1027, App. F).)

Foulkes and co-workers have carefully characterized renal cell uptake of cadmium ion complexed to mercaptoethanol (Cd-ME) in rabbits (Foulkes, 1989; Foulkes and Blanck, 1990). They have done this by a simultaneous bolus injection of [^{109}Cd]ME and [^{3}H]inulin followed by a 40 s occlusion of the aortic artery to allow time for kidney cell uptake. At 40 s, the occlusion is released and blood is immediately sampled from the catheterized renal vein. The ratio of ^{109}Cd to ^{3}H provides a measure of cadmium uptake by the renal cells (inulin is used as an unabsorbed marker). Because only about 10% of plasma is filtered during each pass through the kidney, studies of cadmium uptake from nonfiltered plasma that bathes basolateral renal cell surfaces appear to be relevant. In this system, it has been shown that Cd-MT is not at all extracted from plasma to kidney during the 40 s occlusion, supporting the hypothesis that Cd-MT is reabsorbed at the apical cell surface and not extracted from renal plasma except after glomerular filtration. In contrast, about 50% of cadmium from Cd-ME is extracted during the occlusion (Foulkes and Blanck, 1990). Furthermore, cadmium uptake from Cd-ME is inhibited by zinc, is not subsequently released from the kidney, and is not available to EDTA chelation, indicating that it is intracellular. During transient occlusion, the kidney cortex can be loaded with up to 25 μg Cd/g with no immediate dysfunction! By 1–2 days, a decrease in renal reabsorption of amino acids occurs in response to this mode of cadmium exposure.

2.2.3.3 *Retention and excretion of cadmium by the kidney*

Cadmium is retained by the kidney with a long biological half-life (Friberg *et al.*, 1986a). Cd-MT in plasma that is not reabsorbed in the renal tubules from the glomerular filtrate is excreted intact in urine. Once the renal tubules are damaged by cadmium, urinary cadmium concentrations increase, reabsorption of Cd-MT decreases, and kidney cadmium concentrations fall. The appearance of an abrupt increase in cadmium concentrations in urine is one sign of renal damage from cadmium exposure (Friberg *et al.*, 1986b).

2.2.4 *Liver uptake and transport*

The liver plays a very active and important role in rapidly clearing cadmium ions from the blood. Frazier (1980, 1984) demonstrated that 90% of cadmium intravenously injected as $CdCl_2$ cleared from the plasma within 15 min independent of dose in the range 0.15–1.5 mg/kg. For doses up to 0.6 mg/kg, 50% cleared to tissues by 1–2 min. More than half of the cadmium was deposited in liver, with a liver : plasma concentration ratio of 10–20 : 1. This rapid clearance of cadmium ions from blood to liver serves as an important protective pathway for the body, sparing organs such as kidney and bone from continued exposure to cadmium.

2.2.4.1 *Hepatic uptake of $CdCl_2$*

Recognizing the importance of rapid liver uptake of cadmium ions, a number of investigators have studied the biochemical properties of the liver uptake pathways (reviewed by Shaikh *et al.*, 1995) using primary cultures of rat hepatocytes (Gerson and Shaikh, 1984; Blazka and Shaikh, 1991, 1992) and a human fetal hepatic cell line, WRL-68 (Souza *et al.*, 1996, 1997). Consistent with the *in vivo* data, cadmium ions accumulate with an initial rapid phase lasting about 2 min followed by a sustained slower phase of about 30 min. Blazka and Shaikh (1992) demonstrated that hepatocyte uptake of cadmium ions in the primary cell culture system occurs mainly through the SH-containing transport processes associated with the uptake of zinc. At 3 μM cadmium, this pathway is temperature dependent (80%), inhibited by mercurials (80%), inhibited by zinc (75%), and independent of metabolic inhibitors. Inhibition by other metals, such as copper and iron is much less than that by zinc. During the first 30 min of accumulation, 13% of the cadmium in the hepatocytes is bound to the plasma membrane. Blazka and Shaikh (1991) also showed that, at 3 μM cadmium, 30% of cadmium ion uptake by hepatocytes is blocked by 250 μM verapamil or diltiazem, indicating the participation of calcium ion channels. Participation of these channels appeared to decrease with decreasing cadmium concentration, with no significant effect of calcium-channel blockers at 0.03 μM cadmium. Relating these cadmium concentrations to plasma concentrations *in vivo*, the plasma cadmium concentration at zero time is ~30 μM (~3 μg/ml) after an intravenous dose of 0.15 mg Cd/kg as $CdCl_2$ to rats (Frazier, 1980). This concentration is encountered by the liver at the start of its rapid uptake phase from blood after acute exposure to cadmium ions.

2.2.4.2 *Hepatic uptake of Cd-MT*

Realizing the importance of Cd-MT as a complex that protects cells from cadmium toxicity, investigators began to inquire about the disposition and toxicity of Cd-MT after its injection into the bloodstream (Nordberg *et al.*, 1975; Cherian, 1977). Their finding was surprising in that the Cd-MT complex had almost no affinity for the liver, instead was deposited rapidly and nearly entirely in kidney. The rapid, extensive uptake of cadmium ions by the liver is mirrored exactly by rapid, extensive uptake of Cd-MT by the kidney.

2.2.4.3 *Hepatic retention and excretion of cadmium*

Cadmium is retained by the liver with a long biological half-life (Friberg, 1986a). At environmental levels of exposure, intracellular hepatic cadmium is associated with multiple compartments, including nuclei, high molecular weight proteins and metallothionein, and is not significantly excreted in bile. At higher concentrations, cadmium is excreted from the liver via bile (Cherian and Vostal, 1977; Klaassen and Kotsonis, 1977). Recent progress has demonstrated that this excretion utilizes the pathway for biliary excretion of GSH. First, Gregus and Varga (1985) demonstrated that hepatic GSH depletion decreased biliary cadmium excretion by 97%. Since then, Buchler *et al.* (1996) have cloned the cDNA of the hepatocyte canalicular isoform of the multidrug resistance protein, cMrp. This protein is related to the yeast cadmium factor (YCF1) and is not expressed in the liver of two mutant rat strains, the Eisai hyperbilirubinemic (EHB) rat and the GY/TR-mutant. The EHB rats are characterized by essentially lacking biliary excretion of GSH due to an impaired ATP-driven organic anion transport system (Sugawara *et al.*, 1996, 1997). After cadmium was injected intravenously at 0.1 mg/kg, cadmium excretion in bile reached a maximum of 73.2 ng/15 min in control rats and only 1.8 ng/15 min in EHB rats, while liver cadmium concentrations were slightly higher in EHB rats. These results demonstrate that biliary cadmium excretion occurs via GSH complexation and excretion via canalicular anion transport in rats. It is interesting that this pathway was active for copper excretion but not for zinc excretion through bile (Sugawara *et al.*, 1996).

A second pathway for release of cadmium from liver involves the liver as a continuous source of Cd-MT to the kidney (Jonah and Bhattacharyya, 1989; Chan *et al.*, 1992, 1993). Results have shown that, after parenteral exposure, cadmium deposits at high concentrations in the liver and induces high levels of hepatic MT; Cd-MT in the liver is released to the circulation, deposits in the kidneys, and eventually causes kidney damage (Chan and Cherian, 1993). Some investigators have shown that this liver-to-kidney transfer pathway is enhanced in rats and mice during pregnancy and lactation, even at low levels of cadmium exposure (Bhattacharyya *et al.*, 1982; Chan and Cherian, 1993). Chan *et al.* (1993) provided a dramatic demonstration of this pathway in an experiment in which they transplanted a cadmium-loaded liver into a naive rat and showed high plasma concentrations of Cd-MT followed by dramatic increases in kidney cadmium concentrations and ultimately renal damage. The transplanted livers decreased in cadmium concentration from about 200 to 100 μg/g, plasma Cd-MT concentration rose to 100 μg Cd/L and levels of cadmium in damaged kidneys rose with time to ~200 μg Cd/g. This kidney cadmium concentration is considered by many to be the critical concentration at which renal tubular dysfunction starts in humans continuously exposed to cadmium in an occupational or environmental setting (Friberg, 1986b).

2.2.5 Bone uptake and transport

Bone is a target organ for cadmium toxicity. A single oral dose of cadmium (200 μg) can induce a burst of bone resorption that begins within 24 hours and lasts for approximately 3 days (Wilson and Bhattacharyya, 1997). Lower concentrations of cadmium consumed daily in drinking water (15 μg/L) also cause a transient increase in ^{45}Ca released from bone over a time course of 2–3 weeks (Sacco-Gibson *et al.*, 1992). The time course suggests that cadmium interacts with the bone to cause an upregulation of the resorptive processes that occur during normal cycles of bone remodeling.

2.2.5.1 Cadmium accumulation in bone

Although studies of industrial or environmental exposure to cadmium have demonstrated a slight accumulation of cadmium in the skeleton (Knuuttila *et al.*, 1982; Noda and Kitagawa, 1990; Honda *et al.*, 1997), cadmium does not accumulate to any major extent in bone as does another toxic heavy metal, lead. In *in vitro* studies, cadmium can adsorb onto and incorporate into hydroxyapatite crystals in bone (Christoffersen *et al.*, 1988; Blumenthal *et al.*, 1995). These interactions may change the kinetics of hydroxyapatite formation and dissolution *in vivo*. Young rats that received 10 p.p.m.

cadmium in drinking water from the time of weaning until 5 months later, accumulated 110 ng Cd/g dry weight of bone (Ogoshi *et al.*, 1992). In old rats that did not exhibit the increased remodeling of bones due to growth, 40 p.p.m. cadmium for 7 months deposited 210 ng Cd/g dry weight bone. The long-term (5–7 months) exposure of both age groups caused a lesional effect on bone strength apparent only at the distal end of the femur. In contrast, growing and remodeling bone in young rats is much more susceptible to deficits in mechanical properties, showing a clear response to short-term (1 month) cadmium exposure (Ogoshi *et al.*, 1989).

2.2.5.2 Cadmium uptake into bone cells

Because cadmium is quickly cleared from the blood and does not accumulate to an appreciable extent in the bone matrix, it may interact directly with bone cells to increase bone resorption and/or decrease bone formation, ultimately leading to bone loss. ^{109}Cd injected into rats localizes preferentially in osteoblasts (Bawden and Hammarstrom, 1975). As in many other cell systems, cell–cell interactions between osteoblasts (bone-forming cells) and osteoclasts (bone-resorbing cells) are required for tight coupling of routine bone remodeling. Therefore, osteoblasts may take up cadmium directly and secrete substances that stimulate osteoclastic activity, precluding the necessity for cadmium to be taken up by osteoclasts to induce bone resorption. In fact, it is not known whether cadmium uptake into bone cells is required for its effects on bone, or whether cadmium can bind to extracellular receptors that modulate its effects by triggering intracellular message systems. However, the bone response to cadmium is transient (Sacco-Gibson *et al.*, 1992; Wilson and Bhattacharyya, 1997) and involves an increase in osteoclastic resorptive activity (Wilson *et al.*, 1996) rather than a diminution of a cellular process. Therefore, cadmium may be only necessary as the initial stimulus for a series of intracellular interactions that lead to bone loss.

2.3 Mechanisms of cellular toxicity

2.3.1 Intestinal toxicity

Toxic effects of orally administered low levels of cadmium result from a reduced uptake of essential metals by the intestinal cells. For example, oral cadmium administration decreases iron levels significantly in intestinal mucosa (Elsenhans *et al.*, 1994), resulting in decreased hemoglobin concentration in young rats (Schumann *et al.*, 1996). Dams exposed to cadmium show a decreased intestinal uptake of copper and iron, resulting in the deficiency of these metals in their fetuses (Chmielnicka and Sowa, 1996). At high, cytotoxic concentrations of cadmium, the pathological symptoms in the duodena are severe. Rats treated with a 200 p.p.m. oral dose of cadmium showed enlarged duodena, and there was a significant reduction in the percentage of crypts containing Paneth cells, which were also vacuolated. The epithelial cells covering the villi were swollen and protruded towards the lumenal side (Phillpotts, 1986).

2.3.2 Kidney toxicity

The kidney has long been recognized as a critical target organ for cadmium toxicity. Renal dysfunction in the form of proteinuria, aminoaciduria, glucosuria and phosphaturia has clearly been demonstrated in persons occupationally exposed to cadmium (Friberg *et al.*, 1986b). These results provided the first rational basis for effective restriction of cadmium exposure levels in the workplace. Many investigations have since been directed toward understanding the biochemical and molecular interactions that underlie cadmium-induced renal damage. Here the focus will be on cadmium effects on renal transport pathways, the role of MT in cadmium-induced renal damage, and the influence of GSH and oxidative damage on cadmium-induced renal toxicity.

2.3.2.1 Effects of cadmium on renal transport pathways

Cadmium exposure in animals and humans causes increased urinary excretion of multiple molecular species, including glucose, amino acids, low molecular weight proteins, phosphate, and sometimes calcium (Friberg *et al.*, 1986b). Because the kidney is a critical organ for cadmium toxicity, many investigators have attempted to identify the mechanisms by which cadmium affects renal proximal tubular cell transport pathways, particularly prior to effects on overall cell viability. For example, regarding glucosuria, transport studies utilizing the nonmetabolizable glucose analog, α-methyl-glucoside, have been conducted in pig kidney LLC-PK1 cells and mouse renal cortical tubule cells in primary culture. Results demonstrate that cadmium (0–6 μM $CdCl_2$ for 24 h) inhibits Na^+-dependent glucose transport by up to 50% in primary cell cultures (Blumenthal *et al.*, 1990). No decrease is seen before 3 h, indicating that the effect of cadmium is most likely to be indirect (Blumenthal *et al.*, 1994). Both Blumenthal *et al.* (1996) and Kim and Park (1995) have demonstrated a cadmium-induced decrease in the number of BBMV binding sites for [^{3}H]phlorizin, a competitive inhibitor of glucose for the Na^+–glucose cotransporter. In the primary cell culture system, the effects of cadmium occur with no change in the cells' vital signs, including the ATP/ADP ratio, cellular Na^+ and K^+ concentrations, protein and DNA synthesis, glucose permeability and amino acid transport (Blumenthal *et al.*, 1990). Finally, cadmium decreases the levels of SGLT1 mRNA, the mRNA for the high-affinity isoform of the sodium–glucose cotransporter (Blumenthal *et al.*, 1998) in a dose-dependent manner (0–10 μM). In general, the fall in SGLT1 mRNA precedes the decline in glucose uptake.

Taken together, these results support the hypothesis that cadmium causes glucosuria by decreasing the synthesis of a Na^+–glucose cotransporter in the proximal tubular brush-border membrane, decreasing the abundance of sites for glucose reabsorption from the glomerular filtrate. Adding another dimension to the hypothesis, the results of Kim, Lee and Park suggest that cadmium may also cause a loss of active Na^+–glucose cotransporter sites by direct interaction of the BBMVs with free cadmium (Kim and Park, 1995; Lee *et al.*, 1991). However, the latter direct effect occurs at a much higher cadmium concentration (50 μM incubation for 30 min).

In a second model system, rabbits and rats administered $CdCl_2$ subcutaneously at 2 mg/kg/day developed a marked polyuria and hyposthenuria after 2–3 weeks of daily exposure (Kim *et al.*, 1988, 1990; Lee *et al.*, 1990; Condron *et al.*, 1994; Ahn and Park, 1995). These changes are accompanied *in vivo* by increases in urinary excretion of protein, glucose, urea, calcium, phosphate, chloride and potassium, along with a 19% decrease in surface density of microvillus membrane per unit PTC cell volume. Isolated BBMV preparations from treated rabbits show marked decreases in Na^+-dependent transport of a number of substances, including glucose, amino acids and phosphate. In addition, basolateral membrane vesicles show decreased transport of organic anions (Kim *et al.*, 1988; Lee *et al.*, 1990). This *in vivo* model therefore appears to mimic cadmium-induced renal dysfunction in humans.

With respect to mechanism, for each substance, the transport decrease in the isolated membrane vesicle system can be attributed to a significant decrease in V_{max} with no change in K_m. These results indicate that cadmium may simultaneously, for a number of substances, reduce the effective number of cotransporter units in the renal cortical membranes.

The decreased transport seen in BBMVs prepared from cadmium-treated animals can be reproduced by preincubation of BBMVs from normal animals with 50 μM cadmium for 30–60 min (Kim *et al.*, 1990; Kinne *et al.*, 1995; Park *et al.*, 1997). As in the membranes prepared after *in vivo* treatment, *in vitro*-exposed membranes show a decrease in V_{max} with no change in K_m. In addition, treatment with EDTA after cadmium pretreatment cannot reverse the transport changes, indicating that cadmium has entered the vesicles.

In a model developed by Foulkes, rabbits are exposed to a single pulse of Cd-ME via an intravenous bolus (1 μmol cadmium; 40 μg/kg), immediately followed by a 40-second aortic artery occlusion to provide a short period of renal exposure mainly via the basolateral side of the cells (Foulkes, 1989; Foulkes and Blanck, 1990). In this model, amino acid and organic anion transport remain unchanged for 5–7 h but are significantly decreased by 2–3 days. These results lend support to the hypothesis that the effect of cadmium on renal tubular reabsorption is indirect.

The results of the above studies indicate that:

1 the cadmium-induced polyuria and hyposthenuria after daily subcutaneous cadmium administration to rabbits and rats is an osmotic diuresis induced by proximal tubular rejection of many substances; and
2 the mechanism of impaired renal excretion in cadmium-treated animals is a loss of active cotransporter units in the proximal tubular membranes.

The question remains whether humans exposed over a longer period of time to lower levels of cadmium develop the type of renal tubular dysfunction produced in these short-term animal models.

2.3.2.2 Role of MT in renal effects of cadmium

Induction of MT protein in kidney cells has been shown by many investigators to protect against the renal toxicity of subsequent exposures to cadmium (Friberg *et al.*, 1986b). This finding is shown clearly by the reduced renal toxicity of cadmium in MT transgenic mice that have elevated endogenous levels of MT protein in many organs, including kidney.

In contrast, most current hypotheses prescribe that Cd-MT is a major form of cadmium responsible for cadmium-induced renal damage. This concept arose from the surprising finding that Cd-MT parenterally administered to animals deposits rapidly and exclusively in the kidney and is extremely renal toxic, causing increased urine concentrations of a number of small molecules typical of the human renal disorder caused by cadmium (Nordberg *et al.*, 1975; Cherian, 1977). This important discovery led to a number of questions with respect to the role of Cd-MT in cadmium-induced renal dysfunction:

1 What is the source of Cd-MT to the kidney?
2 By what mechanism does Cd-MT exert its action?
3 Are there substances that protect against Cd-MT renal toxicity?
4 Is Cd-MT the major renal toxic substance, or are there other important renal toxic forms of cadmium?

As described in Sections 2.2.1.3 and 2.2.4.3, both liver and small intestine appear to be important sources of Cd-MT to the kidney. After oral exposure, which is a major route for environmental cadmium exposure in humans, the small intestine serves as a larger source for cadmium transfer to kidney than the liver (Jonah and Bhattacharyya, 1989), and an important form of cadmium moving from intestine to kidney has been postulated by many to be Cd-MT (Nordberg and Nordberg, 1992). This intestine-to-kidney transfer of Cd-MT may explain the puzzle of why kidney damage occurs after long-term dietary cadmium exposure, while hepatic damage predominates after acute high-level parenteral exposure. With respect to mechanism, the rapid lysosomal degradation of MT protein in kidney cells and the local release of high concentrations of cadmium ions are hypothesized to play a central role in the acute renal toxicity of Cd-MT (Bernard *et al.*, 1987; Jin *et al.*, 1987; Dorian *et al.*, 1992a; Vestergaard and Shaikh, 1994). Vestergaard and Shaikh (1994) demonstrated that Cd-MT renal toxicity decreases significantly when its rate of delivery to the circulation is decreased. These results raise the question of whether or when Cd-MT would cause renal damage after its very slow delivery to plasma at levels typical of environmental exposures. One difficulty encountered by investigators trying to study the mechanism of Cd-MT renal toxicity is the finding that cultured renal PTCs do not retain their rapid uptake and exquisite sensitivity to Cd-MT toxicity *in vitro* (Liu *et al.*, 1994a). With respect to protector molecules, acute uptake of Cd-MT by the kidney is blocked by low molecular weight proteins, including $\beta_2\mu G$ (Bernard *et al.*, 1987). In addition, pretreatment by zinc in both MT-normal and MT-knockout mice protects against an acute dose of Cd-MT (Y.P. Liu *et al.*, 1996). This protection in MT-knockout mice indicates that induction of MT protein is not the only way for zinc to protect against Cd-MT toxicity. Experiments with MT-knockout mice have also provided insight into the forms of cadmium that transit from intestine to kidney. For example, recent results demonstrate that cadmium uptake by the kidney early after intestinal absorption is not dependent on the MT status of the mouse (Rajan *et al.*, 1999). These results demonstrate that a form other than Cd-MT must transit from intestine to kidney. The identity of this form and its renal toxicity are not known at this time.

2.3.2.3 Role of GSH and oxidative damage

The hypothesis has developed that GSH protects renal cells from cadmium toxicity early after cellular uptake of cadmium, prior to its complexation by endogenous or induced molecules of metallothionein. GSH is present in kidney cells at millimolar concentrations, is a small molecule that is highly diffusible in cells, and forms a stable complex with cadmium via its SH groups. Vestergaard and Shaikh (1994), for example, demonstrated that rats are significantly more sensitive to the toxicity of $CdCl_2$ and Cd-MT after depletion of cellular stores of GSH. More recently, investigators have begun to focus on an oxidative stress response to cadmium ions and alterations in glutathione status of cells after cadmium exposure (for example, Immenschuh *et al.*, 1995; Ossola and Tomaro 1995; Bagchi *et al.*, 1996, 1997a; Brennan and Schiestl, 1996; Hassoun and Stohs, 1996; Gong and Hart, 1997; Figueiredo-Pereira *et al.*, 1998; Hultberg *et al.*, 1998; Sarkar *et al.*, 1998). In some cell systems, the oxidative stress response to cadmium occurs at low concentrations, such as 0.4–0.6 μM (Hassoun and Stohs, 1996). In addition, Hultberg *et al.* (1998) have shown an increased cellular concentration of GSH and an increased release of GSH into the medium in HeLa cells exposed to 1 μM Cd. Chapter 13 addresses in detail the role of glutathione in metal metabolism, transport and toxicity.

2.3.3 *Liver toxicity*

2.3.3.1 *Toxic effects of $CdCl_2$ in liver cells*

$CdCl_2$ administration by subcutaneous, intraperitoneal, or intravenous routes to experimental animals causes rapid accumulation of cadmium by the liver and acute damage to hepatocytes. Classically, hepatocyte damage both *in vivo* and in hepatocyte cultures is identified by cellular release of liver enzymes and K^+ and ultrastructural changes in mitochondria and endoplasmic reticulum (for example, Groten *et al.*, 1990; Lui *et al.*, 1992, 1994b; Bucio *et al.*, 1995). Recently, however, additional end points have been identified under acute exposure conditions, including induction of genes associated with carcinogenesis (c-*jun*, *p53*) (Zheng *et al.*, 1996), changes in GSH status and oxidative damage (Bucio *et al.*, 1995; Chan and Cherian, 1992), Kupffer cell activation (Sauer *et al.*, 1997a), endothelial cell damage (Liu *et al.*, 1992) and cytoplasmic acidification leading to decreased mitochondrial respiration (Koizumi *et al.*, 1995). Exploring the role of nonparenchymal cells, Sauer *et al.* (1997a) showed that Kupffer cell activation by cadmium enhances the hepatic inflammation response to cadmium, while blocking Kupffer cell function by administering $GdCl_3$ blocks cadmium-induced changes in hepatocyte morphology and the cellular release of alanine aminotransferase. Liu *et al.* (1992) demonstrated that hepatic endothelial cells reflect the differences in sensitivity of two mouse strains to cadmium hepatotoxicity while the parenchymal cells do not. They observed that, during rapid uptake of cadmium by the liver, the first cells encountered are the vascular endothelial cells, which may be damaged before the parenchymal cells. Finally, cytoplasmic acidification is a potentially important early response of hepatocytes to cadmium (Koizumi *et al.*, 1995). Koizumi demonstrated that intracellular acidification from pH 6.8 to 6.6 occurs within the first 10 min of hepatocyte exposure to 5 μM $CdCl_2$. This acidification appears to involve an action of cadmium on HCO_3^-/Cl^- exchange, because it requires medium HCO_3^- and is totally blocked by probenecid, an inhibitor of HCO_3^-/Cl^- exchange. Similar acidification in the absence of cadmium causes a significant (70%) decrease in hepatic mitochondrial respiration *in vitro*, and may explain early cadmium-induced damage to hepatic mitochondria.

Taken together, the above results demonstrate that liver cells show many toxic responses to cadmium under acute exposure conditions *in vivo* and in cell culture.

However, many additional investigations are needed before we understand the extent to which these acute hepatotoxic responses are applicable to humans who are environmentally or occupationally exposed to cadmium.

2.3.3.2 Role of MT in hepatotoxicity of cadmium

Metallothionein is an important hepatic protector molecule in cadmium-exposed animals and cells. MT-null mice, in which the genes for MT-I and MT-II are knocked out, provide the first demonstration that cadmium hepatotoxicity is significantly enhanced in the total absence of hepatic MT (Michalska and Choo, 1993; Masters *et al.*, 1994). One surprise is that MT-null mice are not exquisitely sensitive to cadmium; the hepatotoxicity enhancement factor is only two- to four- fold (Michalska and Choo, 1993; Masters *et al.*, 1994; J. Liu *et al.*, 1996a; Zheng *et al.*, 1996; Conrad *et al.*, 1997). The latter results indicate that the liver has defense mechanisms that come into play in the absence of MT. In another system, Koropatnick and Cherian (1993) identified toxic milk (tx) mice as an interesting model in which MT protects against hepatotoxic responses to cadmium. These mutant mice have hepatic MT concentrations that are 100 times the wild-type levels due to an 80% longer half-life for this protein.

2.3.3.3 Non-MT factors affecting cadmium hepatotoxicity

During the past 5–10 years, many very diverse factors have been identified that significantly modify the hepatotoxicity of cadmium. Table 2.1 provides an overview of a number of these factors and their effects. They include hormones, vitamins, Chinese herbal medicines, sulfhydryl compounds and metals. Each of these factors provides a clue as to the mechanism of cadmium-induced hepatotoxicity, and potentially to its prevention.

Table 2.1 Factors that influence hepatotoxicity of cadmium

Factor	*Increase*	*Decrease*	*Reference*
Vitamin A		X	Sauer *et al.*, (1997b)
Vitamin B_{12}		X	Couce *et al.* (1991)
Ascorbic acid		X	Shukla and Chandra (1989)
Arsenic		X	Hochadel and Waalkes (1997)
Testosterone		X	Shimada *et al.* (1997a)
Progesterone	X		Shiraishi *et al.* (1993); Shimada *et al.*, (1997b)
Heat shock proteins		X	Salminen *et al.* (1996)
Fruit extract		X	Ip *et al* (1996)
Chinese herbal medicines		X	Liu *et al.* (1994b)
Glutathione		X	Chan and Cherian (1992)
Lipoic acid		X	Sumathi *et al.* (1996)
Oleanic acid		X	Liu *et al.* (1995)

2.3.4 *Bone toxicity*

2.3.4.1 *Background*

The skeleton contains more than 99% of the calcium in the body and thus is an important source of calcium for the maintenance of calcium homeostasis when the diet is not sufficient. Factors that regulate the tight coupling of bone resorption and formation in routine remodeling of bone are thus linked to calcium homeostasis. Two hormones, vitamin D and parathyroid hormone (PTH), are the major regulators of calcium homeostasis. Vitamin D stimulates intestinal uptake of calcium and, together with PTH, stimulates osteoblast-mediated bone resorption by osteoclasts. The actions of these hormones are integrally connected to the kidney. PTH induces hydroxylation of 25(OH)-vitamin D to its active form, $1{,}25(OH)_2$-vitamin D, in the kidney. Cadmium-induced renal dysfunction can cause a decrease in the amount of $1{,}25(OH)_2$-vitamin D produced and impair renal calcium reabsorption mechanisms, which impact negatively on calcium homeostasis. Less calcium is absorbed from the diet, less calcium is reabsorbed by the kidney and bone is resorbed to make up for any imbalance. Thus, unchecked renal dysfunction can lead to substantial bone loss and osteomalacia.

Perhaps the best-studied example of cadmium toxicity that impacts on bone is itai-itai ('ouch-ouch') disease (Takeuchi, 1978; Lauwerys, 1979; Nogawa, 1981; Kjellstrom, 1986, 1992; Kido *et al.*, 1989; Noda and Kitagawa, 1990; Tsuritani *et al.*, 1994). Residents of the Jinzu River Basin in the Toyama Prefecture of Japan during the latter part of the Second World War were exposed to cadmium and other pollutants when they used water from the Jinzu River, downstream from a lead and zinc mine, for drinking and rice paddy irrigation. Patients with itai-itai disease experienced bone pain and an increased incidence of fractures, secondary to renal dysfunction. Prolonged cadmium exposure caused proximal renal tubule damage that resulted in inadequate tubular reabsorption of calcium and phosphorus and impaired vitamin D metabolism. Although cadmium was a causal factor of the disease, a number of confounding factors became apparent during the study of the etiology of itai-itai disease. Malnutrition, which was prevalent during the postwar years in Japan, contributed to the severity of the disease. In many cases, high doses of vitamin D were used successfully as a treatment for malnutrition and bone problems. In addition, greater than 95% of itai-itai cases were postmenopausal women with a history of multiple childbirths, although all segments of the population were exposed to cadmium and nutrient deficiencies. The bones of women affected by itai-itai disease may have been more susceptible to cadmium-induced bone loss because of the repeated mobilization of calcium from their skeletons for prenatal ossification and for milk during lactation. The increased turnover of bone during menopause, in combination with cadmium exposure and malnutrition, also added to the stress on the bone that resulted in fracture and osteomalacia.

Although itai-itai patients clearly suffered from cadmium-induced renal damage in addition to bone loss, cadmium can also exert its effects on bone directly, independent of renal toxicity. Bone appears to be a target organ for cadmium damage at lower exposure levels and in a shorter timeframe than found in renal-induced bone loss. A threshold level of approximately 200 μg cadmium per gram of cortex causes renal dysfunction (Friberg *et al.*, 1986b). In studies where the skeletons of mice (Bhattacharyya *et al.*, 1988, 1992, 1993), rats (Ando *et al.*, 1978 and dogs (Sacco-

Gibson *et al.*, 1992) were prelabeled with ^{45}Ca, cadmium induced ^{45}Ca excretion into the feces within 72 hours. Subsequent studies using mice fed a low calcium diet measured stable calcium excretion in feces, and this calcium was demonstrated to be released from bone (Wilson and Bhattacharyya, 1997; Bhattacharyya *et al.*, 1998). These studies indicated that calcium is mobilized from the bone pool starting 8–16 hours after oral cadmium exposure and continuing for several days. A second cadmium dose administered 7 days later also induced calcium release similar in magnitude and duration to the first exposure. The circulating cadmium levels at which a bone response first occurs in mice (Wilson and Bhattacharyya, 1997) and dogs (Sacco-Gibson *et al.*, 1992) is 2–5 μg cadmium/L blood. These levels are comparable to the levels in smokers in the US (2 μg/L, 20 cigarettes or more/day; Elinder *et al.*, 1983) and occupational levels (0.31–22.2 μg/L; Hassler *et al.*, 1983; Chia *et al.*, 1989; Thun *et al.*, 1989; Kawada *et al.*, 1990).

2.3.4.2 *Effects of metallothionein on bone*

MT is a major cadmium-sequestering protein in the kidney and can be induced by cadmium in almost all tissues of the body, including the bone marrow (Tamura and Ohya, 1995). MT is also found in bone in calcifying growth-plate cartilage chondrocytes (Kaji *et al.*, 1986; Litchfield and Sauer, 1996) and is induced by 25-fold by concentrations of cadmium of 0.1 μM or more (Sauer *et al.*, 1998). Although there is no *in vivo* evidence that MT is induced by cadmium in bone cells, MT can be induced by cadmium in osteoblastic cell lines from mouse calvaria (MC3T3-E1; Miyahara *et al.*, 1986) and human osteosarcoma (HOS; Angle *et al.*, 1993) but not from rat osteosarcoma (ROS17/2.8; Angle *et al.*, 1990, 1993; Thomas *et al.*, 1990). The lack of MT induction in ROS 17/2.8 cells cultured in serum-free media by up to 1 μM cadmium exposure was suggested to explain the increased sensitivity of these cells to cadmium. In an *in vivo* study using MT-deficient mice, a single oral cadmium dose (200 μM) induced calcium release from bone (Bhattacharyya *et al.*, 1998) that was twofold higher than for MT-normal mice. The transient nature of the bone response was similar between MT-deficient and MT-normal mice (Fig. 2.3). These data suggest that metallothionein does confer a small amount of protection to the bone against cadmium. However, when cadmium was administered a second time 1 week later, the bone response was similar in duration and magnitude to the first, suggesting that any metallothionein induced in bone cells does not protect against a second exposure to cadmium, as is the case for liver and kidney (Fig. 2.3). Because oral cadmium

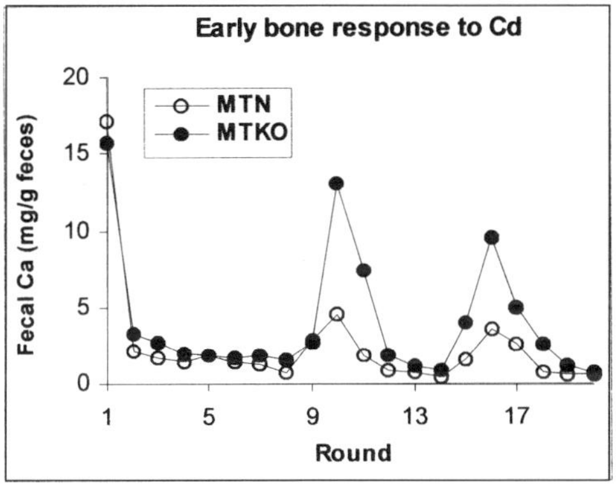

Figure 2.3 Effect of MT status on early bone response to cadmium. The mice received two gavage doses (200 μg as $CdCl_2$) one just prior to collection round 8 (R8) and one just before R14. The duration of each fecal collection round was 24 h except for the following: 72 h, R1–R4; 8 h, R8 and R14; R9 and R15; 48 h, R12 and R20.

exposure can induce metallothionein at several points along the way to bone, such as in the intestine and liver, metallothionein protection may be inferred upstream of the bone cells. Indeed, intestine, liver, and kidney MT levels were elevated 4 days after the last cadmium exposure. However, blood cadmium concentrations 72 hours after each cadmium exposure were not different between MT-deficient and MT-normal mice, suggesting a regulatory point more proximal to bone.

2.3.4.3 *Effects of cadmium on apoptosis*

Apoptosis is a mechanism whereby a stimulus specifically induces cell death in contrast to causing necrosis. Cadmium has been demonstrated to cause apoptosis in liver (Habeebu *et al.*, 1998), testes (Jones *et al.*, 1997 and kidney (Hamada *et al.*, 1991, 1996), all sites of cadmium toxicity. However, cadmium has also been demonstrated to protect against apoptosis in CHO cells when induced by a second heavy metal, chromium (Shimada *et al.*, 1998). This seemingly incongruous finding may help explain the carcinogenicity of cadmium.

In bone, apoptosis may be functioning regularly to restore basal levels of bone cells after the remodeling cycle (Weinstein *et al.*, 1998). However, apoptosis may also be a mechanism by which bone remodeling is uncoupled to cause a net increase or decrease in bone density by selectively decreasing numbers of osteoclasts or osteoblasts. Estrogen, long known as a bone protectant, causes apoptosis of osteoclasts, thus inhibiting bone resorption (Kameda *et al.*, 1997). Estrogen-induced apoptosis is mediated by transforming growth factor-β (TGF-β) in osteoclasts (Hughes *et al.*, 1996). When estrogen is withdrawn using a gonadotropin-releasing hormone (GnRH) analog, the number of osteocytes decreases via apoptosis in human bone (Tomkinson *et al.*, 1997). Because osteocyte viability is thought to be essential for the bone to signal for repair of microdamage and to modulate the effects of mechanical strain, estrogen deficiency could lead to increased bone fragility, even in the absence of net bone loss. Bisphosphonates, an effective treatment for osteoporosis, induce apoptosis in osteoclasts (Hughes *et al.*, 1995; Parfitt *et al.*, 1996; Selander *et al.*, 1996; Luckman *et al.*, 1998), often with the effect of increasing bone mineral density. In glucocorticoid-induced bone disease, characterized by decreased bone formation and osteonecrosis, apoptosis of osteoblasts contributes to decreased bone formation, and apoptosis of osteocytes may contribute to the osteonecrosis (Weinstein *et al.*, 1998). NF-$\kappa\beta$ and tumor necrosis factor-α (TNF-α) mediate apoptosis in osteoblasts (Kitajima *et al.*, 1996).

2.3.4.4 *Cadmium and bone resorption mechanisms*

It is well established that cadmium causes bone loss due to direct action on the bone. For example, in bone organ cultures, 10 nM cadmium caused dramatic demineralization of fetal rat limb bones (Bhattacharyya *et al.*, 1988) and a smaller, but still significant effect was observed at cadmium concentrations of 0.3–50 μM (Miyahara *et al.*, 1980a,b). In intramembranous bone cultures (calvarial or parietal), cadmium at 250–500 nM, but not at 100 nM, caused a two- to threefold increase in the release of ^{45}Ca from prelabeled bones (Suzuki *et al.*, 1989, 1990; Miyahara *et al.*, 1992).

Net bone loss is a result of the uncoupling of bone resorption due to osteoclast activity from bone formation due to osteoblastic activity during normal or upregulated bone turnover. Thus, increased bone resorption and/or decreased bone formation can result in a net loss of bone. Osteoclastic activity involves the fusion of osteoclast precursors to form the typical multinuclear morphology and a change in the cytoskeleton from a dynamic motile configuration into a stationary configuration that creates a ring of tight sealing between the osteoclast and bone. Reactions involved in cadmium-induced bone resorption are shown in Fig. 2.4. Cadmium at 10–100 nM has been demonstrated to stimulate the formation of osteoclast-like multinuclear cells from progenitors in bone

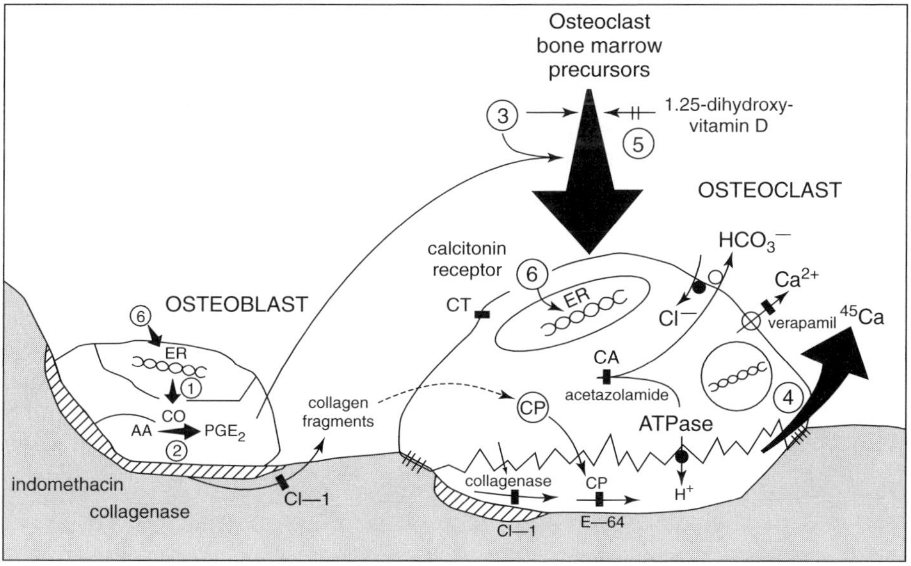

Figure 2.4 Schematic representation of possible mechanisms of cadmium-induced bone resorption and sites of inhibitor actions. Cadmium may cause the osteoblast (left) to interact with the osteoclast (right) to increase bone resorption. The shaded area is mineralized bone; the hatched area is osteoid matrix. Heavy arrows show actions of cadmium. Circled numbers refer to: (1) cadmium-induced increase in osteoblast cyclo-oxygenase (CO) protein and enzyme activity (250–500 nM cadmium), leading to (2) increased osteoblast synthesis of prostaglandin E_2 (PGE_2) from arachidonic acid (AA) (250–500 nM cadmium); (3) cadmium alone (10–90 nM cadmium) or via PGE_2 stimulation (250–500 nM), causes increased formation of osteoclasts from bone marrow precursors, leading to (4) increased osteoclast-mediated release of bone ^{45}Ca (10–500 nM cadmium). Sites of action of pharmacological inhibitors are indicated by heavy bars and are based on data obtained entirely in intramembranous bone systems (calvarial and parietal). Cadmium-induced increases in bone resorption are totally blocked by (left to right): indomethacin, an agent that blocks PGE_2 synthesis, and any agent that blocks the bone-resorbing action of the osteoclasts, such as collagenase inhibitor (CI-1), cysteine proteinase (CP) inhibitor (E-64), carbonic anhydrase (CA) inhibitor (acetazolamide), calcitonin (CT), and the calcium-channel blocker verapamil. Note that cadmium can stimulate formation of osteoclasts directly at concentrations lower than those required for induction of PGE_2. Also, CI-1 inhibits both the collagenase secreted by osteoblasts, which prepares the osteoid matrix for osteoclast activation, and the collagenase secreted by osteoclasts, which degrades collagen matrix in the resorption lacuna. As shown by the upper blocked arrow (5), one report indicates that cadmium itself can inhibit 1,25-dihydroxyvitamin D-stimulated osteoclast formation from bone marrow precursors (≥ 100 nM cadmium).

marrow cultures (Konz *et al.*, 1989; Bhattacharyya, 1991; Miyahara *et al.*, 1991; Wilson *et al.*, 1996) and to increase the numbers of multinuclear osteoclasts from the fusion of mononuclear precursors in osteoclast cultures isolated from neonatal rat bones (Wilson *et al.*, 1996). In contrast, cadmium at 100 nM has also been demonstrated to decrease osteoclast formation when stimulated by 1,25$(OH)_2$-vitamin D (Iwami and Moriyama, 1993), suggesting that cadmium may act via different regulatory pathways in the differentiation of osteoclasts in the presence or absence of osteotropic hormones. An increase in osteoclastic activity also requires the upregulation of proton pumps for acidification and solubilization of the bone mineral, and the upregulation of lysozomal proteinases for collagen matrix degradation. Cadmium-induced bone resorption in endochondral cultures is decreased by a collagenase inhibitor (CI-1), a cysteine proteinase inhibitor (E-64), a carbonic anhydrase inhibitor (acetazolamide) and a calcium-channel blocker (verapamil), all agents that block bone resorption in general. Cadmium-induced PGE_2 release that accompanies the bone resorption is potentiated by the phosphodiesterase inhibitor, isobutylmethylxanthine, suggesting that cAMP may be involved in the bone response to cadmium.

2.3.4.5 Cadmium and bone formation mechanisms

Osteoblasts are responsible for the synthesis and secretion of the collagen matrix that is subsequently mineralized. Cadmium can inhibit bone formation by decreasing osteoblast differentiation from mesenchymal precursors and by increasing osteoblastic activity (Fig. 2.5). Cadmium at 2–20 μM caused a striking inhibition of bone formation in cultured chick embryonic long bones, demonstrated by a decrease in the overall bone length, decreased hydroxyproline synthesis and accumulation into collagen, a release of alkaline phosphatase into the medium, and a degeneration of osteogenic mesenchymal

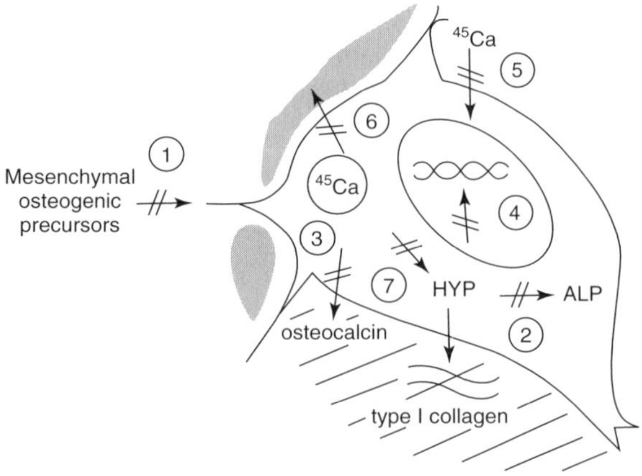

Figure 2.5 Schematic representation of mechanisms of cadmium action on osteoblasts. Shaded areas are mineralized bone, and hatched areas are unmineralized osteoid matrix. Sites of cadmium action are numbered and indicated by arrows blocked by double bars. Cadmium can inhibit bone formation by (1, 2) decreasing osteoblast differentiation from mesenchymal precursors and by (3–7) decreasing multiple osteoblast activities. At lower concentrations (0.1–1 μM), cadmium decreases (2) cellular activity of alkaline phosphatase (ALP), a marker of osteoblast differentiation, and (3) secretion of osteocalcin, a major noncollagenous bone-matrix protein. The DNA content of the osteoblast (4), the cellular accumulation of ^{45}Ca (5), and the calcification (^{45}Ca) of the newly laid down collagen matrix (6) are decreased at 1–2 μM cadmium. Higher concentrations (≥ 2.6 μM) are required to decrease cellular accumulation of hydroxyproline (HYP), a major constituent of type I collagen (7).

cells in the periosteum and osteoblasts around the trabeculae of the bone diaphysis (Miyahara *et al.*, 1978, 1983; Kaji *et al.*, 1988a,b, 1990). Similar results were obtained using osteoblastic cell lines. In a mouse neonatal calvarial cell line (MC3T3-E1), 0.1–1 μM cadmium decreased cellular alkaline phosphatase activity, a marker of osteoblastic differentiation. Cadmium at 1–2 μM caused a decrease in intracellular ^{45}Ca accumulation and a decrease in the extent of calcification of the cultures. Decreases in DNA content and collagen synthesis were only affected at levels of cadmium exposure detrimental to cell viability (>2.7 μM; Miyahara *et al.*, 1988; Iwami and Moriyama, 1993). Cadmium may also be decreasing osteoblastic activity by interfering with intracellular calcium metabolism. In a rat osteosarcoma-derived osteoblastic cell line, ROS17/2.8, 10 nM cadmium increased the flux of ^{45}Ca across the membrane nine-fold (Long, 1997a). This flux was inhibited by calphostin C, an inhibitor of protein kinase C. At higher cadmium concentrations (1–5 μM), intracellular calcium increased twofold, peaking at 40 minutes and returning to baseline by 2 hours (Long, 1997b).

2.3.4.6 Cadmium and the coupling of bone resorption and bone formation

The normal remodeling process is very tightly regulated so that any bone resorption that occurs is closely followed by bone formation, resulting in no net bone loss. Cadmium may act to uncouple this process by both stimulating osteoclastic activity and decreasing osteoblastic activity. However, the levels of cadmium exposure required to stimulate osteoclastic activity (1–500 nM) appear to be lower than those that decrease osteoblast activity (0.1–20 μM) suggesting that differential mechanisms may be occurring, depending on the level of cadmium exposure. Studies with high-level cadmium exposures have demonstrated a disruption of bone formation at the epiphyseal growth plate of rats (Itokawa *et al.*, 1974, 1978; Kimura *et al.*, 1974; Yoshiki *et al.*, 1975; Furata, 1978; Bonner *et al.*, 1981; Iguchi and Sano, 1982). In dogs, 25 p.p.m. but not 10 p.p.m. cadmium in drinking water caused a decrease in activation frequency, radial closure rate, osteoid seams and bone formation rate, all parameters that were reversible after 6 months of no cadmium exposure (Anderson and Danylchuk, 1978, 1979). In contrast, a single, low gavage dose (200 μg $CdCl_2$) caused increased calcium release from bones within 24 hours (Wilson and Bhattacharyya, 1997; Bhattacharyya *et al.*, 1998), a time course that does not allow for the effects of altered bone formation to become apparent.

In many aspects of bone homeostasis, communication between osteoblasts and osteoclasts, whether it is via physical contact (Jimi *et al.*, 1996) or by secretion of soluble cytokines or hormones (Heymann *et al.*, 1998), is essential. Although osteoclasts are the cell type responsible for bone resorption and respond to lower concentrations of cadmium than do osteoblasts, there is no direct evidence that fully differentiated osteoclasts respond directly to cadmium to increase resorptive activity. However, cadmium-induced, osteoblast-mediated increases in osteoclastic activity have been demonstrated in parietal cultures. Prostaglandin (PGE_2) is secreted by osteoblasts in response to cadmium, to signal osteoclasts to increase resorptive activity. An inhibitor of prostaglandin synthesis (indomethacin) blocks this cadmium-induced release of ^{45}Ca into the media (Miyahara *et al.*, 1992).

2.3.4.7 *Estrogen and the bone effect of cadmium*

An important variable in the etiology of itai-itai disease is the connection between cadmium and bone. Because more than 95% of those diagnosed with the disease were postmenopausal, and because the menopause accounts for a substantial amount of the total bone lost in many women over their lifetime, models for the study of itai-itai disease often involve ovariectomy as a way of decreasing estrogen levels *in vivo* (Bhattacharyya *et al.*, 1988, 1992; Sacco-Gibson *et al.*, 1992). Cadmium-induced bone loss is greater in ovariectomized animals than in controls, especially since the effects of cadmium exposure alone on bone at levels relevant to human exposure are subtle.

Estrogen exerts its protective effects on the bone in several ways. It suppresses bone turnover, maintaining a balance between a low rate of bone resorption and formation during the remodeling process. *In vitro*, estrogen has been demonstrated to simulate osteoblast differentiation and to increase deposition of mineral and matrix in mouse bone-marrow cultures (Qu *et al.*, 1998) and in a cultured human osteoblast line (Robinson *et al.*, 1997). Estrogen also decreases the numbers of osteoclasts by blocking early steps in the osteoclast differentiation pathway (Kimble *et al.*, 1996; Sarma *et al.*, 1998) and by inducing apoptosis of mature osteoclasts (Hughes *et al.*, 1996; Kameda *et al.*, 1997; Tompkinson *et al.*, 1997). When estrogen levels decrease during ovariectomy or menopause, cadmium may cause an additive effect, through the mechanisms described, to increase the numbers of osteoclasts available for bone resorption and decrease osteoblastic activity.

Estrogen can exert both genomic and nongenomic effects on bone. Early (regulatory) and late (structural) gene expression are mediated by nuclear activity of the estrogen–receptor complex (Oursler, 1993), whereas more rapid responses such as calcium influx and phospholipid metabolism (Lieberherr *et al.*, 1993; Fiorelli *et al.*, 1996) and mitogen-activated protein kinase activation (Endoh *et al.*, 1997) are triggered by signal-generating cell-surface receptors. Because both osteoblasts and osteoclasts contain estrogen receptors (Eriksen *et al.*, 1988; Komm *et al.*, 1988; Oursler *et al.*, 1991; Hoyland *et al.*, 1997), each cell type can be affected directly by estrogen. Isoforms of the estrogen receptor have been described (Arts *et al.*, 1997; Grandien *et al.*, 1997) and have been found to be tissue specific. In bone cells, the estrogen receptor-α has been localized to osteoblasts and bone-lining cells on trabeculae (Kusec *et al.*, 1998) and to mononuclear osteoclastic precursor cells (Huang *et al.*, 1998), but it is downregulated in multinuclear osteoclasts. In addition, estrogen receptor-α and -β are differentially regulated during osteoblastic differentiation (Arts *et al.*, 1997; Onoe *et al.*, 1997). New estrogen agonist/antagonist ligands that are being developed to treat osteoporosis and cardiovascular disease without affecting uterine and breast tissue (Katzenellenbogen *et al.*, 1997; McDonnell and Norris, 1997) indicate that the mechanism by which the estrogen receptor operates in different tissues is not identical, and this could help explain the mode of action of cadmium in bone. In addition to direct estrogen-receptor effects, cell–cell communication via soluble cytokines associated with bone resorption, such as interleukins (IL-1, IL-6, IL-7, IL-11) and TNF, is downregulated in response to estrogen (Pacifici, 1996, 1997; Miyaura *et al.*, 1997; Cheleuitte *et al.*, 1998).

Although much *in vivo* data exist about the connection between cadmium and estrogen levels, little information is available at the cellular level. In the human breast carcinoma cell line, MCF-7, 1 μM cadmium downregulates the levels of estrogen receptor message and protein by approximately 60% (Garcia-Morales *et al.*, 1994). Cadmium mimics other effects of estrogen, such as increased cell growth and increased progesterone-receptor expression, apparently via the estrogen receptor but independent of estrogen. Many mechanisms of estrogen and cadmium stimulation are similar, such as signal transduction by increased intracellular calcium, phospholipid metabolism and activation of protein kinase C, and by c-jun, c-myc or phorbol ester stimulation of gene transcription. The question of how these mechanisms are differentially regulated when cadmium exerts its greatest effects on bone, during estrogen deficiency, remains to be answered.

2.4 Cadmium-induced gene expression and cell signaling

The effects of cadmium on molecular and cellular processes depend on its exposure concentrations. At low nanomolar to micromolar levels, it causes a transient up-regulation of DNA synthesis (Zglinicki *et al.*, 1992), transcriptional activation (Hidalgo *et al.*, 1976) and cell growth and differentiation in several mammalian cell lines and primary cell cultures (Beyersmann and Hechtenberg, 1997). At higher (millimolar) doses and on prolonged exposure, cadmium produces a variety of cytotoxic effects, including inhibition of DNA, RNA and protein synthesis (Cohen *et al.*, 1991), inhibition of various enzymes, inhibition of DNA repair (Hartwig *et al.*, 1996) and oxidative damage to proteins, lipids and nucleic acids. Examples of oxidative damage include lipid peroxidation and chromosomal aberrations, such as conformational changes, mutations and strand breaks in DNA (Stohs and Bagchi, 1995). Lipid peroxidation is, in turn, a source of reactive oxygen-free radicals. Cd causes cellular oxidative stress through the production of reactive oxygen species (ROS), possibly as a consequence of inflammation. The ROS probably inactivate or decrease the levels of detoxifying antioxidant enzymes such as zinc/copper superoxide dismutase (SOD) and catalase (CAT) (Ossola and Tomaro, 1995), which are scavengers of the free radicals.

At the genetic level, exposure to cadmium initiates the cellular defense mechanism by upregulating the biosynthesis of sulfhydryl compounds such as glutathione (GSH) and metallothionein (MT). Since the induction of these protective metal chelators precedes the cadmium-induced oxidative damage to the cells, the influence of cadmium on gene regulation is not believed to result from oxidative stress (Beyersmann and Hechtenberg, 1997). As a mitogen, cadmium activates transcription factors such as egr-1, c-myc, c-fos and c-jun at the mRNA level (Matsuoka and Call, 1995). It also acts as a weak tumor promoter *in vivo* and causes transformation of cells in culture (Abshire *et al.*, 1996). The mitogenic effects of cadmium are believed to be due to its interference with DNA repair, thereby allowing the accumulation of mutations (Hartwig *et al.*, 1996). Such an inhibition of repair enzymes might lead to tumor promotion and cell growth and, alternatively, to apoptosis. Cadmium is thus implicated in several pathways.

Cadmium can influence cellular signal transduction:

1 at the uptake level, affecting the transport pathways;
2 by influencing secondary messengers; or
3 by affecting gene expression.

The effects of cadmium on gene expression and cellular signal transduction, summarized in Fig. 2.6, arise from its ability to alter cell physiology and influence metabolic pathways. Following is a brief presentation of some of the molecular processes that are influenced by cadmium.

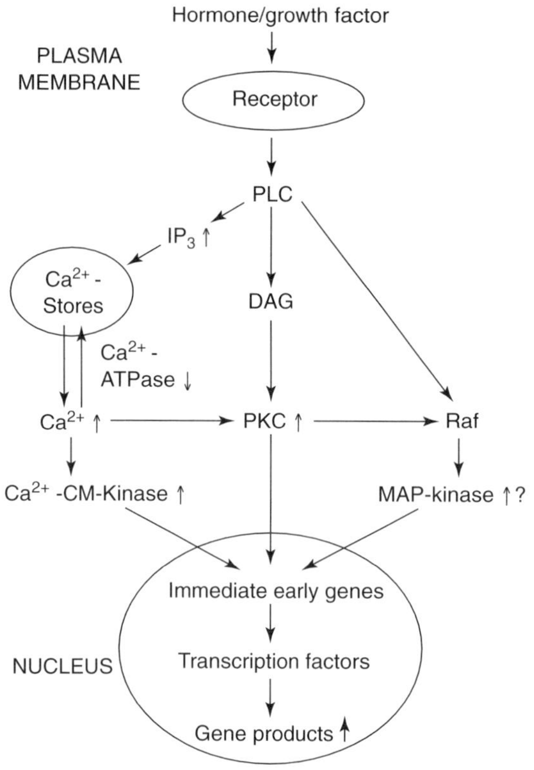

Figure 2.6 Levels of interference by cadmium with cellular signal transduction elements. Depicted are pathways of signaling that are probably affected by cadmium ions immediately or mediated by interlinked steps. Signaling elements known to be altered by cadmium are marked with arrows, indicating the direction of change. DAG, diacylglycerol; CM, calmodulin; IP_3, inositol triphosphate; MAP-kinase, mitogen-activated protein kinase; PKC, protein kinase C; PLC, phospholipase C.

2.4.1 Cd^{2+} modulates Ca^{2+}-requiring protein activities and signaling pathways

A major route of Cd^{2+} entry into the cells is through the voltage-gated (Hinkle *et al.*, 1987) and receptor-mediated (Blazka and Shaikh, 1991) Ca^{2+} channels. Cd^{2+} can easily pass through Ca^{2+} channels because of the similarity in their size (ionic radii: Cd^{2+} = 0.97Å, Ca^{2+} = 0.99Å) and charge. Such a steric fit also allows Cd^{2+} to displace Ca^{2+} from its normal physiological roles, including its binding to and activation of various macromolecules. Thus Cd^{2+} can activate proteins such as calmodulin and enzymes such as protein kinase C (PKC) (Tang and Enger, 1993). Calmodulin regulates cellular processes by activating several enzymes, such as the Ca^{2+}/calmodulin-dependent kinase, phosphodiesterase and the myosin light-chain kinase. PKC is a family of Ser/Thr kinases whose activation is Ca^{2+}- and phospholipid-dependent. These kinases are involved in signal transduction pathways that regulate cell differentiation, cell survival, gene expression, cystoskeletal function and cell–cell interactions. Cd^{2+} can also take the place of Ca^{2+} as a secondary messenger in the activation of calcitropic hormones. At low concentrations (0.1–5 μM), Cd^{2+} interferes with Ca^{2+}-ATPase pumps, leading to an immediate, transient but substantial increase in $[Ca^{2+}]_i$ (Hechtenberg and Beyersmann, 1991). The increase in intracellular free Ca^{2+} results in the formation of inositol triphosphate (IP_3) which triggers the PKC signal cascade (Smith *et al.*, 1989). Substrate phosphorylation by PKC occurs after the translocation of the enzymes from the cytosol to the nuclear membrane and activation by diacylglycerol (DAG) (Rodriguez-Paris *et al.*, 1989). Long (1997b) has shown that Cd^{2+} can activate PKC directly with an activation constant that is 5000-fold smaller than that of Ca^{2+}.

2.4.2 Cd^{2+} can regulate the activities of Zn^{2+}-requiring proteins

Cadmium can compete with heavy metals such as zinc for their binding sites on various proteins and thus modulate their function. Zinc-binding sites, such as the zinc-finger motifs of transcription factors and signal transducers like PKC, or of metalloenzymes like SOD, are all targets for cadmium occupation, leading to nonphysiological regulation of the protein activities. Cadmium is believed to replace zinc in its binding site on Zn/Cu SOD, inhibiting the enzyme and leading to an increased production of ROS (Bauer and Demeter, 1980). In a model that describes the cadmium-induced activation of the PKC-mediated signal transduction, Cd^{2+} substitutes for Zn^{2+} in the regulatory domain of the enzyme thus exposing the putative protein–protein interaction site (Beyersmann *et al.*, 1994).

2.4.3 Cd^{2+} induces the expression of metal-scavenging peptides and proteins

As an early line of defense, cadmium at low concentrations (1–5 μM) induces the expression of several genes, including γ-glutamyl-cysteine synthase (Hatcher *et al.*, 1995), MT (Imbert *et al.*, 1990) and heme oxygenase (HO), also referred to as the heat shock protein HSP32 (Tyrell *et al.*, 1993). The biosynthesis of these molecules, which are metal chelators and antioxidants, is believed to precede cadmium toxicity (Beyersmann and Hechtenberg, 1997).

The MT gene promoter has several metal response elements (MREs: TGCRCNC) (Palmiter, 1987). Cadmium or other heavy metals activate the metal transcription factor-1 (MTF-1) which binds to the MREs, resulting in the transcriptional activation of the *MT* gene (Radtke *et al.*, 1993). The *MT* genes also contain an antioxidant response element (ARE: GTGACNNNGC) (Dalton *et al.*, 1994). Transcription factors such as the activator protein-1 (AP-1: fos/jun heterodimer) have been shown to bind to this element and activate transcription (Friling *et al.*, 1992).

Induction of HSP32 by cadmium might be mediated via a cadmium response element (CdRE: TGCTAGATT) (Takeda *et al.*, 1994) although there are reports that this CdRE may not be functional (Wang and Templeton, 1998). Cadmium or heme activates HSP32 via its AP-1 binding site (Alam *et al.*, 1995). Cadmium-induced expression of the transcription factors could thus activate the HSP.

2.4.4 Cd^{2+}-induced oxidative stress results in the expression of stress proteins

Cd^{2+} at high, cytotoxic concentrations causes an increase in ROS, such as superoxide anions, which cause irreversible damage to various biomolecules (Mikhailova *et al.*, 1997). ROS can also mediate signal transduction by interacting with PKC (Bagchi *et al.*, 1997b). The ROS cause a substantial decrease in the $[GSH]_i$ along with a depletion of other defenses, such SOD and CAT (Ossola and Tomaro, 1995). The decreased level of $[GSH]_i$ results in an increase in the level of the heat shock factor (HSF) and also promotes the binding of HSF to the heat shock element (HSE) consensus sequence (NGAAN) (Morimoto, 1993). This binding results in the transcriptional activation of stress response genes such as HSP70, HSP90 and HSP110 (Ovelgonne *et al.*, 1995). The induction of these HSPs occurs at a cadmium concentration of 10–100 μM. Cadmium might bind the Cys-rich HSF and thereby activate it, although this mechanism has not yet been clearly established. Antioxidants such as *N*-acetyl cysteine (NAC) (Abe *et al.*, 1998), bilirubin or α-tocopherol (Ossola and Tomaro,1995) alter cadmium-induced effects by modulating redox-sensitive cellular components. They prevent the depletion of GSH and thus HSPs are not induced. $CdCl_2$ was also found to activate protein kinase B (PKB) in association with HSP27 in a signaling pathway of various forms of stress (Konishi *et al.*, 1997). The HSP70 has the conserved sequence CGNCCCGG which is homologous to the MRE of the MT genes (Wu *et al.*, 1986) but there is no evidence that MTF-1 or AP-1 regulate its expression.

2.4.5 Cd^{2+} causes transcriptional activation of growth-related genes

Cadmium at 5–50 μM stimulates the expression of proto-oncogenes and genes related to proliferation. Tang and Enger (1993) showed that cadmium induced the expression of c-*fos*, c-*jun* and c-*myc* messages in a dose- and time-dependent manner that was insensitive to cycloheximide. Thus the activation of these genes is believed to be a consequence of direct transcriptional activation and not resulting from the pre-induction of gene-activating growth factors. Cadmium-induced accumulation of c-*myc* transcripts is thought to involve a PKC-mediated pathway (Tang and Enger, 1993). Cadmium induction of c-*fos* is believed to be due to the activation of the erk family of mitogen-activated protein kinases (MAPK), independent of pathways involving PKC or the Ca^{2+}/calmodulin-dependent kinases (Wang and Templeton, 1998). None of these cellular oncogenes has been shown to have a metal response element or a metal-binding site, ruling out a direct interaction of cadmium with their regulatory sequences. So the exact mechanism by which cadmium activates these genes at the transcriptional level is not clearly understood.

Treatment of HeLa cells with 0.2 mM cadmium induced the expression of GADD153 (Luethy and Holbrook, 1994) which is a growth arrest gene and a DNA-damage-inducible member of the CCAAT enhancer-binding family of transcription factors. This induction could be blocked by the addition of antioxidants such as NAC. Cadmium was found to activate the tumor suppressor gene *p53* at 20 μmol/kg (Zheng *et al.*, 1996). It also activates the expression of various cytokines, including the interleukins IL-α, IL-β, IL-6, cell adhesion molecules such as ICAM-1 and growth factors such as TNF-α (Kayama *et al.*, 1995).

In summary, cadmium has varied effects on cellular events, based on the dose of administration and the exposure time. At low concentrations, cadmium promotes cell growth and differentiation, perturbs the cellular homeostasis of Ca^{2+} and Zn^{2+}, activates various signaling pathways, causes transcriptional activation of growth-related genes, stimulates nucleic acid synthesis and induces the production of chelating proteins and peptides. At higher concentrations its effects stem from the oxidative damage that it causes to the macromolecules.

Acknowledgment

Work conducted in the laboratory of Maryka H. Bhattacharyya was supported by NIEHS grants ES07398 and ES04816.

References

Abe, T., Yamamura, K., Gotoh, S., Kashimura, M. and Higashi, K. 1998. Concentration-dependent differential effects of *N*-acetyl-L-cysteine on the expression of HSP70 and metallothionein genes induced by cadmium in human amniotic cells. *Biochim. Biophys. Acta.* **380**, 123–132.

Abel, J., Hohr, D. and Schurek, H.J. 1987. Renal handling of cadmium and cadmium-metallothionein: studies on the isolated perfused rat kidney. *Arch. Toxicol.* **60**, 370–375.

Abshire, M. K., Devor, D. E., Diwan, B. A., Shaughnessy, J. D. Jr. and Waalkes, M. P. 1996. *In vitro* exposure to cadmium in rat L6 myoblasts can result in both enhancement and suppression of malignant progression *in vivo*. *Carcinogenesis* **17**, 1349–1356.

Ahn, D.W. and Park, Y.S. 1995. Transport of inorganic phosphate in renal cortical brush-border membrane vesicles of cadmium-intoxicated rats. *Toxicol. Appl. Pharmacol.* **133**, 239–243.

Alam, J., Camhi, S. and Choi, A. M. 1995. Identification of a second region upstream of the mouse heme oxygenase-1 gene that functions as a basal level and inducer-dependent transcription enhancer. *J. Biol. Chem.* **270**, 11977–11984.

Andersen, O., Nielsen, J. B., Sorensen, J. A. and Scherrebeck, L. 1994. Experimental localization of intestinal uptake sites for metals (Cd, Hg, Zn, Se) in vivo in mice. *Environ. Hlth Perspect. Suppl.* **3**, 199–206.

Anderson, C. and Danylchuk, K.D. 1978. Effect of chronic low-level cadmium intoxication on the Haversian remodeling system in dogs. *Calcif. Tissue Res.* **26**, 143–148.

Anderson, C. and Danylchuk, K.D. 1979. Effect of chronic low-level cadmium intoxication of the Haversian remodeling system in dogs: a reversible phenomenon. *Calcif. Tissue Int.* **27**, 121–126.

Ando, M., Sayato, Y. and Osawa, T. 1978. Studies on the disposition of calcium in bones of rats after continuous oral administration of cadmium. *Toxicol. Appl. Pharmacol.* **46**, 625–632.

Angle, C.R., Thomas, D.J. and Swanson, S.A. 1990. Toxicity of cadmium to rat osteosarcoma cells (ROS 17/2.8): Protective effect of 1,25-dihydroxyvitamin D_3. *Toxicol. Appl. Pharmacol.* **103**, 113–20.

Angle, C.R., Thomas, D.J. and Swanson, S.A. 1993. Osteotoxicity of cadmium and lead in HOS TE 85 and ROS 17/2.8 cells: Relation to metallothionein induction and mitochondrial binding. *Biometals* **6**, 179–84.

Arts, J., Kuiper, G.G., Janssen, J.M., Gustafsson, J.A., Lowik, C.W., Pols, H.A. and van Leeuwen, J.P. 1997. Differential expression of estrogen receptors alpha and beta mRNA during differentiation of human osteoblast SV-HFO cells. *Endocrinology* **138**, 5067–5070.

Bagchi, D., Bagchi, M., Hassoun, E.A. and Stohs, S.J. 1996. Cadmium-induced excretion of urinary lipid metabolites, DNA damage, glutathione depletion, and hepatic lipid peroxidation in Sprague–Dawley rats. *Biol. Trace Elem. Res.* **52**, 143–154.

Bagchi, D., Vuchetich, P.J., Bagchi, M., Hassoun, E.A., Tran, M.X., Tang, L. and Stohs, S.J. 1997a. Induction of oxidative stress by chronic administration of sodium dichromate [chromium VI] and cadmium chloride [cadmium II] to rats. *Free Radic. Biol. Med.* **22**, 471–478.

Bagchi, D., Bagchi, M., Tang, L. and Stohs, S.J. 1997b. Comparative *in vitro* and *in vivo* protein kinase C activation by selected pesticides and transition metals salts. *Toxicol. Lett.* **91**, 31–37.

Bauer, R. and Demeter, I. 1980. Structural properties of the zinc site in Cu, Zn-superoxide dismutase: perturbed angular correlation of gamma ray spectroscopy on the Cu, ^{111}Cd-superoxide dismutase derivative. *Biochem. Biophys. Res. Commun.* **94**, 1296–1302.

Bawden, J.W. and Hammarstrom, L.E. 1975. Distribution of cadmium in developing teeth and bone of young rats. *Scand. J. Dent. Res.* **83**, 179–186.

Bernard, A.M., Ouled, A.A. and Lauwerys, R.R. 1987. The effects of low doses of cadmium-metallothionein on the renal uptake of beta 2-microglobulin in rats. *Toxicol. Appl. Pharmacol.* **15**, 440–445.

Beyersmann, D. and Hechtenberg, S. 1997. Cadmium, gene regulation and cellular signaling in mammalian cells. *Toxicol. Appl. Pharmacol.* **144**, 247–261.

Beyersmann, D., Block, C. and Malviya, A.N. 1994. Effects of cadmium on nuclear protein kinase C. *Environ. Hlth Perspect.* **102**, 177–180.

Bhattacharyya, M.H. 1991. Cadmium-induced bone loss: increased susceptibility in females. *Water Air Soil Pollut.* **57/58**, 665–74.

Bhattacharyya, M.H., Whelton, B.D. and Peterson, D.P. 1982. Gastrointestinal absorption of cadmium in mice during gestation and lactation. II. Continuous exposure studies. *Toxicol. Appl. Pharmacol.* **66**, 368–75.

Bhattacharyya, M.H., Whelton B.D., Stern P.H. and Peterson, D.P. 1988. Cadmium accelerates bone loss in ovariectomized mice and fetal rat limb bones in culture. *Proc. Natl Acad. Sci. USA* **85**, 8761–8765.

Bhattacharyya, M.H., Sacco-Gibson, N.A. and Peterson, D.P. 1992. Cadmium-induced bone loss: increased susceptibility in female beagles after ovariectomy, in *Cadmium in the Human Environment* (eds G. F. Nordberg, R.F.M. Herber and L. Alessio), International Agency for Research on Cancer, Lyon, France, pp. 279–286.

Bhattacharyya, M.H., Peterson, D.P. and Sacco-Gibson, N. 1993. *New, Sensitive Assay for Cadmium in Biological Samples.* Edited Proceedings Seventh International Cadmium Conference, New Orleans, (eds M.E. Cook, S.A. Hiscock, H. Morrow and R. Volpe), Cadmium Association, London; Cadmium Council, Virginia; International Lead Zinc Research Organization, North Carolina, pp. 113–114.

Bhattacharyya, M.H., Blum, C. and Wilson, A.K. 1998. The role of metallothionein in cadmium-induced bone resorption, in *Metallothionein IV*, (ed. C.D. Klaasen), Birkhaüser Verlag, Basel, pp. 473–6.

Blazka, M.E. and Shaikh, Z.A. 1991. Differences in cadmium and mercury uptakes by hepatocytes: Role of calcium channels. *Toxicol. Appl. Pharmacol.* **110**, 355–363.

Blazka, M.E. and Shaikh, Z.A. 1992. Cadmium and mercury accumulation in rat hepatocytes: Interactions with other metal ions. *Toxicol. Appl. Pharmacol.* **113**, 118–125.

Blumenthal, S.S., Lewand, D.L., Buday, M.A., Kleinman, J.G., Krezoski, S.K. and Petering, D.H. 1990. Cadmium inhibits glucose uptake in primary cultures of mouse cortical tubule cells. *Am. J. Physiol.* **258**, F1625-F1633.

Blumenthal, S., Lewand, D., Sochanik, A., Krezoski, S. and Petering, D.H. 1994. Inhibition of Na(+)-glucose cotransport in kidney cortical cells by cadmium and copper: protection by zinc. *Toxicol. Appl. Pharmacol.* **129**, 177–87.

Blumenthal, N.C., Cosma, V., Skyler, D., LeGeros, J. and Walters, M. 1995. The effect of cadmium on the formation and properties of hydroxyapatite *in vitro* and its relation to cadmium toxicity in the skeletal system. *Calcif. Tissue Int.* **56**, 316–322.

Blumenthal, S., Lewand, D., Krezoski, S.K. and Petering, D.H. 1996, Comparative effects of Cd2+ and Cd-metallothionein on cultured kidney tubule cells. *Toxicol. Appl. Pharmacol.* **136**, 220–228.

Blumenthal, S.S., Ren, L., Lewand, D.L. and Petering, D.H. 1998. Cadmium decreases SGLT1 messenger RNA in mouse kidney cells. *Toxicol. Appl. Pharmacol.* **149**, 49–54.

Bonner, F.W., King, L.J. and Parke, D.V. 1981. The acute and subacute effects of cadmium on calcium homeostasis and bone trace metals in the rat. *J. Inorg. Biochem.* **14**, 107–114.

Brennan, R.J. and Schiestl, R.H. 1996. Cadmium is an inducer of oxidative stress in yeast. *Mutat. Res.* **356**, 171–8.

Buchler, M., Konig, J., Brom, M., Kartenbeck, J., Spring, H., Horie, T. and Keppler, D. 1996. cDNA cloning of the hepatocyte canalicular isoform of the multidrug resistance protein, cMrp, reveals a novel conjugate export pump deficient in hyperbilirubinemic mutant rats. *J. Biol. Chem.* **271**, 15091–15098.

Bucio, L., Souza, V., Albores, A., Sierra, A., Chavez, E., Carabez, A. and Gutierrez-Ruiz, M.C. 1995. Cadmium and mercury toxicity in a human fetal hepatic cell line (WRL-68 cells). *Toxicology* **102**, 285–299.

Chan, H.M. and Cherian, M.G. 1992. Protective roles of metallothionein and glutathione in hepatotoxicity of cadmium. *Toxicology* **72**, 281–290.

Chan, H.M. and Cherian, M.G. 1993. Mobilization of hepatic cadmium in pregnant rats. *Toxicol. Appl. Pharmacol.* **120**(2), 308–14.

Chan, H.M., Satoh, M., Zalups, R.K. and Cherian M.G. 1992. Exogenous metallothionein and renal toxicity of cadmium and mercury in rats. *Toxicology* **76**, 15–26.

Chan, H.M., Zhu, L.F., Zhong, R., Grant, D., Goyer, R.A. and Cherian, M.G. 1993. Nephrotoxicity in rats following liver transplantation from cadmium-exposed rats. *Toxicol. Appl. Pharmacol.* **123**, 89–96.

Cheleuitte, D., Mizuno, S. and Glowacki, J. 1998. *In vitro* secretion of cytokines by human bone marrow: effects of age and estrogen status. *J. Clin. Endocrinol. Metab.* **83**, 2043–2051.

Cherian, M.G. 1977. Biliary excretion of cadmium in rat. II. The role of metallothionein in the hepatobiliary transport of cadmium. *J. Toxicol. Environ. Hlth* **2**, 955–961.

Cherian, M.G. 1985. Rat kidney epithelial cell culture for metal toxicity studies. *In Vitro Cell Dev. Biol.* **21**, 505–508.

Cherian, M.G. and Vostal, J.J. 1977. Biliary excretion of cadmium in rat. I. Dose-dependent biliary excretion and the form of cadmium in the bile. *J. Toxicol. Environ. Hlth* **2**(4), 945–54.

Cherian, M.G., Goyer, R.A. and Valberg, L.S. 1978. Gastrointestinal absorption and organ distribution of oral cadmium chloride and cadmium-metallothionein in mice. *J. Toxicol. Environ. Hlth* **4**, 861–8.

Chia, K.S., Ong, C.N., Ong. H.Y. and Endo, G. 1989. Renal tubular function of workers exposed to low levels of cadmium. *Br. J. Ind. Med.* **46**, 165–70.

Chmielnicka, J. and Sowa, B. 1996. Cadmium interaction with essential metals (Zn, Cu, Fe), metabolism metallothionein, ceruloplasmin in pregnant rats and fetuses. *Ecotoxicol. Environ. Safety*, **3**, 277–281.

Christoffersen, J., Christoffersen, M.R., Larsen, R., Rostrup, E., Tingsgaard, P., Andersen, O. and Grandjean, P. 1988. Interaction of cadmium ions with calcium hydroxyapatite crystals: A possible mechanism contributing to the pathogenesis of cadmium-induced bone diseases. *Calcif. Tissue Int.* **42**, 331–339.

Cohen, M., Lotta, D., Coogan, T. and Costa, M. 1991. Mechanisms of metal carcinogenesis: the reactions of metals with nucleic acids, in *Biological Effects of Heavy Metals* (eds E.C. Foulkes), CRC Press, Boca Raton, FL, pp. 19–75.

Condron, R.J., Schroen, C.J. and Marshall, A.T. 1994. Morphometric analysis of renal proximal tubules in cadmium-treated rats. *J. Submicrosc. Cytol. Pathol.* **26**, 51–58.

Conrad, C.C., Walter, C.A., Richardson, A., Hanes, M.A. and Grabowski, D.T. 1997. Cadmium toxicity and distribution in metallothionein-I and -II deficient transgenic mice. *J. Toxicol. Environ. Hlth* **52**, 527–543.

Couce, M.D., Varela, J.M., Sanchez, A., Casas, J.S., Sordo, J. and Lopez-Rivadulla, M. 1991. Effects of vitamin B12 on cadmium toxicity in rats. *J. Inorg. Biochem.* **41**, 1–6.

Dalton, T., Palmiter, R.D. and Andrews, G.K.1994. Transcriptional induction of the mouse metallothionein-I gene in hydrogen peroxide-treated Hepa cells involves a composite major late transcription factor/antioxidant response element and metal response promoter elements. *Nucleic Acids Res.* **22**, 5016–23.

Dawson, D.C. and Ballatori, N. 1995. Membrane transporters as sites of action and routes of entry for toxic metals, in *Toxicology of Metals*, (eds R.A. Goyer and M.G. Cherian), Springer-Verlag, Berlin, pp. 53–76.

Dorian, C. and Klaassen, C.D. 1995. Protection by zinc-metallothionein (ZnMT) against cadmium-metallothionein-induced nephrotoxicity. *Fundam. Appl. Toxicol.* **26**, 99–106.

Dorian, C., Gattone, V.H. II and Klaassen, C.D. 1992a. Renal cadmium deposition and injury as a result of accumulation of cadmium-metallothionein (Cd-MT) by the proximal convoluted tubules – A light microscopic autoradiography study with 109Cd-MT. *Toxicol. Appl. Pharmacol.* **114**, 173–181.

Dorian, C., Gattone, V.H. II and Klaassen, C.D. 1992b. Accumulation and degradation of the protein moiety of cadmium-metallothionein (Cd-MT) in the mouse kidney. *Toxicol. Appl. Pharmacol.* **117**, 242–248.

Dorian, C., Gattone, V.H. II and Klaassen, C.D. 1995. Discrepancy between the nephrotoxic potencies of cadmium-metallothionein and cadmium chloride and the renal concentration of cadmium in the proximal convoluted tubules. *Toxicol. Appl. Pharmacol.* **130**, 161–168.

Elinder, C.-G., Friberg, L., Lind, B. and Jawaid, M. 1983. Cadmium exposure from smoking cigarettes: variations with time and country where purchased. *Environ. Res.* **32**, 220–7.

Elsenhans, B., Kolb, K., Schumann, K. and Forth, W. 1994. The longitudinal distribution of cadmium, zinc, copper, iron and metallothionein in the small-intestinal mucosa of rats after administration of cadmium chloride. *Biol. Trace Elem. Res.* **41**, 31–46.

Elsenhans, B., Strugala, G.J. and Schafer, S.G. 1997. Small-intestinal absorption of cadmium and the significance of mocosal metallothionein. *Hum. Exp. Toxicol.* **16**, 429–434.

Endo, T. and Shaikh, Z.A. 1993. Cadmium uptake by primary cultures of rat renal cortical epithelial cells: Influence of cell density and other metal ions. *Toxicol. Appl. Pharmacol.* **121**, 203–209.

Endo, T., Kimura, O., Hatakeyama, M., Takada, M. and Sakata, M. 1997. Effects of zinc and copper on cadmium uptake by brush border membrane vesicles. *Toxicol. Lett.* **91**, 111–120.

Endo, T., Kimura, O., Sasaya, M., Takada, M. and Sakata, M. 1995. Na(+)- and energy-dependent transport of cadmium into LLC-PK1 cells. *Biol. Pharm. Bull.* **18**, 1689–1693.

Endoh, H., Sasaki, H., Maruyama, K., Takeyama, K., Waga, I., Shimizu, T., Kato, S. and Kawashima H. 1997. Rapid activation of MAP kinase by estrogen in the bone cell line. *Biochem. Biophys. Res. Commun.* **235**, 99–102.

Eriksen, E., Colvard, D., Berg, N., Graham, M., Mann, K., Spelsberg, T.C. and Riggs, B.L. 1988. Evidence of estrogen receptors in normal human osteoblast-like cells. *Science* **241**, 84–7.

Felley-Bosco, E. and Diezi, J. 1991. Cadmium uptake and induction of metallothionein synthesis in a renal epithelial cell line (LLC-PK1). *Arch. Toxicol.* **65**, 160–163.

Figueiredo-Pereira, M.E., Yakushin, S. and Cohen, G. 1998. Disruption of the intracellular sulfhydryl homeostasis by cadmium-induced oxidative stress leads to protein thiolation and ubiquitination in neuronal cells. *J. Biol. Chem.* **273**, 12703–9.

Fiorelli, G., Gori, F., Frediani, U., Franceschelli, F., Tanini, A., Tosti-Guerra, C., Benvenuti, S., Gennari, L., Becherini, L. and Brandi, M.L. 1996. Membrane binding sites and non-genomic effects of estrogen in cultured human pre-osteoclastic cells. *J. Steroid Biochem. Mol. Biol.* **59**, 233–240.

Foulkes, E.C. 1985. Interactions between metals in rat jejunum: Implications on the nature of cadmium uptake. *Toxicology* **37**, 117–125.

Foulkes, E.C. 1989. On the mechanism of cellular cadmium uptake. *Biol. Trace Elem. Res.* **21**, 195–200.

Foulkes, E.C. and Bergman, D. 1993. Inorganic mercury absorption in mature and immature rat jejunum: transcellular and intercellular pathways *in vivo* and everted sacs. *Toxicol. Appl. Pharmacol.* **120**, 89–95.

Foulkes, E.C. and Blanck, S. 1990. Acute cadmium uptake by rabbit kidneys: Mechanism and effects. *Toxicol. Appl. Pharmacol.* **102**, 464–473.

Foulkes, E.C. and McMullen, D.M. 1987. Kinetics of transepithelial movement of heavy metals in rat jejunum. *Am. J. Physiol.*, **253**, G134–G138.

Frazier, J.M. 1980. Cadmium and zinc kinetics in rat plasma following intravenous injection. *J. Toxicol. Environ. Hlth* **6**, 503–518.

Frazier, J.M. 1984. Mathematical model for cadmium kinetics in the isolated perfused rat liver system. *Toxicol. Appl. Pharmacol.* **76**, 426–436.

Friberg, L., Elinder, C.-G., Kjellstrom, T. and Nordberg, G.F. 1986a. *Cadmium and Health: A Toxicological and Epidemiological Appraisal, Volume 1. Exposure, Dose and Metabolism*, CRC Press, Boca Raton.

Friberg, L., Elinder, C., Kjellstrom, T. and Nordberg, G. F. 1986b. *Cadmium and Health: A Toxicological and Epidemiological Appraisal, Volume II. Effects and Responses*, CRC Press, Boca Raton, FL.

Friling, R.S., Bergelson, S. and Daniel, V. 1992. Two adjacent AP-1-like binding sites from the electrophile-responsive element of the murine glutathione S-transferase Ya subunit gene. *Proc. Natl Acad. Sci., USA.* **89**, 668–672.

Furata, H. 1978. Cadmium effects on bone and dental tissues of rats in acute and subacute poisoning. *Experientia* **34**, 1317–1318.

Garcia-Morales, P., Saceda, M., Kenny, N., Kim, N., Salomon, D.S., Gottardis, M.M., Solomon, H.B., Sholler, P.F., Jordan, V.C. and Martin M.B. 1994. Effect of cadmium on estrogen receptor levels and estrogen-induced responses in human breast cancer cells. *J. Biol. Chem.*, **269**, 16896–901.

Gerson, R.J. and Shaikh, Z.A. 1984. Differences in the uptake of cadmium and mercury by rat hepatocyte primary cultures. Role of a sulfhydryl carrier. *Biochem. Pharmacol.* **33**, 199–203.

Gong, Q. and Hart, B.A. 1997. Effect of thiols on cadmium-induced expression of metallothionein and other oxidant stress genes in rat lung epithelial cells. *Toxicology* **119**, 179–91.

Grandien, K., Berkenstam, A. and Gustafasson, J.A. 1997. The estrogen receptor gene: promoter organization and expression. *Int. J. Biochem. Cell. Biol.* **29**, 1343–69.

Gregus, Z. and Varga, F. 1985. Role of glutathione and hepatic glutathione S-transferase in the biliary excretion of methyl mercury, cadmium and zinc: A study with enzyme inducers and glutathione depletors. *Acta Pharmacol. Toxicol. (Copenh)* **56**, 398–403.

Groten, J.P., Sinkeldam, E.J., Luten, J.B. and van Bladaren, P.J. 1990. Comparison of the toxicity of inorganic and liver-incorporated cadmium: A 4-wk feeding study in rats. *Food Chem. Toxicol.* **28**, 435–441.

Groten, J.P., Sinkeldam, E.J., Luten, J.B. and van Bladeren, P.J. 1991. Cadmium accumulation and metallothionein concentrations after 4-week dietary exposure to cadmium chloride or cadmium-metallothionein in rats. *Toxicol. Appl. Pharmacol.* **111**, 504–513.

Groten, J.P., Luten, J.B. and Bladeren, P.J. 1992. Dietary iron lowers the intestinal uptake of cadmium-metallothionein in rats. *Eur. J. Pharmacol.* **228**, 23–28.

Groten, J.P., Koeman, J.H., van Nesselrooij, J.H., Luten, J.B., Fentener van Vlissingen, J.M., Stenhuis, W.S. and van Bladeren P.J. 1994. Comparison of renal toxicity after long-term oral administration of cadmium chloride and cadmium-metallothionein in rats. *Fundam. Appl. Toxicol.* **23**, 544–552.

Habeebu, S.S., Liu, J. and Klaassen, C.D. 1998. Cadmium-induced apoptosis in mouse liver. *Toxicol. Appl. Pharmacol.* **149**, 203–209.

Hamada, T., Nakano, S., Iwai, S., Tanimoto, A., Ariyoshi, K. and Koide, O. 1991. Pathological study on beagles after long-term oral administration of cadmium. *Toxicol. Pathol.* **19**, 138–147.

Hamada, T., Sasaguri, T., Tanimoto, A., Arima, N., Shimarjiri, S., Abe, T. and Sasaguri, Y. 1996. Apoptosis of human kidney 293 cells is promoted by polymerized cadmium-metallothionein. Biochem. *Biophys. Res. Commun.* **219**, 829–834.

Hartwig, A., Schlepegrell, R., Dally, H. and Hartmann, M. 1996. Interference of carcinogenic metal compounds with deoxyribonucleic acid repair process. *Ann. Clin. Lab. Sci.* **26**, 31–38.

Hassler, E., Lind, B. and Pisactor, M. 1983. Cadmium in blood and urine related to present and past exposure. A study of workers in an alkaline battery factory. *Br. J. Ind. Med.* **40**, 420–5.

Hassoun, E.A. and Stohs, S.J. 1996. Cadmium-induced production of superoxide anion and nitric oxide, DNA single strand breaks and lactate dehydrogenase leakage in J774A.1 cell cultures. *Toxicology* **112**, 219–26.

Hatcher, E. L., Chen, Y. and Kang, Y. J. 1995. Cadmium resistance in A549 cells correlates with elevated glutathione content but not antioxidant enzymatic activities. *Free Radic. Biol. Med.* **19**, 805–812.

Hechtenberg, S. and Beyersmann, D. 1991. Inhibition of sarcoplasmic reticulum Ca^{2+} ATPase activity by cadmium, lead and mercury. *Enzyme* **45**, 105–115.

Heymann, D., Guicheux, J., Gouin, F., Passuti, N. and Daculsi, G. 1998. Cytokines, growth factors and osteoclasts. *Cytokine* **10** 155–168.

Hidalgo, H. A., Koppa, V. and Bryan, S. E. 1976. Effect of cadmium on RNA polymerase and protein synthesis in rat liver. *FEBS Lett.* **64**, 159–162.

Hinkle, P.M., Kinsella., P.A. and Osterhoudt, K.C. 1987. Cadmium uptake and toxicity via voltage-sensitive calcium channels. *J. Biol. Chem.* **262**, 16333–16337.

Hochadel, J.F. and Waalkes, M.P. 1997. Sequence of exposure to cadmium and arsenic determines the extent of toxic effects in male Fischer rats. *Toxicology* **116**, 89–98.

Honda, R., Tsuritani, I., Ishizaki, M. and Yamada, Y. 1997. Zinc and copper levels in ribs of cadmium-exposed persons with special reference to osteomalacia. *Environ Res.* **75**, 41–48.

Hoyland, J.A., Mee, A.P., Baird, P., Braidman, I.P., Mawer, E.B. and Freemont, A.J. 1997. Demonstration of estrogen receptor mRNA in bone using in situ reverse-transcriptase polymerase chain reaction. *Bone* **20**, 8–92.

Huang, W.H., Lau, A.T., Daniels, L.L., Fujii, H., Seydel, U., Wood, D.J., Papadimitriou, J.M. and Zheng, M.H. 1998. Detection of estrogen receptor alpha, carbonic anhydrase II and tartrate-resistant acid phosphatase mRNAs in putative mononuclear osteoclast precursor cells of neonatal rats by fluorescence *in situ* hybridization. *J. Mol. Endocrinol.* **20**, 211–219.

Hughes, D.E., Wright, K.R., Uy, H.L., Sasaki, A., Yoneda, T., Roodman, G.D., Mundy, G.R. and Boyce, B.F. 1995. Bisphosphanates promote apoptosis in murine osteoclasts *in vitro* and *in vivo*. *J. Bone Miner Res.* **10**, 1478–87.

Hughes, D.E., Dai, A., Tiffee, J.C., Li, H.H., Mundy, G.R. and Boyce, B.F. 1996. Estrogen promotes apoptosis of murine osteoclasts mediated by TGF-beta. *Nature Medicine* **2**, 1132–6.

Hultberg, B., Andersson, A. and Isaksson, A. 1998. Alterations of thiol metabolism in human cell lines induced by low amounts of copper, mercury or cadmium ions. *Toxicology* **126**, 203–12.

Iguchi, H. and Sano, S. 1982. Effect of cadmium on the bone collagen metabolism of rat. *Toxicol. Appl. Pharmacol.* **62**, 126–136.

Imbert, J., Culotta, V., Furst, P., Gedamu, L. and Hamer, D. 1990. Regulation of metallothionein gene transcription by metals. *Adv. Inorg. Chem.* **8**, 139–164.

Immenschuh, S., Iwahara, S., Satoh, H., Nell, C., Katz, N. and Muller-Eberhard, U. 1995. Expression of the mRNA of heme-binding protein 23 is coordinated with that of heme oxygenase-1 by heme and heavy metals in primary rat hepatocytes and hepatoma cells. *Biochemistry* **34**, 13407–11.

Ip, S.P., Mak, D.H., Li, P.C., Poon, M.K. and Ko, K.M. 1996. Effect of a lignan-enriched extract of *Schisandra chinensis* on aflatoxin B1 and cadmium chloride-induced hepatotoxicity in rats. *Pharmacol. Toxicol.* 78, 413–416.

Itokawa, Y., Abe, T., Tabei, R. and Tanaka, S. 1974. Renal and skeletal lesions in experimental cadmium poisoning. *Arch. Environ. Hlth* **28**, 149–154.

Itokawa, Y., Nishino, K., Takashima, M., Nakuta, T., Kaoto, H., Okamoto, E., Daijo, K. and Kawamura, J. 1978. Renal and skeletal lesions in experimental cadmium poisoning of rats: histology and renal function. *Environ. Res.* **15**, 206–217.

Iwami, K. and Moriyama,. T. 1993. Comparative effect of cadmium on osteoblastic cells and osteoclastic cells. *Arch. Toxicol.* **67**, 352–357.

Jimi, E., Nakamura, I., Amano, H., Taguchi, Y., Tsurukai, T., Tamura, M., Takahashi, N. and Suda, T. 1996. Osteoclast function is activated by osteoblastic cells through a mechanism involving cell-to-cell contact. *Endocrinology* **137**, 2187–90.

Jin, T., Nordberg, G.F. and Nordberg, M. 1986. Uptake of cadmium in isolated kidney cells – influence of binding form and *in vivo* treatment. *J. Appl. Toxicol.* **6**, 397–400.

Jin, T.Y., Leffler, P. and Nordberg, G.F. 1987. Cadmium-metallothionein nephrotoxicity in the rat: Transient calcuria and proteinuria. *Toxicology* **45**, 307–317.

Jonah, M.M. and Bhattacharyya, M.H. 1989. Early changes in the tissue distribution of cadmium after oral but not intravenous cadmium exposure. *Toxicology* **58**, 325–338.

Jones, M.M., Xu, C. and Ladd, P.A. 1997. Selenite suppression of cadmium-induced testicular apoptosis. *Toxicology* **116**, 169–175.

Kaji, T., Yamada, H., Hoshino, T., Miyahara, T., Kozuka, H. and Naruse Y. 1986. A possible mechanism of cadmium–copper interaction in embryonic chick bone in tissue culture. *Toxicol. Appl. Pharmacol.* **86**, 243–252.

Kaji, T., Kawatani, R., Takata, M., Hoshino, T., Miyahara, T., Kozuka, H. and Koizumi, F. 1988a. The effects of cadmium, copper or zinc on formation of embryonic chick bone in tissue culture. *Toxicology* **50**, 303–316.

Kaji, T., Takata, M., Hoshino, T., Miyahara, T., Kozuka, H., Kurashige, Y. and Koizumi, F. 1988b. Role of zinc in protection against cadmium-induced toxicity in formation of embryonic chick bone in tissue culture. *Toxicol. Lett.* **44**, 219–227.

Kaji, T., Takata, M., Miyahara, T., Kozuka, H. and Koisumi, F. 1990. Interaction of zinc with cadmium and copper on ossification of embryonic chick bone in tissue culture. *Arch. Environ. Contam. Toxicol.* **19**, 653–656.

Kameda, T., Mano, H., Yuasa, T., Mori, Y., Miyazawa, K., Shiokawa, M., Nakamaru, Y., Hiroi, E., Hiura, K., Kameda, A. *et al.* 1997. Estrogen inhibits bone resorption by directly inducing apoptosis of the bone-resorbing osteoclasts. *J. Exp. Med.* **186**, 489–95.

Katzenellenbogen, B.S., Montano, M.M., Ekena, K., Herman, M.E. and McInerney, E.M. 1997. William L. McGuire Memorial Lecture. Antiestrogens: mechanisms of action and resistance in breast cancer. *Breast Cancer Res. Treat.* **44**, 23–38.

Kawada, T., Tohyama, C. and Suzuki, S. 1990. Significance of the excretion of urinary indicator proteins for a low level of occupational exposure to cadmium. *Int. Arch. Occup. Environ. Hlth* **62**, 95–100.

Kayama, F., Yoshida, T., Elwell, M.R. and Luster, M.I. 1995. Cadmium-induced renal damage and proinflammatory cytokines: Possible role of IL-6 in tubular epithelial cell regeneration. *Toxicol. Appl. Pharmacol.* **134**, 26–34.

Kello, D., Sugawara, N., Voner, C. and Foulkes, E.C. 1979. On the role of metallothionein in cadmium absorption by rat jejunum *in situ*. **Toxicology 14**, 199–208.

Kido, T., Nogawa, K., Yamada, Y., Honda, R., Tsuritani, I., Ishizaki, M. and Yamaya, H. 1989. Osteopenia in inhabitants with renal dysfunction induced by exposure to environmental cadmium. *Int. Arch. Occup. Environ. Hlth* **61**, 271–276.

Kim, K.R. and Park, Y.S. 1995. Phlorizin binding to renal outer cortical brush-border membranes of cadmium-injected rabbits. *Toxicol. Appl. Pharmacol.* **133**, 244–248.

Kim, K.R., Lee, H.Y., Kim, C.K. and Park, Y.S. 1990. Alteration of renal amino acid transport systems in cadmium-intoxicated rats. *Toxicol. Appl. Pharmacol.* **106**, 102–111.

Kim, Y.K., Choi, J.K., Kim, J.S. and Park, Y.S. 1988. Changes in renal function in cadmium-intoxicated rats. *Pharmacol. Toxicol.* **63**, 342–350.

Kimble, R.B., Srivastava, S., Ross, F.P., Matayoshi, A. and Pacifici, R. 1996. Estrogen deficiency increases the ability of stromal cells to support murine osteoclastogenesis via an interleukin-1 and tumor necrosis factor-mediated stimulation of macrophage colony-stimulating factor production. *J. Biol. Chem.* **271**, 28890–28897.

Kimura, M., Otaki, N., Yoshiki, S., Suzuki, M., Horiuchi, N. and Suda, T. 1974. The isolation of metallothionein and its protective role in cadmium poisoning. *Arch. Biochem. Biophys.* **165**, 340–348.

Kimura, O., Endo, T. and Sakata, M. 1997. Uptake of Cd-metallothionein into LLC-PK1 cells: a comparative study with $CdCl_2$. *Biol. Pharm. Bull.* **20**(2), 158–62.

Kinne, R.K., Schutz, H. and Kinne-Saffran, E. 1995. The effect of cadmium chloride *in vitro* on sodium-glutamate cotransport in brush border membrane vesicles isolated from rabbit kidney. *Toxicol. Appl. Pharmacol.* **135**, 216–221.

Kitajima, I., Nakajima, T., Imamura, T., Takasaki, I., Kawahara, K., Okano, T., Tokioka, T., Soejima, Y., Abeyama, K. and Maruyama, I. 1996. Induction of apoptosis in murine clonal osteoblasts expressed by human T-cell leukemia virus type I tax by NF-κB and TNF-α. *J. Bone Miner. Res.* **11**, 200–10.

Kjellstrom, T. 1986. Effects on bone, on vitamin D, and calcium metabolism, in *Cadmium and Health: A Toxicological and Epidemiological Appraisal, Vol. II, Effects and Responses*, (eds L. Friberg, C. Elinder, T. Kjellstrom and G.F. Nordberg), CRC Press, Boca Raton, FL, pp. 111–158.

Kjellstrom, T. 1992. Mechanism and epidemiology of bone effects of cadmium, in *Cadmium in the Human Environment*, (eds G.F. Nordbert, R.F.M. Herber and L. Allesio), International Agency for Research on Cancer, Publ. 188, pp. 301–310.

Klaassen, C.D. and Kotsonis, F.N. 1977. Biliary excretion of cadmium in the rat, rabbit, and dog. *Toxicol. Appl. Pharmacol.* **41**, 101–12.

Knuuttila, M., Lappalainen, R., Olkkonen, H., Lammi, S. and Alhava, E.M. 1982. Cadmium content of human cancellous bone. *Arch. Environ. Hlth* **37**, 290–294.

Koizumi, T., Yokota, T., Ohmori, S., Kumagai, H. and Suzuki, K.T. 1995. Protective effect of metallothionein on intracellular pH changes induced by cadmium. *Toxicology* **95**, 11–17.

Komm, B.S., Terpening, C., Benz, D., Graeme, K., Gallegos, A. and Korc, M. 1988. Estrogen binding, receptor mRNA, and biologic response in osteoblast-like osteosarcoma cells *Science* **241**, 81–4.

Konishi, H., Matsuzaki, H., Tanaka, M., Takemura, Y., Kuroda, S., Ono, Y. and Kikkawa, U. 1997. Activation of protein kinase B (Akt/RAC-protein kinase) by cellular stress and its association with the heat shock protein Hsp27. *FEBS Lett.* **410**, 493–498.

Konz, R.P., Bhattacharyya, M.H. and Seed, T.M. 1989. *Effects of Cadmium on Formation of Osteoclasts*. Proceedings of the Third National Conference Undergraduate Research, Vol. 2, pp. 411–416.

Koropatnick, J. and Cherian, M.G. 1993. A mutant mouse (tx) with increased hepatic metallothionein stability and accumulation. *Biochem. J.* **296**, 443–449.

Kusec, V., Virdi, A.S., Prince, R. and Triffitt, J.T. 1998. Localization of estrogen receptor-alpha in human and rabbit skeletal tissue. *J. Clin. Endocrinol. Metab.* **83**, 2421–2428.

Lauwerys, R. 1979. Cadmium in man, in *The Chemistry, Biochemistry, and Biology of Cadmium*, (ed. M. Webb), Elsevier/North Holland Biomedical Press, New York, pp. 444–447.

Lee, H.Y., Kim, K.R., Woo, J.S., Kim, Y.K. and Park, Y.S. 1990. Transport of organic compounds in renal plasma membrane vesicles of cadmium intoxicated rats. *Kidney Int.* **37**, 727–735.

Lee, H.Y., Kim, K.R. and Park, Y.S. 1991. Transport kinetics of glucose alanine in renal brush-border membrane vesicles of cadmium-intoxicated rabbits. *Pharmacol. Toxicol.* **69**, 390–5.

Lieberherr, M., Grosse, B., Kachkache, M. and Balsan, S. 1993. Cell signaling and estrogens in female rat osteoblasts: a possible involvement of unconventional nonnuclear receptors. *J. Bone Miner. Res.* **8**, 1365–76.

Lind, Y. and Wicklund-Glynn, A. 1997. The involvement of metallothionein in the intestinal absorption of cadmium in mice. *Toxicol. Lett.* **91**, 179–187.

Litchfield, T.M. and Sauer, G.R. 1996. Metallothionein induction in calcifying growth plate cartilage chondrocytes. *Connect. Tissue Res.* **35**, 189–195.

Liu, J., Kershaw, W.C., Liu, Y.P. and Klaassen, C.D. 1992. Cadmium-induced hepatic endothelial cell injury in inbred strains of mice. *Toxicology* **75**, 51–62.

Liu, J., Liu, Y. and Klaassen, C.D. 1994a. Nephrotoxicity of $CdCl_2$ and Cd-metallothionein in cultured rat kidney proximal tubules and LLC-PK1 cells. *Toxicol. Appl. Pharmacol.* **128**, 264–270.

Liu, J., Liu, Y. and Klaassen, C.D. 1994b. The effect of Chinese hepatoprotective medicines on experimental liver injury in mice. *J. Ethnopharmacol.* **42**, 183–191.

Liu, J., Liu, Y. and Klaassen, C.D. 1995. Protective effect of oleanolic acid against chemical-induced acute necrotic liver injury in mice. *Chung Kuo Yao Li Hsueh Pao* **16**, 97–102.

Liu, J., Liu, Y., Michalska, A.E., Choo, K.H. and Klaassen, C.D. 1996a. Metallothionein plays less of a protective role in cadmium-metallothionein-induced nephrotoxicity than in cadmium chloride-induced hepatotoxicity. *J. Pharmacol. Exp. Ther.* **276**, 1216–1223.

Liu, J., Squibb, K.S., Akkerman, M., Nordberg, G.F., Lipsky, M. and Fowler, B.A. 1996b. Cytotoxicity, zinc protection, and stress protein induction in rat proximal tubule cells exposed to cadmium chloride in primary cell culture. *Ren. Fail.* **18**, 867–882.

Liu, Y.P., Liu, J., Palmiter, R.D. and Klaassen, C.D. 1996, Metallothionein-I-transgenic mice not protected from acute cadmium-metallothionein-induced nephrotoxicity. *Toxicol. Appl. Pharmacol.* **137**, 307–315.

Long, G.J. 1997a. Cadmium perturbs calcium homeostasis in rat osteosarcoma (ROS 17/2.8) cells; a possible role for protein kinase C. *Toxicol. Lett.* **91**, 91–7.

Long, L. J. 1997b. The effect of cadmium on cytosolic free calcium, protein kinase C and collagen synthesis in rat osteosarcoma (ROS17/2.8) cells. *Toxicol. Appl. Pharmacol.* **143**, 189–195.

Luckman, S.P., Hughes, D.E., Coxon, F.P., Graham, R., Russell, G. and Rogers, M.J. 1998. Nitrogen-containing bisphosphonates inhibit the mevalonate pathway and prevent post-translational prenylation of GTP-bindin proteins, including Ras. *J. Bone Miner. Res.* **13**, 581–9.

Luethy, J. D. and Holbrook, N. J. 1994. The pathway regulating GADD153 induction in response to DNA damage is independent of protein kinase C and tyrosine kinases. *Cancer Res.* **54**, 1902S–1906S.

McDonnell, D.P. and Norris, J.D. 1997. Analysis of the molecular pharmacology of estrogen receptor agonists and antagonists provides insights into the mechanism of action of estrogen in bone. *Osteoporos. Int.* **7** (Suppl. 1), S29–S34.

Maitani, T., Waalkes, M.P. and Klassen, C.D. 1984. Distribution of cadmium after oral administration of cadmium–thionein to mice. *Toxicol. Appl. Pharmacol.* **74**, 237–43.

Masters, B.A., Kelly, E.J., Quaife, C.J., Brinster, R.L. and Palmiter, R.D. 1994. Targeted disruption of metallothionein I and II genes increases sensitivity to cadmium. *Proc. Natl Acad. Sci., USA* **91**, 584–588.

Matsuoka, M. and Call, K.M. 1995. Cadmium-induced expression of immediate early genes in LLC-PK1 cells. *Kidney Intl.* **48**, 383–389.

Michalska, A.E. and Choo, K.H. 1993. Targeting and germ-line transmission of a null mutation at the metallothionein I and II loci in mouse. *Proc. Natl Acad. Sci., USA* **90**, 8088–8092.

Mikhailova, M.V., Littlefield, N.A., Hass, B.S., Poirier, L.A. and Chou, M.W. 1997. Cadmium-induced 8-hydroxydeoxyguanosine formation, DNA strand breaks and antioxidant enzyme activities in lymphoblastoid cells. *Cancer Lett.* **115**, 141–148.

Min, K.S., Fujita, Y., Onosaka, S. and Tanaka, K. 1991. Role of intestinal metallothionein in absorption and distribution of orally administered cadmium. *Toxicol. Appl. Pharmacol.* **109**, 7–16.

Min, K.S., Nakatsubo, T., Kawamura, S., Fujita, Y., Onosaka, S. and Tanaka, K. 1992. Effects of mucosal metallothionein in small intestine on tissue distribution of cadmium after oral administration of cadmium compounds. *Toxicol. Appl. Pharmacol.* **113**, 306–310.

Miyahara, T., Kato, T., Nakagawa, S., Kozuka, H., Sakai, T., Nomura, N. and Takayanagi, N. 1978. Influence of poisonous metals on bone metabolism. I. The effect of cadmium on the ossification of chick-embryo tibia in tissue culture. *Eisei Kagaku* **24**, 36–42.

Miyahara, T., Miyokoshi, I., Saito, Y. and Kozuka, H. 1980a. Influence of poisonous metals on bone metabolism. III. The effect of cadmium on bone resorption in tissue culture. *Toxicol. Appl. Pharmacol.* **55**, 477–83.

Miyahara, T., Miyokoshi, M. and Kozuka, H. 1980b. Effect of cadmium on bone resorption in cultured fetal bones. *Bull. Environ. Contam. Toxicol.* **25**, 294–7.

Miyahara, T., Oh, E.Y., Takaine, E. and Kozuka, H. 1983. Interaction between cadmium and zinc, copper, or lead in relation to the collagen and mineral content of embryonic chick bone in tissue culture. *Toxicol. Appl. Pharmacol.* **67**, 41–48.

Miyahara, T., Yamada, H., Ando, R., Nemoto, S., Kaji, T., Mori, M., Kozuka, H., Itoh, N. and Sudo, H. 1986. The effects of cadmium on a clonal osteogenic cell, MC3T3-E1: Inhibition of calcification and induction of metallothionein-like protein by cadmium. *Toxicol. Lett.* **32**, 19–27.

Miyahara, T., Yamada, H., Takeuchi, M., Kozuka, H., Kato, T. and Sudo, H. 1988. Inhibitory effects of cadmium on *in vitro* calcification of a clonal osteogenic cell, MC3T3-E1. *Toxicol. Appl. Pharm.* **96**, 52–9.

Miyahara, T., Takata, M., Miyata, M., Nagai, M., Sugure, A., Kozuka, H. and Kuze, S. 1991. Cadmium stimulated osteoclast-like multinucleated cell formation in mouse bone marrow cell cultures. *Bull. Environ. Contam. Toxicol.* **47**, 283–287.

Miyahara, T., Masakazu, T., Mori-Uchi, S., Miyata, M., Nagai, M., Sugure, A., Matsusta, M., Kozuka, H. and Kuze, S. 1992. Stimulative effects of cadmium on bone resorption in neonatal parietal bone resorption. *Toxicology* **73**, 93–99.

Miyaura, C., Onoe, Y., Inada, M., Maki, K., Ikuta, K., Ito, M. and Suda, T. 1997. Increased B-lymphopoiesis by interleukin 7 induces bone loss in mice with intact ovarian function: similarity to estrogen deficiency. *Proc. Natl Acad. Sci., USA* **94**, 9360–9365.

Morimoto, R. I. 1993. Cells in stress: transcriptional activation of heat shock genes. *Science* **259**, 1409–1410.

Noda, M. and Kitagawa, M. 1990. A quantitative study of iliac bone histopathology on 62 cases with itai-itai disease. *Calcif. Tissue Int.* **47**, 66–74.

Nogawa, K. 1981. Itai-Itai disease and follow-up studies, in *Cadmium in the Environment*, (ed. J.O. Nriagu), John Wiley & Sons, New York, pp. 1–37.

Nordberg, G.F., Goyer, R. and Nordberg, M. 1975. Comparative toxicity of cadmium-metallothionein and cadmium chloride on mouse kidney. *Arch. Pathol.* **99**, 192–197.

Nordberg, M., Jin, T. and Nordberg, G.F. 1992. Cadmium, metallothionein and renal tubular toxicity. *IARC Sci. Publ.* **118**, 293–297.

Ogoshi, K., Moriyama, T. and Nanzai, Y. 1989. Decrease in the mechanical strength of bones of rats administered cadmium. *Arch. Toxicol.* **63**, 320–324.

Ogoshi, K., Nanzai, Y. and Moriyama, T. 1992. Decrease in bone strength of cadmium-treated young and old rats. *Arch. Toxicol.* **66**, 315–320.

Ohta, H. and Cherian, M.G. 1991 Gastrointestinal absorption of cadmium and metallothionein. *Toxicol. Appl. Pharmacol.* **107**, 63–72.

Ohta, H., DeAngelis, M.V. and Cherian, M.G. 1989. Uptake of cadmium and metallothionein by rat everted intestinal sacs. *Toxicol. Appl. Pharmacol.* **101**, 62–69.

Onoe, Y, Miyaura, C., Ohta, H., Nozawa, S. and Suda, T. 1997. Expression of estrogen receptor beta in rat bone. *Endocrinology* **138**, 4509–4512.

Ossola, J.O. and Tomaro, M.L. 1995. Heme oxygenase induction by cadmium chloride: evidence for oxidative stress involvement. *Toxicology* **104**, 141–147.

Oursler, M.H., Osdoby, P., Pyfferoen, J., Riggs, B.L. and Spelsberg, T.C. 1991. Avian osteoclasts as estrogen target cells. *Proc. Natl Acad. Sci., USA* **88**, 6613–17.

Oursler, M.J., Landers, J.P., Riggs, B.L. and Spelsberg, T.C. 1993. Oestrogen effects on osteoblasts and osteoclasts. *Ann. Med.* **25**, 361–71.

Ovelgonne, J.H., Souren, J.E.M., Wiegant, F.A.C. and Van Wijk, R. 1995. Relationship between cadmium-induced expression of heat shock genes, inhibition of protein synthesis and cell death. *Toxicology*, **99**, 19–30.

Pacifici, R. 1996. Estrogen, cytokines, and pathogenesis of postmenopausal osteoporosis. *J. Bone Min. Res.* **11**, 1043–1051.

Pacifici, R. 1997. Idiopathic hypercalciuria and osteoporosis – distinct clinical manifestations of increased cytokine-induced bone resorption. *J. Clin. Endocrinol. Metab.* **82**, 25–31.

Palmiter, R. D. 1987. Molecular biology of metallothionein gene expression. *Experimentia (Suppl.)* **52**, 63–80.

Parfitt, A.M., Mundy, G.R., Roodman, G.D., Hughes, D.E. and Boyce, B.F. 1996. A new model for the regulation of bone resorption, with particular reference to the effects of bisphosphanates. *J. Bone Miner. Res.* **11**, 150–9.

Park, K., Kim, K.R., Kim, J.Y. and Park, Y.S. 1997. Effect of cadmium on Na-P_i cotransport kinetics in rabbit renal brush-border membrane vesicles. *Toxicol. Appl. Pharmacol.* **145**, 255–259.

Phillpotts, C. J. 1986. Histopathological changes in the epithelial cells of rat duodenum following chronic dietary exposure to cadmium, with particular reference to Paneth cells. *Br. J. Exp. Pathol.* **4**, 505–516.

Qu, Q., Perala-Heape, M., Kapanen, A., Dahllund, J., Salo, J., Vaananen, H.K. and Harkonen, P. 1998. Estrogen enhances differentiation of osteoblasts in mouse bone marrow culture. *Bone* **22**, 201–209.

Radtke, F., Heuchel, R., Georgiev, O., Hergersberg, M., Gariglio, M., Bembic, Z. And Schaffner, W. 1993. Cloned transcription of factor MTF-1 activates the mouse metallothionein I promoter. *EMBO J.* **12**, 1355–1362.

Rajan, S.S., Wilson, A.K. and Bhattacharyya, M.H. 1999. Role of metallothionein (MT) in the uptake and distribution of environmental levels of cadmium in mice. *Toxicologist*, **48**, 294.

Robinson, J.A., Harris, S.A., Riggs, B.L. and Spelsberg, T.C. 1997. Estrogen regulation of human osteoblastic cell proliferation and differentiation. *Endocrinology* **138**, 2919–2927.

Rodriguez-Paris, J.M., Shoji, M., Heola, S., Liotta, D., Vogler, W.R. and Kuo, J.F. 1989. Fluorimetric studies of protein kinase C interactions with phospholipids. *Biochem. Biophys. Res. Commun.* **159**, 495–500.

Sacco-Gibson, N., Chaudhury, S., Brock, A., Sickles, A.B., Patel, B., Hegstad, R., Johnston, S., Peterson, D. and Bhattacharyya, M.H. 1992. Cadmium effects on bone metabolism: accelerated resorption in ovariectomized, aged beagles. *Toxicol. Appl. Pharmacol.* **113**, 274–283.

Salminen, W.F. Jr Voellmy, R. and Roberts, S.M. 1996. Induction of hsp70 in HepG2 cells in response to hepatotoxicants. *Toxicol. Appl. Pharmacol.* **141**, 117–123.

Sarkar, S., Yadav, P. and Bhatnagar, D. 1998. Lipid peroxidative damage of cadmium exposure and alterations in autioxidant system in rat erythrocytes: a study with relation to time. *Biometals* **11**, 153–7.

Sarma, U., Edwards, M., Motoyoshi, K. and Flanagan, A.M. 1998. Inhibition of bone resorption by 17beta-estradiol in human bone marrow cultures. *J. Cell. Physiol.* **175**, 99–108.

Sauer, G.R., Nie, D., Wu, L.N. and Wuthier, R.E. 1998. Induction and characterization of metallothionein in chicken epiphyseal growth plate cartilage chondrocytes. *J. Cell. Biochem.* **68**, 110–20.

Sauer, J.M., Waalkes, M.P., Hooser, S.B., Kuester, R.K., McQueen, C.A. and Sipes I.G. 1997a. Suppression of Kupffer cell function prevents cadmium induced hepatocellular necrosis in the male Sprague–Dawley rat. *Toxicology* **121**, 155–164.

Sauer, J.M., Waalkes, M.P., Hooser, S.B., Baines, A.T., Kuester, R.K. and Sipes, I.G. 1997b. Tolerance induced by all-trans-retinol to the hepatotoxic effects of cadmium in rats: Role of metallothionein expression. *Toxicol. Appl. Pharmacol.* **143**, 110–119.

Schumann, K., Friebel, P., Schmolke, G. and Elsenhans, B. 1996. State of iron repletion and cadmium tissue accumulation as a function of growth in young rats after oral cadmium exposure. *Arch. Environ. Contam. Toxicol.* **4**, 483–487.

Selander, K.S., Monkkonen, J., Karhukorpi, E.K., Harkonen, P., Hannuniemi, R. and Vaananen, H.K. 1996. Characteristics of clodronate-induced apoptosis in osteoclasts and macrophages. *Mol. Pharmacol.* **50**, 1127–38.

Shaikh, Z.A., Blazka, M.E. and Endo, T. 1995. Metal transport in cells. Cadmium uptake by rat hepatocytes and renal cortical epithelial cells. *Environ. Hlth Perspect.* **103**, 73–75.

Shimada, H., Bare, R.M., Hochadel, J.F. and Waalkes, M.P. 1997a. Testosterone pretreatment mitigates cadmium toxicity in male C57 mice but not in C3H mice. *Toxicology* **116**, 183–191.

Shimada, H., Hochadel, J.F. and Waalkes, M.P. 1997b. Progesterone pretreatment enhances cellular sensitivity to cadmium despite a marked activation of the metallothionein gene. *Toxicol. Appl. Pharmacol.* **142**, 178–185.

Shimada, H., Shiao, Y.H., Shibata, M. and Waalkes, M.P. 1998. Cadmium suppresses apoptosis induced by chromium. *J. Toxicol. Environ. Hlth* **54**, 159–168.

Shiraishi, N., Barter, R.A., Uno, H. and Waalkes, M.P. 1993. Effect of progesterone pretreatment on cadmium toxicity in the male Fischer (F344/NCr) rat. *Toxicol. Appl. Pharmacol.* **118**, 113–118.

Shukla, G.S. and Chandra, S.V. 1989. Cadmium toxicity and bioantioxidants: status of vitamin E and ascorbic acid of selected organs in rat. *J. Appl. Toxicol.* **9**, 119–122.

Smith, J.B., Dwyer, S.D. and Smith, L. 1989. Cadmium invokes inositol polyphosphate formation and calcium mobilization. *J. Biol. Chem.* **264**, 7115–7118.

Sorensen, J.A., Nielsen, J.B. and Andersen, O. 1993. Identification of the gastrointestinal absorption site for cadmium chloride *in vivo*. *Pharmacol. Toxicol.* **73**, 169–173.

Souza, V., Bucio, L., Jay, D., Chavez, E. and Gutierrez-Ruiz, M.C. 1996. Effect of cadmium on calcium transport in a human fetal hepatic cell line (WRL-68 cells). *Toxicology* **112**, 97–104.

Souza, V., Bucio, L. and Gutierrez-Ruiz, M.C. 1997. Cadmium uptake by a human hepatic cell line (WRL-68 cells). *Toxicology* **120**, 215–220.

Stohs, S.J. and Bagchi, D. 1995. Oxidative mechanisms in the toxicity of metal ions. *Free Radicals Biol. Med.* **18**, 321–336.

Sugawara, N. and Sugawara, C. 1991. Gastrointestinal absorption of Cd-metallothionein and cadmium chloride in mice. *Arch. Toxicol.* **65**, 689–692.

Sugawara, N., Lai, Y.R., Arizono, K. and Ariyoshi, T. 1996. Biliary excretion of exogenous cadmium, and endogenous copper and zinc in the Eisai hyperbilirubinuric (EHB) rat with a near absence of biliary glutathione. *Toxicology* **112**, 87–94.

Sugawara, N., Lai, Y.R., Arizono, K., Kitajima, T. and Inoue, H. 1997. Lack of biliary excretion of Cd linked to an inherent defect of the canalicular isoform of multidrug resistance protein (cMrp) does not abnormally stimulate accumulation of Cd in the Eisai hyperbilirubinemic (EHB) rat liver. *Arch. Toxicol.* **71**, 336–339.

Sumathi, R., Baskaran, G. and Varalakshmi, P. 1996. Relationship between glutathione and DL alpha-lipoic acid against cadmium-induced hepatotoxicity. *Jpn J. Med. Sci. Biol.* **49**, 39–48.

Suzuki, C.A. and Cherian, M.G. 1988. Effects of cadmium-metallothionein on renal organic ion transport and lipid peroxidation in rats. *J. Biochem. Toxicol.* 3, 11–20.

Suzuki, Y., Morita, I., Yamane, Y. and Murota, S. 1989. Cadmium stimulates prostaglandin E_2 production and bone resorption in cultured fetal mouse calvaria. *Biochem. Biophys. Res. Commun.* **158**, 508–13.

Suzuki, Y., Morita, I., Yamane, Y. and Murota, S. 1990. Preventive effect of zinc against cadmium-induced bone resorption. *Toxicology* **62**, 27–34.

Takeda, K., Ishizawa, S., Sato, M., Yoshida, T. and Shibahara, S. 1994. Identification of a *cis*-acting element that is responsible for cadmium-mediated induction of the human heme oxygenase gene. *J. Biol. Chem.* **269**, 22858–22867.

Takeuchi, J. 1978. *Etiology of itai-itai disease*. Proceedings of the First International Cadmium Conference, Metal Bulletin, New York, pp. 222–224.

Tamura, Y. and Ohya, K. 1995. [Study of cadmium-induced metallothionein-like protein in bone marrow of rabbit.] *Kokubyo Gakkai Zasshi* **62**, 404–9.

Tang, N. and Enger, M.D. 1993. Cd^{2+}-induced c-myc mRNA accumulation in NRK-49F cells is blocked by the protein kinase inhibitor H7 but not by HA1004, indicating that protein kinase C is a mediator of the response. *Toxicology* **81**, 155–164.

Thomas, D.J., Angle, C.R. and Swanson, S.A. 1990. Effect of 5-azacytidine on metallothionein inducibility and sensitivity to lethality of cadmium in rat osteosarcoma (ROS 17/2.8) cells. *Toxicology* **65**, 169–178.

Thun, M.J., Osorio, A.M., Schober, A., Hannon, W.H., Lewis, B. and Halperin, W. 1989. Nephropathy in cadmium workers: assessment of risk from airborne occupational exposure to cadmium. *Br. J. Ind. Med.* **46**, 689–97.

Tomkinson, A., Reeve, J., Shaw, R.W. and Noble, B.S. 1997. The deaths of osteocytes via apoptosis accompanies estrogen withdrawal in human bone. *J. Clin. Endocrinol. Metab* **82**, 3128–35.

Tsuritani, I., Honda, R., Ishizaki, M., Yamada, Y., Oshima, K. and Kasuya, M. 1994. Serum bone-type alkaline phosphatase activity in women living in a cadmium-polluted area. *Toxicol. Lett.* **71**, 209–216.

Tyrell, R.M., Applegate, L.A. and Tromvoukis, Y. 1993. The proximal promoter region of the human heme oxygenase gene contains elements involved in stimulation of transcriptional activity by a variety of agents including oxidants. *Carcinogenesis* **14**, 761–765.

Vestergaard, P. and Shaikh, Z.A. 1994. The nephrotoxicity of intravenously administered cadmium-metallothionein: Effect of dose, mode of administration, and preexisting renal cadmium burden. *Toxicol. Appl. Pharmacol.* **126**, 240–247.

Wang, Z. and Templeton, D.M. 1998. Induction of c-fos proto-oncogene in mesangial cells by cadmium. *J. Biol. Chem.* **273**, 73–79.

Weinstein, R.S., Jilka, R.L., Michael Parfitt, A. and Manolagas, S.C. 1998. Inhibition of osteoblastogenesis and promotion of apoptosis of osteoblasts and osteocytes by glucocorticoids. Potential mechanisms of their deleterious effects on bone. *J. Clin. Invest.* **102**, 274–82.

Wilson, A.K. and Bhattacharyya, M.H. 1997. Effects of cadmium on bone: An in vivo model for the early response. *Toxicol. Appl. Pharmacol.* **145**, 68–74.

Wilson, A.K., Cerny, E.A., Smith, B.D., Wagh, A. and Bhattacharyya, M.H. 1996. Effects of cadmium on osteoclast formation and activity in vitro. *Toxicol. Appl. Pharmacol.* **140**, 451–460.

Wu, B.J., Kingston, R.E. and Morimoto, R.I. 1986. Human HSP70 promoter contains at least two distance regulatory domains. *Proc. Natl Acad. Sci., USA* **83**, 629–633.

Yoshiki, S., Yanagisowa, T., Kimura, M., Otaki, N., Suzuki, M. and Suda, T. 1975. Bone and kidney lesions in experimental cadmium intoxication. *Arch. Environ. Hlth* **30**, 559–562.

Zglinicki, T., von Edwall, C., Ostlund, E., Lind, B. Nordberg, M., Ringertz, N.R. and Wroblewski, J. 1992. Very low cadmium concentrations stimulate DNA synthesis and cell growth. *J. Cell Sci.* **103**, 1073–1081.
Zheng, H., Liu, J., Choo, K.H., Michalska, A.E. and Klaasen, C.D. 1996. Metallothionein-I and -II knock-out mice are sensitive to cadmium-induced liver mRNA expression of c-jun and p53. *Toxicol. Appl. Pharmacol.* **136**, 229–235.

3

ARSENIC

Eric Wildfang, Sheila M. Healy and H. Vasken Aposhian

3.1 Introduction

The number and diversity of populations exposed to high levels of inorganic arsenic in their drinking water have increased since the mid-1950s. From the early reports from Taiwan (Tseng, 1977) and Antofagasta, Chile (Borgono *et al.*, 1977), areas with toxic levels of arsenic in their drinking water now include India (Guha Mazumder *et al.*, 1988 and China (Taylor *et al.*, 1989). Yet, the mechanisms by which this metalloid exerts it toxic vascular and carcinogenic effects at the molecular level are still relatively unknown. In this chapter, we have made a point of reviewing some of the older scientific literature since, at a time of ignorance, not only should we be familiar with the most recent findings but also with the less recent.

Arsenic is a naturally occurring, ubiquitous metalloid belonging to the group V(A) elements (N, P, As, Sb and Bi). The toxicity of this environmental toxicant is complex and depends, in part, on its chemical form, dose, route and duration of exposure, degree of accumulation, rate of clearance and animal species. The intricate interplay between all of these variables is only beginning to be unraveled and may hold the key to understanding the mechanisms by which arsenicals produce their toxic and carcinogenic actions.

Arsenic can exist in three possible oxidation states: elemental (0), trivalent (+3, e.g. arsenite; or –3, e.g. arsine) and pentavalent (+5, e.g. arsenate). Because it has multiple and interconvertible oxidation states, arsenic can participate in a number of chemical and biological reactions, including oxidation–reduction reactions, acid–base reactions, covalent interactions with most nonmetals and metals and methylation–demethylation reactions. The complexity and interactions of these reactions have, in part, contributed to the difficulty of understanding its toxicological mechanisms of action.

In general, inorganic forms of arsenic (e.g. arsenite or arsenate) are more toxic than organic forms (e.g., methylarsonate, dimethylarsinate or arsenobetaine). However, elemental arsenic does not appear to demonstrate any significant toxicity. Arsine, the fully reduced gaseous form of arsenic, is considered to be the most toxic arsenical (Squibb and Fowler, 1983). The different chemical forms of arsenic present different mechanisms of actions. The pentavalent arsenates uncouple mitochondrial oxidative phosphorylation (Crane and Lipmann, 1953; Ter Welle and Slater, 1967). Arsenate can competitively substitute for inorganic phosphate in such reactions as endogenous ATP biosynthesis, resulting in an unstable arsenate ester which can be rapidly hydrolyzed. This results in loss of energy-rich ATP, rather than its biochemical conservation.

Table 3.1 Formulae of commonly occurring arsenicals

Arsenite	AsO_2^-, AsO_3^{3-}
Arsenate	$H_2AsO_4^-$, $HAsO_4^{2-}$, AsO_4^{3-}
Arsenous acid	H_3AsO_3
Arsenic acid	H_3AsO_4
Arsenic trioxide	As_2O_3
Arsenic pentoxide	As_2O_5
Gallium arsenide	GaAs
Indium arsenide	InAs
Arsine	AsH_3
Monomethylarsonic acid (MMA)	$CH_3AsO(OH)_2$
Dimethylarsinic acid (DMA)	$(CH_3)_2AsO(OH)$
Trimethylarsine oxide	$(CH_3)_3AsO$
Arsenobetaine	$(CH_3)_3As^+CH_2COO^-\ X^-$
Arsenocholine	$(CH_3)_3As^+CH_2CH_2OH\ X^-$

Trivalent arsenite inhibits a number of enzymes, such as the pyruvate dehydrogenase complex, through its interaction with the sulfhydryls of the enzyme or its cofactors (Webb, 1963). Arsine exposure results in significant hematological disturbances (Fowler and Weissberg, 1974). The toxic actions of other arsenicals and arsenic compounds produced by biotransformation are less understood. In particular, the chemical species or form(s) responsible for the carcinogenic action of arsenic in humans have yet to be identified. The chemical structures of some common arsenicals are presented in Table 3.1.

At present, there exists an urgent need to better understand the mechanisms of action of the toxicity and carcinogenicity of arsenicals. Currently, millions of people around the world are exposed to elevated levels of natural or anthropogenic sources of arsenic known to cause adverse health effects. India (Guha Mazumder *et al.*, 1988; Mudur, 1996), Taiwan (Wu *et al.*, 1989; Chen and Wang, 1990; Chen *et al.*, 1992; Chiou *et al.*, 1995), Mexico (Cebrian *et al.*, 1983) and Chile (Borgono *et al.*, 1977) all have well-defined areas in which populations exhibit the signs of arsenic exposure, except in San Pedro de Atacama in Chile (Aposhian *et al.*, 1997). Chronic arsenic exposure is associated with increased frequencies of liver, lung, bladder, kidney and skin cancers in humans (Taylor *et al.*, 1989; Wu *et al.*, 1989; Jarup and Pershagen, 1991; Bates *et al.*, 1992; Chen *et al.*, 1992; Chiou *et al.*, 1995). An understanding of the factors controlling disposition, metabolism and elimination of arsenicals may help elucidate the mechanisms by which these compounds produce their toxic and carcinogenic effects.

3.2 Organic and inorganic forms of bioavailable arsenic

The biological availability of arsenic is primarily dependent on the chemical and physical properties of the compound (Hindmarsh and McCurdy, 1986). The chemical form of arsenic complicates the determination of its bioavailability since its chemical species can often interconvert, so that the chemical form administered is not necessarily the form at the site of its toxic action or being eliminated. The ability of most animals to metabolize arsenicals further confounds the understanding of which chemical species is actually being absorbed and which is the ultimate toxicant. In most instances, some of the parent compound administered can be detected in biological tissues or fluids.

3.2.1 *Organic arsenicals*

Until recently, alkylated arsenicals, such as monomethylarsonic acid (MMA) and dimethylarsinic acid (DMA), were commonly used in the agricultural industry as insecticides and herbicides (Hindmarsh and McCurdy, 1986; Pershagen, 1986). In solution, these compounds are quite readily absorbed from the gastrointestinal tract (Buchet *et al.*, 1981a; Yamauchi and Yamamura, 1984b; Marafante *et al.*, 1987b; Yamauchi *et al.*, 1988). Following intratracheal instillation of DMA, the extent and rate of lung absorption was greater than that in the gastrointestinal tract (Stevens *et al.*, 1977). Arsanilic acid, a feed additive commonly used in agriculture, is poorly absorbed (<21% of a dose was recovered in the urine over 6 days) from the human gastrointestinal tract following oral administration as a solution or on ingestion of the tissue of poultry that had been fed arsanilic acid (Calesnick *et al.*, 1966).

Marine organisms contain large amounts of organoarsenicals, such as the trimethylarsenic compounds arsenobetaine and arsenocholine. These compounds are easily absorbed following ingestion (Yamauchi *et al.*, 1986; Le *et al.*, 1994). They are not acutely toxic, since they have low biological reactivity and are rapidly excreted from the body (Crecelius, 1977; Buchet *et al.*, 1994). Arsenosugars, the major arsenic species in seaweed, may be further metabolized *in vivo* and have a more prolonged retention time in the human body (Le *et al.*, 1994).

3.2.2 *Inorganic arsenicals*

Inorganic arsenic is the most prevalent form of arsenic in the environment. It is found in soils, plants, animals, fresh and sea water, and the atmosphere (Winship, 1984). Anthropogenic sources have greatly contributed to arsenic contamination of the environment.

Since before the time of the ancient Greeks, various forms of inorganic arsenic have been used for medicinal, as well as homicidal and suicidal, purposes. Hippocrates recommended the paste of arsenic sulfide for the treatment of ulcers (Linsella *et al.*, 1972). In 1786, Thomas Fowler introduced his solution of 1% potassium arsenite for use in the cures of agues, remitting fevers and headaches (Fowler, 1786). Chronic use was associated with arsenic poisoning. Medical folklore also indicates that various inorganic arsenic-containing tonics were consumed in the therapy of acne, psoriasis, sexual disorders, anemia and rheumatism (Winship, 1984). Today, most medicinal uses of inorganic arsenicals have been replaced by safer, more effective therapeutic agents. Arsenicals, however, remain important in the chemotherapy of trypanosomal infections (Flynn and Bowman, 1974; Williamson, 1976; Loiseau *et al.*, 1993) and appear to be efficacious in the treatment of acute promyelocytic leukemia (Chen *et al.*, 1996a; Zhu *et al.*, 1997). Additionally, among practitioners of homeopathic medicine and traditional Chinese medicine, some medicinal applications of arsenicals persist (Espinoza *et al.*, 1996).

Occupational exposure to inorganic arsenic is associated with smelting of nonferrous ores (Vahter *et al.*, 1986; Jarup and Pershagen, 1991), agricultural use (Woolson, 1975), fossil fuel-burning power plants (Weiss *et al.*, 1983), lumber preservation (Baldwin, 1983) and the manufacture of electronics (Willardson, 1983). Inorganic arsenicals used in the manufacture of semiconductor materials for the electronics industry include gallium and indium arsenides.

The primary routes of nonoccupational exposure of humans to inorganic arsenic are via food and water (Pershagen, 1986). Inorganic arsenic is present in both natural and processed foods, predominantly in the pentavalent oxidation state (Winship, 1984). Foods showing the highest concentrations of arsenic include marine products (especially shellfish), poultry and some grain products (Jelinek and Corneliussen, 1977). Humans regularly consuming fish and shellfish show increased excretion of inorganic arsenic (Buchet *et al.*, 1996a). However, it is possible that some arsenic compounds in foods are not measured because of inadequate extraction procedures.

Arsenic is naturally present in all waters, with a mean concentration approaching 1–2 μg/L (Onishi, 1968). It can reach toxic levels (Borgono *et al.*, 1977). Groundwater may be naturally contaminated, due to geologic weathering of arsenic-bearing minerals. Either trivalent or pentavalent arsenic is present, depending upon the depth and chemistry of the aquifer (Hindmarsh and McCurdy, 1986) as well as the season of the year.

The bioavailability of arsenic in soil sources appears to be poor. In dogs given ore-containing soil mixed with food, the mean bioavailability of the inorganic arsenic dose (63 μg/kg) was approximately 8.3% (Groen *et al.*, 1993). Soils near smelters were orally administered to rabbits and demonstrated higher bioavailability, with approximately 24% of the dose being absorbed (Freeman *et al.*, 1993). No dose dependency for soil arsenic bioavailability was observed (Freeman *et al.*, 1993). The oral bioavailability of arsenic from soil and house dust near smelter operations, as determined in cynomolgus monkeys, were 19 and 14%, respectively (Freeman *et al.*, 1995). It was suggested in these latter two studies that the poor bioavailability of soil and dust arsenic may be due to the arsenic existing primarily as insoluble mineral forms (Freeman *et al.*, 1993, 1995). This demonstrates the importance of arsenic speciation in environmental samples. Among children, the hand-to-mouth exposure route may be significant in areas having arsenic-contaminated soil. Styrian mountaineers in Austria used to eat relatively large quantities of inorganic arsenic (the chemical form is uncertain) as they believed that it helped endurance at high altitudes and increased their weight, strength and appetite (Schroeder and Balassa, 1966).

Elemental arsenic does not appear to be toxic upon ingestion, even in substantial amounts, since it is poorly absorbed.

3.3 Disposition of arsenic

3.3.1 Absorption

Variable rates of absorption of arsenicals from the gastrointestinal tract appear to be dependent on chemical form (Fig. 3.1) and dose, although interindividual and interspecies variability cannot be discounted.

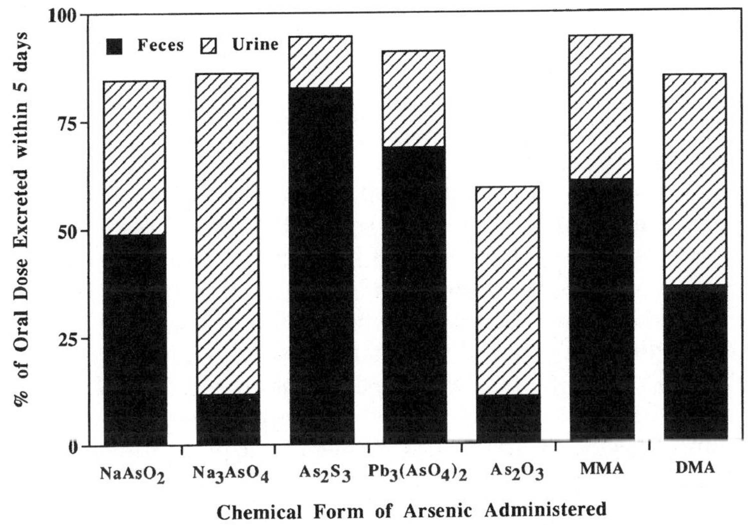

Figure 3.1 Differences in urinary and fecal elimination of total arsenic within 5 days following oral administration of either $Na^{74}AsO_2$ (As III), $Na_3{}^{74}AsO_4$ (As V), $^{74}As_2S_3$ or $Pb_3(^{74}AsO_4)_2$ (2 mg/kg body weight; Marafante and Vahter, 1987); As_2O_3 (4.5 mg/kg body weight; Yamauchi and Yamamura, 1985); MMA or DMA (50 mg/kg body weight; Yamauchi *et al.*, 1988, 1992) in hamsters.

Following ingestion of 63 μg of inorganic arsenic (50 μg arsenite and 13 μg arsenate), a human volunteer excreted 80% of the dose over the following 61 hours, suggesting nearly complete absorption (Crecelius, 1977). Unfortunately, this and a number of other experiments have used only one human. Mice showed a similar percentage of total dose excretion following oral dosing with arsenate (Hughes *et al.*, 1994). Hamsters orally dosed with radiolabeled arsenate exhibited poor absorption, since approximately 70% of the dose was excreted in the feces (Charbonneau *et al.*, 1980). Only a fraction was attributable to biliary elimination. Marine forms (unidentified organoarsenicals) of arsenic appear to be quite readily absorbed (Crecelius, 1977; Charbonneau *et al.*, 1978).

For populations chronically exposed to arsenic via contaminated water, soil, or the burning of high-arsenic-containing coal, dermal absorption represents a limited, yet significant, route of exposure. In rats, cutaneous absorption of arsenite through intact skin occurred (Dutkiewicz, 1977). In a rhesus monkey model, radiolabeled arsenate, in both soil and water formulations, was applied topically to abdominal skin for 24 hours (Wester *et al.*, 1993). Percutaneous absorption of arsenic from the soil was less than 5%, while that from the water was less than 7% of the applied dose (0.6 and 2.1 μg $[^{73}As]$arsenate/cm^2 skin, respectively). The radiolabeled arsenic was detected in the urine for 8 days (Wester *et al.*, 1993).

Studies of dermal absorption of arsenic in humans are relatively sparse. In a cadaver skin model, less than 1% of a topically applied dose of radiolabeled arsenate, in either a soil or water formulation (0.6 and 2.1 μg $[^{73}As]$arsenate/cm^2 skin, respectively), was percutaneously absorbed after 24 h (Wester *et al.*, 1993). Lipophilicity of an arsenic compound may influence dermal absorption. Severe toxicity (peripheral neuropathy, anemia and leukopenia) in a human occurred following dermal exposure to a pesticide spray containing MMA, although inhalation may have contributed to total dose absorbed (Hessl and Berman, 1982).

3.3.2 Distribution

Distribution of arsenicals is highly dependent on chemical form, route of administration, dose and animal species. Different chemical forms of arsenic may be handled quite differently when administered via the same route. Physical characteristics, such as the solubility of the arsenic compound, play an important role, especially following oral dosing (Fig. 3.1) (Marafante and Vahter, 1987). The importance of route of exposure becomes apparent when one observes the disposition of the same chemical form of arsenic in a single animal species, following different routes of administration (Fig. 3.2) (Yamauchi *et al.*, 1988). Finally, the disposition of an arsenical in different animal species can vary greatly (Fig. 3.3) (Vahter, 1994). This is especially evident in rats which, unlike other animal species investigated, sequester a large proportion of the arsenic dose as DMA in erythrocytes (Marafante *et al.*, 1982). For this reason, the National Academy of Sciences (NAS) does not recommend the rat as a model for arsenic metabolism in humans (NAS, 1977).

Hamsters dosed orally with 0.01 μg radiolabeled arsenate had a biphasic elimination profile (Charbonneau *et al.*, 1980). There was an initial-component (98% of dose) half-time of elimination of approximately 7 h, and a second-component (2% of dose) half-time of elimination of 3.8 days. Similar elimination half-times were reported in dogs following 0.3 μg arsenate administered orally, with a 5.9 h first-component (85% of the total dose) half-time of elimination and a 2.4 day second component (14% of the total dose) (Hollins *et al.*, 1979). However, in the dog less than 5% of the dose

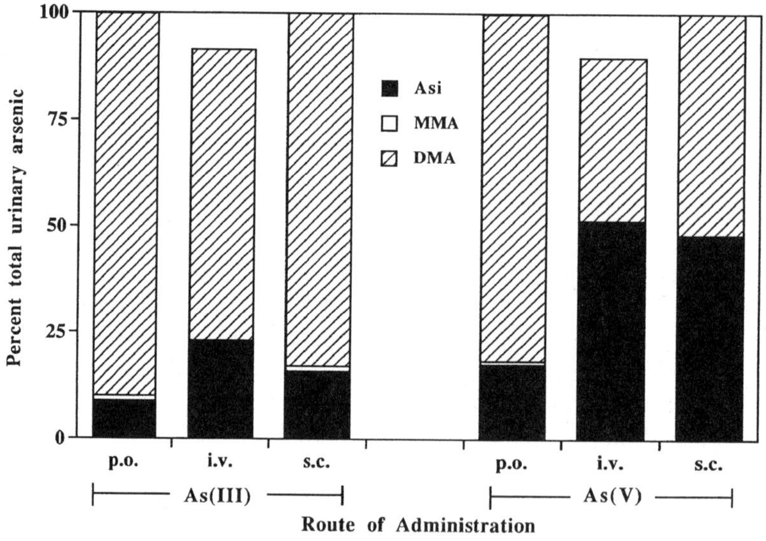

Figure 3.2 Profile of urinary arsenic species from mice administered arsenite or arsenate either orally (p.o., 0.04 mg/kg body weight; Vahter, 1981), intravenously (iv., 0.04 mg/kg body weight; Vahter and Marafante, 1983) or subcutaneously (s.c., 0.4 mg/kg body weight; Vahter, 1981).

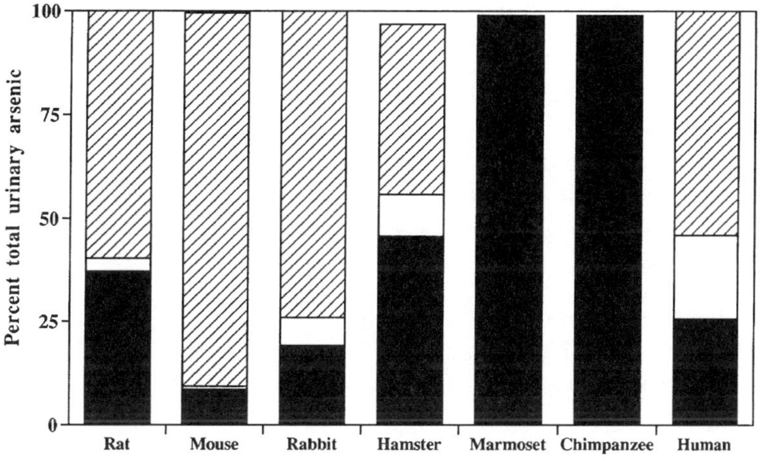

Figure 3.3 Profile of urinary arsenic species of different mammalian species following arsenite administration: rat (0.4 mg/kg, b.w., p.o.; Vahter, 1981), mouse (0.4 mg/kg, b.w., p.o.; Vahter, 1981), rabbit (0.4 mg/kg, b.w., i.v.; Vahter and Marafante, 1983), hamster (5 mg/kg, b.w., p.o.; Hirata *et al.*, 1990), marmoset monkey (0.4 mg/kg, b.w., i.p.; Vahter *et al.*, 1982), chimpanzee (5.8 μg/kg, b.w., i.v.; Vahter *et al.*, 1995b) and human (500 μg p.o.; Buchet *et al.*, 1980). Key as for Fig 3.2.

was recovered in the feces, whereas in the hamster 70% of the dose was found in the feces (Charbonneau *et al.*, 1980). In the hamster, only 6% of fecal arsenic following oral dosing was attributed to biliary excretion, based on results using i.v. dosed hamsters (Charbonneau *et al.*, 1980). This suggests that arsenate is more efficiently absorbed from the gastrointestinal tract of dogs than hamsters.

Pretreatment orally with low doses of arsenate (0.025 or 2.5 mg/L) did not affect the distribution of [^{73}As]arsenate in mice (Hughes and Thompson, 1996). However, in mice acutely dosed with [^{73}As]arsenate (0.5–5000 μg/kg), tissue accumulation was dose-dependent in the bladder, kidney, liver and lung (Hughes *et al.*, 1994).

In Flemish Giant rabbits given intraperitioneal (i.p.) radiolabeled arsenate, peak arsenate activity in the plasma was detected at 30 min and accounted for 75% of the total radioactivity, while arsenite represented an additional 23% (De Kimpe *et al.*, 1996). The latter had a longer clearance half-life than the former. At 45 min, MMA reached a constant 5–10% of the total plasma radiolabeled arsenic. DMA was a constant 60–70% of total plasma arsenic approximately 10 h after arsenate administration (De Kimpe *et al.*, 1996). Total plasma arsenic clearance following i.p. administration of arsenite or arsenate demonstrated biphasic elimination, with a rapid first component (2.0 h) and slower second component (arsenite 42 h; arsenate 58 h) (Bertolero *et al.*, 1981, De Kimpe *et al.*, 1996).

Approximately 92% of an i.p. dose of radiolabeled arsenite (1 μg/kg body weight) was excreted (urine and feces) within 4 days by albino rabbits (Bertolero *et al.*, 1981). Rabbit plasma proteins had little affinity for radiolabeled arsenic metabolites following i.p. administration of arsenate or arsenite (Bertolero *et al.*, 1981; De Kimpe *et al.*, 1996). At 4 and 20 h after i.p. administration of arsenate to Flemish Giant rabbits, the kidneys, liver and lungs had the highest concentration of radiolabeled arsenic (De Kimpe *et al.*,

1996). Similar tissue distribution was observed in albino rabbits administered arsenite i.p. (Bertolero *et al.*, 1981). Inorganic arsenic, MMA and DMA were associated primarily with the soluble tissue fraction of rabbit lung, liver and kidney following i.p. administration of radiolabeled arsenite. DMA was the predominant metabolite (Bertolero *et al.*, 1981). The lungs contained the highest percentage of DMA at all time points examined in rabbits injected i.p. with radiolabeled arsenate or arsenite (Bertolero *et al.*, 1981; De Kimpe *et al.*, 1996). This suggested either a high MMA methylation activity or increased affinity for DMA in the lungs. The importance of pulmonary exhalation of potentially volatile metabolites of inorganic arsenic has not been investigated.

Hamsters dosed intravenously (i.v.) with radiolabeled arsenate (0.01 μg) exhibited a biphasic elimination profile, with an initial-component (65% of dose) half-time of elimination of approximately 9.6 h, and a second-component (35% of dose) half-time of elimination of 4.5 days (Charbonneau *et al.*, 1980). Elimination half-times for dogs following 0.3 μg arsenate administered i.v., were a 5.9 h first-component (85% of the total dose) half-time of elimination and a 2.4 day second component (14% of the total dose) (Hollins *et al.*, 1979). The percentages of the i.v. dose eliminated for both the first and second components were significantly different than those of hamsters dosed orally, apparently due to increased fecal elimination in the latter (Charbonneau *et al.*, 1980). However, the half-times of elimination were not significantly different between hamsters dosed with arsenate i.v. or orally (Charbonneau *et al.*, 1980). Approximately 80% of an i.v dose of arsenate was recovered in the urine of rhesus monkeys over a 7-day period, with 65% being recovered in the initial 24 h (Wester *et al.*, 1993). In dogs administered arsenate i.v., there was biphasic elimination, the first-component half-life being 8.1 h, while that of the second component was 28.4 h (Groen *et al.*, 1993).

Trace levels of arsenic, as both inorganic and organic particulates, may be present in unpolluted air at concentrations approaching 0.01 μg arsenic/m^3 (NAS, 1977). This arsenic generally results from natural weathering processes and biological reduction of arsenicals by microorganisms, especially those present in the soil. Arsenic levels in air can be impacted by anthropogenic sources, such as smelting operations and the burning of arsenic-contaminated fossil fuels. The excretion of urinary arsenic was higher in males than females for residents living near nonferrous smelter operations in Belgium (Buchet *et al.*, 1996b). The area of residence appeared to be a strong determinant of the urinary arsenic output, as demonstrated by an inverse relationship between urinary arsenic excretion and the distance from the nearest smelter.

Information is lacking concerning the disposition of arsenicals in humans in which the dermal route has been confirmed as the only route of exposure. Arsenate was absorbed by human skin *in vitro*, with subsequent accumulation in circulating receptor fluid (Wester *et al.*, 1993). *In vivo*, rhesus monkeys had low percutaneous absorption of radiolabeled arsenate (less than 7% of applied dose) (Wester *et al.*, 1993). Neither the tissue distribution nor the chemical speciation of the urinary arsenic was reported in the study.

Rats administered arsenic subcutaneously, in the form of an implanted pellet containing 2 mg of arsenic trioxide, had highly elevated spleen and erythrocyte arsenic levels for at least 2 months following implantation (Aso and Abiko, 1978). Other tissues, including the kidney, liver, lung, heart, brain and hair, showed less significant elevations (Aso and Abiko, 1978).

3.3.3 Intracellular disposition

Arsenic methylation does not appear to occur in plasma or urine of humans or beagles (Tam *et al.*, 1979b). This suggests that an arsenic species must first be taken into cells, methylated and released back into the bloodstream before being cleared by the kidneys. Using rat liver slices, Georis *et al.* (1990) demonstrated that arsenite uptake was proportional to its concentration in the incubation medium, and that the arsenic concentration usually became more than 10 times greater in the liver slices than in the culture medium. Approximately eightfold less arsenic was taken up when the pentavalent form was incubated with the slices. This suggested that the uptake mechanism of tissue slices probably did not require energy, and that the mechanism was a gradient (facilitated diffusion) uptake (Georis *et al.*, 1990). Similarly, uptake and metabolism of inorganic arsenic in mouse fibroblasts was dose-dependent (Fischer *et al.*, 1985).

3.3.4 Excretion

In humans, urinary arsenic excretion appears to follow first-order kinetics, at least up to 8 days following oral administration (Buchet *et al.*, 1981b). However, the half-time for urinary arsenic excretion increased with increasing dose. There were no dose–response trends in the elimination of arsenate or DMA in mice dosed with 0.5–5000 μg/kg arsenate, although the cumulative urinary elimination of MMA occurred in a dose–response manner (Hughes *et al.*, 1994).

In humans ($n = 3$ or 4), the urinary excretion rate of arsenic metabolites increased as follows: inorganic arsenic ($NaAsO_2$) < DMA < MMA (Buchet *et al.*, 1981a). Following ingestion of 500 μg of either $NaAsO_2$, MMA or DMA, the half-lives were approximately 28 h, less than 4 h, and 11 h, respectively (Buchet *et al.*, 1981a). The total amount of arsenic excreted daily in the urine was related linearly to the daily amount absorbed (Buchet *et al.*, 1981b).

The diet can greatly influence arsenic excretion by the kidneys. Seafood containing different forms of marine arsenicals, which are rapidly filtered, can transiently elevate urinary total arsenic concentration (Buchet *et al.*, 1994). Over 50% of the dose (5 mg total arsenic) from fish containing arsenic was excreted within 2 days of ingestion by human volunteers, after which the excretion rate decreased and eventually became constant (Freeman *et al.*, 1979).

No gender differences in arsenic excretion via the kidneys has been demonstrated in controlled human arsenic challenge studies (Buchet *et al.*, 1984). There were no significant differences in urinary elimination between genders of hamsters dosed i.v. or orally with arsenate (Charbonneau *et al.*, 1980).

In dogs, arsenate underwent net renal tubular reabsorption, apparently in a competitive manner with the phosphate transport mechanism (Ginsburg and Lotspeich, 1963). The major site of arsenic transport was within the proximal tubule (Ginsberg, 1965). The sensitivity of arsenate transport to the rate of urine flow suggested a passive diffusion process. Additionally, Ginsburg (1965) demonstrated intrarenal reduction of arsenate to arsenite.

Biliary excretion of arsenicals appears to be dependent on chemical form, route of administration and species differences. Additionally, enterohepatic circulation of arsenicals may prolong exposure. Arsenic was rapidly excreted into the bile of cannulated rats following i.v. administration of arsenic trichloride (1.0 mg/kg) (Klaassen, 1974). Twenty-five per cent of the dose was recovered in the bile within 2 h. Bile arsenic concentrations were 300- to 600-fold higher than plasma concentrations, which implied an active transport mechanism. However, less than 10% of the administered dose was excreted via the feces over a 7-day period, which suggested the importance of enterohepatic circulation of arsenic in the rat (Klaassen, 1974). Enterohepatic circulation of arsenic was directly established since arsenic was detected in the blood of rats administered arsenic-containing bile intraduodenally (Klaassen, 1974). The biliary arsenic elimination rates in the rabbit and dog were greatly lower, 0.23 and 0.01 μg/min/kg, respectively, than that of the rat (8.4 μg/min/kg) (Klaassen, 1974).

Arsenite and arsenate, when administered intravenously in rats, increased the biliary excretion of endogenous nonprotein thiols, including glutathione, in a dose-dependent manner (Gyurasics *et al.*, 1991a, b). In mice dosed with arsenate (2.5 mg/L) in the drinking water for 28 days, there was a 20–26% decrease in hepatic nonprotein thiols (Hughes and Thompson, 1996). Arsenite was more effective than arsenate in increasing biliary excretion of nonprotein thiols (Gyurasics *et al.*, 1991a). Arsenite was more effectively excreted via the bile than arsenate, although excretion of both species was highest during the 15 min following administration (Cikrt and Bencko, 1974; Cikrt *et al.*, 1980; Gyurasics *et al.*, 1991b). It was proposed that arsenite, as well as arsenate, following reduction to arsenite, formed glutathione complexes with limited stability which were transported into the bile (Gyurasics *et al.*, 1991a).

Fecal elimination of inorganic arsenic and its methylated metabolites generally represents less than 10% of the total arsenic excreted (Ducoff *et al.*, 1948; Hollins *et al.*, 1979; Cikrt *et al.*, 1980), but exceptions have been observed (Charbonneau *et al.*, 1980). The oxidation state of inorganic arsenic does not appear to influence fecal arsenic elimination significantly following intraperitoneal administration (Cikrt *et al.*, 1980), perhaps due to interconversion of the chemical species *in vivo*. Few studies have examined the fecal elimination profile in human subjects successfully (Bettley and O'Shea, 1975).

3.4 Biotransformation

Since the methylated metabolites of inorganic arsenic are less acutely toxic (Lewis and Sweet, 1985), more quickly excreted (Buchet *et al.*, 1981a) and demonstrate less tissue reactivity (Vahter and Marafante, 1983), methylation has been considered to be the primary mechanism of inorganic arsenic biotransformation and detoxification in most animals. Recently, this has been questioned as a detoxification pathway. The methylation pathway is generally believed to follow several sequential steps (Fig. 3.4). Arsenate is reduced to arsenite, from which MMA and DMA are formed enzymatically by the oxidative addition of methyl groups to arsenite and MMA from the methyl donor *S*-adenosylmethionine.

Braman and Foreback (1973) originally demonstrated the presence of methylated arsenicals in human urine. However, an *in vivo* arsenic methylation process for humans was not demonstrated until 4 years later by Crecelius (1977) in a human volunteer who had ingested wine, well water and canned crab meat containing inorganic arsenic. Two methylated metabolites of inorganic arsenic were detected in the urine,

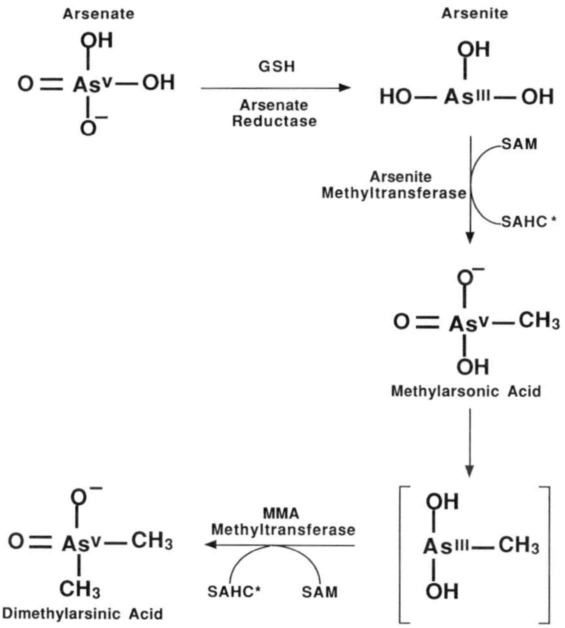

Figure 3.4 Putative pathway of inorganic arsenic biotransformation. Metabolites are those detected in urine, not necessarily those arsenic species present *in vivo*. The trivalent monomethylated species has not been isolated in the urine as yet. SAHC, *S*-adenosylhomocysteine; SAM, *S*-adenosylmethionine; GSH, glutathione.

monomethylarsonic acid (MMA) and dimethylarsinic acid (DMA). Since these early experiments, numerous arsenic challenge studies have been performed on humans and animals, using both inorganic and organic forms of arsenic under a variety of experimental conditions (Marafante *et al.*, 1981; Vahter, 1981; Marafante *et al.*, 1982; Vahter *et al.*, 1982; Yamauchi and Yamamura, 1984a, b; Aposhian, 1997).

In an early human arsenic challenge test, Buchet *et al.* (1980) dosed five volunteers with 3 mg arsenic in the form of $NaAsO_2$. Analysis of urine for total and speciated arsenic demonstrated that 48% of the dose was eliminated over the following 5 days. The biological half-life for total arsenic was approximately 30 h. Inorganic arsenic was excreted mainly during the initial hours of the study. However, this declined rapidly, along with a parallel increased proportion of methylated metabolites, predominantly DMA (Buchet *et al.*, 1980). This effect appears to be accelerated by pretreatment with lipotropic agents, such as methionine, choline or vitamin B_{12} (Buchet *et al.*, 1981a). In volunteers ingesting 500 μg inorganic arsenic as $NaAsO_2$, approximately 20–25% of the excreted amount of arsenic remains in the inorganic form, 15–20% in the monomethylated and 55–65% in the dimethylated form (Buchet *et al.*, 1981a). These values agreed well with those of Crecelius (1977).

It has been suggested that two independent pathways were responsible for the production of MMA and DMA (Buchet *et al.*, 1984; Buchet and Lauwerys, 1988). However, using a purified arsenite and MMA rabbit liver methyltransferase preparation, Zakharyan *et al.* (1995) reported that a single protein appears to perform both methylation steps.

The rate of arsenite methylation has been proposed as one reason for the variation in urinary excretion profiles of inorganic arsenic and its methylated metabolites among different animal species (Wildfang *et al.*, 1998). Partially purified liver arsenic methyltransferase activities demonstrated a higher affinity for arsenite as a substrate than MMA in both the rabbit and hamster (Zakharyan *et al.*, 1996; Wildfang *et al.*, 1998). This could allow accumulation and subsequent urinary excretion of MMA in the early time period following inorganic arsenic exposure. In tissue slices incubated in a low (1 μM) arsenite concentration, DMA formation increased with time, while the MMA concentration remained relatively constant during the incubation period (Georis *et al.*, 1990). These data suggest that, at low doses, arsenite is preferentially methylated until a concentration of MMA is achieved to trigger the second step of methylation, at which point DMA is produced.

Arsenic methylating activity is localized in the tissue cytosol, and is found predominantly in the liver, kidney, lung (Georis *et al.*, 1990; Healy *et al.*, 1998) and testes (Healy *et al.*, 1998). Numerous *in vivo* and *in vitro* studies have elucidated essential cofactors for arsenic methylation. *S*-adenosylmethionine appears to be the primary methyl donor molecule (Buchet and Lauwerys, 1985; Zakharyan *et al.*, 1995). Reduced thiols, principally glutathione (GSH), are required for methylation (Buchet and Lauwerys, 1987; Buchet and Lauwerys, 1988). Cysteine, however, was more active than GSH in the purified rabbit liver arsenite methyltransferase preparation (Zakharyan *et al.*, 1995). Arsenite, rather than arsenate, is the preferred inorganic substrate of the first methylation step (Georis *et al.*, 1990; Zakharyan *et al.*, 1995). MMA is required at up to ten- to fiftyfold higher concentrations than arsenite in order to approach the maximal methylation rate *in vitro* (Georis *et al.*, 1990; Zakharyan *et al.*, 1996).

The specific action of intracellular thiols on arsenic methylation is unclear. Some investigators have suggested that glutathione may act by reducing pentavalent arsenicals prior to the oxidative methylation step (Buchet and Lauwerys, 1987; Buchet and Lauwerys, 1988; Carter, 1995), that glutathione may facilitate trivalent arsenic uptake by cells (Georis *et al.*, 1990), that a glutathione–arsenic complex may be the substrate for the methylating enzymes (Carter, 1995) or that glutathione may function in protecting labile thiol groups (Buchet and Lauwerys, 1998). In addition, glutathione may provide the reducing environment needed by the arsenic methylating protein (Zakharyan *et al.*, 1995). Arsenic methylation activity could not be detected in human liver collected 24 h after death, suggesting labile methylating activities (Buchet and Lauwerys, 1988). It is possible that other cofactors in addition to, or in place of, *S*-adenosylmethionine are required for the methylating activities in humans.

Oral *S*-adenosylmethionine or thiol supplementation following the intraperitioneal administration of $NaAsO_2$ (0.25 or 5 mg/kg) in rats did not affect total arsenic excreted nor the urinary proportions of inorganic arsenic, MMA or DMA (Buchet and Lauwerys, 1987). Induction of glutathione synthesis in rats by oral pretreatment with butylated hydroxytoluene (BHT) did not change total arsenic excretion or the urinary metabolite proportions following oral $NaAsO_2$ administration (Buchet and Lauwerys, 1987). In rat liver slices, addition of glutathione to the incubation medium significantly increased MMA and DMA production (Georis *et al.*, 1990). Glutathione supplementation decreased production of these methylated metabolites in kidney slices (Georis *et al.*, 1990). Glutathione, in 1× and 3× stoichiometric ratios to arsenite concentration, when added to the incubation medium, did not affect arsenic accumulation in rabbit renal cortical slices (Burton *et al.*, 1995). There was no difference of arsenic uptake in rabbit renal slices between arsenite or an arsenic-triglutathione ($As(SG)_3$) species (Burton *et al.*, 1995).

In vivo depletion of glutathione by phorone pretreatment in rats more than doubled the total arsenic excreted within 24 h (Buchet and Lauwereys, 1987). This was due mainly to an increase in the excretion of inorganic arsenic and demonstrated the importance of endogenous glutathione in arsenic methylation (Buchet and Lauwerys, 1987). Glutathione increased arsenic methylation *in vitro* in a dose-dependent manner in rat liver cytosol (Buchet and Lauwerys, 1988) and using purified rabbit liver arsenite methyltransferase (Zakharyan *et al.*, 1995). Other thiol compounds, such as dithiothreitol (DTT), dimercaptosuccinic acid (DMSA) and dimercaptopropanesulfonic acid (DMPS), decreased arsenic methylation with increasing thiol concentration, while mercaptoethanol increased arsenic methylation *in vitro* (Buchet and Lauwerys, 1988). It appears that glutathione (or other intracellular thiols) is necessary for methylation of arsenicals. However, its role in the methylation and cellular uptake of arsenic remains unclear.

Addition of vitamin B_{12} to a rat liver homogenate resulted in increased DMA production, yet had no effect on the amount of MMA (Buchet and Lauwerys, 1985). No effect on arsenite methylation was observed following the addition of NAD, NADP, ATP, succinnate, folate or 5-methyltetrahydrofolate (Buchet and Lauwerys, 1985). However, a human genetic defect, 5,10-methylenetetrahydrofolate reductase deficiency, resulted in severe neurological symptoms in the affected individual after arsenic exposure (Brouwer *et al.*, 1992).

In humans, following repeated oral administration of inorganic arsenic ($NaAsO_2$, five doses over five consecutive days), the combined percentage of MMA and DMA excreted decreased by up to 10% in response to increased dose (Buchet *et al.*, 1981b). This was especially true for the proportion of the dose excreted as DMA following the highest dose (1000 μg $NaAsO_2$). MMA production was nearly linear in response to dose. The fact that methylated metabolites were present in the urine in significant concentration following repeated exposure, even at high doses, indicated that the capacity of the overall methylation pathway had not been saturated. Hopenhayn-Rich *et al.* (1993), after reviewing population exposure studies, concluded that humans do not have a threshold for arsenic methylation. However, others suggest that a sublinear or threshold relationship exists between arsenic ingestion and skin and internal cancers (Rudel *et al.*, 1996).

Following oral arsenite administration, inorganic arsenic, MMA and DMA were the only metabolites found in human urine (Buchet *et al.*, 1981b). After a 500 μg oral dose of MMA, approximately 13% of it was further methylated to DMA by humans over 4 days (Buchet *et al.*, 1981a). But, in the urine of a human given DMA orally, low levels (about 4%) of trimethylarsine oxide (TMAO) were detected (Marafante *et al.*, 1987b). Mice and hamsters metabolized orally administered DMA to TMAO (3.5 and 6.4%, respectively) (Marafante *et al.*, 1987b). Higher urinary concentrations of trimethylarsenic compounds (TMA) have been reported in hamsters given DMA orally (Yamauchi and Yamamura, 1984a). Since no TMA was detected in the feces, it is unlikely that these trimethylarsenic species resulted from gut flora metabolic activity (Yamauchi and Yamamura, 1984; Marafante *et al.*, 1987b).

No trimethylarsenic compounds were detected in rat liver homogenate systems (Buchet *et al.*, 1981b). However, in rats administered DMA orally, trimethylarsine oxide (TMAO) increased in a dose-dependent manner (Chen *et al.*, 1996b). All these findings demonstrate the need for a reliable model system for studying arsenic metabolism.

The small amount of monomethylated arsenic metabolite in the urine of most animals, including humans, following inorganic arsenic exposure is difficult to explain. In the human (Buchet *et al.*, 1981a) and animal *in vivo* (Yamauchi *et al.*, 1988), and *in vitro* (Georis *et al.*, 1990; Zakharyan *et al.*, 1995), MMA is methylated further to DMA. There was about a 4 h and 8 h delay in production of DMA following ingestion of either inorganic arsenic or MMA, respectively, by human volunteers (n = 3 or 4) (Buchet *et al.*, 1981a). This suggests that MMA is the initial methylated metabolite and that it must accumulate before the second step of methylation to DMA can be triggered (Buchet and Lauwerys, 1985). During this interim period, MMA may be filtered at the kidney and excreted in the urine. Since most human studies have not investigated the time course of metabolite excretion in increments of less than 24 h, it is feasible that the MMA values resulted from early elimination in the urine collection period, after which time DMA became the primary excreted metabolite.

Further evidence has been provided by *in vitro* studies using partially purified rabbit and hamster liver arsenite and MMA methyltransferases. Kinetic analysis demonstrated that the K_m of arsenite as a substrate was two orders of magnitude lower than that of MMA^{V} (Zakharyan *et al.*, 1996; Wildfang *et al.*, 1998). This suggested that arsenite may be the favored substrate for methylation, and that methylation of MMA to DMA may not occur to any significant degree until sufficient substrate concentrations (MMA^{III} and MMA^{V}) are achieved. Therefore, MMA formation may be the rate-limiting step of arsenic methylation.

In the human, neither MMA nor DMA appear to undergo *in vivo* demethylation (Buchet *et al.*, 1981a). In rats administered DMA orally, arsenite was detected in the urine and significant concentrations were present in the feces (Chen *et al.*, 1996b). However, demethylation of DMA by intestinal flora could not be ruled out.

Metabolism of arsenicals within the gastrointestinal tract has been insufficiently investigated. Rowland and Davies (1981) investigated arsenic metabolism *in vitro* of different sites of the rat gastrointestinal tract. The contents of rat small intestine and cecum reduced arsenate to arsenite over a period of 2–3 days. Small amounts of MMA and DMA were formed when the contents of the cecum, but not small intestine, were incubated with arsenate for 4 days (Rowland and Davies, 1981). Hall *et al.* (1997) investigated *in vitro* methylation of arsenite and arsenate by mouse cecal contents. MMA was the predominant metabolite, and arsenite appeared to be the preferred methylation substrate.

The metabolism of arsine, the gaseous form of arsenic, by a human accidentally exposed to it has been reported (Apostoli *et al.*, 1997). It was metabolized, and excreted primarily in the urine as both trivalent and pentavalent inorganic arsenic, MMA, DMA and arsenobetaine. The inorganic forms predominated initially. The methylated arsenicals increased during the following 5 days. Although the total dose was not determined, 41 mg total arsenic were excreted over 24 days following exposure (Apostoli *et al.*, 1997). The presence of arsenobetaine in the urine could not be justified solely on dietary intake. Exhalation of arsine or its metabolites were not investigated (Apostoli *et al.*, 1997).

Arsine caused severe hemolysis of rat and dog erythrocytes *in vitro*, lysing 90 and 36% of cells, respectively, within 2 h (Hatlelid *et al.*, 1995). Reactive oxygen intermediates, formed in red blood cells following *in vitro* arsine exposure, were not responsible for the observed damage to hemoglobin (Hatlelid and Carter, 1997). Following arsine gas exposure *in vitro*, rat red blood cells had decreased glutathione concentrations (Blair *et al.*, 1990). This suggested that the mechanism of arsine hemolysis of red blood cells may involve oxidation of sulfhydryl groups in hemoglobin or possibly the red cell membranes (Blair *et al.*, 1990).

Inorganic arsenic inhibited DMA formation *in vivo* (Buchet *et al.*, 1981b) and *in vitro* (Buchet and Lauwerys, 1985, 1988; Georis *et al.*, 1990). Humans chronically exposed to inorganic arsenic in drinking water (0.415 mg/L) demonstrated significant increases in the average proportions of inorganic arsenic (30.6 versus 8.7%) and MMA (11.3 versus 6.8%) in urine, accompanied by a significant decrease in DMA (54.1 versus 78.5%) as compared to nonexposed control subjects (Del Razo *et al.*, 1997). In rat liver slices, inorganic arsenic (5–10 μM) inhibited DMA formation reversibly (Georis *et al.*, 1990). DMA production returned to its normal rate following removal of inorganic arsenic from the incubation medium after 1 h (Georis *et al.*, 1990). MMA production in rat liver cytosol preparations increased linearly up to an arsenite substrate concentration of 0.04 mM (Buchet and Lauwerys, 1987, 1988). DMA production reached a maximum at 0.005 mM and progressively declined with increasing arsenite concentration (Buchet and Lauwerys, 1987, 1988). MMA may have a negative feedback action on the first methylation step (Buchet and Lauwerys, 1988).

Additional inhibitors of arsenic methylation in rat liver cytosol included pyrogallol and sodium diethyldithiocarbamate (decreased MMA and DMA production), mercuric ions (decreased DMA production) and trichloroacetonitrile and selenium (chemical form not indicated; completely prevented arsenic methylation) (Buchet and Lauwerys, 1985). Selenite (12.6 μM), selenate (25 μM) and selenide (>1000 μM) inhibited purified rabbit liver arsenite methyltransferase activity by 50% *in vitro* (Zakharyan *et al.*, 1995). Oral selenite administration to rats slightly affects the proportions of MMA (0.33 μg As/100 g body weight, versus control 0.23 μg) and DMA (2.24 μg/100 g body weight, versus control 2.85 μg) excreted in 24 h, although this is dependent on the route of arsenic administration (Buchet and Lauwerys, 1987). The latter authors speculated that the ability to inhibit specifically one of the two putative methylation activities suggested that methylation may proceed through independent pathways. However, the Aposhian group, using highly purified rabbit liver arsenic methyltransferase, could find no evidence for the two methylation reactions being catalyzed by different proteins (Zakharyan *et al.*, 1995).

The ingestion of fish (ray, cod or plaice) was not associated with an increase in urinary inorganic arsenic (Buchet *et al.*, 1994). However, following the consumption of shellfish (mussels), the excretion of inorganic arsenic was increased, and the amount of urinary DMA was about 25% more than that expected from the inorganic arsenic and its methylated derivatives already present in the shellfish (Buchet *et al.*, 1994, 1996a). This suggested that DMA may be liberated from marine organoarsenicals *in vivo*. Individuals consuming fish and shellfish regularly have an increased urinary excretion of total arsenic as compared to populations ingesting such seafood intermittently (Buchet *et al.*, 1996a).

The health status of the arsenic-exposed individual appears to be important in methylation. Patients diagnosed with liver disease had a significant decrease in urinary excretion of MMA (6.1 versus 12.8%) and increase in urinary excretion of DMA (40.7 versus 24.3%) in the 24 h following intravenous administration of 500 μg $NaAsO_2$, as compared to normal individuals (Buchet *et al.*, 1984). However, the percentage of the total arsenic dose excreted within 24 h was not significantly different. Similar results have been reported among patients with and without liver cirrhosis (Geubel *et al.*, 1988). Monomethylation capacity was restored in two of the cirrhotic patients following liver transplant, suggesting that in humans, the liver is a principal methylating organ (Geubel *et al.*, 1988). Artificially induced liver damage via carbon tetrachloride pretreatment resulted in a significant reduction in DMA production, which was paralleled by a minor decrease in MMA production in rats (Buchet and Lauwerys, 1987).

In addition to methylation, the ability of an animal to interconvert between inorganic forms of arsenic may ultimately be critical in arsenic detoxification. High levels of arsenate were present in the urine of mice administered arsenite (Bencko *et al.*, 1976).

3.4.1 Evidence for arsenic tolerance

Tolerance to the toxic effects following chronic arsenic exposure has been demonstrated *in vivo* (Bencko *et al.*, 1976) and *in vitro* (Fischer *et al.*, 1985). Mice given intramuscular (i.m.) radiolabeled arsenite (1.3 mg/kg) after daily exposure to arsenite (250 mg/L) in their drinking water had higher levels of urinary arsenate than did non-pretreated controls (Bencko *et al.*, 1976). Adaptation of mouse fibroblasts to chronic low-dose (2.5–20 μM) trivalent arsenic resulted in earlier production of the methylated metabolites as compared to nonadapted controls (Fischer *et al.*, 1985).

Evidence of polymorphism in arsenic biotransformation has been suggested in humans (Vahter *et al.*, 1995a) and in guinea-pigs (Healy *et al.*, 1997). Polymorphism of the enzymatic activities responsible for inorganic arsenic methylation may contribute to interindividual variation in arsenic methylation capacity, which may partially explain the greater susceptibility of some exposed populations of humans than others (Del Razo *et al.*, 1997).

3.4.2 Species variation

Arsenic metabolism varies quantitatively, as well as qualitatively, between animal species (Buchet *et al.*, 1981a) as well as within an animal species (Ducoff *et al.*, 1948; Morel *et al.*, 1995). The clearest examples of species variation in arsenic metabolism as measured *in vivo* are the marmoset monkey (Vahter *et al.*, 1982), chimpanzee (Vahter *et al.*, 1995b) and guinea-pig (Healy *et al.*, 1997). These three animal species did not excrete MMA or DMA following inorganic arsenic challenge. Liver arsenite methyltransferase activities are deficient in these species (Aposhian, 1997; Healy *et al.*, 1997). A comparison of species variation of arsenic methylation is presented in Table 3.2. Vahter (1994) has presented an excellent review on species variation in arsenic metabolism.

3.5 Cellular effects of arsenic

The toxicity of arsenic is generally the result of excessive or chronic exposure to inorganic forms of the metalloid, especially those in the trivalent state, such as As_2O_3.

Table 3.2 Diversity of arsenic methylation in mammalian species

Methylators	*Reference*	*Nonmethylators*	*Reference*
Human	Crecelius (1977)	Marmoset monkey	Vahter *et al.* (1982) Zakharyan *et al.* (1996)[b]
Rabbit	Bertolero (1981) Zakharyan *et al.* (1995)[b]	Chimpanzee	Vahter *et al.* (1995b) Zakharyan *et al.*, to be published[b]
Rat	Vahter (1981)	Guinea-pig	Healy *et al.* (1997)[b]
Mouse	Vahter (1981) Healy *et al.* (1998)[b]	Tamarin monkey[a]	Zakharyan *et al.*, (1996)[b]
Dog	Tam *et al.* (1979a, b)		
Hamster	Charbonneau *et al.* (1980) Wildfang *et al.* (1998)[b]		
Rhesus monkey[a]	Zakharyan *et al.* (1996)[b]		

[a] *In vivo* arsenic metabolism data are not currently available for these species.
[b] These references concern the presence or absence of the arsenite methyltransferase as shown by *in vitro* assays.

The general population is exposed to these forms of inorganic arsenic primarily via ingestion of drinking water or foodstuffs and from inhalation of arsenic-containing dusts. The scientific focus on chronic inorganic arsenic exposure has been on its carcinogenic properties. However, long-term exposure to inorganic arsenic may have noncarcinogenic effects on organs such as the lungs, skin, liver, peripheral vascular system, central and peripheral nervous systems, kidneys and the hematopoietic system (WHO, 1981).

3.5.1 Respiration

Noncarcinogenic respiratory effects attributable to arsenic have been reported primarily from investigations dealing with occupational inhalation exposure. Irritant effects on the upper respiratory tract from arsenic exposure during smelter operations (airborne arsenic concentrations ranged from 20 to 2264 $\mu g/m^3$ depending on sampling site) include hoarseness, nasal irritation, rhinitis, pharyngitis and laryngitis (Hine *et al.*, 1977). Lesions of respiratory mucous membranes, including perforation of the nasal septum, were observed among highly exposed individuals (Hine *et al.*, 1977). These effects appear to depend, in part, on the size of the inhaled arsenic particles as well as the chemical form, although the mechanisms by which arsenic causes these effects are unclear.

The absorption, biotransformation and tissue retention of inorganic arsenicals following intratracheal instillation in hamsters were correlated with the *in vitro* and *in vivo* solubilities of the compounds (Marafante and Vahter, 1987). Less than 0.1% of either sodium arsenite or arsenate (soluble arsenicals), as compared to 1.3% of arsenic trisulfide (1.42 μm diameter) or 45% of lead arsenate particles (1.5 μm diameter), was retained in the lungs 3 days after administration.

The influence of arsenic oxidation state and solubility on pulmonary alveolar macrophage (PAM) function was studied by Lantz *et al.* (1994, 1995). Soluble forms of arsenic, either arsenite or arsenate (Lantz *et al.*, 1994), or less soluble forms of arsenic such as arsenic trisulfide or calcium arsenate (Lantz *et al.*, 1995) were instilled intratracheally in rats at 1 mg/kg (as arsenic). Then pulmonary toxicity was assessed by measuring alterations in PAM function *in vivo* and *in vitro*. All of the arsenic compounds caused changes in PAM activity. The investigators suggested that arsenic-induced alterations of PAM function may compromise host defense mechanisms against infections (Lantz *et al.*, 1994, 1995).

The ability of rabbit alveolar macrophages to dissolve lead arsenate and arsenic trisulfide was studied *in vitro* (Marafante *et al.*, 1987a). The solubility of lead arsenate increased, and that of arsenic trisulfide decreased with decreasing pH (Marafante *et al.*, 1987a). Following incubation with macrophages for a 3-day period, 12% more lead arsenate particles were solubilized into the incubation medium than in the absence of the macrophages. However, less of the arsenic trisulfide was solubilized in the presence of macrophages than in their absence (68.8 versus 72.6% of total arsenic added, respectively). The authors concluded that the acidic phagosome environment of the macrophage may have importance in the dissolution of arsenic particles in the lung (Marafante *et al.*, 1987a).

The significance of particulate size has been demonstrated in rats instilled intratracheally with gallium arsenide particulates (8.3 μm mean count diameter) (Webb *et al.*, 1987). The rats developed signs of arsenic intoxication, pulmonary inflammation and pneumocyte hyperplasia. Decreasing the particle mean count diameter to 1.63 μm increased the rate of *in vivo* solubilization and the severity of the pulmonary lesions (Webb *et al.*, 1987).

Inorganic arsenic selectively modulates rat lung microsomal P_{450} monooxygenase activity (Albores *et al.*, 1995). Twenty-four hours after subcutaneous administration of arsenite (75 μmol/kg body weight), lung microsomal P_{450} 1A1 activity was approximately fivefold higher than controls and remained elevated for 48 h after treatment. P_{450} 2B activity remained unchanged. This was apparently not a direct action of arsenite on P_{450} 1A1 activity, as *in vitro* addition of arsenite (up to 1 mM) had no effect on P_{450}-dependent catalysis of control rat lung microsomes (Alborees *et al.*, 1995). The mechanism by which arsenite influences rat lung P_{450} 1A1 activity is not known.

Studies of chronic arsenic inhalation exposure in animals and the effects on the respiratory system are lacking. Lung tissue metabolizes inorganic arsenic to methylated forms (Georis *et al.*, 1990). Arsenic methyltransferase is present in lung cytosol (Healy *et al.*, 1998). The potential toxic actions of these metabolites on the respiratory system need to be investigated. Noncarcinogenic respiratory effects subsequent to the absorption, metabolism and distribution of both inorganic and organic arsenicals from routes of exposure other than inhalation require further investigation.

3.5.2 Neurotoxicity

Clinical and pathologic signs of neurotoxicity subsequent to acute and chronic arsenic exposure have been known for many years (Russell, 1937; Heyman *et al.*, 1956; Hurst, 1959; Jenkins, 1966). Both inorganic (LeQuesne and McLeod, 1977; Feldman *et al.*, 1979) and organic (Hessl and Berman, 1982) forms of arsenic will cause peripheral neuropathy. Inorganic (Itoh *et al.*, 1990; Tripathi *et al.*, 1997) and organic (Hurst, 1959) arsenicals distribute to the central nervous system. In human brain, total arsenic was associated with cerebral white matter and gray matter at a ratio of 2.7, suggesting a nonrandom distribution (Larsen *et al.*, 1979). The arsenic concentration in the brain white matter (Larsen *et al.*, 1979) did not differ significantly from that in the peripheral nerves (Larsen *et al.*, 1972). The choroid plexus appeared to have a protective mechanism since it accumulated arsenic at higher concentrations than that found in either the cerebrospinal fluid or the brain (Zheng *et al.*, 1991).

Neurologic symptoms due to arsenic are primarily those of a peripheral sensory neuritis, predominantly numbness, severe paresthesia of the distal portion of the extremities, diminished sense of touch, pain, heat and cold and symmetrically reduced muscle power (Hindmarsh and McCurdy, 1986; Menkes, 1997). In humans exposed via drinking water to inorganic arsenic (0.18 and 0.21 mg/L) (Southwick *et al.*, 1983), there was slowing of nerve conduction, primarily in sensory nerves. Slowing of motor nerve conduction velocity occurred following a single exposure to inorganic arsenic (LeQuesne and McLeod, 1977).

Following acute inorganic arsenic exposure, peripheral nervous disturbances in the arms and legs are common, manifesting 1–3 weeks after ingestion, with recovery beginning within 2 months after the onset of neurological symptoms (Heyman *et al.*, 1956, LeQuesne and McLeod, 1977). There was severe hearing loss and brain-wave abnormalities among infants exposed to powdered milk contaminated with inorganic pentavalent arsenic (Yamashita *et al.*, 1972). Children living near a power plant that burned coal containing high arsenic levels had hearing loss (Bencko *et al.*, 1977).

Sural nerve biopsies showed axonal degeneration and abnormal myelination in humans following ingestion of arsenic (LeQuesne and McLeod, 1977). *In vitro*, arsenic trioxide (9.6 mM) produced a 50% inhibition of mouse dorsal root ganglion neurite outgrowth (Windebank, 1986).

Arsenite inhibited proliferation of mouse neuroblastoma cells in a dose-dependent manner (median effective concentration, $EC_{50} = 10\ \mu M$) and inhibited lactate dehydrogenase (LDH) and hexosaminidase (HEX) activities by 65% and nearly 100%, respectively, at 0.8 μM (Repetto *et al.*, 1994). Acetylcholinesterase (AChE) activity was inhibited by arsenite in a dose-dependent manner (Repetto *et al.*, 1994). Arsenate was less cytotoxic ($EC_{50} = 54\ \mu M$) to the neuroblastoma cells and had higher EC_{50} values for toxicity indicator enzymes (LDH, HEX and AChE).

Inorganic arsenic produced neurotransmitter disturbances (Kobayashi *et al.*, 1987; Itoh *et al.*, 1990; Nagaraja and Desiraju, 1993). In mouse brain slices, arsenite (0.1 mM) inhibited the synthesis and release of acetylcholine (Kobayashi *et al.*, 1987). Arsenite inhibited (median inhibitory concentration, $IC_{50} = 1.3$ mM) the high-affinity uptake of choline by mouse brain synaptosomes in a dose-dependent manner (Kobayashi *et al.*, 1987). Choline acetyltransferase activity was stimulated by arsenite (0.1 mM), while acetylcholinesterase was dose-dependently inhibited by arsenite ($IC_{50} = 5.4$ mM) (Kobayashi *et al.*, 1987). Arsenite (0.1 mM) did not affect ligand binding at muscarinic acetylcholine receptors (Kobayashi *et al.*, 1987).

In rats administered sodium arsenite (100 p.p.m.) orally for 16 weeks, brain arsenic concentration increased (0.26 versus 0.05 ng/g in controls), while the neurotransmitters dopamine and norepinephrine decreased (Tripathi *et al.*, 1997). Chelation treatment (i.e. meso-2,3-dimercaptosuccinic acid, DMSA; or 2,3-dimercapto-1-propane sulfonate, DMPS) did not reduce brain arsenic levels nor restore arsenic-induced changes in neurotransmitter levels. In arsenite-exposed rats, the glutathione and acetylcholinesterase levels in the brain were decreased, while monoamine oxidase levels increased (Tripathi *et al.*, 1997).

Regional changes in levels of brain neurotransmitters, such as the biogenic amines (e.g. dopamine, serotonin, norepinephrine and their precursor molecules) and amino acids (e.g. glutamate and γ-aminobutyric acid (GABA)), occurred after oral inorganic arsenic administration (Itoh *et al.*, 1990; Nagaraja and Desiraju, 1993). Changes in brain monoamine metabolism did not parallel the arsenic trioxide content of the examined brain regions in mice (Itoh *et al.*, 1990).

Peripheral neuropathy and encephalopathy have been observed in humans exposed to organic arsenicals (Verity, 1995). In a case report, a 16-year-old girl with 5,10-methylenetetrahydrofolate reductase (MTHFR) deficiency exhibited severe neurotoxic symptoms (mental deterioration and paraparesis) following exposure to the pesticide copper acetate arsenite (Brouwer *et al.*, 1992). This suggested that the MTHFR deficiency may have resulted in insufficient *S*-adenosylmethionine levels, which are required for arsenic methylation.

The cellular and molecular mechanisms by which arsenicals cause toxic effects within the nervous system are unclear and represent a significant void in the investigations of arsenic toxicity. Inhibition of the nerve pyruvate oxidase system by arsenic has been proposed as a possible mechanism of action (Hindmarsh and McCurdy, 1986). Since, potentially, arsenic is present in many different chemical forms *in vivo*, the neurotoxic species of arsenic remain to be determined.

3.5.3 Peripheral vascular disease

The peripheral vascular effects of chronic arsenic exposure have been of interest to epidemiologists (Engel and Smith, 1994; Engel *et al.*, 1994). Blackfoot disease is endemic along the south-western coast of Taiwan and has been the most characterized peripheral vascular disease attributed to ingestion of water containing high levels of inorganic arsenic (range 0.01–1.82 mg/L) (TPHD, 1993). Characterized by initial numbness and intermittent claudication, the disease progresses to gangrene and ultimately spontaneous or surgical amputation (TPHD, 1993). Early pathologic findings in Blackfoot disease patients include fibroid degeneration of blood vessels and perivascular small-cell infiltration, followed by thrombotic occlusion and proliferation of the vasa vasorum, ultimately leading to complete occlusion of the affected blood vessels (Yu *et al.*, 1984). Histologically, thromboangiitis obliterans and arteriosclerosis obliterans are observed (TPHD, 1993). Abnormal peripheral microcirculation was observed among seemingly normal individuals living in Blackfoot disease hyperendemic villages of Taiwan (Tseng *et al.*, 1995). They exhibited lower basal perfusion and, after hyperthermic stimulus, lower perfusion in the big toe (Tseng *et al.*, 1995).

It is unclear whether the arsenic in drinking water is solely responsible for such peripheral vascular diseases, since other populations (Zaldivar, 1974; Cebrian *et al.*, 1983) ingesting water containing equivalent amounts of arsenic over a duration similar to those in Taiwan do not present with vascular pathology of the magnitude observed in Blackfoot patients. This suggests the involvement of additional factors besides arsenic in the etiology of Blackfoot disease. A complex mixture of fluorescent substances was discovered in the artesian wells of the Blackfoot-endemic area of Taiwan (TPHD, 1993). The role of these substances in peripheral vascular disease remains unresolved. One of these fluorescent substances, humic acid, inhibited plasma protein kinase C activity in a dose-dependent manner *in vitro* (Yang *et al.*, 1994). However, arsenic trioxide activated protein kinase C. Plasma protein kinase C is an important regulatory molecule of blood coagulation and fibrinolysis (Yang *et al.*, 1994). Defective protein kinase C activity may predispose an individual to thromboembolic complications (Yang *et al.*, 1994).

The cellular and molecular mechanisms for vascular disease caused by arsenic exposure are not known. Using porcine aortic endothelial cells, arsenite elevated transiently both intracellular oxidant formation and nuclear levels of NF-κB in a dose-dependent manner (Barchowsky *et al.*, 1996). Additionally, arsenite activation of NF-κB was attenuated by increasing intracellular thiol pools. Arsenite therefore appears to initiate signaling and translocation of nuclear transcription factors, suggesting that nonlethal levels of arsenite might promote vascular disease by activating the endothelium, rather than by a cytotoxic mechanism (Barchowsky *et al.*, 1996).

3.6 Carcinogenicity

Arsenite is not a mutagen (Rossman, *et al.*, 1980) but may increase mutation frequency by inhibiting the DNA repair enzyme, DNA ligase II (Li and Rossman, 1989). Arsenite decreases incision frequency and impairs ligation repair (Hartwig *et al.*, 1997) and causes hyperploidy (Huang *et al.*, 1995). Sister chromatid exchange induction (Jha *et al.*, 1992) and protein kinase C activation (Liu and Huang, 1997) suggest arsenite as a tumor promotor. The results of these short-term *in vitro* tests suggest genotoxicity.

However, evidence for arsenic carcinogenicity is based on epidemiologic studies. Exposure of humans to arsenic is associated with increased risk of liver, lung, bladder and kidney cancer (Chen *et al.*, 1992; Chiou *et al.*, 1995). Hyperkeratosis and skin cancer have also been related to inorganic arsenic exposure (Neubauer, 1947). However, animal models have failed to provide convincing evidence for arsenic carcinogenicity. In addition, the molecular mechanism of arsenic carcinogenicity in humans is unknown.

The lack of an animal model for carcinogenicity of arsenic in humans has been a critical factor contributing to our ignorance of the mechanism(s) by which inorganic arsenicals cause human cancer. It is still unclear whether trivalent, pentavalent or organic arsenic is the ultimate causative agent. Although oxidative methylation of inorganic arsenic is supposed to reduce its acute toxicity (Aposhian, 1989), there is no unequivocal evidence that methylation decreases its carcinogenicity. Inorganic, not organic, arsenic caused diffuse hyperkeratosis and hyperplasia in an organotypic culture of human keratinocytes (Klimecki *et al.*, 1997). DNA-damaging active oxygen species, superoxide anions and hydroxyl radicals were produced in pulmonary cells after the administration of DMA to mice (Yamanaka *et al.*, 1991). The reduction of molecular oxygen by dimethylarsine is thought to yield the radical O_2^-. Dimethylarsine caused single-stranded DNA breaks in mouse lung cells (Yamanaka *et al.*, 1989). DMA has been implicated as a tumor promoter in bladder, kidney, liver and thyroid carcinogenesis in the rat (Yamamoto *et al.*, 1995). These experiments have been criticized because of the very high DMA dosage. What is of paramount importance is the dose of DMA at its molecular site of action. Oxidative methylation may be a bioactivation pathway for arsenic carcinogenicity.

Recently, two groups have implicated arsenite as an epigenetic carcinogen, and have suggested that perturbed DNA methylation is the basis for arsenite carcinogenicity (Mass and Wang, 1997; Zhao *et al.*, 1997). Both studies presented animal model systems tenable for additional investigation and demonstrated a property of arsenite to modulate DNA structure. The underlying molecular mechanism, however, with respect to interaction and alteration of genetic elements remains unclear.

Because inorganic arsenite is biotransformed via methylation using *S*-adenosylmethionine (SAM) as the methyl donor, inorganic arsenite has been hypothesized to effect DNA methyltransferase activity by influencing cellular SAM concentration. According to Mass and Wang (1997), arsenite increases cellular SAM levels, but according to Zhao *et al.* (1997), arsenite decreases it.

In mammalian and higher plant DNA, a significant fraction of cytosine residues exist in a modified form in which methyl groups are attached to the 5′ carbon atom of the pyrimidine ring (5-MeC). Methyl moieties are added via DNA methyltransferases using SAM as the cofactor. Methylation does not alter the hydrogen bonding or base pairing of the cytosine. Methylation affects gene expression of certain genes in higher eucaryotes. Highly methylated DNA is usually associated with genetically silent regions of the chromosome, while undermethylation is associated with transcriptionally active genes, such as that of metallothionein. Such hypomethylation is thought to alter the structure of DNA so that it is more accessible to activator proteins or so that repressors can dissociate from the gene.

Mass and Wang (1997) exposed human lung adenocarcinoma (A549) cells continuously to low levels of arsenite or arsenate for 2 weeks and observed hypermethylation of the tumor suppressor gene *p53*. At concentrations as low as 0.08 μM arsenite or 30 μM arsenate and in a dose-responsive manner, inorganic arsenic increased resistance to the restriction endonuclease *Hpa*II. This enzyme recognizes and cleaves the CCGG sequence in DNA but it will not cleave methylated CCGG regions. The hypermethylation, a dose-dependent sevenfold increase in 5-MeC, was localized to the promoter region of *p53*. Methylation of cytosine in the promoter region downregulated the transcription of *p53* (Schroeder and Mass, 1997) but arsenic-induced hypermethylation was not restricted to the *p53* gene, since a small increase in total genomic hypermethylation was observed after 2 weeks' exposure to 2 μM arsenite. The authors speculated that arsenite was inhibiting other SAM-dependent methyltransferases, resulting in increased SAM concentrations in the cell. Such a change in SAM concentrations supposedly increased the DNA methyltransferase activity, resulting in hypermethylation of the substrate. Hypermethylation was also observed after exposure to arsenate. No hypermethylation was observed in cells grown in DMA, at concentrations as high as 2000 μM. Since DMA is rarely further methylated in mammals (Marafante *et al.*, 1987b), it would not be expected to influence SAM concentrations or SAM-dependent methyltransferases.

An alternative mechanism of inorganic arsenic carcinogenicity was offered by Mass and Wang (1997). They proposed that the methylation of arsenite would consume SAM, resulting in the depletion of the intracellular SAM pool. The resulting dearth of this cofactor would not allow methylation of cytosine and the maintenance of 5-MeC levels in DNA would be prevented.

This latter hypothesis is supported by the provocative experiments of Zhao *et al.* (1997), showing dose-dependent arsenic-induced DNA hypomethylation. Rat liver epithelial (TRL 1215) cells were transformed after chronic exposure (approximately 13.5 weeks) to low levels of arsenic (0.500 μM), presumably by arsenic-induced global DNA hypomethylation. Inoculation of the arsenic-transformed TRL 1215 cells into Athymic Nude mice resulted in dose-dependent metastatic tumor incidence. When the transformed cells were cultured for an additional 6 weeks in the absence of arsenite, the cells still gave rise to malignant tumors and still maintained their heritable hypomethylated status. However, treatment with arsenite for a shorter period of time (48 h, 1.0 μM), did not induce hypomethylation. These experiments may be the long-awaited beginning of understanding arsenic carcinogenicity.

In contrast to the study by Mass and Wang (1997), *Hpa*II digestion of TRL 1215 genomic DNA was increased threefold after arsenic pretreatment (18 weeks, 0.500 μM) as compared to controls. Consistent with the SAM-depletion hypothesis of arsenic carcinogenesis, there was an approximate fourfold decrease of intracellular SAM in arsenic-transformed cells (Zhao *et al.*, 1997). A 40% decrease in DNA methyltransferase activity was also observed after arsenic treatment. This may be due to decreased SAM concentration, since DNA methyltransferase gene expression was induced by chronic exposure to low levels of arsenite. Another possibility is that arsenite or increased concentrations of *S*-adenosylhomocysteine may also inhibit DNA methyltransferase. Hypomethylation in TRL 1215 cells was also assessed by metallothionein expression. Hypomethylation is associated with activation of *MT* genes as well as the oncogenes c-*myc* and c-*fos* (Rao *et al.*, 1989). In the presence of cadmium as an inducer, the hypomethylated *MT* gene was hyperinducible by arsenic; arsenic-transformed cells had MT protein levels approximately six times greater than untransformed cells. Arsenite-induced hypomethylation increased basal expression and inducibility of MT, which demonstrated altered gene expression by perturbed DNA methylation, hence epigenetic carcinogenesis.

The experiments of Mass and Wang (1997) and Schroeder and Mass (1997) demonstrated that arsenite exposure decreased p53 expression, while the experiments of Salazar *et al.* (1997) showed increased expression of p53 in HeLa and LCL-EBV (lymphoblasts transformed with Epstein–Barr virus) cells upon exposure to 10 μM sodium arsenite. Cancer induction is complex and thus the response to a given carcinogen is a function of the experimental system. This, in part, may reconcile the different mechanisms of epigenetic carcinogenesis detailed in Mass and Wang (1997) and Zhao *et al.* 1997). Both research endeavors offer a viable dose-dependent model of arsenic carcinogenicity worthy of continued study.

One of the key elements in the response of an animal, including humans, would appear to be the biological potential in activating or deactivating exogenous carcinogens. Bettley and O'Shea (1975) suggested that susceptibility to arsenic carcinogenicity among humans may be due to individual metabolic differences in arsenic absorption, storage, or excretion and latency, as evidenced by lower blood and total urinary arsenic levels among carcinoma subjects. The response to chemical carcinogens varies with the species, strain, age and gender of the experimental animal. At present, however, the mechanism of arsenic carcinogenicity is far from being solved. The Waalkes group (Huff *et al.*, 1998) recently suggested more reasonable testing scenarios for arsenic carcinogenicity. If their recommendations are followed, the solution might be found more quickly.

3.7 Mobilization of arsenic in humans by DMPS

This laboratory has had a role in the development of two chelating agents, DMPS (sodium 2,3-dimercapto-1-propane sulfonate, DIMAVAL®) and DMSA (meso-2,3-dimercapto-succinic acid, succimer). Analytical procedures for their study were developed (Maiorino *et al.*, 1986, 1987), their metabolism was investigated (Aposhian *et al.*, 1989; Maiorino and Aposhian, 1989; Rivera *et al.*, 1989; Maiorino *et al.*, 1989, 1991; Hurlbut *et al.*, 1994), and their therapeutic uses honed (Aposhian *et al.*, 1992, 1997; Gonzalez-Ramirez *et al.*, 1995; Maiorino *et al.*, 1996). DMPS and DMSA are chemical analogs of BAL (British anti-Lewisite; dimercaprol; 2,3-dimercapto-1-propanol). BAL has been the drug of choice for the treatment of arsenic, lead and sometimes mercury intoxication in the United States since the late 1940s. However, it has many disadvantages (Klaassen, 1985). Approximately 55% of the patients receiving BAL have some kind of an unpleasant side-effect (Klaassen, 1985).

DMPS was developed in the 1950s in the former Soviet Union by Petrunkin (1956) and became an official drug in the Soviet physician's armamentarium in 1958 (Klimova, 1958). It was not available outside of the Soviet Union until 1978, at which time Heyl, a small pharmaceutical company in Berlin, Germany, specializing in antidotes, announced its synthesis and distribution. Since then, it has had wide use as a chelating agent for both diagnostic and treatment purposes, especially in Germany (Rote Liste, 1995). Its major use has been for mobilizing inorganic mercury in the body (for reviews, see Aposhian 1983; Kemper *et al.*, 1990; Aaseth *et al.*, 1995; Aposhian *et al.*, 1995). In the United States, Investigational New Drug (IND) permission from the FDA is required for its use in research with humans. A Compassionate IND can be obtained quickly from the FDA for its use in emergencies such as life-threatening situations.

DMPS and DMSA are among the most selective and specific chelating agents. Of these two orally useful chelating agents, DMPS has at least three advantages. First, it appears to remain in the body for a longer time than DMSA (Hurlbut *et al.*, 1994). Secondly, it acts more quickly than DMSA (Buchet and Lauwerys, 1989), probably because its distribution is both extracellular and intracellular (Wildenauer *et al.*, 1982; Zheng *et al.*, 1990), whereas DMSA appears to be only extracellular in its distribution (Zheng *et al.*, 1990). Thirdly, preparations of DMPS are available for oral, intravenous or intramuscular use. However, DMSA is available only in capsule form. An intravenous preparation is a distinct advantage when metal toxicity has been so severe that the poisoned patient must be put on dialysis, or when immediate use in the emergency room is necessary.

DMPS has been used to evaluate in a novel manner the arsenic exposure of humans living in two towns in north-eastern Chile (Aposhian *et al.*, 1997). Residents of San Pedro de Atacama drink water containing 593 μg arsenic/L. Those in the control town, Toconao, drink water containing 21 μg arsenic/L. Our hypothesis was that the administration of DMPS (DIMAVAL®) would increase the urinary excretion of arsenic, alter the urinary profile of arsenic species, and thus result in a better indication of the body burden of arsenic and be a better biomarker for arsenic exposure.

The method used to evaluate these subjects was that, after an overnight fast, 300 mg DMPS was given by mouth, and urine was collected at specified time periods. The urines were analyzed for inorganic arsenic, MMA, DMA and total arsenic.

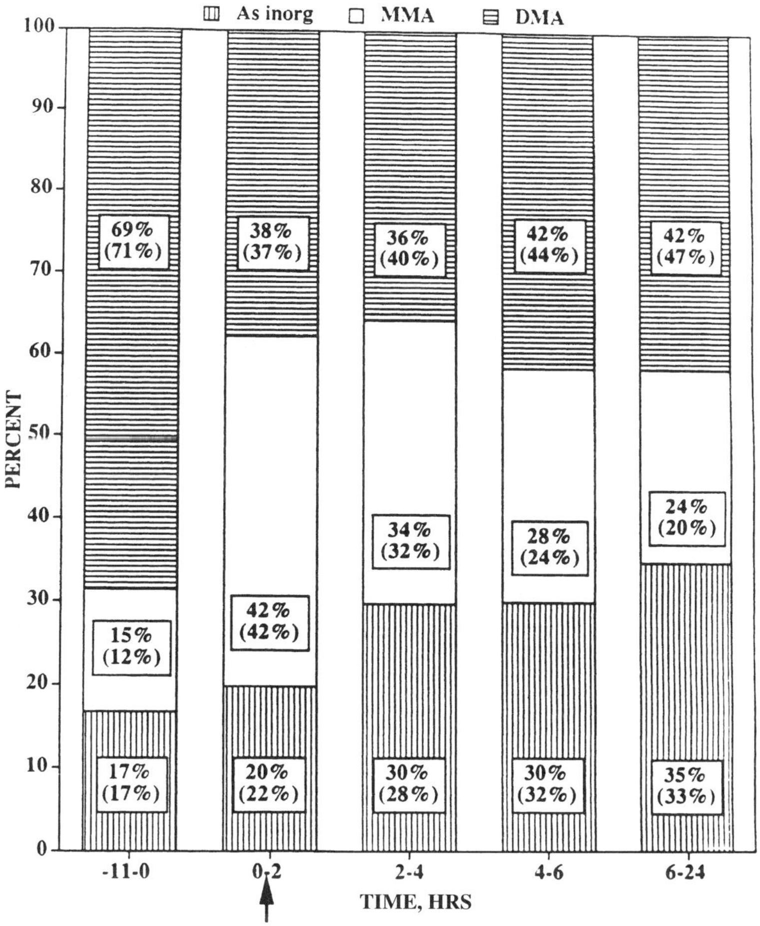

Figure 3.5 Arsenic species expressed as a percentage of urinary total arsenic at time periods before and after DMPS administration. DMPS (300 mg) was given by mouth at zero time (↑). Percentages for the Toconao group are in parentheses.

The exposed subjects had a greater urinary excretion of total arsenic, before and after DMPS administration, than the control subjects (Aposhian *et al.*, 1997). During the 2-hour period following DMPS administration, MMA represented 42%, inorganic arsenic 20–22% and DMA 37–38% of the total urinary arsenic (Fig. 3.5). The usual range of the MMA percentage in human urine has been 10–20%. The percentage MMA increased almost equally for both the arsenic-exposed and control subjects.

These studies (Aposhian *et al.*, 1997) resulted in new and important knowledge about arsenic in humans. They showed that the human body stores substantial amounts of arsenic. DMPS had a highly specific effect on MMA metabolism and urinary excretion. The urinary arsenic concentration after DMPS administration may be more indicative of the body burden of arsenic, since it was greater than that found before DMPS was given.

3.8 Afterthoughts

One is often tempted in a chapter of this type to look ahead and plan or suggest future needs. Inorganic arsenite is a very reactive species in the mammalian body. The isolation of the endogenous molecules, probably proteins, involved in the toxic properties of inorganic arsenic from those involved in its detoxification need to be accomplished and their function(s) clarified. The toxic and detoxifying mechanisms need to be separated and defined. Although this is not a small task, it is possible. Some of this will be accomplished by the introduction of modern molecular biology techniques, as has been done by the Mass and Waalkes groups. The amino-acid sequence of the arsenite methyltransferase needs to be determined since it will allow its gene structure to be unraveled. Nucleotide probes can then be synthesized and used to survey large populations to determine whether genetic differences are responsible for the large variations in the urinary profiles of humans exposed to inorganic arsenite in their drinking water.

It would be inappropriate not to point out the environmental injustices due to drinking water containing high levels of arsenic. It is the economically poor people of West Bengal, Bangladesh, Taiwan and other countries, that usually are subjected to drinking water containing toxic levels of arsenic. The middle class and wealthy can usually use political pressure or spend money to buy purified water. The poor cannot. Water for drinking is essential for life. A lack of it, for only a small number of days, can result in death. So water containing high levels of arsenic has to be consumed by those who have no alternative. One cannot help but wonder whether some investigators are so consumed with doing the best-controlled and scientifically correct investigation that the potential suffering of their subjects is forgotten.

3.9 Summary

For a metalloid having received so much attention for centuries in the lay literature as it has in relation to homicide and suicide, it is surprising that in 1998, at the time of the completion of this chapter, there are still many questions unanswered about arsenic toxicology. For example, much is still unknown about its mechanism of action, biotransformation and carcinogenicity. The many species or chemical forms of arsenic have made the study of its biochemical and toxicological properties one of the most complicated of the heavy metals and metalloids. Inorganic arsenite (As^{III}) has an affinity for sulfhydryl groups of proteins, especially for vicinal sulfhydryl groups. Therefore it can be a potent inhibitor of essential enzymes such as the pyruvate dehydrogenase multienzyme complex.

Distribution of arsenicals is highly dependent on chemical form, route of administration, dose and animal species. Different chemical forms of arsenic may be handled quite differently when administered via the same route. Additionally, arsenic undergoes complex interconversions and metabolism *in vivo*, such that the chemical form of arsenic that one is exposed to may be far different from that being eliminated (Fig. 3.6). Urinary excretion of arsenic metabolites can give information on the degree of exposure of an individual to arsenic and the chemical forms to which one is exposed. In individuals, nonoccupationally exposed and not residing in areas environmentally contaminated by inorganic arsenic, the amount of arsenic (inorganic arsenic + MMA + DMA) excreted is usually below 15 μg/day.

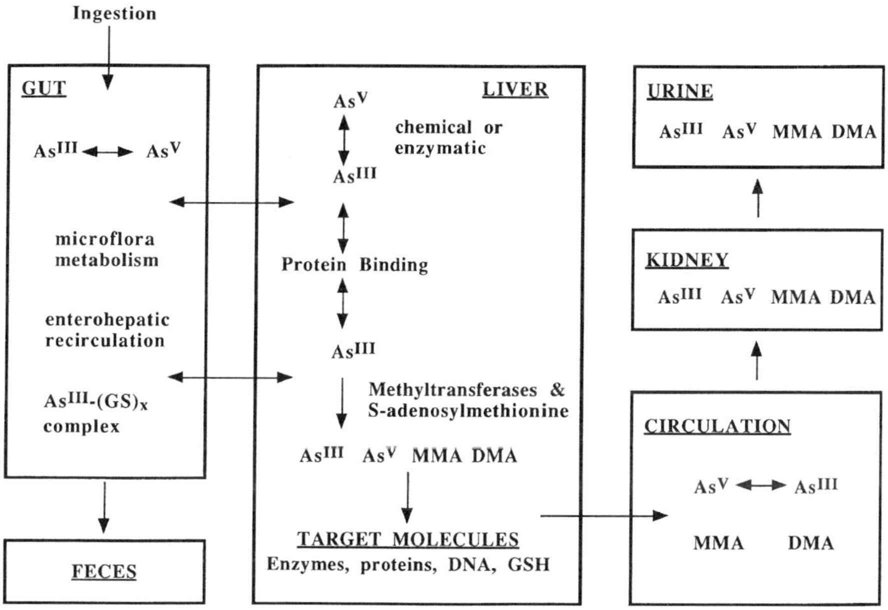

Figure 3.6 The complexity of arsenic's behavior within the body following ingestion of an inorganic form of arsenic (e.g. arsenite or arsenate). While the metabolism and target molecule interactions of these arsenicals are presented as occurring in the liver compartment, this is meant only as a representative target tissue. It is known that other tissues, including the lung, kidney and testes, possess the capabilities to methylate inorganic arsenic, while the lung and skin are certainly target tissues of arsenic toxicity. Also, the reader should notice that oxidation states are not indicated for the methylated arsenicals, MMA and DMA. Current separation and analysis techniques have not allowed differentiation of trivalent and pentavalent forms of these methylation products *in vivo*, or information about their interconvertion.

Methylation has been considered to be the primary mechanism of inorganic arsenic biotransformation and detoxification in most animals, since the methylated metabolites of inorganic arsenic are less acutely toxic, more quickly excreted and demonstrate less tissue reactivity. Recently, this has been questioned as a detoxification pathway. The methylation pathway is generally believed to follow several sequential steps. Arsenate is reduced to arsenite, from which MMA and DMA are formed enzymatically by the oxidative addition of methyl groups to arsenite and MMA from the methyl donor *S*-adenosylmethionine. However, animals show considerable species diversity as to whether they do or do not methylate inorganic arsenite. A number of animal species lack, or are deficient in, arsenite methyltransferase.

Neurologic symptoms due to arsenic are primarily those of a peripheral sensory neuritis, predominantly numbness, severe paresthesia of the distal portion of the extremities, diminished sense of touch, pain, heat and cold and symmetrically reduced muscle power.

Blackfoot disease is endemic along the south-western coast of Taiwan and has been the most characterized peripheral vascular disease attributed to ingestion of water containing high levels of inorganic arsenic (range 0.01–1.82 mg/L). It is unclear whether the arsenic in drinking water is solely responsible for such peripheral vascular diseases, as other populations ingesting water containing equivalent amounts of arsenic over a duration similar to those in Taiwan do not present with vascular pathology of the magnitude observed in Blackfoot patients. This suggests the involvement of additional factors besides arsenic in the etiology of Blackfoot disease.

Arsenic is a known human carcinogen. Since there has not been a suitable animal model, its mechanism of action as a carcinogen is unknown. Recent experiments dealing with the effects of arsenite on methylation of DNA hold a great deal of promise in solving this dilemma. The basis of clinical treatment of arsenic poisoning has been the use of vicinal dithiols such as British anti-Lewisite and, more recently, DMPS (sodium 2,3-dimercapto-1-propane sulfonate, DIMAVAL®).

References

Aaseth, J., Jacobsen, D., Anderson, O. and Wickstrom, E. 1995. Treatment of mercury and lead poisonings with dimercapto succinic acid and sodium dimercaptopropanesulfonate: A review. *The Analyst* **120**, 853–854.

Albores, A., Sinal, C.J., Cherian, M.G. and Bend, J.R. 1995. Selective increase of rat lung cytochrome P_{450} 1A1 dependent monooxygenase activity after acute sodium arsenite administration. *Canadian Journal of Physiology and Pharmacology* **73**, 153–158.

Aposhian, H.V. 1983. DMSA and DMPS – water soluble antidotes for heavy metal poisoning. *Annual Review of Pharmacology and Toxicology* **23**, 193–215.

Aposhian, H.V. 1989. Biochemical toxicology of arsenic. *Reviews in Biochemical Toxicology* **10**, 265–99.

Aposhian, H.V. 1997. Enzymatic methylation of arsenic species and other new approaches to arsenic toxicity. *Annual Review of Pharmacology and Toxicology* **37**, 397–419.

Aposhian, H.V., Maiorino, R.M., Dart, R.C. and Perry, D.F. 1989. Urinary excretion of meso-2,3-dimercaptosuccinic acid in human subjects. *Clinical Pharmacology and Therapeutics* **45**, 520–526.

Aposhian, H.V., Bruce, D.C., Alter, W., Dart, R.C., Hurlbut, K.M. and Aposhian, M.M. 1992. Urinary mercury after administration of 2,3-dimercaptopropane-1-sulfonic acid: Correlation with dental amalgam score. *FASEB Journal* **6**, 2472–2476.

Aposhian, H.V., Gonzalez-Ramirez, D., Maiorino, R.M., Zuniga-Charles, M., Xu, Z.F., Hurlbut, K.M., Junco-Munoz, P., Aposhian, M.M. and Dart, R.C. 1995. Mobilization of heavy metals by newer, therapeutically useful chelating agents. *Toxicology* **97**, 23–38.

Aposhian, H.V., Arroyo, A., Cebrian, M.E., Del Razo, L.M., Hurlbut, K.M., Dart, R.C., Gonzalez-Ramirez, D., Kreppel, H., Speisky, H., Smith, A., Gonsebatt, M.E., Ostrosky-Wegman, P. and Aposhian, M.M., 1997. DMPS-arsenic challenge test: I. Increased urinary excretion of monomethylarsonic acid in humans given dimercaptopropane sulfonate. *Journal of Pharmacology and Experimental Therapeutics* **277**, 938–944.

Apostoli, P., Alessio, L., Romeo, L., Buchet, J.P. and Leone, R. 1997. Metabolism of arsenic after acute occupational arsine intoxication. *Journal of Toxicology and Environmental Health* **52**, 331–342.

Aso, T. and Abiko, Y. 1978. Tissue distribution of arsenic after subcutaneous implantation of arsenic trioxide pellet in rats. *Journal of Toxicological Science* **3**, 109–115.

Baldwin, W.J. 1983. The use of arsenic as a wood preservative, in *Arsenic: Industrial, Biomedical, Environmental Perspectives*, (eds W.H. Lederer and R.J. Fensterheim), Van Nostrand Reinhold Company, New York, pp. 99–110.

Barchowsky, A., Dudek, E.J., Treadwell, M.D. and Wetterhahn, K.E. 1996. Arsenic induces oxidant stress and NF-KB activation in cultured aortic endothelial cells. *Free Radicals in Biology and Medicine* **21**, 783–790.

Bates, M.N., Smith, A.H. and Hopenhayen-Rich, C. 1992. Arsenic ingestion and internal cancers: A review. *American Journal of Epidemiology* **135**, 462–476.

Bencko, V., Benes, B. and Cikrt, M. 1976. Biotransformation of As(III) to As(V) and arsenic tolerance. *Archives of Toxicology*, **36**, 159–162.

Bencko, V., Symon, K., Chladek, V. and Phirt, J. 1977. Health aspects of burning coal with a high arsenic content. II. Hearing changes in exposed children. *Environmental Research* **13**, 386–395.

Bertolero, F., Marafante, E., Edel Rade, J., Pietra, R. and Sabbioni, E. 1981. Biotransformation and intracellular binding of arsenic in tissues of rabbits after intraperitoneal administration of ^{74}As labeled arsenite. *Toxicology* **20**, 35–44.

Bettley, F.R. and O'Shea, J.A. 1975. The absorption of arsenic and its relation to carcinoma. *British Journal of Dermatology* **92**, 563–568.

Blair, P.C., Thompson, M.B., Bechtold, M., Wilson, R.E., Moorman, M.P. and Fowler, B.A. 1990. Evidence for oxidative damage to red blood cells in mice induced by arsine gas. *Toxicology* **63** (1), 25–34.

Borgono, J.M., Vicent, P., Venturino, H. and Infante, A. 1977. Arsenic in the drinking water of the City of Antofagasta: Epidemiological and clinical study before and after the installation of a treatment plant. *Environmental Health Perspectives* **19**, 103–105.

Braman, R.S. and Foreback, C.C. 1973. Methylated forms of arsenic in the environment. *Science* **182**, 1247–1249.

Brouwer, O.F., Onkenhout, W., Edelbroek, P.M., de Kom, J.F.M., de Wolff, F.A. and Peters, A.C.B. 1992. Increased neurotoxicity of arsenic in methylenetetrahydrofolate reductase deficiency. *Clinical Neurology and Neurosurgery* **94**, 307–310.

Buchet, J.P. and Lauwerys, R. 1985. Study of inorganic arsenic methylation by rat liver *in vitro*: Relevance for the interpretation of observations in man. *Archives of Toxicology* **57**, 125–129.

Buchet, J.P. and Lauwerys, R. 1987. Study of factors influencing the *in vivo* methylation of inorganic arsenic in rats. *Toxicology and Applied Pharmacology* **91**, 65–74.

Buchet, J.P. and Lauwerys, R. 1988. Role of thiols in the *in vitro* methylation of inorganic arsenic by rat liver cytosol. *Biochemical Pharmacology* **37**, 3149–3153.

Buchet, J.P. and Lauwerys, R. 1989. Influence of 2,3-dimercaptopropane-1-sulfonate and dimercaptosuccinic acid on the mobilization of mercury from tissues of rats pretreated with mercuric chloride, phenylmercury acetate or mercury vapors. *Toxicology* **54**, 323–333.

Buchet, J.P., Lauwerys, R. and Roels, H. 1980. Comparison of several methods for the determination of arsenic compounds in water and urine. *International Archives of Occupational and Environmental Health* **46**, 11–29.

Buchet, J.P., Lauwerys, R. and Roels, H. 1981a. Comparison of the urinary excretion of arsenic metabolites after a single oral dose of sodium arsenite, monomethylarsonate, or dimethylarsinite in man. *International Archives of Occupational and Environmental Health* **48**, 71–79.

Buchet, J.P., Lauwerys, R. and Roels, H. 1981b. Urinary excretion of inorganic arsenic and its metabolites after repeated ingestion of sodium metaarsenite by volunteers. *International Archives of Occupational and Environmental Health* **48**, 111–118.

Buchet, J.P., Guebel, A., Pauwels, S., Mahiew, P. and Lauwerys, R. 1984. The influence of liver disease on the methylation of arsenite in humans. *Archives of Toxicology* **55**, 151–154.

Buchet, J.P., Pauwels, J. and Lauwerys, R. 1994. Assessment of exposure to inorganic arsenic following ingestion of marine organisms by volunteers. *Environmental Research* **66**, 44–51.

Buchet, J.P., Lison, D., Ruggeri, M., Foa, V. and Elia, G. 1996a. Assessment of exposure to inorganic arsenic, a human carcinogen, due to the consumption of seafood. *Archives of Toxicology* **70**, 773–778.

Buchet, J.P., Staessen, J., Roels, H., Lauwerys, R. and Fagard, R. 1996b. Geographical and temporal differences in the urinary excretion of inorganic arsenic: A Belgian population study. *Occupational and Environmental Medicine* **53**, 320–327.

Burton, C.A., Hatlelid, K., Divine, K., Carter, D.E., Fernando, Q., Brendel, K. and Gandolfi, A.J. 1995. Glutathione effects on toxicity and uptake of mercuric chloride and sodium arsenite in rabbit renal cortical slices. *Environmental Health Perspectives* **103** (Suppl. 1), 81–84.

Calesnick, B., Wase, A. and Overby, L.R. 1966. Availability during human consumption of the arsenic in tissues of chicks fed arsenilic-^{74}As acid. *Toxicology and Applied Pharmacology* **9**, 27–30.

Carter, D.E. 1995. Oxidation–reduction reactions of metal ions. *Environmental Health Perspectives* **103** (Suppl. 1), 17–19.

Cebrian, M.E., Albores, A., Aguilar, M. and Blakely, E. 1983. Chronic arsenic poisoning in the north of Mexico. *Human Toxicology* **2**, 121–133.

Charbonneau, S.M., Spencer, K., Bryce, F. and Sandi, E. 1978. Arsenic excretion by monkeys dosed with arsenic-containing fish or with inorganic arsenic. *Bulletin of Environmental Contamination and Toxicology* **20**, 470–477.

Charbonneau, S.M., Hollins, J.G., Tam, G.K.H., Bryce, F., Ridgeway, J.M. and Willes, R.F. 1980. Whole-body retention, excretion and metabolism of [^{74}As] arsenic acid in the hamster. *Toxicology Letters* **5**, 175–182.

Chen, C.J. and Wang, C.J. 1990. Ecological correlation between arsenic level in well water and age-adjusted mortality from malignant neoplasms. *Cancer Research* **50**, 5470–5474.

Chen, C.J., Chen, C.W., Wu, M.M. and Kuo, T.L. 1992. Cancer potential in liver, lung, bladder, and kidney due to ingested inorganic arsenic in drinking water. *British Journal of Cancer* **66** 888–892.

Chen, G.Q., Zhu, J., Shi, X.G., Ni, J.H., Zhong, H.J., Si, G.Y., Jin, X.L., Tang, W., Li, X.S., Xong, S.M., Shen, Z.X., Sun, G.L., Ma, J., Zhang, P., Zhang, T.D., Gazin, C., Naoe, T., Chen, S.J., Wang, Z.Y. and Chen, Z. 1996a. *In vitro* studies on cellular and molecular mechanisms of arsenic trioxide (As2O3) in the treatment of acute promyelocytic leukemia: As_2O_3 induces NB4 cell apoptosis with downregulation of Bcl-2 expression and modulation of PML-RARalpha/PML proteins. *Blood* **88**, 1052–1061.

Chen, H., Yoshida, K., Wanibuchi, H., Fukushima, S., Inoue, Y. and Endo, G. 1996b. Methylation and demethylation of dimethylarsinic acid in rats following chronic oral exposure. *Applied Organometallic Chemistry* **10**, 741–745.

Chiou, H.Y., Hsueh, Y.M., Liaw, K.F., Horng, S.F., Chiang, M.H., Pu, Y.S., Lin, J.S.N., Huang, C.H. and Chen, C.J. 1995. Incidence of internal cancers and ingested inorganic arsenic: A seven-year follow-up study in Taiwan. *Cancer Research* **55**, 1296–1300.

Cikrt, M. and Bencko, V. 1974. Fate of arsenic after parenteral administration to rats, with particular reference to excretion via bile. *Journal of Hygiene, Epidemiology, Microbiology and Immunology* **18**, 129–136.

Cikrt, M., Bencko, V., Tich, M. and Benes, B. 1980. Biliary excretion of ^{74}As and its distribution in the golden hamster after administration of ^{74}As(III) and ^{74}As(V). *Journal of Hygiene, Epidemiology, Microbiology and Immunology* **24**, 384–388.

Crane, R.K. and Lipmann, F. 1953. The effect of arsenate on aerobic phosphorylation. *Journal of Biological Chemistry* **201**, 235–243.

Crecelius, E.A. 1977. Changes in the chemical speciation of arsenic following ingestion by man. *Environmental Health Perspectives* **19**, 147–150.

De Kimpe, J., Cornelis, R., Mees, L. and Vanholder, R. 1996. Basal metabolism of intraperitoneally injected carrier-free ^{74}As-labeled arsenate in rabbits. *Fundamental and Applied Toxicology* **34**, 240–248.

Del Razo, L.M., Garcia-Vargas, G.G., Vargas, H., Albores, A., Gonsebatt, M.E., Montero, R., Ostrosky-Wegman, P., Kelsh, M. and Cebrian, M.E., 1997. Altered profile of urinary arsenic metabolites in adults with chronic arsenicism. A pilot study. *Archives of Toxicology* **71**, 211–217.

Ducoff, H.S., Neal, W.B., Straube, R.L., Jacobson, L.O. and Brues, A.M. 1948. Biological studies with arsenic[76]. II. Excretion and tissue localization. *Proceedings of the Society for Experimental Biology and Medicine* **69**, 548–554.

Dutkiewicz, T. 1977. Experimental studies on arsenic absorption routes in rats. *Environmental Health Perspectives* **19**, 173–177.

Engel, R.R. and Smith, A.H. 1994. Arsenic in drinking water and mortality from vascular disease: An ecologic analysis in 30 counties in the United States. *Archives of Environmental Health* **49**, 418–427.

Engel, R.R., Hopenhayn-Rich, C., Receveur, O. and Smith, A.H. 1994. Vascular effects of chronic arsenic exposure: A review. *Epidemiologic Reviews* **16**, 184–209.

Espinoza, E.O., Mann, M.J., Bleasdell, B., DeKorte, S. and Cox, M. 1996. Toxic metals in selected traditional Chinese medicinals. *Journal of Forensic Sciences* **41**, 453–456.

Feldman, R.G., Niles, C.A., Kelly-Hayes, M., Sax, D.S., Dixon, W.J., Thompson, D.J. and Landau, E. 1979. Peripheral neuropathy in arsenic smelter workers. *Neurology* **29**, 939–944.

Fischer, A.B., Buchet, J.P. and Lauwerys, R.R. 1985. Arsenic uptake, cytotoxicity, and detoxification studied in mammalian cells in culture. *Archives of Toxicology* **57**, 168–172.

Flynn, I.W. and Bowman, I.B.R. 1974. The action of trypanocidal arsenical drugs on *Trypanosoma brucei* and *Trypanosoma rhodesiense. Comparative Biochemistry and Physiology* **48B**, 261–273.

Fowler, B.A. and Weissberg, J.B. 1974. Arsine poisoning. *New England Journal of Medicine* **291** (22), 1171–1174.

Fowler, T. 1786. *Medical Reports of the Effects of Arsenic in the Cure of Agues, Remitting Fevers and Periodic Headaches*. J. Johnson and W. Brown, London.

Freeman, G.B., Johnson, J.D., Killinger, J.M., Liao, S.C., Davis, A.O., Ruby, M.V., Chaney, R.L., Lovre, S.C. and Bergstrom, P.D. 1993. Bioavailability of arsenic in soil impacted by smelter activities following oral administration in rabbits. *Fundamental and Applied Toxicology* **21**, 83–88.

Freeman, G.B., Schoof, R.A., Ruby, M.V., Davis, A.O., Dill, J.A., Liao, S.C., Lapin, C.A. and Bergstrom, P.D. 1995. Bioavailability of arsenic in soil and house dust impacted by smelter activities following oral administration in cynomolgus monkeys. *Fundamental and Applied Toxicology* **28**, 215–222.

Freeman, H.C., Uthe, J.F., Fleming, R.B., Odense, P.H., Ackman, R.G., Landry, G. and Musial, C. 1979. Clearance of arsenic ingested by man from arsenic contaminated fish. *Bulletin of Environmental Contamination and Toxicology* **22**, 224–229.

Georis, B., Cardenas, A., Buchet, J.P. and Lauwerys, R. 1990. Inorganic arsenic methylation by rat tissue slices. *Toxicology* **63**, 73–84.

Geubel, A.P., Mairlot, M.C., Buchet, J.P. and Lauwerys, R. 1988. Abnormal methylation capacity in human liver cirrhosis. *International Journal of Clinical Pharmacology Research* **7**, 117–122.

Ginsburg, J.M. 1965. Renal mechanism for excretion and transformation of arsenic in the dog. *American Journal of Physiology* **208**, 832–840.

Ginsburg, J.M. and Lotspeich, W.D. 1963. Interrelations of arsenate and phosphate transport in the dog kidney. *American Journal of Physiology* **205**, 707–714.

Gonzalez-Ramirez, D., Maiorino, R.M., Zuniga-Charles, M., Xu, Z., Hurlbut, K.M., Junco-Munoz, P., Aposhian, M.M., Dart, R.C., Diaz Gama, J.H., Echeverria, D., Woods, J.S. and Aposhian, H.V. 1995. Sodium 2,3-dimercaptopropane-1-sulfonate challenge test for mercury in humans. II. Urinary mercury, porphyrins and neurobehavioral changes of dental workers in Monterey, Mexico. *Journal of Pharmacology and Experimental Therapeutics* **272**, 264–274.

Groen, K., Vaessen, H.A.M.G., Kliest, J.J.G., de Boer, J.L.M., van Ooik, T., Timmerman, A. and Vlug, R.F. 1993. Bioavailability of inorganic arsenic from bog ore-containing soil in the dog. *Environmental Health Perspectives* **102**, 182–184.

Guha Mazumder, D.N., Chakraborty, A.K., Ghose, A., Gupta, J.D., Chakraborty, D.P., Dey,

S.B. and Chattopadhyay, N. 1988. Chronic arsenic toxicity from drinking tubewell water in rural West Bengal. *Bulletin of the World Health Organization* **66**, 99–506.

Gyurasics, A., Varga, F. and Gregus, Z. 1991a. Effect of arsenicals on biliary excretion of endogenous glutathione and xenobiotics with glutathione-dependent hepatobiliary transport. *Biochemical Pharmacology* **41**, 937–944.

Gyurasics, A., Varga, F. and Gregus, Z. 1991b. Glutathione-dependent biliary excretion of arsenic. *Biochemical Pharmacology* **42**, 465–468.

Hall, L.L., George, E., Kohan, M.J., Styblo, M. and Thomas, D.J. 1997. *In vitro* methylation of inorganic arsenic in mouse intestinal cecum. *Toxicology and Applied Pharmacology* **147**, 101–109.

Hartwig, A., Gröblinghoff, U.D., Beyersmann, D., Natarjan, A.T., Filon, R. and Mullenders, L.H.F. 1997. Interaction of arsenic (III) with nucleotide excision repair in UV-irradiated human fibroblasts. *Carcinogenesis* **18** 399–405.

Hatlelid, K.M. and Carter, D.E. 1997. Reactive oxygen species do not cause arsine-induced hemoglobin damage. *Journal of Toxicology and Environmental Health* **50**, 463–474.

Hatlelid, K.M., Brailsford, C. and Carter, D.E. 1995. An *in vitro* model for arsine toxicity using isolated red blood cells. *Fundamental and Applied Toxicology* **25**, 302–306.

Healy, S.M., Zakharyan, R.A. and Aposhian, H.V. 1997. Enzymatic methylation of arsenic compounds: IV. *In vitro* and *in vivo* deficiency of the methylation of arsenite and monomethylarsonic acid in the guinea pig. *Mutation Research* **386**, 229–239.

Healy, S.M., Casarez, E.A., Ayala-Fierro, F. and Aposhian, H.V. 1998. Enzymatic methylation of arsenic compounds: V. Arsenite methyltransferase activity in tissues of mice. *Toxicology and Applied Pharmacology* **148**, 65–70.

Hessl, S.M. and Berman, E. 1982. Severe peripheral neuropathy after exposure to monosodium methyl arsonate. *Journal of Toxicology and Clinical Toxicology* **19**, 281–287.

Heyman, A., Pfeiffer, J.B. Jr, Willett, R.W. and Taylor, H.M. 1956. Peripheral neuropathy caused by arsenical intoxication. A study of 41 cases with observations on the effects of BAL (2,3-dimercapto-propanol). *New England Journal of Medicine* **254**, 401–409.

Hindmarsh, J.T. and McCurdy, R.F. 1986. Clinical and environmental aspects of arsenic toxicity. *Critical Reviews in Clinical Laboratory Sciences* **23**, 315–347.

Hine, C.H., Pinto, S.S. and Nelson, K.W. 1977. Medical problems associated with arsenic exposure. *Journal of Occupational Medicine* **19**, 391–396.

Hirata, M., Tanaka, A., Hisanaga, A. and Ishinishi, N. 1990. Effects of glutathione depletion on the acute nephrotoxic potential of arsenite and on arsenic metabolism in hamsters. *Toxicology and Applied Pharmacology* **106**, 469–481.

Hollins, J.G., Charbonneau, S.M., Bryce, F., Ridgeway, J.M., Tam, G.K.H. and Willes, R.F. 1979. Whole body retention and excretion of [^{74}As]arsenic acid in the adult beagle dog. *Toxicology Letters* **4**, 7–13.

Hopenhayn-Rich, C., Smith, A.H. and Goeden, H.M. 1993. Human studies do not support the methylation threshold hypothesis for the toxicity of inorganic arsenic. *Environmental Research* **60**, 161–177.

Huang, R.N., Ho, I.C., Yih, L.H. and Lee, T.C. 1995. Sodium arsenite induces chromosome endoreduplication and inhibits protein phosphatase activity in human fibroblasts. *Environmental and Molecular Mutagenesis* **25**, 188–196.

Huff, J., Chant, P. and Waalkes, M. 1998. Arsenic carcinogenicity testing. *Environmental Health Perspectives* **106**, A170.

Hughes, M.F. and Thompson, D.J. 1996. Subchronic dispositional and toxicological effects of arsenate administered in drinking water to mice. *Journal of Toxicology and Environmental Health* **49**, 177–196.

Hughes, M.F., Menache, M. and Thompson, D.J. 1994. Dose-dependent disposition of sodium arsenate in mice following acute oral exposure. *Fundamental and Applied Toxicology* **22**, 80–89.

Hurlbut, K.M., Maiorino, R.M., Mayersohn, M., Dart, R.C., Bruce, D.C. and Aposhian, H.V. 1994. Determination and metabolism of dithiol chelating agents XVI. Pharmacokinetics of 2,3-dimercapto-propanesulfonate after intravenous administration to human volunteers. *Journal of Pharmacology and Experimental Therapeutics* **268**, 662–668.

Hurst, E.W. 1959. The lesions produced in the central nervous system by certain organic arsenical compounds. *Journal of Pathology and Bacteriology* **77**, 523–534.

Itoh, T., Zhang, Y.F., Murai, S., Saito, H., Nagahama, H., Miyate, H., Saito, Y. and Abe, E. 1990. The effect of arsenic trioxide on brain monoamine metabolism and locomotor activity of mice. *Toxicology Letters* **54**, 345–353.

Jarup, L. and Pershagen, G. 1991. Arsenic exposure, smoking, and lung cancer in smelter workers – A case-control study. *American Journal of Epidemiology* **134**, 545–551.

Jelinek, C.F. and Corneliussen, P.E. 1977. Levels of arsenic in the United States food supply. *Environmental Health Perspectives* **19**, 83–87.

Jenkins, R.B. 1966. Inorganic arsenic and the nervous system. *Brain* **89**, 479–498.

Jha, A.N., Noditi, M., Nilsson, R. and Natarajan, A.T. 1992. Genotoxic effects of sodium arsenite on human cells. *Mutation Research* **284**, 215–221.

Kemper, F.H., Jekat, F.W., Bertram, H.P. and Eckard, R. 1990. New chelating agents, in *Basic Science in Toxicology*, (eds G.M. Volans, J. Sims, F.M. Sullivan and P. Turner), Taylor & Francis, London, pp. 523–546.

Klaassen, C.D. 1974. Biliary excretion of arsenic in rats, rabbits, and dogs. *Toxicology and Applied Pharmacology* **29**, 447–457.

Klaassen, C.D. 1985. Heavy metals and heavy metal antagonists, in *The Pharmacological Basis of Therapeutics*, (eds A.G. Gilman, L.S. Goodman, T.W. Rall and F. Murad), Macmillan, New York, pp. 1605–1627.

Klimecki, W.T., Borchers, A.H., Egbert, R.E., Nagle, R.B., Carter, D.E. and Bowden, G.T. 1997. Effects of acute and chronic arsenic exposure of human-derived keratinocytes in an *in vitro* human skin equivalent system: a novel model of human arsenicism. *Toxicology In Vitro* **11**, 89–98.

Klimova, L,K. 1958. Pharmacology of a new unithiol antidote. *Farmakologiia i Toksikologiia (Moscow)* **21**, 53–59.

Kobayashi, H., Yuyama, A., Ishihara, M. and Matsusaka, N. 1987. Effects of arsenic on cholinergic parameters in brain *in vitro*. *Neuropharmacology* **26**, 1707–1713.

Lantz, R.C., Parliman G., Chen, G.J. and Carter, D.E. 1994. Effect of arsenic exposure n alveolar macrophage function. I. Effect of soluble As(III) and As(V). *Environmental Research* **67**, 183–195.

Lantz, R.C., Parliman, G., Chen, G.J., Barber, D., Winski, S. and Carter, D.E. 1995. Effect of arsenic exposure on alveolar macrophage function. II. Effect of slightly soluble forms of As(III) and As(V). *Environmental Research* **68**, 59–67.

Larsen, N.A., Nielsen, B., Pakkenberg, H., Christoffersen, P., Damsgaard, E. and Heydorn, K. 1972. Neutron activation analysis of arsenic, manganese and selenium concentrations in organs of uraemic and normal persons, in *Nuclear Activation Techniques in the Life Sciences*, IAEA, Vienna, pp. 561–568.

Larsen, N.A., Pakkenberg, H., Damsgaard, E. and Heydorn, K. 1979. Topographical distribution of arsenic, manganese, and selenium in the normal human brain. *Journal of Neurological Sciences* **42**, 407–416.

Le, X.C., Cullen, W.R. and Reimer, K.J. 1994. Human urinary arsenic excretion after one-time ingestion of seaweed, crab, and shrimp. *Clinical Chemistry* **40**, 617–624.

LeQuesne, P.M. and McLeod, J.G. 1977. Peripheral neuropathy following a single exposure to arsenic. Clinical course in four patients with electrophysiological and histological studies. *Journal of the Neurological Sciences* **32**, 437–451.

Lewis, R.L. and Sweet, D.V. (eds.) 1985. *Registry of Toxic Effects of Chemical Substances. 1983–1984 Supplement*, U.S. Department of Health and Human Services, Cincinnati.

Li, J.H. and Rossman, T.G. 1989. Inhibition of DNA ligase activity by arsenite: a possible mechanism of its comutagenesis. *Molecular Toxicology* **2**, 1–9.

Lisella, F.S., Long, K.R. and Scott, H.G. 1972. Health aspects of arsenicals in the environment. *Journal of Environmental Health* **34**, 511–518.

Liu, Y.C. and Huang, H. 1997. Involvement of calcium-dependent protein kinase C in arsenite induced genotoxicity in Chinese hamster ovary cells. *Journal of Cellular Biochemistry* **64**, 423–433.

Loiseau, P.M., Wolf, J.G. and Gayral, P. 1993. Spiroarsoranes of second generation: Evaluation of trypanocidal properties against *Trypanosoma brucei brucei. Tropical Medicine and Parasitology* **44**, 32–36.

Maiorino, R.M. and Aposhian, H.V. 1989. Determination and metabolism of dithiol chelating agents: IV. Urinary excretion of meso-2,3-dimercaptosuccinic acid and mercaptosuccinic acid in rabbits given meso-2,3-dimercaptosuccinic acid. *Biochemical Pharmacology* **38**, 1147–1154.

Maiorino, R.M., Weber, G.L. and Aposhian, H.V. 1986. Fluorometric determination of 2,3-dimercaptopropane-1-sulfonic acid and other dithiols by precolumn derivization with bromobimane and column liquid chromatography. *Journal of Chromatography* **374**, 297–310.

Maiorino, R.M., Barry, T.J. and Aposhian, H.V. 1987. Determination and metabolism of dithiol-chelating agents. Electrolytic and chemical reduction of oxidized dithiols in urine. *Analytical Biochemistry* **160**, 217–226.

Maiorino, R.M., Bruce, D.C. and Aposhian, H.V. 1989. Determination and metabolism of chelating agents: VI. Isolation and identification of the mixed disulfides of meso-2,3-dimercaptosuccinic acid with L-cysteine in human urine. *Toxicology and Applied Pharmacology* **97**, 338–349.

Maiorino, R.M., Dart, R.C., Carter, D.E. and Aposhian, H.V. 1991. Determination and metabolism of dithiol chelating agents: XII. Metabolism and pharmacokinetics of sodium 2,3-dimercaptopropane-1-sulfonate in humans. *Journal of Pharmacology and Experimental Therapeutics* **259**, 808–814.

Maiorino, R.M., Gonzalez-Ramirez, D., Zuniga-Charles, M., Xu, Z.F., Hurlbut, K.M., Aposhian, M.M., Dart, R.C., Woods, J.S., Ostrosky-Wegman, P., Gonsebatt, M.E. and Aposhian, H.V. 1996. Sodium 2,3-dimercaptopropane-1-sulfonate challenge test for mercury in humans. III. Urinary mercury after exposure to mercurous chloride. *Journal of Pharmacology and Experimental Therapeutics* **277**, 938–944.

Marafante, E. and Vahter, M. 1987. Solubility, retention, and metabolism of intratracheally and orally administered inorganic arsenic compounds in the hamster. *Environmental Research* **42**, 72–82.

Marafante, E., Rade, J., Sabbioni, E., Bertolero, F. and Foa, V. 1981. Intracellular interaction and metabolic fate of arsenite in the rabbit. *Clinical Toxicology* **18**, 1335–1341.

Marafante, E., Bertolero, F., Edel, J., Pietra, R. and Sabbioni, E. 1982. Intracellular interaction and biotransformation of arsenite in rats and rabbits. *Science of the Total Environment* **24**, 27–39.

Marafante, E., Lundborg, M., Vahter, M. and Camner, P. 1987a. Dissolution of two arsenic compounds by rabbit alveolar macrophages *in vitro*. *Fundamental and Applied Toxicology* **8**, 382–388.

Marafante, E, Vahter, M., Norin, H., Envall, J., Sandstrom, M., Christakopoulos, A. and Ryhage, R. 1987b. Biotransformation of dimethylarsinic acid in mouse, hamster and man. *Journal of Applied Toxicology* **7**, 111–117.

Mass, M.J. and Wang, L. 1997. Arsenic alters cytosine methylation patterns of the promoter of the tumor suppressor gene *p53* in human lung cells: A model for a mechanism for carcinogenesis. *Mutation Research* **386**, 263–277.

Menkes, J.H. 1997. Man, metals, and minerals, in *Mineral and Metal Neurotoxicology*, (eds M. Yasui, M.J. Strong, K. Ota and M.A. Verity), New York, pp. 5–13.

Morel, G., Cluet, J.L., Telolahy, P., Yang, H.M., Thieffry, N. and de Ceaurriz, J. 1995. Interindividual variability in the urinary excretion of inorganic arsenic metabolites by C57BL/6J mice: Possible involvement of a thiol/disulfide exchange mechanism. *Toxicology Letters* **78**, 111–117.

Mudur, G. 1996. Arsenic poisons 220,000 in India. *British Medical Journal* **313**, 9.

Nagaraja, T.N. and Desiraju, T. 1993. Regional alterations in the levels of brain biogenic amines, glutamate, GABA, and BAD activity due to chronic consumption of inorganic arsenic in developing and adult rats. *Bulletin of Environmental Contamination and Toxicology* **50**, 100–107.

NAS (National Academy of Sciences), 1977. *Medical and Biological Effects of Environmental Pollutants: Arsenic*, NAS, Washington, DC.

Neubauer, O. 1947. Arsenical cancer: A review. *British Journal of Cancer* **1**, 192–251.

Onishi, H. 1968. Arsenic,. in *Handbook of Geochemistry*, (ed. by K.H. Wedepohl), Springer Verlag, New York.

Pershagen, G. 1986. Sources of exposure and biological effects of arsenic, in *Environmental Carcinogens Selected Methods of Analysis*, Vol. 8, (eds I.K. O'Neill, P. Schuller and L. Fishbein), IARC, Lyon, pp. 45–61.

Petrunkin, V.E. 1956. Synthesis and properties of dimercapto derivatives of alkylsulfonic acids. *Ukrainsky Khemichisky Zhurnal* **22**, 603–607.

Rao, P.M., Antony, A., Rajalakshmi, S. and Sarma, D.S.R. 1989. Studies on hypomethylation of liver DNA during early stages of chemical carcinogenesis in rat liver. *Carcinogenesis* **10**, 933–937.

Repetto, G., Sanz, P. and Repetto, M. 1994. Comparative *in vitro* effects of sodium arsenite and sodium arsenate on neuroblastoma cells. *Toxicology* **92**, 143–153.

Rivera, M., Zheng, W., Aposhian, H.V. and Fernando, Q. 1989. Determination and metabolism of dithiol chelating agents: VIII. Metal complexes of meso-dimercaptosuccinc acid. *Toxicology and Applied Pharmacology* **100**, 96–106.

Rossman, T.G., Stone, D., Molina, M. and Troll, W. 1980. Absence of arsenite mutagenicity in *E. coli* and Chinese hamster cells. *Environmental Mutagenesis* **2**, 371–379.

Rote Liste 1995. *Azneimittelverzeichnis des BPI*. Bundesverban der Pharmazeutischen Industrie e.V.; Frankfurt, 12–003, 12–005.

Rowland, I.R. and Davies, M.J. 1981. *In vitro* metabolism of inorganic arsenic by the gastrointestinal microflora of the rat. *Journal of Applied Toxicology* **1**, 278–283.

Rudell, R., Slayton, T.M. and Beck, B. 1996. Implications of arsenic genotoxicity for dose response of carcinogenic effects. *Regulatory Toxicology and Pharmacology* **23**, 87–105.

Russell, D.S. 1937. Changes in the central nervous system following arsphenamine medication. *Journal of Pathology* **45**, 357–366.

Salazar, A.M., Ostrosky-Wegman, P., Menendez, D., Miranda, E., Garcia-Carranca, A. and Rojas, E. 1997. Induction of p53 protein expression by sodium arsenite. *Mutation Research* **381**, 259–265.

Schroeder, H.A. and Balassa, J. 1966. Abnormal trace metals in man: Arsenic. *Journal of Chronic Diseases* **19**, 85–106.

Schroeder, M. and Mass, M.J. 1997. CpG methylation inactivates the transcriptional activity of the promoter of the human p53 tumor suppressor gene. *Biochemical and Biophysical Research Communications* **235**, 403–6.

Southwick, J.W., Western, A.E., Beck, M.M., Whitley, T., Isaacs, R., Petajan, J. and Hansen, C.D. 1983. An epidemiological study of arsenic in drinking water in Millard County, Utah, in *Arsenic: Industrial, Biomedical, Environmental Perspectives*, (eds W.H. Lederer and R.J. Fensterheim), Van Nostrand Reinhold Company, New York, pp. 210–225.

Squibb, K.S. and Fowler, B.A. 1983. The toxicity of arsenic and its compounds, in *Biological and Environmental Effects of Arsenic*, Vol. 6, (ed. B.A. Fowler) Elsevier, New York, pp. 233–269.

Stevens, J.T., Hall, L.L., Farmer, J.D., DiPasquale, L.C., Chernoff, N. and Durham, W.F. 1977. Disposition of ^{14}C and/or ^{74}As-cacodylic acid in rats after intravenous, intratracheal, or peroral administration. *Environmental Health Perspectives* **19**, 151–157.

TPHD (Taiwan Provincial Health Department) 1993. *Blackfoot Disease Control and Prevention in Taiwan Republic of China.*

Tam, G.K.H., Charbonneau, S.M., Lacroix, G. and Bryce, F. 1979a. *In vitro* methylation of ^{74}As in urine, plasma, and red blood cells of human and dog. *Bulletin of Environmental Contamination and Toxicology* **22**, 69–71.

Tam, G.K.H., Charbonneau, S.M., Lacroix, G. and Bryce, F. 1979b. Confirmation of inorganic arsenic and dimethylarsinic acid in urine and plasma of dog by ion-exchange and TLC. *Bulletin of Environmental Contamination and Toxicology* **21**, 371–374.

Taylor, P.R., Qiao, Y.L., Schatzkin, A., Yao, S.X., Lubin, J., Mao, B.L., Rao, J.Y., McAdams, M., Xuan, X.Z. and Li, J.Y. 1989. Relation of arsenic exposure to lung cancer among tin miners in Yunan Province, China. *British Journal of Industrial Medicine* **46**, 881–886.

Ter Welle, H.F. and Slater, E.C. 1967. Uncoupling of respiratory-chain phosphorylation by arsenate. *Biochemica et Biophysica Acta* **143**, 1–17.

Tripathi, N., Kannan, G.M., Pant, B.P., Jaiswal, D.K., Malhotra, P.R. and Flora, S.J.S. 1997. Arsenic-induced changes in certain neurotransmitter levels and their recoveries following chelation in whole rat brain. *Toxicology Letters* **92**, 201–208.

Tseng, C.H., Chong, C.K., Chen, C.J., Lin, B.J. and Tai, T.Y. 1995. Abnormal peripheral microcirculation in seemingly normal subjects living in Blackfoot Disease hyperendemic villages in Taiwan. *International Journal of Microcirculation* **15**, 21–27.

Tseng, W.P. 1977. Effects and dose–response relationships of skin cancer and Blackfoot Disease with arsenic. *Environmental Health Perspectives* **19**, 109–119.

Vahter, M. 1981. Biotransformation of trivalent and pentavalent inorganic arsenic in mice and rats. *Environmental Research* **25**, 286–293.

Vahter, M. 1994. Species differences in the metabolism of arsenic compounds. *Applied Organometallic Chemistry* **8**, 175–182.

Vahter, M. and Marafante, E. 1983. Intracellular interaction and metabolic fate of arsenite and arsenate in mice and rabbits. *Chemico-Biological Interactions* **47**, 29–44.

Vahter, M., Marafante, E., Lindgren, A. and Dencker, L. 1982. Tissue distribution and subcellular binding of arsenic in marmoset monkeys after injection of ^{74}As-arsenite. *Archives of Toxicology* **51**, 65–77.

Vahter, M., Friberg, L., Rahnster, B., Nygren, A. and Nolinder, P. 1986. Airborne arsenic and urinary excretion of metabolites of inorganic arsenic among smelter workers. *International Archives of Occupational and Environmental Health* **57**, 79–91.

Vahter, M., Concha, G., Nermell, B., Nilsson, R., Dulout, F. and Natarajan, A.T. 1995a. A unique metabolism of inorganic arsenic in Native Andean women. *European Journal of Pharmacology* **293**, 455–462.

Vahter, M., Couch, R., Nermell, B. and Nilsson, R. 1995b. Lack of methylation of inorganic arsenic in the chimpanzee. *Toxicology and Applied Pharmacology* **133**, 262–268.

Verity, M.A. 1995. Nervous system, in *Metal Toxicology*, (eds by R.A. Goyer, C.D. Klaassen, and M.P. Waalkes), Academic Press, San Diego, pp. 199–235.

Webb, D.R., Wilson, S.E. and Carter, D.E. 1987. Pulmonary clearance and toxicity of respirable gallium arsenide particulates intratracheally instilled into rats. *American Industrial Hygiene Association Journal* **48**, 660–667.

Webb, J.L. 1963. Arsenicals, in *Enzyme and Metabolic Inhibitors*, Vol. 3, (ed. J.L. Webb), Academic Press, New York, pp. 595–895.

Weiss, C.S., Parks, E.J. and Brinkman, F.E. 1983. Speciation of arsenic in fossil fuels and their conversion process fluids, in *Arsenic: Industrial, Biomedical, Environmental Perspectives*, (eds W.H. Lederer and R.J. Fensterheim), Van Nostrand Reinhold Company, New York, pp. 309–326.

Wester, R.C., Maibach, H.I., Sedik, L., Melendres, J. and Wade, M. 1993. *In vivo* and *in vitro* percutaneous absorption and skin decontamination of arsenic from water and soil. *Fundamental and Applied Toxicology* **20**, 336–340.

WHO (World Health Organization) 1981. *Arsenic*, World Health Organization, Geneva.

Wildenauer, D.B., Reuther, H. and Weger, N. 1982. Interactions of the chelating agent 2,3-dimercaptopropane-1-sulfonate with red blood cells *in vitro*. I. Evidence for carrier mediated transport. *Chemico-Biological Interactions* **42**, 165–177.

Wildfang, E., Zakharyan, R.A. and Aposhian, H.V. 1998. Enzymatic methylation of arsenic compounds: VI. Characterization of hamster liver arsenite and monomethylarsonic acid methyltransferase activities *in vitro*. *Toxicology and Applied Pharmacology* **152**(2), 366–75.

Willardson, R.K. 1983. Arsenic in electronics, in *Arsenic: Industrial, Biomedical, Environmental Perspectives*, (eds W.H. Lederer and R.J. Fensterheim), Van Nostrand Reinhold Company, New York, pp. 72–88.

Williamson, J. 1976. Chemotherapy of African Trypanosomiasis. *Transactions of the Royal Society of Tropical Medicine and Hygiene* **70**, 117–119.

Windebank, A.J. 1986. Specific inhibition of myelination by lead *in vitro*: Comparison with arsenic, thallium, and mercury. *Experimental Neurology* **94**, 203–212.

Winship, K.A. 1984. Toxicity of inorganic arsenic salts. *Adverse Drug Reactions and Acute Poisoning Reviews* **3**, 129–160.

Woolson, E.A. (ed.) 1975. *Arsenical Pesticides*, American Chemical Society, Washington, DC.

Wu, M.M., Kuo, T.L., Hwang, Y.H. and Chen, C.J. 1989. Dose-response relation between arsenic concentration in well water and mortality from cancers and vascular disease. *American Journal of Epidemiology* **130**, 1123–1132.

Yamamoto, S., Konishi, Y., Matsuda, T., Murai, T., Shibata, M.A., Matsui-Yuasa, I., Otani, S., Kuroda, K., Endo, G. and Fukushima, S. 1995. Cancer induction by an organic arsenic compound, dimethylarsinic acid (cacodylic acid) in F344/DuCrj rats after pretreatment with five carcinogens. *Cancer Research* **55**, 1271–1276.

Yamanaka, K., Hasegawa, A., Sawamura, R. and Okada, S. 1989. Dimethylated arsenics induce DNA strand breaks in lung via the production of active oxygen in mice. *Biochemical and Biophysical Research Communications* **165**, 43–50.

Yamanaka, K., Hasegawa, A., Sawamura, R. and Okada, S. 1991. Cellular response to oxidative damage in lung induced by the administration of dimethylarsinic acid, a major metabolite of inorganic arsenic, in mice. *Toxicology and Applied Pharmacology* **108**, 205–213.

Yamashita, N., Doi, M., Nishio, M., Hojo, H. and Tanaka, M. 1972. Current state of Kyoto children by arsenic tainted Morinaga dry milk. *Japanese Journal of Hygiene* **27**, 364–399.

Yamauchi, H. and Yamamura, Y. 1984a. Metabolism and excretion of orally administered dimethylarsinic acid in the hamster. *Toxicology and Applied Pharmacology* **74**, 134–140.

Yamauchi, H. and Yamamura, Y. 1984b. Metabolism and excretion of orally ingested trimethylarsenic in man. *Bulletin of Environmental Contamination and Toxicology* **32**, 682–687.

Yamauchi, H. and Yamamura, Y. 1985. Metabolism and excretion of orally administrated arsenic trioxide in the hamster. *Toxicology* **34**, 113–121.

Yamauchi, H., Kaise, T. and Yamamura, Y. 1986. Metabolism and excretion of orally administered arsenobetaine in the hamster. *Bulletin of Environmental Contamination and Toxicology* **36**, 350–355.

Yamauchi, H., Yamato, N. and Yamamura, Y. 1988. Metabolism and excretion of orally and intraperitoneally administered methylarsonic acid in the hamster. *Bulletin of Environmental Contamination and Toxicology* **40**, 280–286.

Yamauchi, H., Takahashi, K., Yamamura, Y. and Fowler, B.A. 1992. Metabolism of subcutaneous administered indium arsenide in the hamster. *Toxicology and Applied Pharmacology* **116**, 66–70.

Yang, H.L., Tu, S.C., Lu, F.J. and Chiu, H.C. 1994. Plasma protein C activity is enhance by arsenic but inhibited by fluorescent humic acid associated with Blackfoot Disease. *American Journal of Hematology* **46**, 264–269.

Yu, H.S., Sheu, H.M., Ko, S.S., Chiang, L.C., Chien, C.H., Lin, S.M., Tseng, B.R. and Chen, C.S. 1984. Studies on Blackfoot Disease and chronic arsenism in southern Taiwan with special reference to skin lesions and fluorescent substances. *Journal of Dermatology* **11**, 361–370.

Zakharyan, R., Wu, Y., Bogdan, G.M. and Aposhian, H.V. 1995. Enzymatic methylation of arsenic compounds: Assay, partial purification and properties of arsenite methyltransferase and monomethylarsonic acid methyltransferase of rabbit liver. *Chemical Research in Toxicology* **8**, 1029–1038.

Zakharyan, R.A., Wildfang, E. and Aposhian, H.V. 1996. Enzymatic methylation of arsenic compounds: III. The marmoset and tamarin, but not the rhesus, monkeys are deficient in methyltransferases that methylate inorganic arsenic. *Toxicology and Applied Pharmacology* 140, 77–84.

Zaldivar, R. 1974. Arsenic contamination of drinking water and foodstuffs causing endemic chronic poisoning. *Beitrage zur Pathologie* **151**, 384–400.

Zhao, C.Q., Young, M.R., Diwan, B.A., Coogan, T.P. and Waalkes, M.P. 1997. Association of arsenic-induced malignant transformation with DNA hypomethylation and aberrant gene expression. *Proceedings of the National Academy of Sciences of the USA* **94**, 10907–10912.

Zheng, W., Maiorino, R.M., Brendel, K. and Aposhian, H.V. 1990. Determination of dithiol chelating agents: IV. Biliary excretion of dithiols and their interactions with cadmium and metallothionein. *Fundamental and Applied Toxicology* **14**, 598–607.

Zheng, W., Perry, D.F., Nelson, D.L. and Aposhian, H.V. 1991. Choroid plexus protects cerebrospinal fluid against toxic metals. *FASEB Journal* **5**, 2188–2193.

Zhu, J., Koken, M.H.M., Quignon, F., Chelbi-Alix, M.K., Degos, L., Wang, Z.Y., Chen, Z. and de The, H. 1997. Arsenic-induced PML targeting onto nuclear bodies: Implications for the treatment of acute promyelocytic leukemia. *Proceedings of the National Academy of Sciences of the USA* **94**, 3978–3983.

4

CHROMIUM AND NICKEL

Max Costa

4.1 Chromium

4.1.1 Introduction

Chromium (Cr) is abundant in the Earth's crust, with the predominant trivalent form found mostly in nature, and the hexavalent form a by-product of industrial activities (Carson *et al.*, 1986; Costa, 1997). Chromium compounds are used for leather tanning, electroplating additives and production of pigments, catalysts, corrosion inhibitors and as wood preservatives (Costa, 1997). Chromium metal is used widely in the steel industry (Costa, 1997). It is in the first transition element in group 16 and has an extensive range of oxidation states from –2 to +6, however, the most common forms are the trivalent and hexavalent forms (Farrell and Costa, 1997). In most biological reducing environments, hexavalent chromium is converted to trivalent chromium via a mechanism involving various transitional oxidation states, including Cr(V) and Cr(VI) (Farrell and Costa, 1997). In general, these intermediate oxidation states of chromium are extremely unstable, although there are instances where complexes involving these oxidation states can actually be isolated (Farrell and Costa, 1997). Many thousands of compounds have chromium in the +3 oxidation state (Farrell and Costa, 1997). These are characterized by a propensity for a slow ligand exchange reaction and adoption of octahedral coordination geometry (Farrell and Costa, 1997). Trivalent chromium is considered to be a borderline hard acid metal and favors nitrogen and oxygen donors rather than thiol groups (Farrell and Costa, 1997). Hexavalent chromium compounds always exist as oxides or oxohalydes and have very strong oxidizing capabilities (Farrell and Costa, 1997).

4.1.2 Essentiality

A possible role for Cr(III) as an essential element was demonstrated when weaning rats (fed a yeast-based diet) developed progressive impairments in their glucose tolerance (Mertz and Schwarz, 1955). Subsequent experiments demonstrated that chromium deficiencies impaired glucose tolerance (Schwarz and Mertz, 1959). Severe chromium deficiencies caused reductions in rodent growth, longevity, fertility and sperm counts. Evidence that chromium might be essential in humans came from studies of patients undergoing parenteral delivery of fluids containing essential nutrients (Jeejeebhoy *et al.*, 1977; Freund *et al.*, 1979). Omission of chromium from this parenteral regimen yielded two case histories showing patients that had considerable weight loss and hypoglycemia (Jeejeebhoy *et al.*, 1977; Freund *et al.*, 1979). These symptoms were reversed with chromium supplementation of the parenteral fluid (Jeejeebhoy *et al.*, 1977; Freund *et al.*, 1979). The US Department of Agriculture has recommended that a dietary intake of 50–200 μg of chromium per day would be safe and adequate in adults (Freund *et al.*, 1979). At the present time, chromium picolinate is widely advertised as a supplement to enhance athletic performance, muscle mass and as a stabilizer of blood glucose. Chromium bound to the picolinate is extremely stable and it is unclear whether the effects of this compound are due to the chromium or to the chelation of chromium with picolinate. For example, chromium picolinate has been shown to enhance the action of insulin in transporting glucose into cells (Costa, 1997). Thus, one should be cautious about describing these effects as showing the requirement for Cr(III) because chromium picolinate might well be having pharmacologic actions as a complex.

4.1.3 Uptake and distribution of chromium compounds

Internal exposure to chromium can occur through skin absorption, direct ingestion and by inhalation. The absorption of chromium depends upon the oxidation state of the metal and the physical characteristics of the compounds (Costa, 1997). Soluble hexavalent chromium compounds can penetrate the skin more readily than the trivalent forms and uptake is enhanced with increasing pH (Costa, 1997). However, chromium absorption through the skin can be greatly increased by normal abrasions. There are cases of significant internal renal damage following dermal absorption of aberrated skin. Again, by respiration, soluble Cr(VI) compounds are absorbed much better than Cr(III) compounds; however, water-insoluble Cr(III) compounds can be phagocytosed by respiratory cells, whereas water-insoluble compounds cannot enter through the skin (Costa, 1997). By oral administration, approximately 1% of Cr(III) is absorbed, and anywhere from 3 to 6% of Cr(VI) can be absorbed from the GI tract (Costa, 1997). However, recent studies of human volunteers who drank Cr(VI) indicated quite a variation in the level of absorption, with some individuals absorbing more than 25% of an oral dose. The oxidation state of chromium is also a determining factor in its distribution in the body. The trivalent form is primarily bound extracellularly to blood proteins, such as transferrin and microglobulin or serum albumin (Gray and Sterling, 1950; Jeejeebhoy *et al.*, 1977; Freund *et al.*, 1979; Hansen and Stern, 1985), whereas Cr(VI) is transported actively into cells and is bound intracellularly, where it is reduced to Cr(III) and eventually trapped for the life of the cell (Gray and Sterling, 1950; Costa, 1997). However, there is some redistribution of Cr(III) from red blood cells to other sites (Costa, 1997).

Recent studies of human volunteers ingesting Cr(VI) in drinking water showed that substantial amounts enter the red blood cells, and thus Cr(VI) is not entirely reduced to Cr(III) in the GI tract. Chromium accumulates to the highest extent in the kidneys, testes and bones, and to a lesser extent in the heart, pancreas, lungs and brain (Saner, 1980; Costa, 1997). The liver and kidney can account for about 40–50% of the chromium burden in the body. Approximately 80% of a parenteral dose is excreted in the urine and 2–20% in the feces (Saner, 1980; Neiboer and Jusy, 1986). The half-life of chromium in humans has been estimated to be 50–60 days. Some chromium is excreted through the skin in the sweat (Saner, 1980; Neiboer and Jusy, 1986).

4.1.4 Toxicological effects

Again, as discussed above, chromium metal is generally inert and does not exhibit much toxicity. Most of the toxicity of chromium compounds has been attributed to that of Cr(VI). Acutely, the toxicity of hexavalent chromium involves damage to the liver, kidney and brain (US EPA, 1980). The similarity of hexavalent chromate to sulfate and phosphate renders most cells of the body capable of taking up this metal (Costa, 1997). The LD_{50} by parenteral administration is roughly 10–50 mg/kg in animals, whereas the LD_{50} by oral administration is 200–350 mg, indicating that it is 10 to 20 times more toxic by injection than by oral administration. These results show that not all the hexavalent chromate is absorbed, but a substantial portion of it is (US EPA, 1980). The LD_{50} of sodium chromate in humans following ingestion has been reported to be 50 mg/kg, i.e. the level that is equivalent in range to the LD_{50} by intraperitoneal (i.p.) administration in animals; suggesting that humans absorb more hexavalent chromium than experimental animals and thus implying that it is more toxic to humans than animals (NIOSH, 1979). Acute effects in humans of high exposure to near-lethal doses involve gastric distress, nose bleeds, liver damage and yellowing of the tongue and teeth.

Most commonly, humans are exposed to hexavalent chromium through inhalation exposure. Chromate produces perforations of the nasal septum and ulcerations of other skin surfaces since it is a highly corrosive metal (Costa, 1997). Better industrial hygiene practices have greatly lowered the incidence of perforated nasal septums exhibited by workers in the chrome industry (Bidstrup, 1989). Other respiratory effects can occur by prolonged inhalation exposure to Cr(VI)-containing compounds, including hyperemia, chronic rhinitis, congestion of the larynx, polyps of the upper respiratory tract, chronic inflammation of the lung, emphysema, chronic bronchitis and bronchial pneumonia (NIOSH, 1979). Insoluble Cr(VI) compounds seem to have greater retention in the lungs than do the soluble compounds, but soluble chromium compounds, with repeated exposures, can exhibit substantial lung burdens and equivalent effects to the insoluble compounds (NIOSH, 1979; Costa, 1997).

Contact dermatitis is a common problem exhibited by people exposed to hexavalent chromium compounds, but Cr(III) compounds can also cause this problem to some extent. About 10% of all occupationally exposed workers develop allergic contact dermatitis to chromium compounds (Bidstrup, 1989). Among populations not occupationally exposed, the incidence of chromium sensitization is much less. It has been estimated that about 2.2% of the American population is sensitized to chromium by contact in the environment or prolonged contact with leather, which is tanned with chromium (Sheehan *et al.*, 1991). The allergic contact dermatitis depends upon T-lymphocyte activation rather than upon formation of antibodies against chromium-containing allergens (Arfsten *et al.*, 1998).

Cr(VI) ions penetrate cell membranes and undergo intracellular reduction, resulting in Cr(III) ions bound with cellular protein to form Cr(III)–protein complexes. If this damage is severe enough to cause cell death, the damaged cell is engulfed and processed by resident antigen-presenting cells. Chromium complexes are released into the tissue microenvironment. Antigen-presenting cells then present with chromium-modified proteins to naive T lymphocytes and initiate an expansion and proliferation of the vector, and memory lymphocytes that are specific for individual chromium-bearing peptide complexes arise (Arfsten *et al.*, 1998). Sensitization reactions arise from subsequent release of the activated T lymphocyte or lymphokines, which stimulate inflammation and edema (Arfsten *et al.*, 1998). This cascade of events further enhances the chromium-peptide-specific T-lymphocyte proliferation. A final phase is achieved through the continuous renewal of specific memory T lymphocytes (Arfsten *et al.*, 1998).

Chromium-dependent allergic contact dermatitis is associated with an increase in serum IgM and IgA antibodies (reviewed in Cohen and Costa, 1999). Chromium-induced lymphocyte proliferation has been found to be monocyte dependent, but it is not clear whether the monocytes or mature macrophages themselves are involved, or if formation associated with polymorphonuclear leukocytes is affected, contributing to the onset and development of the allergic response (reviewed in Cohen and Costa, 1999). Because inhalation is a primary route of chromium exposure, macrophages have been recovered from the lungs following inhalation exposure of animals to Cr(VI) or Cr(III) (reviewed in Cohen and Costa, 1999). These macrophages display increased chrome-filled cytoplasmic inclusions engulfed in lysosomes. Functionally, these cells display reductions in phagocytic activity, in rates of oxygen consumption and deterioration of other typical macrophage-like functions. Chromium is also cytotoxic to lymphocytes and can acutely, or in cases of chronic high exposure, produce a lessening of the immune response. Thus, it can have both allergenic and immunosuppressive effects (Cohen and Costa, 1999).

Chromium compounds have also been shown to have gastrointestinal effects in humans. Workers exposed to a plating facility with mean concentrations of 4 μg of chromium/m^3 reported stomach pain, duodenal ulcers, gastritis, stomach cramps and frequent indigestion (Costa, 1997). At higher levels of exposure, where the chrome was about 414 μg/m^3, there were 12 cases of chronic tonsillitis, five cases of chronic pharyngitis, and 32 cases of atrophy of the larynx. Ninety-seven workers from a chrome plant exposed to a mixture of insoluble and soluble chromate and dichromate revealed that 10 of the workers had ulcer formation and six had gastritis. Fifteen per cent of the workers exposed to 10 μg of chromium/m^3 for at least 1 year had duodenal ulcers and 9.2% had colitis. Chromium compounds have also been shown to produce liver injury. This effect has been reproduced in experimental animals as well as in humans. There are many documented cases of renal injury following hexavalent chromium exposure, as described in other recent reviews (Farrell and Costa, 1997; Cohen and Costa, 1999).

4.1.5 Carcinogenic effects of chromium

It is obvious from the discussion above that there are no good animal models for the toxicity of hexavalent chromium. The reasons for this are found in the striking differences in the LD_{50} between humans and animals for chromate (NIOSH, 1979). Rodents synthesized ascorbic acid, which is an active extracellular reducer of hexavalent chromium and this may be one basis for the difference in toxicity since humans cannot synthesize ascorbic acid and must rely on the slow reduction accomplished by extracellular or intracellular glutathione. Thus, the absence of a good animal model for chromium toxicity and carcinogenicity makes it difficult to assess the mechanism and extent of chromium toxicity from animal studies. Despite this fact, lung tumors have been induced in Wistar rats exposed to 100 μg/m^3 of sodium dichromate, including two adenomas and one adenocarcinoma, and a malignant tumor of the pharynx was also observed but no tumors were found in the control group (Glaser *et al.*, 1986). The problem with human studies, of course, is the fact that the exposure concentrations are often unknown.

Chromium-induced carcinogenesis has also been studied by different routes of administration. For example, implantation of calcium chromate in the trachea using stainless-steel mesh pellets demonstrated the induction of squamous cell carcinoma and adenocarcinomas in the lungs (IARC, 1990). Similar studies using calcium chromate and zinc chromate implanted in the trachea as pellets were positive for carcinogenicity but negative results were obtained for Cr_2O_3, sodium chromate and lead chromate (IARC, 1990).

However, epidemiologic studies implicate quite strongly the involvement of Cr(VI)-containing compounds in human cancers (IARC, 1990). Cancer mortality rates are 5–40 times higher than expected and the incidence includes lung cancer primarily, but also nasal cancers, laryngeal cancers and, to some extent, gastric cancers (IARC, 1990). Epidemiologic data implicating Cr(VI) compounds as carcinogens are extensive and have been adequately reviewed in previous articles (IARC, 1990; Costa, 1997; Farrell and Costa, 1997). Cr(VI) compounds have also been shown to produce teratogenesis in Golden Syrian hamsters, with a high incidence of resorption and cleft palates in surviving pups exposed to 5–15 mg of Cr_2O_3 per kg on day 8 of gestation (Gale, 1978). However, to date, chromium teratogenesis in humans has not been established, but this is probably because workers exposed to chromate were males.

4.1.6 Genotoxicity and mutagenicity of hexavalent chromium

Hexavalent chromium compounds are positive in perhaps more assays of genotoxicity and mutagenicity than any other known carcinogen (IARC, 1990). The responses in terms of mutations are not very high but are consistent across many different mutational assay systems (IARC, 1990). The hexavalent form of chromium is at least 100 times more active because it is transported into the cell, however, Cr(III) inside the cell is as genotoxic and transforming since the Cr(VI) is eventually reduced to Cr(III) intracellularly. There has been much speculation as to the mechanism of chromium genotoxicity, carcinogenicity and mutagenicity. The reduction of Cr(VI) to Cr(III) generates reactive oxygen species which oxidize DNA bases, and these have been implicated in the mutagenicity and carcinogenicity of hexavalent chromium compounds (Farrell and Costa, 1997). However, oxidative damage to DNA is not very mutagenic and is readily repaired in human cells. It has also been demonstrated that Cr(VI) to Cr(III) reduction generates Cr(III) DNA adducts, involving complexes of proteins, glutathione, cysteine and histidine crosslinked to the phosphate backbone of DNA by the Cr(III) (Zhitkovich *et al.*, 1995, 1996). These complexes are extremely stable because of the unique chemistry of Cr(III) that requires octahedral coordination and forms complexes that are inert to ligand substitution reactions (Zhitkovich *et al.*, 1996). They are poorly repaired and have been shown to be mutagenic in *in vitro* systems (Voitkun *et al.*, 1998). In view of these facts, it is likely that the Cr(III)–DNA adducts play a major role in the genotoxicity, mutagenicity and carcinogenicity of Cr(VI)-containing compounds. Numerous studies have shown that Cr(VI) induces mutations in bacteria and mammalian systems; it produces oxygen radicals and induces a variety of DNA lesions, including strand breaks, alkaline-sensitive sites and DNA–protein crosslinks to

Cr(III), and also affects DNA polymerase activity and fidelity *in vitro*. These studies have been reviewed previously (Hansen and Stern, 1985; Farrell and Costa, 1997). Such effects certainly could contribute to the carcinogenicity of hexavalent chromium compounds and, in fact, hexavalent chromium is unique among the metal carcinogens in that its mechanism of carcinogenic action resembles that of reactive organic carcinogens that form mutagenic DNA adducts. In contrast, many of the other metal carcinogens are not mutagenic and probably do not cause cancer by forming stable and mutagenic DNA adducts.

A model for the uptake and reduction of hexavalent chromium compounds in cellular systems is shown in Fig. 4.1 (Aiyar *et al.*, 1992) illustrating the major events that are responsible for chromium carcinogenicity. Several points of interest should be mentioned with regard to the differences between animal and human systems. First of all, it has recently been shown that human microsomal enzymes have a very high affinity for reduction of Cr(VI) compounds (Myers and Myers, 1998). The affinity is in the micromolar range, or 1000 times higher than the affinity of rodent enzymes for hexavalent chromium (Myers and Myers, 1998). This would indicate that the reduction of Cr(VI) to Cr(III) can be mediated enzymatically in humans whereas, in rodents, it would be mediated only by ascorbic acid and glutathione (Myers and Myers, 1998). Since rodents synthesize ascorbic acid, it is present at high levels both outside and inside cells, and Cr(VI) is rapidly and efficiently reduced to Cr(III) by ascorbic acid, chromium may be less toxic to rodents than to humans. The level of ascorbic acid outside the cell is less in humans but glutathione may participate in some reduction outside and inside the cell in addition to extracellular and intracellular ascorbic acid. There are differences in the effects on DNA of Cr(VI) reduced to Cr(III) by glutathione and ascorbic acid (Farrell and Costa, 1997). The consequences of this enzymatic reduction are uncertain at the present time and more work needs to be done to prove that enzymatic reduction is a significant factor for Cr(VI) reduction in humans.

4.2 Nickel

4.2.1 Introduction

Nickel is widely distributed in nature and makes up about 8.5% of the Earth's core (Carson *et al.*, 1986). The most abundant ores are the oxides and sulfides of nickel. Nickel ore is associated with other metals, such as cobalt, copper, gold, mercury and platinum. Nickel refining involves smelting, which drives out the sulfur, forming nickel oxide. Formation of the volatile gas, nickel carbonyl, and production of nickel metal by reduction represents the Mond method of nickel refining. Nickel is used in the production of stainless steel, as a chemical catalyst in electroplating and production of ceramics pigments, nickel–cadmium batteries and coins (IARC, 1990). Nickel is released into the atmosphere mainly as a result of the combustion of oil. The quantity of nickel depends upon the nature of the oil burned and the concentration of nickel in that fuel. The electrical power industry, which uses mostly oil for generation of electricity, is concerned about newly proposed standards that classify all nickel compounds as carcinogens.

Major health problems have been documented from occupational exposure to nickel and its compounds. Nickel refinery workers have a high incidence of nasal and lung

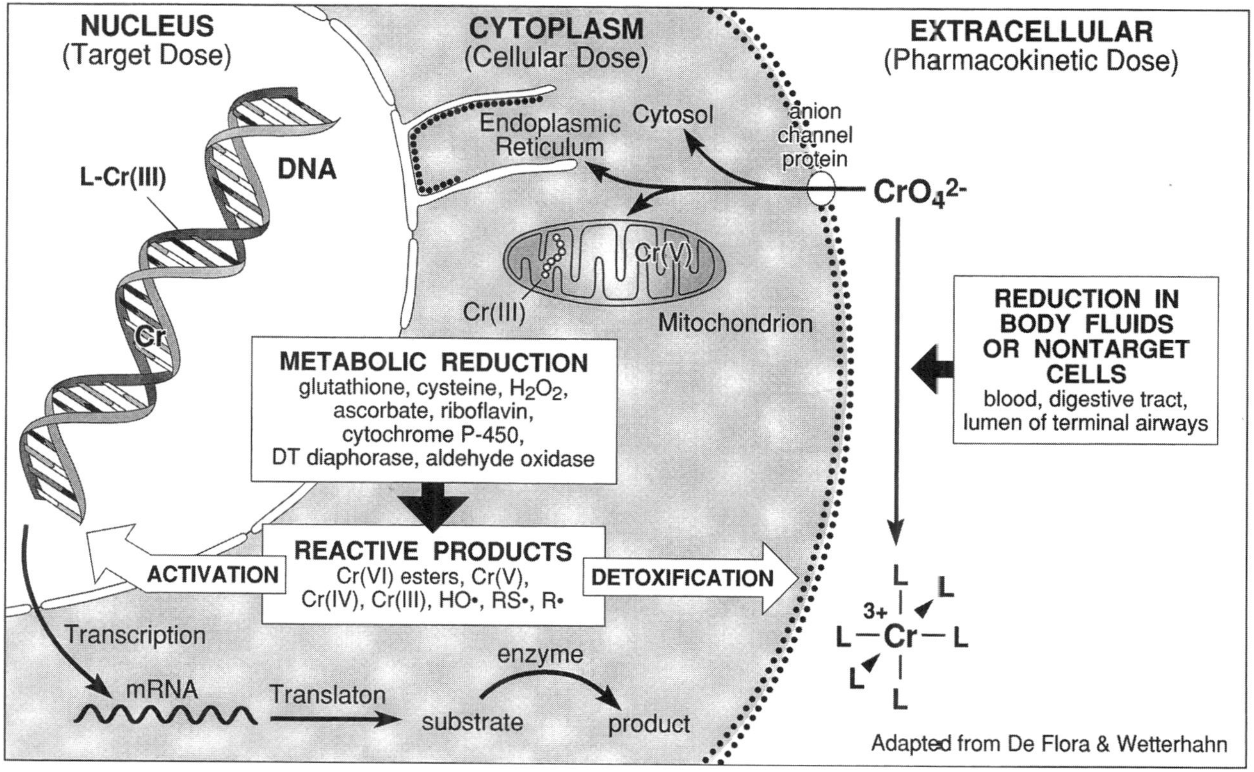

Figure 4.1 Model of chromatic uptake and reduction by cells (adapted from De Flora and Wetterhahn).

cancer as a result of being exposed to particles containing nickel sulfur matte in the crushing operations of nickel refining (IARC, 1990). Nickel carbonyl is one of the most toxic compounds known to man and it produces a very striking pulmonary deterioration and edema that, with sufficient exposure, is lethal. Nickel has also been associated with a high incidence of contact dermatitis (IARC, 1990).

Based upon studies conducted in experimental animals, nickel has been implicated as an essential trace element. Nickel deficiency has resulted in retarded growth, skeletal malformations, anemia and biochemical disorders of carbohydrate, lipid and iron metabolism in rats and fowl (Neilsen and Ollerich, 1974; Schnegg and Kirchgessner, 1975a, b; Kirchgessner and Schnegg, 1976). Nickel has been shown to be an essential element for plant growth and, in particular, it is an essential element of the enzyme urease, as well as being a cofactor in bacterial enzymes, such as hydroxygenases and dehydroxygenases, and F430 coenzymes in microbes (analog to vitamin B_{12}) (Dixon *et al.*, 1980). However, the role of nickel as an essential element in mammalian enzymes has not been established, but all humans are thought to take in more than 200 μg of nickel per day (Anke *et al.*, 1984).

4.2.2 *Carcinogenesis of nickel compounds*

One of the most important long-term health effects from human exposure to nickel compounds is the induction of lung cancer (IARC, 1990). Numerous previous studies have reviewed the epidemiology of nickel carcinogenesis, and it is hypothesized that nickel compounds differ in their carcinogenic activity towards humans, with certain water-insoluble nickel compounds, such as crystalline nickel subsulfide, nickel sulfide and nickel oxide, being considerably more carcinogenic than water-soluble nickel salts (Costa, 1991). In the occupational setting, it is difficult to differentiate exposures to nickel, let alone the species, although recent data suggest that individuals exposed to water-soluble nickel compounds alone also have a higher incidence of cancers, but whether they had exclusive exposures to the water-soluble nickel compounds is not clear (IARC, 1990). In experimental animal studies, the water-soluble nickel salts do not induce tumors, whereas the water-insoluble nickel compounds induce cancers at the site of administration (IARC, 1990). However, there have been studies where animals were exposed to water-insoluble compounds by intramuscular injections and no localized tumors were induced (Kasprzak *et al.*, 1983). More relevant studies have also been conducted where rats and mice have received exposure to nickel subsulfide, green nickel oxide or water-soluble nickel sulfate (NTP, 1994a, b, c). The rats developed a statistically significant higher incidence of lung cancers when exposed to the nickel subsulfide and the green nickel oxide, but did not have an elevated incidence of lung cancer with water-soluble nickel salts. In contrast, mice did not get cancers from any of the nickel compounds. The reason for the lack of cancer induction in mice is not understood but, clearly, there may be species differences in the response to carcinogenic nickel compounds.

The molecular basis for the potent carcinogenic activity of the water-insoluble nickel compounds has been studied using cell culture systems. It has been found that the potent carcinogenic compounds, such as crystalline nickel subsulfide and crystalline nickel sulfide, are phagocytosed by cells that could eventually undergo transformation (Costa and Mollenhauer, 1980). This phagocytosis process yielded very high concentrations of nickel inside the cell, since, following phagocytosis of these particles, they were contained in vacuoles and, with the acidic pH, there was dissolution of nickel into a soluble form that became distributed throughout the cell and entered the nucleus to interact with chromatin (Costa, 1991). Figure 4.2 illustrates the selective phagocytosis, dissolution and delivery of solubilized Ni^{2+} to the nucleus of cells. The potential nickel concentration inside the cell, should 1–4 μm particles totally dissolve, has been estimated to be from the high millimolar to the molar range, indicating the extremely efficient manner by which this phagocytosis process can contribute to elevations of intracellular nickel (Costa, 1991). In contrast, water-soluble nickel salts do not enter cells readily because they compete with the Mg^{2+} transport system and thus must reach millimolar levels to enter. Thus, soluble nickel salts do not give rise to a very high intracellular level of nickel. However, recent findings suggest that soluble nickel salts may have toxic action without entering the cells. This toxic action may involve interactions with Ca^{2+}-sensing sites on the cell membrane (Salnikow *et al.*, 1998; Zhou *et al.*, 1998). It has been known for a number of years that nickel can interact with and block low-voltage L-type Ca^{2+} channels (Kitada, 1994a, b; Zamponi, 1996). It has also been found that nickel can cause the release of intracellular Ca^{2+} without actually entering the cells but by interacting with the cell membrane (Salnikow *et al.*, 1998). This elevation of free intracellular Ca^{2+} induces a variety of responses in the cell. Soluble nickel can distribute to all sites in the body. The potential neurotoxicity of soluble nickel should be considered in view of its ability to block and interact with Ca^{2+} channels.

Once the water-insoluble nickel compound is phagocytosed by the cell, the nickel dissolves inside the cell and the cellular levels of nickel reach high levels (Fig. 4.2). Ni^{2+} is the major nickel ionic species inside the cell. There is a possibility of a Fenton-like reaction with nickel but only following an interaction of Ni^{2+} with certain ligands, such as Gly–Gly–His or the imidazole nitrogen of histidine (Klein *et al.*, 1991a). The binding

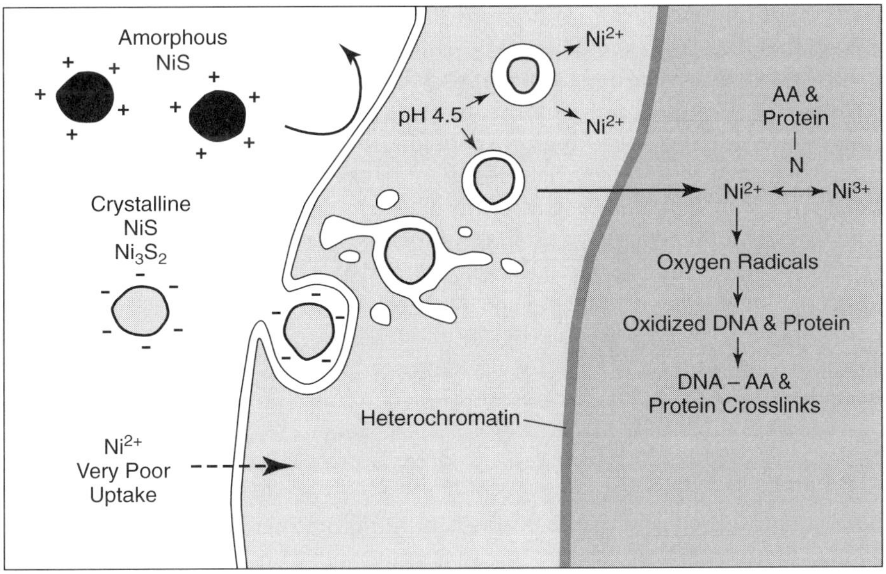

Figure 4.2 Model of nickel-compound phagocytosis and dissolution.

of Ni^{2+} to these ligands diminishes the oxidation potential of Ni^{2+} such that hydrogen peroxide could catalyze nickel oxidation to higher states (Klein *et al.*, 1991a). This generates oxygen radicals but the reaction is dependent on the binding of Ni^{2+} with certain ligands (Klein *et al.*, 1991a).

Nickel has a similar atomic radius to that of magnesium and thus Ni^{2+} will interact at many magnesium-binding sites (Costa, 1991). For example, Ni^{2+} displaces Mg^{2+} from the phosphate backbone of DNA and actually increases chromatin condensation, since Ni^{2+} is more efficient than Mg^{2+} in this regard (Lee *et al.*, 1995). Despite the fact nickel can produce oxidative stress in cells and oxidize DNA bases, nickel compounds that are potently carcinogenic display weak mutagenic activity in standard mutation assays. However, chromosome damage and sister chromatid exchange are significantly and substantially elevated in cells exposed to nickel compounds (Sen and Costa, 1985, 1986). Detectable gene mutations are produced by nickel compounds and these may contribute to its carcinogenesis (IARC, 1990). One of the interesting features of nickel compounds is that they are synergistic with many other agents that produce cancer through mutagenic mechanisms. For example, nickel is synergistic with benzo(a)pyrene, UV and radiation (Costa, 1991). The ability of soluble nickel to inhibit DNA repair activity may represent a molecular basis for this activity, and certainly nickel will inhibit DNA repair, but whether this is the operative mechanism in its carcinogen synergism is unknown. Metal ions, such as Hg^{2+}, that are generally not considered carcinogenic also inhibit DNA repair (Christie *et al.*, 1986).

Our laboratory found that nickel induced *de novo* DNA methylation, particularly in DNA regions that border heterochromatin (Klein *et al.*, 1991b). The mechanism for this has been studied using transgenic cell lines with the bacterial *gpt* gene placed at different sites (Lee *et al.*, 1995). The G12 cell line is very responsive to carcinogenic nickel-induced inactivation of *gpt* and yields a high incidence of 6-thioguanine-resistant colonies. However, the incidence of these 6-thioguanine-resistant colonies is extraordinarily high and therefore unlikely to be caused by mutations. Subsequent work has shown that this high incidence of 6-thioguanine-resistance involves the loss of transcription of the *gpt* gene (Lee *et al.*, 1995). The mechanism for this is thought to be a nickel-induced increase in chromatin condensation of the *gpt* gene in G12, which is located near a large region of heterochromatin. It was hypothesized that nickel displaces magnesium on the phosphate backbone of DNA and as Ni^{2+} is more active than Mg^{2+} in condensing chromatin (Lee *et al.*, 1995), this causes neighboring euchromatin to be drawn into the heterochromatin region (Fig. 4.3). This effect is transient since newly condensed DNA is not methylated, but with Ni^{2+} present, the chromatin remains more condensed (Fig. 4.3). We hypothesize that the newly condensed DNA triggers its own *de novo* methylation, and once the DNA becomes methylated it will be inherited in a methylated state for all subsequent generations in the absence of nickel (Lee *et al.*, 1995). The assumption is made that when the DNA is hypermethylated, it remains in a high-order condensation state. In fact, it was shown that when the *gpt* gene in G12 was relocated to heterochromatin, it was inherited in heterochromatin for all subsequent generations (Lee *et al.*, 1995). This can be attributed to the *de novo* methylation in the coding and flanking regions of this gene. In higher organisms, the 5 position of cytosine is methylated and in some instances this is important in terms of gene expression. Gene expression can be regulated by cytosine methylation and, in general, genes susceptible to such regulation will be expressed if

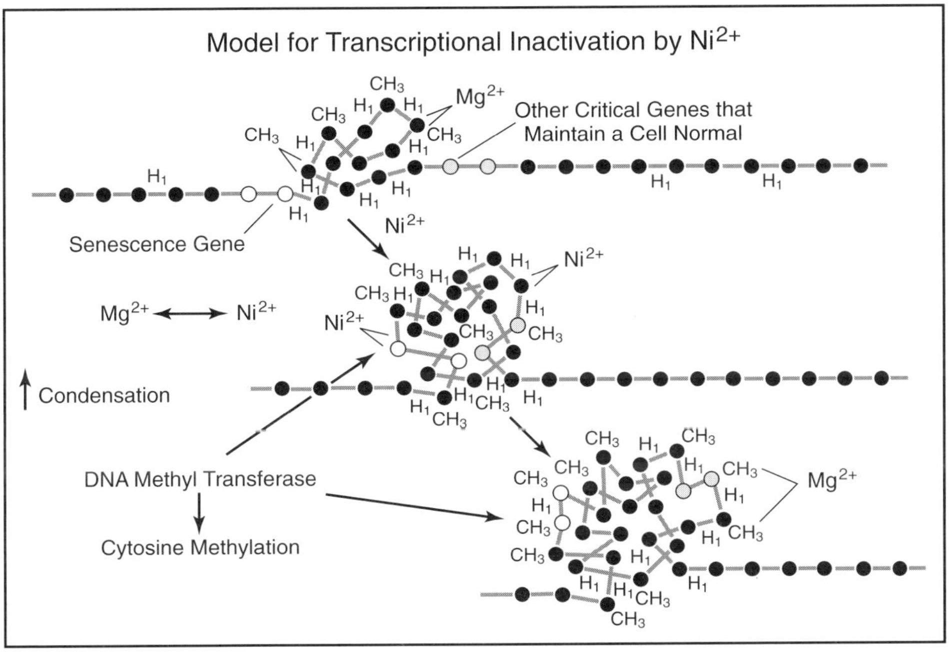

Figure 4.3 Model of the effect of Ni^{2+} on chromatin structure.

the CpG sites in their promoters are not methylated, whereas expression will be diminished if the CpG sites are methylated. CpG sites in the coding region of genes are also methylated, but the significance of these in terms of gene expression is unknown. However, sites of methylated C represents hot spots for mutations in, for example, the p53 coding region where methylated cytosines will be deaminated to thymidine, producing a C → T transition mutation.

The first discovery that nickel could affect DNA methylation came from studies of the inactivation of a senescence gene in nickel-transformed Chinese hamster embryo cells (Klein *et al.*, 1991b). This inactivated senescence gene could be reactivated with the drug 5-aczytidine. Thus, in the X chromosome, nickel induces an inactivation of a senescence gene by inducing *de novo* DNA methylation. It is likely that the same mechanism is operative in this situation as in the transgenic G12 cell line, where the target *gpt* gene was located near heterochromatin. It should be noted that in another cell line such as G10, when the *gpt* gene was placed away from heterochromatin, there was not an inactivation of the transgene. The critical positional dependent inactivation is reminiscent of the positional effect variegation phenomenon in *Drosophila*.

4.2.3 *Other toxicities of nickel*

While the ability of nickel to cause cancer is perhaps its major toxicological effect, other toxicities are also important. Nickel and its compounds are a major cause of dermatosis (IARC, 1990). These effects are not necessarily limited to occupationally exposed individuals since they also occur in the general population, particularly in women, as a result of nickel sensitization among individuals exposed to nickel-containing objects, such as coins and jewelry. Nickel dermatosis often begins on the hand. One study showed that 17 out of 28 patients who ingested 25 mg of nickel and nickel sulfate developed aggravated chronic dermatitis, and that nine of these patients recovered when the dietary intake of nickel was reduced (Kaaber *et al.*, 1978). Nickel-induced allergies have also been associated with dental prostheses (Sunderman, 1986).

Nickel intoxication has also occurred among 23 patients receiving hemodialysis (Webster *et al.*, 1980). These individuals were poisoned from leaching of nickel from nickel-plated tanks used for the preparation of the fluid. Symptoms of poisoning included nausea, vomiting, weakness, headaches and palpitations, which dissipated within a few hours after cessation of hemodialysis. Hypernickelemia has been reported in patients with acute renal failure. The hypernickelemia may contribute to the onset of dialysis-associated hypersensitivity and atherosclerosis (Hopfer *et al.*, 1984; Savory *et al.*, 1984). It was suggested that the permissible upper limit of nickel should be no more than 5 mg/L in common dialysis fluid and 10 mg/L for solutions containing amino acids and albumin (Sunderman, 1983). Higher nickel concentrations have also been observed in patients with acute myocardial infarction and unstable angina pectoris (Leach *et al.*, 1984). Administration of nickel-contaminated intravenous fluids to patients with myocardial infarction could result in hazardous consequences due to increased coronary artery resistance caused by Ni^{2+}. The ability of Ni^{2+} to raise cellular Ca^{2+} levels may be a mechanism for its induction of atherosclerosis and toxic effects on the heart.

Nickel compounds are also known to produce a number of other toxicological effects, including asthma, which is associated with inhalation of nickel-containing mists in the nickel plating industry, and in hypersensitive patients due to non-occupational exposure (Arvidsson and Bogg, 1959; McConnell *et al.*, 1973). Hypertrophic rhinitis and nasal sinusitis have been reported to occur in workers in nickel refineries and electrolysis plants. These conditions are frequently accompanied by nasal polyps and perforation of nasal septa, as well as pulmonary irritations.

In contrast to inorganic nickel compounds, exposure to nickel carbonyl can result in acute poisoning. Nickel carbonyl is a colorless, volatile liquid produced as an intermediate product during the Mond process of nickel refining. Inadvertent formation of nickel carbonyl also occurs during processes that use nickel catalysis, such as petroleum refining, coal gasification and hydrogenation of fats (Sunderman, 1986). Inhalation exposure of nickel carbonyl produces nausea, vertigo, headaches and chest pains. These symptoms disappear a few hours after removal from exposure, however, following a latency period of about 12–36 hours and, in some unusual cases, as long as 5 days, severe pulmonary symptoms develop, with coughing, taxacardio synopsis and profound weakness. There is diffuse pneumonitis, accompanied by either cerebral hemorrhage or pulmonary edema, which are the most likely causes of death should it occur within 4–13 days after exposure. Any recovery from acute poisoning is slow due to pulmonary insufficiency. Measurement of blood nickel concentrations following nickel carbonyl exposure provides a guideline of the severity of poisoning (Sunderman, 1981). The drug diethyldithiocarbonate is used as a chelating agent, but penicillamine and other chelators are also effective in some cases. When diethyldithiocarbonate is used, there is a danger of redistribution of nickel to other organs, such as the brain, so it should be used cautiously. Except for cases of nickel carbonyl exposure, acute nickel poisoning is rare. The major signs of intoxication include gastroenteritis, pulmonary distress, anemia and cachexia. Neurological symptoms, such as tremors and convulsions are associated with acute inorganic nickel poisoning. Only one fatality from acute nickel poisoning has been reported, involving ingestion of approximately 15 g of nickel sulfate (Daldrup *et al.*, 1983). Nickel is known to affect magnesium- and calcium-dependent processes and thus has potential for a wide spectrum of toxic action (Costa, 1991).

Nickel is also teratogenic, as has been well established in animal model systems (Mas *et al.*, 1985). One study showed that when three consecutive generations of rats were administered nickel at 0.5 p.p.m. in the drinking water, there was an increased number of neonatal deaths in all three generations, as well as an increased female to male ratio by the third generation (Schroeder and Mitchener, 1971). Administration of nickel acetate to hamsters at 2–30 mg/kg of body weight on day 8 of gestation increased the number of fetal resorptions and overall embryo mortality, as well as the incidence of unspecified congenital malformations (Ferm and Carpenter, 1968). In addition, in mice malformations were observed from dams exposed to 1.2–6.9 mg/kg as nickel carbonate (i.p. doses) on days 7–11 of gestation (Lu *et al.*, 1976, 1979). Malformations included encephala, cerebral hernia, open eyelids, cleft palate, micromelia, alkalosis of the extremity, clubfoot and skeletal abnormalities. The high incidence of malformations was observed when nickel was injected on day 8 or 9 of gestation. More recent investigations demonstrated that pregnant rats were more susceptible to nickel compared to control groups, as evidenced by the LD_{50} (Mas *et al.*, 1985). Furthermore, teratogenic effects were demonstrated when nickel was administered to dams during organogenesis. While the teratogenic effects of nickel in experimental animals are well established, the dosages are high and may not relate to human exposure. However, there is no information on reproductive or teratogenic effects of nickel in humans, probably because few females have been occupationally exposed to nickel compounds.

References

Aiyar, J., DeFlora, S. and Wetterhahn, K.E. 1992. Reduction of chromium(VI) to chromium(V) by rat liver cytosolic and microsomal fractions: is DT-diaphorase involved? *Carcinogenesis* **13**, 1159–1166.

Anke, M., Groppel, B., Kronemann, H. and Grun, M. 1984. Nickel – an essential element, in *Nickel in the Human Environment*, IARC Scientific Publication, No. 53, (ed. F.W. Sunderman, Jr), International Agency for Research on Cancer, Lyon, pp. 339–365.

Arfsten, D.P., Aylward, L.L. and Karch, N.J. 1998. Immunotoxicology of chromium, in *Immunotoxicology of Environmental and Occupational Metals*, (eds J.T. Zelikoff and P. Thomas), Taylor and Francis, London, pp. 63–92.

Arvidsson, H. and Bogg, A. 1959. Transitory pulmonary infilterations (Leoffler's syndrome) in acute generalized dermatatis. *Acta Derma Venereologica* **39**, 30–34.

Bidstrup, P.L. 1989. Perspective on safety: Personal opinions. *American Industrial Hygiene Association* **50**, 505–509.

Carson, B.C., Ellis, H.V. and McCann, J.L. 1986. *Toxicology and Biological Monitoring of Metals in Humans: Including Feasibility and Need*, Lewis Publishers, Chelsea, Michigan.

Christie, N.T., Cantoni, O., Sugiyama, M., Cattabeni, F. and Costa, M. 1986. Differences in the effects of Hg(II) on DNA repair induced in Chinese hamster ovary cells by ultraviolet or X-rays. *Molecular Pharmacology* **29**, 173–178.

Cohen, M.D. and Costa, M. 1999. Chromium, in *Environmental Toxicants – Human Exposures and their Health Effects*, 2nd edn, (ed. M. Lippmann), Van Nostrand Reinhold, New York, in press.

Costa, M. 1991. Molecular mechanisms of nickel carcinogenesis. *Annual Review of Pharmacology and Toxicology* **31**, 321–337.

Costa, M. 1997. Toxicity and carcinogenicity of Cr(VI) in animal models and humans. *CRC Critical Reviews in Toxicology* **27**, 431–442.

Costa, M. and Mollenhauer, H.H. 1980. Carcinogenic activity of particulate metal compounds is proportional to their cellular uptake. *Science* **209**, 515–517.

Daldrup, T., Haarhoff, K., and Szathmary, S.C. 1983. Teodliche nickel sulfate intoxikation. *Berichte Zur. Gerichlichen Medizin* **41**, 141–144.

Dixon, N.E., Gazzola, C., Asher, C.J., Lee, D.S.W., Blakely, R.L. and Zerner, B. 1980. Jack bean urase (EC 3.5.1.5). II. The relationship between nickel enzymatic activity and the abnormal ultraviolet spectrum. The nickel count of jack beans. *Cancer Journal of Biochemistry* **58**, 474–480.

Farrell, R.P. and Costa, M. 1997. Carcinogenic inorganic chemicals, in *Comprehensive Toxicology, Chemical Carcinogens and Anticarcinogens*, Vol. 12, (eds G.T. Bowden and S.M. Fischer), Elsevier Science Ltd., Pergamon, Oxford, pp. 225–253.

Ferm, V.H. and Carpenter, S. 1968. The teratogenic effects of metals on mammalian embryo. *Advances in Teratology* **5**, 51–75.

Freund, H., Altamian, S. and Fischer, J.E. 1979. Chromium deficiency during total parenteral nutrition. *Journal of the American Medical Association* **241**, 496–498.

Gale, T.T. 1978. Embryotoxic effects of chromium trioxide in hamsters. *Environmental Research* **16**, 101–109.

Glaser, U., Hochrainer, D., Kloppel, H. and Oldiges, H. 1986. Carcinogenicity of sodium dichromate and chromium(VI/III) oxide aerosols inhaled by male Wister rats. *Toxicology* **4**, 219–232.

Gray, S.J. and Sterling, K. 1950. Tagging of red cells and plasma proteins with radioactive chromium. *Journal of Clinical Investigation* **29**, 1604–1613.

Hansen, K. and Stern, R.M. 1985. Welding fumes and chromium compounds in cell transformation assays. *Journal of Applied Toxicology* **5**, 306–314.

Hopfer, S.M., Linden, J.V., Cristomo, C., Catanaotto, F., Galen, M. and Sunderman, F.W. Jr 1984. Hypernickelemia in hemodialysis patients. *Annals of Clinical and Laboratory Science* **14**, 12–13.

IARC (International Agency for Research on Cancer) 1990. *Chromium, Nickel, and Welding, IARC Monographs on the Evaluation of Carcinogenic Risks of Chemicals to Humans*, Vol. 49, World Health Organization, Lyon, pp. 49–256.

Jeejeebhoy, K.N., Chu, R.C., Marliss, E.B., Greenberg, G.R. and Bruce-Robertson, A. 1977. Chromium deficiency glucose intolerance and neuropathy reversed by chromium supplementation in a patient receiving long-term parenteral nutrition. *American Journal of Clinical Nutrition* **30**, 531–538.

Kaaber, K., Veinin, N.K. and Tjell, J.C. 1978. Low nickel diet in the treatment of patients with chronic nickel dermatitis. *British Journal of Dermatology* **98**, 197–210.

Kasprzak, K.S., Gabryel, P., and Jarczewska, K. 1983. Carcinogenicity of nickel(II) hydroxides and nickel(II) sulfate in Wistar rats and its relation to the *in vitro* dissolution rates. *Carcinogenesis* **4**, 275–279.

Kirchgessner, M. and Schnegg, A. 1976. Nickel content of milk from lactating rats fed varying nickel levels. *Archiv. fur Tierernahrung* **11**, 774–746.

Klein, C.B., Frenkel, K. and Costa, M. 1991a. The role of oxidative processes in metal carcinogenesis. *Chemical Research in Toxicology* **4**, 592–604.

Klein, C.B., Conway, K., Wang, X.W., Bhamra, R.K., Lin, S., Cohen, M.D., Annab, L., Barrett, J.C. and Costa, M. 1991b. Senescence of nickel-transformed cells by a mammalian X chromosome: Possible epigenetic control. *Science* **251**, 796–799.

Kitada, Yl. 1994a. A quantitative study of dual action of nickel ions on the taste response to calcium ions of single fibers of the frog glossopharyngeal nerve: inhibition and enhancement by nickel ions. *Chemical Sciences* **19** (5), 401–411.

Kitada, Y. 1994b. Competitive inhibition of the nickel-induced response to choline by calcium ions in single water fibers of the frog glossopharyngeal nerve. *Chemical Sciences* **19** (6), 641–650.

Leach, C.N. Jr, Linden, J., Hopfer, S.M., Chrisostomo, C. and Sunderman, F.W. Jr, 1984. Serum nickel concentrations in patients with unstable angina and myocardial infarction. *Annals of Clinical and Laboratory Science* **14**, 414–415.

Lee, Y-W., Klein, C.B., Kargacin, B., Salnikow, K., Kitahara, J., Dowjat, K., Zhitkovich, A., Christie, N.T. and Costa, M. 1995. Carcinogenic nickel silences gene expression by chromatin condensation and DNA methylatin: A new model for epigenetic carcinogens. *Molecular and Cellular Biology* **15**, 2547–2557.

Lu, C.C., Matsumoto, N. and Iijama, S. 1976. Placental transfer of $NiCl_2$ to fetus in mice. *Teratology* **14**, 245.

Lu, C.C., Matsumoto, N. and Iijama, S. 1979. Teratogenic effects of nickel chloride on embryonic mice and its transfer to embryonic mice. *Teratology* **19**, 137–142.

McConnell, L.H., Fink, J.N., Schlueter, D.P. and Schmidt, M.G. Jr 1973. Asthma caused by nickel sensitivity. *Annals of Internal Medicine* **78**, 888–890.

Mas, A., Holt, D. and Webb, M. 1985. The acute toxicity and teratogenicity of nickel in pregnant rats. *Toxicology* **35**, 47–57.

Mertz, W. and Schwarz, K. 1955. Impaired intravenous glucose tolerance as an early sign of dietary necrotic liver degeneration. *Archives of Biochemistry and Biophysics* **58**, 504–508.

Myers, C. and Myers, J.M. 1998. Iron stimulates the rate of reduction of hexavalent chromium by human microsomes. *Carcinogenesis* **19**(6), 1029–39.

Neiboer, E. and Jusy, A.A. 1986. *Chromium in the Natural and Human Environment*, (eds J.O. Nriagu and E. Neiboer), Wiley Interscience, New York, pp. 21–79.

Neilsen, F.H. and Ollerich, D.A. 1974. Nickel: a new essential trace element. *Federal Proceedings* **33**, 1767–1774.

NIOSH (National Institute for Occupational Safety and Health), 1979. *Registry of Toxic Effects of Chemical Substances* (RTECS), Washington, DC.

NTP (National Toxicology Program) 1994a. *NTP Report on the Toxicology and Carcinogenesis Studies of Nickel Sulfate Hexahydrate* (CAS No. 10101–97–0) in F344/N Rats and $B6C3F_1$ Mice (Inhalation Studies). NTP TR 454, NIH Publication No. 94–3370, US DHHS, NIH, Washington, DC.

NTP (National Toxicology Program) 1994b. *NTP Report on the Toxicology and Carcinogenesis Studies of Nickel Oxide* (CAS No. 1313–99–1) in F344/N Rats and $B6C3F_1$ Mice (Inhalation Studies). NTP TR 451, NIH Publication No. 94–3363, US DHHS, NIH, Washington, DC.

NTP (National Toxicology Program) 1994c. *NTP Report on the Toxicology and Carcinogenesis Studies of Nickel Subsulfide* (CAS No. 12035–72–2) in F344/N Rats and $B6C3F_1$ Mice (Inhalation Studies). NTP TR 453, NIH Publication No. 94–3369, US DHHS, NIH, Washington, DC.

Salnikow, K., Zhou, D., Wang, S. and Costa, M. 1998. Treatment of human lung cells with calcium channel blocker nickel chloride resulted in elevation of intracellular calcium and expression of a novel gene. *The Toxicologist* **43**, 320.

Saner, G. 1980. *Chromium in Nutrition and Disease-Current Topics in Nutrition and Disease*, Vol. 2, Alan R. Liss, New York, pp. 26–29.

Savory, J., Brown, S., Bertholf, R., Ross, R., Savory, M.G. and Wells, M.R. 1984. Serum and lymphocyte nickel and aluminum concentrations in patients with extracorporeal hemodialysis. *Annals of Clinical Laboratory Science* **14**, 413–414.

Schnegg, A. and Kirchgessner, M. 1975a. Essentiality of nickel for growth of animals. *Z. Tierphyiology Tierernah. Futtermittelkunde* **36**, 63–74.
Schnegg, A. and Kirchgessner, M. 1975b. Changes in the hemoglobin content, erythrocyte count and hematocrit in nickel deficiency. *Nutrition Metabolism* **19**, 268–278.
Schroeder, H.A. and Mitchener, M. 1971. Toxic effects of trace elements on the reproduction of mice and rats. *Archives of Environmental Health* **23**, 102–106.
Schwarz, K. and Mertz, W. 1959. Chromium(III) and glucose tolerance factor. *Archives of Biochemistry and Biophysics* **85**, 292–295.
Sen, P. and Costa, M. 1985. Induction of chromosomal damage in Chinese hamster ovary cells by soluble and particulate nickel compounds: Preferential fragmentation of the heterochromatic long arm of the X-chromosome by carcinogenic crystalline NiS particles. *Cancer Research* **45**, 2320–2325.
Sen, P. and Costa, M. 1986. Incidence and localization of sister chromatid exchanges induced by nickel and chromium compounds. *Carcinogenesis* 7, 1527–1533.
Sheehan, P.J., Meyer, D.M., Sauer, M.M. and Paustenbach, D.J. 1991. Assessment of the human health risks posed by exposure to chromium-contaminated soils. *Journal of Toxicology and Environmental Health* 32, 161–201.
Sunderman, F.W. Jr 1981. Recent research on nickel carcinogenesis. *Environmental Health Perspectives* **40**, 131–141.
Sunderman, F.W. Jr 1983. Potential toxicity from nickel contamination of intravenous fluids. *Annals of Clinical and Laboratory Science* **13**, 1–4.
Sunderman, F.W. Jr 1986. Sources of exposure and biological effects of nickel, in *Environmental Carcinogens – Selected Methods of Analysis*, (eds I.K. O'Neill, P. Schuller and L. Fishbein), IARC Publ. No. 71, Lyon, pp. 79–92.
US EPA (US Environmental Protection Agency) 1980. *Health Assessment Document for Chromium*, Environmental Criteria and Assessment Office, Research Triangle Park, NC.
Voitkun, V., Zhitkovich, A. and Costa, M. 1998. Cr(III)-mediated crosslinks of glutathione or amino acids to the DNA phosphate backbone are mutagenic in human cells. *Nucleic Acids Research* **26**(8), 2024–30.
Webster, J.D., Parker, T.F., Alfrey, A.C., Smythe, W.R., Kubo, H., Neal, G. and Hull, A. 1980. Acute nickel intoxication by dialysis. *Annals of Internal Medicine* **92**, 631–633.
Zamponi, G.W., Bourinet, E. and Snutch, T.P. 1996. Nickel block of a family of neuronal calcium channels: Sutbype- and subunit-dependent action at multiple sites. *The Journal of Membrane Biology* **151**, 77–90.
Zhitkovich, A., Voitkun, V. and Costa, M. 1995. Glutathione and free amino acids form stable complexes with DNA following exposure of intact mammalian cells to chromate. *Carcinogenesis* **16** (4), 907–913.
Zhitkovich, A., Voitkun, V. and Costa, M. 1996. Formation of the amino acid–DNA complexes by hexavalent and trivalent chromium *in vitro*: Importance of trivalent chromium and the phosphate group. *Biochemistry* **35**(22), 7275–7282.
Zhou, D., Salnikow, K. and Costa, M. 1998. *Cap43*, a novel gene specifically induced in lungs by Ni^{2+} Compounds. *Cancer Research* **58**, 2182–9.

5

TRANSPORT OF SOFT METALS IN PROCARYOTES

Christopher Rensing and Barry P. Rosen

5.1 Introduction

The most frequent mechanism of resistance to both drugs and metals is extrusion of the toxic compounds from the cell, reducing the intracellular concentration to subtoxic levels (Dey and Rosen, 1995b; Dey and Rosen, 1996). Such extrusion mechanisms are ubiquitous and are found in organisms from every kingdom. In this chapter the mechanisms of resistance to ions of the soft metals zinc, cadmium, copper, silver, nickel, lead, arsenic and antimony in bacteria will be described. Soft-metal ions such as the transition metals Cu(I), Ag(I) and Cd(II) have high polarizing power (a large ratio of ionic charge to the radius of the ion), in contrast to the hard-metal ions of groups I and II such as Na^+ and Ca^{2+}. Some, such as Ag(I) and Pb(II), are also heavy metals, but not all transition metals are '*heavy*'. Zn(II) and the metalloids As(III) and Sb(III) of group V have properties more similar to soft-metal ions and will be considered together with soft metals.

All of these ions are toxic in excess, and bacterial metal resistances probably arose early in evolution due to widespread geochemical sources of metals. Resistance genes to inorganic salts of soft metals, including arsenate, arsenite, antimony, lead, cadmium, nickel, zinc, bismuth and mercury, are found on both plasmids and in chromosomes (Novick and Roth, 1968; Foster *et al.*, 1979; Datta and Hughes, 1983). Although not always the situation, in general plasmid-encoded determinants are strictly resistance systems, while the physiological role of chromosomally encoded systems also includes metal ion homeostasis. In most cases resistance involves extrusion mechanisms. Transport systems for accumulation of the metals also exist. Some of these metal ions are required for growth, in particular copper and zinc. For these there must be a fine balance between uptake and efflux to provide metal ion homeostasis. In the case of metals that are only toxic, uptake may occur through the transport systems for other metals. For example, arsenate is accumulated by phosphate transporters (Rosenberg *et al.*, 1977; Willsky and Malamy, 1980). Thus both routes of uptake and extrusion should be known. Table 5.1 summarizes the known pathways for these transport systems.

Table 5.1 Genes for soft-metal transporters

Metal	*Gene*	*Organism*	*Function*	*Mechanism*	*Reference*
Cd(II), Zn(II), Pb(II)	*cadA*	*Staphylococcus aureus*	Efflux	P-type ATPase	Nucifora *et al.* (1989)
Zn(II), Cd(II), Pb(II)	*zntA*	*Escherichia coli*	Efflux	P-type ATPase	Rensing *et al.* (1997b)
Zn(II), Cd(II), Pb(II)	*zntA*	*Proteus mirabilis*	Efflux	P-type ATPase	Rensing *et al.* (1998)
Cu(I), Ag(I)	*copB*	*Enterococcus hirae*	Efflux	P-type ATPase	Odermatt *et al.* (1992)
Cu(I), Ag(I)	*copA*	*E. hirae*	Probably uptake	P-type ATPase	Solioz *et al.* (1994)
Cu(I), Ag(I)	*ctaA*	*Synechococcus* PCC7942	Possibly uptake, gene disruption produces slight copper resistance	P-type ATPase	Phung *et al.* (1994)
Cu(I), Ag(I)	*pacS*	*Synechococcus* PCC7942	Located in thylakoid membrane, disruption produces copper sensitivity	P-type ATPase	Kanamaru *et al.* (1994)
Co(II), Zn(II), Cd(II)	*czcCBA*	*Alcaligenes eutrophus*	Efflux	CzcA is a proton/cation antiporter and member of the RND family; CzcB is a member of the MFP family; CzcC is a member of the OMF family	Nies (1995), Rensing *et al.* (1997c)
Co(II), Ni(II)	*cnrBA*	*A. eutrophus*	Efflux	CnrA is a proton/cation antiporter and member of the RND family; CnrB is a member of the MFP family	Liesegang *et al.* (1993)

Table 5.1 continued

Metal	*Gene*	*Organism*	*Function*	*Mechanism*	*Reference*
As(III), Sb(III)	*arsAB*	*E. coli*	Efflux	The ArsAB pump is a novel anion translocating ATPase; ArsB can serve as an electrophoretic anion uniporter	Rosen *et al.* (1996), Kuroda *et al.* (1997)
Zn(II)	*znuABC*	*E. coli*	Uptake	ABC transporter	Patzer and Hantke (1998)
Ni(II)	*nik*	*E. coli*	Uptake	ABC transporter	Navarro *et al.* (1993)
Ni(II)	*hoxN*	*A. eutrophus*	Uptake	Unknown	Eitinger *et al.* (1997)
Co(II)	*nhlF*	*Rhodococcus rhodochrous*	Uptake	Unknown	Komeda *et al.* (1997)
Sb(III)	*glpF*	*E. coli*	Uptake	Electroneutral uniporter	Sanders *et al.* (1997)

ATPase, adenosine triphosphatase; RND, resistance, nodulation, division; MFP, membrane fusion protein; OMF, outer membrane factor.

5.2 Zinc, lead and cadmium

5.2.1 Uptake systems

Zinc is essential for all organisms because it plays a critical role in the catalytic activity and structural stability of many proteins. More than 300 zinc-dependent enzymes have been identified (Vallee and Falchuk, 1993). Despite its importance, until recently little was known about the mechanism and regulation of zinc transport in bacteria. A few early studies suggested that bacteria appear to possess a specific energy-dependent zinc transport (Chipley and Edwards, 1972; Bucheder and Broda, 1974). Under conditions of magnesium sufficiency, energy-dependent cadmium uptake in *Escherichia coli* was inhibited by zinc but not by manganese, as in Gram-positive organisms (Laddaga *et al.*, 1985). This might suggest that cadmium and perhaps zinc are taken up by specific transporters unrelated to magnesium system(s).

In *Alcaligenes eutrophus* zinc, magnesium, cobalt, cadmium and nickel appear to be accumulated by a single transport system with low substrate specificity (Nies and Silver, 1989a). When the magnesium concentration in the growth medium was reduced from 1 mM to 0.1 mM, uptake of those divalent metal cations increased about sevenfold; when lowered to 10 μM, uptake increased about twentyfold. Each of the cations competes with each other for transport by this system.

In *E. coli*, energy-dependent uptake of zinc has been reported (Bucheder and Broda, 1974). In that organism, cobalt, manganese, nickel and magnesium ions are transported by the relatively unspecific CorA system (Snavely *et al.*, 1989). However, zinc did not inhibit magnesium uptake by this system, suggesting that it is not a CorA substrate. On the other hand, Zn(II) is a good inhibitor of magnesium transport by MgtA, a P-type ATPase induced under conditions of Mg^{2+} deficiency. In *Salmonella typhimurium* there is a third Mg^{2+} uptake pump, MgtB, a P-type ATPase homologous to MgtA (Snavely *et al.*, 1991). While *E. coli* does not have an MgtB homolog, it has a third magnesium uptake system, MgtE (Townsend *et al.*, 1995). It is reasonable to assume that one or more of these systems ise able to transport Zn(II). *Escherichia coli* may also have a specific zinc-uptake system. When cells were grown in the presence of magnesium, energy-dependent uptake of cadmium was inhibited by zinc but not by manganese, suggesting the existence of a Zn(II)/Cd(II) transporter (Laddaga and Silver, 1985).

Recently a periplasmic zinc-binding protein (PZP1) has been identified in *Haemophilus influenzae* (Lu *et al.*, 1997). A *pzp1*-deficient mutant failed to grow under aerobic conditions and grew poorly under anaerobic conditions. This growth defect was specifically rescued by supplementing the growth medium with high concentrations of zinc. PZP1 has a histidine-rich domain of about 47 amino acids. This domain is extremely rich in potential metal-binding amino acids, including 23 histidines, 10 aspartic acids, and six glutamic acids; it may contain one or more zinc-binding sites. Although potentially it can bind up to five zinc atoms, purified PZP1 was shown to contain an average of 1.6–1.9 zinc atoms/protein molecule. PZP1 has 49.2% identity and 59.4% similarity to a newly identified protein ZnuA (for *zinc uptake*; formerly called YebL) in *E. coli*. The *znuA* gene encodes a 31.4 kDa protein in the *msbB–ruvB* intergenic region precursor (Patzer and Hantke, 1998). ZnuA of *E. coli* has a similar domain structure to PZP1, including a central potential metal-binding domain of 21 amino acid residues and two conserved cysteines in the C-terminal region. ZnuA is organized in an operon similar to that of a typical ATP-binding protein cassette system. ZnuB (formerly *yebI*) is similar to the membrane component of ABC transport systems, while ZnuC (formerly *yebM*) shows strong similarity to the ATPase subunit of ABC transporters (Patzer and Hantke, 1998). These genes were shown to be responsible for high-affinity zinc uptake in zinc-depleted medium. This system is repressed by growth in the presence of 10 μM Zn(II). A repressor of *zurA* termed *zur* (*zinc uptake regulation*), formerly *yibK*, was identified. Zur showed 27% sequence identity with the iron regulator Fur. The Znu proteins belong to a recently defined family of binding-protein dependent divalent cation transport systems, including the Mn(II) transporter Mnt of *Synechocystis* sp. (Bartsevich and Pakrasi, 1995) and a manganese- and a zinc-binding-protein dependent transport system from *Streptococcus pneumoniae* (Dintilhac *et al.*, 1997).

5.2.2 *Efflux systems*

Escherichia coli is intrinsincally resistant to high levels of zinc, lead and cadmium, implying mechanisms that prevent overaccumulation. Recently the *zntA* gene was identified as a zinc- and cadmium-resistance locus, with disruption of *zntA* producing hypersensitivity to all three soft metals (Beard *et al.*, 1997; Rensing *et al.*, 1997b; Rensing *et al.*, 1998b)). The fact that *zntA* is a chromosomal gene suggests that it has a function in normal growth of *E. coli*, probably in zinc homeostasis. ZntA is a member of the superfamily of P-type cation-translocating ATPases but belongs to a subgroup of soft metal ATPases (Lutsenko and Kaplan, 1995; Solioz and Vulpe, 1996; Rosen, 1998) (Fig. 5.1). Representative members of this subfamily include bacterial enzymes such as CadA (Nucifora *et al.*, 1989; Ivey *et al.*, 1992; Tsai *et al.*, 1992), CopA and CopB (Odermatt *et al.*, 1993), CtaA (Phung *et al.*, 1994) and eucaryotic Cu(I)-transporting ATPases such as the proteins associated with Menkes (Vulpe *et al.*, 1993) and Wilson's disease (Bull *et al.*, 1993). As predicted from its sequence (Fig. 5.2), ZntA has been shown to catalyze ATP-coupled accumulation of ^{65}Zn(II) in everted (inside-out) membrane vesicles of *E. coli*, where accumulation into everted membrane vesicles is equivalent to efflux from intact cells (Rensing *et al.*, 1997b). Vesicles prepared from the *zntA*-disrupted strain did not accumulate ^{65}Zn(II), but the transport defect could be complemented by expression of *zntA* on a plasmid. ATP-dependent Zn(II) transport exhibited a K_m of 10^{-5} M and was inhibited by vanadate, a classical inhibitor of P-type ATPases.

The closest homolog to ZntA is a P-type ATPase from *Proteus mirabilis* isolated from a defective ability to swarm (Lai *et al.*, 1998) (Fig. 5.2). The relationship between zinc and swarming is unclear. Zinc may have a direct role, or perhaps high levels of intracellular zinc inhibit some component of the swarming machinery. The gene for this 692-residue ZntA homolog could also complement the zinc- and cadmium-sensitive phenotype of *E. coli*, RW3110, and the protein was therefore called $ZntA_{pm}$, with the *E. coli* homolog termed $ZntA_{ec}$. $ZntA_{pm}$ was as effective as $ZntA_{ec}$ in catalyzing ATP-dependent ^{65}Zn(II) transport in everted membrane vesicles from an *E. coli* strain with a *zntA* disruption (Rensing *et al.*, 1998).

As yet no *zntA* genes have been identified in Gram-positive bacteria. The cadmium-resistance (*cad*) operon, originally identified from the *Staphylococcus aureus* plasmid pI258, has two genes, *cadC*, which encodes a transcriptional repressor, and *cadA*, which encodes a P-type ATPase that is about 30% identical to the two ZntA pumps (Nucifora

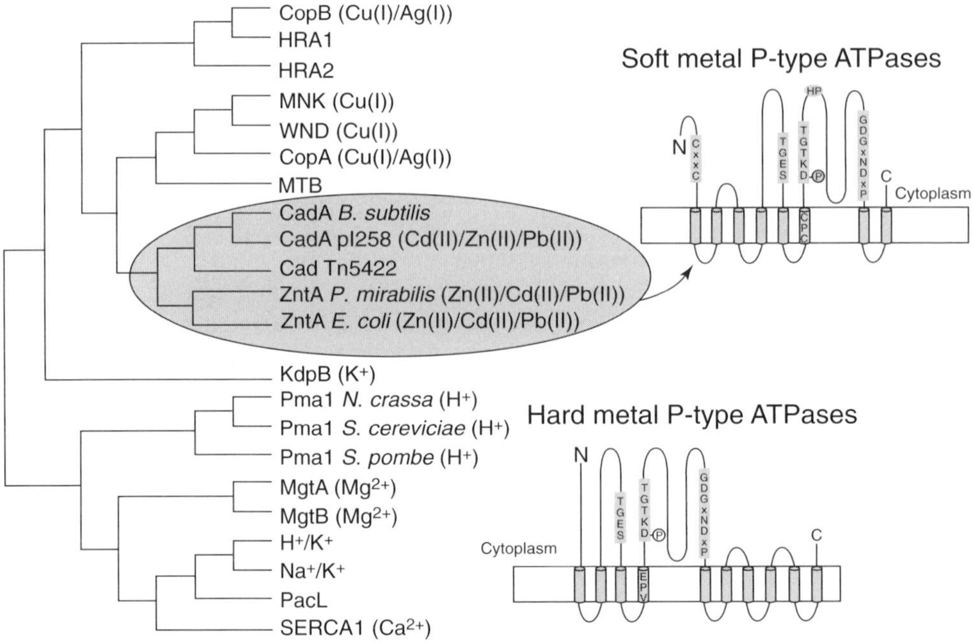

Figure 5.1 Superfamily of P-type ATPases. The superfamily has two branches: soft-metal (top) and hard-metal (bottom) ATPases. Soft-metal ATPases: CopB, *Enterococcus hirae* Cu(I)/Ag(I) pump; HRA1 and HRA2, *Escherichia coli* reading frames; MNK, human Menkes protein; WND, human Wilson protein; CopA, *E. hirae* Cu(I)/Ag(I) uptake pump; MTB, *Mycobacterium tuberculosis* reading frame; CadA, Cd(II) pumps from *Bacillus subtilis*, *Staphylococcus aureus* plasmid pI258 and *S. aureus* transposon Tn5422; ZntA, Zn(II) pumps from *Proteus mirabilis* and *E. coli*; CtaA, *Synechococcus* PCC7942 reading frame; Mtb2. Hard-metal ATPases: KdpB, *E. coli* K^+ uptake pump; Pma1, fungal plasma membrane H^+ pump; MgtA and MgtB, *Salmonella typhimurium* Mg^{2+} uptake pumps; H^+/K^+, mammalian gastric H^+ pump; N^+/K^+, mammalian Na^+ pump; PacL, *Synechococcus* reading frame; SERCA1, mammalian sarcoplasmic reticulum Ca^{2+} pump.

et al., 1989) (Fig. 5.2). CadA has been shown to catalyze ATP-dependent ^{109}Cd(II) uptake in everted membrane vesicles from *B. subtilis* (Tsai *et al.*, 1992). The fact that *cadA* is carried on a plasmid suggests that it is not required for normal physiology of the organism but was acquired as a defense mechanism, just as antibiotic resistance determinants are spread horizontally through gene transfer. CadA homologs have been identified in plasmids and chromosomes from several Gram-positive organisms (Lebrun *et al.*, 1994). The pI258 *cadA* gene could also partially complement the zinc- and cadmium-sensitive phenotypes of an *zntA*-disrupted *E. coli* (Rensing *et al.*, 1998). This clearly demonstrates that the Gram-positive CadA and the Gram-negative ZntA are both functionally and phylogenetically related. However, the preferred substrate of ZntA is zinc, although cadmium is transported at a lower rate. While expression of *zntA* is induced by Zn(II) and Cd(II) (Rensing *et al.*, 1997b), expression of pI258 *cadA* is not induced by Zn(II) but is by Cd(II) (Yoon *et al.*, 1991; Corbisier *et al.*, 1993). This may suggest that physiologically ZntA is involved in zinc homeostasis while CadA is involved in cadmium resistance but not zinc homeostasis (since it would be induced

```
                      10         20         30         40         50
ZNTA-EC     1 MSTPDNHGKK APQFAAFKPL TTVQNANDCC CDGACSSTPT LSENVSGTRY  50
ZNTA-PM     1 ---------- ---------- ---------- ---------- ----------  50
CADA        1 ---------- ---------- ---------- ------MSEQ KVKLMEEEMN  50
                      60         70         80         90        100
ZNTA-EC    51 SWKVSGMDCA ACARKVENAV RQLAGVNQVQ VLFATEKLVV DADNDI----  100
ZNTA-PM    51 ------MDCP TCAQKIETAV LKVVGVKQAK ILFATEKLVV DAD--I----  100
CADA       51 VYRVQGFTCA NCAGKFEKNV KKIPGVQDAK VNFGASKIDV YGNASVEELE  100
                     110        120        130        140        150
ZNTA-EC   101 RAQVESALQK AGYSLRDE-- ----QAAEEP QASR---LKE NLPLITLIVM  150
ZNTA-PM   101 RTDVISAVKS AGFELFDI-- ----SSAGSQ PPAKQSLLKE SSFVIILAIL  150
CADA      101 KAGAFENLKV SPEKLANQTI QRVKDDTKAH KEEKTPFYKK HSTLLFATLL  150
                     160        170        180        190        200
ZNTA-EC   151 MAISWGLEQF NHP---FGQL AFIATTLVGL YPIARQALR- LIKSGSYFAI  200
ZNTA-PM   151 MAISWGAEFI DPQ---AGRA AFIATTLIGL FPIVKKLLR- LIRSGTPFAI  200
CADA      151 IAFGYLSHFV NGEDNLVTSM LFVGSIVIGG YSLFKVGFQN LIRFD--FDM  200
                     210        220        230        240        250
ZNTA-EC   201 ETLMSVAAIG ALFIGATAEA AMVLLLFLIG ERLEGWAASR ARQGVSALMA  250
ZNTA-PM   201 ETLMSVAAIG ALFINATEEA AMVILLFMIG EMLESYAAGR ARRGVSALMA  250
CADA      201 KTLMTVAVIG ATIIGKWAEA SIVVILFAIS EALERFSMDR SRQSIRSLMD  250
                     260        270        280        290        300
ZNTA-EC   251 LKPETATRLR KGEREEVAIN SLRPGDVIEV AAGGRLPADG KLLSPFASFD  300
ZNTA-PM   251 LVPEEAVVIK EGKKQTVPVA QLRPGDIIEI APGRALPTDA ELITAFASFD  300
CADA      251 IAPKEALVRR NGQEIIIHVD DIAVGDIMIV KPGEKIAMDG IIVNGLSAVN  300
                     310        320        330        340        350
ZNTA-EC   301 ESALTGESIP VERATGDKVP AGATSVDRLV TLEVLSEPGA SAIDRILKLI  350
ZNTA-PM   301 ESALTGESVP VERLQGEKVA AGCLSVDRSV QMKVVSEQGQ NAIDRILQLI  350
CADA      301 QAAITGESVP VSKAVDDEVF AGTLNEEGLI EVKITKYVED TTITKIIHLV  350
                     360        370        380        390        400
ZNTA-EC   351 EEAEERRAPI ERFIDRFSRI YTPAIMAVAL LVTLVPPLLF AASWQEWIYK  400
ZNTA-PM   351 EEAEERKAPI ERFVDRFSRY YTPLIMLFSA LVIIIPPLLF AQPWETWFYR  400
CADA      351 EEAQGERAPA QAFVDKFAKY YTPIIMVIAA LVAVVPPLFF GGSWDTWVYQ  400
                     410        420        430        440        450
ZNTA-EC   401 GLTLLLIGCP CALVISTPAA ITSGLAAAAR RGALIKGGAA LEQLGRVTQV  450
ZNTA-PM   401 GLTLLLIGCP CALVISTPAA ITSALAAATK RGALIKGGAA LEQLGTVNTV  450
CADA      401 GLAVLVVGCP CALVISTPIS IVSAIGNAAK KGVLVKGGVY LEKLGAIKTV  450
                     460        470        480        490        500
ZNTA-EC   451 AFDKTGTLTV GKPRVTAIHP -ATGISESEL LTLAAAVEQG ATHPLAQAIV  500
ZNTA-PM   451 ALDKTGTLTE GKPQVTDVIA -NVGFNEKEL LMLASSVEVG SHHPLAKAII  500
CADA      451 AFDKTGTLTK GVPVVTDFEV LNDQVEEKEL FSIITALEYR SQHPLASAIM  500
                     510        520        530        540        550
ZNTA-EC   501 REAQVAELAI --PTAESQRA LVGSGIEAQV NGERVLICAA GKHP------  550
ZNTA-PM   501 NKAQEQQIDV --VEADNRKA LAGKGIEGYL NNQHILVSAP TQLSE--TIP  550
CADA      501 KKAEQDNIPY SNVQVEEFTS ITGRGIKGIV NGTTYYIGSP KLFKELNVSD  550
                     560        570        580        590        600
ZNTA-EC   551 -ADAFTGLIN ELESAGQTVV LVVRNDDVLG VIALQDTLRA DAATAISE--  600
ZNTA-PM   551 LSAQWQQQVA RLEDEGKTVV VVLKEDQFIG VIAMQDTLRN DAIESMKEKK  600
CADA      551 FSLGFENNVK ILQNQGKTAM IIGTEKTILG VIAVADEVRE TSKNVIQK--  600
                     610        620        630        640        650
ZNTA-EC   601 -LNALGVK-G VILTGDNPRA AAAIAGELGL -EFKAGLLPE DKVKAVTEL-  650
ZNTA-PM   601 VLKSMNIN-A VMLTGDNPRA AAAIAQKLGM -DFRAGLLPE DKVTSVMEI-  650
CADA      601 -LHQLGIKQT IMLTGDNQGT ANAIGTHVGV SDIQSELMPQ DKLDYIKKMQ  650
                     660        670        680        690        700
ZNTA-EC   651 NQHAPLAMVG DGINDAPAMK AAAIGIAMGS -GTDVALETA DAALTHNHLR  700
ZNTA-PM   651 SQTHNTMMVG DGINDAPAMK AATIGVAMGS -GTDVALETA DAALTHNRLT  700
CADA      651 SEYDNVAMIG DGVNDAPALA ASTVGIAMGG AGTDTAIETA DIALMGDDLS  700
                     710        720        730        740        750
ZNTA-EC   701 GLVQMIELAR ATHANIRQNI TIALGLKGIF LVTTLLGMTG LWLAVLADTG  750
ZNTA-PM   701 GLPEIIQLSR ATRKIIRENI TIALGLKAIF LVTSLFGITG LWVAVLADSG  750
CADA      701 KLPFAVRLSR KTLNIIKANI TFAIGIKIIA LLLVIPGWLT LWIAILSDMG  750
                     760        770        780        790        800
ZNTA-EC   751 ATVLVTANAL RLLRRR---- ---....... .......... ..........  800
ZNTA-PM   751 ATALVTANAV RLLKVKLKSP HQD....... .......... ..........  800
CADA      751 ATILVALNSL RLMRVK---- -DK....... .......... ..........  800
```

Figure 5.2 Sequence alignment of the ZntA and CadA ATPases. The predicted amino acid sequences from *E. coli* (ZNTA-EC) (Beard *et al.*, 1997; Rensing *et al.*, 1997b), *P. mirabilis* (ZNTA-PM) (Lai *et al.*, 1998; Rensing *et al.*, 1998) and *S. aureus* plasmid pI258 (CADA) (Nucifora *et al.*, 1989) are aligned with identical residues in reverse color.

only when cadmium was present). On the other hand, the regulation of chromosomal *cadA* genes may be different, and the chromosomally encoded CadA pumps may be the zinc exporters of Gram-positive bacteria.

In *Alcaligenes eutrophus* resistance to zinc and cadmium is catalyzed by the gene products of the cobalt–zinc–cadmium (*czc*) resistance operon. The CzcBCA proteins form a membrane-bound complex that catalyzes energy-dependent efflux of Co(II), Zn(II) and Cd(II) (Nies and Silver, 1989b; Nies *et al.*, 1989; Nies, 1995). The data suggest a stoichiometry of 1 : 2 : 1 for CzcA, B and C. It appears that CzcA alone is able to transport zinc and cobalt across the cytoplasmic membrane by acting as a cation/proton antiporter. Expression of only *czcA* confers low-level resistance to zinc and cobalt, suggesting that CzcA may be the catalytic subunit of the exporter (Rensing *et al.*, 1997c). CzcA belongs to the family of RND proteins, with the closest homologs being two cation-RND proteins, CnrA and NccA (Saier *et al.*, 1994). While CzcA is a cytoplasmic membrane protein, CzcB and CzcC are probably located in the periplasm, as judged by the cellular location of the products of translational fusions (Rensing *et al.*, 1997c). CzcB is a member of the membrane fusion protein (MFP) family (Dinh *et al.*, 1994). MFP proteins are hypothesized to span the space between the inner and outer membranes, bringing the outer membrane closer to the central transport protein, as has been proposed for HlyD, a component of the hemolysin ABC transporter HlyB (Pimenta *et al.*, 1996). CzcC belongs to a family of outer membrane factor (OMF) proteins that have recently been suggested to reside in the outer membrane (Paulsen *et al.*, 1997). CzcC is transported across the cytoplasmic membrane, where its leader peptide is processed (Rensing *et al.*, 1997c). Even though *czcC::phoA* translational fusions indicated a periplasmic location, an outer membrane location is more likely; the gene fusion may prevent transfer to its final location. These results suggest a model in which the CzcABC complex spans the inner and outer membranes, with CzcA catalyzing the cation/proton exchange across the inner membrane, CzcC facilitating cation movement across the outer membrane, and CzcB linking the two.

5.3 Copper and silver

5.3.1 Uptake systems

Copper is an essential metal ion. However, cells need to keep the intracellular concentration of free copper ion low, since the free metal ion can facilitate oxidative damage. In bacteria, energy-dependent copper uptake by a specific copper transporter has not been directly demonstrated. The best-studied systems are the copper ATPases of *Enterococcus hirae*, CopA and CopB. While CopB has been shown to be a copper/silver ion efflux pump, CopA is thought to be involved in copper uptake. Cells with a disruption of the *copA* gene are copper-requiring and more silver resistant than the wild type (Odermatt *et al.*, 1993). A close homolog of CopA is CtaA from *Synechococcus* 7942; disruption of the *ctaA* gene rendered the cells more resistant to copper (Phung *et al.*, 1994). Another CopA homolog is PacS from *Synechococcus* 7942. Disruption of *pacS* rendered the cells more sensitive to copper and silver salts. PacS is apparently localized in the thylakoid membrane and is specifically induced by copper and silver (Kanamaru *et al.*, 1994). Eucaryotes possess similar copper ATPases, examples of which are CCC2 in *S. cerevisiae* (Fu *et al.*, 1995) and the proteins related to Menkes (Vulpe *et al.*, 1993) and Wilson's disease (Bull *et al.*, 1993) in humans. These proteins have been shown to be involved in intracellular copper trafficking in all organisms examined, and putative homologs have been found in all sequenced genomes.

Although it is difficult to verify physiologically, the substrate of copper efflux P-type ATPases is probably Cu(I). While in an oxidizing atmosphere Cu(I) rapidly oxidizes to Cu(II), the cytosol of cells is highly reducing, with high concentrations of reduced glutathione or other thiols, so that Cu(I) should be present intracellularly. On the other hand, copper uptake systems such as CopA would be exposed primarily to extracellular Cu(II), so it may be that such systems recognize the oxidized metal. Recently Lutsenko and co-workers cloned the N-terminal sequences from the Menkes and Wilson's disease related proteins and demonstrated that the peptides bound 5–6 nmol of copper/nmol of protein when expressed in *E. coli* (Lutsenko *et al.*, 1997). The copper released from polypeptides reacted with bicinchoninic acid, which reacts with Cu(I) but not Cu(II). These results support, but do not prove, the hypothesis that the reduced metal ion is the pump substrate.

In a parallel study, the N-terminal domain from Wilson's disease related protein has been expressed and purified as a fusion to glutathione *S*-transferase (DiDonato *et al.*, 1997). The fusion protein was able to bind to metal chelate columns charged with various transition metals. The order of binding affinity was Cu(II) >> Zn(II) > Ni(II) > Co(II), with no binding to Fe(II) or Fe(III). Cu(I) binding was also detected by neutron activation. Thus, it is still not clear whether these proteins recognize both oxidized and reduced forms of copper ion.

By multidimensional nuclear magnetic resonance the solution structures of a 72-residue polypeptide corresponding to the fourth metal-binding domain of the Menkes disease related ATPase was recently determined (Gitschier *et al.*, 1998). In the absence of metal, the region containing GMTCXXC, which is a putative metal-binding sequence, was disordered. Binding of Ag(I) induced order to that domain. Ag(I) was bound in a linear bicoordinate manner to the two cysteine residues. It is reasonable to assume that Cu(I) would bind similarly, again suggesting that both Cu(I) and Ag(I) may be substrates of copper pumps.

5.3.2 Efflux systems

The only copper efflux pump that has actually been shown to transport copper is CopB from *Enterococcus hirae*. CopB differs from CopA and its homologs primarily in the N-terminal metal-binding domain. CopA and homologs have one or more N-terminal CXXC metal-binding motifs. In contrast, CopB has a histidine-rich repeat at the N terminus. Disruption of *copB* rendered the cells hypersensitive to copper and silver salts. *In vitro* CopB has been shown to catalyze accumulation of both ^{110}Ag(I) and ^{64}Cu(I) in membrane vesicles prepared from *E. hirae*. These vesicles are presumably everted (inside-out), so uptake corresponds to extrusion from intact cells (Solioz and Odermatt, 1995). In *E. coli*, a homolog of the *E. hirae* CopA termed ORF834 has been identified (Blattner *et al.*, 1997). This *E. coli* homolog, renamed CopA, has been shown to be a Cu(I)-translocating P-type ATPase (Rensing *et al.*, 1999). It is an efflux pump that confers copper ion resistance in *E. coli*.

Another resistance mechanism for copper apparently involves intracellular sequestration. Examples include the copper-resistance determinant (*pco*) from *E. coli* plasmid pRJ1004 and the *cop* resistance determinant from *Pseudomonas syringae* pv. *tomatoe* (Mills *et al.*, 1993; Brown *et al.*, 1995). These two systems are essentially equivalent to each other. The *Pseudomonas cop* operon consists of four structural genes, *copABCD*, controlled by a copper-inducible promoter, most likely under the control of three regulatory genes, *copR* and *copS*, the products of which are homologs to PcoR and PcoS in sequence, plus a third chromosomal gene (Nies and Brown, 1998). The periplasmic *copA* gene product binds 11 Cu(II), while the periplasmic *copC* gene product binds a single Cu(II) ion. The *copB* gene encodes an outer membrane protein, and the *copD* gene encodes an inner membrane protein. The *copA* and *copB* genes are essential to resistance, whereas the *copC* and *copD* genes are only necessary for maximal resistance (Mellano and Cooksey, 1988). The *Pseudomonas cop* system produces bioaccumulation of copper by the bacterial cells (up to 0.3% dry weight) (Cooksey, 1993), and, on copper-supplemented solid media, the bacterial colonies become bright blue. How the outer and inner membrane proteins function is still not understood.

However, the analysis of copper resistance with these systems is complicated by the potential presence of copper efflux pumps such as the *E. coli* CopA gene product (Rensing *et al.*, 1999). For example, if those indeed both encode copper extrusion pumps, then a search for copper hypersensitive mutants might be difficult, requiring simultaneous mutations in both genes. Thus, the plasmid-encoded *pco* determinant of *E. coli* may represent an additional resistance by periplasmic copper sequestration in addition to intrinsic copper efflux. The same might be true for the *cop* determinant from *Pseudomonas syringae*. Indeed, the *cop* resistance determinant of *A. eutrophus* CH34 not only contains genes homologous to *pco* but also contains a gene for a P-type ATPase that is a homolog of PacS and CopA (Nies and Brown, 1998).

Although unrelated to transport systems, genes have been identified in searches for mutations that produce copper-sensitive *E. coli* (Rouch *et al.*, 1989). Some have pleiotropic effects (Gupta *et al.*, 1995). Others are in genes involved in disulfide bond formation. Disruption of *cutA* (*dsbD*) or *dsbC* (Rietsch *et al.*, 1996) makes cells more copper sensitive, perhaps because copper prevents proper folding of essential proteins in those strains. Similarly, disruption of either *dsbA* (Rensing *et al.*, 1997a) and *dsbB* (C. Rensing and B.P. Rosen, unpublished) makes cells more sensitive to cadmium and zinc, perhaps by preventing proper folding of periplasmic proteins.

5.4 Nickel and cobalt

5.4.1 Uptake systems

Both high-affinity and low-affinity bacterial uptake systems for nickel and cobalt have been identified. In *E.coli* nickel is transported at low affinity through the relatively nonspecific magnesium carrier, CorA (Snavely *et al.*, 1989). In addition, a specific nickel transport (Nik) system with high affinity has been reported (Navarro *et al.*, 1993). The Nik transporter is encoded by an operon consisting of five genes. From the deduced amino acid sequences of the genes, the components of the Nik transporter are highly homologous to those of the periplasmic binding protein systems for oligo- and dipeptides and very likely constitute an ATP-binding cassette transport ATPase. Other bacteria use single-polypeptide high-affinity transporters such as HoxN for nickel in *Alcaligenes eutrophus* or NhlF for cobalt in *Rhodococcus rhodochrous*. HoxN is an integral membrane protein that potentially has eight transmembrane helices (Eitinger *et al.*, 1997). A conserved motif (HXXXXDH) is required for high-affinity transport by HoxN and may represent a nickel-binding motif. This motif is also found in NikC, an integral membrane component of the *E. coli* Nik system (Wolfram *et al.*, 1995). However, the cobalt transporter NhlF also contains this motif, and, while cobalt transport by NhlF is inhibited by nickel, nickel transport by HoxN is only slightly inhibited by cobalt (Komeda *et al.*, 1997).

5.4.2 *Efflux systems*

Only two systems have been shown to mediate cobalt and nickel efflux: the cobalt and nickel resistance determinant *cnr* from *Alcaligenes eutrophus* CH34 (Liesegang *et al.*, 1993) and the nickel–cobalt–cadmium resistance determinant *ncc* from *Alcaligenes xylooxidans* (Schmidt and Schlegel, 1994). The proposed genes for the efflux system *cnrCBA* and *nccCBA* are largely homologous to the genes for the three subunits of the much better characterized CzcCBA cation–proton antiporter (Nies *et al.*, 1989; Dong and Mergeay, 1994; Rensing *et al.*, 1997c) and probably have a similar function.

5.5 Arsenic and antimony

5.5.1 *Uptake systems*

Uptake of arsenate, which contains the oxidized form (As(V)) of the semimetal, has been shown to occur by phosphate transport systems. In *Escherichia coli* there are two phosphate transport systems, the high affinity–low velocity Pst (phosphate specific transport) system and the low affinity–high velocity Pit (phosphate transport) system (Willsky and Malamy, 1980), but arsenate is accumulated primarily by the Pit system (Rosenberg *et al.*, 1977) and possibly by the GlpT glycerol-3-phosphate transport system (Elvin *et al.*, 1985).

The routes of entry of arsenite and antimonite, the reduced forms of the metalloids (As(III) and Sb(III)), have not been determined. In an attempt to identify the cellular transporters for arsenite and antimonite, a genetic approach was used (Sanders *et al.*, 1997). As described below, *E. coli* is intrinsically resistant to those metalloid oxyanions because of a chromosomal arsenic resistance (*ars*) operon. A strain of *E. coli* in which the chromosomal *ars* genes were deleted (Carlin *et al.*, 1995) was subjected to random Tn*phoA* mutagenesis, and metalloid-resistant mutants selected. The rationale was that loss of an uptake system could provide resistance. A single mutant resistant to antimonite was isolated, and the transposon was found to be in the *glpF* gene that encodes a polyol transporter (Sanders *et al.*, 1997). Since a single mutation produced resistance, the results suggest that the major route of entry of antimonite into cells of *E. coli* is via the GlpF polyol facilitator. Why should an oxyanion of a metalloid be transported by a protein that recognizes organic alcohols? At neutral pH the major form derived from antimonous acid is the neutral monomeric species $Sb(OH)_3$. Perhaps $Sb(OH)_3$ is recognized by GlpF as the inorganic equivalent of a polyol. However, the disrupted strain was still sensitive to arsenite. It is possible that arsenite is not accumulated by GlpF. A more likely possibility is that it is transported both by

GlpF and another transport system, so that a single mutation would not produce resistance.

5.5.2 *Efflux systems*

Arsenic resistance (*ars*) operons have been identified in many bacteria (Fig. 5.3a) (Xu *et al.*, 1998). In most there are at least three genes, *arsR*, which encodes an As(III)/Sb(III) regulatory protein (Wu and Rosen, 1991), *arsB*, which encodes an As(III)/Sb(III) efflux

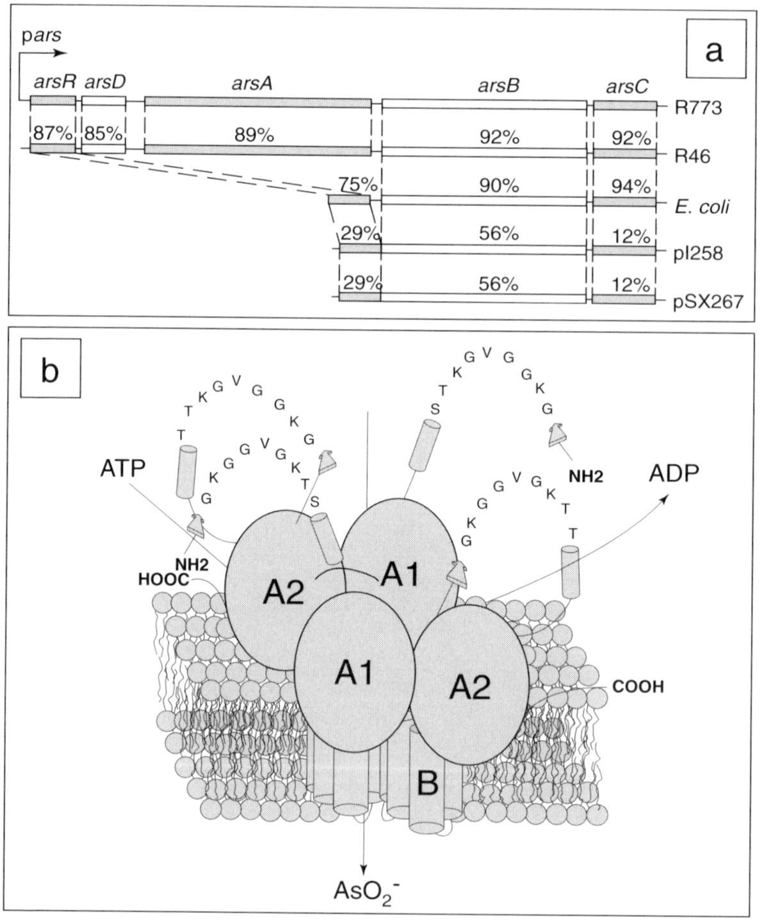

Figure 5.3 Bacterial *ars* operons. (a) In the top line the five genes of the *ars* operon of plasmid R773 are shown, with the direction of transcription indicated by the arrow, starting with the promoter, p_{ars}. Genes are indicated by boxes, with the intergenic spaces as single lines. The genes of homologous *ars* operons of plasmid R46 (Bruhn *et al.*, 1996), the *E. coli* chromosomal operon (Carlin *et al.*, 1995), and staphylococcal plasmids pI258 (Ji and Silver, 1992b) and pSX267 (Rosenstein *et al.*, 1992) are aligned below. The similarities of the gene products to the R773 proteins are given as percentage identity. (b) The products of the *arsA* and *arsB* form a pump that extrudes arsenite or antimonite from the cells, affording metalloid resistance. The pump is postulated to have one ArsB subunit and two ArsA subunits (hence four consensus nucleotide binding sites, two Al and two A2 sites).

protein (Kuroda *et al.*, 1997), and *arsC*, which codes for an arsenate reductase that reduces arsenate (As(V)) to arsenite (As(III)) (Ji and Silver, 1992a; Oden *et al.*, 1994), the substrate of the transporter. In some *ars* operons there are two additional genes, *arsD*, which codes for a second transcriptional repressor (Wu and Rosen, 1993), and *arsA*, the gene for an ATPase that associates with ArsB to form an ATP-driven arsenite (antimonite) extrusion pump (Rosen *et al.*, 1988) (Fig. 5.3b).

The best studied ArsAB pump is that encoded by the *ars* operon of *E. coli* plasmid R773 (Dey *et al.*, 1994a, b). The 583-residue ArsA ATPase has two homologous halves, the N-terminal A1 half and the C-terminal A2 half, each of which contains an ATP-binding motif (Fig. 5.3b) (Chen *et al.*, 1986). ArsA can be purified as a soluble enzyme when *arsA* is expressed in the absence of *arsB*, allowing for examination of its As(III)/Sb(III)-stimulated ATPase activity (Hsu and Rosen, 1989). Mutation in the codons for residues in either nucleotide-binding site led to inactivation of ATPase activity and loss of arsenite resistance and inactivated the enzyme, demonstrating that both sites are required for catalysis (Karkaria *et al.*, 1990; Kaur and Rosen, 1992). From the results of genetic analyses it appears that the A1 and A2 ATP-binding sites interact. When expressed from compatible plasmids, *arsA* genes with single mutations in each site exhibited genetic complementation, suggesting that ArsAs with a defective A1 and a defective A2 site form an active ATPase, even though either alone was inactive (Kaur and Rosen, 1993). This idea was confirmed by biochemical reconstitution of an active ATPase from inactive subunits (Kaur and Rosen, 1994). Recently second-site suppressors of a mutation in the A1 nucleotide-binding site were found in the A2 site (Li *et al.*, 1996; Li and Rosen, 1998). Such intragenic suppression supports a model of interacting A1 and A2 ATP-binding sites. This has led to the hypothesis that ArsA is a dimer with two active catalytic sites, each formed by a pair of interacting A1 and A2 ATP-binding sites (Fig. 5.4a).

As soft metals As(III) and Sb(III) inhibit enzymes by binding to two or three thiol groups. In contrast, ArsA is allosterically activated by

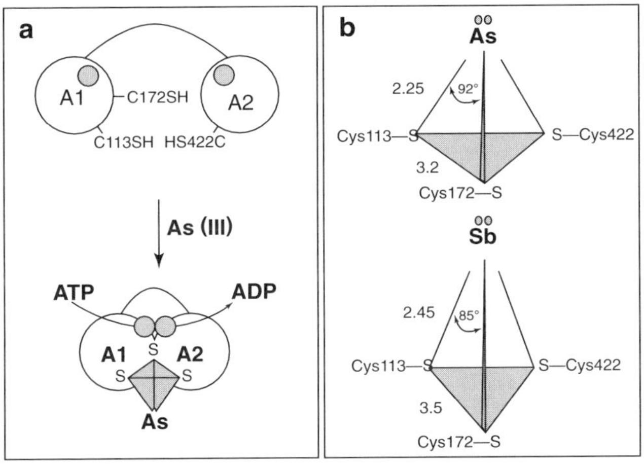

Figure 5.4 Allosteric activation of the ArsA ATPase. (a) ArsA ATPase activity is allosterically regulated by binding of the As(III) or Sb(III) at the allosteric site. This produces a conformational change that brings the A1 and A2 halves of the protein into contact with each other, forming an interface between the two ATP-binding sites. This increases the affinity of the sites for nucleotide, accelerating catalysis. (b) The geometry of the As(III)/Sb(III) binding domain in ArsA is composed of Cys113, Cys172 and Cys422. These residues are located distant from each other in the primary sequence of the protein, and formation of the soft-metal site requires folding of the protein to bring those residues into spatial proximity of each other. The bond angles and distances are postulated from crystallographic analysis of small molecules containing As–S or Sb–S bonds.

As(III) or Sb(III) by coordination of the metal with three residues, Cys113, Cys172 and Cys422, in a novel triganol three-coordinate structure (Fig. 5.4b) (Bhattacharjee *et al.*, 1995; Bhattacharjee and Rosen, 1996). ArsA is the only enzyme ever shown to require arsenic or antimony for activity. Recently the sequence DTAPTGHTIRLL has been identified as conserved in ArsA homologs from eubacteria, archaebacteria, fungi, plants and animals (Fig. 5.5). Although the physiological function of most of these homologs is not known, conservation of this DTAP motif suggests that it has a conserved function. Intrinsic tryptophan fluorescence was used to examine a potential role of the DTAP domain in ATP hydrolysis (Zhou *et al.*, 1995; Zhou and Rosen, 1997). All of the tryptophan codons in *arsA* were changed to other codons, and then the genes for two single tryptophan-containing ArsAs were constructed. One encoded a protein with a single tryptophan residue located at position 141, the N-terminal side of the DTAP

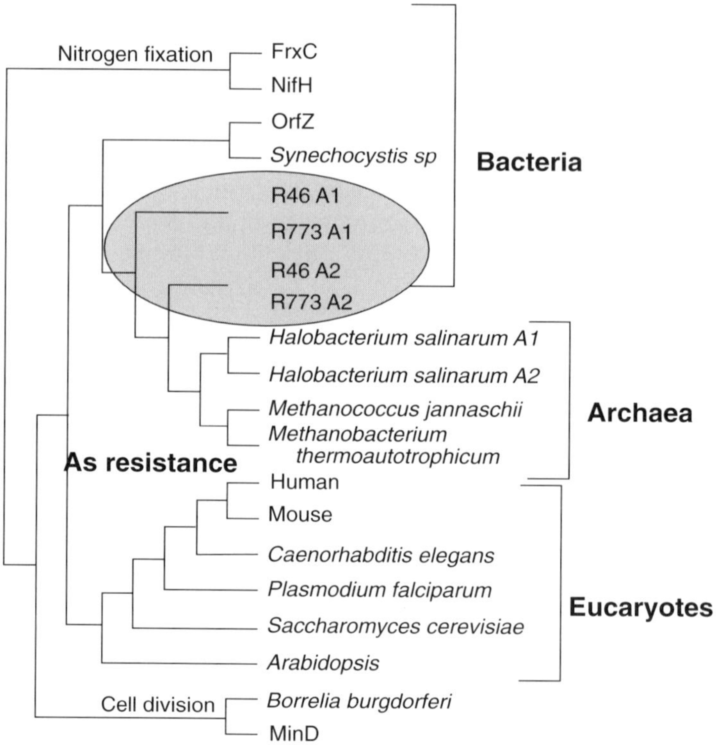

Figure 5.5 The ArsA family. The phylogenetic relationship of the amino-acid sequences of ArsA homologs of procaryotes, archaebacteria and eucaryotes are shown. ArsA homologs: distant relatives of ArsA include proteins involved in nitrogen fixation (Jones *et al.*, 1993) and cell division (de Boer *et al.*, 1991). Homologs proposed to be involved in resistance to As(III)/Sb(III) include the proteins encoded by plasmids R773 and R46, which are the result of gene duplication and fusion, producing A1 and A2 halves, each of which has a consensus nucleotide-binding sequence. The protein from the archaebacterium *H. salinarum* has the same duplicated domain structure. All of the other ArsA homologs have single domain structures with a single nucleotide-binding site and are approximately half the size of the R773 enzyme. Of these single-domain proteins, only the human homolog has been implicated in As(III) resistance (Kurdi-Haidar *et al.*, 1996).

domain. The second engineered gene coded for an ArsA with a single tryptophan residue at position 159, the C-terminal side of the DTAP sequence. These two tryptophans served as intrinsic probes of conformation of the enzyme. The results demonstrated that during ATP hydrolysis the C-terminal end of the DTAP domain moves into a less solvent-exposed environment. With the formation of product, the N-terminal end becomes more accessible to solvent. Thus, this conserved region of ArsA undergoes rotational changes in conformation during the catalytic cycle. The DTAP domain may be involved in transduction of information from binding of metalloid at the allosteric domain to the catalytic domain.

The ArsA–ArsB complex is an obligatorily ATP-coupled primary pump (Fig. 5.6a) (Mobley and Rosen, 1982; Dey *et al.*, 1994a; Dey and Rosen, 1995b). In cells expressing both the *arsA* and *arsB* genes chemical energy was shown to be both necessary and sufficient for extrusion of arsenicals, while electrochemical energy was neither. An *unc* strain of *E. coli* lacking the H^+-translocating ATPase that catalyzes the equilibrium between ATP and the electrochemical proton gradient was used. When starved of endogenous energy reserves, this strain will produce only chemical energy when given glucose in the presence of an inhibitor of respiration such as cyanide. On the other hand, it will generate only electrochemical energy when given a respiratory substrate such as succinate. This allows the establishment of intracellular conditions where either chemical energy or electrochemical energy is available for transport. In such experiments arsenical extrusion was conclusively demonstrated to be linked to chemical energy. While there was a temporal relationship of extrusion with ATP levels, the identity of the compound used as the energy source could not be determined from *in vivo* experiments. However, direct coupling of ATP to $^{73}AsO_2^-$ transport was demonstrated *in vitro* using everted (inside-out) membrane vesicles. Everted membrane vesicles have an orientation opposite to that in intact cells, so, while cells extrude arsenite, vesicles accumulate it. The vesicles were made from *E. coli* cells defective in the *unc* genes, so that coupling of the ArsA–ArsB complex to ATP hydrolysis was direct. Thus the R773 ArsA–ArsB complex

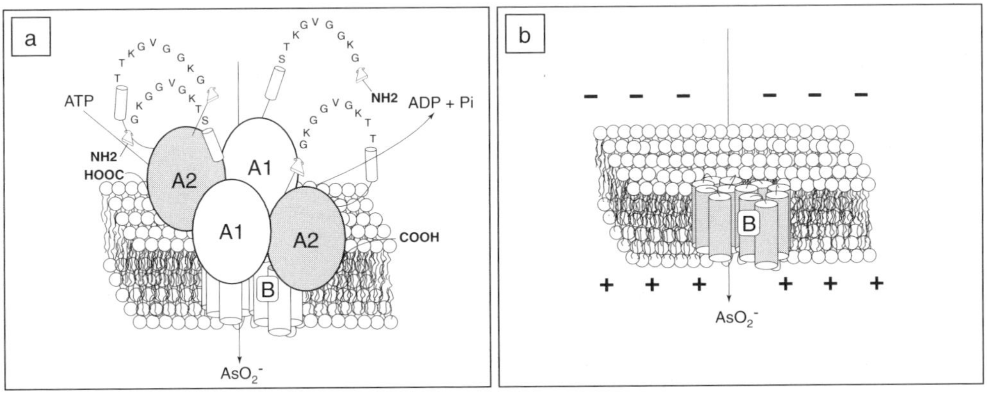

Figure 5.6 Dual energy coupling mechanisms of the Ars oxyanion-translocating system. (a) The ArsA–ArsB complex is an obligatorily ATP-coupled pump for extrusion of oxyanions of As(III) or Sb(III). ArsA is the catalytic subunit, and ArsB is the membrane anchor for ArsA and the anion conducting subunit. (b) ArsB alone functions as a $\Delta\psi$-driven carrier protein in the absence of ArsA, with energy derived from the proton pumping respiratory chain or F_0F_1 ATPase. In contrast, the ArsA–ArsB complex is unable to utilize $\Delta\psi$.

is clearly an obligatorily ATP-coupled pump. ATP could not be replaced with other nucleoside triphosphates nor by the nonhydrolyzable ATP analog, ATPγS. ATP-dependent transport by the ArsA–ArsB complex was insensitive to uncouplers. Transport in vesicles prepared from *unc* strains could not be driven by oxidation of lactate or NADH, demonstrating that electrochemical energy was neither necessary nor sufficient for the Ars pump. As with other transport ATPases, arsenite transport has an absolute requirement for a divalent cation, specifically Mg^{2+} or Mn^{2+}. Neither vanadate, an inhibitor of P-type ATPases, nor azide, an inhibitor of F-type ATPases, inhibited arsenite accumulation.

While the ArsA–ArsB complex is clearly an ATP-coupled pump, ArsB catalyzes arsenite transport coupled to the membrane potential, perhaps as an electrophoretic uniporter (Fig. 5.6b) (Dey and Rosen, 1995a; Kuroda *et al.*, 1997). Thus *arsRBC* operons confer lower resistance to metalloids than *arsRDABC* operons because secondary systems catalyze the formation of smaller concentration gradients than pumps. Since the membrane potential is the driving force, the transported species is most likely an anion such as arsenite rather than the soft-metal form As(III). Consistent with this idea, there are no critical cysteine residues in ArsB. This arsenite carrier has a transmembrane structure with 12 transmembrane spanning sequences, topologically similar to most secondary carriers (Chen *et al.*, 1996). The ability of a single protein to function either as a secondary carrier or subunit of a pump is quite novel and may reflect an evolutionary adaptation to allow for ability to cope with high levels of metalloids (Rosen *et al.*, 1992).

References

Bartsevich, V.V. and H.B. Pakrasi, 1995. Molecular identification of an ABC transporter complex for manganese: analysis of a cyanobacterial mutant strain impaired in the photosynthetic oxygen evolution process. *EMBO Journal* **14**, 1845–1853.

Beard, S.J., R. Hashim, J. Membrillo-Hernandez, M.N. Hughes and R.K. Poole, 1997. Zinc(II) tolerance in *Escherichia coli* K-12: evidence that the *zntA* gene (o732) encodes a cation transport ATPase. *Molecular Microbiology* **25**, 883–891.

Bhattacharjee, H. and B.P. Rosen, 1996. Spatial proximity of Cys113, Cys172, and Cys422 in the metalloactivation domain of the ArsA ATPase. *Journal of Biological Chemistry* **271**, 24465–24470.

Bhattacharjee, H., J. Li, M.Y. Ksenzenko and B.P. Rosen, 1995. Role of cysteinyl residues in metalloactivation of the oxyanion-translocating ArsA ATPase. *Journal of Biological Chemistry* **270**, 11245–11250.

Blattner, F.R., G. Plunkett 3rd, C.A. Bloch, N.T. Perna, V. Burland, M. Riley, J. Collado-Vides, J.D. Glasner, C.K. Rode, G.F. Mayhew, J. Gregor, N.W. Davis, H.A. Kirkpatrick, M.A. Goeden, D.J. Rose, B. Mau and Y. Shao, 1997. The complete genome sequence of *Escherichia coli* K-12. *Science* **277**, 1453–1474.

Brown, N.L., S.R. Barrett, J. Camakaris, B.T. Lee and D.A. Rouch, 1995. Molecular genetics and transport analysis of the copper-resistance determinant (*pco*) from *Escherichia coli* plasmid pRJ1004. *Molecular Microbiology* **17**, 1153–1166.

Bruhn, D.F., J. Li, S. Silver, F. Roberto and B.P. Rosen, 1996. The arsenical resistance operon of IncN plasmid R46. *FEMS Microbiology Letters* **139**, 149–153.

Bucheder, F. and E. Broda, 1974. Energy-dependent zinc transport by *Escherichia coli. European Journal of Biochemistry* **45**, 555-559.

Bull, P.C., G.R. Thomas, J.M. Rommens, J.R. Forbes and D.W. Cox, 1993. The Wilson disease gene is a putative copper transporting P-type ATPase similar to the Menkes gene. *Nature Genetics* **5**, 327–337.

Carlin, A., W. Shi, S. Dey and B.P. Rosen, 1995. The *ars* operon of *Escherichia coli* confers arsenical and antimonial resistance. *Journal of Bacteriology* **177**, 981–986.

Chen, C.M., T.K. Misra, S. Silver and B.P. Rosen, 1986. Nucleotide sequence of the structural genes for an anion pump. The plasmid-encoded arsenical resistance operon. *Journal of Biological Chemistry* **261**, 15030–15038.

Chen, Y., S. Dey and B.P. Rosen, 1996. Soft metal thiol chemistry is not involved in the transport of arsenite by the Ars pump. *Journal of Bacteriology* **178**, 911–913.

Chipley, J.R. and H.M. Edwards Jr, 1972. Cationic uptake and exchange in *Salmonella enteritidis*. *Canadian Journal of Microbiology* **18**, 509–514.

Cooksey, D.A. 1993. Copper uptake and resistance in bacteria. *Molecular Microbiology* **7**, 1–5.

Corbisier, P., G. Ji, G. Nuyts, M. Mergeay and S. Silver, 1993. luxAb gene fusions with the arsenic and cadmium resistance operons of *Staphylococcus aureus* plasmid pI258. *FEMS Microbiology Letters* **110**, 231–238.

Datta, N. and V.M. Hughes, 1983. Plasmids of the same Inc groups in Enterobacteria before and after the medical use of antibiotics. *Nature* **306**, 616–617.

de Boer, P.A., R.E. Crossley, A.R. Hand and L.I. Rothfield, 1991. The MinD protein is a membrane ATPase required for the correct placement of the *Escherichia coli* division site. *EMBO Journal* **10**, 4371–4380.

Dey, S. and B.P. Rosen, 1995a. Dual mode of energy coupling by the oxyanion-translocating ArsB protein. *Journal of Bacteriology* **177**, 385–389.

Dey, S. and B.P. Rosen, 1995b. Mechanisms of drug transport in procaryotes and eucaryotes, in *Drug Transport in Antimicrobial and Anticancer Chemotherapy*, (ed. N. H. Georgopapadakou) Marcel Dekker, New York, pp. 103–132.

Dey, S. and Rosen, B.P. 1996. Heavy metal transporters, in *Transport processes in membrane*, (ed. W. N. K. Konings and H.R. Kaback), Elsevier, Amsterdam, Vol. II, pp. 1–19.

Dey, S., D. Dou and B.P. Rosen, 1994a. ATP-dependent arsenite transport in everted membrane vesicles of *Escherichia coli*. *Journal of Biological Chemistry* **269**, 25442–25446.

Dey, S., D. Dou, L.S. Tisa and B.P. Rosen, 1994b. Interaction of the catalytic and the membrane subunits of an oxyanion-translocating ATPase. *Archives of Biochemistry and Biophysics* **311**, 418–424.

DiDonato, M., S. Narindrasorasak, J.R. Forbes, D.W. Cox and B. Sarkar, 1997. Expression, purification, and metal binding properties of the Nterminal domain from the wilson disease putative copper-transporting ATPase (ATP7B). *Journal of Biological Chemistry* **272**, 33279–33282.

Dinh, T., I.T. Paulsen and M.H. Saier Jr, 1994. A family of extracytoplasmic proteins that allow transport of large molecules across the outer membranes of gram-negative bacteria. *Journal of Bacteriology* **176**, 3825–3831.

Dintilhac, A., G. Alloing, C. Granadel and J.P. Claverys, 1997. Competence and virulence of *Streptococcus pneumoniae*: Adc and PsaA mutants exhibit a requirement for Zn and Mn resulting from inactivation of putative ABC metal permeases. *Molecular Microbiology* **25**, 727–739.

Dong, Q. and M. Mergeay, 1994. Czc/Cnr efflux: a three-component chemiosmotic antiport pathway with a 12-transmembrane-helix protein. *Molecular Microbiology* **14**, 185–187.

Eitinger, T., L. Wolfram, O. Degen and C. Anthon, 1997. A Ni^{2+} binding motif is the basis of high affinity transport of the *Alcaligenes eutrophus* nickel permease. *Journal of Biological Chemistry* **272**, 17139–17144.

Elvin, C.M., C.M. Hardy and H. Rosenberg, 1985. Pi exchange mediated by the GlpT-dependent sn-glycerol-3-phosphate transport system in *Escherichia coli*. *Journal of Bacteriology* **161**, 1054–1058.

Foster, T.J., H. Nakahara, A.A. Weiss and S. Silver, 1979. Transposon A-generated mutations in the mercuric resistance genes of plasmid R100-1. *Journal of Bacteriology* **140**, 167–181.

Fu, D., T.J. Beeler and T.M. Dunn, 1995. Sequence, mapping and disruption of CCC2, a gene that cross-complements the Ca(2+)-sensitive phenotype of *csg1* mutants and encodes a P-type ATPase belonging to the Cu(2+)-ATPase subfamily. *Yeast* **11**, 283–292.

Gitschier, J., B. Moffat, D. Reilly, W.I. Wood and W.J. Fairbrother, 1998. Solution structure of the fourth metal-binding domain from the Menkes copper-transporting ATPase. *Nature Structural Biology* **5**, 47–54.

Gupta, S.D., B.T. Lee, J. Camakaris and H.C. Wu, 1995. Identification of cutC and cutF (nlpE) genes involved in copper tolerance in *Escherichia coli. Journal of Bacteriology* **177**, 4207–4215.

Hsu, C.M. and B.P. Rosen, 1989. Characterization of the catalytic subunit of an anion pump. *Journal of Biological Chemistry* **264**, 17349–17354.

Ivey, D.M., A.A. Guffanti, Z. Shen, N. Kudyan and T.A. Krulwich, 1992. The *cadC* gene product of alkaliphilic *Bacillus firmus* OF4 partially restores Na^+ resistance to an *Escherichia coli* strain lacking an Na^+/H^+ antiporter (NhaA). *Journal of Bacteriology* **174**, 4878–4884.

Ji, G. and S. Silver, 1992a. Reduction of arsenate to arsenite by the ArsC protein of the arsenic resistance operon of *Staphylococcus aureus* plasmid pI258. *Proceedings of the National Academy of Sciences (USA)* **89**, 9474–9478.

Ji, G. and S. Silver, 1992b. Regulation and expression of the arsenic resistance operon from *Staphylococcus aureus* plasmid pI258. *Journal of Bacteriology* **174**, 3684–3694.

Jones, R., P. Woodley, A. Birkmann-Zinoni and R.L. Robson, 1993. The *nifH* gene encoding the Fe protein component of the molybdenum nitrogenase from *Azotobacter chroococcum. Gene* **123**, 145–146.

Kanamaru, K., S. Kashiwagi and T. Mizuno, 1994. A copper-transporting P-type ATPase found in the thylakoid membrane of the cyanobacterium *Synechococcus* species PCC7942. *Molecular Microbiology* **13**, 369–377.

Karkaria, C.E., C.M. Chen and B.P. Rosen, 1990. Mutagenesis of a nucleotide-binding site of an anion-translocating ATPase. *Journal of Biological Chemistry* **265**, 7832–7836.

Kaur, P. and B.P. Rosen, 1992. Mutagenesis of the C-terminal nucleotide-binding site of an anion-translocating ATPase. *Journal of Biological Chemistry* **267**, 19272–19277.

Kaur, P. and B.P. Rosen, 1993. Complementation between nucleotide binding domains in an anion-translocating ATPase. *Journal of Bacteriology* **175**, 351–357.

Kaur, P. and B.P. Rosen, 1994. *In vitro* assembly of an anion-stimulated ATPase from peptide fragments. *Journal of Biological Chemistry* **269**, 9698–9704.

Komeda, H., M. Kobayashi and S. Shimizu, 1997. A novel transporter involved in cobalt uptake. *Proceedings of the National Academy of Sciences (USA)* **94**, 36–41.

Kurdi-Haidar, B., S. Aebi, D. Heath, R.E. Enns, P. Naredi, D.K. Hom and S.B. Howell, 1996. Isolation of the ATP-binding human homolog of the arsA component of the bacterial arsenite transporter. *Genomics* **36**, 486–491.

Kuroda, M., S. Dey, O.I. Sanders and B.P. Rosen, 1997. Alternate energy coupling of ArsB, the membrane subunit of the Ars anion-translocating ATPase. *Journal of Biological Chemistry* **272**, 326–331.

Laddaga, R.A. and S. Silver, 1985. Cadmium uptake in *Escherichia coli* K-12. *Journal of Bacteriology* **162**, 1100–1105.

Laddaga, R.A., R. Bessen and S. Silver, 1985. Cadmium-resistant mutant of *Bacillus subtilis* 168 with reduced cadmium transport. *Journal of Bacteriology* **162**, 1106–1110.

Lai, H.C., D. Gygi, G.M. Fraser and C. Hughes, 1998. A swarming-defective mutant of *Proteus mirabilis* lacking a putative cation-transporting membrane P-type ATPase. *Microbiology* **144**, 1957–1961.

Lebrun, M., A. Audurier and P. Cossart, 1994. Plasmid-borne cadmium resistance genes in

Listeria monocytogenes are similar to *cadA* and *cadC* of *Staphylococcus aureus* and are induced by cadmium. *Journal of Bacteriology* **176**, 3040–3048.

Li, J. and B.P. Rosen, 1998. Steric limitations in the interaction of the ATP binding domains of the ArsA ATPase. *Journal of Biological Chemistry* **273**, 6796–6800.

Li, J., S. Liu and B.P. Rosen, 1996. Interaction of ATP binding sites in the ArsA ATPase, the catalytic subunit of the Ars pump. *Journal of Biological Chemistry* **271**, 25247–25252.

Liesegang, H., K. Lemke, R.A. Siddiqui and H.G. Schlegel, 1993. Characterization of the inducible nickel and cobalt resistance determinant *cnr* from pMOL28 of *Alcaligenes eutrophus* CH34. *Journal of Bacteriology* **175**, 767–778.

Lu, D., B. Boyd and C.A. Lingwood, 1997. Identification of the key protein for zinc uptake in *Hemophilus influenzae*. *Journal of Biological Chemistry* **272**, 29033–29038.

Lutsenko, S. and J.H. Kaplan, 1995. Organization of P-type ATPases: significance of structural diversity. *Biochemistry* **34**, 15607–15613.

Lutsenko, S., K. Petrukhin, M.J. Cooper, C.T. Gilliam and J.H. Kaplan, 1997. N-terminal domains of human copper-transporting adenosine triphosphatases (the Wilson's and Menkes disease proteins) bind copper selectively *in vivo* and *in vitro* with stoichiometry of one copper per metal-binding repeat. *Journal of Biological Chemistry* **272**, 18939–18944.

Mellano, M.A. and D.A. Cooksey, 1988. Nucleotide sequence and organization of copper resistance genes from *Pseudomonas syringae* pv. *tomato*. *Journal of Bacteriology* **170**, 2879–2883.

Mills, S.D., C.A. Jasalavich and D.A. Cooksey, 1993. A two-component regulatory system required for copper-inducible expression of the copper resistance operon of *Pseudomonas syringae*. *Journal of Bacteriology* **175**, 1656–1664.

Mobley, H.L. and B.P. Rosen, 1982. Energetics of plasmid-mediated arsenate resistance in *Escherichia coli*. *Proceedings of the National Academy of Sciences (USA)* **79**, 6119–6122.

Navarro, C., L.F. Wu and M.A. Mandrand-Berthelot, 1993. The *nik* operon of *Escherichia coli* encodes a periplasmic binding-protein-dependent transport system for nickel. *Molecular Microbiology* **9**, 1181–1191.

Nies, D.H. 1995. The cobalt, zinc, and cadmium efflux system CzcABC from *Alcaligenes eutrophus* functions as a cation–proton antiporter in *Escherichia coli*. *Journal of Bacteriology* **177**, 2707–2712.

Nies, D.H. and N.L. Brown, 1998. Two-component systems in the regulation of heavy metal resistance, in *Metal ions in gene regulation*, (eds S. Silver and W. Walden), Chapman & Hall, New York, pp. 77–103.

Nies, D.H. and S. Silver, 1989a. Metal ion uptake by a plasmid-free metal-sensitive *Alcaligenes eutrophus* strain. *Journal of Bacteriology* **171**, 4073–4075.

Nies, D.H. and S. Silver, 1989a. Plasmid-determined inducible efflux is responsible for resistance to cadmium, zinc, and cobalt in *Alcaligenes eutrophus*. *Journal of Bacteriology* **171**, 896–900.

Nies, D.H., A. Nies, L. Chu and S. Silver, 1989. Expression and nucleotide sequence of a plasmid-determined divalent cation efflux system from *Alcaligenes eutrophus*. *Proceedings of the National Academy of Sciences (USA)* **86**, 7351–7355.

Novick, R.P. and C. Roth, 1968. Plasmid-linked resistance to inorganic salts in *Staphylococcus aureus*. *Journal of Bacteriology* **95**, 1335–1342.

Nucifora, G., L. Chu, T.K. Misra and S. Silver, 1989. Cadmium resistance from *Staphylococcus aureus* plasmid pI258 *cadA* gene results from a cadmium-efflux ATPase. *Proceedings of the National Academy of Sciences (USA)* **86**, 3544–3548.

Oden, K.L., T.B. Gladysheva and B.P. Rosen, 1994. Arsenate reduction mediated by the plasmid-encoded ArsC protein is coupled to glutathione. *Molecular Microbiology* **12**, 301–306.

Odermatt, A., H. Suter, R. Krapf and M. Solioz, 1992. An ATPase operon involved in copper resistance by *Enterococcus hirae*. *Annals of the New York Academy of Sciences* **671**, 484–486.

Odermatt, A., H. Suter, R. Krapf and M. Solioz, 1993. Primary structure of two P-type ATPases involved in copper homeostasis in *Enterococcus hirae*. *Journal of Biological Chemistry* **268**, 12775–12779.

Patzer, S.I. and K. Hantke, 1998. The ZnuABC high-affinity zinc uptake system and its regulator Zur in *Escherichia coli*. *Molecular Microbiology* **28**, 1199–1201.

Paulsen, I.T., J.H. Park, P.S. Choi and M.H. Saier Jr, 1997. A family of gram-negative bacterial outer membrane factors that function in the export of proteins, carbohydrates, drugs and heavy metals from gram-negative bacteria. *FEMS Microbiology Letters* **156**, 1–8.

Phung, L.T., G. Ajlani and R. Haselkorn, 1994. P-type ATPase from the cyanobacterium *Synechococcus* 7942 related to the human Menkes and Wilson disease gene products. *Proceedings of the National Academy of Sciences (USA)* **91**, 9651–9654.

Pimenta, A., M. Blight, D. Clarke and I.B. Holland, 1996. The gram-negative cell envelope 'springs' to life: coiled-coil trans-envelope proteins. *Molecular Microbiology* **19**, 643–645.

Rensing, C., Fan, B., Sharma, R., Mitra, B. and Rosen, B.P., 1999. CopA: an *Escherichia coli* Cu(I)-translocating P-type ATPase (submitted).

Rensing, C., B. Mitra and B.P. Rosen, 1997a. Insertional inactivation of *dsbA* produces sensitivity to cadmium and zinc in *Escherichia coli*. *Journal of Bacteriology* **179**, 2769–2771.

Rensing, C., B. Mitra and B.P. Rosen, 1997b. The *zntA* gene of *Escherichia coli* encodes a Zn(II)-translocating P-type ATPase. *Proceedings of the National Academy of Sciences (USA)* **94**, 14326–14331.

Rensing, C., T. Pribyl and D.H. Nies, 1997c. New functions for the three subunits of the CzcCBA cation–proton antiporter. *Journal of Bacteriology* **179**, 6871–6879.

Rensing, C., B. Mitra and B.P. Rosen, 1998a. A Zn(II)-translocating P-type ATPase from *Proteus mirabilis*. *Biochemistry and Cell Biology* **76**, 787–90.

Rensing, C., Sun, Y., Mitra, B. and Rosen, B.P., 1998b. Pb(II)-translocating P-type ATPases. *Journal of Biological Chemistry* **273**, 32614–32617.

Rietsch, A., D. Belin, N. Martin and J. Beckwith, 1996. An *in vivo* pathway for disulfide bond isomerization in *Escherichia coli*. *Proceedings of the National Academy of Sciences (USA)* **93**, 13048–13053.

Rosen, B.P. 1998. The role of efflux in bacterial resistance to soft metals and metalloids, in *Metalloproteins*, (eds S. J. Higgins and D. P. Ballou), Portland Press; London, Vol. 34.

Rosen, B.P., U. Weigel, C. Karkaria and P. Gangola, 1988. Molecular characterization of an anion pump. The *arsA* gene product is an arsenite(antimonate)-stimulated ATPase. *Journal of Biological Chemistry* **263**, 3067–3070.

Rosen, B.P., S. Dey, D. Dou, G. Ji, P. Kaur, M. Ksenzenko, S. Silver and J. Wu, 1992. Evolution of an ion-translocating ATPase. *Annals of the New York Academy of Sciences* **671**, 257–272.

Rosen, B.P., S. Dey and D. Dou, 1996. ATPase, in, *Anion-translocating ATPases*, Jai Press, London, Vol. 5, pp. 241–269.

Rosenberg, H., R.G. Gerdes and K. Chegwidden, 1977. Two systems for the uptake of phosphate in *Escherichia coli*. *Journal of Bacteriology* **131**, 505–511.

Rosenstein, R., A. Peschel, B. Wieland and F. Gotz, 1992. Expression and regulation of the antimonite, arsenite, and arsenate resistance operon of *Staphylococcus xylosus* plasmid pSX267. *Journal of Bacteriology* **174**, 3676–3683.

Rouch, D., J. Camakaris and B.T.O. Lee, 1989. In *Metal Ion Homestasis: Molecular Biology and Chemistry*, (eds D. H. Hamer and D. R. Winge), L.L. Erlbaum Associates, Hillsdale, NJ, pp. 469–77.

Saier, M.H. Jr, R. Tam, A. Reizer and J. Reizer, 1994. Two novel families of bacterial membrane proteins concerned with nodulation, cell division and transport. *Molecular Microbiology* **11**, 841–847.

Sanders, O.I., C. Rensing, M. Kuroda, B. Mitra and B.P. Rosen, 1997. Antimonite is accumulated by the glycerol facilitator GlpF in *Escherichia coli*. *Journal of Bacteriology* **179**, 3365–3367.

Schmidt, T. and H.G. Schlegel, 1994. Combined nickel–cobalt–cadmium resistance encoded by the *ncc* locus of *Alcaligenes xylosoxidans* 31A. *Journal of Bacteriology* **176**, 7045–7054.

Snavely, M.D., J.B. Florer, C.G. Miller and M.E. Maguire, 1989. Magnesium transport in *Salmonella typhimurium*: $^{28}Mg^{2+}$ transport by the CorA, MgtA, and MgtB systems. *Journal of Bacteriology* **171**, 4761–4766.

Snavely, M.D., S.A. Gravina, T.T. Cheung, C.G. Miller and M.E. Maguire, 1991. Magnesium transport in *Salmonella typhimurium*. Regulation of *mgtA* and *mgtB* expression. *Journal of Biological Chemistry* **266**, 824–829.

Solioz, M. and A. Odermatt, 1995. Copper and silver transport by CopB-ATPase in membrane vesicles of *Enterococcus hirae*. *Journal of Biological Chemistry* **270**, 9217–9221.

Solioz, M. and C. Vulpe, 1996. CPx-type ATPases: a class of P-type ATPases that pump heavy metals. *Trends in Biochemical Sciences* **21**, 237–241.

Solioz, M., A. Odermatt and R. Krapf, 1994. Copper pumping ATPases: common concepts in bacteria and man. *FEBS Letters* **346**, 44–47.

Townsend, D.E., A.J. Esenwine, J. George 3rd, D. Bross, M.E. Maguire and R.L. Smith, 1995. Cloning of the *mgtE* Mg^{2+} transporter from *Providencia stuartii* and the distribution of *mgtE* in gram-negative and gram-positive bacteria. *Journal of Bacteriology* **177**, 5350–5354.

Tsai, K.J., K.P. Yoon and A.R. Lynn, 1992. ATP-dependent cadmium transport by the cadA cadmium resistance determinant in everted membrane vesicles of *Bacillus subtilis*. *Journal of Bacteriology* **174**, 116–121.

Vallee, B.L. and K.H. Falchuk, 1993. The biochemical basis of zinc physiology. *Physiological Reviews* **73**, 79–118.

Vulpe, C., B. Levinson, S. Whitney, S. Packman and J. Gitschier, 1993. Isolation of a candidate gene for Menkes disease and evidence that it encodes a copper-transporting ATPase. *Nature Genetics* **3**, 7–13.

Willsky, G.R. and M.H. Malamy, 1980. Characterization of two genetically separable inorganic phosphate transport systems in *Escherichia coli*. *Journal of Bacteroriology* **144**, 356–365.

Wolfram, L., B. Friedrich and T. Eitinger, 1995. The *Alcaligenes eutrophus* protein HoxN mediates nickel transport in *Escherichia coli*. *Journal of Bacteriology* **177**, 1840–1843.

Wu, J. and B.P. Rosen, 1991. The ArsR protein is a *trans*-acting regulatory protein. *Molecular Microbiology* **5**, 1331–1336.

Wu, J. and B.P. Rosen, 1993. The *arsD* gene encodes a second *trans*-acting regulatory protein of the plasmid-encoded arsenical resistance operon. *Molecular Microbiology* **8**, 615–623.

Xu, C., T. Zhou, M. Kuroda and B.P. Rosen, 1998. Metalloid resistance mechanisms in procaryotes. *Journal of Biochemistry (Tokyo)* **123**, 16–23.

Yoon, K.P., T.K. Misra and S. Silver, 1991. Regulation of the *cadA* cadmium resistance determinant of *Staphylococcus aureus* plasmid pI258. *Journal of Bacteriology* **173**, 7643–7649.

Zhou, T. and B.P. Rosen, 1997. Tryptophan fluorescence reports nucleotide-induced conformational changes in a domain of the ArsA ATPase. *Journal of Biological Chemistry* **272**, 19731–19737.

Zhou, T., S. Liu and B.P. Rosen, 1995. Interaction of substrate and effector binding sites in the ArsA ATPase. *Biochemistry* **34**, 13622–13626.

6

METAL TRANSPORT AND METABOLISM

Michael DiDonato and Bibudhendra Sarkar

6.1 Introduction

As one would expect, mutations in the genes encoding proteins that are important for metal metabolism usually result in the manifestation of a severe genetic disorder. In humans, there are several genetic disorders of metal metabolism. The two major genetic disorders of copper metabolism are Menkes and Wilson diseases, which are the results of mutations in the genes *ATP7A* and *ATP7B* respectively, which encode copper-transporting P-type ATPases. For iron metabolism, the major disorders are hemochromatosis, thalassemia and aceruloplasminemia. Disorders that have been attributed to errors in zinc metabolism include acrodermatitis enteropathica and various neurodegenerative conditions. However, little is known about the genetic factors influencing these disorders and therefore they will not be covered in this chapter.

6.2 Genetics of metal transport and metabolism disorders

6.2.1 Copper metabolism disorders

The two major genetic disorders of copper metabolism in humans are Menkes and Wilson diseases. In many ways these two disorders are two sides of the same coin; Wilson disease patients are copper overloaded whereas Menkes disease patients are copper deficient. The recent discovery and cloning of the genes responsible for these disorders has improved our understanding of these diseases, as well as the copper metabolism pathways, immensely (Fig. 6.1). Our understanding of these disorders has also been advanced through the use of animal models and the power of yeast genetics to study orthologous genes.

6.2.1.1 Wilson disease

Wilson disease is an autosomal recessive disorder of copper transport, first described by Kinnear Wilson in 1912 (Wilson, 1912). The disorder results in the toxic accumulation of copper in the various tissues but primarily in the liver, kidney and brain. The disease has a frequency of about 1 in 30 000 in most populations but seems to be more prevalent in Japan, Sardinia and China (Sternlieb, 1990; Reilly *et al.*, 1993).

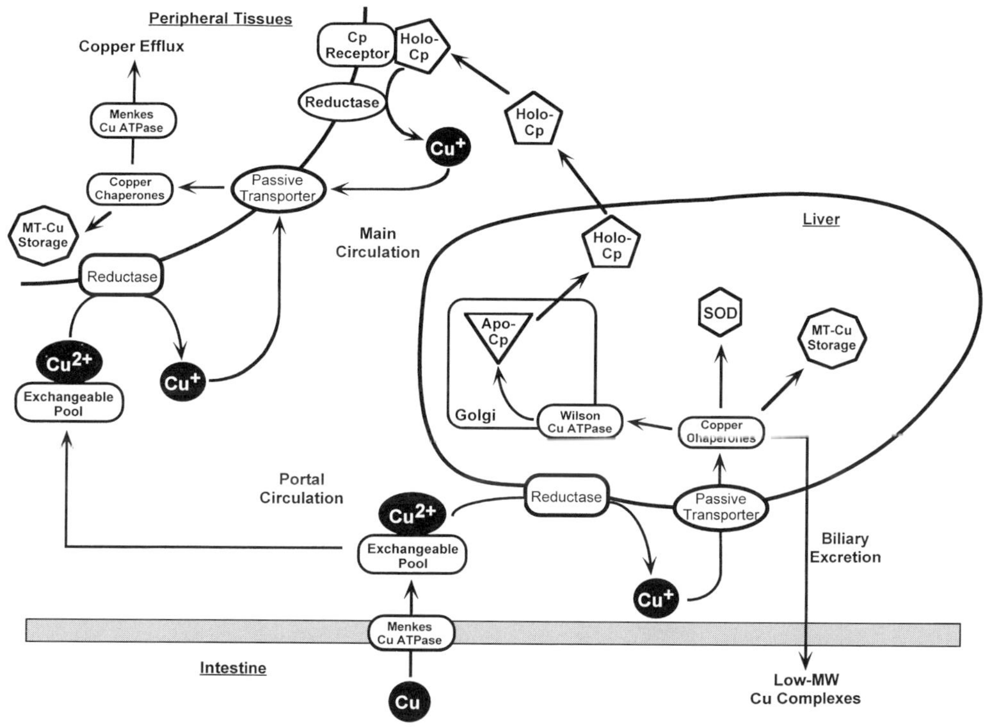

Figure 6.1 Copper transport pathway. Copper is absorbed in the intestine and becomes bound to low molecular weight complexes, amino-acid complexes, and albumin in the Cu(II) form. Prior to uptake, Cu(II) is reduced to Cu(I) by a membrane-bound reductase and enters the cell via a passive transporter. Once in the cell, copper becomes bound to copper chaperones responsible for delivering copper to specific proteins. The Wilson/Menkes ATPase accepts copper from these chaperones and pumps it into the Golgi for incorporation into various proteins such, as ceruloplasmin (Cp). Ceruloplasmin may also play a role in delivering copper to peripheral tissues via cell-surface receptors which internalize the protein. The Wilson disease ATPase may also play a role in the elimination of copper into the bile. (Adapted from DiDonato, M. and Sarkar, B. 1997. *BBA* **1360** (1), 3–16).

Early genetic studies assigned the Wilson disease gene to a single locus by linkage to the esterase D gene (Frydman *et al.*, 1985; Bull *et al.*, 1993) and later studies placed the gene close to a cluster of polymorphic markers in the q14.3 band of chromosome 13 (Bowcock *et al.*, 1987, 1988; Yuzbasiyan-Gurkan *et al.*, 1988). Following these localization studies, the Wilson disease gene was cloned independently by two groups (Bull *et al.*, 1993; Tanzi *et al.*, 1993) and given the designation *ATP7B*. The *ATP7B* gene consists of 22 exons, with exon 21 being expressed in the liver and exon 22 being expressed in the kidney (Petrukhin *et al.*, 1994; Cox, 1995; Thomas *et al.*, 1995). The exons range in size from 77 bp to 2355 bp (Petrukhin *et al.*, 1994). The Wilson disease gene transcript is expressed to a high level in the liver and kidney, with a lower level being detected in the lung and placenta (Bull *et al.*, 1993). The pattern of expression correlates well with the observed clinical and biochemical features of the disease (discussed in a later section). The gene is predicted to encode a copper transporting

P-type ATPase which has many features in common with other cation-transporting P-type ATPases, as well as bacterial heavy-metal transporters (Fig. 6.2). In particular, the protein contains a large N-terminal domain containing six repeats of the GMXCXXC motif thought to be responsible for binding the metal prior to transport. Studies on the recombinant domain have shown that it is able to bind copper as well as other transition metals *in vitro* (DiDonato *et al.*, 1997; Lutsenko *et al.*, 1997).

Mis-sense mutations in the Wilson disease gene have been found in all the exons, with the exception of exons 1–5, which encode the six N-terminal metal-binding motifs, and exon 21, which encodes the C-terminal tail of the protein (Thomas *et al.*, 1995; Shah *et al.*, 1997). The lack of disease mutations in the N-terminal metal-binding region may be a reflection of the redundancy in metal-binding motifs. Many other mutations in the Wilson disease gene, including small insertion/deletions, nonsense, frameshift and splice-site mutations, have been found and are summarized elsewhere (Shimizu *et al.*, 1995; Thomas *et al.*, 1995; Chuang *et al.*, 1996; Loudianos *et al.*, 1996; Nanji *et al.*, 1997). These studies have shown that the most common mis-sense mutation in many

(a)

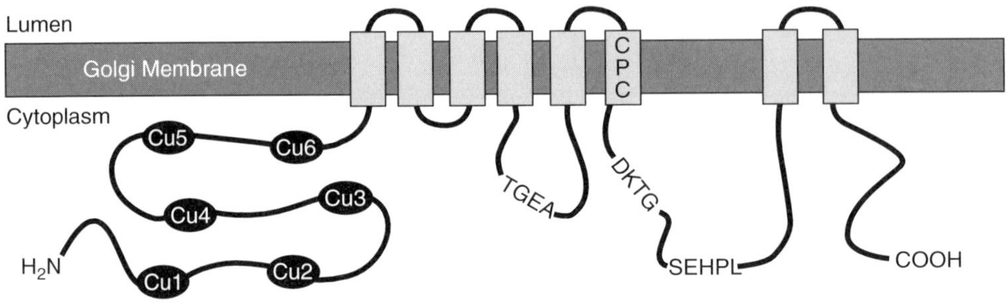

(b)

```
WDCu1   ILGMTCQSCVKSIEDRISNLKGIISMKVSL
WDCu2   VEGMTCQSCVSSIEGKVRKLQGVVRVKVSL
WDCu3   IDGMHCKSCVLNIEENIGQLLGVQSIQVSL
WDCu4   IAGMTCASCVHSIEGMISQLEGVQQISVSL
WDCu5   IKGMTCASCVSNIERNLQKEAGVLSELVAL
WDCu6   ITGMTCASCVHNIESKLTRTNGITYASVAL

ATX1    VV-MTCSGCSGAVNKVLTKLEPDVSKIDIS
HAH1    VD-MTCGGCAEAVSRVLNKL--GGVKYDID
CopA    ITGMTCANCSARIEKELNEQPGVMSATVNL
MerP    ITGMTCDSCAVHVKDALEKVPGVQSADVCT
CadA    VQGFTCANCAGKFEKNVKKIPGVQDAKVNF
```

Figure 6.2 Proposed structure of the Wilson/Menkes copper ATPases. (a) The predicted structure contains domains that are similar to other cation-transporting ATPases as well as other heavy-metal transporters. Cu1–6, N-terminal copper binding motifs; TGEA, transduction motif; CPC, transmembrane motif present in heavy-metal transporters; DKTG, phosphorylation motif; SEHPL, conserved motif found in all heavy-metal transporting ATPases of unknown function. (b) Sequence alignment of metal-binding motifs from the Wilson disease protein (WDCu1–6) and other heavy-metal binding/transporting proteins. ATX1, copper chaperone from *S. cerevisiae*; HAH1, human copper chaperone; CopA, copper-transporting ATPase from *E. hirae*; MerP, mercury-binding protein from *E. coli*; CadA, cadmium-transporting ATPase from *S. aureus*.

populations is a change from histidine at position 1070 to glutamine, which is present in about 38% of homozygous Wilson disease patients. The mutation occurs in a conserved loop motif (SEHPL) which is adjacent to the conserved phosphorylation motif DKTG. Although no specific function has been attributed to this motif, it seems to be necessary for the function of heavy-metal ATPases as it is absolutely conserved between the Wilson disease (ATP7B), Menkes disease (ATP7A) and CopA (bacterial copper ATPase) proteins (Odermatt *et al.*, 1993; Petrukhin *et al.*, 1993; Tanzi *et al.*, 1993). The analogous region in the sarcoplasmic Ca-ATPase has been shown to be the site of interaction for phospholamban (James *et al.*, 1989; Toyofuku *et al.*, 1993). Similarly, for the plasma-membrane Ca-ATPases, this region has been shown to interact with a calmodulin-binding domain (Falchetto *et al.*, 1991). In each case interaction of this region with either phospholamban or calmodulin leads to either a decrease or increase, respectively, in the affinity of the ATPase for its cation (Larsen and Vincenzi, 1979; Larsen *et al.*, 1981). Hence, mutations in this region would be predicted to impair severely the function of the ATPase.

The high frequency of the His1070Gln has permitted a genotype/phenotype analysis of this mutation. In one study, patients homozygous for the His1070Gln mutation were found to have a later average age of onset as compared to the heterozygous patients, 20 years versus 15.4 years (Shah *et al.*, 1997). In addition, all homozygotes were found to have an equal incidence of hepatic and neurologic symptoms. The His1070Gln mutation was also found at high frequency in a study conducted on Mediterranean populations (Figus *et al.*, 1995). This study also found that the His1070Gln mutation is always associated with the same haplotype, in agreement with studies conducted on other populations. The data from this study and others suggest that the His1070Gln mutation is the most common molecular defect in the Wilson disease protein and that it most likely arose from an ancient mutational event (Figus *et al.*, 1995).

6.2.1.2 Menkes disease

Menkes disease is a genetic disorder of copper metabolism characterized by an impairment in the absorption of dietary copper and a severe disturbance in the intracellular transport of copper (Danks *et al.*, 1973; Herd *et al.*, 1987). The disease was first characterized by John Menkes and co-workers in 1962 during the study of five patients from the same family who all died before the age of three (Menkes *et al.*, 1962). Menkes and co-workers noted in their original study that the disease seemed to be confined to males and was inherited in a sex-linked recessive manner. Early genetic work had mapped the disease locus to band Xq13.3 (Verga *et al.*, 1991; Tümer *et al.*, 1992a, b). Following this work, the Menkes disease gene (*ATP7A*) was isolated independently by three groups by positional cloning (Chelly *et al.*, 1993; Mercer *et al.*, 1993; Vulpe *et al.*, 1993). The gene contains 23 exons and spans a genomic region of approximately 150 kb (Dierick *et al.*, 1995; Tümer *et al.*, 1995). The exons range in size from 77 to 726 bp with introns ranging from 196 bp to 60 kb (Dierick *et al.*, 1995). All splice sites conform to the GT–AG rule except for the donor site of intron 9 which precedes an exon that is alternatively spliced (Dierick *et al.*, 1995). After the exon coding for the fifth metal-binding domain (exon 5 in *ATP7A* and exon 3 in *ATP7B*), the coding regions of both the Wilson disease gene (*ATP7B*) and the Menkes disease gene (*ATP7A*) are virtually identical (Tümer *et al.*, 1995). However, significant differences are present in the gene structure at the 3′ end. Whereas the first four metal-binding domains in the Wilson disease gene are contained in one exon, those of the Menkes disease extend over three exons (Dierick *et al.*, 1995).

The Menkes gene encodes a 1500 amino-acid protein which is predicted to be a copper-transporting P-type ATPase (Fig. 6.2) with a high degree of homology to the Wilson disease protein (Vulpe *et al.*, 1993). The overall identity between the two proteins is 57%, and this figure rises to 79% and higher in the phosphatase, transduction–phosphorylation, and ATP-binding domains (Bull *et al.*, 1993). Although the Menkes and Wilson proteins have a high degree of similarity, there are some significant differences, most notably a 78-amino-acid insertion between metal-binding domains 1 and 2 in the Menkes protein which is not present in the Wilson protein (Bull *et al.*, 1993). The implications of this insertion, if any, have yet to be discovered. The Menkes disease protein is expressed in all tissues except the liver, which correlates well with the clinical and biochemical features of the classical Menkes phenotype. Unlike the mutations found in the Wilson disease gene, approximately 20% of the mutations found in the Menkes gene are insertion/deletion mutants of varying sizes (Chelly *et al.*, 1993; Mercer *et al.*, 1993; Vulpe *et al.*, 1993). In a study of 41 unrelated patients presenting the classical severe form of Menkes disease, 19 out of 41 mutations were due to insertion or deletions (Tümer *et al.*, 1997). The two largest of these deletions are a 14 bp deletion in exon 10, and a 10 bp deletion in exon 22. These deletions, as well as the others, would be predicted to alter the reading frame of the gene, leading to the introduction of a stop codon and premature truncation of the protein. Alternatively, these mutations may lead to the skipping of subsequent mutation-bearing exons, which may result in an internally deleted protein that may retain some function, as has been reported for a 2 bp deletion in exon 7 (Das *et al.*, 1994).

The study by Tümer *et al.* (1997) also found that approximately half of the mutations they identified were in exons 7–10 and half of these mutations affected exon 8 specifically. Additionally, 5 of 12 point mutations identified in another series also occur in exon 8 (Das *et al.*, 1994; Tümer *et al.*, 1996). Exon 8 codes for the region of the protein between metal-binding domain 6 and the first transmembrane segment of the protein. The exact role this region plays in the function of the ATPase is still unknown, but it may be important for the precise localization of the metal-binding domain relative to the ATPase core. The importance of this region of the protein is exemplified by the finding that individuals carrying mutations in exon 8 present the classical severe form of Menkes disease (Tümer *et al.*, 1997). In contrast, only two mutations have been found in the corresponding region of the Wilson disease protein, these being a 2 bp deletion and a mis-sense mutation (Figus *et al.*, 1995). These differences in the number of mutations in this region may indicate that these domains play somewhat different roles in the Wilson and Menkes ATPases, despite their high degree of sequence homology.

6.2.2 *Iron metabolism disorders*

6.2.2.1 *Hereditary hemochromatosis*

Hereditary hemochromatosis is an autosomal recessive disorder of iron metabolism, resulting in the abnormal uptake and accumulation of toxic levels of iron, primarily in the parenchymal cells of the liver, heart and pancreas (Bacon, 1997). Current estimates of its frequency indicate that the homozygous genotype occurs in 1 out of every 250 persons, with 1 in 9 being a carrier (McLaren *et al.*, 1995; Witte *et al.*, 1996). Hereditary hemochromatosis is most common in men of northern European descent and it is the most common genetic disorder known in the United States. Early studies indicated that the gene responsible for hereditary hemochromatosis was linked to the human leukocyte antigen (HLA) locus. The HLA-A3 allele was shown to be common in many patients of European origin, being present in 73% of patients in a collective series (Simon *et al.*, 1975, 1980; Shewan *et al.*, 1976; Morris *et al.*, 1977). The question of why this HLA-A3-linked gene has such a high frequency is an intriguing one. Although the exact origin of its high frequency is unknown, it may be that in the past heterozygous individuals had an evolutionary advantage if they absorbed more iron (Motulsky, 1979; Bothwell *et al.*, 1995), or that larger body iron reserves might improve survivability during infancy (Bothwell *et al.*, 1995). Alternatively, the frequency may not be a result of positive selection on the hemochromatosis gene but on the tightly linked HLA-A3 locus (Bothwell *et al.*, 1995). Selection against individuals homozygous for hereditary hemochromatosis would be minimal since the onset of the disorder usually occurs after the reproductive period.

For many years the identification of the gene responsible for hereditary hemochromatosis had eluded investigators. In 1996, using linkage disequilibrium and haplotype analysis, Feder *et al.* identified a gene near the major histocompatibility complex (MHC) on chromosome 6 which contained a mutation present in 83% of individuals homozygous for hereditary hemochromatosis. The gene, termed HLA-H, encodes a 343-amino-acid protein which is very similar to MHC class 1 molecules such as HLA-A2 (Feder *et al.*, 1996). Sequence alignment of HLA-H with other MHC class 1 molecules reveals the presence of several structural features common to these molecules: peptide-binding region, a transmembrane region, an immunoglobulin-like domain, a small cytoplasmic region and a signal sequence (Fig. 6.3). Two mutations have been found in the *HLA-H* gene: the first is a G to A transition, resulting in a change in cysteine at position 282 to a tyrosine (Cys282Tyr); the second results in a change in histidine at position 63 to an aspartate (His63Asp). As mentioned previously, 83% of patients with classical hereditary hemochromatosis were homozygous for the Cys282Tyr mutation with another 4% being compound heterozygotes having Cys282Tyr on one allele and His63Asp on the other allele (Feder *et al.*, 1996).

The discovery of the *HLA-H* gene was an important step forward in not only understanding the disease process, but the iron uptake process in general. It has, perhaps, forced scientists in the field to re-evaluate their understanding of the iron metabolism pathways. Just how is it that an MHC class 1-like molecule is able to regulate the uptake of iron? Although concrete evidence as to the exact function of the HLA-H protein is as yet unavailable, many interesting theories have been proposed. Several studies have shown that a hemochromatosis-like loading of iron is observed in mice where the β_2-microglobulin gene has been knocked out (de Sousa *et al.*, 1994; Rothenberg and Voland, 1996). β_2-microglobulin is a protein that is usually found complexed to MHC class 1 proteins on the surface of most cells. The function of this complex is to display peptides derived from foreign proteins on the cell surface with the purpose of recruiting cytotoxic T cells to remove such cells from the body. If β_2-microglobulin is absent, the complex is destabilized and cell-surface quantities of HLAs are reduced. The Cys282Tyr

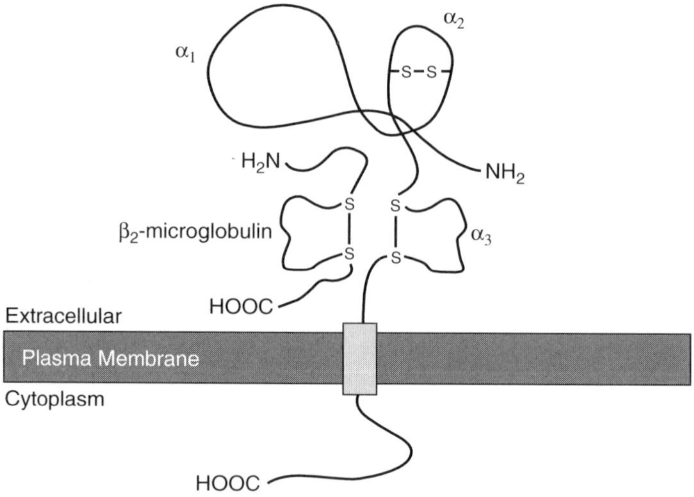

Figure 6.3 Proposed model for the HLA-H protein. The model was constructed by homology to other MHC class 1 proteins. The protein has three extracellular domains which are homologous to the α_1, α_2 and α_3 domains of other MHC class 1 proteins. The protein also has a site for noncovalent interaction with β_2 microglobulin, which is a separate protein. In addition, the protein has a single transmembrane segment and a short cytoplasmic tail. (Adapted from Feder *et al.* (1996).)

mutation common in hemochromatosis patients is very close to the expected site of β_2-microglobulin binding and would be expected to abolish, or at least greatly diminish, this interaction. Further sequence analysis of HLA-H reveals that it contains features common to non-peptide-binding MHC analogs, suggesting that it may not be playing a classical MHC class 1 function (Ueyama *et al.*, 1993; Story *et al.*, 1994; Feder *et al.*, 1996).

Immunohistochemical studies have localized the HLA-H protein to various portions of the gastrointestinal tract with differing expression patterns. In the stomach and colon the staining pattern was polarized and mainly restricted to the basolateral membrane, while the submucosal leukocytes and epithelia of the esophagus showed a nonpolarized staining pattern around the entire plasma membrane (Bacon, 1997; Parkkila *et al.*, 1997). In contrast to these cells, the staining pattern in the small intestine was mainly intracellular and perinuclear. In addition, the staining in the small intestine was restricted to cells in deep crypts and it has been suggested that the HLA-H protein may be involved in predetermining the iron absorption capacity of these cells before they migrate up the villi (Bacon, 1997). This would correlate well with the observation that the primary site of iron absorption is the small intestine (Anderson, 1996).

Based on the above observations it has been proposed that the HLA-H protein is involved in the downregulation of the iron absorption process (Feder *et al.*, 1996; Fleet, 1996). Three possible functions of HLA-H have been proposed in the literature (Feder *et al.*, 1996). First, the HLA-H protein may act as a 'sensor' of body iron concentrations by some form of iron chelate. This would in turn tell the enterocyte that iron stores are high and that iron uptake should be downregulated. Secondly, the HLA-H protein could act as part of a signal transduction pathway, as has been demonstrated in the epidermal growth factor (EGF) receptor systems (Schreiber *et al.*, 1984; Verland *et al.*, 1989; Feder *et al.*, 1996). Finally, it has also been suggested that HLA-H may interact with members of the immune system and indirectly affect iron metabolism in ways that are currently not understood. In order to decipher the exact function of this protein in the iron metabolism pathway, detailed genetic and biochemical studies will have to be carried out. The availability of the β_2-microglobulin knockout mouse will be an invaluable tool in this endeavor.

6.2.2.2 *Thalassemia*

Although the thalassemias are technically not iron metabolism or iron transport disorders, severe iron overload is usually the major complication of the disease. The underlying cause of the disorder is a disturbance in the synthesis of one or more of the globin chains of hemoglobin. There are several different types of hemoglobins, all of which are tetrameric, containing two pairs of different polypeptide chains, and are developmentally regulated. The major hemoglobin present in adults is Hb A ($\alpha_2\beta_2$) which comprises about 97% of the total, with Hb A_2 ($\alpha_2\delta_2$) making up the remainder. Hb F ($\alpha_2\gamma_2$) is the main hemoglobin present in the fetus, with Hb Gower 1 ($\zeta_2\varepsilon_2$), Hb Gower 2 ($\alpha_2\varepsilon_2$), and Hb Portland ($\zeta_2\gamma_2$) being expressed during embryonic development (Weatherall *et al.*, 1995). The six globin genes share a high degree of homology and are thought to have originated from one ancestral gene. The six chains can be divided into two groups, those being α-like (ζ and α, 141 amino acids each) and those being β-like (ε, δ, γ, and β, 146 amino acids each). The structure of human hemoglobin shows that each α-chain makes contact with two β-chains, creating an $\alpha_1\beta_1$ and an $\alpha_1\beta_2$ interface (Baldwin, 1980; Fermi *et al.*, 1984).

The six globin genes are divided into two clusters found on two different chromosomes. The α-like gene cluster is located on chromosome 16 while the β-like gene cluster resides on chromosome 11 (Deisseroth *et al.*, 1976, 1977, 1978). The α-like gene cluster is made up of the ζ gene followed by two copies of the α gene (α_2 and α_1, respectively). In addition, four pseudogenes have been discovered in the cluster which have significant homology to the 'real' genes (Fig. 6.4). These being $\psi\zeta1$, $\psi\alpha1$ and $\psi\alpha2$, which are located between the ζ and $\alpha2$ genes, and $\theta1$ which is located on the 3′ side of the $\alpha1$ gene (Lauer *et al.*, 1980; Proudfoot *et al.*, 1982; Hardison *et al.*, 1986; Marks *et al.*, 1986). The

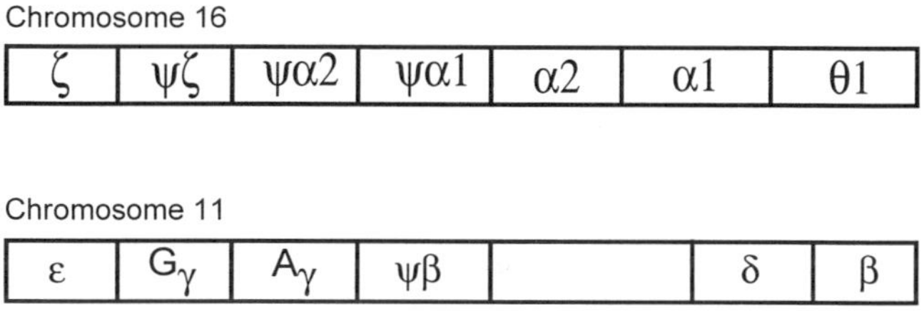

Figure 6.4 Genomic organization of the human globin genes. The α-like gene cluster is located on chromosome 16, while the β-like gene cluster is located on chromosome 11. The α-like gene cluster contain four pseudogenes $\psi\zeta$, $\psi\alpha2$, $\psi\alpha1$ and $\theta1$. The β-like gene cluster contains one pseudogene, $\psi\beta$. All pseudogenes are inactive except for $\theta1$, the function of which is unknown.

β-like gene cluster is composed of the embryonic ε gene followed by the fetal Gγ and Aγ genes, a minor form of the adult δ gene and the β gene. Like the α gene cluster, a pseudogene $\psi\beta$ is found between the Aγ and δ genes and is very similar to the functional β gene. These pseudogenes are inactive, with the exception of $\theta1$, and are thought to be remnants of evolutionary changes in the globin gene cluster. The fact that $\theta1$ is a highly conserved element with no inactivating mutations suggests that it may be an active gene. However, no protein product has yet been isolated and its function remains unknown (Clegg, 1987). Rearrangements within the α and β gene cluster are fairly common and in most cases have no associated clinical condition. The most common of these are individuals with triplicated ζ, γ and α genes, or quadruplicated α and γ arrangements (Higgs *et al.*, 1980; Trent *et al.*, 1981; Winichagoon *et al.*, 1982; Weatherall *et al.*, 1995).

A common feature of all globin genes is the presence of two introns in identical positions relative to the coding sequence. These are of variable length, with the shortest being IVS1 of the α genes (117 bp) and the largest being IVS1 of the ζ gene (1264 bp) (Weatherall *et al.*, 1995). Among the β-like genes there is very little homology between the IVS2 introns except in the case of the Gγ and Aγ genes. In the α-like cluster, IVS1 and IVS2 of the $\alpha2$ and $\alpha1$ genes are highly homologous but share little or no homology with either the ζ or $\psi\zeta$ genes (Weatherall *et al.*, 1995). The flanking regions of both clusters contain sequences that are required for the proper expression of the genes. These include an ATA box, the CCAAT box (about 70–80 bp upstream) and a GC-rich region approximately 100 bp from the cap site (Collins and Weissman, 1984; Nienhuis and Maniatis, 1987). *In vitro* experiments have indicated that mutations in any of these sites will, in most cases, adversely affect the expression of the globin genes (Grosveld *et al.*, 1982; Orkin *et al.*, 1982, 1984).

The six globin genes must be tightly regulated to ensure their expression in both a tissue-specific and developmental-stage-specific manner. Studies have shown that this is accomplished by regulatory sites upstream of each locus. These being the β-globin locus control region (βLCR) located 20 kb upstream of the ε gene and the α-globin hypersensitivity site (αHS-40) located 40 kb upstream of the ζ gene (Grosveld *et al.*, 1987; Stamatoyannopoulos and Nienhuis, 1987; Evans *et al.*, 1990; Hatton *et al.*, 1990; Wilkie *et al.*, 1990). Experiments involving the transfection of human genes into mouse erythroleukemia cells results in the tissue- and developmental-stage-specific expression of the human globin genes at very low levels (Wright *et al.*, 1983; Charnay *et al.*, 1984; Costantini *et al.*, 1985; Townes *et al.*, 1985; Chada *et al.*, 1986; Kollias *et al.*, 1986). These results suggested that regions in and around the globin genes are able to regulate their expression. High-level expression of the human genes was only observed if the βLCR or αHS-40 regions were included in the upstream region of the genes. The βLCR contains four erythroid-specific sites that are hypersensitive to DNase I digest, all of which are required to obtain high-level, tissue-specific expression of the genes from the β-globin locus (Tuan *et al.*, 1985; Forrester *et al.*, 1986; Grosveld *et al.*, 1987; Fraser *et al.*, 1990). Deletions in the βLCR have been shown to abolish expression of the β-globin genes (Driscoll *et al.*, 1989).

The αHS-40 region upstream of the α-globin genes serves the same function as the βLCR (Hatton *et al.*, 1990; Higgs *et al.*, 1990; Wilkie *et al.*, 1990; Jarman *et al.*, 1991). Mutations in this region also result in the lack of expression of the α-globin genes. Analysis of the hypersensitive sites indicates that they contain multiple sites for ubiquitously expressed transcription factors, as well as sites for two erythroid-specific proteins, GATA-1 and NFE-2 (Weatherall *et al.*, 1995). Although it is clear that both the αHS-40 and the βLCR are required for high-level expression from their respective gene clusters, it is not yet clear whether they are involved in the developmental-stage-specific expression of the globin genes.

6.2.2.3 Aceruloplasminemia

Ceruloplasmin is a large, secreted, blue copper glycoprotein that carries 95% of the circulating copper in the body (Frieden, 1986). The crystal structure of holoceruloplasmin has recently been solved and shows that the holoprotein contains six copper atoms, three of which are involved in a trinuclear copper cluster (Zaitseva *et al.*, 1996). The protein is synthesized in the hepatocytes and copper incorporation takes place early in the secretory pathway, possibly in the Golgi or rough endoplasmic reticulum (ER), with *N*-glycosylation not being required for copper loading (Sato and Gitlin, 1991). Although copper availability has no effect on the synthesis or secretion of the protein, apoceruloplasmin lacking oxidase activity is unstable and is rapidly degraded once secreted (Gitlin *et al.*, 1992). Ceruloplasmin has also been shown to contain ferroxidase activity (Roeser *et al.*, 1970; Frieden, 1986; Harris *et al.*, 1995) and because of this it has been proposed to play a role in iron metabolism. The trinuclear copper cluster is responsible for activating oxygen during the catalytic cycle of the protein and residues in both the amino and carboxy termini of the protein provide essential ligands for the trinuclear copper center (Zaitseva *et al.*, 1996). Studies have shown that ceruloplasmin plays a key role in the mobilization and oxidation of iron stores and their subsequent incorporation into transferrin (Osaki and Johnson, 1969). This is supported by the observation that animals deprived of dietary copper have little or no circulating holoceruloplasmin and are unable to release iron into the plasma from tissue stores (Roeser *et al.*, 1970).

The ceruloplasmin gene is located on chromosome 3, spanning approximately 36 kb and consisting of 19 exons (Takahashi *et al.*, 1996). Mutations in this gene have been shown to lead to a condition called aceruloplasminemia, which is characterized by progressive neurodegeneration of the retinal and basal ganglia (Harris *et al.*, 1998). Aceruloplasminemia is inherited in an autosomal recessive manner and thus far five unique mutations have been identified in six separate families: three insertion deletion mutants (Harris *et al.*, 1995, 1996; Okamoto *et al.*, 1996), one nonsense mutation (Daimon *et al.*, 1995; Takahashi *et al.*, 1996) and one splice-site mutation (Yoshida *et al.*, 1995). In all cases these mutations result in a shift in the reading frame, such that residues at the carboxy terminus that are essential for the formation of the trinuclear copper cluster, and hence oxidase activity, are eliminated. This correlates with the finding of little or no oxidase activity in the plasma of affected individuals (Harris *et al.*, 1998).

6.3 Clinical and pathological effects of metal transport and metabolism disorder

6.3.1 Copper metabolism disorders

6.3.1.1 Wilson disease

Wilson disease involves the accumulation of toxic levels of copper in various organs due to a deficiency in the excretion of copper from the body. The age of onset for Wilson disease is variable and can extend from 8–9 years into the mid-50s. The progression of the disease can be divided into three phases. The first phase involves the accumulation of copper in the cytosol of hepatocytes. In the second phase, as more copper is absorbed, the cytosolic concentration rises, causing the induction of large amounts of metallothionein and resulting in the chelation and storage of copper (Hunziker and Sternlieb, 1991). The high levels of copper in the hepatocytes will eventually lead to necrosis, resulting in the release of large amounts of copper into the bloodstream. In phase three, copper begins to accumulate in other organs, such as the brain, kidney and cornea. The sudden rise in serum copper levels results in oxidative damage to erythrocyte membranes, leading to hemolytic anemia (Dobyns *et al.*, 1979; Mowat, 1994). Wilson disease patients can fall into one of three groups: those displaying hepatic symptoms, those displaying neurologic symptoms and those displaying a combination of both.

The deposition of copper in the cornea gives rise to dull-yellow Kayser–Fleischer (K–F) rings which are a diagnostic hallmark of Wilson disease (Scheinberg and Sternlieb, 1984). K–F rings have been found in 95% of all Wilson disease patients and in all patients who display neurologic disorders (Danks, 1995). The deposition of copper in the brain during the latter stages of the disease correlates with the finding that patients with an early age of onset usually display predominantly hepatic symptoms while those with a later age of onset display predominantly neurologic symptoms (Strickland *et al.*, 1973; Vulpe and Packman, 1995). Wilson disease displays extensive clinical heterogeneity. Symptoms of liver dysfunction include jaundice, asymptomatic cirrhosis, subacute hepatitis (non-A or B) and chronic aggressive hepatitis (Mowat, 1994). Neurologic disorders in Wilson disease patients include deteriorating coordination, tremors, dysarthria, drooling, personality changes, dementia, slurring of speech, rigidity and behavioral problems (Mowat, 1994; Strickland and Leu, 1975). The neurologic symptoms in Wilson disease patients can be attributed to cerebellar or extrapyramidal involvement, or a combination of both (Dobyns *et al.*, 1979).

The biochemical findings in patients with Wilson disease (Table 6.1) are consistent with a disturbance in the excretion of copper, primarily a defect in the biliary excretion of copper (Strickland *et al.*, 1973). Urinary copper levels in patients with Wilson disease are increased due to the accumulation of copper in the kidneys, which leads to renal dysfunction in the form of albuminuria and renal rickets (Walshe, 1989). Since incorporation of copper into ceruloplasmin is also impaired in Wilson disease, there is a greater proportion of copper bound to albumin and low-molecular-weight complexes in the serum; however, the overall serum concentration of copper is lowered (Barrow and Tanner, 1988). Ceruloplasmin is a large multicopper oxidase that accounts for the majority of circulating copper in the bloodstream. Its exact function is as yet unknown but it has been implicated in iron metabolism (Roeser *et al.*, 1970; Frieden, 1986; Harris

Table 6.1 Clinical and biochemical features of Wilson and Menkes diseases

	Menkes disease	*Wilson disease*
Laboratory findings	↓ Serum copper levels ↓ Liver copper levels ↑ Intestinal/kidney copper levels	↓ Serum copper levels ↑ Liver copper levels ↑ Urinary copper levels
Defect	Intestinal copper absorption Deficiency of cuproenzymes	Decreased biliary excretion of copper Incorporation of copper into ceruloplasmin
Treatment	Daily Cu-histidine injections	Chelation therapy
Genes	X chromosome/recessive	Chromosome 13/recessive

et al., 1995) because of its ferroxidase activity, as well as copper transport function (Marceau and Aspin, 1973; Hsieh and Frieden, 1975; Campbell *et al.*, 1981; Dameron and Harris, 1987; Hilton *et al.*, 1995). The only effective treatment for Wilson disease is the use of a copper chelator to remove excess copper from the body. Among the chelators most frequently used are D-penicillamine (Walshe, 1956), trientine (Walshe, 1982; Scheinberg *et al.*, 1987) and tetrathiomolybdate (Walshe, 1984; Brewer *et al.*, 1994). Chelation therapy usually lasts for a lifetime and in many patients can lead to significant biochemical and clinical improvement. Side-effects have been observed for many chelators and in the case of D-penicillamine the most common side-effects observed are fever, anorexia, pyridoxine deficiency, and a depression in leukocyte and blood platelet counts (Scheinberg, 1968; Sass-Kortsak, 1975; Mowat, 1994). One alternative to D-penicillamine is the use of oral zinc salts, which lead to the induction of intestinal metallothionein, resulting in the chelation of copper and its excretion in the feces (Hoogenraad *et al.*, 1978, 1979; Brewer *et al.*, 1983; Alexiou *et al.*, 1985; Van Caillie-Bertrand *et al.*, 1985; Lipsky and Gollan, 1987; Yuzbasiyan-Gurkan *et al.*, 1992).

Recent immunohistochemical studies have shown that the Wilson disease ATPase (ATP7B) is localized to the trans-Golgi network under steady-state conditions (Hung *et al.*, 1997; Nagano *et al.*, 1998). Furthermore these studies have shown that the protein undergoes a rapid, copper-dependent relocalization from the Golgi to a cytosolic vesicular compartment as well as to the plasma membrane (Hung *et al.*, 1997; Nagano *et al.*, 1998). The function of the ATPase in the trans-Golgi is most likely to transport copper from the cytoplasm into this compartment for incorporation into copper-containing proteins such as ceruloplasmin. This idea correlates with the finding of low ceruloplasmin levels in patients with Wilson disease. The relocalization of the protein to a vesicular compartment or the plasma membrane in response to elevated copper levels would lead to efflux of copper from the cell either via the bile (through the fusion of vesicles with the canilicular membrane) or across the plasma membrane (Dijkstra *et al.*, 1996; Hung *et al.*, 1997). Recently the Wilson disease protein has also been localized to the mitochondria where it has been suggested to provide copper for incorporation into cytochrome oxidase (Lutsenko and Cooper, 1998).

6.3.1.2 *Menkes disease*

Menkes disease patients affected by the classical severe form of the disorder are unable to absorb copper and hence suffer from the effects of copper deficiency. Unlike Wilson disease, Menkes disease has a very early age of onset and affected individuals usually die before the age of 3 years. In Menkes' original case study (Menkes *et al.*, 1962) all five patients were delivered normally and had normal birthweights. However, it was noted that all patients gained very little weight following birth despite a normal diet and all had hair abnormalities, either from birth or developed soon thereafter. Menkes noted that the hair was coarse and brittle and had an ivory white color due to depigmentation. Upon examination of individual hair strands, several gross abnormalities were noted: they were either twisted (pili torti), fractured at various intervals (trichorrhexis nodosa) or of varying caliber (monilethrix); however, the level of sulfhydryls in the hair were found to be normal (Menkes *et al.*, 1962). The abnormal hair of Menkes disease patients led to the disorder also be referred to as kinky-hair syndrome.

Other case studies in the years following Menkes' initial description of the disorder expanded the list of clinical features to include hypothermia, hyperbilirubinemia, arterial rupture, characteristic facies and bone changes (Danks *et al.*, 1972a). The patients were also found to have extremely low levels of copper in the liver, brain and serum, as well as having extremely low ceruloplasmin levels. Oral administration of copper to correct the problem did not result in an increase in serum copper levels; however, intravenous injection of copper was able to restore ceruloplasmin levels and the level of clearance from the serum was normal (Danks *et al.*, 1972b). This and other evidence pointed to the basic biochemical defect being a severely reduced ability to absorb copper from the intestine and a disturbance in the intracellular transport of copper (Danks *et al.*, 1973; Herd *et al.*, 1987). Cultured fibroblasts from Menkes patients have been shown to accumulate copper in the form of copper metallothionein, and metallothionein induction in cultured muscle cells from these patients is elevated relative to normal cells (Goka *et al.*, 1976; Beratis *et al.*, 1978; Chan *et al.*, 1978; Onishi *et al.*, 1980; Riordan and Jolicoeur-Paquet, 1982; Herzberg *et al.*, 1990).

The clinical symptoms of classical severe Menkes disease (Table 6.1) can be traced back to a deficiency of developmentally important cuproenzymes such as lysyl oxidase, tyrosinase, cytochrome *c* oxidase, dopamine β-hydroxylase, superoxide dismutase and amine oxidase (Menkes, 1988; Danks, 1995). Lysyl oxidase is an essential enzyme needed for the crosslinking of connective tissue, a deficiency in this enzyme would lead to weakened connective tissue and would explain the connective tissue defects such as arterial rupture, observed in Menkes patients (Kuivaniemi *et al.*, 1982; Peltonen *et al.*, 1983; Danks, 1995). Low levels of cytochrome *c* oxidase can lead to temperature instability, while the absence of tyrosinase would explain the hair depigmentation observed in affected individuals (Bankier, 1995). Upon examining the cuproenzyme status in hepatocytes it is found that they have normal activity, although the level of copper in the liver is severely depressed (Phillips *et al.*, 1986). This correlates well with the lack of expression of the Menkes gene in this organ. The only currently available effective treatment for Menkes patients is daily subcutaneous injections of copper-histidine. Patients treated in this way have shown significant clinical and biochemical improvement and can achieve normal levels of ceruloplasmin, dopamine, norepinephrine and serum copper (Sherwood *et al.*, 1989; Kreuder *et al.*, 1993; Sarkar *et al.*, 1993).

As is the case for ATP7B (Wilson disease ATPase), ATP7A (Menkes disease ATPase) has been localized by immunofluorescence studies to the Golgi apparatus under normal copper conditions (Petris *et al.*, 1996; Yamaguchi *et al.*, 1996). The Menkes ATPase would also be expected to be present in the intestinal mucosa, where it would function to transfer copper from the intestine into the bloodstream. In Menkes disease the ATPase is nonfunctional so copper accumulates in the intestinal mucosa as copper metallothionein and is lost from the body when these cells are sloughed off, resulting in a global copper deficiency. It has also been shown that the ATP7A undergoes a reversible, copper-dependent relocalization from the Golgi to the plasma membrane (Petris *et al.*, 1996). Under normal copper conditions ATP7A would be positioned in the Golgi, which would act to provide copper to cuproenzymes as they are synthesized. Under elevated copper conditions the pump is positioned on the plasma membrane where it is able to expel copper from the cell until intracellular copper levels return to normal, then it is recycled back to the Golgi. These observations explain why fibroblasts from Menkes patients have normal copper uptake activity but are not able to expel it to avoid accumulation of copper.

6.3.2 Iron metabolism disorders

6.3.2.1 Hereditary hemochromatosis

As mentioned previously, hereditary hemochromatosis is a disorder of iron metabolism where the uptake of iron from the digestive system is not downregulated, resulting in excessive iron uptake. Since the body has no mechanism for the excretion of iron (besides incidental losses) the level of iron in the body must be regulated at the absorption phase. In hemochromatosis the body continues to absorb high levels of dietary iron regardless of the levels of stored iron in the body. Excess iron is slowly deposited in the liver as hemosiderin with a corresponding increase in plasma iron and plasma transferrin saturation (Bothwell *et al.*, 1995). Since tissue damage only becomes evident once the iron stores have reached 20 to 50 times their normal values (and dietary bioavailable iron is only a few milligrams) the age of onset of the disease is fairly late, between 40 and 60 years of age (Sheldon, 1935; Finch and Finch, 1955; Bothwell *et al.*, 1979; Bothwell *et al.*, 1995). However, cases of neonatal hemochromatosis (Chiu and Davey, 1997) and hemochromatosis in children and young adults have also been described (Perkins *et al.*, 1965; Charlton *et al.*, 1967; Escobar *et al.*, 1987; Haddy *et al.*, 1988). In addition, the disorder seems to be more prevalent in males than in females, presumably due to differences in the rate of iron absorption and loss (Finch and Finch, 1955; Bothwell *et al.*, 1995) Upon examination of iron uptake in hemochromatosis patients the levels appear to be normal, but when compared to the greatly increased stores of iron, the rate of uptake seems unnecessarily high (Fairbanks *et al.*, 1971; Walters *et al.*, 1973; Bothwell *et al.*, 1979; Valberg *et al.*, 1979).

The pathologic and clinical findings in hemochromatosis patients are diverse and include weakness and lethargy, abdominal pain, diabetes mellitus, loss of libido, amenorrhea, dyspnea and various neurologic symptoms (Niederau *et al.*, 1985; Bothwell *et al.*, 1995). Physically many patients display signs of hepatomegaly, skin pigmentation, loss of body hair and splenomegaly (Niederau *et al.*, 1985). However, the onset of clinical manifestations of the disorder are greatly affected by both genetic and environmental factors and the most common symptom of progressive iron overload is the asymptomatic elevation of alanine aminotransferase and aspartate aminotransferase in the liver (McLaren *et al.*, 1995; Witte *et al.*, 1996; McDonnell and Witte, 1997). Histologically, excess iron stores in the form of hemosiderin can be visualized easily in the liver, thyroid and heart where levels may reach 50–100, 25 and 10–15 times the normal values, respectively (Bothwell *et al.*, 1995). An early diagnostic test is to monitor the level of plasma iron and transferrin saturation, since more iron is being delivered to the plasma than can be used by tissues. Repeated measurements of transferrin saturation in the 70% range is usually diagnostic, with typical hemochromatosis patients having values in the 90% range (Bothwell *et al.*, 1995).

Since iron is a redox active metal it most likely exerts its toxic effects through the generation of either hydroxyl radicals via Fenton chemistry (Peters and Seymour, 1976; Bacon and Britton, 1990; Ward *et al.*, 1991) or via the generation of ferryl ions (Gutteridge, 1989). The generation of radical species *in vivo* can then lead to lipid peroxidation, which has been demonstrated both *in vivo* (Bacon *et al.*, 1983; Houglum *et al.*, 1990) and *in vitro* (Link *et al.*, 1985; O'Connell *et al.*, 1985) under iron-overloaded conditions. Lipid peroxidation of intracellular organelle membranes would have deleterious effects on their function and hence on the function of the cell itself (Bacon and Britton, 1989). In addition to lipid peroxidation, the generation of radical species by excess iron stores leads to the irreversible oxidation of ascorbic acid (Bothwell *et al.*, 1995). This can be demonstrated by the administration of large doses of ascorbic acid to patients. If the iron status is normal, most of the ascorbic acid is excreted in the urine; if, however, iron overload is present, the ascorbic acid is oxidized to oxalic acid which then appears in the urine (Lynch *et al.*, 1967). Ascorbic acid deficiency can then lead to a reduction in the amount of iron available for erythropoiesis and there is also some evidence that chronic ascorbic acid deficiency can lead to osteoporosis (Bothwell *et al.*, 1995). Once diagnosed, hereditary hemochromatosis can be treated by weekly venesections of approximately 0.5 L of blood (Bothwell *et al.*, 1979). The amount of iron removed in this way will vary but is usually around 250 mg per 0.5 L of blood (Bothwell *et al.*, 1995). The duration of venesection therapy varies, depending on the amount of iron present, but usually lasts for 2–3 years (Williams *et al.*, 1969). Another treatment option is chelation therapy using deferoxamine; however; this treatment is expensive and the rate of iron removal is slower (Hussain *et al.*, 1977; Propper *et al.*, 1977; Pippard *et al.*, 1978). It has not been widely recommended for the treatment of hereditary hemochromatosis.

6.3.2.2 *Thalassemia*

As mentioned previously, the underlying defect in all thalassemias is the decreased or absent production of one or more globin proteins. The most clinically important forms of thalassemia are α- and β-thalassemia. These can be further subdivided into disorders in which no chains are produced from the affected chromosomes (α^0- and β^0-thalassemia), and those where some chains are synthesized but at a greatly reduced rate (α^+- and β^+-thalassemia). The most common defects leading to α^+-thalassemia are deletions (-$\alpha^{3.7}$ and -$\alpha^{4.2}$) involving one or the other of the duplicated α-globin genes

(Dozy *et al.*, 1979; Higgs *et al.*, 1979). Although much less common, deletion of both α-globin genes, leading to α^0 thalassemia, has also been described (Felice *et al.*, 1984; Higgs *et al.*, 1989; Villegas *et al.*, 1990; Fortina *et al.*, 1991; Waye *et al.*, 1992; Weatherall, 1994). In addition to these gene deletion events, α-thalassemia resulting from mutations in either of the α-globin genes, or in the α-globin regulatory region (αHS-40) is also possible. The picture for the β-thalassemias is much the same. Several deletions that affect the β-globin gene have been reported (Gilman *et al.*, 1984; Padanilam *et al.*, 1984; Thein *et al.*, 1984; Popovich *et al.*, 1986; Diaz-Chico *et al.*, 1987; Anand *et al.*, 1988; Aulehla-Scholz *et al.*, 1989) as well as mutations affecting transcription (Orkin *et al.*, 1983; Antonarakis *et al.*, 1984; Gonzalez-Redondo *et al.*, 1989), splice-site mutations (Antonarakis *et al.*, 1984; Kazazian, 1990) and, in rare cases, deletion of the βLCR (Van der Ploeg *et al.*, 1980; Curtin *et al.*, 1985; Taramelli *et al.*, 1986).

As mentioned previously, thalassemia is not technically a disorder of iron metabolism but patients with the severe form of the disease suffer from iron overload as a side-effect of the disorder. This increase in iron absorption can be explained by looking at the underlying causes of the disorder. The β-globin production in homozygous forms of β-thalassemia is either nonexistent or very low, leading to an excess of α-globin chains in red-cell precursors. Since the α-globin chains are not able to form a viable tetramer, they precipitate into inclusion bodies which can be seen by both light and electron microscopy (Fessas, 1963; Wickramasinghe and Hughes, 1978; Weatherall *et al.*, 1995). The presence of these inclusion bodies leads to the destruction of red-cell precursors, which in turn leads to ineffective erythropoiesis. Mature cells that are released undergo hemolysis due to the presence of excess α-globin chains which, during their degradation, generate free radicals causing extensive membrane damage (Rund and Rachmilewitz, 1995). The combination of ineffective erythropoiesis and hemolysis leads to a general state of anemia and tissue hypoxia. The hypoxic condition stimulates the release of erythropoietin which, in turn, leads to the increased absorption of intestinal iron and the expansion of the bone marrow. Iron loading is further compounded by the fact that blood transfusions are needed to relieve anemia. While this results in a decrease of intestinal iron, the additional input of iron from the transfusion continues to accumulate (Modell and Berdoukas, 1981).

The steady absorption of iron causes it to be deposited in various tissues, starting with the liver and spleen and followed by the myocardium. Deposition of iron in the myocardium usually results in death by intractable cardiac failure (Weatherall *et al.*, 1995). Other clinical conditions which are related to the deposition of iron are hypoparathyroidism and hypogonadism (Modell and Berdoukas, 1981; Bunn and Forget, 1986). It has also been observed that patients with the severe form of thalassemia are more susceptible to bacterial infection, possibly due to the increase in serum iron level which may favor bacterial growth (Modell and Berdoukas, 1981; Bunn and Forget, 1986). It is interesting to note that iron overload is less common in the adult forms of α-thalassemia. This may reflect the fundamental difference between the α- and β-thalassemias. As already discussed, the excess α-globin chains cannot form viable tetramers and precipitate in red-cell precursors, leading to their destruction. The excess β-chains present in α-thalassemia are able to form homotetramers (β_4) which are soluble and do not precipitate in the bone marrow. This is paralleled in the fetal state where excess γ-chains form soluble γ_4 homotetramers. Hence the α-thalassemias are not characterized by a severe degree of ineffective erythropoiesis and hence a milder degree of anemia. The treatment for homozygous β-thalassemias, as in other metal overload disorders, is chelation therapy. The chelator most widely used in the treatment of β-thalassemia is deferoxamine (administered subcutaneously); however, many issues regarding the use of this drug, such as dose-related toxicity and recommended age of initiation, remain unresolved (Rund and Rachmilewitz, 1995). The search for safer, orally adminstered iron chelators has intensified in recent years, leading to clinical trials of many potential new iron chelators such as L1 (Hoffbrand, 1995; Rund and Rachmilewitz, 1995).

6.3.2.3 Aceruloplasminemia

In all cases of aceruloplasminemia studied thus far, the activity of ceruloplasmin has been knocked out due to mutations in the gene which lead to the truncation of the protein. Patients with aceruloplasminemia accumulate iron in various tissues, such as the liver, brain and in the islets of Langerhans with selective loss of pancreatic β-cells. Specifically, magnetic resonance imaging reveals an increase in the levels of iron in the basal ganglia, as well as iron deposits in the microglia and neurons (Harris *et al.*, 1998). These observations are consistent with previous biochemical studies which suggested direct involvement of ceruloplasmin in the iron metabolism pathway. These findings can be explained by examining the function of ceruloplasmin in the context of the movement of iron from one organ to another. In the absence of ceruloplasmin the oxidation of iron from the ferrous to ferric state cannot occur and hence iron cannot be loaded into transferrin for transport to the bone marrow and other organs. This, in turn, leads to an accumulation of ferrous iron in the plasma where it is then rapidly removed from circulation and accumulates in the liver and other organs (Craven *et al.*, 1987; Harris *et al.*, 1998). Evaluation of the hepatic iron and serum ferritin levels in affected individuals shows that they are equivalent to those seen in hemochromatosis patients. However, aceruloplasminemia patients do not suffer from severe anemia since the nonenzymatic oxidation of iron allows some to be loaded into transferrin (Harris *et al.*, 1998).

The clinical features of aceruloplasminemia consist of neurologic disease, diabetes mellitus and retinal degeneration, despite clear evidence of iron overload in the liver (Miyajima *et al.*, 1987; Logan *et al.*, 1994). These observations suggest that ceruloplasmin is playing a critical role in the metabolism of iron in the brain despite the fact that ceruloplasmin is unable to cross the blood–brain barrier, implying that ceruloplasmin gene expression must occur in the brain. This point was clarified by *in situ* hybridization studies on murine and human central nervous system tissue, which demonstrated abundant ceruloplasmin gene expression in astrocytes of the retina and the brain as well as the epithelium of the choroid plexus (Klomp *et al.*, 1996). Further studies of primary cell cultures confirmed the expression, synthesis and secretion of functional ceruloplasmin by astrocytes. Under normal conditions, iron in the brain is oxidized by astrocyte-secreted ceruloplasmin, followed by incorporation into transferrin which is synthesized by oligodendrocytes (Harris *et al.*, 1998). Without ceruloplasmin activity the ferrous iron is taken up by brain parenchymal cells, paralleling the situation in the peripheral tissues and leading to iron overload and cell death. This contention is supported by studies indicating a significant increase in plasma lipid peroxidation in patients with aceruloplasminemia (Miyajima *et al.*, 1996).

It is interesting to note that patients with Wilson disease do not show the symptoms of aceruloplasminemia despite extremely low levels of ceruloplasmin. This suggests that the extrahepatic production of ceruloplasmin is enough to carry out iron metabolism and avoid the effects suffered by aceruloplasminemia patients. As with the treatment of patients with Wilson disease, treatment of aceruloplasminemia involves the use of chelating agents to mediate the removal of iron from various tissues. Among the chelators being evaluated is deferoxamine which has a high affinity for the ferric form of iron, binding in a 1 : 1 complex form. This chelator has been shown to cross the blood–brain barrier and mediate the removal of excess iron in patients afflicted with either inherited or acquired forms of iron overload disorders (Summers *et al.*, 1979). Initial trials using deferoxamine to treat aceruloplasminemia patients have shown a decrease in neurologic symptoms, as well as in amelioration of diabetes and a reduction in body iron stores (Miyajima *et al.*, 1997). Although encouraging, these results must be confirmed by further studies and may provide the basis for treating individuals at risk for aceruloplasminemia, including those who are asymptomatic.

6.4 Conclusions

The cloning of the Wilson and Menkes disease genes and the expression and characterization of their gene products have ushered in a new era in the studies of copper-transport mechanisms. The knowledge of the intracellular localization of transporter proteins and the understanding of their functions will be key to our understanding of how the defective ATPase gene interferes with the transport of copper. These studies will pave the way to developing new therapeutic strategies for the treatment of both Wilson and Menkes diseases. The prevalence of the hemochromatosis gene in the general population presents a considerable risk of excess iron absorption. Consequently, idiopathic hemochromatosis poses a major problem in a situation of iron overload. Individuals afflicted with this condition are treated by performing phlebotomies throughout their life span. If a fetus is suspected to have thalassemia because of its genetic background, recombinant DNA techniques can be used to determine if one or more globin genes are missing from its genome. In spite of these advances, a satisfactory treatment for thalassemia is lacking, other than the use of deferroxamine. Iron and copper are two major redox metals in the biological systems which can cause oxidative damage to tissues. The studies of aceruloplasminemia provide an opportunity to investigate the interplay of copper and iron metabolism and neurodegeneration. We can expect major advances in the area of metal-mediated neuronal injury through future research into the pathogenesis and pathophysiology of aceruloplasminemia.

6.5 References

Alexiou, D., Hatzis, T. and Koutselinis, A. 1985. Maintenance treatment of Wilson's disease with oral zinc. Apropos of a child treated for 4 years. *Archives Françaises de Pédiatrie* **42**, 447–9.

Anand, R., Boehm, C.D., Kazazian, H.H. Jr and Vanin, E.F. 1988. Molecular characterization of a beta zero-thalassemia resulting from a 1.4 kilobase deletion. *Blood* 72, 636–41.

Anderson, G.J. 1996. Control of iron absorption. *Journal of Gastroenterology and Hepatology* **11**, 1030–2.

Antonarakis, S.E., Irkin, S.H., Cheng, T.C., Scott, A.F., Sexton, J.P., Trusko, S.P., Charache, S. and Kazazian, H.H. Jr 1984. beta-Thalassemia in American Blacks: novel mutations in the

'TATA' box and an acceptor splice site. *Proceedings of the National Academy of Sciences, USA* **81**, 1154–8.

Aulehla-Scholz, C., Spiegelberg, R. and Horst, J. 1989. A beta-thalassemia mutant caused by a 300-bp deletion in the human beta-globin gene. *Human Genetics* **81**, 298–9.

Bacon, B.R. 1997. Diagnosis and management of hemochromatosis. *Gastroenterology* **113**, 995–9.

Bacon, B.R. and Britton, R.S. 1989. Hepatic injury in chronic iron overload. Role of lipid peroxidation. *Chemico-Biological Interactions* **70**, 183–226.

Bacon, B.R. and Britton, R.S. 1990. The pathology of hepatic iron overload: a free radical-mediated process? *Hepatology* **11**, 127–37.

Bacon, B.R., Tavill, A.S., Brittenham, G.M., Park, C.H. and Recknagel, R.O. 1983. Hepatic lipid peroxidation *in vivo* in rats with chronic iron overload. *Journal of Clincal Investigation* **71**, 429–39.

Baldwin, J.M. 1980. The structure of human carbonmonoxy haemoglobin at 2.7 Å resolution. *Journal of Molecular Biology* **136**, 103–28.

Bankier, A. 1995. Menkes disease. *Journal of Medical Genetics* **32**, 213–15.

Barrow, L. and Tanner, M.S. 1988. Copper distribution among serum proteins in paediatric liver disorders and malignancies. *European Jornal of Clinical Investigation* **18**, 555–60.

Beratis, N.G., Price, P., Labadie, G. and Hirschhorn, K. 1978. ^{64}Cu metabolism in Menkes and normal cultured skin fibroblasts. *Pediatric Research* **12**, 699–702.

Bothwell, T.H., Charlton, R.W., Cook, J.D. and Finch, C.A. 1979. *Iron Metabolism in Man*, Blackwell Scientific, Oxford.

Bothwell, T.H., Charlton, R.W. and Motulsky, A.G. 1995. Hemochromatosis, in *Metabolic Basis of Inherited Disease*, (eds C.R. Scriver, A.I. Beaudet, W.S. Sly and D. Valle), McGraw-Hill, New York, pp. 2237–69.

Bowcock, A.M., Farrer, L.A., Cavalli-Sforza, L.L., Hebert, J.M., Kidd, K.K., Frydman, M. and Bonne-Tamir, B. 1987. Mapping the Wilson disease locus to a cluster of linked polymorphic markers on chromosome 13. *American Journal of Human Genetics* **41**, 27–35.

Bowcock, A.M., Farrer, L.A., Hebert, J.M., Agger, M., Sternlieb, I., Scheinberg, I.H., Buys, C.H., Scheffer, H., Frydman, M., Chajek-Saul, T. *et al.* 1988. Eight closely linked loci place the Wilson disease locus within 13q14–q21. *American Journal of Human Genetics* **43**, 664–74.

Brewer, G.J., Hill, G.M., Prasad, A.S., Cossack, Z.T. and Rabbani, P. 1983. Oral zinc therapy for Wilson's disease. *Annals of Internal Medicine* **99**, 314–19.

Brewer, G.J., Dick, R.D., Johnson, V., Wang, Y., Yuzbasiyan-Gurkan, V., Kluin, K., Fink, J.K. and Aisen, A. 1994. Treatment of Wilson's disease with ammonium tetrathiomolybdate. I. Initial therapy in 17 neurologically affected patients. *Archives of Neurology* **51**, 545–54.

Bull, P.C., Thomas, G.R., Rommens, J.M., Forbes, J.R. and Cox, D.W. 1993. The Wilson disease gene is a putative copper transporting P-type ATPase similar to the Menkes gene. *Nature Genetics* **5**, 327–37.

Bunn, H.F. and Forget, B.G. 1986. *Hemoglobin: Molecular, Genetic, and Clinical Aspects*, Saunders, Philadelphia.

Campbell, C.H., Brown, R. and Linder, M.C. 1981. Circulating ceruloplasmin is an important source of copper for normal and malignant animal cells. *Biochimica et Biophysica Acta* **678**, 27–38.

Chada, K., Magram, J. and Costantini, F. 1986. An embryonic pattern of expression of a human fetal globin gene in transgenic mice. *Nature* **319**, 685–9.

Chan, W.Y., Garnica, A.D. and Rennert, O.M. 1978. Cell culture studies of Menkes kinky hair disease. *Clinica Chimica Acta* **88**, 495–507.

Charlton, R.W., Abrahams, C. and Bothwell, T.H. 1967. Idiopathic hemochromatosis in young subjects. Clinical, pathological, and chemical findings in four patients. *Archives of Pathology* **83**, 132–40.

Charnay, P., Treisman, R., Mellon, P., Chao, M., Axel, R. and Maniatis, T. 1984. Differences in human alpha- and beta-globin gene expression in mouse erythroleukemia cells: the role of intragenic sequences. *Cell* **38**, 251–63.

Chelly, J., Tumer, Z., Tonnesen, T., Petterson, A., Ishikawa-Brush, Y., Tommerup, N., Horn, N. and Monaco, A. P. 1993. Isolation of a candidate gene for Menkes disease that encodes a potential heavy metal binding protein. *Nature Genetics* **3**, 14–29.

Chiu, M.K. and Davey, A.M. 1997. Neonatal hemochromatosis. *Clinical Pediatrics* **36**, 607–10.

Chuang, L.M., Wu, H.P., Jang, M.H., Wang, T.R., Sue, W.C., Lin, B.J., Cox, D.W. and Tai, T.Y. 1996. High frequency of two mutations in codon 778 in exon 8 of the ATP7B gene in Taiwanese families with Wilson disease. *Journal of Medical Genetics* **33**, 521–523.

Clegg, J.B. 1987. Can the product of the theta gene be a real globin? *Nature* **329**, 465–6.

Collins, F.S. and Weissman, S.M. 1984. The molecular genetics of human hemoglobin, in *Progress in Nucleic Acids Research and Molecular Biology*, (eds W.E. Cohn and K. Moldave), Academic Press, New York, p. 315.

Costantini, F., Radice, G., Magram, J., Stamatoyannopoulos, G., Papayannopoulou, T. and Chada, K. 1985. Developmental regulation of human globin genes in transgenic mice. *Cold Spring Harbor Symposia in Quantum Biology* **50**, 361–70.

Cox, D.W. 1995. Genes of the copper pathway. *American Journal of Human Genetics* **56**, 828–34.

Craven, C.M., Alexander, J., Eldridge, M., Kushner, J.P., Bernstein, S. and Kaplan, J. 1987. Tissue distribution and clearance kinetics of non-transferrin-bound iron in the hypotransferrinemic mouse: a rodent model for hemochromatosis. *Proceedings of the National Academy of Sciences, USA* **84**, 3457–61.

Curtin, P., Pirastu, M., Kan, Y.W., Gobert-Jones, J.A., Stephens, A.D. and Lehmann, H. 1985. A distant gene deletion affects beta-globin gene function in an atypical gamma delta beta-thalassemia. *Journal of Clinical Investigation* **76**, 1554–8.

Daimon, M., Kato, T., Kawanami, T., Tominaga, M., Igarashi, M., Yamatani, K. and Sasaki, H. 1995. A nonsense mutation of the ceruloplasmin gene in hereditary ceruloplasmin deficiency with diabetes mellitus. *Biochemical and Biophysical Research Communications* **217**, 89–95.

Dameron, C.T. and Harris, E.D. 1987. Regulation of aortic Cu,Zn-superoxide dismutase with copper. Caeruloplasmin and albumin re-activate and transfer copper to the enzyme in culture. *Biochemical Journal* **248**, 669–75.

Danks, D.M. 1995. Disorders of Copper Transport, in *Metabolic Basis of Inherited Disease*, (eds C.R. Scriver, A.I. Beaudet, W.S. Sly and D. Valle), New York NY: McGraw-Hill, New York, pp. 2221–35.

Danks, D.M., Campbell, P.E., Stevens, B.J., Mayne, V. and Cartwright, E. 1972a. Menke's kinky hair syndrome. An inherited defect in copper absorption with widespread effects. *Pediatrics* **50**, 188–201.

Danks, D.M., Campbell, P.E., Walker-Smith, J., Stevens, B.J., Gillespie, J.M., Blomfield, J. and Turner, B. 1972b. Menkes' kinky-hair syndrome. *Lancet* **1**, 1100–2.

Danks, D.M., Cartwright, E., Stevens, B.J. and Townley, R.R. 1973. Menkes' kinky hair disease: further definition of the defect in copper transport. *Science* **179**, 1140–2.

Das, S., Levinson, B., Whitney, S., Vulpe, C., Packman, S. and Gitschier, J. 1994. Diverse mutations in patients with Menkes disease often lead to exon skipping. *American Journal of Human Genetics* **55**, 883–889.

Deisseroth, A., Velez, R. and Nienhuis, A.W. 1976. Hemoglobin synthesis in somatic cell hybrids: independent segregation of the human alpha- and beta-globin genes. *Science* **191**, 1262–4.

Deisseroth, A., Nienhuis, A., Turner, P., Velez, R., Anderson, W.F., Ruddle, F., Lawrence, J., Creagan, R. and Kucherlapati, R. 1977. Localization of the human alpha-globin structural gene to chromosome 16 in somatic cell hybrids by molecular hybridization assay. *Cell* **12**, 205–18.

Deisseroth, A., Nienhuis, A., Lawrence, J., Giles, R., Turner, P. and Ruddle, F.H. 1978. Chromosomal localization of human beta globin gene on human chromosome 11 in somatic cell hybrids. *Proceedings of the National Academy of Sciences, USA* **75**, 1456–60.

de Sousa, M., Reimao, R., Lacerda, R., Hugo, P., Kaufmann, S.H. and Porto, G. 1994. Iron overload in beta 2-microglobulin-deficient mice. *Immunology Letters* **39**, 105–11.

Diaz-Chico, J.C., Yang, K.G., Kutlar, A., Reese, A.L., Aksoy, M. and Huisman, T.H. 1987. An approximately 300 bp deletion involving part of the 5′ beta-globin gene region is observed in members of a Turkish family with beta-thalassemia. *Blood* **70**, 583–6.

DiDonato, M., Narindrasorasak, S., Forbes, J.R., Cox, D.W. and Sarkar, B. 1997. Expression, purification, and metal binding properties of the N-terminal domain from the Wilson disease putative copper-transporting ATPase (ATP7B). *Journal of Biological Chemistry* **272**, 33279.

Dierick, H.A., Ambrosini, L., Spencer, J., Glover, T.W. and Mercer, J.F. 1995. Molecular structure of the Menkes disease gene (ATP7A). *Genomics* **28**, 462–9.

Dijkstra, M., Vonk, R.J. and Kuipers, F. 1996. How does copper get into bile? New insights into the mechanism(s) of hepatobiliary copper transport. *Journal of Hepatology* **24**, 109–20.

Dobyns, W.B., Goldstein, N.P. and Gordon, H. 1979. Clinical spectrum of Wilson's disease (hepatolenticular degeneration). *Mayo Clinic Proceedings* **54**, 35–42.

Dozy, A.M., Kan, Y.W., Embury, S.H., Mentzer, W.C., Wang, W.C., Lubin, B., Davis, J.R. Jr and Koenig, H.M. 1979. alpha-Globin gene organisation in blacks precludes the severe form of alpha-thalassaemia. *Nature* **280**, 605–7.

Driscoll, M.C., Dobkin, C.S. and Alter, B.P. 1989. Gamma delta beta-thalassemia due to a *de novo* mutation deleting the 5′ beta-globin gene activation-region hypersensitive sites. *Proceedings of the National Academy of Sciences, USA* **86**, 7470–7474.

Escobar, G.J., Heyman, M.B., Smith, W.B. and Thaler, M.M. 1987. Primary hemochromatosis in childhood. *Pediatrics* **80**, 549–54.

Evans, T., Felsenfeld, G. and Reitman, M. 1990. Control of globin gene transcription. *Annual Review of Cell Biology* **6**, 95–124.

Fairbanks, V.F., Fahey, J.L. and Beutler, E. 1971. *Clinical Disorders of Iron Metabolism*, Grune and Stratton, New York.

Falchetto, R., Vorherr, T., Brunner, J. and Carafoli, E. 1991. The plasma membrane Ca^{2+} pump contains a site that interacts with its calmodulin-binding domain. *Journal of Biological Chemistry* **266**, 2930–6.

Feder, J.N., Gnirke, A., Thomas, W., Tsuchihashi, Z., Ruddy, D. A., Basava, A., Dormishian, F., Domingo, R. Jr, Ellis, M.C., Fullan, A. *et al.* 1996. A novel MHC class I-like gene is mutated in patients with hereditary haemochromatosis. *Nature Genetics* **13**, 399–408.

Felice, A.E., Cleek, M.P., McKie, K., McKie, V. and Huisman, T.H. 1984. The rare alpha-thalassemia-1 of blacks is a zeta alpha-thalassemia-1 associated with deletion of all alpha- and zeta-globin genes. *Blood* **63**, 1253–7.

Fermi, G., Perutz, M.F., Shaanan, B. and Fourme, R. 1984. The crystal structure of human deoxyhaemoglobin at 1.74 Å resolution. *Journal of Molecular Biology* **175**, 159–74.

Fessas, P. 1963. Inclusions of hemoglobin in erythroblasts and erythrocytes of thalassemia. *Blood* **21**, 21.

Figus, A., Angius, A., Loudianos, G., Bertini, C., Dessi, V., Loi, A., Deiana, M., Lovicu, M., Olla, N., Sole, G. *et al.* 1995. Molecular pathology and haplotype analysis of Wilson disease in Mediterranean populations. *American Journal of Human Genetics* **57**, 1318–24.

Finch, S.C. and Finch, C.A. 1955. Idiopathic hemochromatosis, an iron storage disease. A. Iron metabolism in hemochromatosis. *Medicine (Baltimore)* **34**, 381.

Fleet, J.C. 1996. Discovery of the hemochromatosis gene will require rethinking the regulation of iron metabolism. *Nutrition Reviews* **54**, 285–7.

Forrester, W.C., Thompson, C., Elder, J.T. and Groudine, M. 1986. A developmentally stable chromatin structure in the human beta-globin gene cluster. *Proceedings of the National Academy of Sciences, USA* **83**, 1359–63.

Fortina, P., Dianzani, I., Serra, A., Gottardi, E., Saglio, G., Farinasso, L., Piga, A., Gabutti, V. and Camaschella, C. 1991. A newly-characterized alpha-thalassaemia-1 deletion removes the entire alpha-like globin gene cluster in an Italian family. *British Journal of Haematology* **78**, 529–34.

Fraser, P., Hurst, J., Collis, P. and Grosveld, F. 1990. DNaseI hypersensitive sites 1, 2 and 3 of the human beta-globin dominant control region direct position-independent expression. *Nucleic Acids Research* **18**, 3503–8.

Frieden, E. 1986. Perspectives on copper biochemistry. *Clinical Physiology and Biochemistry* **4**, 11–29.

Frydman, M., Bonne-Tamir, B., Farrer, L.A., Conneally, P.M., Magazanik, A., Ashbel, S. and Goldwitch, Z. 1985. Assignment of the gene for Wilson disease to chromosome 13: linkage to the esterase D locus. *Proceedings of the National Academy of Sciences, USA* **82**, 1819–21.

Gilman, J.G., Huisman, T.H. and Abels, J. 1984. Dutch beta 0-thalassaemia: a 10 kilobase DNA deletion associated with significant gamma-chain production. *British Journal of Haematology* **56**, 339–48.

Gitlin, J.D., Schroeder, J.J., Lee Ambrose, L.M. and Cousins, R.J. 1992. Mechanisms of caeruloplasmin biosynthesis in normal and copper-deficient rats. *Biochemical Journal* **282**, 835–9.

Goka, T.J., Stevenson, R.E., Hefferan, P.M. and Howell, R.R. 1976. Menkes disease: a biochemical abnormality in cultured human fibroblasts. *Proceedings of the National Academy of Sciences, USA* **73**, 604–6.

Gonzalez-Redondo, J.M., Stoming, T.A., Kutlar, A., Kutlar, F., Lanclos, K.D., Howard, E.F., Fei, Y.J., Aksoy, M., Altay, C., Gurgey, A. *et al.* 1989. A C–T substitution at nt-101 in a conserved DNA sequence of the promotor region of the beta-globin gene is associated with 'silent' beta-thalassemia. *Blood* **73**, 1705–11.

Grosveld, G.C., de Boer, E., Shewmaker, C.K. and Flavell, R.A. 1982. DNA sequences necessary for transcription of the rabbit beta-globin gene *in vivo*. *Nature* **295**, 120–6.

Grosveld, F., van Assendelft, G.B., Greaves, D.R. and Kollias, G. 1987. Position-independent, high-level expression of the human beta-globin gene in transgenic mice. *Cell* **51**, 975–85.

Gutteridge, J.M. (1989). Iron and oxygen: a biologically damaging mixture. *Acta Paediatrica Scandinavia (Supplement)* **361**, 78–85.

Haddy, T.B., Castro, O.L. and Rana, S.R. 1988. Hereditary hemochromatosis in children, adolescents, and young adults. *American Journal of Pediatrics, Hematology–Oncology* **10**, 23–34.

Hardison, R.C., Sawada, I., Cheng, J.F., Shen, C.K. and Schmid, C.W. 1986. A previously undetected pseudogene in the human alpha globin gene cluster. *Nucleic Acids Research* **14**, 1903–11.

Harris, Z.L., Takahashi, Y., Miyajima, H., Serizawa, M., MacGillivray, R.T. and Gitlin, J.D. 1995. Aceruloplasminemia: molecular characterization of this disorder of iron metabolism. *Proceedings of the National Academy of Sciences, USA* **92**, 2539–43.

Harris, Z.L., Migas, M.C., Hughes, A.E., Logan, J.I. and Gitlin, J.D. 1996. Familial dementia due to a frameshift mutation in the ceruloplasmin gene. *Quarterly Journal of Medicine* **89**, 355–9.

Harris, Z.L., Klomp, L.W.J. and Gitlin, J.D. 1998. Aceruloplasminemia: an inherited neurodegenerative disease with impairment of iron homeostasis. *American Journal of Clinical Nutrition* **67** (supplement), 972S–977S.

Hatton, C.S., Wilkie, A.O., Drysdale, H.C., Wood, W.G., Vickers, M.A., Sharpe, J., Ayyub, H., Pretorius, I.M., Buckle, V.J. and Higgs, D.R. 1990. Alpha-thalassemia caused by a large (62 kb) deletion upstream of the human alpha globin gene cluster. *Blood* **76**, 221–7.

Herd, S.M., Camakaris, J., Christofferson, R., Wookey, P. and Danks, D.M. 1987. Uptake and efflux of copper-64 in Menkes'-disease and normal continuous lymphoid cell lines. *Biochemical Journal* **247**, 341–7.

Herzberg, N.H., Wolterman, R.A., van den Berg, G.J., Barth, P.G. and Bolhuis, P.A. 1990. Metallothionein in Menkes' disease: induction in cultured muscle cells. *Journal of Neurological Sciences* **100**, 50–6.

Higgs, D.R., Pressley, L., Old. J.M., Hunt, D.M., Clegg, J.B., Weatherall, D.J. and Serjeant, G.R. 1979. Negro alpha-thalassaemia is caused by deletion of a single alpha-globin gene. *Lancet* **2**, 272–6.

Higgs, D.R., Old, J.M., Pressley, L., Clegg, J.B. and Weatherall, D.J. 1980. A novel alpha-globin gene arrangement in man. *Nature* **284**, 632–5.

Higgs, D.R., Vickers, M.A., Wilkie, A.O., Pretorius, I.M., Jarman, A.P. and Weatherall, D.J. 1989. A review of the molecular genetics of the human alpha-globin gene cluster. *Blood* **73**, 1081–104.

Higgs, D.R., Wood, W.G., Jarman, A.P., Sharpe, J., Lida, J., Pretorius, I.M. and Ayyub, H. 1990. A major positive regulatory region located far upstream of the human alpha-globin gene locus. *Genes and Development* **4**, 1588–601.

Hilton, M., Spenser, D.C., Ross, P., Ramsey, A. and Mcardle, H.J. 1995. Characterisation of the copper uptake mechanism and isolation of the ceruloplasmin receptor/copper transporter in human placental vesicles. *Biochimica et Biophysica Acta* **1245**, 153–60.

Hoffbrand, A.V. 1995. Iron chelation therapy. *Current Opinion in Hematology* **2**, 153–8.

Hoogenraad, T.U., Van den Hamer, C.J., Koevoet, R. and Korver, E.G. 1978. Oral zinc in Wilson's disease. *Lancet* **2**, 1262.

Hoogenraad, T.U., Koevoet, R. and de Ruyter Korver, E.G. 1979. Oral zinc sulphate as long-term treatment in Wilson's disease (hepatolenticular degeneration). *European Neurology* **18**, 205–11.

Houglum, K., Filip, M., Witztum, J.L. and Chojkier, M. 1990. Malondialdehyde and 4-hydroxynonenal protein adducts in plasma and liver of rats with iron overload. *Journal of Clinical Investigation* **86**, 1991–8.

Hsieh, H.S. and Frieden, E. 1975. Evidence for ceruloplasmin as a copper transport protein. *Biochemical and Biophysical Research Communications* **67**, 1326–31.

Hung, I.H., Suzuki, M., Yamaguchi, Y., Yuan, D.S., Klausner, R.D. and Gitlin, J.D. 1997. Biochemical characterization of the Wilson disease protein and functional expression in the yeast *Saccharomyces cerevisiae*. *Journal of Biological Chemistry* **272**, 21461–6.

Hunziker, P.E. and Sternlieb, I. 1991. Copper metallothionein in patients with hepatic copper overload. *European Journal of Clinical Investigation* **21**, 466–71.

Hussain, M.A., Green, N., Flynn, D.M. and Hoffbrand, A.V. 1977. Effect of dose, time, and ascorbate on iron excretion after subcutaneous desferrioxamine. *Lancet* **1**, 977–9.

James, P., Inui, M., Tada, M., Chiesi, M. and Carafoli, E. 1989. Nature and site of phospholamban regulation of the Ca^{2+} pump of sarcoplasmic reticulum. *Nature* **342**, 90–2.

Jarman, A.P., Wood, W.G., Sharpe, J.A., Gourdon, G., Ayyub, H. and Higgs, D.R. 1991. Characterization of the major regulatory element upstream of the human alpha-globin gene cluster. *Molecular and Cellular Biology* **11**, 4679–89.

Kazazian, H.H. Jr 1990. The thalassemia syndromes: molecular basis and prenatal diagnosis in 1990. *Seminars in Hematology* **27**, 209–28.

Klomp, L.W., Farhangrazi, Z.S., Dugan, L.L. and Gitlin, J.D. 1996. Ceruloplasmin gene expression in the murine central nervous system. *Journal of Clinical Investigation* **98**, 207–15.

Kollias, G., Wrighton, N., Hurst, J. and Grosveld, F. 1986. Regulated expression of human A gamma-, beta-, and hybrid gamma beta-globin genes in transgenic mice: manipulation of the developmental expression patterns. *Cell* **46**, 89–94.

Kreuder, J., Otten, A., Fuder, H., Tumer, Z., Tonnesen, T., Horn, N. and Dralle, D. 1993. Clinical and biochemical consequences of copper-histidine therapy in Menkes disease. *European Journal of Pediatrics* **152**, 828–32.

Kuivaniemi, H., Peltonen, L., Palotie, A., Kaitila, I. and Kivirikko, K.I. 1982. Abnormal copper metabolism and deficient lysyl oxidase activity in a heritable connective tissue disorder. *Journal of Clinical Investigation* **69**, 730–3.

Larsen, F.L. and Vincenzi, F.F. 1979. Calcium transport across the plasma membrane: stimulation by calmodulin. *Science* **204**, 306–9.

Larsen, F.L., Katz, S. and Roufogalis, B.D. 1981. Calmodulin regulation of Ca^{2+} transport in human erythrocytes. *Biochemical Journal* **200**, 185–91.

Lauer, J., Shen, C.K. and Maniatis, T. 1980. The chromosomal arrangement of human alphalike globin genes: sequence homology and alpha-globin gene deletions. *Cell* **20**, 119–30.

Link, G., Pinson, A. and Hershko, C. 1985. Heart cells in culture: a model of myocardial iron overload and chelation. *Journal of Laboratory and Clincal Medicine* **106**, 147–53.

Lipsky, M.A. and Gollan, J.L. 1987. Treatment of Wilson's disease: in D-penicillamine we trust – what about zinc? *Hepatology* **7**, 593–5.

Logan, J. I., Harveyson, K.B., Wisdom, G.B., Hughes, A.E. and Archbold, G.P. 1994. Hereditary caeruloplasmin deficiency, dementia and diabetes mellitus. *Quarterly Journal of Medicine* **87**, 663–70.

Loudianos, G., Dessi, V., Angius, A., Lovicu, M., Loi, A., Deiana, M., Akar, N., Vajro, P., Figus, A., Cao, A. and Pirastu, M. 1996. Wilson disease mutations associated with uncommon haplotypes in Mediterranean patients. *Human Genetics* **98**, 640–2.

Lutsenko, S. and Cooper, M.J. 1998. Localization of the Wilson's disease protein product to mitochondria. *Proceedings of the National Academy of Sciences, USA* **95**, 6004–9.

Lutsenko, S., Petrukhin, K., Cooper, M.J., Gilliam, C.T. and Kaplan, J.H. 1997. N-terminal domains of human copper-transporting adenosine triphosphatases (the Wilson's and Menkes disease proteins) bind copper selectively *in vivo* and *in vitro* with stoichiometry of one copper per metal-binding repeat. *Journal of Biological Chemistry* **272**, 18939–44.

Lynch, S.R., Seftel, H.C., Torrance, J.D., Charlton, R.W. and Bothwell, T.H. 1967. Accelerated oxidative catabolism of ascorbic acid in siderotic Bantu. *American Journal of Clinical Nutrition* **20**, 641–7.

McDonnell, S.M., and Witte, D. 1997. Hereditary hemochromatosis. Preventing chronic effects of this underdiagnosed disorder. *Postgraduate Medicine* **102**, 83–5, 88–91, 94.

McLaren, C.E., Gordeuk, V.R., Looker, A.C., Hasselblad, V., Edwards, C.Q., Griffen, L.M., Kushner, J.P. and Brittenham, G.M. 1995. Prevalence of heterozygotes for hemochromatosis in the white population of the United States. *Blood* **86**, 2021–7.

Marceau, N. and Aspin, N. 1973. The intracellular distribution of the radiocopper derived from ceruloplasmin and from albumin. *Biochimica et Biophysica Acta* **328**, 338–50.

Marks, J., Shaw, J.P. and Shen, C.K. (1986). Sequence organization and genomic complexity of primate theta 1 globin gene, a novel alpha-globin-like gene. *Nature* **321**, 785–8.

Menkes, J. H. 1988. Kinky hair disease: twenty five years later. *Brain Development* **10**, 77–9.

Menkes, J.H., Alter, M., Steigleder, G.K., Weakley, D.R., and Sung, J.H. 1962. A sex-linked recessive disorder with retardation of growth, peculiar hair, and focal cerebral and cerebellar degeneration. *Pediatrics* **29**, 764–79.

Mercer, J.F., Livingston, J., Hall, B., Paynter, J.A., Begy, C., Chandrasekharappa, S., Lockhart, P., Grimes, A., Bhave, M., Siemieniak, D. *et al.* 1993. Isolation of a partial candidate gene for Menkes disease by positional cloning. *Nature Genetics* **3**, 20–5.

Miyajima, H., Nishimura, Y., Mizoguchi, K., Sakamoto, M., Shimizu, T. and Honda, N. 1987. Familial apoceruloplasmin deficiency associated with blepharospasm and retinal degeneration. *Neurology* **37**, 761–7.

Miyajima, H., Takahashi, Y., Serizawa, M., Kaneko, E. and Gitlin, J.D. 1996. Increased plasma lipid peroxidation in patients with aceruloplasminemia. *Free Radicals in Biology and Medicine* **20**, 757–60.

Miyajima, H., Takahashi, Y., Kamata, T., Shimizu, H., Sakai, N. and Gitlin, J.D. 1997. Use of desferrioxamine in the treatment of aceruloplasminemia. *Annals of Neurology* **41**, 404–7.

Modell, C.B. and Berdoukas, V.A. 1981. *The Clinical Approach to Thalassemia*, Grune and Stratton, New York.
Morris, P.J., Vaughan, H., Tait, B.D. and Mackay, I.R. 1977. Histocompatibility antigens (HLA): associations with immunopathic diseases and with responses to microbial antigens. *Australian and New Zealand Journal of Medicine* 7, 616–24.
Motulsky, A.G. 1979. Genetics of hemochromatosis [letter]. *New England Journal of Medicine* **301**, 1291.
Mowat, A.P. 1994. *Liver Disorders in Childhood*, 3rd edn, Butterworth-Heinemann, Cambridge.
Nagano, K., Nakumura, K., Urakami, K.I., Umeyama, K., Uchiyama, H., Koiwai, K., Hattori, S., Yamamoto, T., Matsuda, I. and Endo, F. 1998. Intracellular distribution of the Wilson's disease gene product (ATPase7B) after *in vitro* and *in vivo* exogenous expression in hepatocytes from the LEC rat, an animal model of Wilson's disease. *Hepatology* **27**, 799–807.
Nanji, M.S., Nguyen, V.T.T., Kawasoe, J.H., Inui, K., Endo, F., Nakajima, T., Anezaki, T. and Cox, D.W. 1997. Haplotype and mutation analysis in Japanese patients with Wilson disease. *American Journal of Human Genetics* **60**, 1423–9.
Niederau, C., Fischer, R., Sonnenberg, A., Stremmel, W., Trampisch, H.J. and Strohmeyer, G. 1985. Survival and causes of death in cirrhotic and in noncirrhotic patients with primary hemochromatosis. *New England Journal of Medicine* **313**, 1256–62.
Nienhuis, A.W., and Maniatis, T. 1987. Structure and expression of globin genes in erythroid cells, in *The Molecular Basis of Blood Diseases*, (eds G. Stamatoyannopoulos, A.W. Nienhuis, P. Leder and P.W. Majerus), Saunders, Philadelphia, p. 28.
O'Connell, M.J., Ward, R.J., Baum, H. and Peters, T.J. 1985. The role of iron in ferritin- and haemosiderin-mediated lipid peroxidation in liposomes. *Biochemical Journal* **229**, 135–9.
Odermatt, A., Suter, H., Krapf, R. and Solioz, M. 1993. Primary structure of two P-type ATPases involved in copper homeostasis in *Enterococcus hirae*. *Journal of Biological Chemistry* **268**, 12775–9.
Okamoto, N., Wada, S., Oga, T., Kawabata, Y., Baba, Y., Habu, D., Takeda, Z. and Wada, Y. 1996. Hereditary ceruloplasmin deficiency with hemosiderosis. *Human Genetics* **97**, 755–8.
Onishi, T., Inubushi, H., Tokugawa, S., Muramatsu, M., Nishikawa, K., Suzuki, Y. and Miyao, M. 1980. Abnormal copper metabolism in Menkes cultured fibroblasts. *European Journal of Pediatrics* **134**, 205–10.
Orkin, S.H., Kazazian, H.H. Jr, Antonarakis, S.E., Goff, S.C., Boehm, C.D., Sexton, J.P., Waber, P. G. and Giardina, P.J. 1982. Linkage of beta-thalassaemia mutations and beta-globin gene polymorphisms with DNA polymorphisms in human beta-globin gene cluster. *Nature* **296**, 627–31.
Orkin, S.H., Sexton, J.P., Cheng, T.C., Goff, S.C., Giardina, P.J., Lee, J. I. and Kazazian, H.H. Jr 1983. ATA box transcription mutation in beta-thalassemia. *Nucleic Acids Research* **11**, 4727–34.
Orkin, S.H., Antonarakis, S.E. and Kazazian, H.H. Jr 1984. Base substitution at position -88 in a beta-thalassemic globin gene. Further evidence for the role of distal promoter element ACACCC. *Journal of Biological Chemistry* **259**, 8679–81.
Osaki, S. and Johnson, D.A. 1969. Mobilization of liver iron by ferroxidase (ceruloplasmin). *Journal of Biological Chemistry* **244**, 5757–8.
Padanilam, B.J., Felice, A.E. and Huisman, T.H. 1984. Partial deletion of the 5′ beta-globin gene region causes beta zero-thalassemia in members of an American black family. *Blood* **64**, 941–4.
Parkkila, S., Waheed, A., Britton, R.S., Feder, J.N., Tsuchihashi, Z., Schatzman, R.C., Bacon, B.R., and Sly, W.S. 1997. Immunohistochemistry of HLA-H, the protein defective in patients with hereditary hemochromatosis, reveals unique pattern of expression in gastrointestinal tract. *Proceedings of the National Academy of Sciences, USA* **94**, 2534–9.
Peltonen, L., Kuivaniemi, H., Palotie, A., Horn, N., Kaitila, I. and Kivirikko, K.I. 1983. Alterations in copper and collagen metabolism in the Menkes syndrome and a new subtype of the Ehlers–Danlos syndrome. *Biochemistry* **22**, 6156–63.

Perkins, K.W., McInnes, I.W.S., Blackburn, C.R.B., and Beal, R.W. 1965. Idiopathic hemochromatosis in children. Report of a family. *American Journal of Medicine* **39**, 118.

Peters, T.J. and Seymour, C.A. 1976. Acid hydrolase activities and lysosomal integrity in liver biopsies from patients with iron overload. *Clinical Science and Molecular Medicine* **50**, 75–8.

Petris, M.J., Mercer, J.F., Culvenor, J.G., Lockhart, P., Gleeson, P.A. and Camakaris, J. 1996. Ligand-regulated transport of the Menkes copper P-type ATPase efflux pump from the Golgi apparatus to the plasma membrane: a novel mechanism of regulated trafficking. *EMBO Journal* **15**, 6084–95.

Petrukhin, K., Fischer, S.G., Pirastu, M., Tanzi, R.E., Chernov, I., Devoto, M., Brzustowicz, L.M., Cayanis, E., Vitale, E., Russo, J.J. *et al.* 1993. Mapping, cloning and genetic characterization of the region containing the Wilson disease gene. *Nature Genetics* **5**, 338–43.

Petrukhin, K., Lutsenko, S., Chernov, I., Ross, B.M., Kaplan, J.H. and Gilliam, T.C. 1994. Characterization of the Wilson disease gene encoding a P-type copper transporting ATPase: genomic organization, alternative splicing, and structure/function predictions. *Human Molecular Genetics* **3**, 1647–56.

Phillips, M., Camakaris, J. and Danks, D.M. 1986. Comparisons of copper deficiency states in the murine mutants blotchy and brindled. Changes in copper-dependent enzyme activity in 13-day-old mice. *Biochemical Journal* **238**, 177–83.

Pippard, M.J., Callender, S.T. and Weatherall, D.J. 1978. Intensive iron-chelation therapy with desferrioxamine in iron-loading anaemias. *Clinical Science and Molecular Medicine* **54**, 99–106.

Popovich, B.W., Rosenblatt, D.S., Kendall, A.G. and Nishioka, Y. 1986. Molecular characterization of an atypical beta-thalassemia caused by a large deletion in the 5′ beta-globin gene region. *American Journal of Human Genetics* **39**, 797–810.

Propper, R.D., Cooper, B., Rufo, R.R., Nienhuis, A.W., Anderson, W.F., Bunn, H.F., Rosenthal, A. and Nathan, D.G. 1977. Continuous subcutaneous administration of deferoxamine in patients with iron overload. *New England Journal of Medicine* **297**, 418–23.

Proudfoot, N.J., Gil, A. and Maniatis, T. 1982. The structure of the human zeta-globin gene and a closely linked, nearly identical pseudogene. *Cell* **31**, 553–63.

Reilly, M., Daly, L. and Hutchinson, M. 1993. An epidemiological study of Wilson Disease in the Republic of Ireland. *Journal of Neurology, Neurosurgery, and Psychiatry* **56**, 298–300.

Riordan, J.R. and Jolicoeur-Paquet, L. 1982. Metallothionein accumulation may account for intracellular copper retention in Menkes' disease. *Journal of Biological Chemistry* **257**, 4639–45.

Roeser, H.P., Lee, G.R., Nacht, S. and Cartwright, G.E. 1970. The role of ceruloplasmin in iron metabolism. *Journal of Clinical Investigation* **49**, 2408–17.

Rothenberg, B.E. and Voland, J.R. 1996. beta2 knockout mice develop parenchymal iron overload: A putative role for class I genes of the major histocompatibility complex in iron metabolism. *Proceedings of the National Academy of Sciences, USA* **93**, 1529–34.

Rund, D. and Rachmilewitz, E. 1995. Advances in the pathophysiology and treatment of thalassemia. *Critical Reviews in Oncology–Hematology* **20**, 237–54.

Sarkar, B., Lingertat-Walsh, K. and Clarke, J.T. 1993. Copper-histidine therapy for Menkes disease. *Journal of Pediatrics* **123**, 828–30.

Sass-Kortsak, A. 1975. Wilson's disease. A treatable liver disease in children. *Pediatric Clinics in North America* **22**, 963–84.

Sato, M. and Gitlin, J.D. 1991. Mechanisms of copper incorporation during the biosynthesis of human ceruloplasmin. *Journal of Biological Chemistry* **266**, 5128–34.

Scheinberg, I.H. 1968. Toxicity of penicillamine. *Postgraduate Medical Journal* (Suppl.); 11–24.

Scheinberg, I.H. and Sternlieb, I. 1984. *Wilson's Disease*, Saunders, Philadelphia.

Scheinberg, I.H., Jaffe, M.E. and Sternlieb, I. 1987. The use of trientine in preventing the effects of interrupting penicillamine therapy in Wilson's disease. *New England Journal of Medicine* **317**, 209–213.

Schreiber, A.B., Schlessinger, J. and Edidin, M. 1984. Interaction between major histocompatibility complex antigens and epidermal growth factor receptors on human cells. *Journal of Cell Biology* **98**, 725–31.

Shah, A.B., Chernov, I., Zhang, H.T., Ross, B.M., Das, K., Lutsenko, S., Parano, E., Pavone, L., Evgrafov, O., IvanovaSmolenskaya, I.A. *et al.* 1997. Identification and analysis of mutations in the Wilson disease gene (ATP7B): Population frequencies, genotype–phenotype correlation, and functional analyses. *American Journal of Human Genetics* **61**, 317–28.

Sheldon, J.H. 1935. *Haemochromatosis*, Oxford University Press, London.

Sherwood, G., Sarkar, B. and Kortsak, A.S. 1989. Copper histidinate therapy in Menkes' disease: prevention of progressive neurodegeneration. *Journal of Inherited Metabolic Disease* **12** (Suppl. 2), 393–6.

Shewan, W.G., Mouat, S.A. and Allan, T.M. 1976. Letter: HLA antigens in haemochromatosis. *British Medical Journal* **1**, 281–2.

Shimizu, N., Kawase, C., Nakazono, H., Hemmi, H., Shimatake, H. and Aoki, T. 1995. A novel RNA splicing mutation in Japanese patients with Wilson disease. *Biochemical and Biophysical Research Communications* **217**, 16–20.

Simon, M., Pawlotsky, Y., Bourel, M., Fauchet, R. and Genetet, B. 1975. Idiopathic hemochromatosis associated with HL-A 3 tissular antigen. *Nouvelle Press Medicale* **4**, 1432.

Simon, M., Alexandre, J.L., Fauchet, R., Genetet, B. and Bourel, M. 1980. The genetics of hemochromatosis. *Progress in Medical Genetics* **4**, 135–68.

Stamatoyannopoulos, G. and Nienhuis, A.W. (1987). Hemoglobin switching, in *The Molecular Basis of Blood Diseases*, (eds G. Stamatoyannopoulos, A.M. Nienhuis, P. Leder and P.W. Majerus), Saunders, Philadelphia, p. 66.

Sternlieb, I. 1990. Perspectives on Wilson disease. *Hepatology* **12**, 1234–9.

Story, C.M., Mikulska, J.E. and Simister, N.E. 1994. A major histocompatibility complex class I-like Fc receptor cloned from human placenta: possible role in transfer of immunoglobulin G from mother to fetus. *Journal of Experimental Medicine* **180**, 2377–81.

Strickland, G.T. and Leu, M.L. 1975. Wilson's disease. Clinical and laboratory manifestations in 40 patients. *Medicine* **54**, 113–37.

Strickland, G.T., Frommer, D., Leu, M.L., Pollard, R., Sherlock, S. and Cumings, J.N. 1973. Wilson's disease in the United Kingdom and Taiwan. I. General characteristics of 142 cases and prognosis. II. A genetic analysis of 88 cases. *Quarterly Journal of Medicine* **42**, 619–38.

Summers, M.R., Jacobs, A., Tudway, D., Perera, P. and Ricketts, C. 1979. Studies in desferrioxamine and ferrioxamine metabolism in normal and iron-loaded subjects. *British Journal of Haematology* **42**, 547–55.

Takahashi, Y., Miyajima, H., Shirabe, S., Nagataki, S., Suenaga, A. and Gitlin, J.D. 1996. Characterization of a nonsense mutation in the ceruloplasmin gene resulting in diabetes and neurodegenerative disease. *Human Molecular Genetics* **5**, 81–4.

Tanzi, R.E., Petrukhin, K., Chernov, I., Pellequer, J.L., Wasco, W., Ross, B., Romano, D.M., Parano, E., Pavone, L., Brzustowicz, L.M. *et al.* 1993. The Wilson disease gene is a copper transporting ATPase with homology to the Menkes disease gene. *Nature Genetics* **5**, 344–50.

Taramelli, R., Kioussis, D., Vanin, E., Bartram, K., Groffen, J., Hurst, J. and Grosveld, F.G. 1986. Gamma delta beta-thalassaemias 1 and 2 are the result of a 100 kbp deletion in the human beta-globin cluster. *Nucleic Acids Research* **14**, 7017–29.

Thein, S.L., Old, J.M., Wainscoat, J.S., Petrou, M., Modell, B. and Weatherall, D.J. 1984. Population and genetic studies suggest a single origin for the Indian deletion beta thalassaemia. *British Journal of Haematology* **57**, 271–8.

Thomas, G.R., Forbes, J.R., Roberts, E.A., Walshe, J.M. and Cox, D.W. 1995. The Wilson disease gene: spectrum of mutations and their consequences. *Nature Genetics* **9**, 210–17.

Townes, T.M., Lingrel, J.B., Chen, H.Y., Brinster, R.L. and Palmiter, R.D. 1985. Erythroid-specific expression of human beta-globin genes in transgenic mice. *EMBO Journal* **4**, 1715–23.

Toyofuku, T., Kurzydlowski, K., Tada, M. and MacLennan, D.H. 1993. Identification of regions in the Ca(2+)-ATPase of sarcoplasmic reticulum that affect functional association with phospholamban. *Journal of Biological Chemistry* **268**, 2809–15.

Trent, R.J., Bowden, D.K., Old, J.M., Wainscoat, J.S., Clegg, J.B. and Weatherall, D.J. 1981. A novel rearrangement of the human beta-like globin gene cluster. *Nucleic Acids Research* **9**, 6723–33.

Tuan, D., Solomon, W., Li, Q. and London, I.M. 1985. The 'beta-like-globin' gene domain in human erythroid cells. *Proceedings of the National Academy of Sciences, USA* **82**, 6384–8.

Tümer, Z., Chelly, J., Tommerup, N., Ishikawa-Brush, Y., Tonnesen, T., Monaco, A.P. and Horn, N. 1992a. Characterization of a 1.0 Mb YAC contig spanning two chromosome breakpoints related to Menkes disease. *Human Molecular Genetics* **1**, 483–9.

Tümer, Z., Tommerup, N., Tonnesen, T., Kreuder, J., Craig, I.W. and Horn, N. 1992b. Mapping of the Menkes locus to Xq13.3 distal to the X-inactivation center by an intrachromosomal insertion of the segment Xq13.3–q21.2. *Human Genetics* **88**, 668–72.

Tümer, Z., Vural, B., Tonnesen, T., Chelly, J., Monaco, A.P. and Horn, N. 1995. Characterization of the exon structure of the Menkes disease gene using vectorette PCR. *Genomics* **26**, 437–42.

Tümer, Z., Horn, N., Tonnesen, T., Christodoulou, J., Clarke, J.T. and Sarkar, B. 1996. Early copper-histidine treatment for Menkes disease. *Nature Genetics* **12**, 11–23.

Tümer, Z., Lund, C., Tolshave, J., Vural, B., Tonnesen, T. and Horn, N. 1997. Identification of point mutations in 41 unrelated patients affected with Menkes disease. *American Journal of Human Genetics* **60**, 63–71.

Ueyama, H., Deng, H.X. and Ohkubo, I. 1993. Molecular cloning and chromosomal assignment of the gene for human Zn-alpha 2-glycoprotein. *Biochemistry* **32**, 12968–76.

Valberg, L.S., Ghent, C.N., Lloyd, D.A., Frei, J.V. and Chamberlain, M.J. 1979. Iron absorption in idiopathic hemochromatosis: relationship to serum ferritin concentration in asymptomatic relatives. *Clinical and Investigative Medicine – Médecine Clinique et Experimentale* **2**, 17–22.

Van Caillie-Bertrand, M., Degenhart, H.J., Luijendijk, I., Bouquet, J. and Sinaasappel, M. 1985. Wilson's disease: assessment of D-penicillamine treatment. *Archives of Diseases of the Child* **60**, 652–5.

Van der Ploeg, L.H., Konings, A., Oort, M., Roos, D., Bernini, L. and Flavell, R.A. 1980. Gamma-beta-thalassaemia studies showing that deletion of the gamma- and delta-genes influences beta-globin gene expression in man. *Nature* **283**, 637–42.

Verga, V., Hall, B.K., Wang, S.R., Johnson, S., Higgins, J.V. and Glover, T.W. 1991. Localization of the translocation breakpoint in a female with Menkes syndrome to Xq13.2–q13.3 proximal to PGK-1. *American Journal of Human Genetics* **48**, 1133–8.

Verland, S., Simonsen, M., Gammeltoft, S., Allen, H., Flavell, R.A. and Olsson, L. 1989. Specific molecular interaction between the insulin receptor and a D product of MHC class I. *Journal of Immunology* **143**, 945–51.

Villegas, A., Calero, F., Vickers, M.A., Ayyub, H. and Higgs, D.R. 1990. Alpha thalassaemia in two Spanish families. *European Journal of Haematology* **44**, 109–15.

Vulpe, C.D. and Packman, S. 1995. Cellular copper transport. *Annual Review of Nutrition* **15**, 293–322.

Vulpe, C., Levinson, B., Whitney, S., Packman, S. and Gitschier, J. 1993. Isolation of a candidate gene for Menkes disease and evidence that it encodes a copper-transporting ATPase. *Nature Genetics* **3**, 7–13.

Walshe, J.M. 1956. Penicillamine, a new oral therapy for Wilson disease. *American Journal of Medicine* **21**, 487–95.

Walshe, J.M. 1982. Treatment of Wilson's disease with trientine (triethylene tetramine) dihydrochloride. *Lancet* **1**, 643–7.

Walshe, J.M. 1984. Copper: its role in the pathogenesis of liver disease. *Seminars in Liver Disease* **4**, 252–63.

Walshe, J.M. 1989. Wilson's disease presenting with features of hepatic dysfunction: a clinical analysis of eighty-seven patients. *Quarterly Journal of Medicine* **70**, 253–63.

Walters, G.O., Miller, F.M. and Worwood, M. 1973. Serum ferritin concentration and iron stores in normal subjects. *Journal of Clinical Pathology* **26**, 770–2.

Ward, R.J., Florence, A.L., Baldwin, D., Abiaka, C., Roland, F., Ramsey, M.H., Dickson, D.P., Peters, T.J. and Crichton, R.R. 1991. Biochemical and biophysical investigations of the ferrocene-iron-loaded rat. An animal model of primary haemochromatosis. *European Journal of Biochemistry* **292**, 405–10.

Waye, J.S., Eng, B. and Chui, D.H. 1992. Identification of an extensive zeta-alpha globin gene deletion in a Chinese individual. *British Journal of Haematology* **80**, 378–80.

Weatherall, D.J. (1994). Thalassemia, in *The Molecular Basis of Blood Diseases*, (eds G. Stamatoyannopoulos, A.W. Nienhuis, P. Leder and P.W. Majerus), Saunders, Philadelphia, p. 157.

Weatherall, D.J., Clegg, J.B., Higgs, D.R. and Wood, W.G. 1995. The hemoglobinopathies, in *Metabolic Basis of Disease*, (eds C.R. Scriver, A.I. Beaudet, W.S. Sly and D. Valle), McGraw-Hill, New York, pp. 3417–84.

Wickramasinghe, S.N. and Hughes, M. 1978. Some features of bone marrow macrophages in patients with homozygous beta-thalassaemia. *British Journal of Haematology* **38**, 23–8.

Wilkie, A.O., Lamb, J., Harris, P.C., Finney, R.D. and Higgs, D.R. 1990. A truncated human chromosome 16 associated with alpha thalassaemia is stabilized by addition of telomeric repeat (TTAGGG)n. *Nature* **346**, 868–71.

Williams, R., Smith, P.M., Spicer, E.J., Barry, M. and Sherlock, S. 1969. Venesection therapy in idiopathic haemochromatosis. An analysis of 40 treated and 18 untreated patients. *Quarterly Journal of Medicine* **38**, 1–16.

Wilson, S.A.K. 1912. Progressive lenticular degeneration: A familial nervous disease associated with cirrhosis of the liver. *Brain* **6**, 295–508.

Winichagoon, P., Higgs, D.R., Goodbourn, S.E., Lamb, J., Clegg, J.B. and Weatherall, D.J. 1982. Multiple arrangements of the human embryonic zeta globin genes. *Nucleic Acids Research* **10**, 5853–68.

Witte, D.L., Crosby, W.H., Edwards, C.Q., Fairbanks, V.F. and Mitros, F.A. 1996. Practice guideline development task force of the College of American Pathologists. Hereditary hemochromatosis. *Clinica Chimica Acta* **245**, 139–200.

Wright, S., deBoer, E., Grosveld, F.G. and Flavell, R.A. 1983. Regulated expression of the human beta-globin gene family in murine erythroleukaemia cells. *Nature* **305**, 333–6.

Yamaguchi, Y., Heiny, M.E., Suzuki, M. and Gitlin, J.D. 1996. Biochemical characterization and intracellular localization of the Menkes disease protein. *Proceedings of the National Academy of Sciences, USA* **93**, 14030–5.

Yoshida, K., Furihata, K., Takeda, S., Nakamura, A., Yamamoto, K., Morita, H., Hiyamuta, S., Ikeda, S., Shimizu, N. and Yanagisawa, N. 1995. A mutation in the ceruloplasmin gene is associated with systemic hemosiderosis in humans. *Nature Genetics* **9**, 267–72.

Yuzbasiyan-Gurkan, V., Brewer, G.J., Boerwinkle, E. and Venta, P.J. 1988. Linkage of the Wilson disease gene to chromosome 13 in North-American pedigrees. *American Journal of Human Genetics* **42**, 825–9.

Yuzbasiyan-Gurkan, V., Grider, A., Nostrant, T., Cousins, R.J. and Brewer, G.J. 1992. Treatment of Wilson's disease with zinc: X. Intestinal metallothionein induction. *Journal of Laboratory and Clinical Medicine* **120**, 380–6.

Zaitseva, I., Zaitsev, V., Card, G., Moshkov, K., Bax, B., Ralph, A. and Lindley, P. 1996. The X-ray structure of human serum ceruloplasmin at 3.1 angstrom: Nature of the copper centres. *Journal of Biological and Inorganic Chemistry* **1**, 15–23.

7

THE ROLE OF ION CHANNELS IN THE TRANSPORT OF METALS INTO EXCITABLE AND NONEXCITABLE CELLS

Timothy J. Shafer

7.1 Introduction

Cells employ a variety of mechanisms at the plasma membrane in order to regulate the movement of ions and other molecules into and out of the cell; cotransporters, active and passive transport mechanisms, and ion channels are all important components of cellular ion homeostasis. Of these mechanisms, ion channels have one of the highest rates of transfer of ions across the plasma membrane (Hille, 1992), making them an effective mechanism to alter rapidly the concentration of ions within the cytoplasm. In addition, ion channels are often localized to specific areas of the cell; for example, voltage-gated Na^+ channels are highly localized at the nodes of Ranvier on myelinated axons (Chiu, 1980), and certain types voltage-gated Ca^{2+} channels are found at the presynaptic active zone of nerve terminals (Robitaille *et al.*, 1990; Smith *et al.*, 1993; Issa and Hudspeth, 1994). The localization of ion channels in both excitable (neurons, muscle) and nonexcitable (kidney, liver, endocrine) cells provides a mechanism by which changes in ion concentration can also be localized to specific parts of the cell. Thus, in addition to being a mechanism for transport of ions across the plasma membrane, certain types of ion channels also serve important signaling functions as well. These functions range from the obvious, such as changing membrane potential (Na^+, K^+ and occasionally Ca^{2+} channels) or triggering contraction and secretion (Ca^{2+} channels), to the more subtle, such as generation of second-messenger Ca^{2+} or activation of gene transcription (any type of Ca^{2+} permeable channel).

Among the properties of ion channels is the ability to discriminate for or against, allowing certain ions to move through the channel (see below for a more detailed discussion of ion-channel properties). Some channel types are highly discriminating for

This article has been reviewed by the National Health and Environmental Effects Research Laboratory, US Environmental Protection Agency, and is approved for publication. Approval does not signify that the contents necessarily reflect the views and policies of the Agency, nor does mention of trade names and commercial products constitute endorsement or recommendation for use.

a particular ion, while other channel types are far less specific. For example voltage-sensitive Ca^{2+} channels have a high selectivity for Ca^{2+}, whereas the nicotinic acetylcholine receptor (nAChR) channel is permeable to both Na^+ and K^+, and, in some cases, Ca^{2+}. Regardless of how highly selective an ion channel is for a particular ion, no channel type is permeated exclusively by any one ion, although other permeant ions may be physiologically less relevant or permeate under conditions which are not physiologically relevant. In some cases, differences in the permeability of an ion channel to a series of ions has been used by biophysicists to define the characteristics of the channel and its pore. A classical example of this is voltage-gated Ca^{2+} channels, which are permeable to Sr^{2+} and Ba^{2+} in addition to Ca^{2+}. The relative conductances of these ions are used to distinguish among different types of Ca^{2+} channels (Tsien *et al.*, 1988). The concept that even ion channels that are highly specific for an individual ion are not exclusively permeated by that ion also extends to ionic metals of toxicological interest. As such, ion channels are a route of entry into the cells for many heavy metals, including Cd^{2+} (Taylor, 1988), Pb^{2+} (Cooper and Manalis, 1983; Tomsig and Suszkiw, 1991), Hg^+ (Miyamoto, 1983), CH_3Hg^+ (Atchison, 1987) and Zn^{2+} (Winegar and Lansman, 1990).

This chapter discusses the role of ion channels as routes of entry into the cell for these and other metals. This discussion will focus on routes and mechanisms by which heavy metals permeate ion channels, and will consider briefly the relationships between entry into the cell via ion channels and toxicological consequences. No review or book chapter on this topic could be exhaustive, due to the large number of cell types and ion channels affected by metals. In light of this limitation, the approach taken will be to present an introduction to ion channels and their function, discuss some clear examples of mechanisms by which metals move through ion channels, and provide references for additional resources and reading on the subject of ion channels. Because many of the metals of interest in this book utilize voltage-gated Ca^{2+} channels as a route of entry, this channel type will be discussed as a consistent example in this text.

7.1.1 Driving forces for movement of ions through ion channels

Transport of ions through channels differs from other types of transport in terms of the energy driving the transport. While many types of transport are driven through the hydrolysis of ATP (Na^+, K^+-ATPase, Ca^{2+}, Mg^{2+}-ATPase) and other types of transport take advantage of the energy associated with an ion moving down its electrochemical gradient to 'hitch a ride' into or out of the cell (cotransport systems such as Na^+–Ca^{2+} exchange, amino-acid transporters, etc.), movement of ions through channels is dependent solely on the balance between the electromotive force, determined by the potential difference between the inside and outside of the cell, and the chemical driving force, determined by the intracellular and extracellular concentrations of the ion of interest. The membrane potential at which these forces reach equilibrium can be calculated using the Nernst equation (Nernst, 1888):

$$E_x = \mathrm{R}T/z_x\mathrm{F} \ln [\mathrm{X}]_e/[\mathrm{X}]_i$$

In the above equation, E_x is the equilibrium potential of a positively charged ion; R is the gas constant; T is temperature; z_x is the charge (z) on ion X; F is Faraday's constant;

and $[X]_e$ and $[X]_i$ are the extracellular and intracellular concentrations of (a positively charged) ion X, respectively. In practical terms, when the membrane potential of a cell is at E_x, there will be no net movement of ion X into or out of the cell. Alternatively, the direction of movement of a given ion across the membrane at a given potential can be predicted from the Nernst equation. For example, E_{Ca} is typically greater than +100 mV, thus, at any membrane potential less than +100 mV, Ca^{2+} tends to move into the cell due to the large concentration difference between the external and internal sides of the membrane. Using this equation to calculate E for a divalent metal, the equilibrium potential is reached at +30 mV and +60 mV when concentrations of free metal ion are ten- and 100-fold greater, respectively, in the external compared to internal solution. Thus, for metals such as Cd^{2+}, Pb^{2+}, Hg^{2+}, Ni^{2+} and Zn^{2+}, the Nernst equation predicts that, under most physiological conditions, they would be able to move into the cell by taking advantage of the energy of the electrochemical gradient, similar to the movement of any physiologically relevant ion down its electrochemical gradient through an ion channel.

7.1.2 Structure and function of ion channels

In general terms, most types of ion channels fall into one of two broad classes, which are based on the stimulus that causes the channel to open (gating); ligand-gated (also termed 'receptor operated') or voltage-gated (the terms 'activated' and 'sensitive' are also often used in reference to gating). Other types of ion channels may be gated by mechanical stimuli (stretch-gated) or other stimuli; however, this discussion will focus on voltage- and ligand-gated channels. Although many different stimuli drive the opening and closing of ion channels, ultimately the transition between open and closed states of a channel is a simple reaction for which rate constants can be described. Thus, even in the absence of stimuli that normally control channel gating, ion channels will transition between open and closed states with low probability. Stimuli such as voltage changes or ligand binding greatly increase the probability of channel opening. For example, in excitable cells, the probability of opening of voltage-gated Ca^{2+} channels is very low at resting membrane potentials (–70 mV), but some channels in the population will open briefly and allow Ca^{2+} to enter the cell. Thus even in the absence of excitatory stimuli, Ca^{2+} and potentially other metals, can enter the cell through Ca^{2+} channels, although the amount of entry from any one event would be small. Upon membrane depolarization, the open probability of Ca^{2+} channels increases dramatically, resulting in the opening of the majority of the channels in the population, giving rise to Ca^{2+} currents and changes in intracellular $[Ca^{2+}]$. Once a channel has opened, a number of factors influence the amplitude (size) of current that flows through the channel. Primary among these factors are ion selectivity and membrane potential, but binding of second messengers (Levitan, 1994), steroids (Mermelstein *et al.*, 1996) or other ligands (Scott *et al.*, 1993) also modulate current amplitude in both voltage- and ligand-gated ion channels. Although the general structures of voltage- and ligand-gated ion channels are substantially distinct, families of structurally related proteins comprise both classes of ion channels (reviewed by Barnard, 1996).

The voltage-gated superfamily consists of Na^{+}, K^{+} and Ca^{2+} channels. Voltage-gated ion channels are formed by the assembly of several distinct subunits, denoted by lower-case Greek symbols (e.g. $\alpha\beta\gamma\delta$), where the α (α_1 for Ca^{2+} channels) subunit is the primary

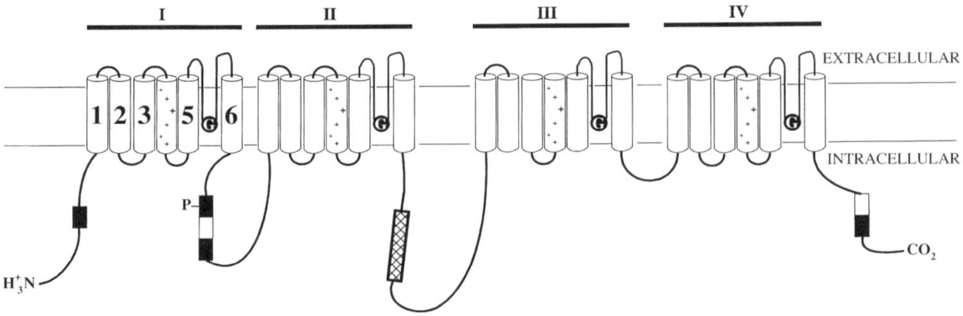

Figure 7.1 Structure of the α_1-subunit of the voltage-gated Ca^{2+} channel. The structure of generic α_1-subunit of the voltage-gated Ca^{2+} channel is depicted in this diagram, and consists of four homologous regions (I–IV), each of which contains six membrane-spanning regions (labeled 1–3, 5–6 in region I). Also depicted are several structural components of the α_1-subunit which have been demonstrated to be important for function of the channel and interaction of the channel with other cellular proteins. The following are illustrated: (1) The S4 region that is involved in voltage sensing (+ symbols); (2) glutamate residues (G) in the pore-loop between the S5 and S6 regions; (3) regions on LI–II, the C and N terminals that are important for interactions with G-protein $\beta\gamma$-subunits (solid boxes); (4) regions on LI–II and the C terminus which are important for binding of Ca^{2+} channel β-subunits (open boxes); (5) a region on LII–III (hatched box) which is important for interaction with exocytotic proteins (also ryanodine receptor binding) (De Waard *et al.*, 1996); and (6) a protein kinase C phosphorylation site (P) which is involved in crosstalk with G proteins (Zamponi *et al.*, 1997). It should be noted that this drawing is not to scale, nor are the locations of different functional regions exact. See the text for references used as a basis for this diagram.

subunit of the ion-channel structure. Recent reviews of K^+ (Jan and Jan, 1997) and Ca^{2+} (De Waard *et al.*, 1996; Reuter, 1996; Randall, 1998) channels, as well as voltage-gated ion channels (Catterall, 1995) are available. A single α-subunit consists of four homologous domains (repeats I–IV) each containing six membrane-spanning segments (S1–S6; Fig. 7.1). Expression of the α-subunit alone is often sufficient to form a functional ion channel, although in many cases, the homomeric α-subunit-derived channel has functional characteristics that differ from those of native channels. Many of the important aspects of voltage-gated ion-channel function have been assigned to the α-subunit, including voltage-sensing, pore formation, ionic selectivity and conductance. The voltage-sensing region of Na^+, K^+ and Ca^{2+} channels resides in the structure of the fourth (S4) membrane-spanning region and is an α-helix typically containing six positively charged lysine and arginine residues (reviewed by Franciolini, 1994). When the potential across the membrane changes, these residues re-align, and the resulting conformational change is thought to control channel gating.

The marriage of electrophysiologic and molecular biological techniques has greatly increased the understanding of ion-channel structure and function over the past two decades. One extremely fruitful approach has been to produce mutations in, or chimeras of, channel subunits (see below), utilizing molecular biological techniques, to express these altered subunits in cells such as oocytes and to measure their functionality using electrophysiologic and pharmacologic techniques. The ultimate result is that, in many cases, different structural components of the channel can be associated with channel functions such as ion selectivity, pore regions and conductance, as well as other channel functions and interactions with other molecules. The intracellular loops that connect the homologous repeats in α-subunits of voltage-gated ion channels appear to be important in interactions with other subunits and other cellular molecules. For example, binding of Ca^{2+}-channel β-subunits to α_1-subunits is dependent on a consensus sequence within the intracellular loop connecting repeats I and II (LI–II; De Waard *et al.*, 1995) although recent evidence also suggests that sequences on the C terminus of the α_1-subunit are also important for β-subunit interactions (Walker *et al.*, 1998).

Many voltage-gated ion channels, including Ca^{2+} channels, are modulated by G proteins, and interact with cytoskeletal or other cellular machinery. Binding of G-protein $\beta\gamma$-subunits to the α_1-subunit of Ca^{2+} channels and inhibition of channel activity is dependent on a consensus sequence, also localized on LI–II (De Waard *et al.*, 1997; Herlitze *et al.*, 1997; Page *et al.*, 1997). Other portions of the α_1-subunit may also be important for interactions with G proteins, as sequences within the amino (Page *et al.*, 1998) as well as C-terminal sites (Qin *et al.*, 1997) also influence G-protein modulation of some Ca^{2+} channel α_1-subunits. Interestingly, the LI–II segment may also play a crucial role in crosstalk between G protein and phosphorylation-dependent modulation of Ca^{2+}-channel activity. Phosphorylation of protein kinase C (PKC) consensus sequences within LI–II blocked inhibition of N and P/Q (see below) Ca^{2+} channel activity by $G\beta\gamma$ (Zamponi *et al.*, 1997). By contrast to LI–II, the LII–III region contains binding sites for exocytotic proteins in α_1-subunits thought to be involved in neurotransmitter release (Sheng *et al.*, 1994; Bezprozvanny *et al.*, 1995; Martin-Moutot *et al.*, 1996; Rettig *et al.*, 1996; Leveque *et al.*, 1998). Considerable attention has also been paid to the amino-acid sequences between S5 and S6, which form short, lipophilic loops (pore loops), and are important for ionic conductance and ionic selectivity properties of channel function (reviewed by MacKinnon, 1995). For voltage-gated Ca^{2+} channels, this segment in the fourth homologous domain (repeat IV) is important in determining single-channel conductance as well as voltage-dependent current activation and inactivation (Yatani *et al.*, 1994a), and this same region in repeat I has been implicated in channel conductance (Dirksen *et al.*, 1997) and nearby regions in voltage-dependent inactivation (Zhang *et al.*, 1994).

Of critical importance for this discussion is how a channel discriminates between ions and 'selects' which ions will pass through the channel. As will be discussed, metal ions, particularly divalent cations, are able to enter cells through channels by virtue of their similarity to a physiologic ion, such as Ca^{2+}. For Ca^{2+} channels, two opposing problems must be solved. First, the channel must selectively allow Ca^{2+} to permeate over much more abundant Na^+ and K^+ ions, which implies that there is a mechanism or binding site which has a high affinity for Ca^{2+} ions in the 'selectivity filter' of the channel. Secondly, in order to carry a current, the flux rate of Ca^{2+} through the channel must be high. Thus, Ca^{2+} ions must exit rapidly from any binding site(s) within a 'selectivity filter'. A number of studies indicate that, like Na^+ and K^+ channels, the amino-acid sequence between S5 and S6 in all four homologous regions of the Ca^{2+} channel is a pore-lining region, and that glutamate residues within this region are critical for binding Ca^{2+} ions (Mikala *et al.*, 1993; Tang *et al.*, 1993; Yang *et al.*, 1993; Yatani *et al.*, 1994b; Ellinor *et al.*, 1995; Parent and Gopalakrishnan, 1995; Bahinski *et al.*, 1997). Each repeat contains one highly conserved glutamate residue which contributes to a negatively charged 'ring' for which Ca^{2+} ions compete. While it is clear that the contributions of the glutamate residues to Ca^{2+} ion binding are not equal (Mikala *et al.*, 1993; Yang *et al.*, 1993; Yatani *et al.*, 1994b; Ellinor *et al.*, 1995), no true consensus has emerged on the structure of the binding site. However, the data indicate that there is a single Ca^{2+} binding site formed by the four glutamate residues, which confers to the pore high selectivity for Ca^{2+} over monovalent cations. The problem of rapid exit of Ca^{2+} ions from this binding site is thought to be overcome by competition between Ca^{2+} ions for occupancy of the site. Some evidence indicates that the glutamate residues may interact with more than one Ca^{2+} ion at a time, which destabilizes the binding site. This, coupled with electrostatic repulsion between Ca^{2+} ions, hastens the exit of Ca^{2+} ions from the binding site and speeds Ca^{2+} ions through the pore in a rapid, single-file manner (Fig. 7.2; Yang *et al.*, 1993).

Although many functional features reside on the α-subunit, the β-,γ- and δ-subunits, often referred to as auxiliary subunits, also have important functional roles (reviewed by Isom *et al.*, 1994). The interaction of α_1-subunits with β-subunits confers upon recombinant channels properties that make them resemble native channels more closely. For example, the β-subunits of voltage-gated Ca^{2+} channels influence the rate of inactivation (Olcese *et al.*, 1994), modulate the rate of voltage-dependent activation (De Waard *et al.*, 1994) and increase the sensitivity of the α_1-subunit to the membrane potential (Cens *et al.*, 1998). In addition, the β-subunits in some cases may be involved in targeting the ion channel to specific subcellular regions or facilitating interaction with other intracellular proteins (reviewed by Sheng and Kim, 1996). For example, coexpression of the α_1- and β-subunits of skeletal muscle Ca^{2+} channels results in higher amounts of α_1-protein being expressed in the membrane than expression of only the α_1-subunit alone (Krizanova *et al.*, 1995). Finally, the β-subunit is also involved in the modulation of Ca^{2+} channel function by G proteins; recent work suggests that the β_3-subunit reduces inhibition of Ca^{2+} channels

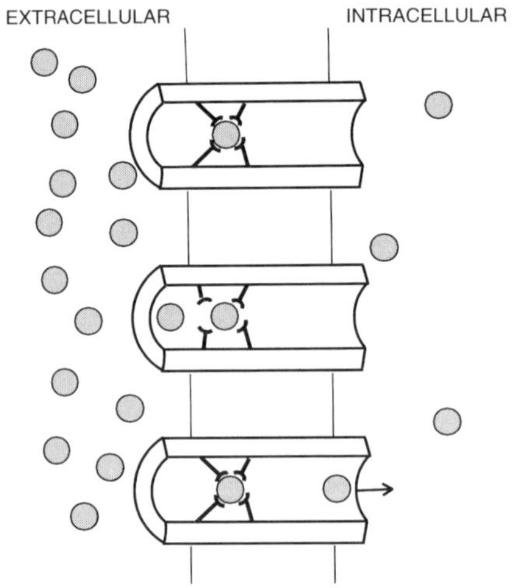

Figure 7.2 Mechanism of permeation of divalent cations through voltage-gated Ca^{2+} channels. Shown in this figure is a schematic representation of the molecular events involved in permeation of a divalent cation (typically Ca^{2+}) through a voltage-gated Ca^{2+} channel pore. A 'cut away' view of the channel pore for three voltage-gated Ca^{2+} channels is presented for simplicity. The top channel illustrates a Ca^{2+} (gray sphere) ion bound stably by four glutamate residues (Y-shaped symbols), which form the Ca^{2+}-binding site and impart Ca^{2+} selectivity to the channel. In the middle channel, a second Ca^{2+} ion has entered the pore region and begins to compete for the glutamate residues – destabilizing binding of the first Ca^{2+} ion. In the bottom channel, the second Ca^{2+} ion has displaced the first Ca^{2+} ion from the glutamate residues, and the first Ca^{2+} ion, in part due to electrostatic repulsion, exits the Ca^{2+} channel pore on the cytoplasmic side. While Ca^{2+} is used as the example in this illustration, Cd^{2+} (and presumably other divalent ions) also permeates the channel by binding to the same glutamate residues. See the text for references used as a basis for this diagram.

containing the α_{1B}-subunits by increasing the dissociation rate of G-protein $\beta\gamma$-subunits from the Ca^{2+} channel α_1 subunit (Roche and Treistman, 1998). The functional roles of the other subunits are also becoming more clear. The presence of the $\alpha_2\delta$-subunit of voltage-gated Ca^{2+} channels increases current amplitude when compared to the α_1-subunit alone (Mori *et al.*, 1991). Increases in the overall expression of α_1 in the membrane, as well as the probability of channel opening, may underlie the increase in Ca^{2+} current observed when the α_1-subunit is coexpressed with the $\alpha_2\delta$-subunit (Shistik *et al.*, 1995).

At least three superfamilies of related gene products give rise to ligand-gated ion channels, and these superfamilies are distinguished by the number of transmembrane segments in each of several subunits that form the channel. The nicotinic acetylcholine (nACh), serotonin ($5HT_3$), γ-aminobutyric acid ($GABA_A$) and glycine receptors all comprise a superfamily in which the channel is formed by a pentamer of subunits, each subunit containing four conserved transmembrane segments. Glutamate-gated ion channels are composed of subunits that contain at least three transmembrane segments and include *N*-methyl-D-aspartate (NMDA), kainate (KA) and α-amino-3-hydroxy-5-methyl-4-isoxazolepropionic acid (AMPA) receptors. Finally, ionotropic ATP receptor subunits have two transmembrane segments. Like the voltage-gated ion channels, ligand-gated ion channels are composed of several different types of subunits. However, the subunits of ligand-gated ion channels appear to combine with different stoichiometries in order to form a functional channel, with each subunit contributing to formation of the pore of the channel. The nACh, $GABA_A$, glycine and $5HT_3$ channels are all formed by a pentamer of subunits and, with the exception of $5HT_3$ and $\alpha7$ nicotinic ACh receptors, each of these channels is composed from several different, but structurally related subunits. For example, a common subunit stoichiometry for $GABA_A$ receptors in the central nervous system is two α-, one β- and two γ-subunits (Mohler *et al.*, 1996). By contrast, the NMDA glutamate receptor appears to be composed from a combination of one NR1 and two NR2 subunits (Sucher *et al.*, 1996). For those who are interested in additional details on individual channel types, a number of excellent reviews are available for the nACh (Albuquerque *et al.*, 1995, 1997; Boyd, 1997), $GABA_A$ (Morrow, 1995; Whiting *et al.*, 1995; Mohler *et al.*, 1996), NMDA (Mori and Mishina, 1995; Sucher *et al.*, 1996), glycine (Rajendra *et al.*, 1997) and $5HT_3$ (Jackson and Yakel, 1995) receptors.

7.1.3 Ion-channel diversity

Ion channels are integral to a diverse array of functions in biological systems; as such, it would be expected that there is considerable diversity in their structure. This is accomplished in part through the existence of multiple isoforms of ion-channel subunits, and is particularly evident in the central nervous system. Over the past decade, the number of physiologically/pharmacologically defined types of voltage-gated Ca^{2+} channels has increased from two to at least six (Pérez-Reyes *et al.*, 1998; Randall, 1998), most of which are thought to consist of a 1 : 1 : 1 : 1 combination of α_1-, β-, α_2- and δ-subunits. Ca^{2+} channels in skeletal muscle have an additional γ-subunit (reviewed by Randall, 1998). A given channel may be formed from any of nine α_1 (α_{1A}–α_{1E}, $\alpha_{1G, H, I}$, α_{1S}) and four β (β_1–β_4) as well as the α_2-,γ- and δ-subunits. Other channel types also display similar diversity in structure (Table 7.1). Among the most diverse is the

Table 7.1 Diversity of subunits comprising voltage- and ligand-gated ion channels[a]

Channel	α-Subunits	β-Subunits	Reference
Na^+	3	1	Catterall (1995)
Ca^{2+}	7[b]	4	De Waard *et al.* (1996)
			Pérez-Reyes *et al.* (1998)
nACh	8	3	Boyd (1997)
$GABA_A$[c]	6	3	Mohler *et al.* (1996)
NMDA[d]	1	4	Sucher *et al.* (1996)

[a] The values provided in this table are the number of genes determined to encode α- or β-subunits, and does not include additional isoforms that may be generated by alternative splicing. For the voltage-gated Ca^{2+} channel, only the α_1-subunit is considered, and for the Na^+ channel, only the β_1-subunit is considered.
[b] The number includes the α_{1S} (skeletal muscle) subunit, which is not expressed in brain.
[c] There are also three different γ- and two different δ-subunits for the $GABA_A$ receptor.
[d] The rat NMDA receptor subunit nomenclature does not include α and β designations, and the values given in these columns reflect the number of NR1 and NR2 subunits, respectively.

ligand-gated $GABA_A$ channel, for which there are six different genes that encode α-subunits, three β genes, three γ- and two δ-subunit genes. Splice variants of ion-channel subunits contribute to additional diversity.

The great diversity of subunit types gives rise to considerable functional and pharmacologic diversity. For example, the $GABA_A$ receptor is modulated by a wide variety of physiological and pharmacological agents, including the benzodiazepine drugs such as diazepam and chlordazepoxide. Sensitivity to benzodiazepine modulation is dependent on the interactions between an $\alpha_{2,\,3\text{ or }5}$- and the γ_2- subunit (Mohler *et al.*, 1996). On occasion, some subunits fulfill distinct functional roles, such as the γ-subunit of the nACh receptor. This subunit is expressed exclusively during development on muscle fibers prior to the formation of the neuromuscular junction. The presence of this subunit confers upon the nAChR a longer open time, which may be important in establishing synaptic connections with motor neurons (reviewed by Kidokoro, 1993).

Because voltage-gated Ca^{2+} channels will be integral to discussions of metal entry into cells in the following sections, some comment is needed here regarding their diversity. The recent description of the gene that encodes the putative α_1-subunit of the T-type voltage-sensitive Ca^{2+} channel (Pérez-Reyes *et al.*, 1998) brings the tally of genes encoding α_1-subunits for Ca^{2+} channels to nine (De Waard *et al.*, 1996; Reuter, 1996; Randall, 1998). At the present time, this correlates well, but not completely, with the number of physiologically and pharmacologically defined Ca^{2+}-channel subtypes: direct assignment of all α_1-subunits to a physiologic channel type has yet to be confirmed. Based on the original nomenclature of Nowycky (Nowycky *et al.*, 1985) these channels have been denoted L-, N-, T-, P-, Q- and R-type. Ca^{2+} channels are distinguished in part on the basis of pharmacological sensitivities to channel blockers, including the dihydropyridines (DHPs, such as nifedipine and nimodopine), peptides derived from the venom of *Conus* sp. fish-hunting snails and the relative sensitivity to block by Cd^{2+} and Ni^{2+}. Biophysical properties, including the voltage at which the channel activates, single-channel conductances and rates of inactivation, provide additional criteria by which voltage-gated Ca^{2+} channels are classified (Table 7.2). The α_1-subunits, which may underlie the differences between these channel types, have also been identified and are designated by subscripts A–E, S (Birnbaumer *et al.*, 1994) and G, H, I (Pérez-Reyes *et al.*, 1998). The L- and T-type channels perhaps represent two extremes of Ca^{2+} channel types. The L-type channel is sensitive to DHPs, inactivates slowly, carries a large current and activates only after strong depolarizations. The L-type Ca^{2+} channel is found in numerous tissue types, including brain, heart and skeletal muscle, where it is abundant. There are some subtle differences between L-type channels in heart, neurons and skeletal muscle. For example, there are slight differences in homology between the α_{1C} (which gives rise to the classical cardiac L-type channel), α_{1D} (classical neuronal and endocrine L-type channel) and α_{1S} (classical skeletal muscle L-type channel) subunits (reviewed by Catterall, 1995). The T-type Ca^{2+} channel is insensitive to dihydropyridines, activates transiently at low (hyperpolarized) voltages and carries a small current. This channel type is found in brain as well as heart tissue. The N, P, Q and R channels are expressed primarily in neural tissue; they are more similar (though not identical) to each other in terms of their biophysical properties and are most often defined by their particular sensitivities to the conotoxins and agatoxins.

Ca^{2+} is a ubiquitous signaling molecule, activating a variety of processes in excitable as well as nonexcitable cells. One factor that adds some specificity to Ca^{2+} signaling is the type of channel through which Ca^{2+} enters the cell. As such, when Ca^{2+} channel function is disrupted by divalent metals, the resulting toxicology will, in part, depend on the channel type that is affected. In general, L-type channels regulate Ca^{2+} entry associated with excitation–contraction processes in muscle cells, but also appear to be important in generation of Ca^{2+} signals for protein phosphorylation and activation of gene expression (Murphy *et al.*, 1991; Finkbeiner and Greenberg, 1996). By contrast, a primary role of the N- (Takahashi and Momiyama, 1993; Wheeler *et al.*, 1994), P- (Charlton and Augustine, 1990; Uchitel *et al.*, 1992; Protti and Uchitel, 1993; Takahashi and Momiyama, 1993; Turner *et al.*, 1993); Q- (Bowersox *et al.*, 1993; Wheeler *et al.*, 1994) and R-type (Wu *et al.*, 1998) Ca^{2+} channels appears to be Ca^{2+} entry associated with neurotransmitter release in the nervous system. Finally, T-type channels are involved in pacemaker activity in heart and neural tissue (reviewed by Huguenard, 1996).

7.2 Ion channels as a route of metal entry into excitable cells

7.2.1 Permeation of Ca^{2+} channels by Cd^{2+} as a model mechanism

Because many of the metals of interest within this book are divalent heavy metals, the permeation of these cations through voltage-gated Ca^{2+} channels will be discussed as a model mechanism by which metals utilize ion channels to enter cells. Interestingly, biophysicists have contributed much to clarifying the mechanisms of heavy-metal movement through Ca^{2+} channels as part of their attempts to study properties of the pore of Ca^{2+} channels. One early model of ion permeation through voltage-gated Ca^{2+} channels in heart presented evidence, based on measurements of Ca^{2+} and Ba^{2+} currents, that at least one and probably two Ca^{2+}-binding sites existed within the pore of Ca^{2+} channels (Hess and Tsien, 1984). The model stated that when the channel is occupied by two Ca^{2+} ions, electrostatic repulsion between ions prevents the channel from being

Table 7.2 Biophysical and pharmacological characteristics of Ca^{2+} channel types (based on Table 1 in Shafer and Atchison 1995, by kind permission of Academic Press)

	T[a]	*N*[a,b]	*L*[a]	*P*[c]	*Q*[d]	*R*[e]
Activation range (mV)	–70	–10	–10	–20 to –30	–40	–40
Inactivation kinetics	Fast $\tau = 20–50$ ms	Moderate $\tau = 95$ ms	Slow $\tau = 0.5$ s	Slow	Moderate $\tau = 116$ ms	Fast $\tau_{1/2} = 12$ ms
Divalent cation sensitivity	$Ni^{2+} > Cd^{2+}$	$Cd^{2+} > Ni^{2+}$	$Cd^{2+} > Ni^{2+}$	$Cd^{2+} > 100$ μM	Cd^{2+}	Ni^{2+}, 66 μM Cd^{2+}, 1 μM
Pharmacological sensitivity	None	ω-CTx-GVIA	DHPs	ω-Aga-IVA FTX	ω-CTx-MVIIC	None
Single-channel conductance (pS in 110 mM Ba^{2+})	8	13	25	10–12 (80 mM Ba^{2+})	15	14
α_1-Subunit[f]	$\alpha_{1G, H, I}$	α_{1B}	$\alpha_{1C, D, S}$	α_{1A}?	α_{1A}?	α_{1E}

[a] Modified from Tsien *et al.* (1988).
[b] Plummer *et al.* (1989).
[c] P-channel data compiled from Llinàs *et al.* (1989), Lin *et al.* (1990).
[d] Q channel data from Sather *et al.* (1992, 1993), Randall *et al.* (1993).
[e] Randall *et al.* (1993), Zhang *et al.* (1994), Ellinor *et al.* (1995).
[f] Birnbaumer *et al.* (1994), Pérez-Reyes *et al.* (1998).

permeated by other cations (thus imparting some selectivity to the channel for Ca^{2+}), and that the repulsion between Ca^{2+} ions also helps propel Ca^{2+} through the channel (Lee and Tsien, 1984). Work by Almers and McCleskey (1984) in frog skeletal muscle also suggested that ions move through Ca^{2+} channel pores in a single-file fashion, binding to multiple sites within the pore. Although the more recent work discussed above has modified this hypothesis, demonstrating that only one divalent cation-binding site exists within the channel pore, the initial observations of single-file movement through the channel pore and electrostatic repulsion appear still to be valid. In addition to Ca^{2+}, the relative permeabilities of Sr^{2+} and Ba^{2+}, as well as monovalent ions, including Na^+, Li^+, K^+ and Cs^+, have been described (Hess *et al.*, 1986). Cd^{2+} had already been demonstrated to be a potent blocker of voltage-gated Ca^{2+} currents in heart (Lansman *et al.*, 1986) and neuronal cells (Byerly *et al.*, 1985). However, it had been noted that block by Cd^{2+} could be relieved at negative membrane potentials, suggesting that Cd^{2+} may move through the channel at negative membrane potentials. Taylor (1988) examined the voltage-dependence of Cd^{2+} block of Ca^{2+} channels in cat sensory neurons to test in part the two-binding-site hypothesis in neuronal cells (Taylor, 1988). Taylor's results also supported the existence of two Ca^{2+}-binding sites within neuronal Ca^{2+} channels. In addition, his data suggested that Cd^{2+} also bound to two sites on the voltage-gated Ca^{2+} channel, and that one of those sites was within the membrane electric field, whereas the other site was probably outside the influence of the membrane electric field (Taylor, 1988). While the two-site model may not be correct for permeation of divalent cations through the Ca^{2+} channel, recent studies using Ca^{2+} channel α_1-subunits with point mutations have confirmed that the four glutamate residues in the pore loops of Ca^{2+} channels (see Fig. 7.2) also bind Cd^{2+} ions (Mikala *et al.*, 1993; Yang *et al.*, 1993; Ellinor *et al.*, 1995). Thus, one of the more significant discoveries (for the purpose of this discussion) from early studies of mechanisms of ion permeation through Ca^{2+} channels was that Cd^{2+} bound to site(s) within the Ca^{2+} channel with high affinity and left slowly, thus suggesting that the ability of Cd^{2+} to block Ca^{2+} currents was a competitive process and that Cd^{2+} could permeate Ca^{2+} channels (Lansman *et al.*, 1986; Taylor, 1988). In fact, Lansman and co-workers (Lansman *et al.*, 1986) demonstrated that multivalent metal ions, such as Ca^{2+}, Sr^{2+}, Ba^{2+}, Cd^{2+}, Mg^{2+}, Co^{2+} and La^{3+} cannot truly be classified as ions that permeate or block Ca^{2+} channels, but rather should be described in terms of the rates at which they enter and exit the pore of the channel; those ions that support measurable currents have higher rates of entry, and those that do not support measurable current, for example Cd^{2+} or Mg^{2+} (Fukushima and Hagiwara, 1985; Hess *et al.*, 1986), have slower rates of entry and exit.

In addition to the studies above, in which it was used as a tool to probe the pore of Ca^{2+}, channels, entry of Cd^{2+} into the cell via voltage-gated Ca^{2+} channels has been demonstrated in a number of other neuronal systems. In PC12 cells (Hinkle and Osborne, 1994), cerebellar granule cells (Usai *et al.*, 1997) and optic nerve preparations (Fern *et al.*, 1996) entry of Cd^{2+} was decreased by antagonists, or increased by agonists, respectively, of L-type Ca^{2+} channels. In PC12 cells, agonists of the L-type Ca^{2+} channel increased the rate of influx and decreased the LC_{50} for Cd^{2+} from 12 to 6 μM, whereas nimodipine, an L-type Ca^{2+} channel antagonist, protected against Cd^{2+} cytotoxicity (Hinkle and Osborne, 1994). In addition, in PC18 cells, which lack functional Ca^{2+} channels, the LC_{50} for Cd^{2+} cytotoxicity was higher (15 μM), and not sensitive to Ca^{2+} channel antagonists (Hinkle and Osborne, 1994). In the optic nerve, Cd^{2+} reduced the area of the compound action potential (CAP) in a manner that was facilitated by depolarization and could be blocked by nifedipine (Fern *et al.*, 1996). In addition, entry of Cd^{2+} via Ca^{2+} channels provided only a route of Cd^{2+} entry into the neuron; Cd^{2+} block of Ca^{2+} channels was not responsible for reductions in the CAP, as pharmacologic block of Ca^{2+} channels did not affect the CAP. Ultimately, disrupted mitochondrial respiration secondary to Cd^{2+} entry through Ca^{2+} channels accounted for the toxicity of Cd^{2+} in this preparation (Fern *et al.*, 1996). Direct measurement of Cd^{2+} entry using fura-2 (Hinkle and Osborne, 1994; Usai *et al.*, 1997) also demonstrates that this ion permeates voltage-gated Ca^{2+} channels.

7.2.2 Entry of other metals through voltage-gated Ca^{2+} channels

Voltage-gated Ca^{2+} channels represent a route of entry into the cell for a wide variety of divalent and trivalent (Lansman, 1990) metals. The general theory that di- and trivalent cations move through the Ca^{2+} channel has been confirmed numerous times in a wide variety of cells, including neurons (Taylor, 1988; Swandulla and Armstrong, 1989; Chow, 1991; Thévenod and Jones, 1992), heart (Hess and Tsien, 1984; Hess *et al.*, 1986; Lansman *et al.*, 1986) and myotubes (Lansman, 1990), such that entry of metals through voltage-gated Ca^{2+} channels must be considered a potential route of entry for any cell containing Ca^{2+} channels. A number of metals in addition to Cd^{2+} have also been examined and found to have similar blocking effects and are likely to permeate Ca^{2+} channels, including Zn^{2+} (Winegar and Lansman, 1990; Wakamori *et al.*, 1998), Fe^{2+}, Co^{2+} (Winegar *et al.*, 1991) and Ni^{2+} (Winegar *et al.*, 1991; Zamponi *et al.*, 1996; Wakamori *et al.*, 1998). In general, the rates of entry and exit of ions influence their effects and relative ability to block Ca^{2+} channels. The entry and exit rates of divalent and trivalent ions are directly related to the rate at which water molecules are replaced by pore ligands as ions enter the pore, and electrostatic repulsions between ions, respectively (Lansman, 1990; Winegar and Lansman, 1990; Winegar *et al.*, 1991).

The mechanism of permeation of Ca^{2+} channels by cations appears to be similar for L-, N-, P-, Q- and R-type Ca^{2+} channels, and probably involves similar processes for the low voltage-activated T-type channel. However, differences clearly exist; the relative potency of block of Ca^{2+} channels by Ni^{2+} and Cd^{2+} has long been used to establish the identity of Ca^{2+} channel types. The relative potency of block of T- and L-type channels by a number of heavy metals has been compared in neuroblastoma cells. L-type channels were preferentially blocked by Cd^{2+} while La^{3+} blocked both L- and T-type channels with high potency. In addition, the L- and T-type channels differed in their sensitivities to block by a series of metals. The relative potencies for block of the T-type channel were $La^{3+} << Ni^{2+} < Cd^{2+} < Co^{2+}$, whereas for the L-type channel the relative potencies were $La^{3+} < Cd^{2+} << Ni^{2+} < Co^{2+}$ (Narahashi *et al.*, 1987). A second group has examined differences in Ni^{2+} effects on $\alpha_{1A,B,\ C\ and\ E}$-subunits, as well as the role of β-subunits in block by Ni^{2+} (Zamponi *et al.*, 1996). In addition to blocking the pore of the channel, Ni^{2+} also inhibited the gating of the channel. Thus, there presumably are structural differences between the Ca^{2+}-channel types which could account for these differences, yet potential structural differences between the different Ca^{2+}-channel α_1-subunits, which may account for differences in heavy-metal effects on Ca^{2+} channels, have not yet been examined in detail. In part, this may be due to the lack of a definitive α_1-subunit for the T-type Ca^{2+}-channel, and perhaps the recent cloning and characterization of the α_{1G}-subunit (Pérez-Reyes *et al.*, 1998) will facilitate such experiments. A systematic study of the influence of structural differences on heavy-metal permeation, such as the above-mentioned studies of the structure of the Ca^{2+} binding site in the channel pore, might provide important toxicological information regarding differences in the accumulation of heavy metals into different cell types.

While the work discussed above demonstrated that metal ions, including heavy metals, could move through voltage-gated Ca^{2+} channels, often the physiologic conditions required the use of extremely strong hyperpolarizations or depolarizations in order to measure exit rates of some metals from the pore. Theoretically, these experiments suggest that divalent heavy-metal cations might utilize voltage-gated Ca^{2+} channels to cross the cell membrane; and in fact, this has been demonstrated in a number of preparations. Many of the initial observations that metals utilized Ca^{2+} channels for transport across the cell membrane were indirect, but none the less clearly demonstrated this phenomenon. The neuromuscular junction is a well-characterized model synapse (Atchison, 1988; Atchison and Spitsbergen, 1994; Shafer and Atchison, 1995), which has been utilized extensively for study of mechanisms of neurotoxic heavy metals. Evoked endplate potentials (EPPs) recorded from the motor endplate are the result of synchronous release of acetylcholine (ACh) from the nerve terminal following depolarization by an action potential (typically by electrical stimulation of the motor nerve), whereas spontaneous miniature endplate potentials (MEPPs) are recorded at the motor endplate as a result of the release of a single quanta (vesicle) of ACh from the presynaptic terminal (reviewed by Atchison, 1988). The frequency of MEPPs ($MEPP_f$) is related to the intracellular [Ca^{2+}] in the presynaptic terminal, and is increased *in vitro* by a number of divalent heavy metals, including Pb^{2+} and methylmercury (MeHg) (Atchison and Narahashi, 1982, 1984). Treatments that increase Ca^{2+} entry into the nerve terminal, particularly the Ca^{2+}-channel activator Bay K 8644, decreased the latency of the increase in $MEPP_f$ induced by MeHg, suggesting that MeHg may enter the nerve terminal via Ca^{2+} channels (Atchison, 1987). Other data support this conclusion. Whole-cell patch–clamp recordings from PC12 cells suggest that MeHg enters the pore of Ca^{2+} channels, as MeHg blocks N- and L- type currents in this cell line, and block is facilitated by increasing the frequency of channel opening (Shafer and Atchison, 1991). In primary cerebellar granule-cell cultures, blockers of N-, L- and P-type Ca^{2+} channels delay CH_3Hg^+-induced increases in intracellular Ca^{2+} concentration (Marty and Atchison, 1997), also confirming that MeHg enters cells through voltage-gated Ca^{2+} channels. Block of L-type Ca^{2+} channels by flunarizine protects against MeHg toxicity *in vivo* and cytotoxicity *in vitro*, although the overall distribution of MeHg *in vivo* was not altered by flunarizine, which suggests that the protective effect of this drug may not have been due to prevention of MeHg entry into cells (Sakamoto *et al.*, 1996).

Additional evidence that mercurials could enter the cell through voltage-sensitive Ca^{2+} channels comes from comparisons of the potency and efficacies of block of synaptosomal Ca^{2+} influx by mercurials of differing charge and lipophilicity (Hewett and Atchison, 1991). Only positively charged mercurials ($EtHg^+$, $MeHg^+$, Hg^{2+}) were able to block influx of $^{45}Ca^{2+}$ through voltage-sensitive Ca^{2+} channels in synaptosomes, and this block was voltage-dependent, suggesting that these ions interact within the pore of the channel (Hewett and Atchison, 1991). Neutral mercurials, such as dimethylmercury (Me_2Hg) and *p*-chloromercuribenzoate (PCMB) or negatively charged mercurials such as *p*-chloromercuriphenylsulfonate ($PCMBS^-$) did not block Ca^{2+} channel-mediated influx into synaptosomes (Hewett and Atchison, 1991). These data suggest that only positively charged mercurials can utilize voltage-gated Ca^{2+} channels as a route of entry into the cell, but do not rule out other mechanisms, such as passive diffusion by dissolving in the lipid bilayer, for the neutral or negatively charged mercurials. Presumably, the highly lipophilic nature of Me_2Hg was largely responsible for the recent tragic death of a researcher working with this compound (Nierenberg *et al.*, 1998).

Pb^{2+} enters bovine adrenal medullary cells through voltage-gated Ca^{2+} channels; its influx is increased by depolarization and blocked by organic Ca^{2+} channel blockers (Simons and Pocock, 1987). Tomsig and Suszkiw (1991) used fura-2, a divalent cation-sensitive fluorescent probe (typically used for measurement of intracellular Ca^{2+}) to monitor the entry of Pb^{2+} into bovine chromaffin cells. Depolarization (and hence activation of Ca^{2+} channels) clearly increased Pb^{2+} entry into these cells, and this component of Pb^{2+} entry was enhanced and decreased by agonists (Bay K 8644) and antagonists (nifedipine) of the L-type voltage-gated Ca^{2+} channel (Tomsig and Suszkiw, 1991).

Consider the earlier discussions of interactions between Ca^{2+} channels α_1-subunits and neurotransmitter release mechanisms, as well as the ability of Pb^{2+} to increase $MEPP_f$, which is related to increases in intraterminal [Ca^{2+}] (reviewed by Atchison, 1988). A number of studies indicate that Pb^{2+} can trigger directly vesicle release from nerve terminals (Toth *et al.*, 1987; Shao and Suszkiw, 1991) and bovine adrenal chromaffin cells (Tomsig and Suszkiw, 1993). Pb^{2+} is also two orders of magnitude more effective at triggering exocytosis than is Ca^{2+} (Shao and Suszkiw, 1991). In addition, this effect is not limited to Pb^{2+}, as Cd^{2+}, La^{3+}, Co^{2+} and Zn^{2+} also can activate exocytotic mechanisms (Tomsig and Suszkiw, 1996). Thus, the actions of Pb^{2+} and other heavy metals on synaptic transmission, increases in spontaneous release and decreases in evoked release (Cooper and Manalis, 1984; Cooper *et al.*, 1984), are potentially both related to interactions of heavy metals with Ca^{2+} channels at the active zone. Under resting conditions, small amounts of Pb^{2+} (or other heavy metals) entering the nerve terminal as a result of the probabilistic opening of Ca^{2+} channels at the active zone have direct access to exocytotic release mechanisms, and may trigger spontaneous release of individual vesicles of neurotransmitter, contributing to the increases in $MEPP_f$. When the nerve terminal is depolarized, block of Ca^{2+} channels by Pb^{2+} decreases dramatically the total amount of Ca^{2+} entry, thus decreasing synchronous release of acetylcholine and the EPP.

7.2.3 Metal movement into excitable cells through other types of ion channels

While the movement of divalent heavy metals through voltage-gated Ca^{2+} channels has been studied intensely, the role of other types of ion channels in heavy metal entry into cells has not been widely examined. NMDA channels, in addition to allowing the flux of Na^+ and K^+, are readily permeable to Ca^{2+} (reviewed by Ozawa, 1996). In addition, certain subunits of AMPA (Ozawa, 1996) and nACh (Boyd, 1997) receptors are also highly permeable to Ca^{2+}. Several divalent cations that block voltage-gated Ca^{2+}

channels also block NMDA receptor, such as Pb^{2+} (Ujihara and Albuquerque, 1992), Cd^{2+} and Zn^{2+} (Mayer *et al.*, 1989). Like voltage-gated Ca^{2+} channels, the NMDA receptor has a single cation-binding site in the pore of the channel, and cations move through this channel in a single-file manner (Zarei and Dani, 1994). However, under normal physiologic conditions, the channel is blocked by Mg^{2+} bound within the pore. Activation of the receptor in conjunction with strong depolarization of the neuronal membrane clears Mg^{2+} from the channel and current can flow (Mayer and Westbrook, 1984). In addition to Mg^{2+}, Co^{2+}, Mn^{2+} and Ni^{2+} also block NMDA channels in a voltage-dependent manner, suggesting that these ions also may move through the ion pore (Mayer and Westbrook, 1987). In cerebellar granule cells, activation of glutamatergic pathways increased Cd^{2+} entry (Usai *et al.*, 1997). However, by contrast to voltage-gated Ca^{2+} channels, Cd^{2+}, Hg^{2+} and Zn^{2+} appear to bind to a site near the extracellular surface of the NMDA receptor (Mayer *et al.*, 1989). Hence, it is not clear from this study whether these metals permeate the channel. In addition, Cd^{2+} had little effect on currents gated by kainate and quisqualate (Mayer *et al.*, 1989). This suggests that glutamatergic-stimulated Cd^{2+} entry occurred through non-L-type voltage-gated Ca^{2+} channels subsequent to depolarization of the membrane by activation of non-NMDA glutamate receptors, but this hypothesis has not yet been investigated.

Zn^{2+} enters excitable cells via both Ca^{2+} channels and glutamatergic receptor-gated ion channels, including the NMDA and AMPA receptor. Winegar and Lansman (1990) demonstrated that Zn^{2+} interacts with voltage-gated Ca^{2+} channels in a manner similar to other divalent heavy metals, suggesting that it can enter cells via this route. Entry of Zn^{2+} through voltage-gated Ca^{2+} channels can be blocked by antagonists of L-type Ca^{2+} channels in neurons (Weiss *et al.*, 1993; Sensi *et al.*, 1997) and heart cells (Atar *et al.*, 1995). In neurons, Zn^{2+} entry via Ca^{2+} channels may, under conditions of high Zn^{2+} concentrations, contribute to Zn^{2+}-dependent cytotoxicity. By contrast, in GH_3 cells, Zn^{2+} entry via Ca^{2+} channels was directly linked to activation of gene transcription (Atar *et al.*, 1995). In addition to Ca^{2+} channels, other channel types also mediate Zn^{2+} entry into neurons. This includes both NMDA (Koh and Choi, 1994) and AMPA receptors (Yin and Weiss, 1995). In perhaps the most comprehensive study of routes of Zn^{2+} entry into neurons, four pathways were demonstrated, including voltage-gated Ca^{2+} channels, NMDA- and AMPA- or kainate-gated channels, and reversed Na^{+}–Ca^{2+} exchange mechanisms (Sensi *et al.*, 1997).

The blocking action of Pb^{2+} has been studied on both NMDA (Alkondon *et al.*, 1990; Uteshev *et al.*, 1993; Ishihara *et al.*, 1995a; Yamada *et al.*, 1995) and nACh (Oortgiesen *et al.*, 1990, 1995; Ishihara *et al.*, 1995b; Nagata *et al.*, 1997; Zwart, 1997) receptors, and this metal is a potent blocker of these channel types in addition to Ca^{2+} channels. However, actual transport of Pb^{2+} into the cell via the NMDA and nACh receptor channels has not been examined in any detail. Given the permeability of the NMDA receptor channel and some nACh channels to Ca^{2+}, it is likely that some of these channels are permeable to Pb^{2+} as well.

7.3 Ion channels as a route of metal entry into nonexcitable cells

Divalent metals exhibit a broad array of toxicities, including renal toxicity (Hg^{2+}, Cd^{2+}), carcinogenicity (Ni^{2+}, Cd^{2+}, Pb^{2+}), altered heme synthesis and hematological function (Pb^{2+}) and teratogenicity (Pb^{2+}, Cd^{2+}), which are discussed in detail in other chapters of this book. As in excitable cells, a number of mechanisms are available for metals to enter nonexcitable cells, including ion channels. Once again, sticking with the example of Ca^{2+} channels, entry of toxic metals into target cells does utilize ion channels, and *in vitro* a number of heavy metals rely on Ca^{2+} channels as a major mechanism of entry into the cell. By contrast to neural tissue, the diversity of voltage-gated Ca^{2+} channels in nonexcitable tissues is not as great. While expression of other subtypes cannot be ruled out, only the L- and T-type channels appear to be expressed readily in non-neural tissue. The involvement of the L-type Ca^{2+} channels has been investigated more extensively than that of the T-type channel as a mechanism for metal entry, probably in large part due to the greater capacity of the L-type channel to allow Ca^{2+} entry into the cell, as it carries a much larger unitary current and remains open for longer time than the T-type channel (reviewed by Tsien *et al.*, 1988). Not surprisingly, the metals that permeate L-type channels in nonexcitable cells are the same as those that permeate L-type channels in excitable cells (Cd^{2+}, Pb^{2+}, Hg^{2+}, Ni^{2+}) and utilize the same mechanisms.

In general, studies that have focused on the interactions between Ca^{2+} channels and metals in nonexcitable cells, as compared to those in excitable cells, have emphasized the role of the ion channel as a mechanism of entry, rather than using various metals to study the function of the ion channel. One benefit of this more toxicological focus is that the size of the contribution of channel-mediated entry of the metal compared to other routes of entry is often quantified to some degree, and these studies demonstrate that channel-mediated influx of heavy metals can contribute a substantial fraction to the overall entry of metal. This has been demonstrated for several different metals and tissues. For example, Ni^{2+} utilizes L-type Ca^{2+} channels in immortalized human kidney epithelial cells, and movement of Ni^{2+} through this channel may contribute up to 75–80% of the total entry (Refsvik and Andreassen, 1995). By contrast, approximately 20% of Ni^{2+} entry into hepatocytes appears to be mediated by the L-type Ca^{2+} channels, as high concentrations of nicardipine (a DHP) reduced Ni^{2+} content by that amount only when Ca^{2+} was not present (Funakoshi *et al.*, 1997). However, activation of Ca^{2+} channels by vasopressin did increase Ni^{2+} content by 24% (Funakoshi *et al.*, 1997), suggesting that it enters the cell through Ca^{2+} channels when they are gated by endogenous agonists.

Cd^{2+} also enters hepatic cells via L-type Ca^{2+} channels and, like Ni^{2+}, these channels represent a significant pathway for Cd^{2+} entry. Using WRL-68 cells, a human hepatic cell line, Souza and co-workers (Souza *et al.*, 1997) demonstrated that approximately 35% of Cd^{2+} entry utilizes Ca^{2+} channels. The data generated in this cell line are very similar to those of earlier studies using primary cultures of rat hepatocytes, in which verapamil, and other L-type Ca^{2+}-channel antagonists, decreased the rate of Cd^{2+} entry, and decreased overall Cd^{2+} entry by 30% (Blazka and Shaikh, 1991). Interestingly, entry of Hg^{2+} was not affected by verapamil (Blazka and Shaikh, 1991). In hepatocytes, Cd^{2+} and Hg^{2+} utilize very different routes of entry, as other divalent cations, including Ca^{2+}, Cu^{2+}, Zn^{2+} and Fe^{2+}, compete with Cd^{2+} but not Hg^{2+} for entry into the hepatocyte (Blazka and Shaikh, 1992). Of these, Zn^{2+} was the most effective at blocking Cd^{2+} entry, suggesting that Cd^{2+} in addition to ion channels, uses the same pathways as Zn^{2+} to enter the hepatocyte (Blazka and Shaikh, 1992). Finally, Cd^{2+} permeation through L-type Ca^{2+} channels in HIT-T15 cells, a transformed pancreatic β-cell line, accounts for the majority of Cd^{2+} into the cell, as low micromolar concentrations of nimodopine completely inhibited Cd^{2+} entry (Usai *et al.*, 1997).

In kidney, which is sensitive to Cd^{2+} toxicity, different mechanisms appear to underlie the route of Cd^{2+} entry into proximal versus distal convoluted tubule cells. The L-type Ca^{2+}-channel blocker verapamil had no effect on Cd^{2+} entry into rat glomerular mesangial cells nor into LLC-PK_1 cells (Templeton, 1990), a proximal tubule cell line. A comparison of mechanisms of Cd^{2+} entry into primary cultures of hepatocytes and renal cortical epithelial cells also supports a difference in mechanisms by which Cd^{2+} enters cells; in hepatocytes, transport is mediated by temperature-sensitive processes such as ion channels and carriers, whereas in confluent renal cortical epithelial cells, transport is not energy or temperature dependent and occurs largely by passive diffusion (Shaikh *et al.*, 1995). While the greater overall amount of Cd^{2+} entry into proximal convoluted tubule cells probably underlies much of the increased sensitivity of these cells to Cd^{2+} toxicity, entry of Cd^{2+} into proximal tubule cells appears not to be mediated by ion channels, particularly voltage-gated Ca^{2+} channels.

By contrast, Cd^{2+} entry into renal distal tubule cells does appear to be mediated by Ca^{2+} channels. Parathyroid hormone (PTH) is a physiologic regulator of Ca^{2+} reabsorption in renal tissue, particularly distal convoluted tubules, where it acts to depolarize the cell and stimulate Ca^{2+} entry through L-type Ca^{2+} channels (Gesek and Friedman, 1992). In Mandin–Darby canine kidney cells, which have a distal tubule cell phenotype, PTH-stimulated Cd^{2+} entry was increased by 100% in the presence of the L-type Ca^{2+}-channel agonist Bay K 8644 and this effect was blocked by nifedipine, suggesting that L-type Ca^{2+} channels mediate PTH-stimulated Cd^{2+} entry in MDCK cells (Flanagan and Friedman, 1991). Similarly, under unstimulated conditions, Cd^{2+} entry into mouse distal convoluted tubule cells is not mediated by ion channels (Friedman and Gesek, 1994). However, when stimulated with PTH, Cd^{2+} entry into the cell increased and was sensitive to Bay K 8644 and nifedipine, suggesting that L-type Ca^{2+} channels mediated a component of Cd^{2+} entry into distal convoluted tubule cells (Friedman and Gesek, 1994).

Cells with secretory functions may also accumulate metals through voltage-gated Ca^{2+} channels linked to release of secretory vesicles. In pancreatic β-cells, glucose-stimulated Ca^{2+} influx and insulin release were augmented by Bay K 8644 and inhibited by Cd^{2+}, but not Ni^{2+} (Plasman *et al.*, 1990), suggesting that Cd^{2+} could possibly enter these cells via L-type Ca^{2+} channels. In melanotrophs, Mn^{2+}, Co^{2+}, La^{3+} and Cd^{2+}, but not Ni^{2+}, quenched fura-2 fluorescence signals, indicating that they readily enter the cell, likely through voltage-gated Ca^{2+} channels (Shibuya and Douglas, 1992). In another secretory cell line derived from pituitary (GH_4C_1 cells), Hinkle and co-workers had previously demonstrated that Cd^{2+} enters through voltage-gated L-type Ca^{2+} channels, and that toxicity of Cd^{2+} could be increased or decreased by Ca^{2+}-channel agonists and antagonists, respectively (Hinkle *et al.*, 1987).

7.4 Toxicological consequences of metal entry through ion channels

As stated earlier in the chapter, the toxicological consequences of metal entry through ion channels will depend in part on the type of ion channel used for entry into the cell, and partly on its location. Some of the studies discussed above demonstrated that the use of ion-channel agonists or antagonists increased or decreased, respectively, the toxicity of metals in the cells being examined. Thus, in some cases, the roles of ion channels in metal toxicity have been demonstrated (Atchison, 1987; Marty and

Atchison, 1997; Hinkle and Osborne, 1994; Fern *et al.*, 1996). However, studies demonstrating a role for ion channels in metal toxicity in which sequential mechanisms of events are clearly determined are not plentiful. Thus, in the future, one research need in this area is to associate more directly entry of metals through ion channels with downstream toxicological events.

As discussed above for Ca^{2+} channels and release proteins, ion channels do not 'float' freely in the membrane, rather they are associated with a variety of intracellular proteins, including rapsyn, gephyrin and PSD-95 proteins. These ion-channel-associated proteins not only serve to organize ion channels within the membrane, but may also provide a framework for interaction of ions such as Ca^{2+} with signaling proteins such as calmodulin, protein kinases and nitric oxide synthase (reviewed by Sheng, 1996; Kornau *et al.*, 1997). Activation of these second messengers by Ca^{2+} influx may result in transient modifications of cellular function. Alternatively, the importance of Ca^{2+}-conducting ion channels for Ca^{2+}-dependent regulation of nuclear events is clear; these channels provide a conduit for Ca^{2+} activation of gene transcription (reviewed by Ghosh and Greenberg, 1995; Lerea, 1997; Hardingham *et al.*, 1998) through pathways that include mitogen-activated protein (MAP) kinases (Xia *et al.*, 1996), cyclic AMP-responsive elements and serum-responsive elements (Johnson *et al.*, 1997). For example, activation of voltage-gated Ca^{2+} channels in PC12 cells upregulates the expression of the immediate early genes c-*fos* and egr1 (Ginty *et al.*, 1991). As in many biological systems, bidirectional regulation may occur, as c-*fos* and c-*jun* can be involved in regulation of Ca^{2+}-channel expression (Cavalié *et al.*, 1994).

It is well established that a number of heavy metals disrupt Ca^{2+}-dependent signaling processes in a variety of cells. The heavy metals Pb^{2+} (Markovac and Goldstein, 1988; Long *et al.*, 1994; Tomsig and Suszkiw, 1995) and Cd^{2+} (Long, 1997) activate PKC at nanomolar concentrations and inhibit PKC activity at micromolar concentrations (Saijoh *et al.*, 1988; Murakami *et al.*, 1993; Tomsig and Suszkiw, 1995). Pb^{2+} also alters other signaling pathways, including calmodulin (Kern and Audesirk, 1995), c-*fos* (Kim *et al.*, 1997) and cyclic GMP phosphodiesterase (Srivastava *et al.*, 1995). Cd^{2+} also induces the expression of immediate early genes in a number of different cell types (Beyersmann and Hechtenberg, 1997). Thus, entry of metals through Ca^{2+} conducting ion channels provides heavy metals with direct access to any number of different signaling pathways in which they can substitute for or disrupt the normal physiologic actions of Ca^{2+}. Metal-induced disruption of Ca^{2+} homeostasis and signaling is hypothesized to be a major mechanism underlying metal toxicity.

7.5 Summary and future research needs

To enter the cell, divalent, inorganic heavy metals, and to some extent monovalent metals, utilize ion channels that have substantial permeability to Ca^{2+}. Some other exceptions, for example the monovalent, organic metal MeHg, can also utilize ion channels as a route of entry. Ion channels are widely utilized as a pathway for divalent metal entry into both excitable cells such as neurons and nonexcitable cells such as liver or kidney cells, but may not be a 'universal' pathway. For example, entry of Cd^{2+} into the proximal tubule may not depend on entry through ion channels.

On exit from Ca^{2+}-conducting channels, divalent metals such as Cd^{2+} or Pb^{2+} may be delivered directly to cellular signaling proteins whose function they are capable of disrupting or inappropriately activating. This potential sequence of events is likely to be involved in the toxicity of heavy metals; block of Ca^{2+} channels in a number of preparations discussed above decreased metal entry and delayed initiation of, or protected against, metal toxicity.

However, more sophisticated studies are needed regarding the role of ion channels as a conduit for metal entry and toxicity. Many of the studies discussed in this chapter utilized *in vitro* preparations to examine metal entry into cells, and were designed so that metal ions were delivered to the cell only as the free metal ion. In the *in vivo* exposure situation, free metal concentrations in biological fluids are kept extremely low, due to binding of metals to serum proteins and complexation with organic molecules such as cysteine. For example, the concentration of free Pb^{2+} in plasma is approximately 10^{-12} M when total blood Pb^{2+} content is 10 μg/dL (reviewed by Simons, 1997). Thus a major question to be resolved is whether ion channels play a significant role *in vivo* in the entry of metals into the cell. At least one study mentioned above suggests that Ca^{2+} channels may not be important for MeHg entry into cells following *in vivo* exposure (Sakamoto *et al.*, 1996). However, this organic metal may have access into the cell through routes not available to inorganic metals. Some indirect evidence suggests that ion channels do play a role in metal toxicity *in vivo*, as a number of different studies have found that expression of Ca^{2+}-conducting ion channels, such as the voltage-gated Ca^{2+} channel (Ruis *et al.*, 1984; Singh and Jiang, 1997) and NMDA receptor-activated channel (reviewed by Guilarte, 1997), is altered following *in vivo* exposure to Pb^{2+}. Such responses may represent a compensatory response to, or an undesirable consequence of, Pb^{2+} exposure and need further investigation.

Although this chapter has focused on movement of metals through voltage-gated Ca^{2+} channels as they cross the plasma membrane, it should be recognized that there are a number of other types of channels which may also serve as pathways for metal permeation into cells, but for which heavy-metal entry has not been examined in any detail. In particular, Ca^{2+} release-activated Ca^{2+} (CRAC) channels need to be considered. These channels are particularly important in Ca^{2+} entry and regulation in nonexcitable cells, where they are responsible for Ca^{2+} entry following depletion of intracellular Ca^{2+} stores (reviewed by Parekh and Penner, 1997). Electrophysiologic characterization of CRAC channels indicates that their permeability to Ca^{2+}, Mg^{2+}, Ba^{2+} and monovalent cations is similar to that of voltage-dependent Ca^{2+} channels (Hoth, 1995; Lepple-Weinhues and Cahalan, 1996; Kerschbaum and Cahalan, 1998). La^{3+} blocks CRAC current at low micromolar concentrations (Aussel *et al.*, 1996; Fasolato and Nilius, 1998), suggesting that it, and perhaps other heavy metals, may also permeate CRAC channels.

Finally, it should be noted that there are many different types of ion channels present on intracellular organelles such as the mitochondria, endoplasmic reticulum and nuclear envelope. Although the intracellular concentration of many metal ions is also kept extremely low (Simons, 1997), little information exists concerning the intracellular movement and compartmentalization of metal ions. Recently, divalent cation-permeable ion channels have been found in the nuclear envelope (Longin *et al.*, 1997). If these channels have selectivity and permeation properties similar to, or turn out to be less selective than, voltage-gated Ca^{2+} channels, then they could also allow heavy metals to enter the nucleus, and disrupt processes there. Such studies have not yet been performed.

Acknowledgments

I would like to thank Dr William D. Atchison of Michigan State University and Dr Marga Oortgiesen of Duke University for their helpful comments and suggestions regarding this chapter.

References

Albuquerque, E.X., Pereira, E.F.R., Castro, N.G. and Alkondon, M. 1995. Neuronal nicotinic receptors: function, modulation and structure. *Seminars in Neuroscience* 7, 91–101.

Albuquerque, E.X., Pereira, E.F.R., Alkondon, M., Schrattenholz, A. and Maelicke, A. 1997. Nicotinic acetylcholine receptors on hippocampal neurons: distribution on the neuronal surface and modulation of receptor activity. *Journal of Receptor and Signal Transduction Research* **17**, 243–66.

Alkondon, M., Costa, A.C.S., Radhakrishnan, V., Aronstam, R.S. and Albuquerque, E.X. 1990. Selective blockade of NMDA-activated channel currents may be implicated in learning deficits caused by lead. *Federation of European Biochemical Societies Letters* **261**, 124–30.

Almers, W. and McCleskey, E.W. 1984. Non-selective conductance in calcium channels in frog muscle: calcium selectivity in a single-file pore. *Journal of Physiology (London)* **353**, 585–608.

Atar, D., Backx, P.H., Appel, M.M., Gao, W.D. and Marban, E. 1995. Excitation–transcription coupling mediated by zinc influx through voltage-dependent Ca^{2+} channels. *Journal of Biological Chemistry* **270**, 2473–7.

Atchison, W.D. 1987. Effects of activation of sodium and calcium entry on spontaneous release of acetylcholine induced by methylmercury. *Journal of Pharmacology and Experimental Therapeutics* **241**, 131–9.

Atchison, W.D. 1988. Effects of neurotoxicants on synaptic transmission: lessons learned from electrophysiological studies. *Neurotoxicology and Teratology* **10**, 393–416.

Atchison, W.D. and Narahashi, T. 1982. Methylmercury-induced depression of neuromuscular transmission in the rat. *NeuroToxicology* **3**, 37–50.

Atchison, W.D. and Narahashi, T. 1984. Mechanism of action of lead on neuromuscular junctions. *NeuroToxicology* **5**, 267–84.

Atchison, W.D. and Spitsbergen, J.M. 1994. The neuromuscular junction as a target for toxicity, in L.W. Chang (Ed.) *Principles of Neurotoxicology*, (ed. L.W. Chang), Marcel Dekker, New York, pp. 265–307.

Aussel, C., Marhaba, R., Pelassy, C. and Breittmayer, J.P. 1996. Submicromolar La^{3+} concentrations block the calcium release-activated channel, and impair CD69 and CD25 expression in CD3- or thapsigargin-activated Jurkat cells. *Biochemical Journal* **313**, 909–13.

Bahinski, A., Yatani, A., Mikala, G., Tang, S., Yamamoto, S. and Schwartz, A. 1997. Charged amino acids near the pore entrance influense the ion-conduction of the channel. *Molecular and Cellular Biochemistry* **166**, 125–34.

Barnard, E.A. 1996. The transmitter-gated channels: a range of receptor types and structures. *Trends in Pharmacological Sciences* **17**, 305–9.

Beyersmann, D. and Hechtenberg, S. 1997. Cadmium, gene regulation and cellular signalling in mammalian cells. *Toxicology and Applied Pharmacology* **144**, 247–61.

Bezprozvanny, I., Sheller, R.H. and Tsien, R.W. 1995. Functional impact of syntaxin on gating of N-type and Q-type calcium channels. *Nature (London)* **378**, 623–7.

Birnbaumer, L., Campbell, K.P., Catterall, W.A., Harpold, M.M., Hofmann, F., Horne, W.A., Mori, Y., Schwartz, A., Snutch, T.P., Tanabe, T. and Tsien, R.W. 1994. The naming of voltage-gated Ca^{2+} channels. *Neuron* **13**, 505–6.

Blazka, M.E. and Shaikh, Z.A. 1991. Differences in cadmium and mercury uptakes by hepatocytes: role of calcium channels. *Toxicology and Applied Pharmacology* **110**, 355–63.

Blazka, M.E. and Shaikh, Z.A. 1992. Cadmium and mercury accumulation in rat hepatocytes: interactions with other metals. *Toxicology and Applied Pharmacology* **113**, 118–25.

Bowersox, C., Ko, C-P., Sugiura, Y., Li, C.Z., Fox, J., Hoffman, B.B. and Miljanich, G. 1993. Omega-conopeptide SNX-230 (MVIIC) blocks calcium channels in mouse neuromuscular junction nerve terminals. *Neuroscience Abstracts* **19**, 1478.

Boyd, R.T. 1997 The molecular biology of neuronal nicotinic acetylcholine receptors. *Critical Reviews in Toxicology* **27**, 299–318.

Byerly, L., Chase, P.B. and Stimers, J.R. 1985. Permeation and interaction of divalent cations in calcium channels of snail neurons. *Journal of General Physiology* **85**, 491–518.

Catterall, W.A. 1995. Structure and function of voltage-gated ion channels. *Annual Review of Biochemistry* **64**, 493–531.

Cavalié, A., Beringer, B., Haas, C.A., García, D.E., Lindholm, D., and Lux, H.D. 1994. Constitutive upregulation of calcium channel currents in rat phaeochromocytoma cells: role of c-*fos* and c-*jun*. *Journal of Physiology (London)* **479**,11–27.

Cens, T., Dalle, C. and Charnet, P. 1998. Expression of the β subunit modulates surface potential sensing by calcium channels. *Pflügers Archives–European Journal of Physiology* **435**, 865–7.

Charlton, M.P. and Augustine, G.J. 1990. Classification of presynaptic calcium channels at the squid giant synapse: neither T-, L- nor N-type. *Brain Research* **525**, 133–9.

Chiu, S.Y. 1980. Asymmetry currents in the mammalian myelinated nerve. *Journal of Physiology (London)* **309**, 499–519.

Chow, R.H. 1991. Cadmium block of squid calcium currents. Macroscopic data and a kinetic model. *Journal of General Physiology* **98**, 751–70.

Cooper, G.P. and Manalis, R.S. 1983. Influence of heavy metals on synaptic transmission: a review. *NeuroToxicology* **4**, 69–84.

Cooper G.P. and Manalis, R.S. 1984. Cadmium: effects on transmitter release at the frog neuromuscular junction. *European Journal of Pharmacology* **99**, 251–6.

Cooper, G.P., Suszkiw, J.B. and Manalis, R.S. (1984). Presynaptic effects of heavy metals, in *Cellular and Molecular Neurotoxicology*, (ed. T. Narahashi), Raven Press, New York, pp. 1–21.

De Waard, M., Pragnell, M. and Campbell, K.P. 1994. Ca^{2+} channel regulation by a conserved β subunit domain. *Neuron* **13**, 495–503.

De Waard, M., Witcher, D.R., Pragnell, M., Liu, H. and Campbell, K.P. 1995. Properties of the alpha 1-beta anchoring site in voltage-dependent Ca^{2+} channels. *Journal of Biological Chemistry* **270**, 12056–64.

De Waard, M., Gurnett, C. and Campbell, K.P. 1996. Structural and functional diversity of voltage-activated calcium channels, in *Ion Channels*, (ed. T. Narahashi), Plenum Press, New York, Vol. 4, pp. 41–87.

De Waard, M., Liu, H., Walker, D., Scott, V.E.S., Gurnett, C.A. and Campbell, K.P. 1997. Direct binding of G-protein $\beta\gamma$ complex to voltage-dependent calcium channels. *Nature (London)* **385**, 446–50.

Dirksen, R.T., Nakai, J., Gonzalez, A., Imoto, K. and Beam, K.G. 1997. The S5–S6 linker of repeat I is a critical determinant of L-type Ca^{2+} channel conductance. *Biophysical Journal* **73**, 1402–9.

Ellinor, P.T., Yang, J., Sather, W., Zhang, J-F. and Tsien, R.W. 1995. Ca^{2+} channel selectivity at a single locus for high-affinity Ca^{2+} interactions. *Neuron* **15**, 1121–32.

Fasolato, C. and Nilius, B. 1998. Store depletion triggers the calcium release-activated calcium current (I_{CRAC}) in macrovascular endothelial cells: a comparison with Jurkat and embryonic kidney cell lines. *Pflüger's Archives* **436**, 69–74.

Fern, R., Black, J.A., Ransom, B.R. and Waxman, S.G. 1996. Cd^{2+}-induced injury in CNS white matter. *Journal of Neurophysiology* **76**, 3264–73.

Finkbeiner, S. and Greenberg, M.E. 1996. Ca^{2+}-dependent routes to ras: mechanisms for neuronal survival, differentiation and plasticity. *Neuron* **16**, 233–6.

Flanagan, J.L. and Friedman, PA. 1991. Parathyroid hormone-stimulated cadmium accumulation in Mandin–Darby canine kidney cells. *Toxicology and Applied Pharmacology* **109**, 241–50.

Franciolini, F. 1994. The S4 segement and gating of voltage-dependent cationic channels. *Biochimica et Biophysica Acta* **1197**, 227–36.

Friedman, P.A. and Gesek, F.A. 1994. Cadmium uptake by kidney distal convoluted tubule cells. *Toxicology and Applied Pharmacology* **128**, 257–63.

Funakoshi T., Inoue, T., Shimada, H. and Kojima, S. 1997. The mechanisms of nickel uptake by rat primary hepatocyte cultures: role of calcium channels. *Toxicology* **124**, 21–6.

Fukushima, Y. and Hagiwara, S. 1985. Current carried by monovalent cations through Ca^{2+}, channels in mouse neoplastic B lymphocytes. *Journal of Neurophysiology* **358**, 255–85.

Gesek, F.A. and Friedman, P.A. 1992. On the mechanism of parathyroid hormone stimulation of calcium uptake by mouse distal convoluted tubule cells. *Journal of Clinical Investigation* **90**, 749–58.

Ghosh, A. and Greenberg, M.E. 1995. Calcium signaling in neurons: molecular mechanisms and cellular consequences. *Science* **268**, 239–47.

Ginty, D.D., Glowaka, D., Bader, D.S., Hidaka, H. and Wagner, J.A. 1991. Induction of immediate early genes by Ca^{2+} influx requires cAMP-dependent protein kinase in PC12 cells. *Journal of Biological Chemistry* **266**, 17454–8.

Guilarte, T.R. 1997. Glutamatergic system and developmental lead neurotoxicity. *NeuroToxicology* **18**, 665–72.

Hardingham, G.E., Cruzalegui, F.H., Chawla, S. and Bading, H. 1998. Mechanisms controlling gene expression by nuclear calcium signals. *Cell Calcium* **23**, 131–4.

Herlitze, S., Hockerman, G.H., Scheuer, T. and Catterall, W.A. 1997. Molecular determinants of inactivation and G-protein modulation in the intracellular loop connecting domains I and II of the calcium channel α_{1A} subunit. *Proceedings of the National Academy of Sciences of the United States of America* **94**, 1512–6.

Hess., P. and Tsien, R.W. 1984. Mechanism of ion permeation through calcium channels. *Nature (London)* **309**, 453–6.

Hess, P., Lansman, J.B. and Tsien, R.W. 1986. Calcium channel selectivity for divalent and monovalent cations. Voltage and concentration dependence of single channel current in ventricular heart cells. *Journal of General Physiology* **88**, 293–319.

Hewett, S.J. and Atchison, W.D. 1991. Effects of charge and lipophilicity on mercurial-induced reduction of $^{45}Ca^{2+}$ uptake in isolated nerve terminals of the rat. *Toxicology and Applied Pharmacology* **113**, 267–73.

Hille, B. 1992. *Ionic Channels of Excitable Membranes*, 2nd edn, Sinauer Associates, Sunderland, MA.

Hinkle, P.M. and Osborne, M.E. 1994. Cadmium toxicity in rat pheochromocytoma cells: studies on the mechanism of uptake. *Toxicology and Applied Pharmacology* **124**, 91–8.

Hinkle, P.M., Kinsella, P.A. and Osterhoudt, K.C. 1987. Cadmium uptake and toxicity via voltage-sensitive calcium channels. *Journal of Biological Chemistry* **262**, 16333–7.

Hoth, M. 1995. Calcium and barium permeation through calcium release-activated calcium (CRAC) channels. *Pflüger's Archives* **430**, 315–22.

Huguenard, JR. 1996. Low threshold calcium currents in central nervous system neurons. *Annual Review of Physiology* **58**, 329–48.

Ishihara, K., Alkondon, M., Montes, J.G. and Albuquerque, E.X. 1995a. Nicotinic responses in acutely dissociated rat hippocampal neurons and the selective blockade of fast-desensitizing nicotinic currents by lead. *Journal of Pharmacology and Experimental Therapeutics* **273**,1471–82.

Ishihara, K., Alkondon, M., Montes, J.G. and Albuquerque, E.X. 1995b. Ontogenically related properties of *N*-methyl-D-aspartate receptors in rat hippocampal neurons and the age-specific sensitivity of developing neurons to lead. *Journal of Pharmacology and Experimental Therapeutics* **273**, 1459–70.

Isom, L.L., De Jongh, K.S. and Catterall, W.A. 1994. Auxiliary subunits of voltage-gated ion channels. *Neuron* **12**, 1183–94.

Issa, N.P. and Hudspeth, A.J. 1994. Clustering of Ca^{2+} channels and Ca^{2+} activated K^+ channels at fluorescently labeled presynaptic active zones of hair cells. *Proceedings of the National Academy of Sciences of the United States of America* **91**, 7578–82.

Jackson, M.B. and Yakel, J.L. 1995. The 5-HT_3 receptor channel. *Annual Review of Physiology* **57**, 447–68.

Jan, L.Y. and Jan, Y.N. 1997. Voltage-gated and inwardly rectifying potassium channels. *Journal of Physiology (London)* **505**, 267–82.

Johnson, C.M., Hill, C.S., Chawla, S., Treisman, R. and Bading, H. 1997. Calcium controls gene expression via three distinct pathways that can function independently of the ras/mitogen-activated protein kinases (ERKs) signaling cascade. *Journal of Neuroscience* **17**, 6189–202.

Kern, M. and Audesirk, G. 1995. Inorganic lead may inhibit neurite development in cultured rat hippocampal neurons through hyperphosphorylation. *Toxicology and Applied Pharmacology* **134**, 111–23.

Kerschbaum, H.H. and Cahalan, M.D. 1998. Monovalent permeability, rectification, and ionic block of store-operated calcium channels in Jurkat T lymphocytes. *Journal of General Physiology* **111**, 521–37.

Kidokoro, Y. 1993. Developmental changes of transmitter-gated channels. *Japanese Journal of Pharmacology* **43**, 727–43.

Kim, K., Annadata, M., Goldstein, G.W. and Bressler, J.P. 1997. Induction of c-*fos* mRNA by lead in PC12 cells. *International Journal of Developmental Neuroscience* **15**, 175–82.

Koh, J.-Y. and Choi, D.W. 1994. Zinc toxicity on cultured cortical neurons: involvement of *N*-methyl-D-aspartate receptors. *Neuroscience* **60**, 1049–57.

Kornau, H.-C., Seeburg, P.H. and Kennedy, M.B. 1997. Interaction of ion channels and receptors with PDZ domain proteins. *Current Opinion in Neurobiology* **7**, 368–73.

Krizanova, O., Varadi, M., Schwartz, A. and Varadi, G. 1995. Coexpression of skeletal muscle voltage-dependent calcium channel α_1 and β cDNAs in mouse Ltk^- cells increases the amount of α_1 protein in the cell membrane. *Biochemical and Biophysical Research Communications* **211**, 921–7.

Lansman, J.B. 1990. Blockade of current through single calcium channels by trivalent lanthanide cations. Effect of ionic radius on the rates of ion entry and exit. *Journal of General Physiology* **95**, 679–96.

Lansman, J.B., Hess, P. and Tsien, R.W. 1986. Blockade of current through single calcium channels by Cd^{2+}, Mg^{2+}, and Ca^{2+}. *Journal of General Physiology* **88**, 321–47.

Lee K.S. and Tsien, R.W. 1984. High selectivity of calcium channels in single dialysed heart cells of the guinea-pig. *Journal of Physiology (London)* **354**, 253–72.

Lepple-Wienhues, A. and Cahalan, M.D. 1996. Conductance and permeation of monovalent cations through depletion-activated Ca^{2+} channels (I_{CRAC}) in Jurkat T cells. *Biophysical Journal* **71**, 787–94.

Lerea, L.S. 1997. Glutamate receptors and gene induction: signalling from receptor to nucleus. *Cellular Signalling* **9**, 219–26.

Leveque, C., Pupier, S., Marqueze, B., Geslin, L., Kataoka, M., Takahashi, M., De Waard, M. and Seagar, M. 1998. Interaction of cysteine string proteins with the α_{1A} subunit of the P/Q-type calcium channel. *Journal of Biological Chemistry* **273**, 13488–92.

Levitan, I.B. 1994. Modulation of ion channels by protein phosphorylation and dephosphorylation. *Annual Review of Physiology* **56**,193–212.

Lin, Y.W., Rudy, B. and Llinás, R.R. 1990. Funnel-web spider venom and a toxin fraction block calcium current expressed from rat brain mRNA in *Xenopus* oocytes. *Proceedings of the National Academy of Sciences of the United States of America* **87**, 4538–42.

Llinás, R., Sugimori, M., Lin, J.-W. and Cherksey, B. (1989). Blocking and isolation of a calcium channel from neurons in mammals and cephalopods utilizing a toxin fraction (FTX) from funnel-web spider poison. *Proceedings of the National Academy of Sciences of the United States of America* **86**,1689–93.

Long, G.J. 1997. The effect of cadmium on cytosolic free calcium, protein kinase C, and collagen synthesis in rat osteosarcoma (ROS 17/2.8) cells. *Toxicology and Applied Pharmacology* **143**, 189–95.

Long, G.J., Rosen, J.F. and Schanne, F.A.X. 1994. Lead activation of protein kinase C from rat brain. *Journal of Biological Chemistry* **269**, 834–7.

Longin, A.-S., Mezin, P., Favier, A. and Verdetti, J. 1997. Presence of zinc and calcium permeant channels in the inner membrane of the nuclear envelope. *Biochemical and Biophysical Research Communications* **235**, 236–241.

MacKinnon, R. 1995. Pore loops: an emerging theme in ion channel structure. *Neuron* **14**, 889–92.

Markovac, J. and Goldstein, G.W. 1988. Picomolar concentrations of lead stimulate brain protein kinase C. *Nature (London)* **334**, 71–3.

Martin-Moutot, N., Charvin, N., Leveque, C., Sato, K., Nishiki, T., Kozaki, S., Takahashi, M. and Seagar, M. 1996. Interaction of SNARE complexes with P/Q-type calcium channels in rat cerebellar synaptosomes. *Journal of Biological Chemistry* **271**, 6567–70.

Marty, M.S. and Atchison, W.D. 1997. Pathways mediating Ca^{2+} entry in rat cerebellar granule cells following *in vitro* exposure to methyl mercury. *Toxicology and Applied Pharmacology* **147**, 319–30.

Mayer, M.L. and Westbrook, G.L. 1984. Voltage-dependent block by Mg^{2+} of NMDA responses in spinal cord neurons. *Nature (London)* **309**, 261–3.

Mayer, M.L. and Westbrook, G.L. 1987. Permeation and block of *N*-methyl-D-aspartic acid receptor channels by divalent cations in mouse cultured central neurones. *Journal of Physiology (London)* **394**, 501–27.

Mayer, M.L., Vyklicky, L. and Westbrook, G.L. 1989. Modulation of excitatory amino acid receptors by group IIB metal cations in cultured mouse hippocampal neurons. *Journal of Physiology (London)* **415**, 329–50.

Mermelstein, P.G., Becker, J.B. and Surmeier, D.J. 1996. Estradiol reduces calcium currents in rat neostriatal neurons via a membrane receptor. *Journal of Neuroscience* **16**, 595–604.

Mikala, G., Bahinski, A., Yatani, A., Tang, S. and Schwartz, A. 1993. Differential contribution by conserved glutamate residues to an ion-selectivity site in the L-type Ca^{2+} channel pore. *Federation of European Biochemical Societies Letters* **335**, 265–9.

Miyamoto, M.D. 1983. Hg^{2+} causes neurotoxicity at an intracellular site following entry through Na and Ca channels. *Brain Research* **267**, 375–9.

Mohler H., Fritschy, J.M., Lüscher, B., Rudolph, U., Benson, J. and Benke, D. 1996. The $GABA_A$ receptors. From subunits to diverse functions, in *Ion Channels*, (ed. T. Narahashi), Plenum Press, New York, Vol. 4, pp. 89–113.

Mori, H. and Mishina, M. 1995. Structure and function of the NMDA receptor channel. *Neuropharmacology* **34**, 1219–37.

Mori, Y., Fredrich, T., Kim, M.S., Mikami, A., Nakai, J., Ruth, P., Bosse, E., Hofmann, F., Flockerzi, V., Furuichi, T., Mikoshiba, K., Imoto, K., Tanabe, T. and Numa, S. 1991. Primary structure and functional expression from complementary DNA of a brain calcium channel. *Nature (London)* **350**, 398–402.

Morrow, A.L. 1995. Regulation of $GABA_A$ receptor function and gene expression in the central nervous system. *International Review of Neurobiology* **38**, 1–41.

Murakami, K., Feng, G and Chen, S.G. 1993. Inhibition of brain protein kinase C subtypes by lead. *Journal of Pharmacology and Experimental Therapeutics* **264**, 757–61.

Murphy, T.H., Worley, P.F. and Baraban, J.M. 1991. L-type voltage-sensitive calcium channels mediate synaptic activation of immediate early genes. *Neuron* **7**, 625–35.

Nagata, K., Haung, C.-S., Song, J.-H. and Narahashi, T. 1997. Lead modulation of the neuronal nicotinic acetylcholine receptor in PC12 cells. *Brain Research* **754**, 21–7.

Narahashi, T., Tsunoo, A. and Yoshii, M. 1987. Characterization of two types of calcium channels in mouse neuroblastoma cells. *Journal of Physiology* **383**, 231–49.

Nernst, W. 1888. Zur kinetik der lösung befindlichen körper: theorie der diffusion. *Z. Phys. Chem.* 613–37.

Nierenberg, D.W., Nordgren, R.E., Chang, M.B., Siegler, R.W., Blayney, M.B., Hochberg, F., Toribara, T.Y., Cernichiari, E. and Clarkson, T. 1998. Delayed cerebellar disease and death after accidental exposure to dimethlymercury. *New England Journal of Medicine* **338**, 1672–6.

Nowycky, M.C., Fox, A.P. and Tsien, R.W. 1985. Three types of neuronal calcium channels with different calcium agonist sensitivity. *Nature (London)* **316**, 440–3.

Olcese, R., Qin, N., Schneider, T., Neely, A., Wei, X., Stephani, E. and Birnbaumer, L. 1994. The amino terminus of a calcium channel β subunit sets the rates of channel inactivation independently of the subunit's effect on activation. *Neuron* **13**, 1433–8.

Oortgiesen, M., Van Kleef, R.G.D.M., Bajnath, R.B. and Vijverberg, H.P.M. 1990. Nanomolar concentrations of lead selectively block neuronal nicotinic acetlycholine responses in mouse neuroblastoma cells. *Toxcology and Applied Pharmacology* **103**, 165–74.

Oortgiesen M., Zwart, R., van Kleef, R.G.D.M. and Vijverberg, H.P.M. 1995. Subunit-dependent action of lead on neuornal nicotinic acetylcholine receptors expressed in *Xenopus* oocytes. *Clinical and Experimental Pharmacology and Physiology* **22**, 364–5.

Ozawa, S. 1996. Permeation of calcium through glutamate receptor channels. *Seminars in Neuroscience* **8**, 261–9.

Page, K.M., Stephens, G.J., Berrow, N.S. and Dolphin, A.C. 1997. The intracellular loop between domains I and II of the B-type calcium channel confers aspects of G-protein sensitivity to the E-type calcium channel. *Journal of Neuroscience* **17**, 1330–8.

Page, K.M., Canti, C., Stephens, G.J., Berrow, N.S. and Dolphin, A.C. 1998. Identification of the amino terminus of neuronal Ca^{2+} channel α_1 subunits α_{1B} and α_{1E} as an essential determinant of G-protein modulation. *Journal of Neuroscience* **18**, 4815–4824.

Parekh, A.B. and Penner, R. 1997. Store depletion and calcium influx. *Physiological Reviews* **77**, 901–30.

Parent, L. and Gopalakrishnan, M. 1995. Glutamate substitution in repeat IV alters divalent and monovalent cation permeation in the heart Ca^{2+} channel. *Biophysical Journal* **69**, 1801–13.

Pérez-Reyes, E., Cribbs, L.L., Daud, A., Lacerda, A.E., Barclay, J., Williamson, M.P., Fox, M., Rees, M. and Lee, J.-H. 1998. Molecular characterization of a neuronal low-voltage-activated T-type calcium channel. *Nature (London)* **39**, 896–900.

Plasman, P.O., Hermann, M., Herchuelz, A. and LeBrun, P. 1990. Sensitivity to Cd^{2+} but resistance to Ni^{2+} of Ca^{2+} inflow into rat pancreatic islets. *American Journal of Physiology* **258**, E529–E533.

Plummer, M.R., Logothetis, D.E. and Hess, P. 1989. Elementary properties and pharmacological sensitivities of calcium channels in mammalian peripherial neurons. *Neuron* **2**,1453–63.

Protti, D.A. and Uchitel, O.D. 1993. Transmitter release and presynaptic Ca^{2+} currents blocked by the spider toxin ω-Aga-IVA. *NeuroReport* **5**, 333–6.

Qin, N., Platano, D., Olcese, R., Stefani, E. and Birnbaumer, L. 1997. Direct interaction of $G\beta\gamma$-binding domain of the Ca^{2+} channel α_1 subunit is responsible for channel inhibition by G protein coupled receptors. *Proceedings of the National Academy of Sciences of the United States of America* **94**, 8866–71.

Rajendra, S., Lynch, J. and Schofield., P.R. 1997. The glycine receptor. *Pharmacology and Therapeutics* **7**, 121–46.

Randall, A.D. 1998. The molecular basis of voltage-gated Ca^{2+} channel diversity: Is it time for T? *Journal of Membrane Biology* **16**, 207–13.

Randall, A.D., Wendland, B., Schweizer, F., Miljanich, G., Adams, M.E. and Tsien, R.W. 1993.

Five pharmacologically distinct high voltage-activated Ca^{2+} channels in cerebellar granule cells. *Neuroscience Abstracts* **19**, 1478.

Refsvik, T. and Andreassen, T. 1995. Surface binding and uptake of nickel (II) in human epithelial kidney cells: modulation by ionomycin, nicardipine and metals. *Carcinogenesis* **16**,1107–12.

Rettig, J., Sheng, Z-H., Kim, D.K., Hodson, C.D., Snutch, T.P. and Catterall, W.A. 1996. Isoform-specific interaction of the α_{1A} subunits of brain Ca^{2+} channels with the presynaptic proteins syntaxin and SNAP-25. *Proceedings of the National Academy of Sciences of the United States of America* **93**, 7363–8.

Reuter, H. 1996. Diversity and function of presynaptic calcium channels in the brain. *Current Opinion in Neurobiology* **6**, 331–7.

Robataille, R., Adler, E.M. and Charlton, M.P. 1990. Strategic location of calcium channels at transmitter release sites of frog neuromuscular synapses. *Neuron* **5**, 773–9.

Roche, J.P. and Treistman, S.N. 1998. Ca^{2+} channel β_3 subunit enhances voltage-dependent relief of G-protein inhibition induced by muscarinic receptor activation and $G\beta\gamma$. *Journal of Neuroscience* **18**, 4883–90.

Ruis, R.A., Lucchi, L., Govoni, S. and Trabucchi, M. 1984. *In vivo* chronic lead exposure alters [^{3}H]nitrendipine binding in rat striatum. *Brain Research* **322**, 180–3.

Saijoh, K., Inoue, Y., Ratsuyama, H. and Sumino, K. 1988. The interaction of cations with activity of soluble protein kinase C from mouse brain. *Pharmacology and Toxicology* **63**, 221–4.

Sakamoto, M., Ikegami, N. and Nakano, A. 1996. Protective effects of Ca^{2+} channel blockers against methylmercury toxicity. *Pharmacology and Toxicology* **78**, 193–9.

Sather, W.A., Tanabe, T., Mori, Y., Adams, M.E., Miljanich, G., Numa, S. and Tsien, R.W. 1992. Contrasts between cloned BI and cardiac L-type Ca^{2+} channels expressed in *Xenopus* oocytes. *Neuroscience Abstracts* **18**, 10.

Sather, W.A., Tanabe, T., Zhang, J.F., Mori, Y., Adams, M.E. and Tsien, R.W. 1993. Distinctive biophysical and pharmacological properties of class A (BI) calcium channel α_1 subunits. *Neuron* **11**, 291–303.

Scott, R.H., Sutton, K.G. and Dolphin, A.C. 1993. Interactions of polyamines with neuronal ion channels. *Trends in Neuroscience* **16**, 153–60.

Sensi, S., Canzoniero, L.M.T., Yu, S.P., Ying, H.S., Koh, J-Y., Kerchner, G.A. and Choi, D.W. 1997. Measurement of intracellular free zinc in living cortical neurons: routes of entry. *Journal of Neuroscience* **17**, 9554–64.

Shafer, T.J. and Atchison, W.D. 1991. Methylmercury blocks N- and L-type Ca^{++} channels in nerve growth factor-differentiated pheochromocytoma (PC12) cells. *Journal of Pharmacology and Experimental Therapeutics* **258**, 149–57.

Shafer, T.J. and Atchison, W.D. 1995. Electrophysiological methods for analysis of effects of neurotoxicants on synaptic transmission, in *Neurotoxicology: Approaches and Methods*, (eds L.W. Chang and W. Slikker Jr), Academic Press, New York, pp. 157–81.

Shaikh, Z.A., Blazka, M.E. and Endo, T. 1995. Metal transport in cells: cadmium uptake by rat hepatocytes and renal cortical epithelial cells. *Environmental Health Perspectives* **103**, 73–5.

Shao, Z. and Suszkiw, J.B. 1991. Ca^{2+}-surrogate action of Pb^{2+} on acetylcholine release from rat brain synaptosomes. *Journal of Neurochemistry* **56**, 568–74.

Sheng, M. 1996. PDZs and receptor/channel clusturing: rounding up the latest suspects. *Neuron* **17**, 575–8.

Sheng, M. and Kim, E. 1996. Ion channel associated proteins. *Current Opinion in Neurobiology* **6**, 602–8.

Sheng, Z-H., Rettig, J., Takahashi, M. and Catterall, W.A. 1994. Identification of a syntaxin-binding site on N-type calcium channels. *Neuron* **13**, 1303–13.

Shibuya, I. and Douglas, W.W. 1992. Calcium channels in rat melanotrophs are permeable to manganese, cobalt, cadmium, and lanthanium, but not to nickel: evidence provided by fluorescence changes in fura-2-loaded cells. *Endocrinology* **131**, 1936–41.

Shistik, E., Ivanina, T., Puri, T., Hosey, M. and Dascal, N. 1995. Ca^{2+} current enhancement by α_2/δ and β subunits in *Xenopus* oocytes: contribution of changes in channel gating and α_1 protein level. *Journal of Physiology (London)* **489**, 55–62.

Simons, T.J.B. 1997. Interactions between lead and metals in cellular toxicity, in *Mineral and Metal Toxicology*, (eds B. Yasui, K. Ota, M.J. Strong and M.A. Verity), CRC Press, Boca Raton, pp. 243–51.

Simons, T.J.B. and Pocock, G. 1987. Lead enters bovine adrenal medullary cells through calcium channels. *Journal of Neurochemistry* **48**, 383–9.

Singh, A.K. and Jiang, Y. 1997. Developmental effects of chronic low level lead exposure on voltage-gated calcium channels in brain synaptosomes obtained from the neonatal and the adult rats. *Comp. Biochemistry and Physiology* **118C**, 75–81.

Smith, S.J., Buchanan, J., Osses, L.R., Charlton, M.P. and Augustine, G.J. 1993. The spatial distribution of calcium signals in squid presynaptic terminals. *Journal of Physiology (London)* **472**, 573–93.

Souza, V., Bucio, L. and Gutierrez-Ruiz, M.C. 1997. Cadmium uptake by a human hepatic cell line (WRL-68 cells). *Toxicology* **120**, 215–20.

Srivastava, D., Hurwitz, R.L. and Fox, D.A. 1995. Lead- and calcium-mediated inhibition of bovine rod cGMP phosphodiesterase: interactions with magnesium. *Toxicology and Applied Pharmacology* **134**, 43–52.

Sucher, N.J., Awobuluyi, M., Choi, Y-B. and Lipton, S.A. 1996. NMDA receptors: from genes to channels. *Trends in Pharmacological Sciences* **17**, 348–55.

Swandulla, D. and Armstrong, C.M. 1989. Calcium channel block by cadmium in chicken sensory neurons. *Proceedings of the National Academy of Sciences of the United States of America* **86**, 1736–40.

Takahashi, T. and Momiyama, A. 1993. Different types of calcium channels mediate central synaptic transmission. *Nature (London)* **366**, 156–8.

Tang, S., Mikala, G., Bahinski, A., Yatani, A., Varadi, G. and Schwartz, A. 1993. Molecular localization of ion selectivity sites within the pore of a human L-type cardiac calcium channel. *Journal of Biological Chemistry* **268**, 13026–9.

Taylor, W.R. 1988. Permeation of barium and cadmium through slowly inactivating calcium channels in cat sensory neurons. *Journal of Physiology (London)* **407**, 433–52.

Thévenod, F. and Jones, S.W. 1992. Cadmium block of calcium current in frog sympathetic neurons. *Biophysical Journal* **63**, 162–8.

Templeton, D.M. 1990. Cadmium uptake by cells of renal orgin. *Journal of Biological Chemistry* **265**, 21764–70.

Tomsig, J.L. and Suszkiw, J.B. 1991. Permeation of Pb^{2+} through calcium channels: fura-2 measurements of voltage and dihydropyridine-sensitive Pb^{2+} entry in isolated bovine chromaffin cells. *Biochimica et Biophysica Acta* **1069**, 197–200.

Tomsig, J.L. and Suszkiw, J.B. 1993. Intracellular mechanism of Pb^{2+}-induced norepinephrine release from bovine chromaffin cells. *American Journal of Physiology* **265**, C1630–C1636.

Tomsig, J.L. and Suszkiw, J.B. 1995. Multisite interactions between Pb^{2+} and protein kinase C and its role in norepinephrine release from bovine adrenal chromaffin cells. *Journal of Neurochemistry* **64**, 2667–73.

Tomsig, J.L. and Suszkiw, J.B. 1996. Metal selectivity of exocytosis in alpha-toxin-permeabilized bovine chromaffin cells. *Journal of Neurochemistry* **66**, 644–50.

Toth, G.P., Cooper, G.P. and Suszkiw, J.B. 1987. Effects of divalent cations on acetylcholine release from digitonin-permeabilized rat cortical synaptosomes. *NeuroToxicology* **8**, 507–16.

Tsien, R.W., Lipscombe, D., Madison, D.V., Bley, K.R. and Fox, A.P. 1988. Multiple types of neuronal calcium channels and their selective modulation. *Trends in Neurosciences* **11**, 431–8.

Turner, T.J., Adams, M.E. and Dunlap, K. 1993. Multiple Ca^{2+} channel types coexist to regulate synaptosomal neurotransmitter release. *Proceedings of the National Academy of Sciences of the United States of America* **90**, 9518–22.

Uchitel, O.D., Protti, D.A., Sanchez, V., Cherksey, B.D., Sugimori, M. and Llinàs, R. 1992. P-type voltage-dependent calcium channel mediates presynaptic calcium influx and transmitter release in mammalian synapses. *Proceedings of the National Academy of Sciences of the United States of America* **89**, 3330–3.

Ujihara, H. and Albuquerque, E.X. 1992. Developmental change of the inhibition by lead of NMDA-activated currents in cultured hippocampal neurons. *Journal of Pharmacology and Experimental Therapeutics* **263**, 868–75.

Usai, C., Barberis, A., Moccagatta, L. and Marchetti, C. 1997. Pathways of cadmium uptake in excitable mammalian cells: a microspectrofluorimetric study. *European Journal of Histochemistry* **41** (Suppl. 2), 189–90.

Uteshev, V., Büsselberg, D. and Haas, H.L. 1993. Pb modulates the NMDA-receptor-channel complex. *Naunyn-Schmiedeberg's Archives of Pharmacology* **347**, 209–13.

Wakamori, M., Strobeck, M., Niidome, T., Teremoto, T., Imoto, K. and Mori, Y. 1998. Functional characterization of ion permeation pathway in the N-type Ca^{2+} channel. *Journal of Neurophysiology* **79**, 622–34.

Walker, D., Bichet, D., Campbell, K.P. and De Waard, M. 1998. A beta 4 isoform-specific interaction site in the carboxyl-terminal region of the voltage-dependent Ca^{2+} channel alpha 1A subunit. *Journal of Biological Chemistry* **273**, 2361–7.

Weiss, J.H., Hartley, D.M., Koh, J.-Y., and Choi, D.W. 1993. AMPA receptor activation potentiates zinc neurotoxicity. *Neuron* **10**, 43–9.

Wheeler, D.B., Randall, A. and Tsien, R.W. 1994. Roles of N-type and Q-type Ca^{2+} channels in supporting hippocampal synaptic transmission. *Science* **264**, 107–11.

Whiting, P.J., McKernan, R.M. and Wafford, K.A. 1995. Structure and pharmacology of vertebrate $GABA_A$ receptor subtypes. *International Review of Neurobiology* **38**, 95–138.

Winegar, B.D. and Lansman, J.B. 1990. Voltage-dependent block by zinc of single dihydropyridine-sensitive calcium channels in mouse C2 myotubules. *Journal of Physiology (London)* **425**, 563–78.

Winegar, B.D., Kelly, R. and Lansman, J.B. 1991. Block of current through single calcium channels by Fe, Co, and Ni. Location of the transition metal binding site in the pore. *Journal of General Physiology* **97**, 351–67.

Wu, L.-G. Borst, J.G.G. and Sakmann, B. 1998. R-type Ca^{2+} currents evoke transmitter release at a rat central synapse. *Proceedings of the National Academy of Sciences of the United States of America* **95**, 4720–5.

Xia, Z., Dudek, H., Miranti, C.K. and Greenberg, M.E. 1996. Calcium influx via the NMDA receptor induces immediate early gene transcription by a MAP kinase/ERK-dependent mechanism. *Journal of Neuroscience* **16**, 5425–36.

Yamada, Y., Ujihara, H., Sada, H. and Ban, T. 1995. Pb^{2+} reduces current from NMDA receptors expressed in *Xenopus* oocytes. *Federation of European Biochemical Societies Letters* **377**, 390–2.

Yang, J., Ellinor, P.T., Sather, W.A., Zhang, J.-F. and Tsien, R.W. 1993. Molecular determinants of Ca^{2+} selectivity and ion permeation in L-type Ca^{2+} channels. *Nature (London)* **366**, 158–61.

Yatani, A., Bahinski, A., Wakamori, M., Tang, S., Mon, Y., Kobayshi, T. and Schwartz, A. 1994a. Alteration of channel characteristics by exchange of pore-forming regions between two structurally related Ca^{2+} channels. *Molecular and Cellular Biochemistry* **140**, 93–102.

Yatani, A., Bahinski, A., Mikala, G., Yamamoto, S. and Schwartz, A. 1994b. Single amino acid substitutions within the ion permeation pathway alter single-channel conductance of the human L-type cardiac Ca^{2+} channel. *Circulation Research* **75**, 315–23.

Yin, H.Z. and Weiss, J.H. 1995. Zn^{2+} permeates Ca^{2+} permeable AMPA/kainate channels and triggers selective neural injury. *NeuroReport* **6**, 2553–6.

Zamponi, G.W., Bourinet, E. and Snutch, T.P. 1996. Nickel block of a family of neuronal calcium channels: subtype- and subunit-dependent action at multiple sites. *Journal of Membrane Biology* **151**, 77–90.

Zamponi, G.W., Bourinet, E., Nelson, D., Nargeot, J. and Snutch, T.P. 1997. Crosstalk between G proteins and protein kinase C mediated by the calcium channel α_1 subunit. *Nature (London)* **385**, 442–6.

Zarei, M.M. and Dani, J.A. 1994. Ionic permeability characteristics of the *N*-methyl-D-aspartate receptor channel. *Journal of General Physiology* **103**, 231–48.

Zhang, J.-F., Ellinor, P., Aldrich, R.W. and Tsien, R.W. 1994. Molecular determinants of voltage-dependent inactivation in calcium channels. *Nature (London)* **372**, 97–100.

Zwart, R., VanKleef, R.G.D.M., Van Hooft, J.A., Oortgiesen, M. and Vijverberg, H.P.M. 1997. Cellular aspects of persistent neurotoxicants: effects of Pb^{2+} on neuronal nicotinic acetylcholine receptors. *NeuroToxicology* **18**, 709–18.

8

RESPONSES OF THE RESPIRATORY TRACT TO CADMIUM

Beth A. Hart

8.1 Introduction

Cadmium is a known industrial and environmental pollutant and a contaminant of tobacco smoke. Because of its widespread use, interest in its toxicity and biological effects has grown. The chief route of cadmium exposure is via the respiratory system. Not surprisingly, cadmium has been implicated as a causative agent in the development of a variety of pulmonary diseases, including emphysema, interstitial fibrosis and cancer. However, the mechanisms responsible for these diverse cadmium-induced pathologies are not currently known.

Investigations related to the pulmonary effects of cadmium have, in general, relied on two major methods for delivering cadmium to the lung: (1) intratracheal instillation and (2) inhalation. Administration of the cadmium compound by inhalation is by far the most physiologically relevant method. However, unlike intratracheal instillation, this technique requires the use of inhalation exposure facilities which are costly to construct and to operate. As a consequence, comparatively few cadmium inhalation studies have been performed, especially those that address mechanistic questions. Moreover, the vast majority of the inhalation studies that have been conducted report mainly about the effects of a single, acute cadmium exposure, as opposed to chronic effects employing lower concentrations.

The chemical form of the cadmium compound (chloride, oxide, sulfate, sulfide) and its physical form (aqueous solution, dust or fume) are also important considerations and can have a significant bearing on the results obtained. Most of the cadmium pulmonary studies have been conducted with rats, but strain differences and interspecies differences have been reported that can make extrapolations of research results to humans more difficult.

Pulmonary studies involving cadmium are further complicated by the complexity of the lung with its more than 40 different cell types. Cadmium-induced changes occurring in a less-abundant cell type may potentially be masked and go undetected if the analysis is performed using tissue from the whole lung. Moreover, cell isolation and purification protocols are currently available for only a few of the diverse cell types present in the lung, and maintenance of these cells in culture for extended periods is, in general, not possible. Even when studies with cells in culture can be conducted, the researcher must always be concerned whether results obtained *in vitro* accurately reflect the conditions as they exist *in vivo*. This is particularly true for an agent such as cadmium that has the potential to elicit an inflammatory reaction whereby cells from the blood are recruited into the lung. Such infiltrating inflammatory cells are usually in an activated state and, as a result, possess the capability of releasing a plethora of cytokines as well as reactive oxygen and nitrogen species that can contribute substantially to the disease process.

This chapter summarizes what is currently known about the pulmonary response of the lung to cadmium and spans the period from the early 1970s to the present. Results from both *in vivo* and *in vitro* studies are presented. Attempts have been made to include, wherever possible, mechanistic studies, regardless of whether they come from *in vivo* or *in vitro* studies. Hypotheses related to the potential role that epigenetic phenomena may play in cadmium-induced lung carcinogenesis are also presented. It is hoped that they will generate discussion and promote ideas for experiments that will lead to a better understanding of this disease process.

8.2 Disposition of inhaled cadmium

8.2.1 Deposition, clearance, retention and translocation

Several studies have been performed to assess the retention kinetics of cadmium deposited in the lung (Oberdorster *et al.*, 1979; Hadley *et al.*, 1980; Hart, 1986; Rusch *et al.*, 1986; Hirano *et al.*, 1989a,b; Klimisch, 1993). Data from these clearance studies often appear to disagree, which may be due, in part, to differences in experimental design (for example, method of administration, dose and frequency of exposures, chemical and physical form of cadmium, and length of time over which the clearance and translocation measurements were conducted).

Early studies (Oberdorster *et al.*, 1979) compared the pulmonary clearance kinetics of inhaled cadmium chloride, a highly soluble form, with cadmium oxide, a less soluble one. Rats exposed once to an aerosol of cadmium chloride had a higher lung burden of cadmium than animals exposed to cadmium oxide. Although the half-life for long-term pulmonary clearance was the same for both compounds (70 days), the initial phase of clearance was faster for cadmium oxide.

In a subsequent study, the deposition and clearance of cadmium chloride was compared with that of cadmium sulfide (Klimisch, 1993). In these experiments, animals were exposed 6 hours/day for 10 days to either cadmium chloride, generated as an aerosol from an aqueous solution, or to cadmium sulfide, generated as a dry dust to avoid problems of possible photoreduction. Although both groups were exposed to the same concentration of each test substance, more than twice as much cadmium was deposited in the lungs of animals exposed to cadmium chloride than to cadmium sulfide. Differences also existed between the short- and long-term clearance of these compounds. Despite greater aqueous solubility of cadmium chloride, only 10% was cleared rapidly after 10 days of exposure, compared to almost 40% for cadmium sulfide. After 3 months, 45% of the total lung deposit of cadmium chloride remained in the lung, compared to only 20% for cadmium sulfide. The calculated half-life for cadmium chloride (87 days) was twice that for cadmium sulfide (42 days).

These results indicate strongly that the pulmonary deposition and clearance of cadmium is highly dependent upon its chemical form. It is important to differentiate clearly between the chemical solubility of a cadmium compound in water and its actual solubility in a biological system such as the lung (Oberdorster, 1990). Although cadmium oxide and cadmium sulfide are both rather insoluble compounds in water, compared to cadmium chloride, particles of cadmium oxide can be readily solubilized in the lung (Hadley *et al.*, 1980), presumably by the acid milieu of phagolysosomes in alveolar macrophages (Lundborg *et al.*, 1984). On the other hand, cadmium sulfide particles are not readily solubilized in this manner.

Cadmium deposition can also vary as a function of species and strain. Oberdorster *et al.* (1994) reported that Balb/c mice retained approximately twice as much cadmium in the lungs than Fisher 344 rats. In contrast, McKenna *et al.* (1997) found similar percentages of inhaled cadmium deposited in the lungs of C57 mice and Wistar rats. However, the values were less than the percentage of pulmonary cadmium retained by DBA mice. These interspecies and interstrain differences could be attributed to variations in ventilation rate per unit body weight, pulmonary retention half-time of cadmium, pulmonary airway geometry, and nasal filtering capabilities (McKenna *et al.*, 1997).

The mechanisms of pulmonary cadmium clearance vary with the time following exposure and with the chemical and biological nature of the cadmium compound inhaled. The early phase of clearance is due primarily to mucocilliary clearance and secondarily to alveolar clearance of phagocytosed material carried onto the mucocilliary escalator or to the lymphatic system (Klimisch, 1993). These mechanisms can continue to function during long-term clearance, particularly for cadmium sulfide which has limited chemical and biological solubility. After mucocilliary clearance and absorption from the gastrointestinal tract occurs, cadmium can also be translocated to other organs, such as the liver and the kidney (Rusch *et al.*, 1986). The translocation of cadmium from the lung to either the liver or the kidney is likely to involve a protein-bound cadmium form. This mechanism may play a much more prominent role in long-term clearance, especially for cadmium chloride and cadmium oxide which have high solubilization rates (Oberdorster, 1990; Klimisch, 1993).

Hart (1986) showed that repeated cadmium inhalation exposure (3 hours/day, 5 days/week for 6 weeks) of rats increased the long-term cadmium burden in the liver and kidney, as well as in the lung. The magnitude of cadmium accumulation in each organ occurred in direct relationship to the frequency of exposures. On a per-gram, dry-weight basis, lungs of cadmium-exposed animals accumulated approximately six times more cadmium than the kidney and 12 times more cadmium than the liver.

In humans, smoking cigarettes is a major source of inhalation cadmium exposure. Tobacco plants acquire cadmium by absorption from soil and accumulate it in the leaves, which are then used for manufacturing cigarettes. A significant amount of cadmium in the tobacco is transferred to the smoke during the burning process, and the cadmium in the smoke is s subsequently inhaled by the smoker (Wagner and Yeargan, 1986). It has been known for some time that the tissue content of cadmium is significantly higher for smokers than for nonsmokers (Paakko *et al.*, 1989). Using well-defined smoke generation and exposure conditions, Gairola and Wagner (1991) confirmed that significant accumulation of cadmium occurs in the lung and kidney of rodents given long-term inhalation exposure to mainstream cigarette smoke.

8.2.2 Uptake and accumulation in lung cells

Inhaled cadmium is dissolved in epithelial lining fluid and is likely to be associated with proteins. Cadmium bound to protein could be exchanged at specific sites on cell membranes. Alternatively, the metal-bound protein may be transported intact across cell membranes. Soluble components present in the alveolar space may also facilitate cadmium uptake by stimulating phagocytosis and endocytosis (Sibille and Reynolds, 1990).

Significantly more cadmium is accumulated by alveolar macrophages than by type II alveolar epithelial cells following repeated exposure of animals to cadmium aerosols (Hart *et al.*, 1995a). It is not clear whether this reflects differences in pulmonary particle deposition or a more efficient mechanism for metal uptake. Both cell classes are topographically located within the alveolus and, as a consequence, are readily exposed to inhaled cadmium. However, type II cells are attached to the basal membrane of the epithelium, whereas alveolar macrophages are able to move freely along the entire surface of the alveolar epithelium (Dubois, 1986; Schlesinger, 1989). The kinetics of cadmium uptake in long-term experiments are likely to be more complex due to the induction of metal-binding proteins.

In vitro transport studies report that the uptake of ionized cadmium by alveolar macrophages is time, dose, temperature and energy dependent, indicating that uptake does not occur by passive diffusion but rather by an active transport mechanism (Cox and Waters, 1978; Hart, 1978). The participation of calcium channels in cadmium transport in lung epithelial cells is suggested by the finding that verapamil treatment decreases metal uptake (Kang *et al.*, 1990). Cadmium uptake can also be suppressed in these cells by the inclusion of *N*-acetyl cysteine in the culture medium (Gong and Hart, 1997), presumably because cadmium forms complexes with the sulfhydryl group on this compound and is prevented from entering the cell. Extracellular glutathione would likely have a similar effect. Pre-incubation of lung fibroblasts with *N*-methylmaleimide also inhibits cadmium uptake, which implies that surface sulfhydryl groups participate in the uptake mechanism (Ochi, 1991).

8.3 Inflammatory response of the lung to cadmium

8.3.1 Analysis of bronchoalveolar lavage fluid

Pulmonary lavage represents an important tool for assessing the response of the lung to cadmium. In this technique, the bronchoalveolar region of the respiratory tract is washed or lavaged with an isotonic salt solution. The cellular component of the recovered bronchoalveolar lavage fluid can be characterized with regard to total cell number and the percentages of the major inflammatory cell types (macrophages, lymphocytes and polymorphonuclear leukocytes). The fluid component of the lavagate can be analyzed for protein and cellular enzymes to determine the magnitude of lung injury.

It is well documented that inhalation of cadmium, administered either as an aerosol of a cadmium salt or in cadmium-enriched cigarette smoke, provokes an intense inflammatory response which is characterized by an influx of polymorphonuclear leukocytes and macrophages (Strauss *et al.*, 1976; Henderson *et al.*, 1979; Hart, 1986; Grose *et al.*, 1987a; Hirano *et al.*, 1989b; Gairola, 1989; Gavett and Oberdorster, 1994; Oberdorster *et al.*, 1994; McKenna *et al.*, 1997). This inflammatory cell response is accompanied by alterations in the chemical composition of the lavage fluid. The presence of increased levels of protein in the lavage fluid is suggestive of a loss of epithelial cell integrity after cadmium exposure, whereas frank cellular damage is reflected by elevations in lactic dehydrogenase (Henderson *et al.*, 1979; Hart, 1986; McKenna *et al.*, 1997). Such damage may be the result of direct interaction of cadmium with the pulmonary cells or, alternatively, may be caused by reactive oxygen species, proteases or cytokines released by newly recruited inflammatory cells or by the resident macrophages that have become activated (Henderson, 1989). Arachidonic acid, released from lipid membranes in response to cadmium exposure (Nelson *et al.*, 1994), is taken up in the alveolar epithelium and metabolized by lipoxygenase pathways to form leukotrienes. Some members of the leukotriene family are extremely potent chemoattractants that may provide the stimulus for the rapid recruitment of polymorphonuclear leukocytes and macrophages into the lungs of cadmium-exposed animals (Henderson, 1989).

The intensity of the pulmonary inflammatory response depends upon the chemical form of cadmium. Inhalation exposure to cadmium chloride or cadmium oxide elicits similar increases in bronchoalveolar lavage protein and polymorphonuclear leukocytes (Grose *et al.*, 1987a) whereas exposure to cadmium sulfide causes little cytological or biochemical change (Oberdorster *et al.*, 1985). This difference is no doubt caused by the chemical and biological insolubility of cadmium sulfide.

Oberdorster *et al.* (1994) found that pulmonary inflammation in response to cadmium inhalation was greater in mice (Balb/c) than in rats (Fisher 344). McKenna and colleagues (1997) subsequently showed that the degree of inflammation also varies in mice strains. The influx of polymorphonuclear leukocytes into the alveolar space was more pronounced in C57 mice than in either Wistar rats or DBA mice.

8.3.2 Alterations in alveolar macrophage function

In vitro effects of cadmium on the pulmonary macrophage have been studied by several investigators. Coin and Stevens (1986) showed that cadmium caused dose-dependent decreases in cell viability that were significantly affected by temperature, medium composition and time of exposure. The release of lysosomal enzymes following cadmium exposure paralleled cell death. Scanning electron microscopic studies revealed cadmium-induced alterations in the ruffled membrane surface typically observed in control macrophages. *In vitro* exposure to cadmium also caused dose-dependent decreases in the percentage of macrophages capable of phagocytosis and reductions in the total number of opsonized yeast particles engulfed by individual cells (Greenspan and Morrow, 1984; Coin and Stevens, 1986). Presumably all of these effects could be directly or indirectly caused by inhibition of oxygen metabolism and ATPase activity by cadmium (Mustafa *et al.*, 1971).

Cadmium is also capable of modifying the phagocytic ability of macrophages *in vivo*. Gardner *et al.* (1977) and Bouley *et al.* (1977) found that acute exposure of animals to an aerosol of cadmium markedly reduced the ability of the lung to defend itself against microbial insults. Clearance kinetics of inhaled bacteria were impaired and there was an increased death rate in the cadmium-exposed group. Using a combined *in vitro–in vivo* approach, Greenspan and Morrow (1984) also showed that an acute cadmium exposure inhibited phagocytosis by macrophages obtained by lavage.

On the other hand, repeated inhalation exposure of animals using considerably lower cadmium oxide concentrations (Johansson *et al.*, 1983; Hart, 1986) and long-term exposure to cadmium-enriched cigarette smoke (Gairola, 1989) appear to produce stimulatory effects in alveolar macrophages. The number of macrophages recovered in lavage fluid of chronically exposed animals are increased. Moreover, these macrophages exhibit characteristics compatible with metabolic activation, including:

1 differences in cell size and number of inclusion bodies;
2 increased oxidative activity after stimulation with *Escherichia coli* (Johansson *et al.*, 1983);
3 enhanced ability to phagocytose latex particles and immunoglobulin-coated sheep red blood cells (Gairola, 1989); and
4 increased secretion of an interleukin-1-like cytokine (Hart *et al.*, 1996b).

8.3.3 Lipid peroxidation

Lipid peroxidation is one of the hallmarks of cadmium exposure (Cross *et al.*, 1979; Vincent *et al.*, 1989). In this process polyunsaturated fatty acids become oxidatively degraded to a variety of products, including ketones, aldehydes, alkenes and alkanes. Comparatively little information is available on the nature of mechanisms responsible for the generation of free radicals and the ensuing lipid peroxidation in lung tissue following acute exposure to cadmium compounds. Unlike other bioactive metals, cadmium displays a weak redox potential. This suggests that free radicals must be generated indirectly, possibly as a consequence of intracellular damage (Vincent *et al.*, 1989) or following the inflammatory response and subsequent proliferative repair process (Kobrle *et al.*, 1991). Research aimed at elucidating mechanisms of cadmium-induced lipid peroxidation in the lung have focused on:

1 possible disturbances of antioxidant defense (Manca *et al.*, 1991);
2 alterations in respiratory chain components responsible for the reduction of oxygen and changes in mitochondrial membrane integrity (Mustafa *et al.*, 1971);
3 inflammatory processes resulting in the release of oxidants by activated cells during the respiratory burst (Cross *et al.*, 1979; Manca *et al.*, 1994); and
4 the participation of metals such as iron and copper in the reductive transfer of electrons to molecular oxygen (Vincent *et al.*, 1989).

It is possible that all of these mechanisms may contribute, to a greater or lesser extent, to the oxidative changes in the lung induced by cadmium. However, some mechanisms may play a more important role in the early phases of toxicity, whereas others may occur as a consequence of the primary damage. The most plausible events responsible for cadmium-induced lipid peroxidation include:

1 depletion of glutathione reserves by cadmium or by oxidants released from inflammatory cells (Cross *et al.*, 1979; Manca *et al.*, 1994); and
2 alterations in the homeostasis and pulmonary bioavailability of redox-active endogenous metals such as Fe^{2+} and Cu^{2+} (Vincent *et al.*, 1989).

Lipid-derived breakdown products, such as malondialdehyde, could then diffuse into the nucleus, interact directly with DNA, and potentially cause heritable mutations. These possibilities deserve further study (Howden and Faux, 1996).

8.4 Proliferative response of the alveolar epithelium

Cell renewal is necessary to maintain the integrity of the lung in response to injury. Several investigators have demonstrated that cadmium inhalation results in a substantial cell proliferative response in the lung. It is characterized by a doubling in lung DNA content and an increased tritiated thymidine incorporation which is believed to be primarily associated with type II alveolar epithelial cells (Palmer *et al.*, 1975). The overall pattern of cell proliferation following cadmium inhalation is suggestive of that seen with oxygen, ozone and nitrogen dioxide exposure, at least with respect to type II cell activity. It is not clear what triggers this response or whether the mechanism is the same for each of these diverse agents. The wave of cellular proliferation and tritiated thymidine incorporation coincides with substantial lipid accumulation, in the lung tissue itself and in bronchoalveolar fluid. Pulmonary elevations also occur in the activity of glucose 6-phosphate dehydrogenase, a key enzyme in the pentose phosphate pathway (Hayes *et al.*, 1976; Amanuma and Suzuki, 1987; Boudreau *et al.*, 1988). Stimulation of this pathway could result from an increased demand for ribose in RNA and DNA synthesis during type II cell proliferation, as well as for NADPH, required as a reductant in lipid biosynthesis. The increased phosphatidyl choline content of lung homogenates and bronchoalveolar lavage fluid from cadmium-exposed animals is thought to be caused by increased production and release of surfactant by the hyperplastic and hypertrophied type II cells as opposed to cell lysis (Amanuma and Suzuki, 1987; Boudreau *et al.*, 1988).

8.5 Cadmium-induced lung diseases

8.5.1 Emphysema

Administration of cadmium to rodents by aerosol or intratracheal instillation has been shown to produce emphysema, characterized by distention of the air spaces distal to the terminal bronchiole and by destruction of alveolar septa (Lewis, 1973; Karlinsky and Snider, 1978). Release of proteolytic enzymes by inflammatory cells has often been cited as a potential cause. However, other factors could be involved, particularly those related to collagen metabolism which is critical in determining lung structure. Chambers *et al.* (1994) found that subtoxic concentrations of cadmium chloride selectively decrease lung fibroblast procollagen production and increase the proportion of newly synthesized procollagen that is degraded. The effects of cadmium on procollagen degradation are of particular interest, because this process plays an important role in regulating collagen production and deposition. Intracellular degradation is known to increase if defective procollagen is produced. Procollagen undergoes post-translational modifications, including glycosylation, hydroxylation and triple helix formation. Thus, cadmium-induced inhibition of any one of these processes would cause formation of unstable procollagen. This, in turn, would increase intracellular procollagen degradation (Bienkowski 1989) and limit collagen accumulation.

In other experiments, fibroblasts that were made resistant to cadmium by continuous exposure, were shown to have lower levels of lysyl oxidase mRNA, decreased intra-cellular pro-protein production, and less mature enzyme secreted into the culture medium compared to cadmium-sensitive cells. Mature lysyl oxidase catalyzes a reaction that initiates crosslinking of collagen. Thus, low levels of this enzyme would promote procollagen instability and diminish collagen accumulation (Li *et al.*, 1995). A similar decrease in lysyl oxidase activity is also observed in lungs of animals repeatedly exposed to cadmium aerosols (Sampson *et al.*, 1984).

8.5.2 Pulmonary fibrosis

Interstitial pulmonary fibrosis is another possible outcome of acute cadmium exposure. In the evolution of this disease, damaged type I epithelial cells are replaced with type II cells. Thickening of the alveolar wall also occurs due to the combined effects of fibroblast proliferation and increased accumulation of connective tissues products, particularly type I collagen (Snider *et al.*, 1988; Frankel *et al.*, 1991; Driscoll *et al.*, 1992). Development of fibrosis is thought to be caused by a battery of mediators released by activated macrophages. Driscoll *et al.* (1992) demonstrated that fibronectin production by alveolar macrophages is significantly increased for weeks after intratracheal instillation of cadmium. By acting together with other growth factors, fibronectin can stimulate fibroblasts to proliferate (Bitterman *et al.*, 1983). Another mediator capable of promoting fibrogenesis is transforming growth factor-β, a positive effector of lysyl oxidase gene expression in lung fibroblasts (Boak *et al.*, 1995). The fibrogenic response in cadmium-exposed animals is characterized by an increased expression of type I and type III procollagen mRNA (Frankel *et al.*, 1991), as well as increased lysyl oxidase activity (Sampson *et al.*, 1984).

8.5.3 Lung cancer

Epidemiological studies have shown that the incidence of lung cancer is significantly increased among workers who are occupationally exposed to cadmium via the respiratory route. Examples of these occupational settings include the manufacture of nickel–cadmium batteries, the smelting and refining of zinc, lead and copper ores, and welding with metal solders. Additional pulmonary cadmium exposure occurs in smokers (IARC, 1993). Aerosols of cadmium oxide, chloride, sulfate and sulfide have also been tested for their carcinogenic effects in rodents. All cadmium compounds tested, regardless of their solubility, appeared to have a similar carcinogenic potential in rat lung (Takenaka *et al.*, 1983; Sanders and Mahaffey, 1984; Oldiges *et al.*, 1989; Glaser *et al.*, 1990; Heinrich, 1992). The malignant lung tumors observed in rats exposed to inhaled cadmium contained numerous cells, similar to type II alveolar cells and these cells originated from epithelial cells in the bronchiolar/alveolar region (Takenaka *et al.*, 1983; Oberdorster, 1986).

Studies using V79 lung fibroblasts have been performed to establish possible mechanisms for the genotoxic action of cadmium. Cadmium chloride has been shown to induce DNA single-strand breaks, DNA–protein crosslinks (Ochi and Ohsawa, 1983) and chromosomal aberrations (Ochi *et al.*, 1984) in these cells. Experiments applying free-radical scavengers support the involvement of oxygen free-radicals in this process (Snyder, 1988). The relevance of these results is somewhat questionable because of the high cadmium concentrations required to cause these changes.

Although cadmium is weakly mutagenic when administered alone, it can enhance the co-mutagenicity of ultraviolet light in the hypoxanthine phosphoribosyl transferase gene in V79 cells (Hartwig and Beyersmann, 1989). These results suggested that the carcinogenic action of cadmium might be related to its ability to interfere with DNA repair mechanisms. It has subsequently been confirmed that DNA repair processes can be disturbed using low, noncytotoxic concentrations of cadmium compounds (Hartwig, 1994, 1995). An investigation of the total incision frequency during the repair of DNA damage induced by ultraviolet light revealed a dose-dependent inhibition of the incision frequency at concentrations as low as 1 μM cadmium (Hartwig *et al.*, 1996). Cadmium, at a concentration as low as 0.5 μM, also inhibited the repair of oxidative DNA lesions induced by visible light (Dally and Hartwig, 1997). Cadmium appears to interfere with two steps in the repair process. It disrupts DNA–protein interactions involved in damage recognition and also impairs ligation capacity, perhaps by inactivating essential sulfhydryl groups (Hartwig *et al.*, 1996). Since DNA is continuously damaged by exogenous and endogenous sources, an impaired repair capacity might not only be relevant for enhancing effects in combination with other DNA-damaging agents, but might also account for the carcinogenic action of cadmium itself. A reduced repair capacity, particularly toward oxidative DNA damage, can enhance the background level of the lesions *in vivo* and this might increase the risk of cancer.

Cadmium-induced oxidative damage to DNA could also be caused indirectly from the intense inflammatory response that cadmium inhalation elicits. An association between inflammation-related and neoplastic events has been documented in many tissues, including the lung (Borm and Driscoll, 1996). Inflammatory cells that infiltrate the lung in response to cadmium inhalation exposure would be expected to release reactive oxygen species into the surrounding tissue. Oxidants such as superoxide anions and hydroxyl radicals can, in turn, initiate lipid peroxidation. This process produces lipid radicals which then react with oxygen and propagate a chain reaction, ultimately producing a vast array of highly reactive hydroperoxides and aldehydes. All of these agents, including cadmium itself, can be damaging to DNA. Mismatched bases can be introduced into DNA molecules during repair replication. Cycles involving DNA damage, repair, misrepair and hence mismatch acquisition can occur in growing cells as well as in resting ones. Mismatch acquisition is an important source of new mutations, which are more likely to occur in cases where DNA repair mechanisms may be compromised. Both processes can contribute to cancer development (MacPhee, 1995, 1996). Epigenic mechanisms, responsible for tumor promotion and progression in cadmium-induced lung carcinogenesis, are possible, and are detailed in subsequent sections of this chapter.

8.6 Cadmium-induced biochemical and molecular changes

8.6.1 Alterations in enzymatic and nonenzymatic antioxidants

Cadmium exposure produces changes in the antioxidant protective system of the lungs. The effects observed are dependent upon cadmium dosage, method of cadmium administration, time after exposure and whether the animals were exposed to cadmium only once or repeatedly. Salovsky *et al.*, (1992) reported that a single intratracheal instillation of cadmium acetate (500 μg/kg) provoked a decrease in catalase in the earliest stage after exposure, particularly by day 5. Superoxide dismutase activity was also decreased but, unlike catalase, its activity was only affected by cadmium exposure toward the end of a 30-day observation period. Glutathione peroxidase activity and the level of reduced glutathione, on the other hand, were significantly elevated from day 5 to day 15. Results from an intratracheal study using a significantly lower cadmium dosage (Cross *et al.*, 1979), reported elevations in catalase, superoxide dismutase and glutathione peroxidase-associated enzymes within 24 hours after cadmium instillation (90 μg/kg) and a return to baseline values by day 7.

Grose *et al.* (1987a, b) compared the pulmonary effects of inhaled cadmium chloride and cadmium oxide. At the highest concentration (4.5 mg cadmium/m^3), the lung lesions were more severe in response to cadmium oxide than to cadmium chloride, but the pulmonary biochemical responses were similar. Total glutathione peroxidase activity and non-selenium-dependent glutathione peroxidase activity were reduced immediately after a 2 hour cadmium exposure, whereas selenium-dependent glutathione peroxidase activity was unaffected. This response was followed 3 days later by an increase in total and selenium-dependent glutathione peroxidase activities and by a decrease in the activity of the non-selenium-dependent form. Glutathione transferase, glutathione reductase, and glucose 6-phosphate dehydrogenase showed no initial change but all three enzyme activities were maximally elevated 3 days after exposure (Grose *et al.*, 1987a) and remained elevated for the next 2 weeks (Boudreau *et al.*, 1988).

In a similar fashion, animals repeatedly exposed by inhalation to an atmosphere containing 1.6 mg cadmium/m^3 (3 hours/day, 5 days/week) for 1–4 weeks, showed a twofold increase in pulmonary catalase, glutathione peroxidase, glutathione reductase and superoxide dismutase in the first week of exposure. These enzyme activities remained at the elevated levels for the next 3 weeks of exposure (Hart *et al.*, 1989b, 1990). In addition to these enzymatic changes, repeated inhalation of cadmium also resulted in more than a fourfold increase in pulmonary levels of nonprotein thiols, which suggests that enzymes involved in glutathione synthesis (γ-glutamylcysteine synthetase) or in glutathione transport (γ-glutamyl transpeptidase) may also be increased in activity. It is not currently known whether any of these biochemical alterations are due to gene activation or whether they result from changes in lung cellularity.

In general, the biochemical changes in antioxidant levels observed in the lung following cadmium exposure are similar to those that occur in response to oxygen, ozone or nitrogen dioxide. This suggests that cadmium acts by disturbing the cellular redox potential, thus rendering the cell vulnerable to oxidative damage. Support for this comes from recent work by Yang and co-workers (Yang *et al.*, 1997). By using the fluorescent probe 2′,7′-dichlorofluorescin diacetate, these investigators demonstrated a significant increase in production of reactive oxygen species by cadmium-treated lung fibroblasts.

8.6.2 Induction of metallothionein

The ability of the lung to retain cadmium is believed to be due to the induction of metallothioneins (MTs) which serve to chelate metals. These metal-binding, thiol-rich proteins are present in extremely low basal levels in the lungs of unexposed animals but increase significantly following intratracheal or inhalation cadmium exposure. Pulmonary MT induction has been demonstrated in a variety of species, including Syrian hamsters (Benson and Henderson, 1980), rabbits (Post *et al.*, 1982), rats (Oberdorster and Koerdel, 1981; Hart, 1986; Hart *et al.*, 1996a; Kenaga *et al.*, 1996; McKenna *et al.*, 1998) and mice (Oberdorster *et al.*, 1994; Kenaga *et al.*, 1996; McKenna *et al.*, 1998). Two isoforms of MT (MT-1 and MT-2) have been isolated in near-equal proportions from lungs of cadmium-exposed rats by ion exchange chromatography (Hart, 1986; Hart *et al.*, 1996a). Cadmium inhalation exposure also induces MT production in the liver and to a lesser extent in the kidney of these animals. The amount of MT that accumulates is directly proportional to the cadmium burden of each organ (Hart, 1986).

Zinc-thionein is also induced in the lung following aerosol exposure to zinc (Oberdorster and Koerdel, 1981; Cosma *et al.*, 1992) whereas copper-thionein accumulates in the lungs of animals exposed to increased oxygen tension (Hart *et al.*, 1989a, 1990; Piedboeuf *et al.*, 1994). Although the inducer of MT during hyperoxic exposure is not known for certain, there are several likely candidates. Hyperoxic exposure has been shown to cause significant elevations in ceruloplasmin, an acute-phase, copper-containing, plasma protein (Hart *et al.*, 1989a) that is synthesized in the liver in response to oxidant stress and then released into the circulation (Cousins, 1985). Elevations of circulating levels of ceruloplasmin may lead to an increased pulmonary uptake of copper which, in turn, could promote pulmonary metallothionein synthesis (Winge *et al.*, 1981; Cousins, 1985). Other likely MT inducers include cytokines (Liu *et al.*, 1991), such as interleukin-1, produced by activated macrophages (Hart *et al.*, 1996b), and reactive oxygen species (Dalton *et al.*, 1994; Brookens *et al.*, 1994). These agents could act independently or synergistically to induce MT production.

MT protein that accumulates in the lung in response to cadmium inhalation is preceded by a rapid, upregulation of MT gene expression (Hart *et al.*, 1996a; McKenna *et al.*, 1998). MT-1 mRNA and MT-2 mRNA levels rise above control levels within 15 min following the completion of a 3 hour cadmium inhalation exposure, reach maximal hybridization at 2 hours, and are still significantly elevated at 96 hours. The lag time between the maximal MT mRNA increase and the maximal MT protein increase (48 hours) has been observed by others and is likely due to stability and translation efficiencies of MT mRNAs and turnover rate of MT protein (Klaassen and Lehman-McKeeman, 1989; Piedboeuf *et al.*, 1994; Hart *et al.*, 1996a; McKenna *et al.*, 1998). The concentration of cadmium in the aerosol is directly proportional to lung cadmium burden, MT protein accumulation and MT mRNA levels (Hart *et al.*, 1996a). Based upon the time course and dose–response results, it would seem reasonable to expect that daily inhalation exposures to cadmium would result in a progressive accumulation of MT transcripts as well as MT protein. This appears to be true for MT protein but not for MT mRNAs. Although pulmonary mRNA levels in animals exposed to cadmium aerosols on five consecutive occasions are higher that MT mRNA levels in animals exposed only once, the incremental increase in MT mRNAs following a single exposure is significantly greater in animals with no prior exposure to cadmium (Hart *et al.*, 1996a). Further studies are required to determine whether this difference in response of cadmium-pretreated animals resides in redox-mediated alterations in transcription factors governing MT gene expression. Differential intra- and interspecies inducibility of MT genes has also been been reported (Oberdorster *et al.*, 1994; McKenna *et al.*, 1998), prompting the prediction that animals with low degrees of pulmonary MT inducibility are likely to exhibit a greater susceptibility for cadmium-

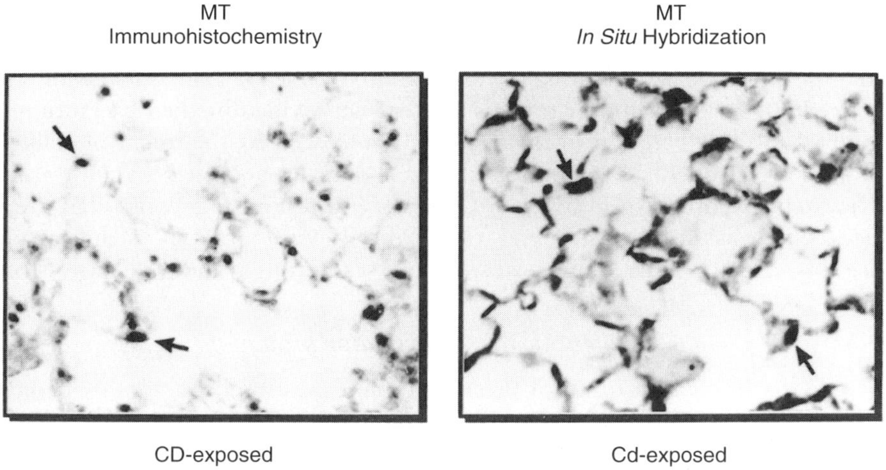

Figure 8.1 Pulmonary localization of metallothionein (MT) protein (by immunohistochemistry) and MT mRNA (by *in situ* hybridization). Arrows specify MT-positive cells. (Reprinted from Hart, B.A., Gong, Q. and Eneman, J.D. 1996. Pulmonary metallothionein expression in rats following single and repeated exposure to cadmium aerosols. *Toxicology* **112**, 205–18, with permission from Elsevier Science.)

induced lung injury. However, it is important to realize that metallothionein expression is not the only factor determining the sensitivity to cadmium.

Using immunohistochemistry and *in situ* hybridization techniques (Fig. 8.1), MT has been localized primarily within the alveolar compartment (Hart *et al.*, 1985, 1990, 1996a; Hart and Garvey, 1986; McKenna *et al.*, 1997) and in some interstitial cells (Hart and Keating, 1980; Hart *et al.*, 1985, 1995c). Isolation of purified populations of alveolar macrophages and type II alveolar epithelial cells from cadmium-exposed animals (Hart *et al.*, 1995a) reveals that both MT mRNA and MT protein are expressed in both cell types, albeit in greater amounts in macrophages. However, not all macrophages or type II cells are positive for MT and the degree of MT immunostaining and MT transcript level varies in each cell population. MT protein is localized in both the cytoplasm and nucleus of immunopositive cells (Hart *et al.*, 1995a,b, 1996a,b).

8.6.3 Thiol-modulation of cadmium-induced gene expression

Recent studies (Gong and Hart, 1997) have also shown that cadmium exposure coordinately upregulates expression of glutathione *S*-transferase α (GST), heme oxygenase-1 (HO) and MT-1 in lung epithelial cells. These genes code for proteins that all function, albeit by strikingly different mechanisms, in a protective capacity by increasing cellular resistance to oxidants (Sato and Bremner, 1993; Vile *et al.*, 1994; Choi and Alam, 1996). GST is involved in the detoxification of lipid and nucleotide hydroperoxides, and it also catalyzes the conjugation of glutathione to electrophilic xenobiotics, including mutagens, carcinogens and other toxic compounds (Daniel, 1993). HO, on the other hand, degrades heme to biliverdin, which subsequently is converted to bilirubin, a known antioxidant, and to an increase in ferritin, a protein that limits the availability of iron in the Fenton reaction that generates hydroxyl radicals

(Choi and Alam, 1996). Of particular interest is the finding that the pro-oxidant property of cadmium, reflected in its demonstrated ability to lower intracellular glutathione levels, plays an important role in regulating the expression of MT, GST and HO genes (Gong and Hart, 1997). Further support for this concept is provided by the observation that the cadium-induced expression of these genes is enhanced by treatment of cells with buthionine sulfoxamide, an agent that lowers glutathione levels by inhibiting γ-glutamylcysteine synthetase activity (γ-GCS), the rate-limiting controlled step in glutathione synthesis. In contrast, pretreatment of cells with *N*-acetyl cysteine, an agent that raises intracellular glutathione levels, suppresses cadmium-induced expression.

8.6.4 Potential role of activator protein-1

It is possible that induction of all of these pulmonary resistance factors by cadmium may be mediated, in whole or in part, by a common transcription factor. A likely candidate is activator protein-1 (AP-1). It exists as a heterodimer, consisting of the oncoproteins Fos and Jun, or as a homodimer of Jun proteins (Angel and Karin, 1991). AP-1 expression and binding activity can be increased by a variety of agents, including cadmium, oxidants and cytokines, as well as by decreased levels of reduced glutathione (Angel and Karin, 1991; Bergelson *et al.*, 1994a,b). AP-1 is thought to mediate the transcriptional activity of HO-1 (Alam *et al.*, 1994), GST-α (Bergelson *et al.*, 1994b) and γ-GCS (Rahman *et al.*, 1996). There is a good possibility that all of these cadmium-inducible genes could be upregulated in a similar fashion, but this remains to be established. AP-1 is also known to play a key role in cell proliferation and differentiation (Angel and Karin, 1991). Thus, AP-1 expression might serve a dual role in pulmonary cadmium adaptation, both by mediating resistance-factor induction and by stimulating type II cell proliferation. This hypothetical mechanism is presented in Fig. 8.2.

8.7 Cadmium adaptation and cross-resistance to oxidants

8.7.1 Adaptive process

It has been shown unequivocally that pulmonary adaptation develops in rats repeatedly exposed, via inhalation, to cadmium concentrations well within the carcinogenic range (Hart, 1986). The process of pulmonary adaptation to cadmium, shown in Fig. 8.3, develops over a 4–6 week period. Initial pulmonary injury is evident during the first 2 weeks of cadmium exposure (1.6 mg cadium/m^3, 3 hours/day, 5 days/week) and is characterized by the release of cellular enzymes into the alveolar space and an intense inflammatory response involving polymorphonuclear leukocytes and activated macrophages. After 4–6 weeks of repeated inhalation treatment, airway alterations diminish. This occurs in spite of progressive increases in lung cadmium burden (Hart, 1986). The histology of the cadmium-adapted lung resembles that of the air control, with the notable exception that type II alveolar epithelial cells are increased threefold in number and fourfold in volume. The absence of interstitial fibrosis suggests that lysyl oxidase activity is at or near control levels, as observed in another study in which animals were repeatedly exposed to cadmium (Sampson *et al.*, 1984).

As discussed previously in this chapter, a variety of defense mechanisms are substantially increased in the cadmium-adapted lung. In addition to a significant elevation in MT protein that exhibits an increasing linear trend in response to repeated cadmium

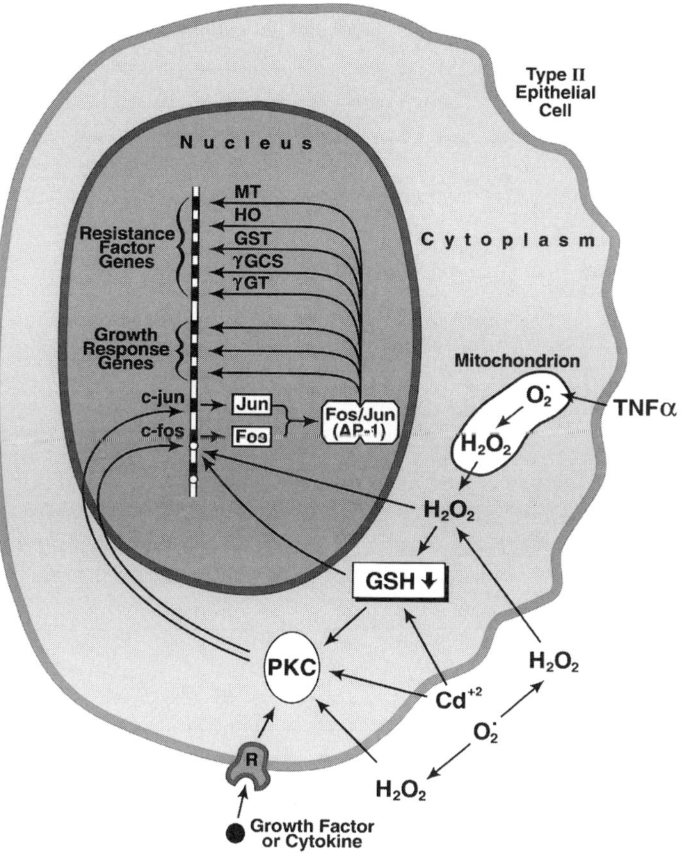

Figure 8.2 This illustrates the potential role of activator protein-1 (AP-I) in mediating cadmium-induced oxidant resistance and type II alveolar cell proliferation. MT, metallothionein; HO, heme oxygenase-1; GST, glutathione *S*-transferase; γGCS, γ-glutamyl-cysteine synthetase; γGT, γ-glutamate transferase; TNFα, tumor necrosis factor-α; H_2O_2, hydrogen peroxide; $O_2^{\cdot}$, superoxide anion; GSH, glutathione; PKC, protein kinase C; Cd^{+2}, cadmium ion; R, growth factor or cytokine receptor. (Illustration prepared by Gary Nelson, University of Vermont Department of Medical Photography.)

exposure (Hart, 1986; Hart *et al.*, 1996a), the cadmium-adapted lung also possesses increased activities of several antioxidant enzymes (glutathione peroxidase, glutathione reductase and manganese superoxide dismutase) and significant elevations in reduced pulmonary glutathione levels, which could be due to the induction of γ-glutamyl cysteine synthetase and γ-glutamyl transpeptidase (Hart *et al.*, 1989b, 1990). *In vitro* studies by Gong and Hart (1997) suggest that cadmium is also likely to activate the pulmonary genes for members of the glutathione *S*-transferase family and for heme oxygenase-1.

8.7.2 Cadmium-induced resistance to cadmium and cross-resistance to oxidants

Animals pretreated by repeated inhalation exposure to cadmium aerosols (Hart, 1986) exhibit pulmonary tolerance to a subsequent acute cadmium challenge (Hart *et al.*,

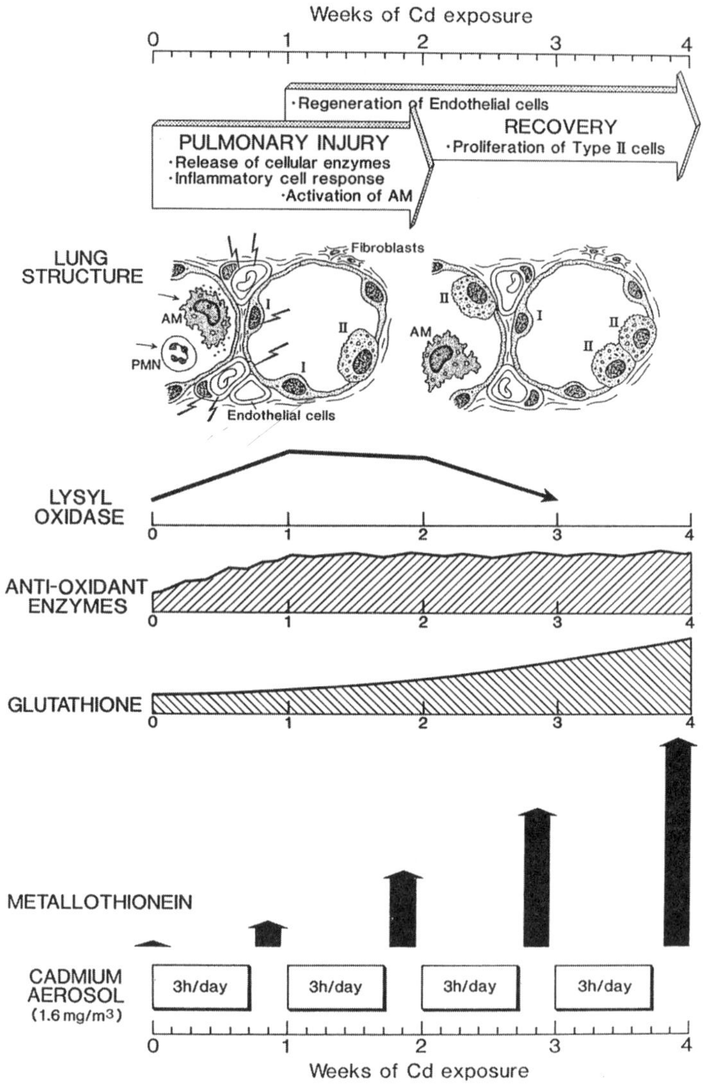

Figure 8.3 This illustration depicts the sequence of cytological and biochemical events that occur as the lung adapts to repeated inhalation of cadmium (1.6 mg cadmium/m^3, 3 hours/day, 5 days/week). (Illustration by Gary Nelson, University of Vermont, Department of Medical Photography.)

1989b). Buckley and Bassett (1987) used an isolated perfused lung model to demonstrate that lungs from cadmium-pretreated animals had an increased ability to maintain levels of glutathione during hydroperoxide challenge. Hart *et al.* (1990) subsequently reported that cadmium-adapted animals show collateral tolerance to hyperoxia (Hart *et al.*, 1990). Cadmium pretreated animals were still alive after 8 continuous days of exposure to 100% oxygen whereas air-pretreated control animals succumbed to hyperoxia after 54–62 hours. Extensive remodeling occurred in the lungs of cadmium-pretreated animals in response to the hyperoxic challenge, which was similar to structural changes observed in animals exposed to 85% oxygen for 5–7 days (Crapo *et al.*, 1980). The enhanced oxidant resistance of cadmium pretreated animals is likely due to their increased capability to repair structural damage by proliferation of alveolar lining cells and to their augmented enzymatic and nonenzymatic antioxidants, including the thiol-rich protein MT that has known free-radical scavenging capabilities (Thornalley and Vasak, 1985; Sato and Bremner, 1993).

Hart *et al.* (1995b) proposed that alveolar type II cells from lungs of cadmium-adapted rats are more resistant to oxidant injury than cells from nonadapted animals. To test this hypothesis, type II cells were isolated and purified from cadmium-adapted and nonadapted animals and then were challenged *in vitro* by exposure to hydrogen peroxide. This experimental model may be a particularly useful and relevant system for studying mechanisms of pulmonary oxidant resistance. The results of this study clearly showed differences in the oxidative stress response of cells from the two groups of animals. Hydrogen peroxide was significantly less cytotoxic to cadmium-adapted type II pneumocytes than to nonadapted cells. Moreover, greater amounts of hydrogen peroxide could be removed from the extracellular milieu by adapted cells than by non-adapted ones. Cadmium-induced oxidant resistance has also been demonstrated in lung fibroblasts (Mello-Filho and Meneghini, 1988; Hart *et al.*, 1995c), but in these studies cells were adapted to cadmium *in vitro* as opposed to *in vivo*.

Similar to cadmium-induced changes observed in whole lung (Hart *et al.*, 1989b, 1990, 1996a), type II cells isolated from cadmium-adapted animals have increased levels of glutathione reductase and glutathione peroxidase as well as increased MT mRNA and protein (Hart *et al.*, 1995a,b). Immunocytochemical procedures have been used by Hart *et al.* (1995c) to determine the percentage of type II cells that express MT, to subclassify positive cells according to their MT content by the intensity of immuno-staining, and to verify the subcellular localization of MT. As shown in Fig. 8.4, type II cells isolated from animals exposed more frequently to cadmium have a significantly greater percentage of darkly staining cells and more intense nuclear MT staining than cells from animals exposed on fewer occasions. The subcellular distribution of MT is likely to be critical in determining its ability to scavenge free-radicals generated during oxidant stress. Chubatsu and Meneghini (1993) examined the oxidant sensitivity of V79 Chinese hamster fibroblasts transfected with a plasmid for the MT gene. Only transfectants that expressed MT in both the nucleus and the cytoplasm were protected against hydrogen peroxide-induced DNA single-strand breaks (Chubatsu and Meneghini, 1993; Schwarz *et al.*, 1994).

8.8 Potential role of adaptation in cadmium carcinogenesis

8.8.1 Genetic disease or epigenetic process?

In the short run, these cadmium-adaptive mechanisms would appear to be protective for the lung. However, the emergence of a resistant phenotype in pulmonary stem cells capable of division (type II alveolar epithelial cells) could contribute to the carcinogenic process. Scientists have debated for years the issue of whether cancer is fundamentally a genetic disease or whether it could be an epigenetic process. A large component of epigenetic events could be considered under the general category of adaptive responses (Farber and Rubin, 1991). It cannot be denied that genomic changes are crucial to

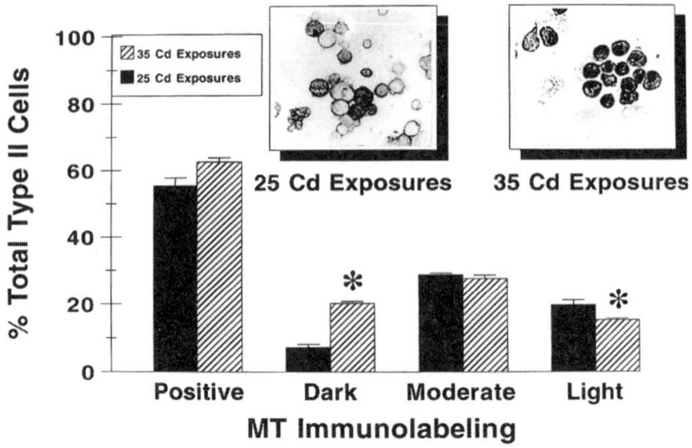

Figure 8.4 Metallothionein (MT) immunolabeling in type II alveolar epithelial cells isolated from animals exposed to 1.6 mg cadmium/m^3 for 5 or 7 weeks (3 hours/day, 5 days/week). Each bar represents the mean ±S.E.M. where $N = 3$. Paired t-tests were used to evaluate the difference in immunolabeling between the two exposure groups. (*indicates significantly different at $P < 0.05$). The inset depicts typical microscopic fields from preparations of the two exposure groups (no counterstain, original magnification × 540). The reaction product appears as black or gray. Cells negative for MT are almost transparent when viewed by light microscopy. (Reprinted from Hart, B.A., Eneman, J.D., Gong, Q. and Durieux-Lu, C-C. 1995. Increased oxidant resistance of alveolar epithelial type II cells isolated from rats following repeated exposure to cadmium aerosols. *Toxicology Letters* 81, 131–9, with permission from Elsevier Science.)

carcinogenesis. Clearly, mutations, gene rearrangements, translocations and other forms of genomic disorganization are common in cancer development. However, given the acceptance of some form of genomic change in most forms of cancer, we are still faced with whether adaptive responses could provide a selective advantage for an initiated cell to survive and to eventually undergo clonal expansion and tumor development (Cerutti, 1989; Farber and Rubin, 1991; Rubin, 1994).

A cell can react to adequate doses of a carcinogen in two ways: either the cell dies or it sustains damage and survives due to adaptive mechanisms. In the latter case, the cell's metabolic, genetic and biochemical properties, and its sensitivity to damaging agents, may undergo change. Resistance to the initial agent and to a number of other harmful stimuli is an example of such an adaptive change. If the resistant cell has become initiated, perhaps due to defective or inhibited DNA repair mechanisms, the cell may still remain dormant for long periods until exposure to a promoting agent. This would lead to clonal expansion of the nascent tumor (Monceviciute-Eringiene, 1996).

8.8.2 Cadmium-induced resistance and suppression of apoptosis

Apoptosis is a mechanism whereby the body can delete abnormal or damaged cells without wholesale tissue destruction. It normally occurs following the arrest of a cell that has been genetically altered beyond its ability to repair. Many tumors commonly possess nonfunctional gene products vital for cell cycle arrest and the onset of apoptosis. The extensive investigations related to proteins like p53 and Bcl-2 point to the importance that resistance to apoptosis plays in tumor development (Hockenbury *et al.*, 1993; Wiseman and Halliwell, 1996). Neoplastic growth results from an imbalance between cell replication and cell death, such that a net increase in cell number occurs. Thus, suppression of apoptosis (cell death) might be considered an important mediator in cancer development (Carson and Ribeiro, 1993; Wright *et al.*, 1994). Known apoptotic agents include reactive oxygen and nitrogen species (Buttke and Sandstrom, 1994; Sakamoto *et al.*, 1996) and cytokines such as tumor necrosis factor (Wright *et al.*, 1994). Antioxidants play an important role in protecting normal cells from oxidant injury. Until recently, there has been an expectation that 'more is better' with regard to antioxidant protection. However, an antioxidant imbalance, in presumptive sites of cadmium-induced tumor formation (i.e. type II epithelial cells) (Takenaka *et al.*, 1983; Oberdorster, 1986), may result in the suppression of apoptosis. This, in turn, would provide a selective advantage to these cells and could lead to promotion and progression of tumor growth. Recent studies suggest that overexpression of MT potentiates tumor cell growth, whereas downregulation of MT in these malignant cells induces growth arrest and apoptosis, thus eliciting an antiproliferative effect (Abdel-Mageed and Agrawal, 1997). Developing tumors with elevated MT levels (Hart *et al.*, 1993) and increased resistance to apoptosis would make many conventional cancer therapies ineffective (Cerutti, 1994; Cerutti *et al.*, 1994).

8.8.3 Hypothesis to explain the role of cadmium adaptation in lung carcinogenesis

A working hypothesis outlining the mechanisms that may be responsible for cadmium-induced lung carcinogenesis is illustrated in Fig. 8.5. It is proposed that adaptation to cadmium results in the emergence of a lung epithelial cell subpopulation with superior defenses (MTs, GSTs, γ-GCS, γ-GT, HO-1) that increase resistance to cadmium- and oxidant-induced cell death. This resistant phenotype is due to cadmium-induced alterations in gene expression that are mechanistically linked to cellular redox status, as determined by glutathione levels. Some cells bearing this resistant phenotype are pre-neoplastic. Initiation likely occurred prior to adaptation and was caused by an inhibition of DNA repair processes. These resistant, pre-neoplastic cells are capable of increased proliferation, when exposed to tumor promoters, and are more resistant to apoptosis by oxidants and cytokines, such as tumor necrosis factor-α (TNF-α) and transforming growth factor-β (TGF-β). As a consequence, selective clonal expansion occurs that ultimately results in tumor formation and malignancy. Future research, conducted using *in vivo* and *in vitro* models, is needed to test the various components of this hypothesis. Such studies have important medical applications in understanding environmentally induced lung cancers, mechanisms of tumor promotion, and resistance of primary carcinomas to chemotherapy.

8.9 Summary

From the information presented in this chapter and in the literature, several approaches and potential areas for future research emerge:

1 The results of single, acute inhalation exposures to cadmium provide important clues about the response of the lung to cadmium, especially at the molecular level.

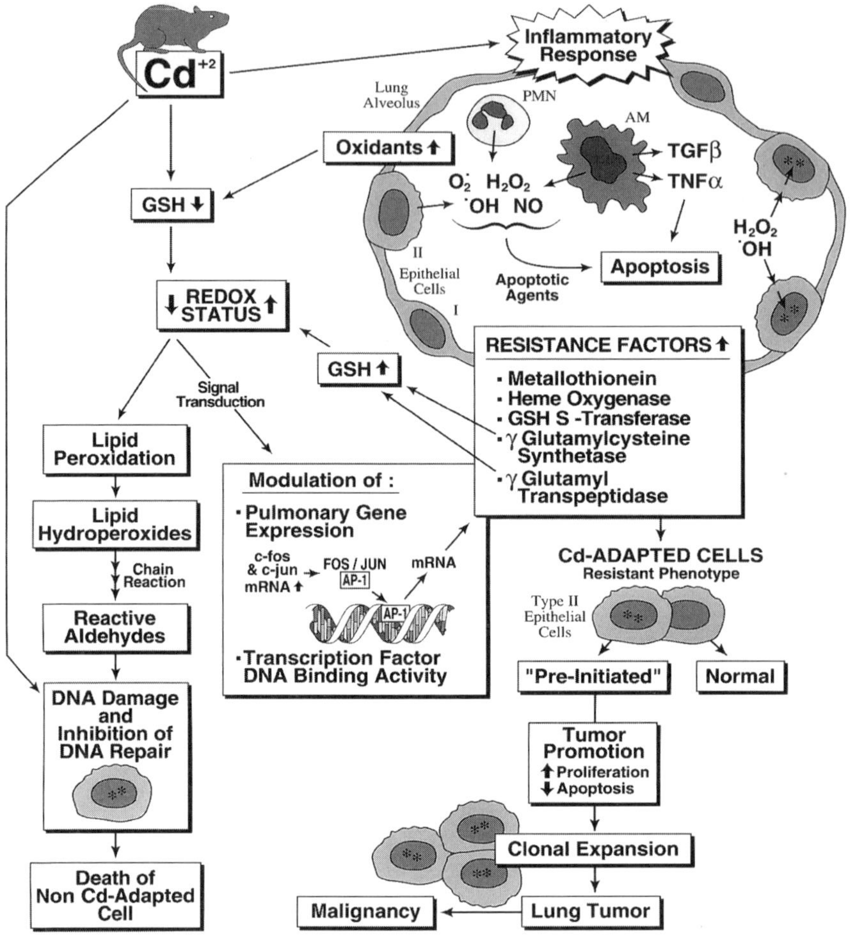

Figure 8.5 This illustration depicts a working hypothesis to explain how repeated inhalation of cadmium can cause lung cancer. Enhancement of endogenous mutations through the inhibition of DNA repair, by cadmium and oxidants, is proposed as the initiating event. Reactive oxygen and nitrogen species may be generated as a consequence of inflammatory reactions induced by cadmium. Tumor promotion and progression involve (a) the induction of multiple resistance factors that convey a selective advantage to initiated cells by suppressing normal apoptotic processes, and (b) oncoproteins that stimulate initiated cells with superior antioxidant defenses to divide. (Illustrated by Gary Nelson, University of Vermont, Department of Medical Photography.)

However, more emphasis must also be placed on performing experiments involving the long-term consequences of repeated inhalation at lower doses.

2 The pulmonary responses of different rat strains should be compared. Cadmium-induced biochemical and molecular alterations should be correlated with final outcome (i.e. development of lung tumors).
3 Greater emphasis should be placed on the use of *in vivo–in vitro* models.
4 Better methods must be developed for the long-term culturing of type II alveolar epithelial cells which are the presumptive sites of cadmium-induced tumor

formation. This technology is essential in order to test whether type II cells from cadmium-adapted animals are more likely to become transformed if grown *in vitro* in the presence of tumor promoters.

5 Although metallothionein is clearly an important pulmonary protectant, it is not the only resistance factor that is induced in response to cadmium exposure. The role of glutathione and the activation of genes involved in its synthesis and transport must not be underestimated.

Acknowledgments

The preparation of this paper was supported, in part, by NIH grant ES-08991. The author expresses thanks to her laboratory colleagues, the staff at the Dana Medical Library, and Gary Nelson, medical illustrator at the University of Vermont College of Medicine.

References

Abdel-Mageed, A.B. and Agrawal, K.C. 1997. Antisense down-regulation of metallothionein induces growth arrest and apoptosis in human breast carcinoma cells, *Cancer Gene Therapy* **4**, 199–207.

Alam, J., Cai, J. and Smith, A. 1994. Isolation and characterization of the mouse heme oxygenase-1 gene, *Journal of Biological Chemistry* **269**, 1001–9.

Amanuma, K. and Suzuki, K.T. 1987. Effect of intratracheal instillation of cadmium chloride on phospholipids in alveolar wash fluid. *Toxicology* **44**, 321–8.

Angel, P. and Karin, M. 1991. The role of Jun, Fos, and the AP-1 complex in cell proliferation and transformation. *Biochimica and Biophysica Acta* **1072**, 129–57.

Benson, J.M. and Henderson, R.F. 1980. Isolation and characterization of a low-molecular-weight cadmium-binding protein from Syrian hamster lung, *Toxicology and Applied Pharmacology* **55**, 370–7.

Bergelson, S., Pinkus, R. and Daniel, V. 1994a. Intracellular glutathione levels regulate Fos/Jun induction and activation of glutathione S-transferase gene expression. *Cancer Research* **54**, 36–40.

Bergelson, S., Pinkus, R. and Daniel, V. 1994b. Induction of AP-l (Fos/Jun) by chemical agents mediates activation of glutathione S-transferase and quinone reductase gene expression, *Oncogene* **9**, 565–71.

Bienkowski, R.S. 1989. Intracellular degradation of newly synthesized collagen, in *Current Trends in the Study of Intracellular Protein Degradation*, (eds S. Grisolia and E. Knecht), Springer International, Bilbao, Spain, Vol. II, pp. 423–43.

Bitterman, P.B., Rennard, S.I., Adelberg, S. and Crystal, R. 1983. The role of fibronectin as a growth factor for fibroblasts. *Journal of Cell Biology* **97**, 1932.

Boak, A.M., Roy, R., Berk, J., Taylor, L., Polgar, P., Goldsterin, R.H. and Kagan, H.M. 1995. Regulation of lysyl oxidase expression in lung fibroblasts by transforming growth factor-β and prostaglandin E_2. *American Review of Respiratory Cell and Molecular Biology* **11**, 751–5.

Borm, P.KJ. and Driscoll, K. 1996. Particles, inflammation and respiratory tract carcinogenesis. *Toxicology Letters* **88**, 188–99.

Boudreau, J., Vincent, R., Nadeau, D., Trottier, B., Fournier, M., Krzystyniak, K. and Chevalier, G. 1988. Toxicity of inhaled cadmium chloride: Early responses of the antioxidant and surfactant systems in rat lungs. *Journal of Toxicology and Environmental Health* **23**, 241–56.

Bouley, G., Dubreuil, A., Despaux, N. and Boudene, C. 1977. Toxic effects of cadmium microparticles on the respiratory system: An experimental study on rats and mice. *Scandinavian Journal of Work, Environment, and Health* **3**, 116–21.

Brookens, M., Hickens, R., Lazo, J. and Pitt, B. 1994. Partially reduced oxygen species directly regulate metallothionein gene expression. *American Journal of Respiratory Critical Care Medicine* **149** (4), 570.

Buckley, B.J. and Bassett, D.J.P. 1987. Glutathione redox status of control and cadmium oxide-exposed rat lungs during oxidant stress. *Journal of Toxicology and Environmental Health* **22**, 287–99.

Buttke, T.M. and Sandstrom, P.A. 1994. Oxidative stress as a mediator of apoptosis. *Immunology* **15**, 7–10.

Carson, D.A. and Ribeiro, J.M. 1993. Apoptosis and disease. *Lancet* **341**, 1251–4.

Cerutti, P. 1989. Response modification in carcinogenesis. *Environmental Health Perspective* **81**, 39–43.

Cerutti, P. 1994. Oxy-radicals and cancer. *Lancet* **344**, 862–3.

Cerutti, P., Ghost, R. and Amstad, P. 1994. Role of cellular antioxidant defenses in oxidant carcinogenesis. *Environmental Health Perspective* **102** (suppl. 10), 123–30.

Chambers, R.C., McAnulty, R.J., Shock, A., Campa, J.S., Newman-Taylor, A.J. and Laurent, G.J. 1994. Cadmium selectively inhibits procollagen production and proliferation. *American Journal of Physiology* **267**, L300–8.

Choi, A.M.K. and Alam, J. 1996. Heme oxygenase-1: function, regulation, and induction of a novel stress-inducible protein in oxidant-induced lung injury. *American Journal of Respiratory Cell and Molecular Biology* **15**, 9–19.

Chubatsu, L.S. and Meneghini, R. 1993. Metallothionein protects DNA from oxidative damage. *Biochemical Journal* **291**, 193–8.

Coin, P.G. and Stevens, J.B. 1986. Toxicity of cadmium chloride *in vitro*: Indices of cytotoxicity with the pulmonary alveolar macrophage. *Toxicology and Applied Pharmacology* **82**, 140–50.

Cosma, G., Fulton, H., DeFeo, T. and Gordon, T. 1992. Rat lung metallothionein and heme oxygenase gene expression following ozone and zinc oxide exposure. *Toxicology and Applied Pharmacology* **117**, 75–80.

Cousins, R.J. 1985. Absorption, transport, and hepatic metabolism of copper and zinc: Special reference to metallothionein and ceruloplasmin. *Physiological Reviews* **65**, 238–308.

Cox, C.C. and Waters, M.D. 1978. Isolation of a soluble cadmium-binding protein from pulmonary alveolar macrophages. *Toxicology and Applied Pharmacology* **46**, 385–94.

Crapo, J.D., Barry, B.E., Foscue, H.A. and Shelburne, J.S. 1980. Structural and biochemical changes in rat lungs occurring during exposures to lethal and adaptive doses of oxygen. *American Review of Respiratory Diseases* **122**, 123–43.

Cross, C.E., Omaye, S., Rifas, D., Hasegawa, G.K. and Reddy, K.A. 1979. Biochemical effects of intratracheal instillation of cadmium chloride on rat lung. *Biochemical Pharmacology* **28**, 381–8.

Dally, H. and Hartwig, A. 1997. Induction and repair inhibition of oxidative DNA damage by nickel (II) and cadmium (II) in mammalian cells. *Carcinogenesis* **18**, 1021–6.

Dalton, T., Palmiter, R.D. and Andrews, G. 1994. Transcriptional induction of the mouse metallothionein-1 gene in hydrogen peroxide-treated Hepa cells involves a composite major late transcription factor/antioxidant response element and metal response promoter elements. *Nucleic Acid Research* **22**, 5016–23.

Daniel, V. 1993. Glutathione S-transferases: gene structure and regulation of expression. *Critical Reviews in Biochemistry and Molecular Biology* **28**, 173–207.

Driscoll, K.E., Maurer, J.K., Poynter, J., Higgins, J., Asquith, T. and Miller, N.S. 1992. Stimulation of rat alveolar macrophage fibronectin release in a cadmium chloride model of lung injury and fibrosis. *Toxicology and Applied Pharmacology* **116**, 30–7.

Dubois, R.M. 1986. The alveolar macrophage. *Thorax* **40**, 321–7.
Farber, E. and Rubin, H. 1991. Cellular adaptation in the origin and development of cancer. *Cancer Research* **51**, 2751–61.
Frankel, F.R., Steeger, J.R., Damiano, V.V., Sohn, M., Oppenheim, D. and Weinbaum, G. 1991. Induction of unilateral pulmonary fibrosis in the rat by cadmium chloride. *American Journal of Respiratory Cell and Molecular Biology* **5**, 385–94.
Gairola, C.G. 1989. Cadmium-enriched cigarette smoke-induced cytological and biochemical alterations in rat lungs. *Journal of Toxicology and Environmental Health* **27**, 317–29.
Gairola, C.G. and Wagner, G.J. 1991. Cadmium accumulation in the lung, liver, and kidney of mice and rats chronically exposed to cigarette smoke. *Journal of Applied Pharmacology* **11**, 355–8.
Gardner, D.E., Miller, F.J., Illing, J.W. and Kirtz, J.M. 1977. Alterations in bacterial defense mechanisms of the lung induced by inhalation of cadmium. *Bulletin European de Physiopathologie Respiratoire* **13**, 157–74.
Gavett, S.H. and Oberdorster, G. 1994. Cadmium chloride and cadmium metallothionein induced pulmonary injury and recruitment of polymorphonuclear leukocytes. *Experimental Lung Research* **20**, 517–37.
Glaser, U., Hochrainer, D., Otto, F.J. and Oldiges, H. 1990. Carcinogenicity and toxicity of four cadmium compounds inhaled by rats. *Toxicological and Environmental Chemistry* **27**, 153–62.
Gong, Q. and Hart, B.A. 1997. Effect of thiols on cadmium-induced expression of metallothionein and other oxidant stress genes in rat lung epithelial cells. *Toxicology* **119**, 179–91.
Greenspan, B.J. and Morrow, P.E. 1984. The effects of *in vitro* and aerosol exposures to cadmium on phagocytosis by rat pulmonary macrophages. *Fundamental and Applied Toxicology* **4**, 48–57.
Grose, E.C., Richards, J.H. Jaskot, R.H., Menache, M.G., Graham, J.A. and Dauterman, W.C. 1987a. A comparative study of the effects of inhaled cadmium chloride and cadmium oxide: Pulmonary response. *Journal of Toxicology and Environmental Health* **21**, 219–32.
Grose, E.G., Richards, J.H., Jaskot, R.H., Menache, M.G., Graham, J. A. and Dauterman, W.C. 1987b. Glutathione peroxidase and glutathione transferase activity in rat lung and liver following cadmium inhalation. *Toxicology* **44**, 171–9.
Hadley, J.G., Conklin, A.W. and Sanders, C.L. 1980. Rapid solubiliztion and translocation of ^{109}CdO following pulmonary deposition. *Toxicology and Applied Pharmacology* **54**, 156–60.
Hart, B.A. 1978. Transport of cadmium by the alveolar macrophage. *Journal of the Reticuloendothelial Society* **24**, 363–75.
Hart, B.A. 1986. Cellular and biochemical response of the rat lung to repeated inhalation of cadmium. *Toxicology and Applied Pharmacology* **82**, 281–91.
Hart, B.A. and Garvey, J.S. 1986. Detection of metallothionein in bronchoalveolar cells and lavage fluid following repeated cadmium inhalation. *Environmental Research* **50**, 269–78.
Hart, B.A. and Keating, R.F. 1980. Distribution and accumulation of cadmium in human lung fibroblasts. *Chemico-Biological Interactions* **29**, 67–83.
Hart, B.A., Cherian, M.G. and Angel, A. 1985. Cellular localization of metallothionein in the lung following repeated cadmium inhalation. *Toxicology* **37**, 171–9.
Hart, B.A., Voss, G.W. and Garvey, J.S. 1989a. Induction of pulmonary metallothionein following oxygen exposure. *Environmental Research* **50**, 269–78.
Hart, B.A., Voss, G.W. and Willean, C.L. 1989b. Pulmonary tolerance to cadmium following cadmium aerosol pretreatment. *Toxicology and Applied Pharamacology* **101**, 447–60.
Hart, B.A., Voss, G.W., Shatos, M.A. and Doherty, J. 1990. Cross-tolerance to hyperoxia following cadmium aerosol pretreatment. *Toxicology and Applied Pharmacology* **103**, 255–70.
Hart, B.A., Voss, G.W. and Vacek, P.M. 1993. Metallothionein in human lung carcinoma. *Cancer Letters* **75**, 121–8.

Hart, B.A., Gong, Q., Eneman, J.D. and Durieux-Lu, C-C. 1995a. *In vivo* expression of metallothionein in rat alveolar macrophages and type II epithelial cells following repeated cadmium aerosol exposures. *Toxicology and Applied Pharmacology*, **133**, 82–90.

Hart. B.A., Eneman, J.D., Gong, Q. and Durieux-Lu, C.-C., 1995b. Increased oxidant resistance of alveolar epithelial type II cells isolated from rats following repeated exposure to cadmium aerosols. *Toxicology Letters* **81**, 131–9.

Hart, B.A., Prabhu, R.M., Eneman, J.D., Durieux-Lu, C.-C., Janssen, A.M.L. and Borm, P.J.A. 1995c. Oxidant resistance of cadmium-adapted human lung fibroblasts. *Toxicology* **98**, 1–13.

Hart, B.A., Gong, Q. and Eneman, J.D. 1996a. Pulmonary metallothionein expression in rats following single and repeated exposure to cadmium aerosols. *Toxicology* **112**, 205–18.

Hart, B.A., Gong, Q., Eneman, J.D., Durieux-Lu, C.-C., Kimberly, P. and Hacker, M.P. 1996b. Increased oxidant resistance of alveolar macrophages isolated from rats repeatedly exposed to cadmium aerosols. *Toxicology* **107**, 163–75.

Hartwig, A. 1994. Role of DNA repair inhibition in lead- and cadmium-induced genotoxicity. *Environmental Health Perspectives* **102** (Suppl. 3), 45–50.

Hartwig, A. 1995. Current aspects in metal genotoxicity. *BioMetals* **8**, 3–11.

Hartwig, A. and Beyersmann, D. 1989. Comutagenicity and inhibition of DNA repair by metal ions in mammalian cells. *Biological Trace Element Research* **21**, 359–65.

Hartwig, A., Schlepegrell, R., Dally, H. and Hartmann, M. 1996. Interaction of carcinogenic metal compounds with deoxyribonucleic acid repair processes. *Annals of Clinical and Laboratory Science* **26**, 31–8.

Hayes, J.A., Snider, G.L. and Palmer, K.C. 1976. The evolution of biochemical damage in the rat lung after acute cadmium exposure. *American Review of Respiratory Disease* **113**, 121–30.

Heinrich, U. 1992. Pulmonary carcinogenicity of cadmium by inhalation in animals, in *Cadmium in the Human Environment: Toxicity and Carcinogenicity*, (eds G.F. Nordberg, R.F.M. Herber and L. Alessio), International Agency for Research on Cancer, Lyon, pp. 405–13.

Henderson, R.F. 1989. Bronchoalveolar lavage: A tool for assessing the heath status of the lung, in *Concepts in Inhalation Toxicology*, (eds R.O. McClellan and R.F. Henderson), Hemisphere Publishing Corporation, New York, pp. 415–44.

Henderson, R.F., Rebar, A.H., Pickrell, J.A. and Newton, G.J. 1979. Early damage indicators in the lung: Biochemical and cytological response of the lung to inhaled metal salts. *Toxicology and Applied Pharmacology* **50**, 123–36.

Hirano, S., Tsukamoto, N., Kobayashi, E. and Suzuki, K.T. 1989a. Toxicity of cadmium oxide instilled into the rat lung. I. Metabolism of cadmium oxide in the lung and its effects on essential elements. *Toxicology* **55**, 15–24.

Hirano, S., Tsukamoto, N., Higo, S. and Suzuki, K.T. 1989b. Toxicity of cadmium oxide instilled into the rat lung. II. Inflammatory responses in broncho-alveolar lavage fluid. *Toxicology* **55**, 25–35.

Hockenbury, D., Oltval, Z.N., Yin, S-M., Milliman, C.L. and Korsmyer, S.J.H. 1993. Bcl-2 functions in an antioxidant pathway to prevent apoptosis. *Cell* **75**, 241–51.

Howden, P.J. and Faux, S.P. 1996. Fibre-induced lipid peroxidation leads to DNA adduct formation in *Salmonella typhimurium* RA104 and rat lung fibroblasts. *Carcinogenesis* **17**, 413–19.

IARC 1993. *Beryllium, cadmium, mercury, and exposures in the glass manufacturing industry*, International Agency for Research on Cancer Monographs on the Evaluation of the Carcinogenic Risks to Humans. Vol. 48. IARC Scientific Publications, Lyon, France, pp. 119–238.

Johansson, A., Camner, P., Jarstrand, C. and Wiernick, A. 1983. Rabbit alveolar macrophages after inhalation of soluble cadmium, cobalt, and copper: a comparison with the effects of soluble nickel. *Environmental Research* **31**, 340–54.

Kang, Y.-J., Liu, M.-S. and Enger, M.D. 1990. Diamide reduces cadmium accumulation by human lung carcinoma A549 cells. *Toxicology* **62**, 53–8.

Karlinsky, J.B. and Snider, G.L. 1978. Animal models of emphysema. *American Review of Respiratory Diseases* **117**, 1109–33.

Kenaga, C., Cherian, M.G., Cox, C. and Oberdorster, G. 1996. Metallothionein induction and pulmonary responses to inhaled cadmium chloride in rats and mice. *Fundamentals of Applied Pharmacology* **30**, 204–12.

Klaassen, C.D. and Lehman-McKeeman, L.D. 1989. Regulation of the isoforms of metallothionein. *Biological Trace Element Research* **21**, 119–29.

Klimisch, H.-J., 1993. Lung deposition, lung clearance and renal accumulation of inhaled cadmium chloride and cadmium sulfide in rats. *Toxicology* **84**, 103–24.

Kobrle, V., Hurych, J., Cikft, M., Jones, M.M. 1991. Sodium bis(hydroxyethyl)-dithiocarbamate reduces acute lung tissue damage induced by cadmium in rats. *Fundamentals of Applied Toxicology* **16**, 733–41.

Lewis, G.P. 1973. Centrilobular emphysema experimentally induced by cadmium chloride aerosol. *American Review of Respiratory Diseases* **108**, 40–8.

Li, W., Chou, I.-N., Boak, A. and Kagan, H.M. 1995. Downregulation of lysyl oxidase in cadmium-resistant fibroblasts. *American Journal of Respiratory Cell and Molecular Biology* **13**, 418–25.

Liu, J., Liu, Y.-P., Sendelbach, L.E. and Klaassen, C.D. 1991. Endotoxin induction of hepatic metallothionein is mediated through cytokines. *Toxicology and Applied Pharmacology* **109**, 235–40.

Lundborg, M., Lind, B. and Camer, P. 1984. Ability of rabbit alveolar macrophages to dissolve metals. *Experimental Lung Research* **7**, 11–22.

McKenna, I.M., Waalkes, M.P., Chen, L.C. and Gordon, T. 1997. Comparison of inflammatory lung responses in Wistar rats and C57 and DBA mice following acute exposure to cadmium oxide fumes. *Toxicology and Applied Pharmacology* **146**, 196–206.

McKenna, I.M., Gordon, T., Chen, L.C., Anver, M.R. and Waalkes, M.P. 1998. Expression of metallothionein protein in the lungs of Wistar rats and C57 and DBA mice exposed to cadmium oxide fumes. *Toxicology and Applied Pharmacology* 3, 169–78.

MacPhee, D.G. 1995. Mismatch repair, somatic mutations, and the origins of cancer. *Cancer Research* **55**, 5489–92.

MacPhee, D.G. 1996. Mismatch repair as an important source of new mutations in non-dividing cells. *Experientia* **52**, 357–63.

Manca, D., Ricard, A.C., Trottier, B. and Chevalier, G. 1991. Studies on lipid peroxidation in rat tissues following administration of low and moderate doses of cadmium chloride. *Toxicology* **67**, 303–23.

Manca, D., Richard, A.C., Tra, H.V. and Chevalier, G. 1994. Relation between lipid peroxidation and inflammation in the pulmonary toxicity of cadmium. *Archives of Toxicology* **68**, 364–9.

Mello-Filho, A.C. and Meneghini, R. 1988. V79 Chinese hamster cells rendered resistant to high cadmium concentrations also become resistant to oxidative stress. *Biochemical Journal* **256**, 475–9.

Monceviciute-Eringiene, E. 1996. Cancer and malignant resistance of cells as phenomena of adaptation to damaging factors. *Medical Hypotheses* **46**, 459–62.

Mustafa, M.G., Cross, C.E. and Tyler, W.S. 1971. Interference of cadmium ion with oxidative metabolism of alveolar macrophages. *Archives of Internal Medicine* **127**, 1050–8.

Nelson, J.M., Duane, P.G., Rice, K.L. and Niewoehner, D.E. 1994. Cadmium ion-induced alterations of phospholipid metabolism in endothelial cells. *American Journal of Respiratory Cell and Molecular Biology* **5**, 328–36.

Oberdorster, G. 1986. Airborne cadmium and carcinogenicity of cadmium. *Scandinavian Journal of Work and Environmental Health* **12**, 523–37.

Oberdorster, G. 1990. Lung clearance mechanisms of soluble and insoluble cadmium compounds, in *Environmental Hygiene*, (eds N.H. Seemayer and W. Hadnagy), Springer-Verlag, Heidelberg, Vol. II, pp. 221–4.

Oberdorster, G. and Koerdel, W. 1981. Metallothionein content in lung, liver, and kidney after chronic cadmium oxide and zinc oxide inhalation in rats, in *Heavy Metals in the Environment*, CEP Consultants Ltd, Edinburgh, pp. 502–5.

Oberdorster, G., Baumert, H.-P., Hochrainer, D. and Stoeber, W. 1979. The clearance of cadmium aerosols after inhalation exposure. *American Industrial Hygiene Association Journal* **40**, 443–50.

Oberdorster, G., Guth, D.J., Lee, Y.H. and Mavis, R.D. 1985. Lung toxicity of cadmium: Importance of the chemical form, in *Heavy Metals in the Environment*, (ed. T.D. Lekkas), CEP Consultants Ltd, Edinburgh, pp. 571–4.

Oberdorster, G., Cherian, M.G. and Baggs, R.B. 1994. Correlation between cadmium-induced pulmonary carcinogenicity, metallothionein expression, and inflammatory processes: A species comparison. *Environmental Health Perspectives* **102** (Suppl. 3), 257–63.

Ochi, T. 1991. Cadmium-resistant V79 cells with decreased accumulation of cadmium. *Chemico-Biological Interactions* **78**, 207–21.

Ochi, T. and Ohsawa, M. 1983. Induction of 6-thioguanine-resistant mutants and single strand scissions of DNA by cadmium chloride in cultured Chinese hamster cells. *Mutation Research* **111**, 69–78.

Ochi, T., Mogi, M., Watanabe, M. and Ohsawa, M. 1984. Induction of chromosomal aberrations in cultured Chinese hamster cells by short-term treatment with cadmium chloride. *Mutation Research* **137**, 103–9.

Oldiges, H., Hochrainer, D. and Glaser, U. 1989. Long-term inhalation study with Wistar rats and four cadmium compounds. *Toxicological and Environmental Chemistry* **19**, 217–22.

Paakko, P., Kokkone, P., Anttila, S. and Kalliomaki, P.L. 1989. Cadmium and chromium as markers of smoking in human lung tissue. *Environmental Research* **49**, 197–207.

Palmer, K.C., Snider, G.L. and Hayes, J.A. 1975. Cellular proliferation induced in the lung by cadmium aerosol. *American Review of Respiratory Disease* **112**, 173–9.

Piedboeuf, B., Johnson, C.J., Watkins, R.H., Hudak, B.B., Lazo, J.S., Cherian, M.G. and Horowitz, S. 1994. Increased expression of tissue inhibitor of metalloproteinases and metallothionein in murine lungs after hyperoxic exposure. *American Journal of Respiratory and Cell Molecular Biology* **11**, 165–72.

Post, C.T., Squibb, K.S., Fowler, B.A., Gardner, D.E., Illing, J. and Hook, G.E.R. 1982. Production of low molecular weight cadmium binding proteins in rabbit lung following exposure to cadmium chloride. *Biochemical Pharmacology* **31**, 2969–75.

Rahman, I., Smith, C.A.D., Lawson, M.F., Harrison, D.J. and MacNee, W. 1996. Induction of γ-glutamylcysteine synthetase by cigarette smoke is associated with AP-1 in human alveolar epithelial cells. *Federation of European Biochemical Societies* **396**, 21–5.

Rubin, H. 1994. Epigenic nature of neoplastic transformation, in *Developmental Biology and Cancer*, (eds G.M. Hodges and C. Rowlatt), CRC Press, Boca Raton, Florida, pp. 61–84.

Rusch, G.M., O'Grodnick, J.S. and Reinhart, W.E. 1986. Acute inhalation study in the rat of comparative uptake, distribution, and excretion for different cadmium containing materials. *American Industrial Hygiene Association* **47**, 754–60.

Sakamoto, T., Repasky, W.T., Uchia, K., Hirata, A. and Hirata, F. 1996. Modulation of cell death pathways to apoptosis and necrosis of hydrogen peroxide treated rat thymocytes by lipocortin I. *Biochemical and Biophysical Research Communications* **220**, 643–7.

Salovsky, P., Shopova, V., Dancheva, V. and Marev, R. 1992. Changes in antioxidant lung protection against single intra-tracheal cadmium acetate instillation in rats. *Human and Experimental Toxicology* **11**, 217–22.

Sampson, C.E., Chichester, C., Hayes, J.A. and Kagan, H.M. 1984. Alterations in collagen biosynthesis and in metallothionein in lungs of rat acutely or repeatedly exposed to cadmium chloride aerosol. *American Review of Respiratory Diseases* **129**, 619–24.

Sanders, C.L. and Mahaffey, J.A. 1984. Carcinogenicity of single and multiple intratracheal instillations of cadmium oxide in the rat. *Environmental Research* **33**, 227–33.

Sato, M. and Bremner, I. 1993. Oxygen free radicals and metallothionein. *Free Radicals in Biology and Medicine* **14**, 325–37.

Schlesinger, R.B. 1989. Deposition and clearance of inhaled particles, in *Concepts in Inhalation Toxicology*, (eds R.O. McClellan and R.F. Henderson), Hemisphere Publishing Corporation, New York, pp. 163–92.

Schwartz, M.A., Lazo, J.S., Yalowich, J.C., Reynolds, I., Kagan, V.E., Tyurin, V., Kim, Y-M., Watkines, S.C. and Pitt, B.R. 1994. Cytoplasmic metallothionein overpression protects NIH 3T3 cells from *tert*-butyl hydroperoxide toxicity. *Journal of Biological Chemistry* **269**, 15238–43.

Sibille, Y. and Reynolds, H.Y. 1990. Macrophages and polymorphonculear neutrophils in lung defense and injury. *American Review of Respiratory Diseases* **141**, 471–501.

Snider, G.L., Lucey, E.C., Faris, B., Jung-Legg, Y., Stone, P.J. and Franzblau, C. 1988. Cadmium chloride-induced air space enlargement with interstitial fibrosis is not associated with destruction of lung elastin. *American Review of Respiratory Diseases* **137**, 918–23.

Snyder, R.D. 1988. Role of active oxygen species in metal-induced DNA strand breakage in human diploid fibroblasts. *Mutation Research* **193**, 237–46.

Strauss, R.H., Palmer, K.C. and Hayes, J.A. 1976. Acute lung injury induced by cadmium aerosol: Evolution of alveolar cell damage. *American Journal of Pathology* **84**, 561–78.

Takenaka, S., Oldiges, H., Konig, H., Hochrainer, D. and Oberdoster, G. 1983. Carcinogenicity of four cadmium aerosols in Wistar rats. *Journal of the National Cancer Institute* **70**, 367–71.

Thornalley, P.J. and Vasak, M. 1985. Possible role for metallothionein in protection against radiation-induced oxidative stress. Kinetics and mechanism of its reaction with superoxide and hydroxyl radicals. *Biochimica and Biophysica Acta* **827**, 36–44.

Vile, G.F., Basu-Modak, S., Waltner, C. and Tyrrell, R.M. 1994. Heme oxygenase-1 mediates an adaptive response to oxidative stress in human skin fibroblasts. *Proceedings of the National Academy of Science* **91**, 2607–10.

Vincent, R., Boudreau, J., Nadeau, D., Fournier, M., Krzystynia, K.; Trottier, B. and Chevalier, G. 1989. Lipid peroxidation in rat lungs following an acute inhalation exposure to cadmium chloride. *Journal of Aerosol Medicine* **2**, 349–63.

Wagner, G.J. and Yeargan, R. 1986. Variation in cadmium accumulation and tissue distribution of cadmium in tobacco. *Plant Physiology* **82**, 274–9.

Winge, D.R.R., Geller, B.L. and Garvey, J.S. 1981. Isolation of copper-thionein from rat liver. *Archives of Biochemistry and Biophysics* **208**, 160–6.

Wiseman, H. and Halliwell, B. 1996. Damage to DNA by reactive oxygen and nitrogen species: role in inflammatory disease in progression to cancer. *Biochemical Journal*, **313**, 17–29.

Wright, S.C., Zhong, J. and Larrick, J. 1994. Inhibition of apoptosis as a mechanism of tumor promotion. *Federation of American Societies for Experimental Biology* **8**, 654–60.

Yang, C.-F., Shen, H.-M., Shen, Y., Shuang, Z.-X. and Ong, C.-N. 1997. Cadmium-induced oxidative cellular damage in human fetal lung fibroblasts (MRC-5 cells). *Environmental Health Perspectives* **105**, 712–16.

9

MERCURY: MOLECULAR INTERACTIONS AND MIMICRY IN THE KIDNEY

Rudolfs K. Zalups

9.1 Introduction

Among metals, mercury is unique in that it is found in the environment in several physical and chemical forms. At room temperature, elemental (or metallic) mercury exists as a liquid. Due its high vapor pressure, this form of mercury can be, and is, released into the environment as mercury vapor. Mercury also exists as a cation having an oxidation state of 1+ (mercurous) or 2+ (mercuric). In occupational and environmental settings, the most common cationic form of mercury encountered is the mercuric form, which may have a valence of 1+ or 2+, depending on whether the mercuric ion is covalently bonded to a carbon atom of an organic side group, such as an alkyl group. With respect to the organic forms of mercury, methylmercury is the most frequently encountered organic mercuric compound in the environment. This particular organic mercuric compound occurs mainly as the result of conversion of inorganic forms of mercury by microorganisms in soil and water.

Due to industrialization and changes in the environment during the twentieth century, humans and animals are exposed to numerous chemical forms of mercury, including elemental mercury vapor (Hg^0), inorganic mercurous (Hg^+) and mercuric (Hg^{2+}) compounds and organic mercuric ($R\text{-}Hg^+$ or $R\text{-}Hg\text{-}R\cdot$, where R represents any organic ligand) compounds. Inasmuch as mercury is found in so many places, it is nearly impossible for most humans to avoid exposure to some form(s) of mercury on a regular basis.

Elemental mercury and inorganic and organic mercury-containing compounds cause toxic effects in a number of systems and organs, depending on the chemical form of mercury, the level of exposure, the duration of exposure and the route of exposure. The kidneys are the primary target organs where inorganic mercury is taken up, accumulated, and expresses toxicity. Organic mercuric compounds also express toxic effects in the kidneys, although they have more diffuse systemic distributions and affect other target organs, including hematopoietic and neural tissues (Clarkson, 1972; Ballatori, 1991). Differences in transport of inorganic and organic forms of mercury are likely responsible for the differences in tissue and organ distribution, pattern of biological effect and toxic potency (Ballatori, 1991; Zalups and Lash, 1994).

When considering the biological activity of mercuric ions that have gained entry into the body of humans or other mammals, one must take into account the bonding characteristics of these ions. Although mercuric ions will bind to numerous nucleophilic groups on molecules, they have a greater predilection to bond to reduced sulfur atoms, especially those on small endogenous thiol-containing molecules, such as glutathione, cysteine, homocysteine and metallothionein. In the presence of an excess of a thiol-containing molecule, mercuric ions have a high propensity toward linear II coordination with two of these molecules. For example, in a situation where there are twice as many molecules of glutathione to inorganic mercuric ions in aqueous solution, there will be a tendency for each inorganic mercuric ion to be bonded to the sulfur atoms of two molecules of glutathione in a linear II coordinate covalent complex (Rabenstein, 1989). Organic mercurials, such as methylmercury, tend to form 1 : 1 complexes with thiol-containing molecules.

In spite of the extremely high affinity of inorganic mercury and methylmercury towards sulfhydryl groups, the complexes formed between some thiol-containing molecules and mercuric ions appear to be labile under certain circumstances in the living organism. In fact, thiol- or other nucleophilic-exchange is about the only cogent explanation that can account for the biological effects and interactions of mercury-containing compounds in various compartments of the body. For example, most of the mercuric ions present in plasma soon after exposure to inorganic mercury are bound to sulfhydryl-containing proteins, such as albumin (Friedman, 1957; Mussini, 1958; Cember *et al.*, 1968; Lau and Sarkar, 1979). Yet, these mercuric ions do not appear to remain bound to these proteins for very long, inasmuch as there is a rapid decrease in the plasma burden of mercury concurrent with a rapid rate of uptake of inorganic mercury in the kidneys and liver. Since the current body of evidence (to be discussed) indicates that mercuric *S*-conjugates of small endogenous thiols (such as cysteine and glutathione) are likely to be the transportable forms of mercury in the kidneys, it must be that the mercuric ions are transferred from the plasma proteins to these small molecular weight thiols by some form of complex ligand-exchange mechanisms that are currently unknown. It should also be mentioned that the effectiveness of thiol-containing pharmacologic agents, such as penicillamine, *N*-acetylpenicillamine, *meso*-2,3-dimercaptosuccinic acid (DMSA), 2,3-dimercapto-1-propanesulfonic acid (DMPS), dithioerythritol and dithiothreitol, in reversing or protecting against toxic effects of mercury-containing compounds can only be explained by the ability of these agents to remove mercury from endogenous ligands, thereby forming new thiol–mercury complexes. Although mercury can bind to other cellular nucleophiles, the affinity constant for thiolate anions is of the order of 10^{15}–10^{20} whereas affinity constants for oxygen- or nitrogen-containing ligands, such as carbonyl groups or amino groups, are about ten orders of magnitude lower (Ballatori, 1991). Hence, it is reasonable, in most cases, to consider biological effects of inorganic mercury or organic mercury in terms of their interactions with thiol-containing residues.

Inorganic mercury exhibits extremely steep dose–response relationships in a variety of renal systems, including *in vivo* treatment of rats and rabbits (Zalups and Diamond, 1987; Zalups *et al.*, 1988; Zalups and Lash, 1990; Zalups, 1991c), renal cortical slices from rabbits (Ruegg *et al.*, 1987), isolated segments of proximal tubules from rabbits (Barfuss *et al.*, 1990; Zalups *et al.*, 1993a), freshly isolated proximal tubular cells from rats (Lash and Zalups, 1992), and primary cultures of renal cortical cells from rats

(Smith *et al.*, 1987). In all these various renal systems, a threshold effect is generally observed, in that no cellular necrosis is observed up to a certain dose. Above that dose, however, cellular necrosis rapidly progresses, and in some systems an all-or-none response is observed. This does not mean that subtoxic doses of mercury do not have biochemical or physiological effects. One possible explanation for the threshold effect and the subsequent steep dose–response curve is that endogenous ligands bind mercury and may act as a buffer to prevent functional changes from occurring. Above a certain dose or concentration of mercury, the buffer becomes saturated, allowing additional mercury to bind to critical nucleophilic groups in the cell, thereby causing functional impairment. Intracellular thiol-containing proteins such as metallothionein or low-molecular-weight thiols, in particular glutathione, likely function in such a capacity.

It has become increasingly clear that in order to understand the nephropathy induced by mercury, and to find therapeutic regimens to treat this nephropathy, it is essential to understand the mechanisms involved in the uptake, intracellular interactions and cellular elimination of mercury in the target cells, namely the epithelial cells lining the proximal tubule. Moreover, besides the chemical properties of mercury-containing compounds and intracellular buffering capacity of both target and nontarget organs, other factors must be considered in understanding the biochemical and molecular mechanisms of action of mercury-containing compounds. Susceptibility to the injurious effects of mercury may be modified by a number of intracellular and extracellular factors. These very factors are the theoretical basis for most of the currently employed therapeutic strategies. Physiological or pathological alterations in cellular function, particularly in the kidney and liver, may also play important roles in modifying susceptibility to mercury-induced renal injury. Consideration of these factors can provide clues that will aid in understanding the basic mechanisms of mercury-induced renal cellular injury.

One of the primary aims of this chapter is to present information on the molecular interactions between mercuric ions and various target molecules in mechanisms involved in the renal tubular transport, accumulation, excretion and toxicity of mercury. Particular attention will be paid to the role of molecular mimicry and the species of mercury involved in the renal (proximal) tubular uptake and transport of mercuric ions.

9.2 Renal disposition and transport of mercury

The kidneys are the primary targets where mercuric ions accumulate in humans and other mammals exposed to elemental or inorganic forms of mercury (reviewed in Zalups and Lash, 1994). Renal uptake of mercury *in vivo* is very rapid. Recent findings indicate that as much as 50% of a low dose of inorganic mercury is present in the kidneys of rats within a few hours of exposure (Zalups, 1993a). Significant amounts of mercury also accumulate in the kidneys after exposure to organic forms of mercury, but to a lesser extent than that which occurs after exposure to inorganic or elemental forms of mercury.

9.2.1 *Intrarenal distribution and localization of mercury*

Within the kidneys, both inorganic and organic forms of mercury accumulate primarily in the cortex and outer stripe of the outer medulla (Zalups and Lash, 1994).

Histochemical and autoradiographic data from mice and rats (Taugner *et al.*, 1966; Hultman *et al.*, 1985; Magos *et al.*, 1985; Hultman and Enestrom, 1986, 1992; Rodier *et al.*, 1988; Zalups, 1991b), and tubular microdissection data from rats and rabbits (Zalups and Barfuss, 1990; Zalups, 1991a), indicate that the accumulation of inorganic mercury that occurs in the renal cortex and outer stripe of the outer medulla occurs mainly along the convoluted and straight segments of the proximal tubule. Although the segments of the proximal tubule appear to be the primary sites where mercuric ions are taken up and accumulate, there remains the possibility that some level of uptake and transport of mercury may occur along other segments of the nephron and collecting duct. It is interesting that deposits of presumed inorganic mercury have also been found along segments of proximal tubules in the kidneys of rats and mice treated with organic forms of mercury (Magos *et al.*, 1985; Rodier *et al.*, 1988). Additional findings indicate that a significant fraction of the mercury in the kidneys of animals exposed to methylmercury is in the inorganic form (Gage, 1964; Norseth and Clarkson, 1970a,b; Omata *et al.*, 1980; Zalups *et al.*, 1992), suggesting that organic mercury is oxidized to inorganic mercury prior to or after it enters the renal tubular epithelial cells. Furthermore, there is evidence that intracellular conversion of methylmercury to inorganic mercury can occur (Dunn and Clarkson, 1980). However, the mechanism for this conversion is currently unknown.

9.2.2 Mechanisms involved in the proximal tubular uptake and transport of mercury

Until recent years, very little was known about the mechanisms by which inorganic or organic forms of mercury gain entry into the epithelial cells lining the nephron and collecting duct. In order to define the mechanisms involved in the renal transport of mercury, it was necessary to first determine which segments of the nephron are involved in the uptake, transport and accumulation of the various forms of mercury. Several different sets of findings from numerous studies indicate quite conclusively that the epithelial cells lining the three segments of the proximal tubule are the primary sites where mercuric ions are taken up and transported within the kidneys (reviewed in Zalups and Lash, 1994). It should be stressed, however, that there are currently insufficient data to exclude the possibility that other segments of the nephron may also, to a minor extent, take up, transport and accumulate various forms of mercury.

Extensive investigations on the uptake and transport of inorganic mercury indicate that at least two distinct mechanisms are involved in the uptake of inorganic mercuric ions by proximal tubular epithelial cells. One of the mechanisms is localized on the luminal membrane (Zalups, 1995; Zalups and Minor, 1995) and the other is localized on the basolateral membrane (Zalups and Barfuss, 1993, 1995a; Zalups, 1995; Zalups and Minor, 1995).

9.2.2.1 Mechanisms involved in the luminal uptake of mercury

There is a strong body of evidence linking the luminal uptake of inorganic mercury to the activity of γ-glutamyltransferase, which, in the kidney, is localized predominantly in the luminal membrane of proximal tubular epithelial cells. The function of this enzyme is to cleave catalytically the γ-glutamylcysteine bond in molecules of glutathione (that are present in the proximal tubular lumen). Much of the aforementioned evidence comes from *in vivo* experiments, in which inhibition of renal γ-glutamyltransferase, by pretreatment with AT 125 (acivicin), caused profound effects on the renal disposition of administered mercury. More specifically, pretreatment with acivicin has been shown to cause significant decreases in the renal uptake and accumulation of mercury and significant increases in the urinary excretion of mercury in mice (Tanaka *et al.*, 1990; Tanaka-Kagawa *et al.*, 1993) and rats (Berndt *et al.*, 1985; de Ceaurriz *et al.*, 1994; Zalups, 1995) administered inorganic mercury or in mice administered methylmercury (Tanaka-Kagawa *et al.*, 1993) or exposed to mercury vapor (Kim *et al.*, 1995). Enhanced urinary excretion of glutathione has also been documented in acivicin-pretreated rats that were subsequently injected with inorganic mercury (Berndt *et al.*, 1985). We (Zalups and Barfuss) also have unpublished data indicating that the luminal uptake of mercuric ions is significantly decreased in isolated S2 segments of the proximal tubule of the rabbit when these segments are perfused through the lumen with mercuric conjugates of glutathione, following the inhibition of γ-glutamyltransferase with acivicin (Cannon *et al.*, 1999). Collectively, these data indicate that a significant fraction of the mercuric ions taken up by proximal tubular epithelial cells is accomplished by a luminal absorptive mechanism dependent on the actions of γ-glutamyltransferase, which presumably involves the cleavage of the γ-glutamylcysteine bond in molecules of glutathione bonded to mercuric ions.

As mentioned previously in this chapter, mercuric ions, under the appropriate conditions, have an extremely high propensity to form linear II coordinate covalent complexes with two molecules of small nonprotein thiols, such as glutathione or cysteine (Rabenstein, 1989). The implications of the findings in the above-mentioned studies indicate that at least some pool of the mercuric ions present in the lumen of the proximal tubule is in the form of some type of mercuric *S*-conjugate of glutathione, prior to being taken up. Although it is not known exactly where these mercuric conjugates of glutathione are formed prior to arriving in the lumen of the proximal tubule, one must consider the possibility that some of these mercuric *S*-conjugates of glutathione are formed outside the kidneys and then enter into the lumen of the proximal tubule via glomerular filtration. There are two reasons to suspect that this may occur. First of all, the formation of mercuric conjugates of glutathione in the plasma (after exposure to mercuric compounds) is possible because the concentration of this nonprotein thiol in plasma has been estimated to be approximately 10 μM (Lash and Jones, 1985a), which provides a sufficiently large pool of glutathione to form conjugates with mercuric ions in plasma. Secondly, the size and shape of these conjugates are such that they should pass through the glomerular filtration barrier unimpeded. It is also possible that some mercuric conjugates of glutathione are actually formed in the lumen of the proximal tubule or are secreted into the lumen from within proximal tubular epithelial cells after being formed within the cells. Inasmuch as the renal γ-glutamyltransferase and the renal tubular uptake and accumulation of mercury (Zalups and Barfuss, 1990; Zalups, 1991a,b) are both localized mainly in segments of the proximal tubule, the luminal absorptive uptake of mercury almost certainly occurs along these segments of the nephron.

It also appears that the liver plays some role in the renal uptake and transport of mercury. Evidence for this hypothesis comes from recent dispositional studies. In one study, specific depletion of hepatic glutathione, prior to the administration of inorganic mercury, was shown to cause a significant diminution in the renal uptake and accumulation of inorganic mercury (Tanaka *et al.* 1990). In other studies, it was demonstrated that biliary ligation prior to the administration of inorganic mercury causes a decrease in the renal tubular uptake and accumulation of inorganic mercury (Zalups and Barfuss, 1996a; Zalups 1998a). Together, these findings tend to link some aspect of the liver, presumably some pool of glutathione synthesized in the liver, to a fraction or component of the renal uptake of inorganic mercury. Additional studies are clearly needed to better determine the role of the liver in the renal tubular uptake of mercury.

Considering that the uptake of mercuric ions in the lumen of the proximal tubule is linked to the activity of γ-glutamyltransferase and the presence of mercuric *S*-conjugates of glutathione in the tubular lumen, the actual luminal uptake of mercuric ions into proximal tubular cells would appear to involve the transport of some product formed as the result of the actions of the γ-glutamyltransferase. One such product might be a mercuric conjugate of cysteinylglycine, which could be transported by a sodium-dependent small peptide transport system in the luminal membrane (Silbernagl, 1992). However, since the activity of membrane dipeptidases, such as cysteinylglycinase, is very high in the proximal tubule, it seems more likely that the actual species of mercury transported at the luminal membrane is a mercuric conjugate of L-cysteine via one or more of the amino-acid transport systems. It should also be stressed that there is *in vitro* evidence indicating that sequential enzymatic degradation of glutathione to cysteinylglycine and then to cysteine is possible while molecules of glutathione are bonded to a mercuric ion (Naganuma *et al.* 1988).

As mentioned above, mercuric conjugates of cysteine, presumably in the form of a dicysteinyl–mercury complex, appear to be the most likely species of inorganic mercury transported at the luminal membrane of proximal tubular cells. There are several different, recent lines of evidence supporting this hypothesis. One line of evidence comes from numerous sets of *in vivo* data showing that the renal uptake and accumulation of inorganic mercury (Zalups and Barfuss, 1995b) and the renal tubular injury induced by inorganic mercury (Zalups and Barfuss, 1996b) are increased in animals when the inorganic mercury is administered as a mercuric conjugate of cysteine. Another line of evidence comes from recent *in vitro* data from isolated brush-border membrane vesicles indicating that mercuric ions gain entry into brush-border membrane vesicles far more readily when they are exposed to mercuric conjugates of cysteine than mercuric conjugates of glutathione or even mercuric chloride (Zalups and Lash, 1997a). By far the most convincing evidence (not yet published) comes from recent studies in our laboratories (Cannon *et al.*, 1999). In brief, the findings from one of these studies indicate that mercuric conjugates of cysteine, presumably in the form of a single mercuric ion bonded to the sulfur atoms of two molecules of cysteine in a linear II coordinate covalent complex, are taken up at the luminal membrane of proximal tubular cells by known amino-acid transporters. More importantly, our data indicate strongly that some of the same transport systems involved in the luminal uptake of the amino acid cystine are involved in the luminal uptake of mercuric conjugates of cysteine. Additional data (not yet published) indicate that the luminal uptake of these mercuric conjugates of cysteine involves at least two separate amino-acid transport systems, with one of these systems being sodium dependent and the other being sodium independent. Thus, it appears that some of the absorptive luminal transport of a dicysteinyl–mercuric complex in the proximal tubule occurs by a mechanism of molecular mimicry, whereby the mercuric conjugate of cysteine mimics the amino acid cystine at the site of the transporter(s) that are responsible for the luminal uptake of cystine. A diagrammatic summary of the known and putative mechanisms involved luminal transport of inorganic mercury is presented in Fig. 9.1. In addition, Fig. 9.2 shows similarities, in a three-dimensional space-filled rendering, between the molecule cystine and the complex formed by two molecules of cysteine bonded to an inorganic mercuric ion.

Molecular mimicry is not a novel concept. In a recent review, Clarkson (1993) discusses the concept that mercury, and other metals, form complexes with biological molecules that mimic structurally endogenous molecules. For example, the complex formed between methylmercury and cysteine can mimic the amino acid methionine, which is thought to be a mechanism by which methylmercury gains entry into the central nervous system. Another example is that complexes formed between inorganic mercury or methylmercury and glutathione can mimic conjugates of glutathione or glutathione disulfide. The implication of a di-cysteinyl-mercury complex mimicking the amino acid cystine does, however, appear to be a novel addition to the purported species of molecules that are mimicked in the process of the transport of a toxic metal into an epithelial cell. It should be stressed that the mechanisms involved in the basolateral uptake of inorganic and organic forms of mercury also likely involve some form of molecular mimicry.

9.2.2.2 Mechanisms involved in the basolateral uptake of mercury

In addition to the uptake of mercury that occurs at the luminal membrane of proximal tubular cells, there is also uptake of mercuric ions at the basolateral membrane of these cells. Convincing evidence for the involvement of a basolateral mechanism came from a recent study in which the uptake and disposition of administered inorganic mercury were studied in rats in which glomerular filtration had been reduced to negligible levels in one or both kidneys by pretreating with mannitol in combination with ureteral ligation (Zalups and Minor, 1995). The findings from that study indicate clearly that a significant fraction of the renal uptake of mercury is linked to a basolateral mechanism. Additional data from other recent studies indicate that the primary mechanism involved in the basolateral uptake of inorganic mercury is linked to the activity of the organic anion transport system (Zalups and Lash, 1994; Zalups, 1995, 1997, 1998b; Zalups and Barfuss, 1995a, 1998a,b). Evidence for the involvement of the organic anion transport system comes primarily from *in vivo* experiments in which pretreatment with *p*-aminohippurate (which is a specific inhibitor of the renal organic anion transporter; Roch-Ramel *et al.*, 1992) was shown to inhibit significantly the renal tubular uptake of administered inorganic mercury in both normal rats and in rats whose ureters had been ligated prior to the administration of inorganic mercury (Zalups and Lash, 1994; Zalups, 1995, 1997, 1998b,c; Zalups and Barfuss, 1995a, 1996a). There are also data implicating the organic anion transporter in the basolateral uptake of organic mercuric compounds. These data show that the renal uptake, accumulation (Tanaka *et al.*, 1992) and toxicity (Ban and de Ceaurriz, 1988) of methylmercury are reduced significantly in mice pretreated with probenecid, which is another competitive inhibitor of the transport of organic anions in renal proximal tubules (Roch-Ramel *et al.*, 1992).

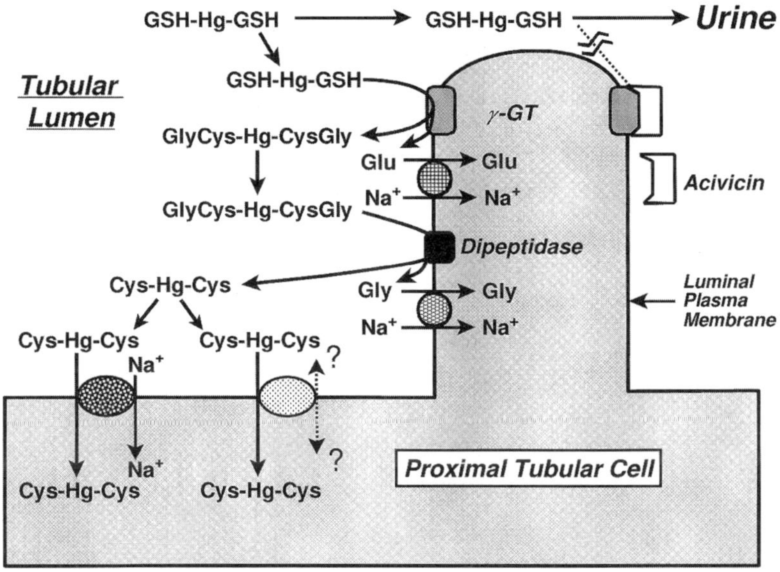

Figure 9.1 Scheme of likely mechanisms involved in the luminal uptake of mercuric ions in proximal tubular epithelial cells. Recent findings provide strong support for the existence of the luminal mechanisms displayed in this figure. Findings from several different *in vivo* and *in vitro* experiments indicate that near complete inhibition of renal γ-glutamyltranspeptidase (with acivicin) causes decreased luminal uptake of inorganic mercury and enhanced urinary excretion of mercury and glutathione (GSH). According to the presented scheme, the mechanism of action of the γ-glutamyltranspeptidase in the proximal tubular uptake of mercury probably involves the catalytic cleavage of the γ-glutamylcysteine bond on molecules of GSH bound to inorganic mercury (that could be bound to another molecule of GSH or some other nonprotein thiol, such as cysteine). After the action of the γ-glutamyltranspeptidase, the resulting cysteinylglycine remains bound to inorganic mercury and can potentially enter two pathways. One pathway involves the transport of a mercuric *S*-conjugate of cysteinylglycine into the proximal tubular cell by the sodium-dependent dipeptide transport system, and the second pathway involves the catalytic cleavage of the cysteinylglycinyl bond on molecules of cysteinylglycine bonded to a mercuric ion by the dipeptidase cysteinylglycinase located on the luminal membrane. Due to the abundance of dipeptidase activity on the luminal membrane of proximal tubular epithelial cells, it seems unlikely at present that any appreciable amount of transport of a mercuric conjugate of cysteinylglycine would occur under normal circumstances. Therefore, it seems more likely that the majority of the mercuric conjugates of cysteinylglycine enter the pathway in which the cysteinylglycine bond is cleaved by the dipeptidase. After the actions of the dipeptidase, cysteine from the cysteinylglycine remains bonded to the mercuric ion. At least some of the resulting mercuric *S*-conjugates of cysteine then appear to enter the proximal tubule by a mechanism involving molecular mimicry via transporters involved in the luminal absorption of cystine. Current evidence from our laboratories (Zalups and Barfuss) indicates that at least two transporters are involved in the luminal uptake of mercuric *S*-conjugates of cysteine. One of these transporters appears to be sodium dependent and the other sodium independent. It should be pointed out that sequential enzymatic degradation of GSH to cysteine while the sulfhydryl of cysteine is bonded to a mercuric ion has been demonstrated *in vitro*. The scheme presented in this figure is meant to summarize *some* of the current lines of thinking regarding the mechanisms involved in the luminal uptake of inorganic mercury, and it is in no way meant to exclude other possibilities, for they may very well exist. It is also possible, if not likely, that the same mechanisms involved in the proximal tubular uptake of inorganic mercury are also involved in the uptake of organic forms of mercury.

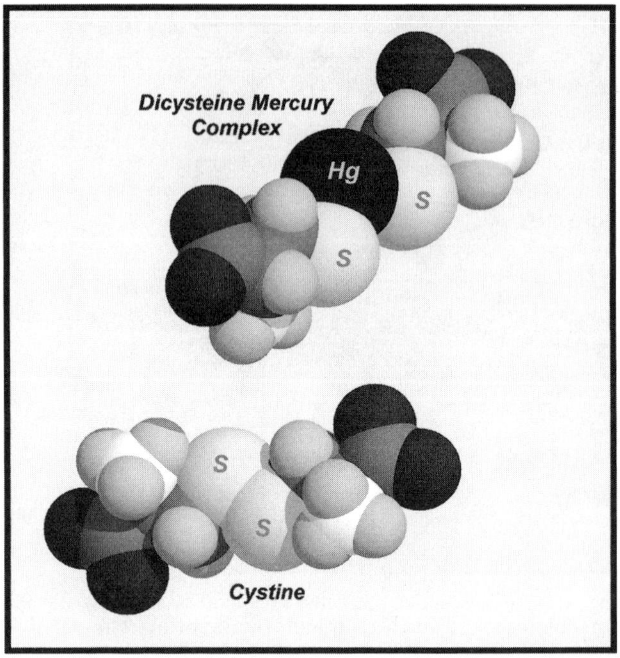

Figure 9.2 Three-dimensional space-fill rendering of the molecule cystine and the complex formed by two molecules of cysteine bonded to an inorganic mercuric ion. According to the current line of evidence, the complex formed by two molecules of cysteine bonded to an inorganic mercuric ion may be transported into proximal tubular epithelial cells at the luminal membrane by a mechanism involving molecular mimicry of cystine.

More recently, it has been shown that pretreatment with small aliphatic dicarboxylic acids, such as succinic, glutaric or adipic acid, also inhibits the renal (basolateral) uptake of inorganic mercury (Zalups and Barfuss, 1998b). In recent reviews on renal organic anion transport (Pritchard and Miller, 1993; Dantzler, 1996), evidence is provided that part of the organic anion transporter can be driven by an organic anion/dicarboxylic acid exchange. Some of this evidence indicates that intracellular generation of α-ketoglutarate (from normal metabolic processes) creates an intracellular chemical gradient favoring the movement of this dicarboxylic acid out of the cell. Moreover, it appears that α-ketoglutarate exits proximal tubular cells at the basolateral membrane by exchanging with organic anions. There is additional evidence indicating that α-ketoglutarate (and other dicarboxylic acids) enters back into the cell across the basolateral membrane via a symport involving the cotransport of sodium. This symport is driven by the sodium gradient generated by the (Na^+, K^+)-stimulated ATPase. Although it is not exactly clear by which mechanism the administration of succinic, glutaric or adipic acid inhibits the renal tubular uptake of inorganic mercury, it seems likely that an excess of any of these three dicarboxylic acids in the extracellular compartment creates competition for the sodium-dependent entry of α-ketoglutarate at the site of the dicarboxylic acid cotransporter situated in the basolateral membrane. This would likely cause a decrease in the intracellular concentration of α-ketoglutarate which, in turn, would decrease the chemical gradient favoring the movement of α-ketoglutarate out of the proximal tubular epithelial cell in exchange for the uptake of an organic anion from the plasma. The net result would be a decreased rate of uptake of organic anions, and a decreased rate of uptake of mercuric conjugates, presumably of cysteine or glutathione, that are transported at this site. Since dicarboxylic acids are themselves organic anions, an excess of these molecules in the extracellular fluid likely

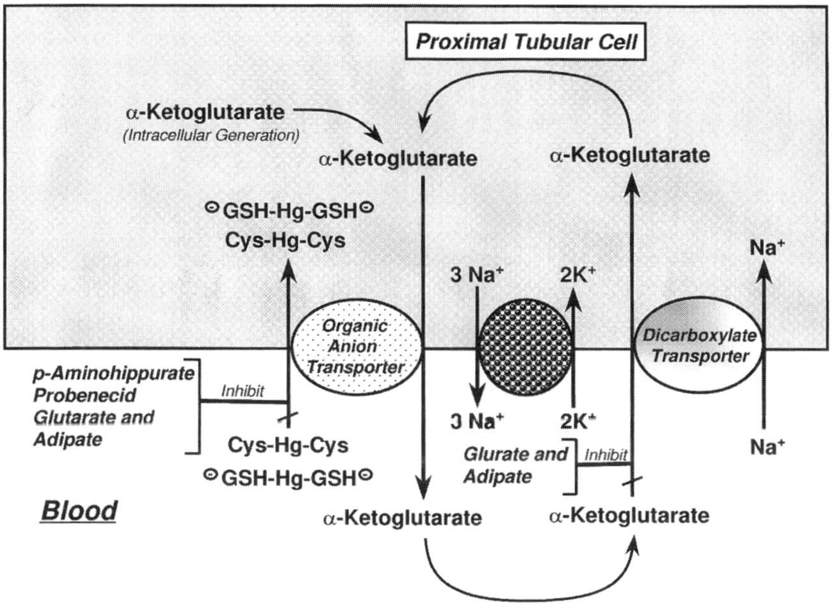

Figure 9.3 Diagram outlining the putative roles of both the organic anion and dicarboxylic acid transport systems in the basolateral uptake of inorganic mercury along the proximal tubule. On the basis of the current line of evidence regarding the organic anion transport system, it appears that intracellular generation of α-ketoglutarate (as a result of normal metabolic processes) creates a chemical gradient favoring the movement of this dicarboxylic acid out of the cell. Moreover, it appears that α-ketoglutarate exits proximal tubular cells at the basolateral membrane by exchanging with organic anions. Once in the extracellular compartment, α-ketoglutarate enters back into the cell across the basolateral membrane via a symport involving the cotransport of sodium, which is driven by the sodium gradient generated by the Na^+, K^+-ATPase. According to the scheme presented, inorganic mercury enters into proximal tubular epithelial cells (presumably as a conjugate of glutathione (GSH) or cysteine (Cys)) via the organic anion transport system in exchange for intracellular α-ketoglutarate. Evidence supporting this notion comes from the fact that the basolateral uptake of mercury can be inhibited by *p*-aminohippurate (PAH), probenecid and dicarboxylic acids (particularly glutarate). The scheme presented also shows that succinate, glutarate and adipate compete at the site of the dicarboxylic acid transporter for the basolateral uptake of α-ketoglutarate. It appears that both glutarate and adipate, but not succinate, can influence the basolateral uptake of mercury by theoretically acting at two sites. Experimental evidence indicates that these two dicarboxylic acids can compete for the uptake of substrates transported by the organic anion transporter and other dicarboxylic acids (such as α-ketoglutarate) transported by the sodium–dicarboxylic acid symport system. Succinate, on the other hand, appears to influence the basolateral uptake of mercury only by competing for the dicarboxylic acid transporter, which would cause a decrease in the intracellular concentration of α-ketoglutarate, resulting in a decrease in the activity of the organic anion transporter.

creates competition with whatever form of mercury is putatively transported by the organic anion transporter, and thus, results in decreased uptake of mercury at the basolateral membrane. There is evidence that adipate and glutarate, but not succinate or malonate, can exchange with intracellular α-ketoglutarate at the site of the organic anion transport system (Pritchard and Miller, 1993). Figure 9.3 presents some of these features in a putative model of the mechanisms involved in the basolateral uptake of inorganic mercury.

It should be stressed that after exposure to inorganic mercury, virtually all of the mercuric ions that end up in systemic circulation are bound to the cellular components of blood and various endogenous molecular ligands. Of the mercuric ions present in plasma, the majority of them are bound to albumin and other large proteins (Friedman, 1957; Cember *et al.*, 1968; Lau and Sarkar, 1979; Zalups and Lash, 1994). Since the organic anion transporter does not transport large proteins, small molecular conjugates of mercury are likely transported. As mentioned above, two conjugates of inorganic mercury that likely form in the body and that may be transported by the organic anion transport system are mercuric conjugates of glutathione and cysteine. It is established that formation of mercuric conjugates of glutathione and cysteine does occur systemically within certain organs, such as in the liver (Ballatori and Clarkson, 1985). A fraction of these conjugates, as well as those that may have formed in blood, are likely carried in the plasma to the kidneys. A link between the liver and some aspects of the renal uptake of inorganic mercury has already been demonstrated in our laboratories (Zalups and Barfuss, 1996a). The formation of mercuric conjugates of glutathione and cysteine in blood also seems probable, due to the fact that concentrations of glutathione and cysteine in the plasma (of rats) are around 10 μM (Lash and Jones, 1985a). As alluded to earlier, these concentrations of glutathione and cysteine in plasma provide a significantly large enough pool of nonprotein thiols to which mercuric ions can bind. Recent unpublished data from our laboratory also indicate that biliary ligation causes a marked reduction in the pool of both reduced glutathione and cysteine in the plasma within one hour after ligation, and that this reduction is linked to the decreased uptake of mercury in the kidneys.

We have also recently provided additional evidence linking the organic anion transport system to the uptake of inorganic mercury in the form of mercuric conjugates of thiol-containing molecules. While investigating the renal uptake of mercury when administered as a mercuric conjugate of cysteine, homocysteine or *N*-acetylcysteine, we demonstrated that virtually all of the renal tubular uptake of mercuric conjugates of *N*-acetylcysteine occurs at the basolateral membrane, and that the majority of this transport can be inhibited by pretreatment with *p*-aminohippurate. The negative charge on the molecules of *N*-acetylcysteine likely promotes the rapid transport of mercuric conjugates of *N*-acetylcysteine into proximal tubular cells at the site of the organic anion transporter. However, additional evidence indicates that this same negative charge prevents or impedes the uptake of these conjugates when they are present in the lumen (Zalups and Barfuss, 1998b), and thus promotes the rapid urinary excretion of mercury when it is in this form.

Since there is a net negative charge on molecules of glutathione, and since there is evidence for the transport of organic conjugates of glutathione across renal basolateral membranes (Lash and Jones, 1985b), it seems plausible that mercuric conjugates of glutathione are also transported at the basolateral membrane. Findings from a recent study in which mercuric conjugates of glutathione or cysteine were administered to rats that had undergone bilateral ureteral ligation support this contention. Despite the fact that cysteine has a net neutral charge at physiological pH, there is *in vitro* evidence indicating that some organic conjugates of cysteine are in fact taken up at the basolateral membrane of proximal tubular epithelial cells by the organic anion transporter (Lash and Anders, 1989; Dantzler *et al.*, 1995) One line of recent evidence indicating that mercuric conjugates of cysteine may also be taken up at the basolateral membrane comes from *in vitro* data obtained with basolateral-membrane vesicles isolated from the kidneys of rats. These data show that rates of association and transport of inorganic mercury in basolateral-membrane vesicles tends to be greater when the vesicles are exposed to mercuric conjugates of cysteine than when they are exposed to mercuric chloride (Zalups and Lash, 1997a). The hypothesis that basolateral uptake of mercuric conjugates of cysteine occurs is also supported indirectly by data obtained from rats whose ureters had been ligated and who were treated with inorganic mercury plus cysteine (Zalups, 1998b). These data show that bilateral ureteral ligation causes a reduction of approximately 50% in the renal burden of mercury in both control rats treated with a low, nontoxic dose of inorganic mercury alone and in rats co-administered this dose of inorganic mercury with a fourfold greater amount L-cysteine. These findings also show that the renal uptake of mercury is enhanced in the animals treated with inorganic mercury plus cysteine, relative to animals treated with inorganic mercury alone, while the relative intrarenal distribution of mercury remains similar in both groups of rats. The most reasonable explanation for these findings is that by injecting mercuric conjugates of cysteine, more of these conjugates, than normally are formed when inorganic mercury is administered alone, are made available at the site of the organic anion transporter (and possibly other basolateral transporters) to promote the uptake of mercury. Pretreatment with *p*-aminohippurate (PAH) also causes a significant decrease in the renal uptake of mercury in rats (whose ureters had been ligated) administered inorganic mercury plus cysteine (Zalups, 1998c). This finding further indicates that the organic anion transporter is likely involved in the basolateral uptake of mercuric conjugates of cysteine.

Although current experimental evidence is tending to point to mercuric conjugates of cysteine and glutathione being involved in the luminal and basolateral uptake of inorganic mercury along the proximal tubule after exposure to mercuric chloride, it is clear that other thiols, especially homologs of cysteine, can influence significantly the manner in which inorganic mercury is handled in the kidneys (Zalups and Barfuss, 1998b). For example, our data indicate that when inorganic mercury is bound to homocysteine, there is some uptake of mercury at the luminal membrane, but basolateral uptake of mercury predominates. In addition, our findings indicate that when inorganic mercury is bound to *N*-acetylcysteine, uptake of mercury occurs almost exclusively at the basolateral membrane. These patterns of uptake and accumulation of mercury are clearly different from those detected in rats treated with inorganic mercury alone or in combination with cysteine.

Findings from a couple of *in vitro* studies indicate that glutathione can be taken up by the basolateral membrane of proximal tubular epithelial cells as an intact tripeptide, by a mechanism that is sodium dependent (Lash and Jones, 1983, 1984). The findings in one study demonstrate that the uptake of glutathione in isolated vesicles of basolateral membranes from proximal tubules can be blocked by probenecid in a concentration-dependent manner (Lash and Jones, 1984), indicating that glutathione is taken up across the basolateral membrane of proximal tubular epithelial cells by the organic anion transport system. Additional findings from another study indicate that *S*-conjugates of glutathione, such as *S*-(1,2-dichlorovinyl)glutathione, are also transported by a probenecid-sensitive mechanism localized in the basolateral membrane of proximal tubular epithelial cells (Lash and Jones, 1985b). Since *S*-conjugates of glutathione and mercury are probably present in blood plasma, and since they mainly possess a net negative charge, it seems plausible (based on the findings from this and other studies) that some inorganic (and perhaps organic) mercury may gain entry into proximal tubular epithelial cells as a result of the transport of an *S*-conjugate of glutathione and mercury by the organic anion transport system.

One must also consider that mercuric ions are transported into proximal tubular epithelial cells across the basolateral membrane as *S*-conjugates of cysteine. Recent findings indicate that organic *S*-conjugates of cysteine (such as *S*-(1,2-dichlorovinyl)-L-cysteine) are taken up by proximal tubular epithelial cells by a sodium-dependent, probenecid- and PAH-sensitive organic anion transport system (Lash and Anders, 1989). Since both reduced glutathione and cysteine are in the plasma of rats at concentrations of approximately 10 μM (Lash and Jones, 1985b), there appears to be a large-enough pool of these ligands to which inorganic or organic mercuric ions can bind in the plasma. Formation of these conjugates in the plasma can be explained simply on the basis of the affinity of mercuric ions for the free sulfhydryl groups on these two ligands (especially the one on glutathione; Rabenstein, 1989).

Approximately 40% of the dose is normally taken by the total renal mass of rats during the initial hour after the intravenous injection of a nontoxic dose of inorganic mercury (Zalups and Lash, 1994; Zalups and Barfuss, 1995a, 1998a,b; Zalups, 1997) the current line of evidence indicates that approximately 50% of this renal burden of mercury can be attributed to a basolateral mechanism.

9.2.3 *Intracellular distribution of mercury*

Once inorganic mercuric ions gain entry in proximal tubular cells, it appears that they distribute throughout all intracellular pools (Madsen, 1980; Omata *et al.*, 1980; Baggett and Berndt, 1985; Houser and Berndt, 1988). Cellular fractionation studies using the renal cortex from rats treated with mercuric chloride acutely or chronically indicate that mercury distributes in nuclear, lysosomal, mitochondrial, brush border and supernatant fractions, with the nuclear fraction containing the greatest amount of mercury among all organelle fractions (Madsen, 1980; Madsen and Hansen, 1980). Similar findings have also been obtained in other studies using the renal cortex of normal and uninephrectomized rats treated with mercuric chloride (Baggett and Berndt, 1985; Houser and Berndt, 1988). In these studies, the cytosolic fraction was found to contain the greatest content of mercury. Interestingly, the relative specific content of mercury increased to the greatest extent in the lysosomal fraction when rats were made proteinuric with an aminoglycoside (Madsen, 1980) or when rats were treated chronically with mercuric chloride (Madsen and Hansen, 1980). Increases in the lysosomal content of mercury may reflect the fusion of primary lysosomes with endocytotic or cytosolic vesicles containing complexes of inorganic mercury bound to proteins.

9.3 Urinary excretion of mercury

Urinary and fecal excretion of mercury are the principal means by which humans and other mammals eliminate mercury from the body after exposure to elemental, organic or inorganic forms of mercury. In general, a greater fraction of a dose of mercury is excreted in the feces than in the urine (Rothstein and Hayes, 1960; WHO, 1991; Zalups *et al.*, 1987, 1988, 1991a, 1992, 1993a; Zalups and Lash, 1994). In rats, it has been shown that more than twice as much inorganic mercury is excreted in the feces than in the urine during the initial few days after exposure to a nontoxic dose of mercuric chloride (Rothstein and Hayes, 1960; Zalups *et al.*, 1987, 1988; Zalups and Lash, 1994). Generally less than 10% of the dose is excreted in the urine during this time. In one study, rats injected intravenously with a nontoxic dose of mercuric chloride excreted about 20% of the dose of mercury in the urine and 30% of the dose in the feces by the end of 54 days following injection (Rothstein and Hayes, 1960). Even less mercury is excreted in the urine after exposure to organic forms of mercury. In a recent study, both normal and uninephrectomized rats excreted only about 3% of the dose of mercury in the urine by the end of the initial 7 days following the intravenous injection of a low dose of methylmercury (Zalups *et al.*, 1992). By contrast, greater than 15% of the administered dose was excreted in the feces by these animals during the same period of time. In a recent study where seven male adult human subjects were administered a tracer amount of ^{203}Hg-labeled methylmercury, the cumulative fecal excretion of mercury over 70 days was much greater than the cumulative urinary excretion of mercury (Smith *et al.*, 1994). In fact, about 30% of the dose was excreted in the feces in 70 days while only about 4% of the dose was excreted in the urine.

At present, very little is known about the mechanisms involved in the urinary excretion of mercury. It is clear that inorganic and organic forms of mercury are taken up very avidly by proximal tubular epithelial cells. What is not clear, however, is to what extent filtered mercury is taken up, and whether any mercury is added to the luminal fluid through a secretory mechanism. There are several experimental maneuvers that cause the urinary excretion of mercury, mainly by decreasing the luminal absorption of mercury along the proximal tubule (or the rest of the nephron). For example, as mentioned earlier in this chapter, when renal γ-glutamyltransferase is inhibited, the urinary excretion of inorganic mercury increases tremendously, which actually serves as evidence of decreased luminal absorption and transport of mercury in the proximal tubule. Moreover, when inorganic mercury is administered as a mercuric conjugate of *N*-acetylcysteine (Zalups and Barfuss, 1998b), 2,3-dimercaptopropane-1-sulfonate (DMPS; Zalups, 1993c), 2,3-dimercaptosuccinic acid (DMSA; Zalups 1993c) or metallothionein, urinary excretion of mercury increases greatly due to the lack of luminal uptake of these mercuric conjugates. In general, it appears that when mercuric ions are bound to organic ligands possessing a net negative charge, the mercuric conjugates of these molecules are taken up poorly at the luminal membrane, and in turn, are excreted in the urine. More studies are clearly needed to better define the factors and mechanisms involved in the urinary excretion of mercury and mercury-containing compounds.

9.4 Molecular interactions and effects of mercury in renal epithelial cells

9.4.1 Effects of mercury on intracellular thiol metabolism and status

One major effect of mercury on intracellular thiol status is induction of, and binding to, metallothionein (Piotrowski *et al.*, 1974). Metallothioneins are a group of small proteins, having an approximate molecular weight of 6000–7000 Da, that contain multiple sulfhydryl groups and have the capacity to bind numerous metal ligands, including inorganic mercury, cadmium, zinc, copper, silver and platinum. Administration of a single, daily, nontoxic dose of mercuric chloride over several days has been shown to cause a near doubling in the concentration of metallothionein in the renal cortex or outer stripe of the outer medulla in rats (Zalups and Cherian, 1992a). Induction of synthesis of metallothionein in kidney has also been demonstrated in rats exposed to elemental mercury vapor over the course of several days (Cherian and Clarkson, 1976). The increase appears to be tissue-selective, as changes in hepatic metallothionein synthesis have not been demonstrated. Mercury vapor is converted into inorganic mercury, which is predominantly (>98%) recovered in kidney, suggesting that the induction of metallothionein by exposure to elemental mercury may be mediated by inorganic mercury. The induction of metallothioneins in the kidney by inorganic mercury likely first involves increased transcription of *MT1* and *MT2* genes via interaction of zinc-dependent metal transcription factor(s) (MTFs) and *cis*-acting DNA elements termed metal-responsive elements (MREs) resident in the promoter region of the metallothionein genes. Subsequently, there is post-transcriptional control of the translation of new messenger RNA into metallothionein I and II proteins (see Chapter 18 for additional details on induction of metallothionein by metals).

Inorganic and organic forms of mercury also cause changes in intracellular glutathione status and metabolism. These effects are observed acutely after short-term, single treatments, and are concentration dependent. Several reports, using both *in vivo* and *in vitro* systems, have demonstrated increases in intracellular contents of glutathione in renal tubular epithelial cells after administration of relatively low, toxic or subtoxic doses of either methylmercury (Woods *et al.*, 1992) or inorganic mercury (Chung *et al.*, 1982; Fukino *et al.*, 1986; Siegers *et al.*, 1987; Zalups and Lash, 1990; Lash and Zalups, 1992, 1993). At higher doses of inorganic mercury, however, decreases in renal content of glutathione, which are often substantial, are observed (Fukino *et al.*, 1984; Zalups and Lash, 1990; Lash and Zalups, 1992, 1993).

Dose-dependent effects of inorganic mercury on renal glutathione status have been demonstrated in male Sprague–Dawley rats that received one of several nontoxic or nephrotoxic intravenous doses of mercuric chloride (Zalups and Lash, 1990). At the level of the whole kidney, or in samples derived from the cortex or the outer stripe of the outer medulla, the subtoxic or the moderately nephrotoxic dose of mercuric chloride produced significant increases in the renal concentration of glutathione. This effect was most marked in the outer stripe of the outer medulla, where the concentration of glutathione increased by as much as 85%. There is toxicological significance in this finding since the outer stripe of the outer medulla is mainly where tubular injury induced by mercury is localized. At the highest nephrotoxic dose of mercuric chloride, the concentrations of glutathione in the renal cortex and outer stripe of the outer medulla were similar to those in controls.

Since the cellular content of glutathione is under feedback control, the large increases in the renal content of glutathione observed after treatment with inorganic mercury suggest that subtoxic or moderately toxic doses of inorganic mercury induce synthesis of glutathione via γ-glutamylcysteine synthetase (γ-GCS), which is the rate-limiting enzyme involved in the biosynthesis of glutathione. More recent data from the laboratories of Lash and Zalups (1993) support this hypothesis. They found that the activity of γ-glutamylcysteine synthetase was increased in renal proximal epithelial cells isolated from rats treated with inorganic mercury relative to that in proximal tubular cells isolated from control rats. Further support for the hypothesis that mercuric ions induce γ-glutamylcysteine synthetase in renal epithelial cells comes from a study by Woods *et al.* (1992), who showed that the mRNA for γ-glutamylcysteine synthetase increased (by 4.4 fold) in kidneys of male Fischer 344 rats treated with methylmercury hydroxide for 3 weeks. Therefore, both inorganic and organic forms of mercury, at subnephrotoxic doses, appear to induce synthesis of glutathione via the activity of γ-glutamylcysteine synthetase.

In addition to causing upregulation of γ-glutamylcysteine synthetase, inorganic mercury also alters, in a dose-dependent manner, the activity of other glutathione-dependent enzymes. The effects of inorganic mercury on these enzymes differ, depending on whether a subtoxic, a moderately toxic or a highly toxic dose is administered. Subtoxic doses of mercuric chloride apparently cause increases in the activities of glutathione disulfide (GSSG) reductase and glutathione peroxidase in isolated epithelial cells from both proximal tubular and distal tubular regions of the rat nephron (Lash and Zalups, 1993). In contrast, marked decreases in the activities of renal GSSG reductase and glutathione peroxidase have been observed in male rats treated chronically with a relatively high dose of mercuric chloride by one group of investigators (Addya *et al.*, 1984). They also found apparent adaptive increases in catalase activity. Similarly, others have found significant decreases in the activity of GSSG reductase after the administration of highly nephrotoxic doses of mercuric chloride (Chung *et al.*, 1982; Gstraunthaler *et al.*, 1983; Fukino *et al.*, 1984). While two groups of investigators found small (20–35%), but statistically significant, decreases in the activity of glutathione peroxidase (Chung *et al.*, 1982; Gstraunthaler *et al.*, 1983), another group did not see any change in the activity of this enzyme (Fukino *et al.*, 1984). It is important to keep in mind, however, that it becomes nearly impossible to interpret *in vivo* data obtained from renal tissue in which there has been extensive cellular injury and death, inasmuch as decreases in the content of an enzyme or molecule of interest (in samples of renal tissue) can be accounted for simply by the release and excretion of the cytoplasmic contents during the process of necrosis.

9.4.2 Role of lipid peroxidation and oxidative stress in mercury-induced renal cellular injury

Findings from several studies suggest that an important mechanism involved in renal cellular injury induced by either *in vivo* or *in vitro* exposure to inorganic or organic forms of mercury involves production of oxidative stress. The high affinity of mercuric ions for binding to thiols naturally suggests that the ensuing depletion of intracellular thiols (especially glutathione) either directly causes or predisposes proximal tubular cells to oxidative stress. Furthermore, other cellular antioxidants, including ascorbic acid and vitamin E, have been reported to be depleted in the kidneys of rats treated with mercuric chloride (Fukino *et al.*, 1984). Activity of several antioxidant enzymes also appears to be markedly diminished after *in vivo* exposure of rats to nephrotoxic doses of mercuric chloride. For example, it has been reported that administration of mercuric chloride to male Sprague–Dawley rats caused marked decreases in activity of superoxide dismutase, catalase, glutathione peroxidase, and glutathione disulfide reductase in the renal cortex (Gstraunthaler *et al.*, 1983).

Decreases in activities of these protective enzymes would be expected to increase the susceptibility of renal epithelial cells to oxidative injury. There has been some disagreement as to whether mercury itself causes oxidative injury or whether it merely makes the renal epithelial cells more sensitive to agents that produce oxidative stress. Fukino *et al.* (1984) found that thiobarbiturate reactants, which indicate occurrence of lipid peroxidation, were markedly increased in renal cortical homogenates from rats 12 hours after a subcutaneous injection of a nephrotoxic dose of mercuric chloride. Gstraunthaler *et al.* (1983) observed increases in formation of malondialdehyde in renal cortical homogenates obtained from mercuric chloride-treated rats (relative to homogenates generated from control rats treated only with cumene hydroperoxide). Since the two groups of rats were administered equivalent doses of mercuric chloride, concentration-dependence cannot be invoked to explain the difference in observed responses. Based on these findings, it appears that inorganic mercury can enhance the ability of other agents to induce lipid peroxidation.

There are intimate relationships between maintenance of normal renal function, renal cellular content of glutathione, cellular redox status, and the generation of ATP in mitochondria. These relationships served as an impetus for Lund *et al.* (1991) to investigate the role of mercury-induced oxidative stress in mitochondria of renal epithelial cells as a mechanism of mercury-induced renal cellular injury. More specifically, they investigated the effects of inorganic mercury on hydrogen peroxide production by isolated renal cortical mitochondria from rats. Depending on the supply and coupling-site-specificity of respiratory substrates, variable increases in the formation of hydrogen peroxide were observed; incubation of isolated mitochondria with 30 nmol mercuric chloride/mg protein increased formation of hydrogen peroxide fourfold at the ubiquinone–cytochrome *b* region and twofold at the NADH dehydrogenase region. Additionally, iron-dependent lipid peroxidation was increased 3.5-fold at the NADH dehydrogenase region and by 25% at the ubiquinone–cytochrome *b* region. Intra-mitochondrial glutathione was decreased in a time- and concentration-dependent manner by mercuric chloride. In fact, a 12 nmol/mg protein concentration of mercuric chloride completely depleted mitochondrial glutathione within 30 min, suggesting that targeting of mitochondrial glutathione by mercury may be responsible for the intra-mitochondrial oxidative stress. More recently, Lund *et al.* (1993) have demonstrated that production of hydrogen peroxide, depletion of glutathione and lipid peroxidation increased in mitochondria isolated from renal cortical homogenates of rats treated *in vivo* with mercuric chloride after addition of appropriate respiratory substrate. These findings support *in vitro* data and lead one to suggest that mercury-induced oxidative stress within mitochondria is an important mechanism involved in renal tubular injury induced by mercury.

9.4.3 Effects of mercury on renal mitochondrial function

As described above, Lund *et al.* (1991, 1993) have demonstrated that inorganic mercury interferes with respiratory function, leading to enhanced production of hydrogen peroxide, particularly at coupling site II of the electron transport chain. Furthermore, the demonstration of oxidative stress in mitochondria isolated from kidneys after *in vivo* treatment of rats with mercuric chloride indicates that an oxidative stress localized to the mitochondria may be responsible for mercury-induced inhibition of various energy-dependent processes in the kidneys.

In an earlier series of studies, Weinberg *et al.* (1982a,b) compared the effects of mercuric chloride on mitochondrial function *in vitro* after either *in vivo* or *in vitro* treatment with mercuric chloride. When mitochondria were isolated from male Sprague–Dawley rats and then treated *in vitro* with inorganic mercury (Weinberg *et al.*, 1982a), a marked uncoupling of respiration (i.e. increase in state 4 rate of oxygen consumption) and a significant decrease in the rate of substrate-stimulated respiration (i.e. state 3 respiration) were observed. Additionally, atractyloside-insensitive ADP uptake and both basal and Mg^{2+}-activated oligomycin-sensitive ATPase activities were markedly increased by inorganic mercury. These *in vitro* effects occurred with a threshold concentration of mercuric chloride of 2 nmol/mg protein. Similarly, if renal cortical mitochondria were isolated from rats treated with mercuric chloride *in vivo* (5 mg/kg, subcutaneously), the most prominent effects were inhibition of ADP uptake and decreases in the rates of state 3 and uncoupler-stimulated respiration (Weinberg *et al.*, 1982b). These effects were not attributed to interaction of mercury with mitochondria during the isolation procedure. However, with both *in vivo* and *in vitro* treatment, inorganic mercury was not readily washed out of mitochondria, suggesting complex formation between mercuric ions and mitochondrial thiols, possibly protein sulfhydryl groups.

Chavez and Holguin (1988) and Chavez *et al.* (1991) also found uncoupling of mitochondrial respiration after either *in vivo* or *in vitro* treatment with mercuric chloride. Consistent with this finding, they found that inorganic mercury induced calcium efflux from mitochondria, oxidation of pyridine nucleotides and a collapse of the membrane potential. In support of the mechanism of mercury-induced mitochondrial injury involving complex formation between mercuric ions and mitochondrial sulfhydryl groups, Chavez and Holguin (1988) reported that inorganic mercury bound to mitochondrial protein in a concentration-dependent pattern, with saturation at approximately 9 nmol mercuric chloride/mg protein.

Jung *et al.* (1989) employed ATP depletion by different chemical agents in microdissected nephron segments to localize the nephron site specificity of injury. They found that 1 μM mercuric chloride produced a significant depletion of intracellular ATP content only in S2 segments; nephron segments derived from the other portions of the proximal tubule (i.e. S1 or S3) or other regions of the nephron, such as the distal convoluted tubule or the medullary thick ascending limb of the loop of Henle, were not as sensitive to ATP depletion after incubation with inorganic mercury. This pattern agrees with histopathological data, which demonstrates that the pars recta of the proximal tubule is the primary target of inorganic mercury. These data also support the conclusion that renal mitochondria are early intracellular targets of inorganic mercury. This is logical considering the extremely high content of sulfhydryl-containing proteins in both the mitochondrial matrix and the mitochondrial inner membrane.

In a recent study, Zalups *et al.* (1993b) studied the accumulation and toxicity of inorganic mercury in suspensions of isolated segments of renal proximal tubules from the rabbit. Incubation of isolated proximal tubules with mercuric chloride, in the absence of extracellular thiols, caused a marked time- and concentration-dependent inhibition of nystatin-stimulated oxygen consumption, demonstrating mitochondrial toxicity in an intact *in vitro* renal cellular model. Furthermore, inhibition of oxygen consumption by mercuric chloride preceded the development of irreversible cellular injury, as assessed by release of lactate dehydrogenase (LDH) from the tubular segments, suggesting that inhibition of cellular energetics is a critical component of the nephrotoxic response to inorganic mercury.

9.4.4 Effects of mercury on intracellular distribution of calcium ions

Inorganic mercury induces efflux of calcium ions from renal mitochondria of rats both *in vivo* and *in vitro* (Chavez and Holguin, 1988; Chavez *et al.*, 1991). The importance of maintaining appropriate intracellular concentrations of calcium for proper cellular function is well documented, suggesting that the prominent effects of mercury on mitochondrial calcium status may play an important part in the acute nephropathy induced by mercury.

Smith *et al.* (1987) used primary cultures of renal tubular cells from rabbits that were mostly of proximal tubular origin as an *in vitro* model system to study the effects of inorganic mercury on the intracellular distribution of ionic calcium. They employed the fluorescent dye fura-2 to quantitate the cytosolic content of free ionized calcium. Treatment of cells with low (2.5–10 μM) concentrations of mercuric chloride produced two to tenfold increases in the intracellular content of calcium. In contrast to this, exposure of cells to higher (25–100 μM) concentrations of mercuric chloride produced an initial, rapid ten to twelvefold increase in intracellular calcium, which quickly returned to levels that were about twice those in control cells. This was subsequently followed by a second, more gradual increase in the intracellular content of calcium that was dependent on the presence of extracellular calcium. Cytotoxicity was also associated with this phase of increase in the intracellular content of calcium and was similarly dependent on the presence of extracellular calcium. The increases in the cytosolic content of calcium that were independent of extracellular concentrations of calcium were primarily due to release of intracellular calcium ions from nonmitochondrial intracellular stores, presumably derived from the endoplasmic reticulum. The subsequent decrease in intracellular calcium content may be due to buffering processes, such as uptake though the microsomal Ca^{2+},Mg^{2+}-ATPase or through the mitochondrial uniporter. The dependence of the slow, late phase of increase in cytosolic calcium content on extracellular calcium with higher concentrations of inorganic mercury suggests that nonlethal effects of inorganic mercury in renal cells are associated with redistribution of intracellular stores of calcium, but that toxic effects are associated with mercury-induced changes in permeability of the plasma membrane.

9.4.5 Alterations in plasma membrane (Na^+, K^+)-stimulated ATPase induced by mercury

Cellular plasma membranes contain a large number of proteins containing sulfhydryl groups that are critical for enzymatic activity and membrane structure (Rothstein, 1970). Among these is the (Na^+, K^+)-stimulated ATPase located on the basolateral membrane of epithelial cells in both the proximal and distal regions of the nephron, which is markedly inhibited by alkylation or oxidation of its sulfhydryl group. Anner and colleagues (Anner and Moosmayer, 1992; Anner *et al.*, 1992; Imesch *et al.*, 1992) have conducted a detailed series of studies on the interaction between mercury-containing compounds and purified and reconstituted ATPase protein from the renal outer medulla of the rat, rabbit and sheep. To determine the molecular details of the interaction between mercury-containing compounds and the (Na^+, K^+)-stimulated ATPase, studies had to be done with purified and reconstituted enzyme rather than intact renal epithelial cells or renal tubules.

Anner *et al.* (1992) showed that a number of mercury-containing compounds, including mercuric chloride, mersalyl, and *p*-mercuribenzene-sulfonic acid, potently inhibited the activity of the ATPase by binding to a site distinct from that at which the cardiac glycosides (e.g. digoxin and ouabain) bind. The binding of inorganic mercury was concentration dependent and was modulated by addition of mercury chelators, such as EDTA or 2,3-dimercapto-1-propanesulfonic acid, indicating that inorganic mercury binding to the enzyme is reversible.

Imesch *et al.* (1992) showed that the inactivation of the (Na^+, K^+)-stimulated ATPase by mercuric chloride (0.1–100 μM) apparently loosens the interaction between the α- and β-subunits of the ATPase molecule, thereby altering the sensitivity of the enzyme to extracellular drugs, hormones and antibodies.

Moreover, Anner and Moosmayer (1992) showed that the binding of inorganic mercury to the (Na^+, K^+)-stimulated ATPase molecule occurs primarily at the cytosolic surface. Binding of mercury was closely correlated with inhibition of uptake of ^{86}Rb, indicating that the metal-binding site is critical to the active transport function of the ATPase.

An important extension of these studies will be to design experiments to investigate effects of mercury on (Na^+, K^+)-stimulated ATPase function in more intact renal systems, such as isolated perfused tubule segments or isolated cells. Considering the potency of the interaction and the fact that the plasma membrane is a very early target site for mercury, it is likely that this interaction will be important in the mechanism of mercury-induced renal cellular injury. It is also likely that sulfhydryl groups on other membrane proteins, particularly those in the epithelial cells lining the proximal tubule, interact with mercury and may play a role in the nephropathy induced by mercury.

9.4.6 Influence of mercury on heme metabolism

Exposure to mercury *in vivo* has also been shown to induced porphyrinuria (Woods *et al.*, 1990a). The porphyrinogenic properties of mercury-containing compounds have been initially attributed to metal-induced alterations in the regulation of enzymes involved in heme biosynthesis or degradation in target tissues. However, since the magnitude of porphyrin excretion during prolonged exposure to either methylmercury or inorganic mercury is greater than can be accounted for by changes in heme metabolism alone, Woods *et al.* (1990a,b) needed to invoke alternative biochemical mechanisms to explain their findings. They showed that mercuric ions promoted free-radical-mediated oxidation of reduced porphyrins. The mechanism involved depletion or interference of normal antioxidants in renal epithelial cells, such as endogenous thiols like glutathione. Furthermore, the ability of inorganic mercury and glutathione to react with endogenously produced reactive oxygen metabolites from both hepatic and renal mitochondria from rats was correlated with porphyrinogen oxidation.

An important clinical application of this effect of mercury is illustrated in a study by Bowers *et al.* (1992), who evaluated patterns of urinary excretion of porphyrin in male Fischer 344 rats as a diagnostic tool to assess exposure to inorganic mercury or methylmercury. Evaluation of the urinary excretion of porphyrins is a noninvasive method that can be applied to human populations suspected of being exposed to mercury-containing compounds (Woods *et al.*, 1993).

9.4.7 Expression of stress proteins after exposure to mercury

Various environmental stimuli, including toxic chemicals, increase the synthesis of a class of proteins known as stress proteins. Goering *et al.* (1992) evaluated the effect of a nephrotoxic dose of mercuric chloride (1 mg/kg) on patterns of protein synthesis in the kidneys of male Sprague–Dawley rats. Enhanced *de novo* synthesis of 70 and 90 kDa molecular mass proteins were detected as early as 2 hours after exposure to inorganic mercury, and maximal increases in protein levels were observed at 4–8 hours post-treatment. By 16 hours after injection, rates of synthesis of the stress proteins decreased back towards basal levels. Changes in protein expression also occurred in liver, but were of smaller magnitudes and were not observed until 16–24 hours post-injection.

Goering *et al.* (1992) concluded that alterations in expression of stress proteins precede overt renal injury and are target-organ-specific, suggesting that they may serve as biomarkers of renal injury. Furthermore, once the biological function(s) of these proteins are identified, a more complete understanding of the early effects of mercury can be obtained.

9.5 Renal toxicity of mercury

All forms of mercury are nephrotoxic *in vivo* (Magos and Clarkson, 1977; Zalups and Lash, 1994), although the inorganic forms of mercury are far more acutely nephrotoxic (Magos and Clarkson, 1977; Zalups and Lash, 1994). With organic mercuric compounds, multiple exposures to relatively large amounts of these compounds are generally required to induce renal injury (Chang *et al.*, 1973; Magos *et al.*, 1985; McNeil *et al.*, 1988). By contrast, renal injury induced by inorganic mercury is generally expressed fully during the initial 24 hours following exposure, and can be induced in rats with a single dose of as little as 1.5 μmol mercury/kg (Zalups and Diamond 1987; Zalups *et al.*, 1988). It should be pointed out, however, that rats tend to be more vulnerable to the nephrotoxic effects of inorganic mercury than New Zealand white rabbits or several strains of mice (Zalups, unpublished findings). We have also found some strain differences in the severity of the nephropathy induced by inorganic mercury in rats (Zalups, unpublished findings).

In rats, the oral LD_{50} for inorganic mercury, in the form of mercuric chloride, has been reported to be in the range 25.9–77.7 mg/kg (Kostial *et al.*, 1978). A slightly lower range of doses of inorganic mercury (10–42 mg/kg), in the form of mercuric chloride, has been estimated to be fatal in humans (Gleason *et al.*, 1957). In one study of human poisoning with mercuric chloride, nine patients died after ingesting a single dose of inorganic mercury ranging from 29 mg/kg to greater than 50 mg/kg (Troen *et al.*, 1951; WHO, 1991). It should be pointed out that death due to ingestion of a single dose of inorganic mercury is generally due to multiple effects. In addition to acute renal failure, cardiovascular collapse, shock and severe gastrointestinal damage and bleeding are contributing causes of death.

9.5.1 Site of tubular injury induced by mercury

It is well established that the pars recta (straight segment) of the proximal tubule (particularly the portion at the junction of the cortex and outer medulla) is the segment of the nephron that is most vulnerable to the toxic effects of both inorganic and organic forms of mercury (Cuppage and Tate, 1967; Gritzka and Trump, 1968; Fowler, 1972; Ganote *et al.*, 1974; Zalups and Diamond, 1987; Zalups *et al.*, 1988; Zalups, 1991a). Depending on the severity of the nephropathy induced by mercury, cellular injury and necrosis can occur along the entire length of the pars recta, from just underneath the capsule to the junction of the outer and inner stripes of the outer medulla. Convoluted portions of proximal tubules and sometimes distal segments of the nephron can be involved when the nephropathy is very severe. However, there is no direct evidence indicating that mercury has direct *in vivo* toxic effects on segments distal to the proximal tubule, and thus, the involvement of these segments may be due to secondary effects elicited by severe damage to the pars recta of proximal tubules. Further research is clearly needed to clarify this issue.

It is interesting that, in contrast to the effects of mercury *in vivo*, all three segments (S1, S2 and S3) of the proximal tubule (of the rabbit) become intoxicated with either inorganic mercury or methylmercury when the mercury-containing compounds are perfused through the lumen of each of these segments *in vitro* (Barfuss *et al.*, 1990; Zalups *et al.*, 1991a; Zalups and Barfuss, 1993). The differences between the *in vivo* and *in vitro* findings are somewhat perplexing since all segments of the proximal tubule accumulate mercury under both experimental conditions. Another interesting difference between the *in vivo* and *in vitro* situation is that, *in vitro*, organic mercury (specifically methylmercury) is more toxic to proximal tubular epithelial cells than inorganic mercury. This has been demonstrated in primary cultures of proximal tubular epithelial cells (Aleo *et al.*, 1987, 1992) and in isolated perfused segments of the proximal tubule (Zalups and Barfuss, 1993).

9.5.2 Markers of renal cellular injury and impaired renal function

A number of methods have been used to detect renal tubular injury induced by mercury. One noninvasive method that has been employed frequently is measuring the urinary excretion of a number of cellular enzymes (Ellis *et al.*, 1973; Planas-Bohne, 1977; Stroo and Hook, 1977; Kirschbaum, 1979; Buchet *et al.*, 1980; Price, 1982; Stonard *et al.*, 1983; Gottelli *et al.*, 1985; Zalups and Diamond, 1987). The rationale for using the urinary excretion of cellular enzymes as an indicator of renal cellular injury is based on the close association between renal cellular necrosis and enzymuria. After renal epithelial cells have undergone cellular necrosis, most, if not all, of the contents of the necrotic epithelial cells, including numerous cellular enzymes, are released into the tubular lumen and are excreted in the urine. The usefulness of any particular cellular enzyme as a marker of renal cellular injury or necrosis depends on the stability of the enzyme in urine, whether the enzyme or the activity of the enzyme is greatly influenced by the toxicant that is being studied and the subcellular localization of the enzyme relative to the subcellular site of injury.

During the early stages of the nephropathy induced by mercury, prior to tubular necrosis, cells along the proximal tubule undergo a number of degenerative changes and begin to lose some of their luminal (brush-border) membrane (Zalme *et al.*, 1976). Evidence from several studies shows that the urinary excretion of the brush-border enzymes, alkaline phosphatase and γ-glutamyltransferase (GGT), increases during the nephropathy induced by mercury-containing compounds (Price, 1982; Gottelli *et al.*,

1985; Zalups *et al.*, 1988; 1991b). When tubular injury becomes severe and necrosis of tubular epithelial cells is apparent, then the urinary excretion of a number of other intracellular enzymes, such as lactate dehydrogenase (LDH), aspartate aminotransferase (AST), alanine aminotransferase (ALT) and *N*-acetyl-D-glucosaminidase (NAG), increases (Planas-Bohne, 1977; Zalups and Diamond, 1987; Zalups *et al.*, 1988, 1991b; WHO, 1991; ATSDR, 1994).

After a significant number of proximal tubules have become functionally compromised by the toxic effects of mercury, the capacity for the reabsorption of filtered plasma solutes and water is greatly diminished. As a consequence of this diminished absorptive capacity, there is an increase in the urinary excretion of both water and a number of plasma solutes, such as glucose, amino acids, albumin and other plasma proteins (Price, 1982; Zalups and Diamond, 1987; Diamond, 1988; Zalups *et al.*, 1988). In a recent study of workers exposed to mercury vapor, it was demonstrated that increased urinary excretion of Tamm–Horsfall glycoprotein and tubular antigens and decreased urinary excretion of prostaglandin E_2 and F_2 and thromboxane B_2 can also be used as indices of renal pathology induced by mercury (Cardenas *et al.*, 1993).

In two reports, the urinary excretion of mercury (factored by the total renal mass) was demonstrated (in normal and uninephrectomized rats) to correlate very closely with the level of injury in pars recta segments of proximal tubules during the acute nephropathy induced by low toxic doses of inorganic mercury (Zalups and Diamond, 1987; Zalups *et al.*, 1988). In these reports, the urinary excretion of mercury was shown to correlate with the histopathological scoring of injury to the pars recta of proximal tubules and increased urinary excretion of albumin, total protein and the cellular enzymes LDH, GGT and NAG (Zalups and Diamond, 1987). Overall, it appears that as the level of renal injury increases in the kidneys, there is a corresponding increase in the urinary excretion of mercury. Other nephrotoxic agents have also been shown to decrease the retention of mercury in the kidney and to increase the excretion of mercury in the urine (Clarkson and Magos, 1967; Magos and Stoychev, 1969; Trojanowska *et al.*, 1971), presumably by causing the release of mercury from renal epithelial cells undergoing necrosis and decreased uptake. Although the urinary excretion of mercury appears to correlate well with the level of acute renal injury induced by mercuric chloride, there does not appear to be a close correlation between the severity of renal injury and the renal concentration or content of mercury (Zalups and Diamond, 1987; Zalups *et al.*, 1988).

When renal tubular injury becomes severe during the nephropathy induced by mercury, the concentration of creatinine in plasma increases due to a decrease in glomerular filtration rate (GFR) (Barenberg *et al.*, 1968; McDowell *et al.*, 1976; Zalups *et al.*, 1991b). The mechanisms responsible for the decreased GFR are not known at present, but are likely complex and involve are number of factors. In addition to causing decreases in GFR, mercury causes the fractional excretion of sodium and potassium to increase (McDowell *et al.*, 1976). These functional changes likely reflect a significant decrease in the number of functioning nephrons, inasmuch as similar changes occur in rats and mice when their total renal mass has been reduced surgically by approximately 75% (Zalups *et al.*, 1985; Zalups, 1989; Zalups and Henderson, 1992). As part of the severe nephropathy induced by mercury, blood urea nitrogen (BUN) also increases as plasma creatinine increases, due to the significant decrease in glomerular filtration rate. Thus, the measure of plasma creatinine or BUN or both may be used as an indicator of impaired renal function induced by mercury (McDowell *et al*., 1976). However, it is preferable to use the clearance of creatinine or inulin over measurement of BUN as an index of renal function. BUN can be elevated by some nonrenal causes, and therefore is not as sensitive an indicator of renal function. After exposure to high doses of mercury, an oliguric or anuric acute renal failure ensues. The factors that lead to acute renal failure are complex, involving multiple systems. Clearly, further research is needed to better understand the mechanisms involved in the induction of acute renal failure following exposure to mercury.

9.5.3 Mercury-induced renal autoimmunity

There is evidence from studies with rabbits, inbred Brown-Norway rats, and a cross between Brown-Norway and Lewis rats that multiple exposures to inorganic mercury can lead to the production of antibodies against the glomerular basement membrane and results in an immunologically mediated membranous glomerular nephritis (Bigazzi, 1988, 1992). This glomerular nephropathy is characterized by the binding of antibodies to the glomerular basement membrane, followed by the deposition of immune complexes in the glomerulus (Sapin *et al*., 1977; Druet *et al*., 1978; Roman-Franco *et al*., 1978). There is also evidence from studies with several strains of both mice and rats that repeated exposures to inorganic mercury can lead to the deposition of immune complexes in the mesangium and along the glomerular basement membrane, which leads to an immune-complex glomerulonephritis (Weening *et al*., 1981; Enestrom and Hultman, 1984; Bigazzi, 1988; Hultman and Enestrom, 1992). Whether mercury can induce an autoimmune glomerulonephritis in humans is not clear at present. It should be pointed out that a great majority of the cases of glomerulonephritis in humans are classified as idiopathic. It is possible that some of the idiopathic forms of glomerulonephritis could be induced by exposure to mercury or other environmental or occupational toxicants.

The autoimmunity induced by mercury likely reflects some complex effects of mercuric ions on cell signaling and gene-expression events in immune cells, such as in monocytes and lymphocytes. For example, we have recently demonstrated that exposure of human monocytes to low, nontoxic doses of inorganic mercury causes a rapid suppression of activation of signaling events that are normally induced in these cells by lipopoylsaccharide or phorbol ester (Koropatnick and Zalups, 1997). See Chapter 18 for additional details on some of the effects of mercuric ions on cell signaling and gene expression in human leukocytes.

9.6 Factors that modifiy the renal toxicity of mercury

9.6.1 Influence of intracellular thiols on the renal accumulation and toxicity of mercury

Two major intracellular thiols, glutathione and metallothionein, appear to be important in regulating the renal accumulation of mercury and ultimately, the renal susceptibility to mercury-induced renal cellular injury. It is likely that other molecules within cells, including the large supply of nonmetallothionein, protein sulfhydryls, play some role in the renal cellular accumulation and toxicity of mercury.

Intracellular concentrations of glutathione can be manipulated readily within a relatively brief period of time. Several investigators have employed diethyl maleate to conjugate glutathione, thereby lowering the amount of intracellular glutathione available to interact with mercuric ions. Berndt and colleagues (Berndt *et al.*, 1985; Baggett and Berndt, 1986; Zalups and Lash, 1997b) and Johnson (1982) showed that depletion of intracellular glutathione or nonprotein thiols was accompanied by decreases in the renal accumulation of inorganic mercury in animals treated with mercuric chloride. In the studies by Berndt and colleagues (Berndt *et al.*, 1985; Baggett and Berndt, 1986), depletion of intracellular glutathione also resulted in increased severity of renal injury induced by treatment with mercuric chloride. Zalups and Lash (1990) also found a close correlation between intrarenal concentrations of glutathione and accumulation of inorganic mercury. There are some conflicting findings on the effects of diethyl maleate from the laboratory of Girardi and Elias (1991), who have reported increases in renal accumulation of inorganic mercury in mice treated with diethyl maleate. In a more recent study, Zalups and Lash (1997b) demonstrated, in rats, that acute depletion of glutathione in the kidneys and liver by treatment with diethyl maleate causes significant decreases in the renal uptake and accumulation of mercury during the initial hour after the administration of a low, nontoxic dose of mercuric chloride. Interestingly, while the renal accumulation of mercury decreases following treatment with diethyl maleate, the net hepatic accumulation of mercury increases. Thus, depletion of renal and hepatic glutathione has mixed effects on the disposition of mercury.

In other experiments, Tanaka-Kagawa *et al.* (1993) lowered the intracellular content of glutathione in the kidneys of mice by administration of buthionine sulfoximine, which is a potent inhibitor of the intracellular synthesis of glutathione, and then inhibited degradation of glutathione with the potent inhibitor of γ-glutamyltransferase, acivicin, and observed no changes in accumulation of either inorganic mercury or methylmercury, when compared with control animals. Zalups and Lash (1997b) have also shown in rats that acute depletion of renal glutathione with buthionine sulfoximine does not affect the early aspects of the accumulation of inorganic mercury in the kidneys. However, when the disposition of mercury is studied 24 hours after treatment with inorganic mercury, then pretreatment with buthionine sulfoximine does have a signifcant effect on the disposition of mercury in the kidneys (unpublished findings).

In studies where acivicin was used to inhibit γ-glutamyltransferase, Berndt *et al.* (1985) and Zalups (1995) showed in rats, and Tanaka *et al.* (1990) showed in mice, that the urinary excretion of glutathione and inorganic mercury increased after inhibition of glutathione degradation. Tanaka-Kagawa *et al.* (1993) also found that the urinary excretion of inorganic mercury increased while the renal accumulation of either inorganic mercury or methylmercury decreased.

Tanaka *et al.* (1990) also found that when mice were pretreated with 1,2-dichloro-4-nitrobenzene, to deplete the hepatic content of glutathione, prior to injection of mercuric chloride, there was a marked reduction in the renal accumulation of mercury and a significant decrease in the level of renal cellular injury induced by inorganic mercury. These findings tend to suggest that hepatically synthesized glutathione and the activity of γ-glutamyltransferase are involved in the renal uptake of mercury. Additional findings from a set of recent studies in which bile flow was either diverted or prevented from entering the small intestine of rats demonstrates that some aspect of hepatic function is linked to a component of the renal uptake and accumulation of mercury (Zalups, 1998a; Zalups and Barfuss, 1996a).

Increases in the intracellular contents of glutathione and other nonprotein thiols can be achieved by several means. Girardi and Elias (1991, 1993) treated mice with *N*-acetylcysteine and found that intracellular accumulation of inorganic mercury was lower in both kidney and liver as compared with animals possessing normal intracellular thiol status. Inasmuch as transport of inorganic mercury with glutathione has been established in liver (Ballatori, 1991) and higher intracellular contents of glutathione would be expected to provide increased numbers of ligands for binding to inorganic mercury, the seemingly paradoxical results of Girardi and Elias (1991) and the discrepancies described above suggest that the intrarenal and intracellular disposition of mercury-containing compounds must be regulated by a more complex array of factors than the availability of thiol ligands on glutathione.

Additional experiments by Tanaka-Kagawa *et al.* (1993), involving changes in intracellular metallothionein levels, may provide some clarification of the contradictory reports of glutathione-depletion and the accumulation of mercury. These investigators reported that induction of renal metallothionein with $Bi(NO_3)_3$ diminished the ability of acivicin to decrease intrarenal accumulation of either inorganic mercury or methylmercury. They interpreted this as indicating that inorganic mercury or methylmercury that is bound to ligands other than metallothionein in renal cells can be readily secreted into the tubular lumen with intracellular glutathione. Other studies (Fukino *et al.*, 1984, 1986; Zalups and Cherian, 1992a,b) have documented increases in intrarenal mercury accumulation and decreases in nephrotoxicity induced by either organic or inorganic mercury after induction of renal metallothionein. Thus, it appears that there is a complex interplay between protein and nonprotein thiols that ultimately determines the renal disposition of mercury.

9.6.2 Modulation of renal accumulation and toxicity of mercury by extracellular thiols

While manipulation of intracellular thiols is sometimes employed therapeutically to alter the accumulation of mercury and to modulate effects of mercury once it enters target sites, administration of thiol-containing compounds can be applied prior to, or simultaneously with, mercury-containing compounds to alter the pharmacokinetics and pharmacodynamics of mercury. 2,3-Dimercaptopropane-1-sulfonate (DMPS) and *meso*-2,3-dimercaptosuccinic acid (DMSA) are becoming two of the more commonly used metal chelators employed as antidotes for mercury poisoning, and their chemical and pharmacological properties have been reviewed recently (Aposhian, 1983; Aposhian and Aposhian, 1990). Examples of some of their most distinguishing features are that, in contrast to the earlier chelator dimercaprol (also known as British anti-Lewisite or BAL), DMPS and DMSA are fairly nontoxic, are very water soluble, are not very lipid soluble, and are effective if administered orally. The two compounds are quite versatile, being capable of chelating arsenic, lead, cadmium and mercury. However, they differ in potency and specificity; DMSA is generally the more effective of the two in chelating organic mercury, whereas DMPS is the more effective of the two in chelating inorganic mercury (Zalups, 1993b). Additional extracellular thiol reagents that have been employed clinically for removal of methylmercury are D-penicillamine and

N-acetyl-DL-penicillamine (Aposhian, 1983). Some of the reported variabilities in effectiveness and potency of the various chelators of mercury may be attributed to species differences, routes of administration and doses of chelators given.

Zalups *et al.* (1991b) demonstrated dose-dependent protection in rats from the nephropathy induced by inorganic mercury. Their data suggest that the protective effects of DMPS may be attributed to decreases in the renal mercury burden and increases in urinary excretion of mercury. Furthermore, Maiorino *et al.* (1991) demonstrated a high correlation between effectiveness of DMPS and urinary excretion of both inorganic mercury and DMPS in humans. In a recent study (Zalups, 1993b) the same dose of DMPS or DMSA, administered to rats 24 hours after the animals were given an intravenous nontoxic dose of mercuric chloride, was shown to reduce the renal burden of mercury significantly during the subsequent 24 hours after treatment with the respective chelator. Treatment with DMPS caused a reduction in the renal burden of mercury by more than 80%, while DMSA caused a reduction in the renal burden of mercury by about 50%. These findings indicate that DMPS is more effective (on a per mole basis) in reducing the renal burden of mercury after exposure to inorganic mercury. The kinetics involved in the rapid reduction of the renal burden of mercury after treatment with DMPS or DMSA appear to indicate that transport of both of these chelating agents by the epithelial cells along the proximal tubule is involved in the reduction of the renal tubular burden of mercury. It is well established that both organic anions, such as sulfonates and dicarboxcylic acids, such as succinic acid, are transported by proximal tubular epithelial cells.

In a recent mechanistic study using isolated perfused proximal tubular segments, we provide data indicating that DMPS is taken up rapidly at the basolateral membrane by the *p*-aminohippuric acid-dependent organic anion transport system (Zalups *et al.*, 1998). The findings also show that once inorganic mercury binds to DMPS, the mercuric conjugates are not taken up readily at either the luminal or basolateral membranes. These particular findings are somewhat puzzling inasmuch as it had been thought that a mercuric conjugate of DMPS might be taken up at the basolateral membrane by the organic anion transport system (Zalups, 1993b). Perhaps the most important findings from this study are those indicating that DMPS can extract accumulated inorganic mercury from proximal tubular cells while it is being transported from the basolateral to the luminal side of these cells.

Additional support for the hypothesis that transport of DMPS and intracellular chelation of mercury occur along segments of the proximal tubule after treatment with DMPS comes from the study by Klotzbach and Diamond (1988). With isolated perfused kidneys from male Long–Evans rats, they showed that DMPS undergoes net tubular secretion by a kinetically saturable process that is inhibited by *p*-aminohippurate and probenecid. They also found that DMPS produced a dose-dependent decrease in the retention of inorganic mercury and an increase in urinary excretion of inorganic mercury. Furthermore, both effects were blocked by probenecid, suggesting that the mechanism of protection by DMPS is by chelation of inorganic mercury within the proximal tubular cell. Many investigators have observed that DMPS is readily oxidized in perfusates or in plasma to the disulfide form. To enable interaction with metals, DMPS is reduced back to the dithiol form within proximal tubular cells by a glutathione-dependent thiol–disulfide exchange reaction (Klotzbach and Diamond, 1988; Stewart and Diamond, 1988).

Other low-molecular-weight thiols have been employed experimentally to modulate nephrotoxicity of mercury. Because of its prominence as the primary intracellular, nonprotein thiol, exogenous glutathione is a logical choice to try as a modulatory agent. Work by Jones and colleagues (Aw *et al.*, 1991) has demonstrated that oral administration of glutathione can significantly increase the content of glutathione in the lung, kidney, heart, brain, small intestine and skin, but not in liver, under conditions where glutathione is depleted. This suggests that glutathione taken orally may supplement cellular glutathione in some tissues under certain toxicological or pathological conditions. A large body of data has been accumulated, both *in vivo* and in *in vitro* systems showing that exogenous glutathione can protect against mercury-induced renal injury. Zalups *et al.* (1991a) perfused isolated rabbit proximal tubules with 18.4 M mercuric chloride and various thiols, including glutathione or cysteine. Both thiols, present in the perfusate at fourfold higher concentrations than inorganic mercury, either prevented or significantly decreased the extent of tubular injury. An ultrafiltrate of rabbit plasma was similarly protective. The mechanism of protection by glutathione, cysteine or plasma ultrafiltrate appeared to involve decreased movement of inorganic mercury across the luminal membrane, as measured by the disappearance flux of inorganic mercury from the lumen which was significantly decreased in the presence of these thiols or thiol-containing compounds. Houser and Berndt (1988) administered glutathione monoethyl ester to rats and found that both renal cortical accumulation of inorganic mercury and the severity of mercury-induced renal injury were diminished.

The protective effects of exogenous glutathione and DMPS have also been demonstrated in suspensions of isolated proximal tubular cells from rats (Lash and Zalups, 1992). Proximal tubular cells were incubated for 15 minutes in an extracellular buffer containing bovine serum albumin and various concentrations of glutathione or DMPS. Cells were then incubated for an additional 1 hour in the presence of 250 μM mercuric chloride, which was found to be the threshold concentration of inorganic mercury that produced cellular injury under these incubation conditions. Glutathione provided concentration-dependent protection from mercury-induced cytotoxicity, as assessed by decreases in total cellular lactate dehydrogenase activity. A glutathione concentration of 500 μM, or twice that of inorganic mercury, was required to completely protect the proximal tubular cells. DMPS, in contrast, provided complete protection against 250 μM mercuric chloride at a concentration (175 μM) that was less than that of inorganic mercury. This difference in behavior between glutathione and DMPS likely arises from differences in chemistry and renal handling of the two compounds.

In contrast to the results described above, Tanaka *et al.* (1990) found that co-administration of glutathione and mercuric chloride to mice caused the renal content of mercury to increase relative to that in mice that were given mercuric chloride alone. These investigators concluded that transport of inorganic mercury to the kidney may occur as a mercury–glutathione complex and that the simultaneous presence of glutathione enhances uptake of mercury. Zalups and Barfuss (1995b,c) have observed similar effects in rats co-administered a nontoxic dose of inorganic mercury with glutathione or cysteine. Consistent with these findings, Miller and Woods (1993) showed recently that complexes of glutathione and Hg^{2+}, or GSSG and Hg^{+}, promoted uroporphyrinogen oxidation and catalyzed decomposition of hydrogen peroxide, indicating that mercury–glutathione (or other thiol) complexes likely contribute to mercury-induced toxicity. Some of these results have also been confirmed in rats in our laboratory (unpublished findings). We have very recent data from rats indicating that when a toxic 2.0 μmol/kg dose of mercuric chloride is co-administered with a 2 : 1 mole ratio of glutathione or cysteine, the nephropathy induced by the inorganic mercury is made more severe. Resolution of the marked contrast between these findings and those described above will require a detailed mechanistic description of the renal transport of inorganic mercury. Although advances have been made in the understanding of mechanisms of renal transport of mercury, the role of thiols in the renal cellular uptake of mercury is still unclear.

Although DMPS and DMSA are dithiols and are highly effective protective agents against mercury-induced injury, less definitive results have been obtained with two other dithiols, dithioerythritol and dithiothreitol. On the one hand, Barnes *et al.* (1980) observed, in rats, evidence of protection against morphologic lesions and losses of activities of key marker enzymes for plasma membrane and mitochondria induced by mercury with dithiothreitol. Weinberg *et al.* (1982a) provided evidence of protection for isolated renal mitochondria from mercuric chloride-induced dysfunction by dithioerythritol, but only if the dithiol was added *in vitro* simultaneously with mercuric chloride; when the dithiol agent was added *in vitro* after the rats had been treated with mercuric chloride *in vivo*, no protection or reversal of toxicity was observed. To complicate further the understanding of how dithiols interact with mercury-containing compounds in biological systems, Chavez and Holguin (1988) reported that addition of dithiothreitol to renal mitochondria isolated from rats that had been treated with inorganic mercury actually increased the degree of mitochondrial injury induced by mercury. They suggested that the dithiol made additional sulfhydryl-sensitive sites available for interaction with mercury, thereby enhancing the toxic response. In the same study (Chavez and Holguin, 1988), the investigators also reported that the monothiol 2-mercaptoethanol also enhanced mercuric chloride-induced mitochondrial injury, although higher concentrations than those of the dithiol were required to reproduce the effect.

Chavez *et al.* (1991) have also reported that the angiotensin-converting-enzyme inhibitor, captopril (1-(3-mercapto-2-methyl-1-oxopropyl)-1-proline), was an effective protective agent both *in vivo* and *in vitro* against mercuric chloride-induced mitochondrial injury and morphological damage.

Besides low-molecular-weight thiols, administration of inorganic mercury complexed to the small sulfhydryl-containing protein metallothionein, although not providing protection against the toxicity induced by inorganic mercury, altered the renal site of injury (Chan *et al.*, 1992). Whereas the primary target of renal injury due to mercuric chloride is the pars recta (S2, S3) segments of the proximal tubule, the primary target of renal injury due to mercury–metallothionein appears to be the pars convoluta and early pars recta (S1, S2) segments of the proximal tubule. Intrarenal accumulation and urinary excretion of inorganic mercury in rats was also greater when mercury was administered with metallothionein than when mercury was administered by itself.

9.6.3 Effects of reduced nephron number and compensatory renal growth on mercury

9.6.3.1 Accumulation and renal toxicity

Reduction in the number of functioning nephrons, which can occur as a consequence of aging, renal disease, or surgical removal of renal tissue, has profound effects on renal cellular function and consequently, on the renal handling of exogenous chemicals and on the susceptibility of renal tissue to chemically induced injury (Meyer *et al.*, 1991). The remnant renal tissue undergoes compensatory growth, which is predominantly (i.e. >85%) due to cellular hypertrophy, particularly in segments of the proximal tubule. One of the more prominent changes in renal function that occurs as a result of compensatory hypertrophy includes marked increases in mitochondrial metabolism, which may lead to an enhanced susceptibility of renal tissue to oxidative stress (Nath *et al.*, 1990).

Many animal studies have shown that rats that have undergone a reduction in renal mass, such as unilateral nephrectomy, are more susceptible to the nephropathy induced by inorganic mercury (Houser and Berndt, 1986, 1988; Zalups *et al.*, 1988; Zalups and Lash, 1990; Lash and Zalups, 1992, 1993). The biochemical changes that occur as a consequence of reduced renal mass and compensatory hypertrophy are retained *in vitro* when proximal tubular cells are isolated from rats (Lash and Zalups, 1992, 1993). Furthermore, the enhanced susceptibility of rats to the nephrotoxic effects of inorganic mercury is also retained *in vitro*. In the absence of exogenous thiols in the extracellular incubation medium, isolated proximal tubular cells from unilaterally nephrectomized rats in which compensatory renal growth has occurred (NPX) exhibit irreversible cellular injury at significantly lower concentrations of mercuric chloride than isolated proximal tubular cells from sham-operated (SHAM) rats.

Although the mechanism(s) for the enhanced susceptibility of proximal tubular cells from NPX rats to injury induced by mercury are not well characterized, it appears that enhanced accumulation of mercury is a contributing factor. Findings from studies with both mercuric chloride (Zalups and Diamond, 1987; Zalups *et al.*, 1988; Zalups and Lash, 1990; Zalups, 1991b) and methylmercuric chloride (Zalups *et al.*, 1992) indicate that greater amounts of mercury, on a per gram tissue basis, accumulate in the remnant kidney of NPX rats than in the kidneys of SHAM or control rats. Moreover, the findings indicate that greatest increase in the accumulation of mercury occurs in the outer stripe of the outer medulla, and specifically in pars recta segments of proximal tubules (Zalups, 1991b), which coincides with the site where the toxicity of mercury is expressed in the kidney. Other factors, such as changes in intrarenal handling of mercury, are also probably involved in changing the cellular response to mercury exposure. Some of the altered accumulation of mercury that occurs in the remnant kidney is probably related to alteration in the renal concentrations of intracellular thiols. Recent findings show that both the intracellular metabolism of glutathione (Zalups and Lash, 1990) and metallothionein (Zalups and Cherian, 1992 a,b) are altered significantly after renal mass is reduced following unilateral nephrectomy and compensatory renal growth.

9.7 Conclusion

Toxicology of heavy metals encompasses a large field of research and is of interest to many because of the widespread environmental distribution of these toxicants. This is particularly true of mercury-containing compounds. Significant advances have been made during the past decade, in part due to the development and validation of various *in vitro* biological preparations, including isolated and perfused microdissected segments of the nephron and isolated cellular suspensions and culture techniques. Principal areas that were discussed include renal accumulation and transport of mercury, molecular interaction of mercury in renal epithelial cells, renal excretion and toxicity of mercury, and factors influencing the renal toxicity of mercury.

9.7.1 Renal accumulation and transport of mercury

In summary, the kidneys are the primary site for accumulation of various forms of mercury. Inorganic and organic mercury accumulate primarily in the renal cortex and outer stripe of the outer medulla. Most of the accumulation of mercury that occurs in the cortex and outer stripe of the outer medulla occurs mainly along the three segments (S1, S2 and S3) of the proximal tubule. At present, there appears to be at least two primary mechanisms involved in the uptake of mercuric ions by proximal tubular epithelial cells. One of the mechanisms is localized on the luminal plasma membrane and involves the activity of the γ-glutamyltransferase. Moreover, it appears that the primary species being transported at the luminal membrane is a mercuric conjugate of cysteine. The uptake of some of these conjugates appears to occur at the site of both sodium-dependent and sodium-independent amino-acid transporters via a mechanism of molecular mimicry whereby the mercuric conjugates of cysteine mimic the amino acid cystine. On the basolateral membrane, uptake of mercury appears to involve the organic anion transport system. Likely species of mercury that are transported at this site include mercuric conjugates of glutathione, cysteine, homocysteine and *N*-acetylcysteine. It is not known at present whether other segments of the nephron and collecting duct play a significant role in the renal uptake, transport, accumulation and excretion of mercurials.

To understand the mechanisms involved in the tubular uptake of mercury, one must consider and understand the molecular interactions that occur between the various forms of mercury and sulfhydryl-containing molecules that are present in various compartments in the body. To further emphasize this point, one only needs to view the current body of evidence on the renal tubular transport of mercury, which shows overwhelmingly that mercury is likely to be cotransported into renal tubular (proximal) epithelial cells with thiol-containing compounds. However, further work is needed to establish the relationship between the renal cellular uptake of mercury and the interactions and relationships between mercurous and mercuric ions and cysteine, GSH, metallothioneins, albumin and other ligands (containing sulfhydryl groups) in the different compartments of the body.

9.7.2 Molecular interactions with mercury in renal epithelial cells

Although many of the biological effects of mercury-containing compounds in renal tissue can be attributed to the binding of mercury to plasma membrane or intracellular thiols, many findings are inconsistent with this being the sole mechanism of action. It has been established, however, that through alterations in intracellular thiol status, mercury can promote oxidative stress, lipid peroxidation, mitochondrial dysfunction and changes in heme metabolism.

Conner and Fowler (1993) have attempted explain some of the biological effects of mercury-containing compounds in renal tissue. In their scheme, after mercuric or methylmercuric ions enter proximal tubular epithelial cells by transport across either the brush-border or basolateral membrane, they interact with thiol-containing compounds, principally GSH and metallothionein. They propose that the early of effects of mercury include alterations in membrane permeability to calcium ions and inhibition of mitochondrial function. Moreover, they propose that through unknown signaling mechanisms, mercury induces the synthesis of glutathione, various glutathione-dependent enzymes, metallothionein and several stress proteins. It is important to realize that the inductive effects of mercuric or methylmercuric ions occur primarily at subtoxic or moderately toxic doses of mercury; at higher doses, cellular injury occurs and biosynthetic processes are inhibited. While this model accounts for some of the data in the literature, little mechanistic information regarding mercury transport across luminal and basolateral membranes of renal tubular epithelial cells or certain biochemical effects of mercury in renal tubular epithelial cells, such as oxidative stress, is provided. Hence, while this model explains some effects of mercury and illustrates how some of these processes may interact to produce renal cellular injury, a more thorough integrated model is needed to explain the biochemical mechanism for mercury-induced renal cellular injury.

9.7.3 Renal toxicity of mercury

Despite compelling histopathological data implicating the pars recta of the proximal tubule as the primary target site that is adversely affected by mercury, other segments of the nephron may also be involved in the nephropathy induced by mercurials. To date, however, little work has been performed on *in vitro* systems derived from nephron segments other than the proximal tubules, so it is not clear how susceptible other renal tubular cell populations are to direct exposure to mercury or what role injury in these segments plays in the overall toxic response in the kidney.

Although intrarenal accumulation of mercury *per se* would logically appear to be the parameter that one would characterize and expect to correlate with the severity of renal injury induced by mercury, it appears that urinary excretion of mercury correlates much more closely to the severity of renal injury. This is an important observation because it can potentially provide a noninvasive means for assessing the severity of the acute nephropathy induced by mercury. An additional noninvasive mechanism to monitor exposure to mercury and severity of renal injury induced by mercury is measurement of the urinary excretion of a number of plasma solutes and renal cellular enzymes.

9.7.4 Factors influencing the renal toxicity of mercury

As a means to both better understand the mechanism of action of mercury-containing compounds and to develop antidotes and treatments for mercury-poisoning, factors, both intracellular and extracellular, have been investigated as tools to alter the disposition and metabolism of mercury. The intracellular content of glutathione or metallothionein can be manipulated by a number of techniques to thereby affect the intracellular distribution of mercury-containing compounds. Similarly, exogenous glutathione or metallothionein can influence the renal accumulation and toxicity of mercury. Additionally, other chelators, most notably DMPS and DMSA, have been employed clinically as antidotes to either reverse or prevent toxic effects of mercury-containing compounds.

Physiological or pathological processes can dramatically alter the renal handling of, and cellular responses to, mercury. One such process that has received considerable attention is reduced renal mass and compensatory renal growth. Justification for studying this model lies in the fact that a large population of individuals in our society have reduced renal mass (caused by a whole host of factors) and these individuals may be at a greater risk of becoming intoxicated by mercurials or other nephrotoxic compounds. Hence, an understanding of how compensatory renal growth and progressive losses in the number of functioning nephrons modify the handling and toxicity of mercury in the remnant renal tissue is a very relevant and important issue.

9.7.5 Future directions

Much work is still needed to delineate the precise biochemical mechanisms involved in the renal tubular uptake, transport and handling of mercury-containing compounds and to understand how mercury alters biochemical processes within renal tubular epithelial cells. Continued use and further validation of *in vitro* models such as microdissected perfused tubules from defined nephron segments, freshly isolated renal tubular epithelial cells and primary cultures of renal tubular epithelial cells, will hopefully enable many of the unanswered questions regarding mercury transport and toxicity to be addressed. With these methodologies it should also be possible to resolve some of the conflicting data in the literature, particularly regarding the interaction of thiols and mercury-containing compounds.

Although a great deal of new information has been gathered over the past several years regarding the renal tubular handling and toxicity of the various forms of mercury, much more experimental research is needed to gain a better understanding of the precise mechanisms involved in the transport of mercury along the nephron and the specific mechanisms involved in the nephropathy induced by the different forms of mercury.

References

Addya, A., Chakravarti, K., Basu, A., Santra, M., Haldar, S. and Chatterjee, G.C. 1984. Effects of mercuric chloride on several scavenging enzymes in rat kidney and influence of vitamin E supplementation. *Acta Vitaminology and Enzymology* **6**, 103–107.

Aleo, M.D., Taub, M.L., Olson, J.R., Nickerson, P.A. and Kostyniak, P.J. 1987. Cellular uptake and response of primary cultures of rabbit renal proximal tubule cells exposed to mercuric chloride and methylmercury chloride, in *In Vitro Toxicology: Approaches to Validation*, (ed. A.M. Goldberg), Mary Ann Liebert, Inc., New York, pp. 211–26.

Aleo, M.D., Taub, M.L. and Kostyniak, P.J. 1992. Primary cultures of rabbit renal proximal tubule cells. III. Comparative cytotoxicity of inorganic and organic mercury. *Toxicology and Applied Pharmacology* **112**, 310–317.

Anner, B.M. and Moosmayer, M. 1992. Mercury inhibits Na-K-ATPase primarily at the cytosolic side. *American Journal of Physiology* **262**, F843–F848.

Anner, B.M., Moosmayer, M. and Imesch, E. 1992. Mercury blocks Na-K-ATPase by a ligand-dependent and reversible mechanism. *American Journal of Physiology* **262**, F830–F836.

Aposhian, H.V. 1983. DMSA and DMPS – Water soluble antidotes for heavy metal poisoning. *Annual Review of Pharmacology and Toxicology* **23**, 193–215.

Aposhian, H.V. and Aposhian, M.M. 1990. *Meso*-2,3-Dimercaptosuccinic acid: Chemical, pharmacological and toxicological properties of an orally effective metal chelating agent. *Annual Review of Pharmacology and Toxicology* **30**, 279–306.

ATSDR 1994. *Toxicological Profile for Mercury*. US Department of Health and Human Services, Public Health Service, Agency for Toxic Substance and Disease Registry, TP-93/10.

Aw, T.Y., Wierzbicka, G., and Jones, D.P. 1991. Oral glutathione increases tissue glutathione *in vivo*. *Chemico-Biological Interactions* **80**, 89–97.

Baggett, J. McC. and Berndt, W.O. 1985. The effect of potassium dichromate and mercuric chloride on urinary excretion and organ and subcellular distribution of [^{203}Hg]mercuric chloride in rats. *Toxicology Letters* 29, 115–21.

Baggett, J. McC. and Berndt, W.O. 1986. The effect of depletion of nonprotein sulfhydryls by diethyl maleate plus buthionine sulfoximine on renal uptake of mercury in the rat. *Toxicology and Applied Pharmacology* **83**, 556–62.

Ballatori, N. 1991. Mechanisms of metal transport across liver cell plasma membranes. *Drug Metabolism Reviews* **23**, 83–132.

Ballatori, N. and Clarkson, T.W. 1985. Biliary secretion of glutathione and of glutathione-metal complexes. *Fundamental and Applied Toxicology* **5**, 816–31.

Ban, M. and de Ceaurriz, J. 1988. Probenecid-induced protection against hexachloro-1,3-butadiene and methylmercury toxicity to the mouse kidney. *Toxicology Letters* **40**, 71–6.

Barenberg, R.L., Solomon, S., Papper, S. and Anderson, R. 1968. Clearance and micropuncture study of renal function in mercuric chloride treated rats. *Journal of Laboratory and Clinical Medicine* **72**, 473–84.

Barfuss, D.W., Robinson, M.K. and Zalups, R.K. 1990. Inorganic mercury transport in the proximal tubule of the rabbit. *Journal of the American Society of Nephrology* **1**, 910–17.

Barnes, J.L., McDowell, E.M., McNeil, J.S., Flamenbaum, W. and Trump, B.F. 1980. Studies on the pathophysiology of acute renal failure. IV. Protective effect of dithiothreitol following administration of mercuric chloride in the rat. *Virchows Archive. B Cell Pathology* **32**, 201–32.

Berndt, W.O., Baggett, J. McC., Blacker, A. and Houser, M. 1985. Renal glutathione and mercury uptake by kidney. *Fundamental and Applied Toxicology* **5**, 832–9.

Bigazzi, P.E. 1988. Autoimmunity induced by chemicals. *Clinical Toxicology* **26**, 125–56.

Bigazzi, P.E. 1992. Lessons from animal models: The scope of mercury-induced autoimmunity. *Clinical Immunology and Immunopathology* **65**, 81–4.

Bowers, M.A., Aicher, L.D., Davis, H.A. and Woods, J.S. 1992. Quantitative determination of porphyrins in rat and human urine and evaluation of urinary porphyrin profiles during mercury and lead exposures. *Journal of Laboratory and Clinical Medicine* **120**, 272–81.

Buchet, J.P., Roels, H., Bernard, A. and Lauwerys, R. 1980. Assessment of renal function of workers exposed to inorganic lead, cadmium or mercury vapor. *Journal of Occupational Medicine* **22**, 741–50.

Cannon, V.T., Barfuss, D.W. and Zalups, R.K. 1999. Molecular homology and the luminal transport of HG^{++} in the renal proximal tubule. *Journal of the American Society of Nephrology*. In press.

Cardenas, A., Roels, H., Bernard, A.M., Barbon, R., Buchet, J.P., Lauwreys, R.R., Rosello, J., Hotter, G., Mutti, A., Franchini, I. *et al.* 1993. Markers of early renal changes induced by industrial pollutants. I. Application to workers exposed to mercury vapour. *British Journal of Industrial Medicine* **50**, 17–27.

Cember, H., Gallagher, P. and Faulkner, A. 1968. Distribution of mercury among blood fractions and serum proteins. *American Industrial Hygiene Association Journal* **29**, 233–7.

Chan, H.M., Satoh, M., Zalups, R.K. and Cherian, M.G. 1992. Exogenous metallothionein and renal toxicity of cadmium and mercury in rats. *Toxicology* **76**, 15–26.

Chang, L.W., Ware, R.A. and Desnoyers, P.A. 1973. A histochemical study on some enzymes changes in the kidney, liver and brain after chronic mercury intoxication in the rat. *Food and Cosmetic Toxicology* **11**, 283–6.

Chavez, E. and Holguin, J.A. 1988. Mitochondrial calcium release induced by Hg^{2+}. *Journal of Biological Chemistry* **283**, 3582–7.

Chavez, E., Zazueta, C., Osornio, A., Holguin, J.A. and Miranda, M.E. 1991. Protective behavior of captopril on Hg^{++}-induced toxicity on kidney mitochondria. *In vivo* and *in vitro* experiments. *Journal of Pharmacology and Experimental Therapeutics* **256**, 385–90.

Cherian, M.G. and Clarkson, T.W. 1976. Biochemical changes in rat kidney on exposure to elemental mercury vapor: Effect on biosynthesis of metallothionein. *Chemico-Biological Interactions* **12**, 109–20.

Chung, A.-S., Maines, M.D. and Reynolds, W.A. 1982. Inhibition of the enzymes of glutathione metabolism by mercuric chloride in the rat kidney: Reversal by selenium. *Biochemical Pharmacology* **31**, 3093–100.

Clarkson, T.W. 1972. The pharmacology of mercury compounds. *Annual Review of Pharmacology and Toxicology* **12**, 375–406.

Clarkson, T.W. 1993. Molecular and ionic mimicry of toxic metals. *Annual Review of Pharmacology and Toxicology* **32**, 545–71.

Clarkson, T.W. and Magos, L. 1967. The effect of sodium maleate on the renal disposition and excretion of mercury. *British Journal of Pharmacology and Chemotherapy* **31**, 560–7.

Conner, E.A. and Fowler, B.A. 1993. Mechanisms of metal-induced nephropathy, in *Toxicology of the Kidney*, 2nd edn, (eds J.B. Hook and R.S. Goldstein), Raven Press, New York, pp. 437–57.

Cuppage, F.E. and Tate, A. 1967. Repair of the nephron following injury with mercuric chloride. *American Journal of Pathology* **51**, 405–29.

Dantzler, W.H. 1996. Comparative aspects of renal organic anion transport. *Cellular Physiology and Biochemistry* **6**, 28–38.

Dantzler, W.H., Evans, K.K. and Wright, S.H. 1995. Kinetics of interactions of *para*-aminohippurate, probenecid, cysteine conjugates and *N*-acetylcysteine conjugates with basolateral organic anion transporter in isolated rabbit proximal renal tubules. *Journal of Pharmacology and Experimental Therapeutics* **272**, 663–72.

de Ceaurriz, J., Payan, J.P., Morel, G. and Brondeau, M.T. 1994. Role of extracellular glutathione and γ-glutamyltranspeptidase in the disposition and kidney toxicity of inorganic mercury in rats. *Journal of Applied Toxicology* **14**, 201–6.

Diamond, G.L. 1988. Biological monitoring of urine for exposure to toxic metals, in *Biological Monitoring of Toxic Metals*, (eds T.W. Clarkson, L. Friberg, G.F. Nordberg, and P.R. Sager), Plenum Publishing, New York, pp. 515–29.

Druet, P., Druet, E., Potdevin, F. and Sapin, C. 1978. Immune type glomerulonephritis induced by $HgCI_2$ in the Brown-Norway rat. *Annals of Immunology (Institute Pasteur)* **129C**, 777–92.

Dunn, J.D. and Clarkson, T.W. 1980. Does mercury exhalation signal demethylation of methylmercury? *Health Physics* **38**, 411–14.

Ellis, B.G., Price, R.G. and Topham, J.C. 1973. The effect of tubular damage by mercuric chloride on kidney function and some urinary enzymes in the dog. *Chemico-Biological Interactions* **7**, 101–13.

Enestrom, S. and Hultman, P. 1984. Immune-mediated glomerulonephritis induced by mercuric chloride in mice. *Experientia* **40**, 1234–40.

Fowler, B.A. 1972. The morphological effects of dieldrin and methyl mercuric chloride on pars recta segments of the rat kidney proximal tubules. *American Journal of Pathology* **69**, 163–78.

Friedman, H.L. 1957. Relationship between chemical structure and biological activity in mercurial compounds. *Annals of the New York Academy of Science* **65**, 461–70.

Fukino, H., Hirai, M., Hsueh, Y.M. and Yamane, Y. 1984. Effect of zinc pretreatment on mercuric chloride-induced lipid peroxidation in the rat kidney. *Toxicology and Applied Pharmacology* **73**, 395–401.

Fukino, H., Hirai, M., Hsueh, Y.M., Moriyasu, S. and Yamane, Y. 1986. Mechanism of protection by zinc against mercuric chloride toxicity in rats: Effects of zinc and mercury on glutathione metabolism. *Journal of Toxicology Environmental Health* **19**, 75–89.

Gage, J.C. 1964. Distribution and excretion of methyl and phenylmercury salts. *British Journal of Industrial Medicine* **21**, 197–202.

Ganote, C.E., Reimer, K.A. and Jennings, R.B. 1974. Acute mercuric chloride nephrotoxicity: An electron microscopic and metabolic study. *Laboratory Investigations* **31**, 633–47.

Girardi, G. and Elias, M.M. 1991. Effectiveness of *N*-acetylcysteine in protecting against mercuric chloride-induced nephrotoxicity. *Toxicology* **67**, 155–64.

Girardi, G. and Elias, M.M. 1993. Effect of different renal glutathione levels on renal mercury disposition and excretion in the rat. *Toxicology* **81**, 57–67.

Gleason, M.N., Gosselin, R.E. and Hodge, D.C. 1957. *Clinical Toxicology of Commercial Products*, Williams and Wilkins, Baltimore, p. 154.

Goering, P.L., Fisher, B.R., Chaudhary, P.P. and Dick, C.A. 1992. Relationship between stress protein induction in rat kidney by mercuric chloride and nephrotoxicity. *Toxicology and Applied Pharmacology* **113**, 184–91.

Gottelli, C.A., Astolfi, E., Cox, C., Cernichiari, E. and Clarkson, T.W. 1985. Early biochemical effects of organic mercury fungicide on infants: 'Dose makes the poison'. *Science* **227**, 638–40.

Gritzka, T.L. and Trump, B.F. 1968. Renal tubular lesions caused by mercuric chloride: Electron microscopic observations: Degeneration of the pars recta. *American Journal of Pathology* **52**, 1225–77.

Gstraunthaler, G., Pfaller, W. and Kotanko, P. 1983. Glutathione depletion and *in vitro* lipid peroxidation in mercury or maleate-induced acute renal failure. *Biochemical Pharmacology* **32**, 2969–72.

Houser, M.T. and Berndt, W.O. 1986. The effect of unilateral nephrectomy on the nephrotoxicity of mercuric chloride in the rat. *Toxicology and Applied Pharmacology* **83**, 506–15.

Houser, M.T. and Berndt, W.O. 1988. Unilateral nephrectomy in the rat: Effects on mercury handling and renal cortical subcellular distribution. *Toxicology and Applied Pharmacology* **93**, 187–94.

Hultman, P. and Enestrom, S. 1986. Localization of mercury in the kidney during experimental acute tubular necrosis studied by the cytochemical silver amplification method. *British Journal of Experimental Pathology* **67**, 493–503.

Hultman, P. and Enestrom, S. 1992. Dose-response studies in murine mercury-induced autoimmunity and immune-complex disease. *Toxicology and Applied Pharmacology* **113**, 199–208.

Hultman, P., Enestrom, S. and von Schenck, H. 1985. Renal handling of inorganic mercury in mice. The early excretion phase following a single intravenous injection of mercuric chloride studied by the silver amplification method. *Virchows Archive [Cell Pathology]* **49**, 209–24.

Imesch, E., Moosmayer, M. and Anner, B.M. 1992. Mercury weakens membrane anchoring of Na-K-ATPase. *American Journal of Physiology* **262**: F837–F842.

Johnson, D.R. 1982. Role of renal cortical sulfhydryl groups in development of mercury-induced renal toxicity. *Journal of Toxicology and Environmental Health* **9**, 119–26.

Jung, K.Y., Uchida, S. and Endou, H. 1989. Nephrotoxicity assessment by measuring cellular

ATP content. I. Substrate specificities in the maintenance of ATP content in isolated rat nephron segments. *Toxicology and Applied Pharmacology* **100**, 369–82.

Kim, C.Y., Watanabe, C., Kasanuma, Y. and Satoh, H. 1995. Inhibition of gamma-glutamyltranspeptidase decreases renal deposition of mercury after mercury vapor exposure. *Archives of Toxicology* **69**, 722–4.

Kirschbaum, B.B. 1979. Alanine aminopeptidase excretion after mercuric chloride renal failure. *Biochemical Medicine* **21**, 220–5.

Klotzbach, J.M. and Diamond, G.L. 1988. Complexing activity and excretion of 2,3-dimercapto-1-propane sulfonate in rat kidney. *American Journal of Physiology* **254**, F871–F878.

Koropatnick, J.D. and Zalups, R.K. 1997. Effect of non-toxic mercury, zinc or cadmium pretreatment on the capacity of human monocytes to under lipopolysaccharide-induced activation. *British Journal or Pharmacology* **120**, 797–806.

Kostial, K., Kello, D., Jugo, S. *et al.* 1978. Influence of age on metal metabolism and toxicity. *Environmental Health Perspectives* **25**, 81–6.

Lash, L.H. and Anders, M.W. 1989. Uptake of nephrotoxic *S*-conjugates by isolated rat renal proximal tubular cells. *Journal of Pharmacology and Experimental Therapeutics* **248**, 531–7.

Lash, L.H., and Jones, D.P. 1983. Transport of glutathione by renal basal-lateral membrane vesicles. *Biochemistry Biophysics Research Communications* **112**, 55–60.

Lash, L.H., and Jones, D.P. 1984. Renal glutathione transport: characteristics of the sodium-dependent system in the basal-lateral membrane. *Journal of Biological Chemistry* **259**, 14508–14.

Lash, L.H. and Jones, D.P. 1985a. Distribution of oxidized and reduced forms of glutathione and cysteine in rat plasma. *Archives of Biochemistry and Biophysics* **240**, 583–92.

Lash, L.H. and Jones, D.P. 1985b. Uptake of the glutathione conjugate *S*-(1,2-dichlorovinyl) glutathione by renal basal-lateral membrane vesicles and isolated kidney cells. *Molecular Pharmacology* **28**, 278–82.

Lash, L.H. and Zalups, R.K. 1992. Mercuric chloride-induced cytotoxicity and compensatory hypertrophy in rat kidney proximal tubular cells. *Journal of Pharmacology and Experimental Therapeutics* **261**, 819–29.

Lash, L.H. and Zalups, R.K. 1993. Activities of enzymes involved in renal cellular glutathione metabolism after uninephrectomy in the rat. *Archives of Biochemistry and Biophysics* **309**, 129–38.

Lau, S. and Sarkar, B. 1979. Inorganic mercury (II)-binding components in normal human blood serum. *Journal of Toxicology and Environmental Health* **5**, 907–16.

Lund, B.-O., Miller, D.M. and Woods, J.S. 1991. Mercury-induced H_2O_2 production and lipid peroxidation *in vitro* in rat kidney mitochondria. *Biochemical Pharmacology* **42** (Suppl.), S181–S187.

Lund, B.-O., Miller, D.M. and Woods, J.S. 1993. Studies on Hg(II)-induced H_2O_2 formation and oxidative stress *in vivo* and *in vitro* in rat kidney mitochondria. *Biochemical Pharmacology* **45**, 2017–24.

McDowell, E.M., Nagle R.B., Zalme, R.C., McNeil, J.S., Flamenbaum, W. and Trump, B.F. 1976. Studies on the pathophysiology of acute renal failure. I. Correlation of ultrastructure and function in the proximal tubule of the rat following administration of mercuric chloride. *Virchows Archive [Cell Pathology]* **22**, 173–96.

McNeil, S.I., Bhatnagar, M.K. and Turner, C.J. 1988. Combined toxicity of ethanol and methylmercury in rat. *Toxicology* **53**, 345–63.

Madsen, K.M. 1980. Mercury accumulation in kidney lysosomes of proteinuric rats. *Kidney International* **18**, 445–53.

Madsen, K.M. and Hansen, J.C. 1980. Subcellular distribution of mercury in the rat kidney cortex after exposure to mercuric chloride. *Toxicology and Applied Pharmacology* **54**, 443–53.

Magos, L. and Clarkson, T.W. 1977. Renal injury and urinary excretion, in *Handbook of Physiology, Section 9, Renal Physiology*, (ed. D.H.K. Lee) American Physiological Society, Bethesda, pp. 503–12.

Magos, L. and Stoychev, T. 1969. Combined effect of sodium maleate and some thiol compounds on mercury excretion and redistribution in rats. *British Journal of Pharmacology* **35**, 121–6.

Magos, L., Brown, A.W., Sparrow, S., Bailey, E., Snowden, R.T., and Skipp, W.R. 1985. The comparative toxicology of ethyl- and methyl-mercury. *Archives of Toxicology* **57**, 260–7.

Maiorino, R.M., Dart, R.C., Carter, D.E. and Aposhian, H.V. 1991. Determination and metabolism of dithiol chelating agents. XII. Metabolism and pharmacokinetics of sodium 2,3-dimercaptopropane-1-sulfonate in humans. *Journal of Pharmacology and Experimental Therapeutics* **259**, 808–14.

Meyer, T.W., Scholey, J.V. and Brenner, B.M. 1991. Nephron adaptation to renal injury, in *The Kidney*, 4th edn, (eds B.M. Brenner and F.C. Rector Jr), W.B. Saunders, Philadelphia, pp. 1871–908.

Miller, D.W. and Woods, J.S. 1993. Redox activities of mercury–thiol complexes. Implications for mercury induced porphyria and toxicity. *Chemical–Biological Interactions* **88**, 23–35.

Mussini, E. 1958. Bonds of mercurial diuretics to blood proteins. *Boll. Soc. Ital. Biol. Sper.* **34**, 1588–90.

Naganuma, A., Oda-Urano, N., Tanaka, T. and Imura, N. 1988. Possible role of hepatic glutathione in transport of methylmercury into mouse kidney. *Biochemical Pharmacology* **37**, 291–6.

Nath, K.A., Croatt, A.J. and Hostetter, T.H. 1990. Oxygen consumption and oxidant stress in surviving nephrons. *American Journal of Physiology* **258**, F1354–F1362.

Norseth, T. and Clarkson, T.W. 1970a. Studies on the biotransformation of ^{203}Hg-labeled methylmercury chloride in rats. *Archives of Environmental Health* **21**, 717–27.

Norseth, T. and Clarkson, T.W. 1970b. Biotransformation of methylmercury salts in the rat studied by specific determination of inorganic mercury. *Biochemical Pharmacology* **19**, 2775–83.

Omata, S., Sato, M., Sakimura, K. and Sugano, H. 1980. Time-dependent accumulation of inorganic mercury in subcellular fractions of kidney, liver, and brain or rats exposed to methylmercury. *Archives of Toxicology* **44**, 231–41.

Piotrowski, J.K., Trojanowska, B., Wisniewska-Knypl, J.M. and Bolanowska, W. 1974. Mercury binding in the kidney and liver of rats repeatedly exposed to mercuric chloride. *Toxicology and Applied Pharmacology* **27**, 11–19.

Planas-Bohne, F. 1977. The effect of mercuric chloride on the excretion of two urinary enzymes in the rat. *Archives of Toxicology* **37**, 219–25.

Price, R.G. 1982. Urinary enzymes, nephrotoxicity and renal disease. *Toxicology* **23**, 99–134.

Pritchard, J.B. and Miller, D.S. 1993. Mechanisms mediating renal secretion of organic anions and cations. *Physiological Reviews* **73**, 765–88.

Rabenstein, D.L. 1989. Metal complexes of glutathione and their biological significance, in *Glutathione: Chemical, Biochemical and Medical Aspects*, Vol. 3, *Coenzymes and Cofactors*, (eds D. Dophin, O. Auramovibc and R. Poulson), Wiley, New York, pp. 147–86.

Roch-Ramel, F., Besseghir, K. and Murer, H. 1992. Renal excretion and tubular transport of organic anions and cations, in *Handbook of Physiology, Section 8, Renal Physiology*, (ed. E.E. Windhager), Oxford University Press, New York, pp. 2189–262.

Rodier, P.M., Kates, B. and Simons R. 1988. Mercury localization in mouse kidney over time: Autoradiography versus silver staining. *Toxicology and Applied Pharmacology* **257**, 235–45.

Roman-Franco, A.A., Turiello, M., Albini, B., Ossi, E., Milgrom, F. and Andres, G.A. 1978. Anti-basement membrane antibodies and antigen-antibody complexes in rabbits injected with mercuric chloride. *Clinical Immunology and Immunopathology* **9**, 464–81.

Rothstein, A. 1970. Sulfhydryl groups in membrane structure and function. *Current Topics in Membrane Transport* **1**, 135–76.

Rothstein, A. and Hayes, A.D. 1960. The metabolism of mercury in the rat studied by isotope techniques. *Journal of Pharmacology and Experimental Therapeutics* **130**, 166–76.

Ruegg, C.E., Gandolfi, A.J., Nagle, R.B. and Brendel, K. 1987. Differential patterns of injury to the proximal tubule of renal cortical slices following *in vitro* exposure to mercuric chloride, potassium dichromate, or hypoxic conditions. *Toxicology and Applied Pharmacology* **90**, 261–73.

Sapin, C., Dreut, E. and Dreut, P. 1977. Induction of anti-glomerular basement membrane antibodies in the brown norway rat by mercuric chloride. *Clinical and Experimental Immunology* **28**, 173–9.

Siegers, C.-P., Schenke, M. and Younes, M. 1987. Influence of cadmium chloride, mercuric chloride, and sodium vanadate on the glutathione-conjugating enzyme system in liver, kidney, and brain of mice. *Journal of Toxicology and Environmental Health* **22**, 141–8.

Silbernagl, S. 1992. Tubular transport of amino acids and small peptides, in *Handbook of Physiology, Section 8, Renal Physiology*, (ed. E.E. Windhager) Oxford University Press, New York, pp. 1938–76.

Smith, J.C., Allen, P.V., Turner, M.D., Most, B., Fisher, H.L. and Hall, L.L. 1994. The kinetics of intravenously administered methyl mercury in man. *Toxicology and Applied Pharmacology* **128**, 251–6.

Smith, M.W., Ambudkar, I.S., Phelps, P.C., Regec, A.L. and Trump, B.F. 1987. $HgCI_2$-induced changes in cytosolic Ca^{2+} of cultured rabbit renal tubular cells. *Biochimie Biophysics Acta* **931**, 130–42.

Stewart, J.R. and Diamond, G.L. 1988. *In vivo* tubular secretion and metabolism of the disulfide of 2,3-dimercaptopropane-1-sulfonate. *Drug Metababolism and Disposition* **16**, 189–95.

Stonard, M.D., Chater, B.V., Duffield, D.B., Nevitt, A.L., O'Sullivan, J.J. and Steele, G.T. 1983. An evaluation of renal function in workers occupationally exposed to mercury vapor. *International Archives of Occupational and Environmental Health* **52**, 177–89.

Stroo, W.E. and Hook, J.B. 1977. Enzymes of renal origin in urine as indicators of nephrotoxicity. *Toxicology and Applied Pharmacology* **39**, 423–34.

Tanaka, T., Naganuma, A. and Imura, N. 1990. Role of γ-glutamyltranspeptidase in renal uptake and toxicity of inorganic mercury in mice. *Toxicology* **60**, 187–98.

Tanaka, T., Naganuma, A. and Imura, N. 1992. Routes for renal transport of methylmercury in mice. *European Journal of Pharmacology* **228**, 9–14.

Tanaka-Kagawa, T., Naganuma, A. and Imura, N. 1993. Tubular secretion and reabsorption of mercury compounds in mouse kidney. *Journal of Pharmacology and Experimental Therapeutics* **264**, 776–82.

Taugner, R., Winkel, K. and Iravani, J. 1966. Zur Lokalization der Sublimatanreicherung in der Rattenneire. *Virchows Archive Pathology Anatomy and Physiology* **340**, 369–83.

Troen, P., Kaufman, S.A. and Katz, K.H. 1951. Mercuric bichloride poisoning. *New England Journal of Medicine* **244**, 459–63.

Trojanowska, B., Piotrowski, J.K. and Szendzikowski, S. 1971. The influence of thioacetamide on the excretion of mercury in rats. *Toxicology and Applied Pharmacology* **18**, 374–86.

Weening, J.J., Hoedemaker, P.J. and Bakker, W.W. 1981. Immunoregulation and anti-nuclear antibodies in mercury-induced glomerulopathy in the rat. *Clinical and Experimental Immunology* **45**, 64–71.

Weinberg, J.M., Harding, P.G. and Humes, H.D. 1982a. Mitochondrial bioenergetics during the initiation of mercuric chloride-induced renal injury. I. Direct effects of *in vitro* mercuric chloride on renal cortical mitochondrial function. *Journal of Biological Chemistry* **257**, 60–7.

Weinberg, J.M., Harding, P.G. and Humes, H.D. 1982b. Mitochondrial bioenergetics during the initiation of mercuric chloride-induced renal injury. II. Functional alterations of renal cortical mitochondria isolated after mercuric chloride treatment. *Journal of Biological Chemistry* **257**, 68–74.

WHO 1991. *Environmental Health Criteria 118: Inorganic Mercury* World Health Organization, Geneva.

Woods, J.S., Calas, C.A., Aicher, L.D., Robinson, B.H. and Mailer, C. 1990a. Stimulation of porphyrinogen oxidation by mercuric ion. I. Evidence of free radical formation in the presence of thiols and hydrogen peroxide. *Molecular Pharmacology* **38**, 253–60.

Woods, J.S., Calas, C.A. and Aicher, L.D. 1990b. Stimulation of porphyrinogen oxidation by mercuric ion. Promotion of oxidation from the interaction of mercuric ion, glutathione, and mitochondria-generated hydrogen peroxide. *Molecular Pharmacology* **38**, 261–6.

Woods, J.S., Davis, H.A. and Baer, R.P. 1992. Enhancement of γ-glutamylcysteine synthetase mRNA in rat kidney by methyl mercury. *Archives of Biochemistry and Biophysics* **296**, 350–3.

Woods, J.S., Martin, M.D., Naleway, C.A. and Echeverria, D. 1993. Urinary porphyrin profiles as a biomarker of mercury exposure: Studies on dentists with occupational exposure to mercury vapor. *Journal of Toxicology and Environmental Health* **40**, 239–50.

Zalme, R.C., McDowell, F.M., Nagle, R.B., McNeil, J.S., Flamenbaum, W. and Trump, B.F. 1976. Studies on the pathophysiology of acute renal failure. II. A histochemical study of the proximal tubule of the rat following administration of mercuric chloride. *Virchows Archive [Cell Pathology]* **22**, 197–216.

Zalups, R.K. 1989. Effect of dietary K^+ and 75% nephrectomy on the morphology of principal cells in CCDs. *American Journal of Physiology* **257**, F387–F396.

Zalups, R.K. 1991a. Autometallographic localization of inorganic mercury in the kidneys of rats: Effect of unilateral nephrectomy and compensatory renal growth. *Experimental and Molecular Pathology* **54**, 10–21.

Zalups, R.K. 1991b. Method for studying the *in vivo* accumulation of inorganic mercury in segments of the nephron in the kidneys of rats treated with mercuric chloride. *Journal of Pharmacological Methods* **26**, 89–104.

Zalups, R.K. 1991c. Renal accumulation and intrarenal distribution of inorganic mercury in the rabbit: Effect of unilateral nephrectomy and dose of mercuric chloride. *Journal of Toxicology and Environmental Health* **33**, 213–28.

Zalups, R.K. 1993a. Early aspects of the intrarenal distribution of mercury after the intravenous administration of mercuric chloride. *Toxicology* **79**, 215–28.

Zalups, R.K. 1993b. Influence of 2,3-dimercaptopropane-1-sulfonate (DMPS) and meso-2,3-dimercaptosuccinic acid (DMSA) on the renal disposition of mercury in normal and uninephrectomized rats exposed to inorganic mercury. *Journal of Pharmacology and Experimental Therapeutics* **267**, 791–800.

Zalups, R.K. 1995. Organic anion transport and action of γ-glutamyltranspeptidase in kidney linked mechanistically to renal tubular uptake of inorganic mercury. *Toxicology and Applied Pharmacology* **132**, 289–298.

Zalups, R.K. 1997. Enhanced renal outer medullary uptake of mercury associated with uninephrectomy: Implication of a luminal mechanism. *Journal of Toxicology and Environmental Health* **50**, 173–94.

Zalups, R.K. 1998a. Intestinal handling of mercury in the rat: Implication of intestinal secretion of inorganic mercury following biliary ligation or cannulation. *Journal of Toxicology and Environmental Health* **53**, 615–36.

Zalups, R.K. 1998b. Basolateral uptake of inorganic mercury in the kidney. *Toxicology and Applied Pharmacology* **150**, 1–8.

Zalups, R.K. 1998c. Basolateral uptake of mercuric conjugates of *N*-acteylcysteine and cysteine in the kidney involves the organic anion transport system. *Journal of Toxicology and Environmental Health* **54**, 101–17.

Zalups, R.K. and Barfuss, D.W. 1990. Accumulation of inorganic mercury along the renal proximal tubule of the rabbit. *Toxicology and Applied Pharmacology* **106**, 245–53.

Zalups, R.K. and Barfuss, D.W. 1993. Transport and toxicity of methylmercury along the proximal tubule of the rabbit. *Toxicology and Applied Pharmacology* **121**, 176–85.

Zalups, R.K. and Barfuss, D.W. 1995a. Pretreatment with *p*-aminohippurate inhibits the renal uptake and accumulation of injected inorganic mercury in the rat. *Toxicology* **103**, 23–35.

Zalups, R.K. and Barfuss, D.W. 1995b. Accumulation and handling of inorganic mercury in the kidney after co-administration with glutathione. *Journal of Toxicology and Environmental Health* **44**, 385–99.

Zalups, R.K. and Barfuss, D.W. 1995c. Renal disposition of mercury in rats after intravenous injection of inorganic mercury and cysteine. *Journal of Toxicology and Environmental Health* **44**, 401–13.

Zalups, R.K. and Barfuss, D.W. 1996a. Diversion or prevention of biliary outflow from the liver diminishes the renal uptake of injected inorganic mercury. *Drug Metabolism and Disposition* **24**, 480–6.

Zalups, R.K. and Barfuss, D.W. 1996b. Nephrotoxicity of inorganic mercury co-administered with L-cysteine. *Toxicology* **109**, 15–29.

Zalups, R.K. and Barfuss, D.W. 1998a. Small aliphatic dicarboxylic acids inhibit renal uptake of administered mercury. *Toxicology and Applied Pharmacology* **148**, 183–93.

Zalups, R.K. and Barfuss, D.W. 1998b. Participation of mercuric conjugates of cysteine, homocysteine, and *N*-acetylcysteine in mechanisms involved in the renal tubular uptake of inorganic mercury. *Journal of the American Society of Nephrology* **9**, 551–61.

Zalups, R.K. and Cherian, M.G. 1992a. Renal metallothionein metabolism after a reduction of renal mass. I. Effect of unilateral nephrectomy and compensatory renal growth on basal and metal-induced renal metallothionein metabolism. *Toxicology* **71**, 83–102.

Zalups, R.K. and Cherian, M.G. 1992b. Renal metallothionein metabolism after a reduction of renal mass. II. Effect of zinc pretreatment on the renal toxicity and intrarenal accumulation of inorganic mercury. *Toxicology* **71**, 103–17.

Zalups, R.K. and Diamond, G.L. 1987. Mercuric chloride-induced nephrotoxicity in the rat following unilateral nephrectomy and compensatory renal growth. *Virchows Archive B* **53**, 336–46.

Zalups, R.K. and Henderson, D.A. 1992. Cellular morphology in outer medullary collecting duct: Effect of 75% nephrectomy and K^+ depletion. *American Journal of Physiology* **263**, F1119–F1127.

Zalups, R.K. and Lash, L.H. 1990. Effects of uninephrectomy and mercuric chloride on renal glutathione homeostasis. *Journal of Pharmacology and Experimental Therapy* **254**, 962–70.

Zalups, R.K. and Lash, L.H. 1994. Advances in understanding the renal transport and toxicity of mercury. *Journal of Toxicology and Environmental Health* **42**, 1–44.

Zalups, R.K. and Lash, L.H. 1997a. Binding of mercury in renal brush-border and basolateral membrane-vesicles: Implication of a cysteine conjugate of mercury involved in the luminal uptake of inorganic mercury. *Biochemical Pharmacology* **53**, 1889–900.

Zalups, R.K. and Lash, L.H. 1997b. Depletion of glutathione in the kidney and the renal disposition of administered inorganic mercury. *Drug Metabolism and Disposition* **25**, 516–23.

Zalups, R.K. and Minor, K.H. 1995. Luminal and basolateral mechanisms involved in the renal tubular uptake of inorganic mercury. *Journal of Toxicology and Environmental Health* **46**, 73–100.

Zalups, R.K., Stanton, B.A., Wade, J.B. and Giebisch, G. 1985. Structural adaptation in initial collecting tubule following reduction in renal mass. *Kidney International* **27**, 636–42.

Zalups, R.L., Klotzbach, J.M. and Diamond, G.L. 1987. Enhanced accumulation of inorganic mercury in renal outer medulla after unilateral nephrectomy. *Toxicology and Applied Pharmacology* **89**, 226–36.

Zalups, R.K., Cox, C. and Diamond, G.L. 1988. Histological and urinalysis assessment of nephrotoxicity induced by mercuric chloride in normal and uninephrectomized rats, in *Biological Monitoring of Toxic Metals*, (eds T.W. Clarkson, L. Friberg, G.F. Nordberg and P.R. Sager), Plenum Publishing, New York, pp. 531–45.

Zalups, R.K., Robinson, M.K. and Barfuss, D.W. 1991a. Factors affecting inorganic mercury transport and toxicity in the isolated perfused proximal tubule. *Journal of the American Society of Nephrology* **2**, 866–78.

Zalups, R.K., Gelein, R.M. and Cernichiari, E. 1991b. DMPS as a rescue agent for the nephropathy induced by mercuric chloride. *Journal of Pharmacology and Experimental Therapeutics* **256**, 1–10.

Zalups, R.K., Barfuss, D.W. and Kostyniak, P.J. 1992. Altered intrarenal accumulation of mercury in uninephrectomized rats treated with methylmercury chloride. *Toxicology and Applied Pharmacology* **115**, 174–82.

Zalups, R.K., Cherian, M.G. and Barfuss. D.W. 1993a. Mercury–metallothionein and the renal accumulation and handling of mercury. *Toxicology* **83**, 61–78.

Zalups, R.K., Knutson, K.L. and Schnellmann, R.G. 1993b. *In vitro* analysis of the accumulation and toxicity of inorganic mercury in segments of the proximal tubule isolated from the rabbit kidney. *Toxicology and Applied Pharmacology* **119**, 221–7.

Zalups, R.K., Parks, L., Cannon, V.T. and Barfuss, D.W. 1998. Mechanisms of action of 2,3-dimercaptopropane-1-sulfonate and the transport, disposition, and toxicity of inorganic mercury in isolated perfused segments of rabbit proximal tubules. *Molecular Pharmacology* **54**, 353–63.

10
TRANSPORT OF METALS IN THE NERVOUS SYSTEM

Michael Aschner and Laura E. Kerper

10.1 Introduction

The concept of a blood–brain barrier (BBB) was advanced by the German bacteriologist, Paul Ehrlich (1906). He noted that when injected intravenously, certain dyes stained all tissues and organs, with the exception of the central nervous system (CNS). His interpretation of these results was that the injected dyes had low affinity for the CNS, and, therefore, did not stain the brain parenchyma. However, Ehrlich was challenged by one of his students who asserted that his mentor's original conclusion was erroneous (Goldmann, 1913). To prove his point, Goldmann injected a dye (trypan blue) directly into the cerebrospinal fluid (CSF) of the spinal cord. The dye readily stained the brain parenchyma, but it did not enter the vasculature nor did it stain peripheral organs and tissues, prompting Goldmann to suggest that a physical barrier separated the brain compartment from the systemic circulation. The term 'blood–brain barrier' (*Bluthirnschranke*) was originally coined by Lewandowsky in 1900, after demonstrating that the dye, Prussian blue, injected into the peripheral vascular bed was excluded from the CNS parenchyma.

The BBB is a specialized structure responsible for the maintenance of the neuronal microenvironment. It plays a pivotal role in tissue homeostasis, vasotonus regulation, fibrinolysis and coagulation, the vascularization of normal and neoplastic tissues, and blood cell activation and migration during physiological (development and repair) and pathological processes (reviewed by Risau and Wolfburg, 1990; Risau, 1994, 1995). Regulation of blood–brain exchange of nutrients and macromolecules is accomplished by individual endothelial cells. These cells form the functional barrier, and they are continuously linked to each other by tight junctions (or zonulae occludentes) formed by their outer membrane leaflets. This specialization effectively and actively compartmentalizes the CNS and vasculature bed, and therefore negates oncotic and osmotic forces that govern blood–tissue exchange in peripheral tissues and organs. A pivotal function of the endothelial cells is to regulate the selective transport and metabolism of substances from blood to brain, as well as their transport in the opposite direction. Because of the existence of tight junctions between endothelial cells, nonspecific paracellular ionic leakage across the brain capillaries is kept to a minimum. It should be noted, however, that not all of brain vasculature is endowed with these specialized morphologic features. In contrast to the zonulae occludentes junctions of tight barrier areas, endothelial cells in the circumventricular organs possess maculae occludentes junctions. These junctions only partially occlude the gaps between juxtaposed endothelial cells, and, therefore, the barrier is not as tight or 'seamless' as in the zonulae occludentes-type junctions, and diffusion across the capillaries is more prevalent. From the toxicologic point of view, those areas that lack the zonulae occludentes-type junctions represent sites for potential neurotoxin transport or leakage into the CNS, because passage into the brain parenchyma is less restricted.

A number of factors determine transport across the endothelium of the BBB (Rapoport, 1976). For most organic solutes and macromolecules, permeability is largely dependent upon their lipophilicities. Certain molecules needed for brain metabolism, however, penetrate the BBB much more readily than one would predict solely based on their lipid solubilities (Oldendorf, 1970). Hydrophilic solutes and macromolecules are believed to cross the barrier via active transport carrier mechanisms or facilitated diffusion. Many of these carriers are symmetrically localized on both the luminal (blood-facing) and abluminal (brain-facing) membranes, while others have an asymmetric distribution (Goldstein and Betz, 1986). For example, the carriers for the essential neutral amino acids such as tryptophan, which are necessary for neurotransmitter synthesis, can be found on both the luminal and abluminal membranes. In contrast, the carrier for the inhibitory amino acid glycine is located only on the abluminal endothelial membrane. This asymmetric distribution of the glycine transporter facilitates its removal from the CNS, thus maintaining its CNS concentration at a low level. Similarly, the abluminal membrane is more densely populated than the luminal membrane with the enzyme ATPase (Goldstein and Betz, 1986). This enzyme forms the basis of the Na^+, K^+-ATPase pump that simultaneously transports Na^+ out of the endothelium into the brain, while pumping K^+ in the opposite direction. Like glycine, K^+ has a potent effect on the transmission of nerve impulses and neuron firing. Therefore, this asymmetric distribution functions to maintain the concentration of K^+ in the extracellular fluid at relatively low levels, preventing the repetitive depolarization of neurons. To date, an excess of 20 transport systems have been localized to the BBB endothelium. They function to facilitate and regulate the influx (as well as efflux) of hormones, vitamins, peptides and organic nutrients to the CNS. In addition to the ion transporters for Na^+ and K^+, a number of other ion-transporter systems (Ca^{2+}-ATPase) and numerous channels have also been localized on the BBB capillaries. These function to control and regulate the ionic homeostasis of the brain extracellular fluid (Smith *et al.*, 1997).

The permeability values of the BBB for many polar, hydrophilic compounds are generally 100–1000 times lower than those reported for peripheral capillaries, reflecting the restrictive transport properties of the BBB (Smith, 1990). While the half-times of transport for polar, hydrophilic compounds in peripheral tissues range from seconds to minutes, their half-times for transport and interstitial equilibration across the BBB range from hours to days (Smith *et al.*, 1997). The BBB is also characterized by a high electrical resistance, approximating 2000 ohm × cm^2, indicating a low conductance even to small ions (Risau and Wolfburg, 1990). The combination of this low passive permeability and the expression of facilitated and active transport systems on the endothelial cells provides for a highly regulated environment and facilitates the maintenance of internal milieu that is optimal for neural function.

The development of the unique properties of the brain microvasculature is a consequence of tissue-specific interactions between endothelial cells of extraneural origin and developing neuronal cells. Invading capillaries express BBB characteristics in response to inductive signals that originate from the surrounding neural tissue (Farrell and Risau, 1994; Risau, 1995). In the mammalian CNS, brain capillaries develop from solid cords of endothelial cells. These cords develop a slit-like lumen, the latter progressively increasing its caliber (Jacobson, 1987). As the endothelial cell layer becomes thicker, newly formed endothelial cell cords that sprout from more mature capillaries invade it. These cords are separated from juxtaposed neurons by a basement membrane, and they progressively become ensheathed by resident astrocytes. The most rapid capillary sprouting corresponds to the period of glial cell proliferation and neuronal dendritic development and arborization, suggesting a permissive role of glia (and astrocytes in particular) in BBB development. The density of CNS capillaries in various brain areas is thought to be correlated directly with the metabolic activity and oxygen consumption in each region.

Of particular importance to the development of the BBB are the astrocytes, which are thought to provide trophic factors for barrier induction and maintenance (reviewed by Aschner, 1998). Until about four decades ago, it was thought that the restrictive barrier was formed structurally by the astrocytic foot-processes, since the presence of juxtaposed astrocytes was the most obvious distinguishing feature between brain capillaries (nonfenestrated) and all other capillaries in the periphery (of the fenestrated type). However, advances in electron microscopy in the 1950s showed that the barrier proper resided in the zonulae occludentes between the endothelial cells, and not the astrocytes themselves, since there was free passage of markers between the astrocytic end-feet processes (reviewed by Goldstein and Betz, 1986). An interesting feature of the BBB is the extremely high density of intermembranous particle assemblies in astrocytic membranes facing the blood capillaries (Landis and Reese, 1982). The exact nature of these systems remains unknown, but it is thought that they comprise K^+ channels, since a high K^+ conductance has been found on the astrocytic membranes (Newman, 1986).

Many metals (calcium, copper, magnesium, manganese, iron, zinc, cobalt and molybdenum) are essential and required for optimal CNS function. They play important roles in brain function as catalysts, second messengers and gene modulators. Many of them also regulate enzyme and receptor activity and, accordingly, each must be supplied to the CNS at an optimal level, because both deficiency and excess can result in aberrant CNS function. The first step in the uptake of these metals into the CNS is their transport across the BBB. A series of active or receptor-mediated transport systems inherent to the BBB vasculature serve to control the transport of these metals into the brain, maintaining their optimal concentrations. Other metals that are nonessential (i.e. have no known functional attributes), such as mercury and lead, can also gain access to the CNS (Fig. 10.1). Despite the specialization of the BBB and its 'built-in' safeguards, this highly restrictive barrier is incapable of preventing the exchange of nonessential metals from the blood to the brain, since their transport is often governed by the same physiologic properties that govern the exchange of nutrients, therapeutic agents or hormones. Wetterhahn-Jennette (1981) and Clarkson (1993) have detailed elegantly in their respective reviews how the structural similarities of many of these nonessential metals to endogenous ligands lead to transport across the BBB via 'molecular mimicry'; namely, a metal complex is formed that mimics the behavior of normal endogenous substrates, utilizing transport systems inherent to the BBB to gain access to the CNS.

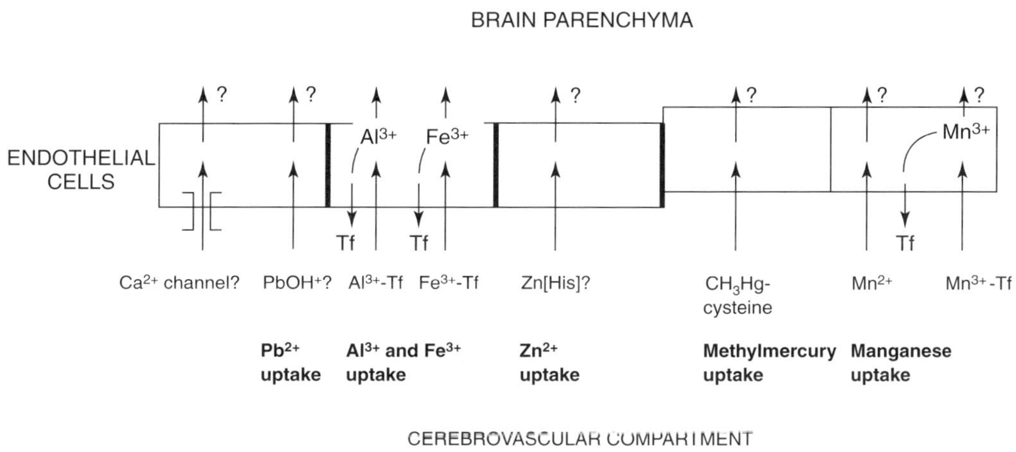

Figure 10.1 Mechanisms of metal transport at the blood–brain barrier. For details on each specific metal, refer to text. ? denotes unknown mechanism.

This review is not intended to be an exhaustive account of the transport of all metals in the CNS. Rather, it intends to provide examples that illustrate mechanisms of transport, addressing transport systems for both essential and nonessential metals. Readers interested in the morphological, evolutionary and embryological considerations of the BBB are referred to reviews by Bradbury (1997), Neuwlet (1989), Risau and Wolfburg (1990), Abbott (1991), Risau (1994, 1995) and Farrell and Risau (1994).

10.2 Transport of various essential metals across the BBB

10.2.1 Iron

The concentration of iron (Fe) in the CNS is quite high, and in certain regions, such as the basal nuclei (globus pallidus, putamen, etc.), iron reaches concentrations similar to those present in liver tissue (Bradbury, 1997). Elevated concentrations of iron within the CNS are necessary due to its requirement as a cofactor in certain enzymes along the mitochondrial electron transfer chain, such as aconitase, NADH–ubiquinone reductase (Cpx-I), and succinate dehydrogenase (succinate–ubiquinone reductase, SDH, complex-II). Aconitase, an enzyme present in mitochondria and cytoplasm, is responsible for the conversion of citrate to isocitrate in the tricarboxylic acid cycle. It contains a 4Fe–4S cubane cluster which serves as the enzymatic binding site for citrate. Cpx-I is present in the inner membrane of mitochondria and catalyzes the transfer of two electrons from NADH to the small, lipid-soluble molecule ubiquinone, referred to as coenzyme Q. This reaction is the first step in the respiratory chain that delivers electrons to the energy storage molecule, ATP. The enzyme consists of 16–24 nonheme Fe atoms in 5–8 Fe–S centers. SDH is a second enzyme involved in the mitochondrial electron transfer chain. This enzyme has 8 Fe–S bonds in three centers. In addition, iron has an important role as a cofactor in enzymes involved in the synthesis of catecholamines and myelin.

In the plasma, iron circulates predominantly as the trivalent ion (Fe^{3+}) complexed with transferrin (Tf), a glycoprotein with an approximate molecular weight of 80 kDa

(Huebers and Finch, 1987). Transport of iron across the luminal membrane of the capillary endothelium occurs by receptor-mediated endocytosis of ferric transferrin (for recent references see Ueda *et al.*, 1993; Broadwell *et al.*, 1996; Descamps *et al.*, 1996; Bradbury, 1997; van Gelder *et al.*, 1997), resulting in an initial linear uptake of iron into adult rat brain at an average rate relative to serum of about 3.3×10^{-3} ml/g brain/h. The receptor is a disulfide-linked, integral membrane glycoprotein with a molecular weight of approximately 180 000 Da (Huebers and Finch, 1987). Its affinity for apo-transferrin is approximately two orders of magnitude lower than its affinity for diferric transferrin (K_d 2–7 nM; Smith *et al.*, 1997).

Once transported across the luminal endothelial membrane, most of the iron is separated from Tf by acidification within endosomes (Huebers and Finch, 1987). Pulse–chase experiments in a culture model of the BBB (Descamps *et al.*, 1996) suggest that approximately 10% of Tf is recycled to the luminal side of the cells, whereas the majority of Tf is transcytosed to the abluminal side. The iron is transported across the endothelial abluminal membrane into the brain interstitial fluid by a currently unknown mechanism. Bradbury (1997) has postulated that iron is absorbed from the vesicular system into cytoplasm and transported across the abluminal endothelial cell membrane into the interstitial fluid as one or more species of low molecular weight. The same author has suggested that the transported species is ionic Fe^{2+}. Once within the brain interstitial fluid, iron binds with unsaturated Tf (Bradbury, 1997). The brain is endowed with a machinery for its own synthesis of Tf, which occurs in oligodendrocytes and the choroid plexus (Bloch *et al.*, 1985, 1987; Connor and Fine, 1987; Zheng, 1996). Tf-bound iron is subsequently transported from the interstitial fluid across cell membranes into neurons and glia. Iron leaves the CNS via the bulk outflow of CSF through arachnoid villi and other channels (Bradbury, 1997). Normally, 70–80% of CNS iron is found in the myelin fraction; histochemically it is localized in oligodendrocytes (Dwork *et al.*, 1988; Connor and Menzies, 1990).

While most results clearly show that the Fe–Tf complex is transcytosed across brain capillary endothelial cells by a receptor-mediated pathway, the potential for other transport systems should not be overlooked. The melanotransferrin (MTf) molecule, also called p97, is a melanoma cell-specific molecule, with a high level of sequence homology to Tf and lactoferrin. It is somewhat unusual since it occurs predominantly as a membrane-bound, glycosylphosphatidyl-inositol (GPI)-anchored molecule. The potential for this molecule to carry iron across the BBB was recently suggested by Rothenberger *et al.* (1996). These authors have noted a remarkable similarity between, yet differential distribution of, MTf and Tf. In all brain tissues, both MTf and the Tf receptor were found to localize highly to capillary endothelium, while Tf itself was shown to localize predominantly to glial cells. The authors suggest that GPI-anchored MTf may potentially provide a novel route for iron uptake independent of Tf and its receptor.

10.2.2 Zinc

The transport of zinc (Zn) into various brain regions of the rat was recently studied by Buxani-Rice *et al.* (1994). Radiolabeled Zn^{2+} ($^{65}Zn^{2+}$] was buffered with either histidine or bovine serum albumin prior to short cerebrovascular perfusion. In both cases the flux of $^{65}Zn^{2+}$ into the CNS appeared linear with time through 90 seconds of perfusion.

In the presence of albumin, the brain flux of $^{65}Zn^{2+}$ in the parietal cortex appeared to follow Michaelis–Menten kinetics, with an apparent K_m of 16 nM and a V_{max} of 44 nmol/kg/min. $^{65}Zn^{2+}$ flux into the CNS was enhanced by increased L-histidine concentrations (1 and 1000 nM). The combined effect of histidine and Zn^{2+} on CNS uptake was best described by a function of [$ZnHis^{+}$], and was not affected by either L-arginine or L-phenylalanine. These authors concluded, therefore, that a specific transport system for Zn^{2+} resides in the plasma membranes of the brain endothelium, and that the enhancement of Zn^{2+} flux into the CNS in the presence of L-histidine is associated with the diffusion of [$ZnHis^{+}$] product 'across unstirred layers "ferrying" zinc to and from transport sites' (Buxani-Rice *et al.*, 1994).

10.2.3 Aluminum

Like iron, aluminum (Al) is believed to be transported predominantly across the endothelial BBB as a Tf conjugate (Radunovic, 1994; Smith *et al.*, 1997). However, since the binding of aluminum (3+ oxidation state) to Tf is weaker than that of Fe^{3+}, it is also likely to circulate in plasma as a conjugate with other small molecular weight compounds, such as citrate (see below; Martin *et al.*, 1987; Allen *et al.*, 1995; Ackley and Yokel, 1997). The Al–Tf conjugate binds to the Tf receptor with a K_d approximating 13 nM (Roskams and Connor, 1990), only slightly higher than that for the Fe–Tf conjugate. Tf-mediated transport of aluminum into the CNS is relatively slow, with a BBB transfer constant of 3×10^{-7} ml/s/g (Radunovic, 1994). In addition to the above transporter, recent studies by Allen *et al.* (1995) have alluded to the presence of an efficient, energy-dependent carrier for aluminum that removes the metal efficiently from brain extracellular fluid, either into brain cells or blood. A follow-up microdialysis study by the same group (Ackley and Yokel, 1997) suggests that small molecular weight complexes of Al–citrate may be important in carrying aluminium across the BBB, and that the monocarboxylic acid transporter predominantly located at the BBB, rather than at neuronal or glial cell membranes, maintains a steady-state aluminum citrate brain-to-blood-ratio (Ackley and Yokel, 1997).

10.2.4 Manganese

In the Earth's crust, in water, and in atmospheric particulate matter, manganese (Mn) exists abundantly in a number of physical and chemical forms. Because its outer electron shell can donate up to seven electrons, manganese can assume 11 different oxidation states. Of environmental importance are Mn^{2+}, Mn^{4+} and Mn^{7+}. In living tissue, manganese has been found as Mn^{2+}, Mn^{3+} and Mn^{4+} (Archibald and Tyree, 1987). The ability of manganese to assume valence states ranging all the way from 3– to 7+ in combination with other elements is extremely important, emphasizing the potential for this element to act either as a pro-oxidant or antioxidant in biological media. Manganese also tends to form tight complexes with other substances (Keen, 1995). Chemically and biochemically, it shares a number of similarities with iron:

1 both are transition elements adjacent to each other in the periodic table;
2 both carry similar valent charges (2+ and 3+) in biological environments;
3 both have similar ionic radii;

4 both bind Tf strongly (Aschner and Aschner, 1990; Suarez and Eriksson, 1993; Ueda *et al.*, 1993); and
5 both accumulate preferentially within mitochondria (Gavin *et al.*, 1990).

Within the CNS, approximately 80% of manganese accumulates in astrocytes (Wedler and Denman, 1984).

Although only a small portion of systemic manganese is transported into the CNS, it has a defined role both in CNS homeostasis and pathology. Its presence is critical throughout the life span of the mammalian organism. It serves as a necessary constituent of many metalloproteins, including the mitochondrial enzymes superoxide dismutase and pyruvate carboxylase, as well as the glial-specific enzyme glutamine synthetase (for extensive reviews see Prohaska, 1987; Wedler, 1993).

Measurements of the initial rate of manganese transport into rat brain show that the transport rate is saturable (Aschner and Aschner, 1990; Murphy *et al.*, 1991; Rabin *et al.*, 1993; Ingersoll *et al.*, 1995). Murphy *et al.* (1991) have determined the *in vivo* BBB transport rate constant for Mn^{2+} and found a value of 3.2×10^{-4} ml/s/g in the parietal cortex. However, since manganese is thought to circulate in plasma as a conjugate with Tf, it has been suggested that Tf may also carry it into the CNS (Aschner and Aschner, 1990). When complexed with Tf, manganese is exclusively present in the trivalent oxidation state, with two metal ions tightly bound to each transferrin molecule. A bicarbonate-binding site is activated for each bound Mn ion, a step requiring carbonic anhydrase, the enzyme that converts CO_2 and H_2O to bicarbonate. In plasma, oxidizing agents such as ceruloplasmin may play a role in the oxidation of Mn^{2+} to Mn^{3+} (Gibbons *et al.*, 1976).

The role of Tf in manganese transport across the BBB is favored by several lines of evidence. In the rat brain, the return of manganese to control levels is significantly slower when it is administered beyond days 18–20 of life (Rhenberg *et al.*, 1981), a developmental period that coincides with the immunocytochemical localization of Tf (Connor and Fine, 1987) and carbonic anhydrase (Sapirstein, 1983) in the brain. There is additional theoretical and experimental evidence that Tf functions prominently in manganese transport across the BBB. In the absence of iron, the binding sites of transferrin can accommodate a number of other metals, including gallium, copper, chromium, cobalt, vanadium, aluminum, terbium and plutonium, raising the possibility that Tf functions *in vivo* as a transport agent for many of these metals. At normal plasma iron concentrations (0.9–2.8 μg/ml), normal iron-binding capacity (2.5–4 μg/ml), and normal Tf concentration in plasma (3 mg/ml), with two metal-ion-binding sites per molecule, only 30% are occupied by Fe^{3+}. Thus Tf has available 50 μmol of unoccupied Mn^{3+} binding sites per liter. Accordingly, analogous to the situation with aluminum, manganese may not need to displace iron in order to bind to Tf.

Examination of the distribution of Tf receptors in relationship to manganese accumulation in the brain is intriguing. Manganese concentrations are highest in the pallidum, thalamic nuclei and substantia nigra. The areas with the highest Tf receptor distribution are not identical to those that concentrate manganese. However, manganese-accumulating areas are efferent to areas of high Tf receptor density, suggesting that perhaps these sites may accumulate manganese through neuronal transport. For example, the manganese-rich areas of the ventral pallidum, globus pallidus and substantia nigra receive input from the nucleus accumbens and the caudate putamen

– two areas abundantly rich in Tf receptor. Indeed, recent studies by Sloot and Gramsbergen (1994) confirmed axonal transport of manganese, and that Mn taken up by nerve cells in the striatum is subsequently transported into the substantia nigra by way of anterograde axonal transport. The presence of Tf receptor in the neuropil may provide the anatomic substrate for the 'internalization' of Mn^{3+}–Tf conjugates into neurons. Other studies (Suarez and Eriksson, 1993) have corroborated the internalization of a Mn–Tf complex in SHSY5Y neuroblastoma cells. The Tf receptor on SHSY5Y cells can bind to and internalize an Mn–Tf complex as efficiently as an Fe–Tf complex. Accordingly, and analogous to the transport of iron (Roberts *et al.*, 1993), the evidence supports a manganese-transport model in which manganese-loaded Tf is taken up by receptor-mediated endocytosis at the luminal membrane of brain capillaries. The manganese then dissociates from the Tf complex, presumably in endosomal compartments, and is transcytosed by unknown mechanisms, while the Tf is retroendocytosed.

10.2.5 Copper

Copper is a trace element necessary for the normal function of several important enzymes, yet copper homeostasis is still poorly understood. Copper is required for infant growth, host defense mechanisms, bone strength, red and white cell maturation, iron transport, cholesterol and glucose metabolism, myocardial contractility and brain development (Walshe, 1995; Olivares and Uauy, 1996; Tumer and Horn, 1996). The mechanism for maintaining balance of the metal in the body is controlled by at least two genes (Walshe, 1995; Tumer and Horn, 1996), and disruption of one gene on the X chromosome leads to defective copper absorption, with consequent deficiency of available copper at the cellular level, commonly referred to as Menkes disease (an X-linked recessive disorder). The Menkes locus is mapped to Xq13.3 and the defective gene has been isolated (Tumer and Horn, 1996). Menkes disease is associated with abnormalities of collagen formation and brain maturation, culminating in early death. Disruption of the other copper-controlling gene, located on chromosome 13, is associated with Wilson's disease, manifested by excess copper accumulation in the body. This occurs initially in the liver, leading to cirrhosis, and subsequently in the CNS, leading to destruction of the motor control center and, frequently, changes in personality. Copper excess can also cause renal damage, osteoarticular changes and joint pains. Two human genes defective in Wilson's disease have also been isolated and, like their counterpart in Menkes disease, the predicted gene products are P-type ATPases. Both genes exhibit extensive sequence similarity and appear to be members of a new class of ATP-driven copper pumps involved in the regulation of cellular copper homeostasis (Solioz *et al.*, 1994).

Although increased brain copper uptake in Wilson's disease has been known for three decades (Oldendorf and Kitano, 1965), little is known about the cellular mechanisms associated with the transport of copper across the BBB. Evidence for uptake of nonceruloplasminic copper in the brain, either as free ion or amino acid conjugates, was provided by Chutkow (1978), but the properties and kinetics of copper transport across the BBB are yet to be determined.

10.3 Transport of various nonessential metals across the BBB

10.3.1 Lead

Neurotoxicity from lead (Pb) exposure is of great concern, especially since lead is ubiquitous in the environment and even very low concentrations in blood can cause deleterious neurological effects (reviewed by Goyer, 1993). Lead is known to cross the BBB and exert direct effects on the brain, but there have been few investigations into the mechanisms of transport.

Because the BBB excludes transport of most proteins from blood to brain, it is generally assumed that the transported species of lead is either free Pb^{2+} or a low molecular weight complex of lead. Free Pb^{2+} in human serum has been calculated to be approximately 1/5000 of total lead (Al-Modhefer *et al.*, 1991); under conditions of normal lead exposure in humans that would be roughly equal to 1×10^{-12} M free lead. Al-Modhefer *et al.* (1991) studied lead-binding characteristics in human serum, and report that approximately 60% of serum lead was bound to low molecular weight sulfhydryl-containing compounds, mainly cysteine. The remaining 40% was bound to proteins, mainly albumin. These authors found no significant binding to transferrin, ruling out transferrin-mediated transport across the BBB, as has been reported for a number of other metals (described above).

With an intravenous infusion system in rats, Bradbury and Deane (1986) found that uptake of ^{203}Pb into brain was linear with time up to 4 h, at both tracer concentrations and at 500 μM in the infusate. They also reported that the permeability of lead at the BBB was at least 24 times that of calcium. The authors suggested that these data indicate a rate limitation at the BBB and a large reservoir of lead-binding capacity within the brain which reduces the amount of lead available for back transport from brain to blood (Bradbury and Deane, 1993).

These authors also measured brain uptake of ^{203}Pb after short cerebrovascular perfusion (Deane and Bradbury, 1990). In these experiments ^{203}Pb uptake was linear over a concentration range of 0.1–4 μM and was not affected by the addition of 2.5 mM calcium, 50 μM D600 (a calcium-channel blocker), 10 mM $BaCl_2$, 1 mM bicarbonate, 50 μM 4,4′-diisothiocyanostilbene-2,2′-disulfonic acid (DIDS, an inhibitor of the Cl^-/HCO_3^- anion transport) or by the removal of sodium from the perfusate. The bicarbonate and the DIDS were added because a bicarbonate-dependent uptake of lead via anion transport had been reported previously in red blood cells (Simons, 1986).

^{203}Pb entry into brain was increased in these studies (Deane and Bradbury, 1990) by addition of 2 mM sodium vanadate, 2 μM carbonyl cyanide 4-(trifluoromethoxy)-phenylhydrazone (FCCP) or 2 μM stannic chloride, all inhibitors of Ca-ATPase. ^{203}Pb uptake into brain was decreased by addition of 0.5 mM *p*-chloromercuribenzene-sulfonate (PCMBS, a blocker of sulfhydryl groups), albumin, cysteine, or EDTA (substances that bind lead), or by decreasing pH of perfusate from 7.4 to 6.56.

The authors concluded that lead transport across the BBB was by passive diffusion of Pb^{2+} or a simple inorganic complex of lead, and further hypothesized that the transported species was $PbOH^+$ because of the pH effects on transport. They also suggested that lead was pumped out of brain endothelial cells back to blood by a Ca-ATPase, based on the effects of the Ca-ATPase inhibitors mentioned above. Ca-ATPase has been localized to both the luminal and abluminal membranes of brain endothelial cells (Nag, 1987), and this mechanism has also been reported for lead efflux from human red blood cells (Simons, 1988, 1993b).

In an *in vitro* system, Kerper and Hinkle (1997a) found evidence for another mechanism of lead transport across the BBB. These investigators used primary cultures of bovine brain capillary endothelial cells loaded with a fluorescent calcium-indicating dye, indo-1. This dye changes its fluorescence characteristics when bound to calcium, but it has a higher affinity for lead, and fluorescence can be measured at excitation and emission wavelengths where calcium binding has no effect but lead decreases fluorescence. In this system, lead uptake into brain endothelial cells was increased by depletion of intracellular calcium stores, analogous to store-operated calcium transport. Most eucaryotic cells have voltage-independent plasma-membrane calcium channels that open in response to depletion of calcium pools within the sarcoplasmic or endoplasmic reticulum (ER) (Putney and Bird, 1993; Berridge, 1995; Clapham, 1995). In a physiologic system these pools can be depleted in response to hormone or agonist binding to a G-protein-coupled receptor at the plasma membrane and subsequent intracellular signaling events that ultimately generate inositol trisphosphate. Inositol trisphosphate binds to a receptor on the endoplasmic membrane that is a calcium channel, and calcium leaves the ER and enters the cytosol. This event signals the plasma membrane calcium channels to open, and calcium enters the cell from the extracellular medium. Kerper and Hinkle (1997a) found that lead uptake could also be stimulated by depleting the ER of calcium. Whether this uptake was through calcium channels or other store-operated cation channels in the plasma membrane remains unclear.

10.3.2 Mercury

Mercury (Hg) exists in biological systems in both inorganic and organic forms. The most common organic form of mercury in biological systems is methylmercury (CH_3Hg^+), and this is more toxic to the nervous system than inorganic forms. Therefore, most of the literature concerning mercury transport in the nervous system is about methylmercury specifically, and we will focus on methylmercury in this part of the chapter.

Methylmercury is a potent neurotoxin, as evidenced by a number of tragic outbreaks of poisoning in the past few decades (Bakir *et al.*, 1973; Harada, 1995). It is especially damaging to the developing nervous system, causing decreased cell numbers and impaired cell migration in the brain (Reuhl and Chang, 1979). This can occur at blood concentrations that cause no apparent symptoms in adults. At higher levels of exposure, methylmercury is also neurotoxic to adults.

Upon ingestion, methylmercury distributes readily throughout the body. Transport into brain is rapid, with methylmercury reaching brain tissue within 5 min after intravenous injection (Thomas and Smith, 1979; Hirayama 1985), and within 4 h after ingestion (Sager *et al.*, 1982). It has been widely held that the high lipid solubility of methylmercury compounds accounts for the ease with which they are transported across the BBB. While it is true that methylmercury-halide compounds (e.g. CH_3HgCl) are lipophilic, such compounds account for only a negligible fraction of methylmercury in the body. Methylmercury has a very high affinity for sulfhydryl (SH) groups, and this high affinity governs its reactions in the body (Hughes, 1957). It is bound almost exclusively to proteins and low molecular weight sulfhydryl-containing compounds in blood and other tissues. Hirayama (1980, 1985), found that intravenous injection of CH_3HgCl together with cysteine, a sulfhydryl-containing amino acid, increased methylmercury uptake in brain. This increased uptake was inhibited by the large, neutral amino acid, phenylalanine, providing evidence that transport of methylmercury across the BBB was associated with the large, neutral amino-acid carrier, or L system.

Aschner and Clarkson (1987) measured the effects of infused L-cysteine and L- and D-leucine on the distribution of methylmercury in pregnant and nonpregnant rats, and found that infusion of L-cysteine caused an increase in the amount of methylmercury in brain as compared to saline-infused control rats. Infusion of L-leucine reduced methylmercury uptake into the brains of the nonpregnant rats only, and D-leucine had no effect on either group compared to saline. In a second study (Aschner and Clarkson, 1988a), this group also noted that infusion of L-cysteine into pregnant rats caused an increase over saline-infused controls in brain methylmercury uptake in the unborn pups as well as in the dams. Infusions of L-leucine or glutathione (GSH) caused decreased uptake of methylmercury in the brains of both pups and dams. Taken together, these two studies present further evidence of involvement of amino acids, and especially of L-cysteine, in brain uptake of methylmercury. These authors suggested that the CH_3Hg–cysteine complex acts as an amino-acid analog, similar in structure to methionine, and is transported by the L-system carrier for neutral amino acids.

In support of this theory, this group (Aschner and Clarkson, 1988b; Kerper *et al.*, 1992) also showed that when a mixture of CH_3HgCl and L-cysteine was injected into rats via the rapid carotid perfusion technique, and allowed one pass through the brain, the uptake of methylmercury into brain was higher than in controls, which were perfused with CH_3HgCl alone, and also higher than uptake with the D isomer of cysteine, demonstrating stereoselectivity of transport. Furthermore, the increased uptake due to L-cysteine was inhibited by co-injection of other large, neutral amino acids, but not by injection of the acidic amino acid, L-aspartic acid. These characteristics are all consistent with L-system transport.

Similar effects were seen by this group using an *in vitro* system with measurement of methylmercury uptake into isolated bovine brain capillary segments (Aschner and Clarkson, 1989). Uptake of methylmercury into the brain capillaries was enhanced by adding L-cysteine, but not D-cysteine or L-methionine. Again, the increased uptake with L-cysteine was abolished by the inclusion of L-methionine in the incubation solution. Mokrzan *et al.* (1995) used primary cultures of bovine brain capillary endothelial cells to show stereoselective uptake of CH_3Hg–L-cysteine that was inhibited by L-system substrates but not by substrates for the acidic amino-acid carrier.

Since methylmercury has a high affinity for thiol groups, it is assumed that a mixture of methylmercury and cysteine, in the absence of competing thiols, forms a stable complex at a 1 : 1 molar ratio. Although the CH_3Hg–thiol bonds are thermodynamically very stable, there is a rapid exchange of methylmercury among competing thiol groups (Rabenstein and Fairhurst, 1975). In a mixture of thiols, such as would exist in plasma, some of the methylmercury would presumably be bound to cysteine and some to other thiols. The concentration of free cysteine in plasma is approximately 10 μM (Bannai, 1984), and by itself may not account for the relatively rapid uptake of methylmercury into brain after exposure by intravenous injection or ingestion. Although GSH concentration in plasma is also approximately 10 μM, the turnover rate is high; it is therefore constantly being made and metabolized (McIntyre and Curthoys, 1980). The GSH complex with methylmercury may be a source of CH_3Hg–cysteine in the plasma available for transport across the BBB.

The mechanism for methylmercury efflux from brain endothelial cells has also been investigated. Kerper *et al.* (1996) found that efflux of methylmercury from primary cultures of bovine brain capillary endothelial cells was mediated by GSH. Depletion of intracellular GSH inhibited efflux from these cells. Incubation of cells with glutathione *S*-conjugates or GSH analogs also inhibited methylmercury efflux, presumably by competing for the same transport systems. Large, neutral amino acids had no effect on transport, nor did depletion of intracellular ATP. The authors concluded that methylmercury transport from capillary endothelial cells into brain may occur by complexation with GSH and transport across the cell membrane by an ATP-independent GSH conjugate transport system. GSH has also been implicated in the efflux of methylmercury from other cell types, including hepatocytes (Dutczak and Ballatori, 1994), astrocytes (Fujiyama *et al.*, 1994), renal cells (Yasutake *et al.*, 1989) and PC12 cells (Miura and Clarkson, 1993).

10.3.3 Cadmium

While the main target organs for cadmium poisoning are liver, kidney and lung, there have been some reports of neurotoxicity due to cadmium exposure in both humans and animals (reviewed by Arvidson, 1983). In experimental animals, cadmium is preferentially accumulated by liver and kidney, with small amounts reaching the brain (Chan and Cherian, 1993; Choudhuri *et al.*, 1996). Cadmium was found in the brains of newborn rats, born to mothers injected with ^{109}Cd during pregnancy (Lucis *et al.*, 1972). When cadmium in the nervous system was measured after administration to rats, the greatest accumulation was seen in peripheral ganglia, and CNS levels were low. Within the CNS and associated structures, most of the cadmium was located in the hypophysis, meninges, choroid plexus and pineal gland, structures that do not have a BBB (Arvidson and Tjälve, 1986). Thus cadmium entry to the brain is greatly limited by the BBB. In developing mice, distribution of cadmium to the brain decreased with increasing age of the mice (Choudhuri *et al.*, 1996). This was probably due to two factors – development of the BBB and expression of brain metallothionein, which is low in newborns and increases with age.

10.4 Non-blood–brain barrier transport

10.4.1 Neuronal transport

10.4.1.1 Lead

Lead has long been known to interfere with cellular calcium processes and to block calcium transport into cells (reviewed by Simons, 1993a). In recent years, lead has also been shown to enter various cell types through calcium channels. Neurons have several types of voltage-sensitive calcium channels that open in response to depolarization of the cell membrane (reviewed by Tsien *et al.*, 1987). Simons and Pocock (1987) used bovine adrenal medullary cells to show entry of lead through voltage-sensitive calcium channels. Bovine adrenal medullary cells are not neurons, but are derived from neural crest tissue, have voltage-sensitive calcium channels, and in primary culture have characteristics of sympathetic postganglionic cells. In this model, lead uptake was increased by depolarization of the cells with 77 mM KCl to approximately four times that of control cells. BAY K 8644, an agonist of voltage-sensitive calcium channels, also increased lead uptake. Uptake of lead was inhibited by addition of excess Ca^{2+} to the extracellular medium, and by methoxyverapamil, a calcium-channel antagonist. These results were confirmed by Tomsig and Suszkiw (1991). These authors used fura-2, a calcium-indicating dye that also binds lead, to measure lead entry into bovine adrenal medullary cells. Again, lead uptake was stimulated by depolarization with excess KCl and by BAY K 8644, and was inhibited by nifedipine, an antagonist for voltage-sensitive calcium channels. These investigators also noted that, unlike calcium, lead uptake was sustained over several minutes, indicating the absence of the channel inactivation that normally occurs after voltage-dependent calcium uptake. A continuous voltage-independent uptake of lead was also observed in cells under resting conditions; the nature of this uptake was not investigated.

It should be noted that lead uptake through voltage-sensitive calcium channels does not occur universally. In primary cultures of rat dorsal root ganglion neurons, Domann *et al.* (1997) found no evidence of lead uptake, either voltage-dependent or -independent. Lead did, however, inhibit the voltage-dependent uptake of calcium in these cells.

Lead may also be transported in neurons by retrograde axonal transport. In a study by Baruah *et al.* (1981), ^{203}Pb that was injected into the rat sciatic nerve reached the spinal cord within 9 days after injection. In this way it may be possible for lead to enter the central nervous system without crossing the BBB.

10.4.1.2 Mercury

There is surprisingly little information about the transport of mercury (organic or inorganic) in neurons. Hare and Atchison (1995) and Marty and Atchison (1997) observed that a disruption of Ca^{2+} regulation by methylmercury during depolarization in NG108-15 neuroblastoma cells and rat cerebellar granule cells was delayed by the voltage-sensitive calcium-channel blocker, nifedipine, while voltage-sensitive calcium-channel activators had the reverse effect. This group suggested that while methylmercury blocks calcium entry through voltage-sensitive calcium channels, it may also be able to enter the cells through these channels.

Arvidson (1987) presented evidence of retrograde axonal transport of inorganic mercury as a way for mercury to enter the brain while bypassing the BBB. He injected ^{203}Hg into the tongues of rats, and after 2 weeks found ^{203}Hg in the hypoglossal nuclei in the lower brainstem of the injected rats.

Efflux of methylmercury was studied in PC12 cells, a line derived from a rat pheochromocytoma that has neuronal characteristics when cultured with nerve growth factor (Miura and Clarkson, 1993). In this investigation a subline of PC12 cells was cloned that was resistant to both methylmercury and inorganic mercury toxicity. This subline took up methylmercury more slowly, accumulated lower levels, and had a faster rate of methylmercury efflux than the parent line. It also had four times as much intracellular glutathione (GSH). The authors suggested that the higher levels of GSH were responsible for the increased resistance to methylmercury. This was probably due in part to increased efflux of methylmercury as a GSH complex.

10.4.1.3 Cadmium

Cadmium transport in neurons has not been studied, extensively probably because the brain is not a major target organ of cadmium toxicity. There is evidence in another excitable cell type, the GH4C1 pituitary cell line, that cadmium enters through voltage-sensitive L-type calcium channels (Hinkle *et al.*, 1987). Whether this mode of entry also occurs in neuronal cells is not currently known.

Cadmium, like lead and mercury, has also been observed to undergo retrograde axonal transport (Arvidson, 1985). ^{109}Cd that was injected into the tongue of rats was found in the hypoglossal nuclei of the lower brainstem 2 weeks after injection. Thus cadmium may also be able to reach the brain without crossing the BBB.

10.4.1.4 Manganese

Manganese is widely used as a calcium surrogate to study cation transport through calcium channels in neurons and other cell types. It enters cells through both voltage-sensitive (Anderson, 1983; Drapeau and Nachshen, 1984; Lansman *et al.*, 1986) and voltage-insensitive (Hallam and Rink, 1985; Jacob, 1990) calcium channels. Like other divalent cations such as Sr^{2+}, Ba^{2+}, Cd^{2+} and Mg^{2+}, Mn^{2+} is transported through calcium channels more slowly than Ca^{2+}, but is an effective blocker of Ca^{2+} transport (Nachshen and Blaustein, 1982; Nachshen, 1984).

Sloot and Gramsbergen (1994) observed anterograde axonal transport of manganese in rat basal ganglia. Other mechanisms of neuronal manganese uptake are not currently known.

10.4.2 Neuroglial cell transport

10.4.2.1 Lead

Astrocytes accumulate high amounts of lead compared to neurons and the surrounding medium (Holtzman *et al.*, 1984; Tiffany-Castiglioni, 1986). At the same time, they are more resistant to lead toxicity than other CNS cells (Holtzman *et al.*, 1987; Wu and Tiffany-Castiglioni, 1987). In one study, rat astrocytes in culture accumulated lead to at least 55 times the concentration in the culture medium (Tiffany-Castiglioni, 1986). Astrocytes that are exposed to high amounts of lead contain lead inclusion bodies in the cytoplasm and nucleus (Tiffany-Castiglioni *et al.*, 1989). This sequestration of lead by astrocytes may serve as a mechanism to protect the brain from lead-induced damage. In fact, developing rats become resistant to lead encephalopathy as their astrocytes develop the ability to sequester lead (Holtzman *et al.*, 1984).

Astrocytes contain high levels of GSH compared to other cells (Raps *et al.*, 1989), and this may account in part for their sequestration and relative resistance to lead toxicity. In cultured rat astrocytes, lead has been shown to increase intracellular levels of GSH after several days of exposure (Legare *et al.*, 1993). There are many ways that GSH can protect a cell from metal toxicity (these will be discussed below), but one of the most important may be direct chelation. This can serve dual purposes: binding to GSH keeps the metal from binding to other cell components where it may cause more harm, and may also facilitate efflux of the metal from the cell (as described below for methylmercury). Whether GSH facilitates efflux of lead from astrocytes is not known.

Since these cells tend to sequester lead, this mechanism may not be as important for lead as it is for methylmercury (Fujiyama *et al.*, 1994).

Very little is known about mechanisms of lead uptake into astrocytes and other glial cells. Unlike neurons, glial cells do not have voltage-sensitive calcium channels. Kerper and Hinkle (1997b) showed that uptake in C6 rat glioma cells was increased by depletion of intracellular calcium stores, indicating transport via a store-operated cation channel. A basal component of lead uptake was also observed in these cells; the mechanism of this uptake was not determined.

10.4.2.2 Mercury

Methylmercury transport across the BBB as a complex with L-cysteine, via the neutral amino-acid system L, was described above. Work by Aschner *et al.* (1990a; 1991) suggests that the uptake of methylmercury by astrocytes occurs by a similar mechanism. When methylmercury is added to the media as the L-cysteine conjugate, uptake by astrocytes exhibits the kinetic criteria of a specific transport system. Saturation kinetics, substrate specificity and inhibition, and *trans*-stimulation were demonstrated in the presence of this SH-containing amino acid. Cysteine-mediated uptake of methylmercury was inhibited by the co-administration of L-methionine, and 2-aminobicyclo-(2, 2, 1)-heptane-2-carboxylic acid, a specific substrate for the neutral amino-acid transport-system L. 2-Methylaminoisobutyric acid was ineffective in inhibiting the uptake of the methylmercury–cysteine conjugate. Preloading of the astrocytes with glutamate was moderately effective in *trans*-stimulating the uptake of methylmercury–cysteine conjugates, while in the absence of cysteine, uptake of [^{203}Hg]methylmercury was unchanged. These results indicate the presence in astrocytes of a neutral amino-acid carrier transport-system L, capable of selectively mediating methylmercury–cysteine uptake.

Additional studies indicate that the putative methylmercury transport system in astrocytes can also mediate methylmercury transport in the net efflux direction (Aschner *et al.*, 1991). Metal exchange is not required for efflux since efflux of [^{203}Hg]CH_3HgCl occurs in methylmercury-free buffer. The apparent stimulatory effects of extracellular methylmercury–L-cysteine conjugates on the rate and amount of ^{203}Hg efflux are clearly shown to reflect displacement of [^{203}Hg]CH_3HgCl from intracellular pools via a specific effect on the transport system itself. The simplest interpretation is that the efflux observed represents transport by the same neutral amino-acid system L that facilitates methylmercury uptake (Aschner *et al.*, 1990b).

Studies on the relationship between intracellular GSH levels and the efflux of methylmercury from astrocytes were recently carried out by Fujiyama *et al.* (1994). Treatment of astrocytes with 2-oxothiazolidine-4-carboxylic acid (OTC), resulted in a significant increase of cellular levels of GSH, and OTC also increased the efflux of preloaded methylmercury. GSH isopropyl ester, a second GSH enhancer, also facilitated methylmercury efflux from the astrocytes. Additional studies by the same group revealed that methylmercury was exclusively released as a methylmercury–GSH conjugate, and that efflux occurred via the probenecid-sensitive organic acid transport system, since it could be significantly abolished in the presence of probenecid. Accordingly, conjugation with GSH may also represent a major pathway for methylmercury release from astrocytes.

10.4.2.3 Cadmium

There is very little information about transport of cadmium in glial cells. In C6 rat glioma cells, cadmium uptake was observed with intracellular fluorescent calcium indicator dyes that also bind cadmium (Hinkle *et al.*, 1992). These cells accumulated cadmium much more slowly than excitable GH3 cells that have voltage-sensitive calcium channels. The mechanism of uptake in C6 cells was not determined.

10.4.2.4 Manganese

In view of the preferential accumulation of brain manganese in astrocytes, recent studies were directed to determine the transport mechanisms of manganese in these cells. The time-dependent uptake of ^{54}Mn into cultured astrocytes was studied by Aschner *et al.* (1992). The initial rate (1 min) of manganese uptake by astrocytes is concentration-dependent. The apparent kinetic constants are $K_m = 0.30 \pm 0.03$ μM and $V_{max} = 0.30 \pm 0.02$ nmol manganese/min/mg protein. It was found that both Ca^{2+} and Mn^{2+} compete for the same transport systems in astrocyte plasma membranes which may include Ca^{2+} channels, Na^+/Ca^{2+} exchange, or both. Co^{2+}, Zn^{2+} and Pb^{2+} do not have any effect on the initial rate of ^{54}Mn uptake, suggesting that only Ca^{2+} and Mn^{2+} share the same transport system. Transport of ^{54}Mn into astrocytes is also independent of the presence of $[Na^+]_o$.

The kinetic data showing saturation kinetics, competition by related substrates (i.e. Ca^{2+}) and a probable counter transport mechanism exhibit the accepted criteria of a transport process facilitated by a specific membrane transport protein. The relatively high capacity (V_{max}) of 0.30 nmol/mg protein/min and the apparent K_m are consistent with an efficient uptake of Mn^{2+} by astrocytes. Metal exchange is not obligatory for manganese efflux. However, increasing $[Mn^{2+}]_o$ has a marked effect on the initial rate of release of manganese. The simplest interpretation of these data is that extracellular manganese *trans*-stimulates the efflux of intracellular manganese, most consistent with uptake and efflux occurring on the same transport system or a separate exchange system. The biphasic kinetic profiles of manganese efflux suggest several types of intracellular manganese pools or different transport systems (Aschner *et al.*, 1992). The fast phase of efflux (up to 10 min) must occur from a pool that is in rapid equilibrium with the extracellular medium. This pool can be assumed to contain all intracellular complexes whose dissociation rate constants are higher than the rate constant for efflux. A second pool must be in slow equilibrium with the rapid equilibrium pool to account for the slower phase of manganese release between 10 and 120 min. Finally, a third pool of manganese (~25–30%) is unavailable for efflux, representing irreversibly bound manganese. This pool may account for the principal concentrative mechanism for manganese in astrocytes and may represent manganese uptake into mitochondria.

10.4.3 Choroid plexus transport

The choroid plexus manufactures and maintains the chemical composition of the CSF. It has active transport mechanisms to take up nutrients from blood, and it also transports substances into blood from the CSF (Mihorat, 1976). The epithelial cells of the choroid plexus have tight junctions, restricting passage of substances into the CSF (Spector, 1989). There is evidence that the choroid plexus sequesters heavy metals

(Zheng *et al.*, 1991), thereby limiting their concentrations in the CSF and brain parenchyma. Zheng *et al.* (1991) measured lead, cadmium, mercury and arsenic in rats or rabbits 24 h after intraperitoneal injection, and found concentrations in the choroid plexus to be many times higher than in CSF, blood or brain. In addition, there was evidence of an energy-requiring mechanism for uptake of metals from the CSF into the choroid plexus. Thus, the choroid plexus may serve to protect the brain from heavy metal exposure.

10.5 Glutathione as an effector of metal distribution

Glutathione (GSH) is the most abundant sulfhydryl-containing nonprotein peptide in many cells (Meister, 1988). It protects cells against metal toxicity in a number of ways. It protects against oxidative damage in the cell, removing reactive oxygen species such as hydrogen peroxide, scavenging free-radicals, and maintaining the intracellular concentration of reduced sulfhydryl groups (Meister and Anderson, 1983). It can also protect the cell by directly chelating heavy metals, which have high affinities for the SH group of the cysteine moiety. This chelation of metals can serve two purposes: it prevents the metals from binding to other, more vulnerable cell components, and in some cases it also may facilitate efflux of the metal from the cell.

There are several lines of evidence that GSH is involved with distribution and toxicity of heavy metals in the brain. Choi *et al.* (1996) increased resistance to methylmercury toxicity in mice by increasing the GSH concentrations in liver and brain. GSH levels were increased by oral or intraperitoneal administration of GSH-glycoside, which entered cells and apparently was metabolized to GSH. Miura and Clarkson (1993) developed a subline of PC12 cells that were resistant to methylmercury toxicity; this subline had four times the concentration of GSH of the parent line. Legare *et al.* (1993) showed that astrocytes that were exposed to lead in culture had increased concentrations of GSH after several days; this may be a protective mechanism against metal toxicity. Astrocytes have higher amounts of GSH than neurons (Slivka *et al.*, 1987; Raps *et al.*, 1989), and are more resistant to lead (Holtzman *et al.*, 1984; Wu and Tiffany-Castiglioni, 1987). Astrocytes also tend to sequester lead (Holtzman *et al.*, 1984; Tiffany-Castiglioni *et al.*, 1987), methylmercury (Oyake *et al.*, 1966; Garman *et al.*, 1975) and manganese (Aschner *et al.*, 1992). Whether or not GSH is involved in this sequestration remains to be investigated.

As noted above, GSH may also afford protection against heavy-metal toxicity in the nervous system by mediating efflux of metals from cells. There is direct evidence that this is the case for methylmercury. Efflux of methylmercury mediated by GSH has been shown in neuron-like PC12 cells (Miura and Clarkson, 1993), astrocytes (Fujiyama *et al.*, 1994) and brain capillary endothelial cells (Kerper *et al.*, 1996). Efflux probably occurs after formation of a methylmercury–glutathione complex, via a glutathione or glutathione conjugate transport system. Whether or not glutathione is involved in efflux of other heavy metals in the nervous system has yet to be determined.

Acknowledgments

This paper was partially supported by grants from the National Institute of Environmental Health Services ES 07331, AA 11617, and the US Environmental Protection Agency R-819210 to MA, and National Research Service Award ES05714 to LEK.

References

Abbott, N.J. 1991. Glial–neuronal interactions. *Annals of the New York Academy of Sciences* **633**.

Ackley, D.C. and Yokel, R.A. 1997. Aluminum citrate is transported from brain into blood via the monocarboxylic acid transporter located at the blood–brain barrier. *Toxicology* **120**, 89–97.

Allen, D.D., Orvig, C. and Yokel, R.A. 1995. Evidence for energy-dependent transport of aluminum out of brain extracellular fluid. *Toxicology* **98**, 31–9.

Al-Modhefer, A.J.A., Bradbury, M.W.B. and Simons, T.J.B. 1991. Observations of the chemical nature of lead in human blood serum. *Clinical Sciences* **81**, 823–9.

Anderson, A. 1983. Mn ions pass through calcium channels. *Journal of General Physiology* **81**, 805–27.

Archibald, F.S. and Tyree, C. 1987. Manganese poisoning and the attack of trivalent manganese upon catecholamines. *Archives of Biochemistry and Biophysics* **256**, 638–50.

Arvidson, B. 1983. Cadmium toxicity and neural cell damage, in *Neurobiology of the Trace Elements*, (eds I.E. Dreosti and R.M. Smith), Humana Press, Clifton, NJ, pp. 51–78.

Arvidson, B. 1985. Retrograde axonal transport of cadmium in the rat hypoglossal nerve. *Neuroscience Letters* **62**, 45–9.

Arvidson, B. 1987. Retrograde axonal transport of mercury. *Experimental Neurology* **98**, 198–203.

Arvidson, B. and Tjälve, H. 1986. Distribution of ^{109}Cd in the nervous system of rats after intravenous injection. *Acta Neuropathol.* **69**, 111–16.

Aschner, M. 1998. Blood–brain barrier: Physiological and functional considerations, in *Handbook of Developmental Neurotoxicology*, (eds W. Slikker, Jr and L.W. Chang), Academic Press, San Diego, 339–51.

Aschner, M. and Aschner, J.L. 1990. Manganese transport across the blood–brain barrier: relationship to iron homeostasis. *Brain Research Bulletin* **24**, 857–60.

Aschner, M. and Clarkson, T.W. 1987. Distribution of mercury 203 in pregnant and non-pregnant rats following systemic infusions with thiol-containing amino acids. *Teratology* **36**, 321–8.

Aschner, M. and Clarkson, T.W. 1988a. Distribution of mercury 203 in pregnant rats and their fetuses following systemic infusions with thiol-containing amino acids and glutathione during late gestation. *Teratology* **38**, 145–55.

Aschner, M. and Clarkson, T.W. 1988b. Uptake of methylmercury in the rat brain: effects of amino acid. *Brain Research* **462**, 31–9.

Aschner, M. and Clarkson, T.W. 1989. Methylmercury uptake across bovine brain capillary endothelial cells *in vitro*: the role of amino acid. *Pharmacology and Toxicology* **64**, 293–7.

Aschner, M., Eberle, N.B., Goderie, S. and Kimelberg, H.K. 1990a. Methylmercury uptake in rat primary astrocyte cultures: the role of the neutral amino acid transport system. *Brain Research* **524**, 221–8.

Aschner, M., Eberle, N.B., Miller, K. and Kimelberg, H.K. 1990b. Interactions of methylmercury with rat primary astrocyte cultures: Inhibition of rubidium and glutamate uptake and induction of swelling. *Brain Research* **530**, 245–50.

Aschner, M., Eberle, N. and Kimelberg, H.K. 1991. Interaction of methylmercury with rat primary astrocyte cultures: methylmercury efflux. *Brain Research* **554**, 10–14.

Aschner, M., Gannon M. and Kimelberg, H.K. 1992. Manganese (Mn) uptake and efflux in cultured rat astrocytes. *Journal of Neurochemistry* **58**, 730–5.

Bakir, F., Damluji, S.F., Amin-Zaki, L., Murtadha, M., Khalidi, A., al-Rawi, N.Y., Tikriti, S., Dahahir, H.I., Clarkson, T.W., Smith, J.C. and Doherty, R.A. 1973. Methylmercury poisoning in Iraq. *Science* **181**, 230–41.

Bannai, S. 1984. Transport of cystine and cysteine in mammalian cells. *Biochimica et Biophysica Acta* **779**, 289–306.

Baruah, J.K., Rasool, C.G., Bradley, W.G. and Munsat, T.L. 1981. Retrograde axonal transport of lead in rat sciatic nerve. *Neurology* **31**, 612–16.

Berridge, M.J. 1995. Capacitative calcium entry. *Biochemical Journal* **312**, 1–11.

Bloch, B., Popovici, T., Levin, M.J., Tuil, D. and Kahn, A. 1985. Transferrin gene expression visualized in oligodendrocytes of the rat brain using 'in situ' hybridization and immunohistochemistry. *Proceedings of the National Academy of Sciences, USA* **82**, 6706–10.

Bloch, B., Popovici, T., Chouham, S., Levin, M.J., Tuil, D. and Kahn, A. 1987. Transferrin gene expression in choroid plexus of the adult rat brain. *Brain Research Bulletin* **18**, 573–6.

Bradbury, M.W.B. 1997. Transport of iron in the blood–brain–cerebrospinal fluid system. *Journal of Neurochemistry* **69**, 443–54.

Bradbury, M.W.B. and Deane, R. 1986. Rate of uptake of lead-203 into brain and other soft tissues of the rat at constant radiotracer levels in plasma. *Annals New York Academy of Sciences* **481**, 142–60.

Bradbury, M.W.B. and Deane, R. 1993. Permeability of the blood–brain barrier to lead. *Neurotoxicology* **14**, 131–6.

Broadwell, R.D., Baker-Cairns, B.J., Friden, P.M., Oliver, C. and Villegas, J.C. 1996. Transcytosis of protein through the mammalian cerebral epithelium and endothelium. III. Receptor-mediated transcytosis through the blood–brain barrier of blood-borne transferrin and antibody against the transferrin receptor. *Experimental Neurology* **142**, 47–65.

Buxani-Rice, S., Ueda, F. and Bradbury, M.W. 1994. Transport of zinc-65 at the blood–brain barrier during short cerebrovascular perfusion in the rat: its enhancement by histidine. *Journal of Neurochemistry* **62**, 665–72.

Chan, H.M. and Cherian, M.G. 1993. Mobilization of hepatic cadmium in pregnant rats. *Toxicology and Applied Pharmacology* **120**, 308–14.

Choi, B.H., Yee, S. and Robles, M. 1996. The effects of glutathione glycoside in methyl mercury poisoning. *Toxicology and Applied Pharmacology* **141**, 357–64.

Choudhuri, S., Liu, W.L., Berman, N.E.J. and Klaassen, C.D. 1996. Cadmium accumulation and metallothionein expression in brain of mice at different stages of development. *Toxicology Letters* **84**, 127–33.

Chutkow, J.G. 1978. Evidence for uptake of nonceruloplasminic copper in the brain: effect of ionic copper and amino acids. *Proceedings of the Society for Experimental Biology and Medicine* **158**, 113–16.

Clapham, D.E. 1995. Calcium signaling. *Cell* **80**, 259–68.

Clarkson, T.W. 1993. Molecular and ionic mimicry of toxic metals. *Annual Reviews of Pharmacology and Toxicology* **32**, 545–71.

Connor, J.R. and Fine, R.E. 1987. Development of transferrin-positive oligodendrocytes in the rat central nervous system. *Journal of Neuroscience Research* **17**, 51–9.

Connor, J.R. and Menzies, S.L. 1990. Altered cellular distribution of iron in the central nervous system of myelin deficient rats. *Neuroscience* **34**, 265–71.

Deane, R. and Bradbury, M.W.B. 1990. Transport of lead-203 at the blood–brain barrier during short cerebrovascular perfusion with saline in the rat. *Journal of Neurochemistry* **54**, 905–14.

Descamps, L., Dehouck, M.P., Torpier, G. and Cecchelli, R. 1996. Receptor-mediated transcytosis of transferrin through blood–brain barrier endothelial cells. *American Journal of Physiology* **270**, H1149–H1158.

Domann, R., Wunder, L. and Büsselberg, D. 1997. Lead reduces depolarization-induced calcium entry in cultured DRG neurons without crossing the cell membrane: Fura-2 measurements. *Cellular and Molecular Neurobiology* **17**, 305–14.

Drapeau, P. and Nachshen, D.A. 1984. Manganese fluxes and manganese-dependent neurotransmitter release in presynaptic nerve endings isolated from rat brain. *Journal of Physiology (London)* **348**, 493–510.

Dutczak, W.J. and Ballatori, N. 1994. Transport of the glutathione–methylmercury complex across liver canalicular membranes on reduced glutathione carriers. *Journal of Biological Chemistry* **269**, 9746–51.

Dwork, A.J., Schon, E.A. and Herbert, J. 1988. Non identical distribution of transferrin and ferric iron in human brain. *Neuroscience* **27**, 333–45.

Ehrlich, P. 1906. Über die Beziehungen von chemische Constitution, Vertheilung, und pharmakologisher Wirkung. *Collected Studies in Immunity. Repr. and Trans.*, Wiley, New York, pp. 567–95.

Farrell, C.L. and Risau, W. 1994. Normal and abnormal development of the blood–brain barrier. *Microscopy Research and Technique* **27**, 495–506.

Fujiyama, J., Hirayama, K. and Yasutake, A. 1994. Mechanism of methylmercury efflux from cultured astrocytes. *Biochemical Pharmacology* **47**, 1525–30.

Garman, R.H., Weiss, B. and Evans, H.L. 1975. Alkylmercurial encephalopathy in the monkey; a histopathologic and autoradiographic study. *Acta Neuropathologiae* **32**, 61–74.

Gavin, C.E., Gunter, K.K. and Gunter, T.E. 1990. Manganese and calcium efflux kinetics in brain mitochondria. Relevance to manganese toxicity. *Biochemical Journal* **266**, 329–34.

Gibbons, R.A., Dixon, S.N., Hallis, K., Russell, A.M., Sansom, B.F. and Symonds, H.W. 1976. Manganese metabolism in cows and goats. *Biochimica et Biophysica Acta* **444**, 1–10.

Goldmann, E.E. 1913. Vitalfärbung am Zentralnervensystem. *Abh. Preuss. Akad. Wiss. Phys.-Math.* **1**, 1–60.

Goldstein, G.A. and Betz, A.L. 1986. The blood–brain barrier. *Scientific American* **255**, 74–83.

Goyer, R.A. 1993. Lead toxicity: Current concerns. *Environmental Health Perspectives* **100**, 177–87.

Hallam, T.J. and Rink, T.J. 1985. Agonists stimulate divalent cation channels in the plasma membrane of human platelets. *FEBS Letters* **186**, 175–9.

Harada, M. 1995. Minamata disease: Methylmercury poisoning in Japan caused by environmental pollution. *Critical Reviews in Toxicology* **25**, 1–24.

Hare, M.F. and Atchison, W.D. 1995. Nifedipine and tetrodotoxin delay the onset of methylmercury-induced increase in $[Ca^{2+}]_i$ in NG108-15 cells. *Toxicology and Applied Pharmacology* **135**, 299–307.

Hinkle, P.M., Kinsella, P.A. and Osterhoudt, K.C. 1987. Cadmium uptake and toxicity via voltage-sensitive calcium channels. *Journal of Biological Chemistry* **262**, 16333–7.

Hinkle, P.M., Shanshala, E.D. II and Nelson, E.J. 1992. Measurement of intracellular cadmium with fluorescent dyes. *Journal of Biological Chemistry* **267**, 25553–9.

Hirayama, K. 1980. Effects of amino acids on brain uptake of methylmercury. *Toxicology and Applied Pharmacology* **55**, 318–23.

Hirayama, K. 1985. Effects of combined administration of thiol compounds and methylmercury chloride on mercury distribution in rats. *Biochemical Pharmacology* **34**, 2030–2.

Holtzman, D., De Vries, C., Nguyen, H., Olson, J. and Bensch, K. 1984. Maturation of resistance to lead encephalopathy: Cellular and subcellular mechanisms. *Neurotoxicology* **5**, 97–124.

Holtzman, D., Olson, J.E., De Vries, C. and Bensch, K. 1987. Lead toxicity in primary cultured cerebral astrocytes and cerebellar granular neurons. *Toxicology and Applied Pharmacology* **89**, 211–25.

Huebers, H.A. and Finch, C.A. 1987. The physiology of transferrin and transferrin receptors. *Physiology Reviews* **67**, 520–82.

Hughes, W.H. 1957. A physicochemical rationale for the biological activity of mercury and its compounds. *Annals of the New York Academy of Sciences* **65**, 454–60.

Ingersoll, R.T., Montgomery, E.B. and Aposhian, H.V. 1995. Central nervous system toxicity of manganese. I. Inhibition of spontaneous motor activity in rats after intrathecal administration of manganese chloride. *Toxicology and Applied Pharmacology* **27**, 106–13.

Jacob, R. 1990. Agonist-stimulated divalent cation entry into single cultured human umbilical vein endothelial cells. *Journal of Physiology (London)* **421**, 55–77.

Jacobson, M. 1987. Histogenesis and morphogenesis of the central nervous system, in *Developmental Neurobiology*, (ed. M. Jacobson), Plenum Press, New York, pp. 57–114.

Keen, C.L. 1995. Overview of manganese toxicity, in *Proceedings of the Workshop on the Bioavailability and Oral Toxicity of Manganese*, (ed. S. Velazquez), EPA Liaison, US EPA, Environmental Criteria and Assessment Office, pp. 3–11.

Kerper, L.E. and Hinkle, P.M. 1997a. Lead uptake in brain capillary endothelial cells: Activation by calcium store depletion. *Toxicology and Applied Pharmacology* **146**, 127–33.

Kerper, L.E. and Hinkle, P.M. 1997b. Cellular uptake of lead is activated by depletion of intracellular calcium stores. *Journal of Biological Chemistry* **272**, 8346–52.

Kerper, L.E., Ballatori, N. and Clarkson, T.W. 1992. Methylmercury transport across the blood–brain barrier by an amino acid carrier. *American Journal of Physiology* **262**, R761–R765.

Kerper, L.E., Mokrzan, E.M., Ballatori, N. and Clarkson, T.W. 1996. Methylmercury efflux from brain capillary endothelial cells is modulated by intracellular glutathione but not ATP. *Toxicology and Applied Pharmacology* **141**, 526–31.

Landis, D. and Reese, T.S. 1982. Regional organization of astrocytic membranes in cerebellar cortex. *Neuroscience* 7, 937–50.

Lansman, J.B., Hess, P. and Tsien, R.W. 1986. Blockade of current through single calcium channels by Cd^{2+}, Mg^{2+}, and Ca^{2+}. *Journal of General Physiology*, **88**, 321–47.

Legare, M.E., Barhoumi, R., Burghardt, R.C. and Tiffany-Castiglioni, E. 1993. Low-level lead exposure in cultured astroglia: Identification of cellular targets with vital fluorescent probes. *Neurotoxicology* **14**, 267–72.

Lewandowsky, M. 1900. Zur Lehre der Cerebrospinalflussigkeit. *Z. Klin. Med.* **40**, 480–94.

Lucis, O.J., Lucis, R. and Shaikh, Z.A. 1972. Cadmium and zinc in pregnancy and lactation. *Archives of Environmental Health* **25**, 14–22.

McIntyre, T.M. and Curthoys, N.P. 1980. The interorgan metabolism of glutathione. *International Journal of Biochemistry* **12**, 545–51.

Martin, R.B., Savory, J., Brown, S., Bertholf, R.L. and Wills, M.R. 1987. Transferrin binding of Al^{3+} and Fe^{3+}. *Clinical Chemistry* **33**, 405–7.

Marty, M.S. and Atchison, W.D. 1997. Pathways mediating Ca^{2+} entry in rat cerebellar granule cells following *in vitro* exposure to methyl mercury. *Toxicology and Applied Pharmacology* **147**, 319–30.

Meister, A. 1988. Glutathione metabolism and its selective modification. *Journal of Biological Chemistry* **263**, 7205–8.

Meister, A. and Anderson, M. 1983. Glutathione. *Annual Reviews of Biochemistry* **52**, 711–60.

Mihorat, T.H. 1976. Structure and function of the choroid plexus and other sites of cerebrospinal fluid formation. *International Reviews of Cytology* **47**, 225–89.

Miura, K. and Clarkson, T.W. 1993. Reduced methylmercury accumulation in a methylmercury-resistant rat pheochromocytoma PC12 cell line. *Toxicology and Applied Pharmacology* **118**, 39–45.

Mokrzan, E.M., Kerper, L.E., Ballatori, N. and Clarkson, T.W. 1995. Methylmercury–thiol uptake into cultured brain capillary endothelial cells on amino acid system L. *Journal of Pharmacology and Experimental Therapeutics* **272**, 1277–84.

Murphy, V.A., Wadhwani, K.C., Smith, Q.R. and Rapoport, S.I. 1991. Saturable transport of manganese (II) across the rat blood–brain barrier. *Journal of Neurochemistry* **57**, 948–54.

Nachshen, D.A. 1984. Selectivity of the Ca binding site in synaptosome Ca channels. *Journal of General Physiology* **83**, 941–67.

Nachshen, D.A. and Blaustein, M.P. 1982. Influx of calcium, strontium, and barium in presynaptic nerve endings. *Journal of General Physiology* **79**, 1065–87.

Nag, S. 1987. Ultrastructural localization of calcium-activated adenosine triphosphatase (Ca^{2+}-ATPase) in cerebral endothelium. *Laboratory Investigations* **57**, 52–6.

Neuwlet, E.A. (ed.) 1989. In *Implications of the Blood–Brain Barrier and its Manipulation*, Plenum Publishing, New York.

Newman, E. 1986. High potassium conductance in astrocyte endfeet. *Science* **233**, 453–4.

Oldendorf, W.H. 1970. Measurement of brain uptake of radiolabeled substances using a tritiated water internal standard. *Brain Research* **24**, 372–6.

Oldendorf, W.H. and Kitano, M. 1965. Increased brain radiocopper uptake in Wilson's disease. *Archives of Neurology* **13**, 533–40.

Olivares, M. and Uauy R. 1996. Copper as an essential nutrient. *American Journal of Clinical Nutrition* **63**, 791S–796S.

Oyake, Y., Tanaka, M., Kubo, H. and Cichibu, H. 1966. Neuropathological studies on organic mercury poisoning with special reference to the staining and distribution of mercury granules. *Advances in Neurological Sciences* **10**, 744–50.

Prohaska, J.R. 1987. Function of trace elements in brain metabolism. *Physiology Reviews* **67**, 858–901.

Putney, J.W. and Bird, G.St.J. 1993. The inositol phosphate–calcium signaling system in nonexcitable cells. *Endocrinology Reviews* **14**, 610–31.

Rabenstein, D.L. and Fairhurst, M.T. 1975. Nuclear magnetic resonance studies of the solution chemistry of metal complexes. XI. The binding of methylmercury by sulfhydryl-containing amino acids and by glutathione. *Journal of the American Chemical Society* **97**, 2086–92.

Rabin, O., Hegedus, L., Bourre, J.M. and Smith, Q.R. 1993. Rapid brain uptake of manganese (II) across the blood–brain barrier. *Journal of Neurochemistry* **61**, 509–17.

Radunovic, A. 1994. Transport of aluminium, compared with 67-Ga and 59-Fe, into brain and other tissues of the young adult rat, Ph.D. thesis.

Rapoport, S.I. (ed.) 1976. Permeability and osmotic properties of the blood–brain barrier, in *Blood–Brain Barrier in Physiology and Medicine*, Raven Press, New York, pp. 87–127.

Raps, S.P., Lai, J.C.K., Hertz, L. and Cooper, A.J.L. 1989. Glutathione is present in high concentrations in cultured astrocytes but not in cultured neurons. *Brain Research* **493**, 398–401.

Reuhl, K.R. and Chang, L.W. 1979. Effects of methylmercury on the development of the nervous system: a review. *Neurotoxicology* **1**, 21–55.

Rhenberg, G.L., Hein, F.J., Carter, D.S., Linko, S.R. and Laskey, W.J.. 1981. Chronic ingestion of Mn_3O_4 by young rats: tissue accumulation, distribution and depletion. *Journal of Experimental Toxicology and Environmental Health* 7, 263–72.

Risau W. 1994. Molecular biology of blood–brain barrier ontogenesis and function. *Acta Neurochirurgica – Supplementum* **60**, 109–12.

Risau, W. 1995. Differentiation of endothelium. *FASEB Journal* **9**, 926–33.

Risau, W. and Wolfburg, H. 1990. Development of the blood–brain barrier. *Trends in Neurological Sciences* **13**, 174–8.

Roberts, R.L., Fine, R.E. and Sandra, A. 1993. Receptor-mediated endocytosis of transferrin at the blood–brain barrier. *Journal of Cell Sciences* **104**, 521–32.

Roskams, A.J. and Connor, J.R. 1990. Aluminum access to the brain: A role for transferrin and its receptor. *Proceedings of the National Academy of Sciences, USA* **87**, 9024–7.

Rothenberger, S., Food, M.R., Gabathuler, R., Kennard, M.L., Yamada, T., Yasuhara, O., McGeer, P.L. and Jefferies, W.A. 1996. Coincident expression and distribution of melanotransferrin and transferrin receptor in human brain capillary endothelium. *Brain Research* **712**, 117–21.

Sager, P.R., Doherty, R.A. and Rodier, P.M. 1982. Effects of methylmercury on developing mouse cerebellar cortex. *Experimental Neurology* 77, 179–93.

Sapirstein, V.S. 1983. Development of membrane-bound carbonic anhydrase and glial fibrillary acidic protein in normal and quaking mice. *Developmental Brain Research* **6**, 13–19.

Simons, T.J.B. 1986. The role of anion transport in the passive movement of lead across the human red cell membrane. *Journal of Physiology, (London)* **378**, 287–312.

Simons, T.J.B. 1988. Active transport of lead by the calcium pump in human red blood cells. *Journal of Physiology, (London)* **405**, 105–13.

Simons, T.J.B. 1993a. Lead–calcium interactions in cellular lead toxicity. *Neurotoxicology* **14**, 77–86.

Simons, T.J.B. 1993b. Lead transport and binding by human erythrocytes *in vitro*. *Pflügers Archives* **423**, 307–13.

Simons, T.J.B. and Pocock, G. 1987. Lead enters bovine adrenal medullary cells through calcium channels. *Journal of Neurochemistry* **48**, 383–9.

Slivka, A., Mytilineou, C. and Cohen, G. 1987. Histochemical evaluation of glutathione in the brain. *Brain Research* **409**, 275–84.

Sloot, W.N. and Gramsbergen, J.B. 1994. Axonal transport of manganese and its relevance to selective neurotoxicity in the rat basal ganglia. *Brain Research* **657**, 124–32.

Smith, Q.R. 1990. Regulation of metal uptake and distribution within the brain, in *Nutrition and the Brain*, Vol. 8, (eds R.J. Wurtman and J.J. Wurtman), Raven Press, New York, pp. 25–74.

Smith, Q.R., Rabin, O. and Chikhale, E.G. 1997. Delivery of metals to brain and the role of the blood–brain barrier, in *Metals and Oxidative Stress*, (ed. J.R. Connor), Plenum Press, New York, pp. 113–29.

Solioz, M., Odermatt A. and Krapf, R. 1994. Copper pumping ATPases: common concepts in bacteria and man. *FEBS Letters* **346**, 44–7.

Spector, R. 1989. Micronutrient homeostasis in mammalian brain and cerebrospinal fluid. *Journal of Neurochemistry* **53**, 1667–74.

Suarez, N. and Eriksson, H. 1993. Receptor-mediated endocytosis of a manganese complex of transferrin into neuroblastoma (SHSY5Y) cells in culture. *Journal of Neurochemistry* **61**, 127–31.

Thomas, D.J. and Smith, J.C. 1979. Partial characterization of a low-molecular weight methylmercury complex in rat cerebrum. *Toxicology and Applied Pharmacology* **47**, 547–56.

Tiffany-Castiglioni, E. 1986. Cellular targets of lead neurotoxicity: *in vitro* models. *Toxicology* **42**, 303–15.

Tiffany-Castiglioni, E., Zmudzki, J., Wu, J.-N. and Bratton, G.R. 1987. Effects of lead treatment on intracellular iron and copper concentrations in cultured astroglia. *Metabolic Brain Disease* **2**, 61–79.

Tiffany-Castiglioni, E., Sierra, E.M., Wu, J.-N. and Rowles, T.K. 1989. Lead toxicity in neuroglia. *Neurotoxicology* **10**, 417–44.

Tomsig, J.L. and Suszkiw, J.B. 1991. Permeation of Pb^{2+} through calcium channels: Fura-2 measurements of voltage- and dihydropyridine-sensitive Pb^{2+} entry in isolated bovine chromaffin cells. *Biochimica et Biophysica Acta* **1069**, 197–200.

Tsien, R.W., Hess, P., McCleskey, E.W. and Rosenberg, R.L. 1987. Calcium channels: Mechanisms of selectivity, permeation, and block. *Annual Reviews of Biophysics and Biophysical Chemistry* **16**, 265–90.

Tumer, Z. and Horn, N. 1996. Menkes disease: recent advances and new insights into copper metabolism. *Annals of Medicine* **28**, 121–9.

Ueda, F., Raja, K.B., Simpson, R.J., Trowbridge, I.S., and Bradbury, M.W.B. 1993 Rate of ^{59}Fe uptake into brain a cerebrospinal fluid and the influence thereon of antibodies against the transferrin receptor. *Journal of Neurochemistry* **60**, 106–13.

van Gelder, W., Cleton-Soeteman, M.I., Huijskes-Heins, M.I., van Run, P.R. and van Eijk, H.G. 1997. Transcytosis of 6.6-nm gold-labeled transferrin: an ultrastructural study in cultured porcine blood–brain barrier endothelial cells. *Brain Research* **746**, 105–16.

Walshe, J.M. 1995. Copper: not too little, not too much, but just right. *Journal of the Royal College of Physicians of London* **29**, 280–8.

Wedler, F.C. 1993. Biological significance of manganese in mammalian systems, in *Progress in Medicinal Chemistry*, Vol. 30, (eds G.P. Ellis and D.K. Luscombe), Elsevier Science Publishers, Amsterdam, pp. 89–133.

Wedler, F.C. and Denman, R.B. 1984. Glutamine synthetase: the major Mn(II) enzyme in mammalian brain. *Current Topics in Cell Regulation* **24**, 153–69.

Wetterhahn-Jennette, K. 1981. The role of metals in carcinogenesis: Biochemistry and metabolism. *Environmental Health Perspectives* **40**, 233–52.

Wu, J.-N., and Tiffany-Castiglioni, E. 1987. Reduction by lead of hydrocortisone-induced glycerol phosphate dehydrogenase activity in cultured rat oligodendroglia. *In Vitro Cell Development and Biology* **23**, 765–74.

Yasutake, A., Hirayama, K. and Inoue, M. 1989. Mechanism of urinary excretion of methylmercury in mice. *Archives of Toxicology* **63**, 479–83.

Zheng, W. 1996. Choroid plexus and metal toxicity, in *Toxicology of Metals*, (ed. L.W. Chang), CRC Press, Boca Raton, pp. 605–22.

Zheng, W., Perry, D.F., Nelson, D.L. and Aposhian, H.V. 1991. Choroid plexus protects cerebrospinal fluid against toxic metals. *FASEB Journal* **5**, 2188–93.

11

TRANSPORT OF METALS IN THE GASTROINTESTINAL SYSTEM AND KIDNEYS

Gary L. Diamond

11.1 Overview of functional organization of transporting epithelia

The transporting epithelia of the renal tubule and small intestine are characterized by layers of anatomically and biochemically polarized cells that are connected to each other by tight junctions, and which rest on a basement membrane. The basic features of this organization are depicted in Fig. 11.1. The cell membranes are anatomically polarized into an apical (luminal, mucosal) region, which faces the tubule or intestinal lumen, and a basolateral (serosal) region. The tight junction forms the borders between the apical and basolateral membranes. Biochemical polarity is reflected in distinct enzymes and transport proteins in the apical and basolateral membranes. On the basolateral membrane, these include transport proteins that are common to most cells, such as Na^+,K^+-ATPase, Ca^{2+}-ATPase, Na^+/Ca^{2+}-exchanger, Cl^-/HCO_3^--exchanger, sodium-independent glucose and amino-acid carriers and barium-sensitive K^+ channels. The apical membrane contains transport proteins that are largely unique to epithelial cells; these include Na^+-coupled glucose, phosphate and amino-acid carriers, H^+-ATPase and amiloride-sensitive Na^+ channels. The apical membrane also is enriched in the activities of certain peptidases and hydrolases.

The tight junction is a region of particularly close association of the plasma membranes of adjacent epithelial cells. The junction serves as a permeability barrier between adjacent cells and, possibly, as a barrier to the diffusion of membrane proteins between the apical and basolateral membranes and, thus, may play an important role in maintaining the polarity of the cell (Dragsten *et al.*, 1981). The permeability properties of tight junctions vary considerably. Epithelia that exhibit low tight-junction permeability to ions tend to exhibit relatively high electrical resistance (low conductance) and relatively high transepithelial electrical potential differences (PD) (e.g. 30–150 mV). Such epithelia are classified as 'tight epithelia'; examples include the mammalian renal distal tubule and colon. By contrast, 'leaky epithelia' tend to exhibit high tight-junction permeability to ions, relatively low electrical resistance and low transepithelial electrical PD (e.g. 0–10 mV). The mammalian renal proximal tubule and small intestine fall into this category. In leaky epithelia, a large fraction of solute transport occurs between adjacent epithelial cells across the tight junctions; this is referred to as the paracellular

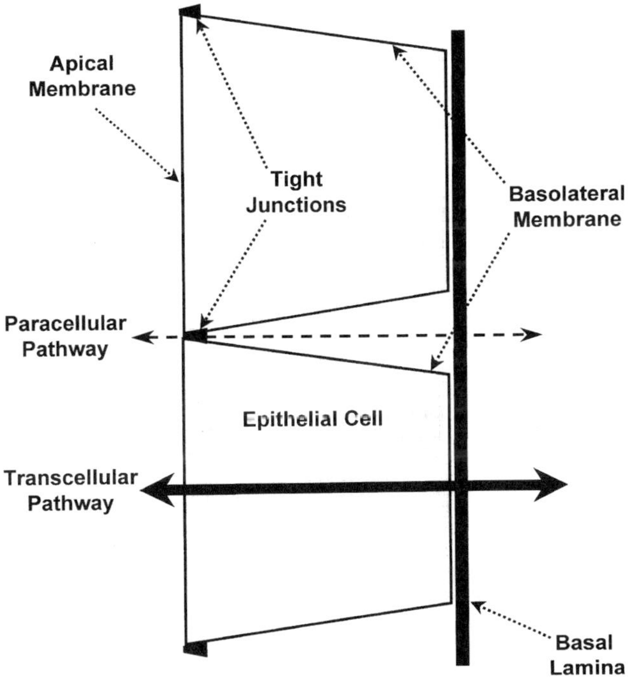

Figure 11.1 Diagram depicting some of the anatomic features of transporting epithelia. The tight junction separates the membrane into two domains; the apical (mucosal) membrane, which faces the intestinal lumen or renal tubule lumen, and the basolateral membrane (serosal). Transport can occur between cells, through the tight junction (paracellular pathway), or through the cells (transcellular pathway).

(shunt) pathway. In the renal proximal convoluted tubule, approximately one- to two-thirds of sodium absorption is thought to occur through the paracellular pathway (Fromter *et al.*, 1973; Rector, 1983). By contrast, solute transport in tight epithelia is predominantly across the cell, or transcellular. Because the lipid bilayer of plasma membranes constitutes a formidable barrier to diffusion of charged species (Al-Awqati *et al.*, 1986), this necessitates specialized transport proteins in the apical and basolateral cell membranes to facilitate transcellular transport of ions.

11.1.1 Paracellular pathway

It is evident from the above discussion, that two potential pathways exist for the transepithelial transport of metals across epithelia: a paracellular pathway and a transcellular pathway (Fig. 11.1). The relative magnitudes of each will depend on the permeability properties of the tight junctions, the physical and chemical properties of the metal species (or complex), and the electrochemical potential gradient for the transported species across the tight junction. The relative paracellular permeabilities (P) of K^+ and Cl^- (relative to Na^+) in small intestinal epithelia of a variety of species has been shown to be approximately: P_K, 1.3 > P_{Na}, 1.0 > P_{Cl}, 0.4, which suggests that the paracellular pathway is selectively permeable to cations (Powell, 1981). In the rabbit ileum, the relative permeabilities of the alkali-metal cations are similar: P_{Cs}, 1.6 > P_{Rb}, 1.4 > P_K, 1.2 > P_{Na}, 1.0 > P_{Li}, 0.8, and generally reflect their relative mobilities in free solution; on the other hand, the permeability of the tetraethylammonium ion is negligible (Armstrong, 1987). The ionic radius of tetraethylammonium is approximately 4 Å (10^{-10} m) , whereas the radius of the Cs^+ ion is approximately 1.7 Å (Evans, 1994). These observations are consistent with a model of the tight junction as an anionic channel having an average diameter of approximately 10–15 Å (Naftalin and Tripathi, 1985). On this basis alone, one would predict relatively high permeabilities of leaky epithelia to the ions Pb^{2+} (1.2 Å), Ca^{2+} (1.0 Å), Cd^{2+} (0.8 Å), Hg^{2+} (0.7 Å) and Zn^{2+} (0.6 Å), whereas, large complexes of these ions with organic ligands, such as amino acids, nonprotein sulfhydryls (NPSH), peptides and proteins, would be expected to have relatively low permeabilities.

11.1.2 Transcellular pathway

The transcellular pathway consists of the apical and basolateral membranes and the intervening cytosolic space (Fig. 11.1). In general, the diffusion of ions of chemical species through the lipid bilayer of plasma membranes is greatly limited by the energy barrier associated with the solvation of the ion in the hydrophobic environment of the membrane (Al-Awqati *et al.*, 1986). The same applies to neutral hydrophilic non-electrolyte species. For essential electrolytes that have been rigorously studied, this energy barrier is surmounted by the investment of cellular energy in the maintenance and function of specialized membrane transfer pathways, such as channels or carriers. For reasons that are discussed below, unambiguous evidence for channel-mediated or carrier-mediated transport of the heavier metals ions (e.g. Cd^{2+}, Pb^{2+}, Hg^{2+}) has been difficult to procure.

The specialized mechanisms for solute transport eluded to above fall into two general categories, facilitated diffusion and pumps. Pumps couple an exothermic chemical reaction directly to the translocation of solute across the membrane. Examples of this are the basolateral membrane Na^+,K^+-ATPase and Ca^{2+}-ATPase, and the apical membrane H^+-ATPase. In each case, the translocation of one or more ions across the membrane is coupled directly to the hydrolysis of ATP. Facilitated diffusion mechanisms include channels and carriers. Channels are generally thought to be membrane-spanning hydrophilic pores, whereas carriers are thought to consist of binding proteins that move through, or reorient, within the membrane to shuttle the transported solute or solutes across the membrane. In either case, the facilitating mechanism serves primarily to decrease the energy barrier for solvation of the solute in the membrane, enabling the solute to move across the membrane down an electrochemical potential gradient for the transported solutes (Parsegian, 1969).

Ion pumps are critical for epithelial transport because they provide a mechanism for generating electrochemical potential gradients across the apical and basolateral membranes and across the epithelium. These gradients can be utilized to drive diffusion (facilitated or otherwise) of other solutes at other locations in the cell. A typical small intestinal epithelial cell has an electrical PD across the apical membrane of approximately –35 mV (intracellular-negative) and –40 mV across the basolateral membrane (Fig. 11.2). Typical intracellular and extracellular sodium concentrations in the cell

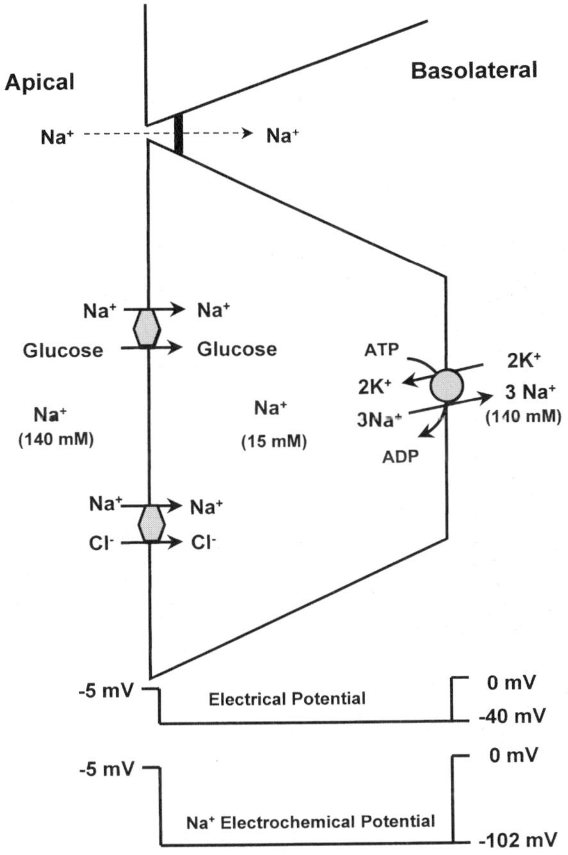

Figure 11.2 Diagram showing selected features of Na^+ transport in a typical small intestinal cell, an example of a leaky epithelium. Activity of the Na^+,K^+-ATPase on the basolateral membrane contributes to the maintenance of an electrochemical PD for Na^+ of approximately –100 mV (intracellular-negative) across the apical and basolateral membranes. The energy stored in the electrochemical gradient can be used to transport other solutes, such as Cl^- or glucose, by coupling their transport to the transport of Na^+ down its electrochemical gradient.

are approximately 15 mM and 140 mM, respectively (Armstrong, 1987). Based on this information, along with some simplifying assumptions, the electrochemical PD for sodium across the apical and basolateral membranes and epithelium can be estimated from equation 11.1:

$$
\begin{aligned}
\Delta\mu &= RT\ln\frac{C_i}{C_o} + zF\Delta\Psi \\
\Delta\mu_{\text{apical}} &= -9280 \text{ J/mol } (-97 \text{ mV}) \\
\Delta\mu_{\text{basolateral}} &= -9769 \text{ J/mol } (-102 \text{ mV}) \\
\Delta\mu_{\text{epithelium}} &= -480 \text{ J/mol } (-5 \text{ mV})
\end{aligned}
\qquad (11.1)
$$

where $\Delta\mu$ is the electrochemical PD between the interior and exterior of the cell (J/mol); R is the universal gas constant (8.135 J/deg/mol); T is temperature (K); C_i and C_o are the concentrations, in this case of Na^+, inside and outside the cell, respectively; z is the charge of the transported ion; F is the Faraday constant (95 600 C/mol); and $\Delta\Psi$ is the electrical PD between the interior and exterior of the cell.

Equation 11.1 predicts a substantial electrochemical PD for sodium across both the apical and basolateral membranes and the direction of this difference is intracellular-negative. Therefore, transport of Na^+ into the cell from the lumen of the intestine, across the apical membrane, involves movement of Na^+ down an electrochemical potential gradient favoring diffusion of Na^+ into the cell, whereas transport of Na^+ out of the cell, across the basolateral membrane, occurs against an electrochemical gradient. Several additional observations are relevant:

1 Extrusion of sodium from the cell across the basolateral membrane is an 'active' process in that it occurs against an electrochemical gradient for Na^+.
2 The active extrusion of Na^+ across the basolateral membrane, which is mediated by the Na^+,K^+-ATPase, results in the translation of some of the chemical energy stored in ATP into an electrochemical PD for Na^+ across the apical and basolateral membranes.
3 The energy stored in the electrochemical PD for Na^+ can be used to drive the transport of Na^+ and other ions across both membranes, either by a direct coupling to Na^+ transport or by indirect coupling to the electrical potential difference. The examples shown in Fig. 11.2 are Na^+–Cl^- and Na^+–glucose cotransport at the apical membrane.
4 Net transfer of Na^+ from the intestinal lumen across the epithelium occurs against a small electrochemical PD (lumen-negative); thus, it appears to be an 'active' (energy-requiring) process.

11.1.3 Characterization of epithelial transport mechanisms for metals

The preceding discussion referred to several possible ways to categorize the epithelial transport of metals and their complexes:

- active or passive;
- passive diffusion, facilitated diffusion or pump-driven; and
- paracellular or transcellular.

Ultimately, discerning active from passive mechanisms requires that measurements be made of the activities (or, at the very least, concentrations) of the specific metal species and complexes on opposite sides of the membrane or epithelium of interest, so that the magnitude of the electrochemical gradients for each species can be estimated. This has been achieved for calcium, as discussed below, but not for most of the metals of toxicological interest (e.g. cadmium, lead, mercury). Success with calcium has been achieved because the ligand chemistry of the Ca^{2+} ion in physiological solutions has been rigorously characterized and because of the development of ion-specific electrodes and fluorescent dyes for quantifying Ca^{2+} ion concentrations in biological fluids (Friedman and Gesek, 1995). The ligand reactions of other metals are not understood as well and, until recently, direct methods for quantifying ion concentrations have not been available for the so-called heavy metals (Hinkle *et al.*, 1992; Kerper and Hinkle, 1997a,b). For the most part, studies of the transport of metals in kidney and intestine have relied on radioisotope tracer techniques. These techniques can provide information about concentrations and rates of transport of the metal in all forms present in the system being studied, but they cannot alone yield estimates of electrochemical gradients for the individual metal species that may be participating in transport.

In the absence of quantitative information on metal speciation in the system being studied, a variety of less direct approaches have been used to characterize transport mechanisms of metals. Examples of these can be found in the discussions of cadmium, lead and mercury. They include the following types of assessments:

1. Is the isotope transport rate or isotope concentration gradient dependent on metabolic energy?
2. Is transport capacity limited (saturable)?
3. Does the transport process show chemical specificity?
4. Is there evidence for physiologic regulation of transport (e.g. induction or hormone sensitivity)?

These assessments can yield insights regarding the possible involvement of facilitated mechanisms in the transport of the metal, but have yet to yield a satisfactory mechanistic characterization of the transport of the metals of greatest toxicological interest.

The following sections summarize the experimental basis for our current understanding of transport mechanisms for calcium, lead, mercury and cadmium, and in doing so, illustrate many of the above points. Relative to our knowledge about the heavier metals, we can luxuriate in information about the mechanism of transport of calcium. For this reason calcium is described in some detail as an example of a divalent cation for which epithelial transport processes are reasonably well characterized. Calcium also serves to illustrate how all of the major categories of transport mechanisms can be involved in the epithelial transport of a single ion species, and this may be relevant to other metals.

11.2 Calcium transport

11.2.1. Magnitude of transepithelial transport in kidney

The calcium concentration in plasma is approximately 2.5 mM. Approximately 45% is bound to protein, mainly to albumin and globulins, 45% exists as free calcium ion (Ca^{2+}) and 10% is complexed with low-molecular-weight anions (e.g. bicarbonate, phosphate, sulfate and citrate). The ratio of the calcium concentrations in tubular fluid (TF) of Bowman's space (glomerular filtrate) and plasma (P) ($[TF/P]_{Ca}$) in the hamster and rat is 0.55–0.65, indicating that approximately 55–65% of plasma calcium is filtered, an amount equivalent to all of the unbound fraction in plasma (Lassiter *et al.*, 1963; Harris *et al.*, 1974; Le Grimellec *et al.*, 1975). Approximately 10% of the ultrafilterable calcium is thought to be complexed; thus, the calcium and Ca^{2+} ion concentrations of the glomerular ultrafiltrate are approximately 1.4 mM and 1.1 mM, respectively.

In rats and rabbits, approximately 60–70% of the filtered calcium is absorbed in the proximal tubule, another 20% in the loop of Henle, and most of the remaining filtered load in the distal, connecting and collecting tubules (Sutton and Dirks, 1975). In human beings this would amount to a tubular absorption rate of approximately 250 mmol/day at a glomerular filtration rate of 180 L/day.

11.2.2 Evidence for active and passive mechanisms of transport

Active transport of calcium appears to occur to some degree in all segments of the renal tubule, however, the relative contribution of active and passive mechanisms is not the same in each segment. Measurements of bidirectional fluxes of calcium across the *in situ* perfused rat proximal tubule have shown that peritubular–lumen backflux is approximately three times greater than net lumen-to-peritubular flux, suggesting a relatively high epithelial permeability to calcium (Murayama *et al.*, 1972). The epithelial calcium permeability of the rat proximal convoluted tubule *in vivo* is approximately 3×10^{-4} cm/s (Frick *et al.*, 1965) and approximately $8–9 \times 10^{-5}$ cm/s in the isolated perfused rabbit proximal convoluted tubule (Friedman *et al.*, 1981; Ng *et al.*, 1984), comparable to that of NaCl (Bomsztyk *et al.*, 1984), which has a high, relatively large, passive absorption component in the proximal tubule. The high permeability of the proximal tubule to calcium has made it somewhat difficult to quantify the active absorption component in the *in situ* free-flowing tubule. Free-flow micropuncture studies of the rat nephron have shown that the electrical potential across the proximal tubule ranges from approximately –2 mV (lumen-negative) in the early proximal convoluted tubule to +3 mV (lumen-positive) in the late proximal convoluted tubule (Koeppen and Giebisch, 1983). Electrochemical equilibrium across an epithelium that is freely permeable to Ca^{2+} would be achieved at these electrical potential differences and corresponding transepithelial chemical gradients for Ca^{2+} of approximately 0.8–1.2. The observed ratio of concentrations of calcium in tubular fluid and glomerular ultrafiltrate (UF) ($[TF/UF]_{Ca}$) along segments of the proximal convoluted tubule that are accessible to micropuncture (S1 and S2 segments) falls within this range; approximately 1.1–1.2 in the rat under a variety of conditions, including NaCl-induced diuresis (Lassiter *et al.*, 1963; Le Grimellec *et al.*, 1973; Harris *et al.*, 1974; Sutton and Dirks, 1975). Under conditions of mannitol-induced diuresis, the $[TF/UF]_{Ca}$ in the proximal convoluted tubule drops to slightly below one, consistent with isosmotic absorption (Lassiter *et al.*, 1963; Durante and Watson, 1967; Wong *et al.*, 1979).

An active component in the proximal convoluted tubule has been detected using *in situ* perfusion techniques in which the composition of the lumen and peritubular fluid can be more rigorously controlled. If the peritubular vasculature is perfused and the flow of tubular fluid in the tubule lumen stopped (stop-flow conditions), a static head electrochemical potential gradient for calcium (and other ions) develops across the epithelium. Under these conditions, net epithelial absorption of calcium is zero and unidirectional peritubular-to-lumen flux of calcium (reflecting passive backflux) provides an estimate of unidirectional lumen-to-peritubular active transport of calcium. Based on such measurements, a sodium-dependent, ouabain-sensitive active component of calcium absorption was observed in the rat proximal convoluted tubule. By analogy to sodium, the active component may account for 20–30% of total calcium absorption in the proximal convoluted tubule (Ullrich *et al.*, 1976).

In the distal convoluted tubule, connecting and collecting tubule, active calcium absorption is more readily apparent, since it can be shown to occur against both a lumen-to-peritubular chemical potential gradient ($[TF/UF]_{Ca}$ = 0.3–0.5) and lumen-negative electrical potential gradient (0 to –50 mV, lumen-negative) (e.g. Lassiter *et al.*, 1963; Boulpaep and Seely, 1971; Wright, 1971; Costanzo and Windhager, 1978). The larger electrochemical gradients for Ca^{2+} that are sustainable in the distal segments of the tubule are indicative of a lower calcium permeability relative to that of the proximal convoluted tubule (Costanzo and Windhager, 1978). Absorption of calcium in the cortical thick ascending limb of the loop of Henle, distal convoluted tubule and connecting tubule (and possibly other segments of the nephron) is regulated by parathyroid hormone (PTH), calcitonin and 1,25-dihydroxy-vitamin D_3 ($1,25(OH)_2D_3$) (Friedman and Gesek, 1995).

11.2.3 Apical membrane transport of calcium – evidence for channels and carriers

Intracellular Ca^{2+} concentration is approximately 2×10^{-7} M in kidney epithelial cells (Friedman and Gesek, 1995), whereas the Ca^{2+} concentration in tubular fluid is approximately 1×10^{-3} M. The resulting chemical concentration gradient of 2×10^{-4} M across the apical membrane of the tubule cell, in combination with an electrical potential gradient that ranges from –20 to –75 mV (intracellular-negative) (Allen and Barratt, 1985; Yoshitomi *et al.*, 1989; Gesek and Friedman, 1992b), creates an electrochemical potential difference of approximately –135 to –190 mV (intracellular-negative) that favors entry of Ca^{2+} into the cell from the tubular lumen. The exact nature of the mechanisms that facilitate transfer of calcium through the apical membrane have not been completely characterized, however, two types of mechanisms have been proposed: Ca^{2+} conductance channels and carrier-mediated transfer.

Evidence for Ca^{2+} channels in the apical membranes of renal tubule cells is derived mainly from two types of observations: sensitivity of hormone-stimulated (e.g. PTH) calcium influx to dihydropyridine Ca^{2+} channel blockers, and direct measurement of Ca^{2+} (and/or Ba^{2+}) conductance in membrane patches. Saunders and Isaacson (1990) described Ca^{2+} channel activity (detected as Ba^{2+} conductance) in cell-detached, inside-out patches from apical membranes of proximal straight tubules, thick ascending limb, distal convolute tubules and cortical collecting ducts of the rabbit. Channel activity was voltage-sensitive; open-state probabilities increased with increasing electrical potential. Activity was depressed by verapamil and nifedipine, and increased by BAY K 8644. A Ca^{2+} channel has been detected in cell-attached patches of apical membrane of rabbit distal convoluted tubules (Poncet *et al.*, 1992). In cell-detached, inside-out patches, channel activity was blocked by cytoplasmic lanthanum, nifedipine and verapamil.

PTH has been shown to increase the transcellular transport of calcium in a variety of tubule preparations, including cortical thick ascending limbs, distal convoluted tubules and connecting tubules (Costanzo and Windhager, 1978, 1980; Shareghi and Stoner, 1978; Bourdeau and Burg, 1979; Lajeunesse *et al.*, 1994). In mouse distal convoluted tubules, PTH hyperpolarizes epithelial cells and stimulates a calcium influx pathway which is increased by hyperpolarizing stimuli and is sensitive to dihydropyridines, decreasing in response to nifedipine and increasing in response to BAY K 8644 (Bacskai and Friedman, 1990; Gesek and Friedman, 1992b). In cell-attached patches of apical membranes of mouse distal convoluted tubule cells grown in culture, Ba^{2+} conductance channels have been observed after stimulation of the cells with PTH (Matsunaga *et al.*, 1994). The PTH-stimulated channel activity was depressed by nifedipine and increased by BAY K 8644. The channels also exhibited a voltage sensitivity; the open-state probability increased as the membrane electrical potential difference was increased from –50 to –90 mV (intracellular-negative). Thus, PTH-stimulated calcium influx appears to be associated with the induction (or activation) by PTH of a voltage-sensitive Ca^{2+} channel in the apical membrane, which could operate within the range of the transmembrane electrical potentials observed in unstimulated and PTH-stimulated tubule cells. The PTH-stimulated conductance pathway in the apical membrane of the mouse distal convoluted tubule cells has been characterized as being distinct from L-type Ca^{2+} channels found in other cells, in that the channels exhibit longer open times (seconds); a lower conductance (2 pS in 110 mM Ba^{2+}); activation by hyperpolarizing membrane potentials; greater Ca^{2+} than Ba^{2+} conductance; and insensitivity to millimolar concentrations of Zn^{2+}, Ni^{2+}, Cd^{2+}, Co^{2+} or verapamil (Gesek and Friedman, 1992a,b; Matsunaga *et al.*, 1994).

cAMP-stimulated and PTH-stimulated Ca^{2+} conductance pathways in the apical membrane of rabbit connecting tubules have been described. The transcellular transport and calcium influx in rabbit connecting tubule epithelial cells are both stimulated by cAMP (Imai, 1981; Bourdeau and Lau, 1989, 1992; Bourdeau and Eby, 1990; Shimizu *et al.*, 1990). In cell-attached patches from freshly isolated tubules, cAMP increased Ca^{2+} channel activity (measured as Ba^{2+} conductance). In cell-detached, inside-out patches, open-state probability was decreased by exposure of the cytoplasmic surface to nifedipine and increased by combined exposure to BAY K 8644 and protein kinase A catalytic subunit (Tan and Lau, 1993). Maximal channel activity was observed at an electrical potential difference of –70 mV (cytoplasmic-negative) and decreased with decreasing or increasing electrical potential. The slope conductance extrapolated to the reversal potential was 25 pS. These characteristics are more similar to L-type channels. An apical membrane Ca^{2+} channel that is activated by PTH in combination with increased lumen perfusion pressure in the isolated perfused rabbit connecting tubule has also been reported (Taniguchi *et al.*, 1994).

Evidence for carrier-mediated mechanisms of calcium transfer across the apical membrane of the renal tubule is less robust than that for channels. Brunette *et al.* (1992) described a saturable calcium-uptake mechanism in apical-membrane vesicles prepared from rabbit proximal and distal tubules. The mechanism was inhibited by Na^+, either with a *trans*-Na^+ gradient or in the absence of a Na^+ gradient, suggesting that uptake was not mediated by Na^+/Ca^{2+} exchange. Calcium uptake was greater under alkaline conditions, either with a *cis*-H+ gradient or in the absence of a proton gradient (i.e. $pH_{in} = pH_{out}$). A H^+/Ca^{2+}-exchange mechanism in the apical membrane has also been proposed; however, definitive evidence for its existence has not been established (Friedman and Gesek, 1995). In apical-membrane vesicles from rabbit small intestine, binding (and/or uptake) of calcium is enhanced by an inward-negative electrical potential, suggesting the existence of some type of direct or indirect rheogenic mechanism that transfers calcium into the vesicle, where it binds to the interior (cytosolic side) of the membrane (Merrill *et al.*, 1986).

11.2.4 Basolateral membrane calcium transport – evidence for carriers and pumps

Efflux of calcium from renal tubule cells across the basolateral membrane occurs against a steep electrochemical gradient for Ca^{2+} (approximately –180 mV, inside negative) and, thus, is an active process. The predominant active calcium transport mechanism is a plasma membrane Ca^{2+}-ATPase. The basolateral Na^+,K^+-ATPase may also contribute to the active extrusion of Ca^{2+} by establishing a favorable electrochemical Na^+ gradient for carrier-mediated Na^+/Ca^{2+}-exchange.

Ca^{2+}-ATPase activity occurs in all tubule segments, however, activity is highest in the distal convoluted tubules and cortical collecting ducts (Doucet and Katz, 1982; Brunette *et al.*, 1987). The enzyme is located predominantly, if not exclusively, in the basolateral membrane (Gmaj *et al.*, 1979, 1983; Borke *et al.*, 1987; Brunette *et al.*, 1987). The high-affinity form of the enzyme (calmodulin-activated) has a dissociation constant for Ca^{2+} of approximately $0.3–0.7 \times 10^{-6}$ M (Doucet and Katz, 1982; Brunette *et al.*, 1990), which enables the enzyme to be active at physiologic intracellular Ca^{2+} concentrations in renal tubule epithelial cells (Mandel and Murphy, 1984; Bourdeau and Lau, 1989; Kubota *et al.*, 1990; Gesek and Friedman, 1992a,b). The Ca^{2+}-ATPase activity of the basolateral membrane from proximal and distal convoluted tubule appears to be stimulated by PTH (Levy *et al.*, 1986; Tsukamoto *et al.*, 1992).

In addition to the Ca^{2+}-ATPase which extrudes Ca^{2+} from the cell, the basolateral membranes of renal tubule cells contain a Na^+/Ca^{2+}-exchanger which directly couples the exchange transfer of Ca^{2+} and Na^+ across the basolateral membrane. Evidence for the Na^+/Ca^{2+}-exchanger, in renal tubule epithelial cells derives from the following types of observations:

1. sodium dependence of Ca^{2+} transfer in *in situ* or isolated perfused tubules (Ullrich *et al.*, 1976; Friedman *et al.*, 1981; Lorenzen *et al.*, 1984; Yang *et al.*, 1988);
2. coupling of Ca^{2+} and Na^+ fluxes in isolated basolateral membranes from renal proximal (Gmaj *et al.*, 1979, 1985; Jayakumar *et al.*, 1984; Van Heeswijk *et al.*, 1984; Scoble *et al.*, 1985; Dominguez *et al.*, 1989, 1991; Hanaoka *et al.*, 1993) and distal convoluted tubules (Ramachandran and Brunette, 1989; Gesek and Friedman, 1995);
3. immunoreactivity of basolateral membrane protein to antibodies against the exchanger protein (Dominguez *et al.*, 1992; Bourdeau *et al.*, 1993; Reilly *et al.*, 1993); and
4. demonstration of mRNA for the exchanger in renal proximal (Dominguez *et al.*, 1992) and distal convoluted tubule cells (Yu *et al.*, 1992).

The Na^+/Ca^{2+}-exchanger will transport Ca^{2+} into or out of the cell, depending on the direction and relative strengths of the electrochemical potentials for Na^+ and Ca^{2+} across the basolateral membrane and the stoichiometry of the exchanger. Since the electrochemical potential for Ca^{2+} normally exceeds that of Na^+, exchange stoichiometries of $1Na^+ : 1Ca^{2+}$ or $2Na^+ : Ca^{2+}$ would favor Ca^{2+} entry into the cell in exchange for Na^+ extrusion. However, a $3Na^+ : 1Ca^{2+}$ stoichiometry, which has been observed for the cell membrane Na^+/Ca^{2+}-exchanger of cardiac tissue (Reeves and Hale, 1984), would favor net extrusion of Ca^{2+} in exchange for Na^+ entry. Participation of the Na^+/Ca^{2+}-exchanger in Ca^{2+} extrusion from the cell is further suggested from studies in which Na^+/Ca^{2+}-exchange activity has been observed to increase in response to stimulation from PTH, which increases net reabsorptive transfer of calcium (Costanzo and Windhager, 1980; Jayakumar *et al.*, 1984; Scoble *et al.*, 1985; Hanai *et al.*, 1986; Bouhtiauy *et al.*, 1991; Hanaoka *et al.*, 1993).

11.2.5 Intracellular transfer of calcium – evidence for involvement of binding proteins

Renal and intestinal epithelial cells contain 1,25$(OH)_2D_3$-inducible Ca^{2+}-binding proteins (calbindin-D) (Gross and Kumar, 1990). A 9 kDa protein (calbindin-D_{9K}) and a 28 kDa protein (calbindin-D_{28K}) are the predominant forms in mammalian intestine and kidney. Calcium binding is cooperative, with a stoichiometry of two moles Ca^{2+}/mole protein for calbindin-D_{9K} (Linse *et al.*, 1987, 1991; Akke *et al.*, 1995) and three moles Ca^{2+}/mole protein for calbindin-D_{28K} (Fullmer and Wasserman, 1981; Gross *et al.*, 1988). The dissociation constant (K_d) for Ca^{2+} binding is approximately 10^{-7}–10^{-8} M.

Several observations suggest that calbindin-D is involved in Ca^{2+} transport or the regulation of transport. Calbindin-D is induced by 1,25$(OH)_2D_3$ in the small intestine and kidney (Gross and Kumar, 1990). The highest concentrations in both tissues occur in absorptive cells (Thomasset *et al.*, 1986), in regions where calcium transport is stimulated by 1,25$(OH)_2D_3$; the duodenum (Taylor and Wasserman, 1967; Moriuchi *et al.*, 1975; Perret *et al.*, 1985) and renal distal convoluted and connecting tubules (Taylor *et al.*, 1982; Bindels *et al.*, 1991a,b). The concentration of calbindin-D is highest at the duodenal villus tips, where Ca^{2+} transport activity is highest, and lowest in the villus crypts (Taylor and Wasserman, 1967; Corradino and Taylor, 1985; Van Corven *et al.*, 1985). Changes in concentrations of calbindin-D, alkaline phosphatase and Ca^{2+}-ATPase activity along the villus-crypt axis in response to 1,25$(OH)_2D_3$ are related (Bikle *et al.*, 1984; Van Corven *et al.*, 1985). Induction of calbindin-D precedes, or is concurrent with, the increased calcium absorption that occurs in rat duodenum in response to vitamin D (Corradino and Wasserman, 1971; Wasserman *et al.*, 1982; Bishop *et al.*, 1983, 1984; Taylor, 1983).

The exact role that calbindin-D plays in Ca^{2+} transport is not understood. Bronner *et al.* (1986) estimated the rate of diffusion of Ca^{2+} across the rat intestinal epithelial cell and compared it to the measured maximal rate (V_m) of transepithelial flux of Ca^{2+} across the rat *in situ* isolated duodenal loop. The V_m exceeded the diffusion rate by a factor 70, suggesting that under certain conditions intracellular diffusion of Ca^{2+} could be a rate-limiting step in the absorptive transfer of calcium. Kretsinger *et al.* (1982) and Feher (1983, 1984) have suggested that limitations in the rate of diffusion of Ca^{2+} (mol/s/cell) from the apical to basolateral membrane might be overcome by a much larger diffusion rate of calbindin-D. In vitamin-D replete rats, the concentration of calbindin-D is approximately 0.4 mM (Ueng *et al.*, 1979; Thomasset *et al.*, 1982) in intestinal cells. The high concentration of calbindin-D, coupled with the high affinity constant (e.g. 10^{-6} M), would result in a 60-fold 'amplification' of the calcium diffusion rate, assuming a K_d of 3×10^{-7} M, a cytosolic Ca^{2+} concentration of 2×10^{-7} M, and that the diffusion coefficient for calbindin-D was approximately 1/14 that of Ca^{2+} (i.e. square root of the molecular weight ratio).

Calbindin-D may have more direct effects on Ca^{2+} transport mechanisms. Bouhtiauy *et al.* (1994a,b) showed that calbindin-D_{28K}, but not calbindin-D_{9K}, increases calcium uptake by apical-membrane vesicles prepared from rabbit distal tubules, whereas calbindin-D_{9K}, but not calbindin-D_{28K}, stimulates ATP-dependent calcium transport across basolateral membrane vesicles. Calbindin-D_{9K} also stimulates basolateral membrane Ca^{2+}-ATPase activity in the rat duodenum (Walters, 1989; Walters *et al.*, 1990). The plasma membrane Ca^{2+}-ATPase of rat duodenum and bovine Ca^{2+}-ATPase, have a calbindin-D_{9K} binding site (Ghijsen *et al.*, 1986; James *et al.*, 1991), suggesting a possible mediating function of calbindin-D_{9K} on Ca^{2+}-ATPase.

11.2.6 Model of epithelial calcium transport

In summary, the model of calcium transfer across the renal distal convoluted tubule shown in Fig. 11.3 illustrates the basic features of calcium transport across epithelia as they are currently understood. Similar mechanisms appear to be active in other types of transporting epithelia, including the renal proximal tubule and small intestine, although the relative contributions of the various transfer pathways vary, as may the degree of hormonal regulation. The main features of the model are:

1. transfer across the epithelium against an electrochemical gradient for Ca^{2+} involving paracellular and transcellular transfer pathways;
2. facilitated mechanisms for Ca^{2+} entry into the cell across the apical membrane and down a steep electrochemical gradient for Ca^{2+};
3. active extrusion of Ca^{2+} from the cell across the basolateral membrane, against a steep electrochemical gradient; and
4. energy transduction by direct coupling of Ca^{2+} transport to ATP hydrolysis (Ca^{2+}-ATPase) and indirect coupling through the utilization of the membrane sodium gradient (and possibly other ion gradients) produced by the Na^{+},K^{+}-ATPase.

Bronner *et al.* (1986) proposed a kinetic model for intestinal calcium absorption in the vitamin-D replete rat, based on observations made using the *in situ* perfused rat duodenum (Pansu *et al.*, 1981, 1983a,b; Bronner *et al.*, 1986). The model represents transepithelial transport of calcium as the sum of nonsaturable and saturable components, consistent with the evidence of a capacity-limited active transcellular pathway(s) and a paracellular 'shunt' pathway (equation 11.2):

$$J^{net}_{m \rightarrow s} = \frac{V_m C_m}{K_m + C_m} + (K_p C_m) \qquad (11.2)$$

where *J* is the net rate of transepithelial (i.e. mucosal-to-serosal m → s) transport of calcium; V_m (22 μmol/h/g wet tissue) and K_m (3.9 μM) are the apparent maximum rate and the half-saturation constant for capacity-limited transepithelial transport, respectively; K_p is the coefficient for capacity-unlimited transepithelial transport (0.16/h); and C_m is the mucosal calcium concentration (see inset panel of Fig. 11.4). The model predicts that at a dietary calcium intake of approximately 70 mg/day, typical for rats consuming a 0.5% calcium diet (Zornitzer and Bronner, 1971), approximately 88% (62 mg/day) will be absorbed by the saturable pathway.

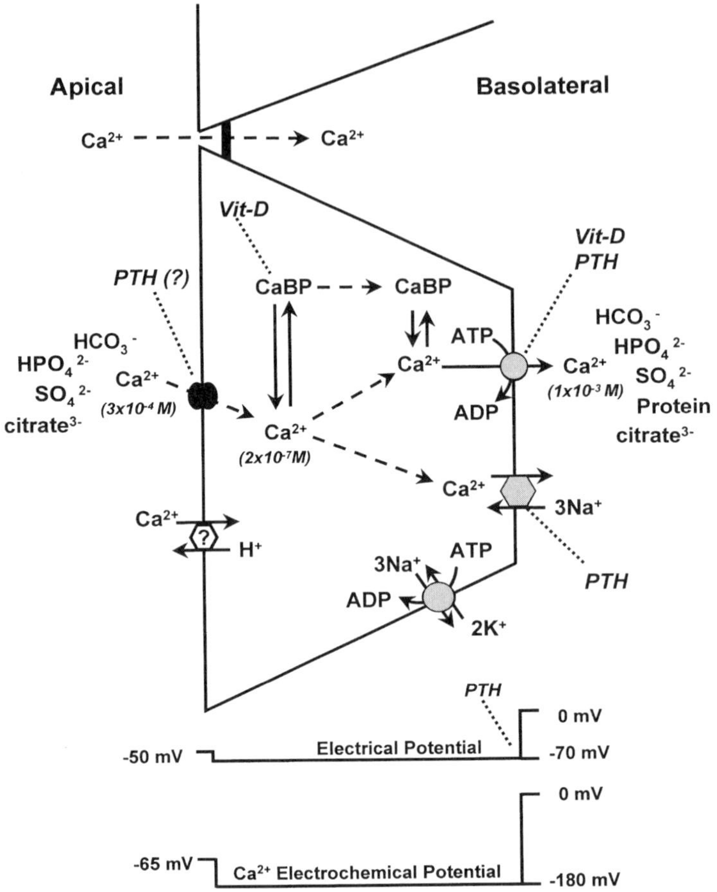

Figure 11.3 Diagram of the major features of Ca^{2+} transport in epithelial cells as reflected in our current understanding of transport in the renal distal convoluted tubule epithelium. In this model, the Ca^{2+}-ATPase on the basolateral membrane (as well as other intracellular Ca^{2+} buffering and storage mechanisms) maintain a steep electrochemical PD for Ca^{2+} (intracellular-negative) gradient across the basolateral and apical membranes. Ca^{2+} can move down this gradient to enter the cell through Ca^{2+} channels in the apical membrane, and possibly by other facilitated mechanisms (e.g. Ca^{2+}/H^+-exchange). The Ca^{2+}/Na^+-exchanger in the basolateral membrane may function in the direction of Ca^{2+} extrusion or Ca^{2+} entry. With the assumed stoichiometry of $3Na^+/Ca^{2+}$, and at the electrochemical PD shown, extrusion would be favored energetically. Calcium-binding proteins (CaBP), such as the calbindins, may also participate in transport if intracellular diffusion of Ca^{2+} is rate limiting (see text for discussion). Mechanisms that appear to be regulated by parathyroid hormone (PTH) or vitamin D (Vit-D) are indicated.

11.3 Lead transport

Our understanding of the mechanisms of lead transport in epithelia has been limited by our inability to measure electrochemical gradients for the numerous lead species that exist in physiologic systems. This limitation derives, in part, from the complexity of the ligand reactions of Pb^{2+} and lack of tools for quantifying Pb^{2+} and its complexes. The

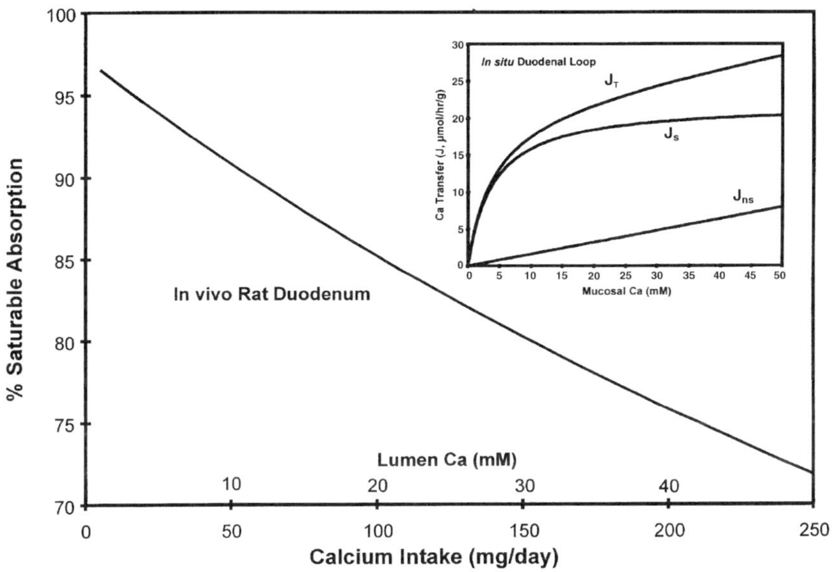

Figure 11.4 Saturable absorption of calcium as a percentage of total calcium absorbed in the rat duodenum predicted by the kinetic model of Bronner *et al.* (1986). Percent saturable absorption is displayed as a function of lumen calcium concentration or equivalent calcium intakes, calculated assuming a daily water intake of 25 ml and a fivefold dilution of ingesta in the duodenum of the rat (Bronner *et al.*, 1986; Powell, 1987). The inset shows the transport rates of the saturable and nonsaturable pathways as a function of duodenal calcium concentration.

first quantification of Pb^{2+} concentrations in cells was made only recently from measurements of the fluorescence quench of the Ca^{2+}-binding dye, indo-1 (Kerper and Hinkle, 1997a,b). The intracellular concentration of Pb^{2+} in 301 cells, a kidney cell line, exposed to 1 μM lead nitrate was estimated to be approximately 10–20 μM. Such techniques have not been applied to studies of lead transport in epithelia. All of our current understanding of lead transport mechanisms derives from isotope tracer studies. These studies have described various aspects of the kinetics of membrane or transepithelial transport of lead, but have provided little insight as to the form of lead that is being transported and, therefore, about the molecular mechanisms involved. However, some general concepts of importance have emerged from these studies and are summarized here.

11.3.1 Lead transport in small intestine

Lead transport in the small intestine has been studied in the *in situ* perfused mouse intestine, the *in situ* ligated chicken intestine, and in *in vitro* isolated segments of rat intestine (Cikrt, 1970; Flanagan *et al.*, 1979; Aungst and Fung, 1981; Mykkänen and Wasserman, 1981; Barton, 1984; Fullmer and Rosen, 1990). Aungst and Fung (1981) examined lead isotope transport across everted segments of rat small intestine. Net rates of transepithelial transport of lead increased and then appeared to approach a limiting rate as mucosal lead concentration was increased, suggesting a capacity limitation in the transepithelial transport of lead. The relationship between net transepithelial transport and mucosal lead concentration conformed to the kinetic model shown in equation 11.2 where: *J* is the net rate of transepithelial (i.e. mucosal-to-serosal, m → s) transport of lead; V_m (1.44×10^{-5} μmol/min) and K_m (0.59 μM) are the apparent maximum rate and the half-saturation constant for capacity-limited transepithelial transport, respectively; K_p is the coefficient for capacity-unlimited transepithelial transport (1.3×10^{-5} L/day), and C_m is the mucosal lead concentration (0.5–50 mM). The model represents net transepithelial lead transport as the sum of nonsaturable and saturable pathways, conceptually analogous to kinetic models proposed for Ca^{2+} transfer across the rat small intestine (Bronner *et al.*, 1986) (see Fig. 11.4). Net transepithelial lead transfer was shown to be at least partially dependent on metabolic energy, as transfer rates were depressed 40% and 50% under anoxic conditions or by sodium fluoride, respectively.

Barton (1984) used metabolic inhibitors and various other manipulations of *in vitro* segments of rat intestine to show that mucosal uptake of lead and transepithelial transport of lead involved distinct mechanisms. The observations that dissociated the two processes included the following:

1 Net transepithelial transfer was much more sensitive to anoxia and metabolic inhibitors than net mucosal uptake of lead; for example, under anoxic conditions, net mucosal-to-serosal transfer was depressed by 98% under anoxic conditions, whereas net mucosal uptake (sum of mucosal-to-serosal transfer and tissue lead uptake) was depressed by only 20%.
2 Net mucosal-to-serosal transfer of lead was depressed in duodenal segments isolated from rats that had received a 1 mg oral dose of $FeCl_2$ or a dose of bacterial endotoxin, whereas mucosal uptake was not affected by either treatment.
3 Addition of 1 mM $FeCl_2$ to the mucosal bathing fluid had no effect on mucosal uptake but depressed mucosal-to-serosal transfer of lead by 80%.
4 Mucosal-to-serosal lead transfer and mucosal uptake were highest in duodenal segments compared to segments of jejunum or ileum; however, mucosal uptake was only 40% lower in jejunum and ileum, whereas mucosal-to-serosal transfer was approximately 80% lower in jejunum and ileum.

In addition to the above observations, Barton (1984) observed serosal fluid/mucosal fluid lead concentration ratios ($[SF/MF]_{Pb}$) as high as six in the duodenal segment incubations, which were decreased to less than 0.1 under anoxic conditions or in the presence of metabolic inhibitors. These observations are consistent with a metabolically dependent, capacity-limited process for transport of lead from the intestinal epithelium to the serosal fluid space. Evidence that transepithelial transport of lead is capacity limited is further supported by studies conducted in the *in situ* ligated intestinal loop of the chicken (Mykkänen and Wasserman, 1981) and the *in situ* perfused intestinal loop of the mouse (Flanagan *et al.*, 1979).

The mechanism(s) that confer a limited capacity to cell-to-serosal fluid lead transport have not been elucidated; however, some insight as to their possible nature has been provided by studies of the effects of calcium and iron on lead absorption. Transepithelial transport of lead across the *in situ* ligated intestinal loop of the chicken is enhanced under conditions of dietary calcium deficiency (Mykkänen and Wasserman, 1981) and depressed under conditions of vitamin D deficiency; the latter depression of lead transport is relieved by administration of cholecalciferol or $1{,}25(OH)_2D_3$ (Mykkänen and Wasserman, 1982). This 'vitamin D-associated' component of lead absorption was observed to be most pronounced in the duodenum, relative to the chick jejunum or ileum, and reflected primarily an enhancement of the serosal transfer of lead from the epithelium, not enhancement of uptake from the lumen. This is similar to the effects of vitamin D status on transepithelial transport of calcium (Mykkänen and Wasserman, 1982), which is thought to be mediated through stimulation of basolateral membrane Ca^{2+}-ATPase activity or induction of calbindin D (Bronner *et al.*, 1986), or both. In the mouse, iron deficiency or anemia have been shown to increase transepithelial transport of lead across the *in situ* lumen-perfused duodenal loop (Flanagan *et al.*, 1979). Most of the increased total uptake of lead from the duodenal lumen was accounted for by an increase in serosal transfer of iron. Thus, interactions between iron and lead also appear to involve either intracellular transfer or basolateral transfer mechanisms.

The above observations on lead transport kinetics and lead–calcium and lead–iron interactions yield a highly incomplete description of lead transport in the small intestine, yet one that is nevertheless consistent with mechanisms that could be inferred from our understanding of the transepithelial transport of Ca^{2+} and other elemental cations (Fig. 11.5). The kinetics of transepithelial transport of lead can be described as the sum of nonsaturable and saturable components, consistent with the possibility of a capacity-limited transcellular pathway and a paracellular 'shunt' pathway. A similar model has been proposed for intestinal calcium absorption in the rat, based on observations made using the *in situ* perfused rat duodenum (Pansu *et al.*, 1981, 1983a,b; Bronner *et al.*, 1986) (Fig. 11.4). The saturable, and perhaps rate-limiting, mechanisms appear

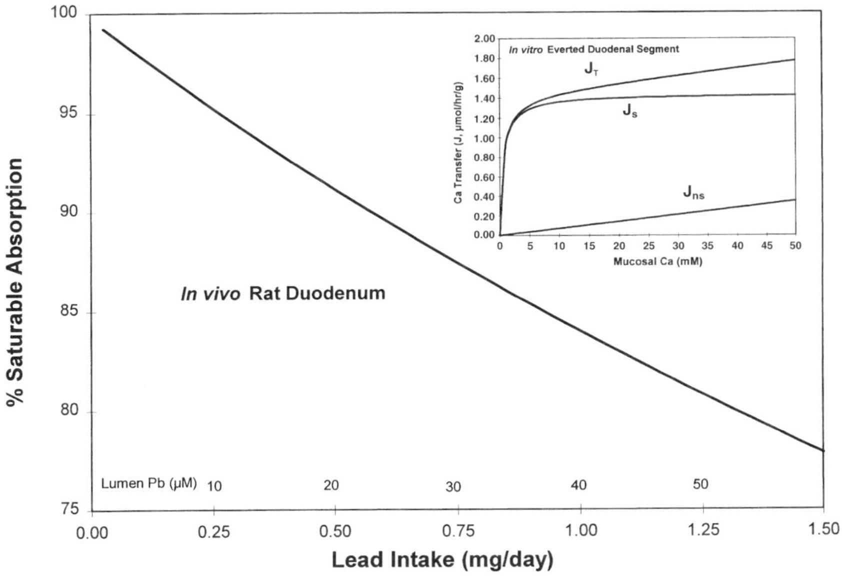

Figure 11.5 Saturable absorption of lead as a percentage of total lead absorbed in the rat duodenum predicted by the kinetic model of Aungst and Fung (1981) (see Fig. 11.4 caption for explanation).

to be primarily associated with transfer of lead from the epithelium to the serosal fluid space and are, at least partially, dependent on metabolic energy. By analogy to other divalent cations, saturable transport mechanisms for Pb^{2+} may exist within the apical and basolateral membranes and within the intestinal epithelial cell. For calcium, these are thought to represent membrane carriers on the basolateral membrane (e.g. Ca^{2+}-ATPase, Ca^{2+}/Na^{+}-exchange) or facilitated diffusion pathways in the apical membrane (e.g. Ca^{2+} channel) and intracellular Ca^{2+}-binding proteins (Bronner *et al.*, 1986; Gross and Kumar, 1990; Friedman and Gesek, 1995).

A comparison of the kinetic models of lead and calcium absorption in the small intestine is revealing (Table 11.1). The duodenal concentrations predicted to result in 20% nonsaturable absorption are 50 μM lead and 30 mM calcium; 83 times the K_m for saturable lead absorption (0.6 μM) and only eight times the K_m for saturable calcium absorption (3.9 mM). If the nonsaturable component of lead and calcium absorption reflects primarily transfer through the paracellular pathway, then the models suggest the possibility of a lower permeability of the paracellular pathway to lead. This cannot be explained by the larger ionic radius of the Pb^{2+} ion (Pb^{2+}, 1.2 Å; Ca^{2+}, 1.0 Å) since the paracellular pathway is highly permeable to larger cations such as Cs^{+} (1.7 Å) and Rb^{+} (1.6 Å) (Armstrong, 1987; Evans, 1994). Alternatively, the relatively small, nonsaturable pathway for lead may reflect the lower permeability of a chemical species of lead other than Pb^{2+}, which is more restricted from nonsaturable pathways available to calcium and other inorganic cations. Direct evidence for paracellular transfer of lead is lacking. Lead has been detected histochemically in the lateral interstitial spaces of transporting epithelia (Morton *et al.*, 1985); however, this by itself is not sufficient evidence for paracellular transfer, as lead could appear in this region if it were transported across the basolateral membrane adjoining the lateral interstitial space. Diaz and Monreal (1995) examined the permeability of myelin lipid liposomes to lead and detected a rheogenic diffusion pathway which they attributed to the lead hydroxide ion, $Pb(OH^{+})$. Whether or not diffusion of lead across intact cell membranes contributes to the nonsaturable pathway for lead remains an open question.

11.3.2 Lead transport in the kidney

Much less is known about the transport of lead in the kidney. Measurements of the renal clearance of radioisotopic lead in dogs have shown that lead can be absorbed from and secreted into the tubule depending on the acid–base status of the animal (Vander

Table 11.1 Predicted relationship between relative magnitudes of saturable and nonsaturable absorption pathways for lead and calcium in rats

	Duodenal concentration	*% Saturable*	*% Nonsaturable*
Lead	0.6 μM (K_m)	99.4	0.6
	50 μM	80	20
Calcium	3.9 mM (K_m)	97	3
	30 mM	80	20

Saturable and nonsaturable absorption of lead and calcium were estimated based on the kinetic models of Aungst and Fung (1981) and Bronner *et al.* (1986), respectively. See text for a description of the models.

et al., 1977; Victery *et al.*, 1979a). Net reabsorption was observed during acidosis induced by hydrogen chloride infusion and normal conditions (sodium chloride infusion), whereas net secretion was observed during alkalosis induced by sodium bicarbonate infusion. Further evidence for bidirectional transfer of lead was provided by observations of the effects of ureteral ligation on lead uptake by the kidney. Glomerular filtration ceases after ligation of the ureter during mannitol-induced diuresis (Malvin and Wilde, 1973), therefore lead uptake observed in ligated kidneys (stop-flow) can be assumed to have resulted primarily from transfer of lead to the kidney from the plasma (basolateral) side of the tubule, rather than from the glomerular filtrate or luminal side of the tubule. The difference between uptake observed during free-flow (ureter-open) and stop-flow conditions provides an estimate of the luminal uptake component. Lead uptake under stop-flow conditions in the dog was shown to be approximately 50% of the uptake that occurs under free-flow conditions, suggesting that approximately half of the lead uptake by the kidney may occur at the basolateral side of the tubule (Victery *et al.*, 1979b). During alkalosis induced by sodium bicarbonate infusion, the magnitude of the basolateral component was twice that observed during normal conditions (sodium chloride infusion) and was estimated to contribute approximately 30% to total lead uptake. The estimated luminal uptake component was four times the filtered load of lead that was observed under free-flow conditions, suggesting that a major component of luminal uptake may represent lead that was secreted into the tubular lumen. These observations are consistent with clearance measurements, which showed net tubular secretion of lead in alkalosis and net reabsorption under normal conditions (Vander *et al.*, 1977; Victery *et al.*, 1979a).

The mechanisms of secretory and absorptive transfer of lead in the kidney have not been characterized. Victery *et al.* (1979b) examined the kinetics of urinary excretion of lead, sodium and inulin during the transition from stop-flow to free-flow conditions in the dog. The concentration profiles of urine excreted immediately after removal of the ureter ligation reflect the profile achieved along the tubule lumen during stop-flow in combination with any further processing of the tubular fluid (e.g. absorption or secretion) that occurs after release of the ureter ligature, the latter is ideally minimized by intense free-flow diuresis (Malvin and Wilde, 1973). Stop-flow analysis indicated that lead was absorbed from the tubular lumen at a site distal to the proximal tubule (i.e. co-located with the sodium concentration minimum) and in the proximal tubule (i.e. co-located with the sodium concentration plateau). Alkalosis induced by sodium bicarbonate infusion enhanced the proximal absorption component.

Uptake of lead in the rabbit renal cortical slice appears to be at least partially dependent on metabolic energy, in that it is depressed under anoxic conditions and by metabolic inhibitors (e.g. dinitrophenol, sodium cyanide, sodium iodoacetate), but not by the Na^+,K^+-ATPase inhibitor, oubain (Vander *et al.*, 1979). Lead uptake, but not *p*-aminohippurate (PAH) uptake, was decreased when slices were incubated with $SnCl_4$ and when slices were incubated in the presence of 20 g/L bovine serum albumin. Cysteine or citrate increased lead uptake by approximately 20% (Vander and Johnson, 1981).

11.3.3 Apical membrane transport mechanisms

Specific mechanisms of transport of lead across the apical membrane have not been elucidated, although there have only been few studies designed to examine such mechanisms in small intestine or other 'leaky' transporting epithelia. Victery *et al.* (1984) studied interactions between lead and isolated apical-membrane vesicles of rat renal proximal tubules. Incubation of the isolated membranes with Pb^{2+} resulted in saturable binding of lead to the membranes which was insensitive to transient electrochemical Na^+,K^+, H^+ or amino-acid gradients. Binding was blocked by Sn^{2+}, Sn^{4+}, La^{3+}, Fe^{2+}, Fe^{3+} or Cu^{2+}, but was not affected by Mg^{2+}, Zn^{2+}, Cd^{2+} or Hg^{2+}, and most notably, Ca^{2+}. The extensive binding to the apical-membrane vesicles may have hindered detection of carrier-mediated transport of lead in this preparation, if such mechanisms were to exist. The same can be said for studies of calcium transport in isolated apical-membrane vesicles (Gmaj *et al.*, 1979). However, in apical-membrane vesicles from rabbit small intestine, binding (or uptake, or both) of calcium is enhanced by an inward-negative electrical potential, suggesting the existence of some type of direct or indirect rheogenic mechanism that transfers calcium into the vesicle where it binds to the interior (cytosolic side) of the membrane (Merrill *et al.*, 1986). Thus, to the extent that lead transport has been studied in isolated apical-membrane vesicles, the characteristics of lead binding seem to be different from those of calcium.

The possibility of lead crossing cell membranes through cation channels may be relevant to lead transport across the apical membrane of the small intestine and kidney. Lead has been shown to enter cells through voltage-gated L-type Ca^{2+} channels in bovine adrenal medullary cells (Simons and Pocock, 1987; Tomsig and Suszkiw, 1991) and through store-operated Ca^{2+} channels in pituitary GH_3, glial C_3, human embyronic kidney and bovine brain capillary endothelial cells (Kerper and Hinkle, 1997a,b). Evidence that such mechanisms play an important role in the transport of lead in epithelia is lacking, although there is growing evidence suggesting an important role of cation channels in apical-membrane transport of Ca^{2+} in the kidney (Friedman and Gesek, 1995).

11.3.4 Model of epithelial lead transport

In summary, a detailed mechanistic model of the epithelial transport of lead is beyond our current understanding. Evidence is in support of the existence of saturable and nonsaturable pathways in the duodenum and that the saturable pathway involves some mechanism associated with serosal transfer, presumably transport mechanisms within the cell or within the basolateral membrane (Fig. 11.6). Lead transfer out of the lumen, across the cell and across the basolateral membrane may involve the lead ion (Pb^{2+}) and/or complexes with one or more ligands. At the apical membrane, lead or its complexes may enter the cell via membrane carriers or channels. Transport as a complex with proteins via endocytosis (Fowler *et al.*, 1994) or as a complex with amino acids is also possible. Although direct evidence for transport of an organic lead complex has not been provided, that such a mechanism is possible is evident by the recent report of transport of a cysteine–mercury complex in the renal proximal tubule (Zalups and Lash, 1997b) and a cysteine–methylmercury complex via an amino-acid transport mechanism in brain endothelial cells (Mokrzan *et al.*, 1995). Once in the cell, lead will interact with a variety of intracellular ligands, including calcium binding proteins (Fullmer *et al.*, 1985) and high-affinity lead-binding proteins (Goering and Fowler, 1984; Fowler and DuVal, 1991; Quintanilla-Vega *et al.*, 1995). Transfer across the cell or basolateral membrane (or both) involves a mechanism(s) that may be sensitive to vitamin D and

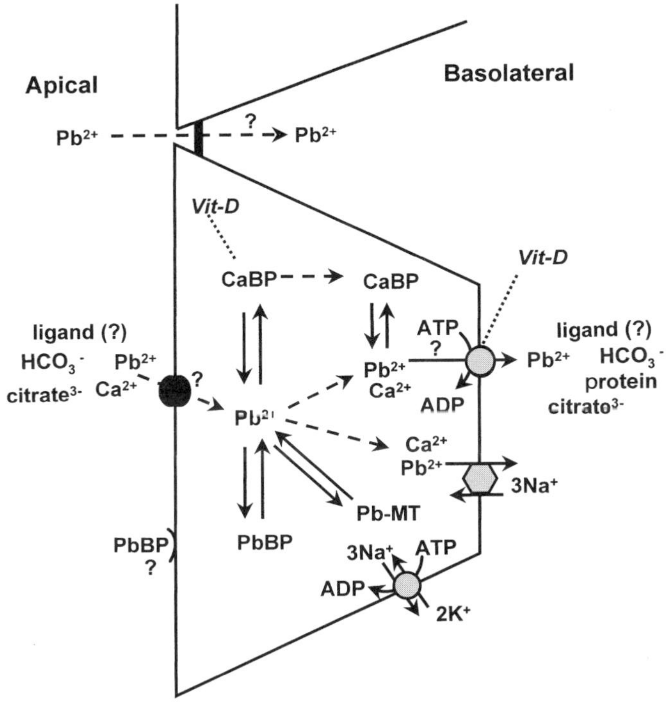

Figure 11.6 Hypothetical model of lead transport in epithelia based on our current understanding of transport in the small intestine. In this model, lead crosses the apical membrane, possibly through Ca^{2+} channels or by endocytosis of a lead-binding protein (PbBP). In the epithelial cell, lead may displace Ca^{2+} from calcium-binding proteins (CaBP) such as the calbindins, or bind to other proteins, including metallothionein (Pb-MT). Lead may exit the cell across the basolateral membrane by displacing Ca^{2+} from the Ca^{2+}-ATPase or the Ca^{2+}/Na^{+}-exchanger. Possible sites of interaction with vitamin D (Vit-D) are indicated. Other mechanisms of membrane transport of lead have not been elucidated, although facilitated mechanisms are suggested from kinetic analyses of radioisotope tracer studies. A paracellular pathway is suggested from the kinetic analyses of lead isotope tracer studies in small intestine (see Fig. 11.5). Electrochemical gradients for Pb^{2+} and other species of lead and lead complexes have not been quantified.

iron status. The vitamin D-associated mechanism may be a Ca^{2+}-ATPase, which is one of the primary calcium-transport mechanisms induced by vitamin D, or possibly the induction of calbindin-D. An alternative transport mechanism is the Ca^{2+}-Na^{+}-exchanger, which does not appear to be regulated by vitamin D (Bronner *et al.*, 1986). The nature of the basolateral iron-associated mechanism has not been elucidated.

11.4 Cadmium transport

11.4.1 Ligand reactions of cadmium

The physiologic disposition of cadmium is influenced substantially by the formation of complexes with protein sulfhydryls and nonprotein sulfhydryl (NPSH) ligands. Cadmium binds to plasma albumin with an apparent dissociation constant of approximately 10^{-4} M (Rao and Lal, 1958; Verma *et al.*, 1982); because of the relatively high concentration of albumin in plasma (approximately 1 mM), under most circumstances the cadmium–albumin complex (Cd–albumin) will be the main form of cadmium in plasma. Cadmium forms highly stable complexes with sulfhydryls; the dissociation constant for the cadmium–cysteine complex (Cd–cys) is approximately 10^{-10} M (Andersen, 1984) and for metallothionein, approximately 10^{-20}–10^{-25} M (Kagi and Vallee, 1961). Thus, cadmium in the plasma will be delivered to the kidney predominantly in the form of complexes with albumin and possibly other high-molecular-weight (HMW) proteins, as the cadmium–metallothionein complex (Cd–MT), or as low-molecular-weight (LMW) complexes with NPSH compounds. The concentration of sulfhydryl complexes of cadmium in the glomerular filtrate will greatly exceed that of Cd–albumin because of the relatively low glomerular sieving coefficient of albumin. Only under conditions in which sulfhydryl concentrations become limiting in the formation of Cd–sulfhydryl complexes will forms of free inorganic cadmium begin to predominate. Jumarie *et al.* (1997) estimated the following concentrations at chemical equilibrium in a HEPES-, phosphate-buffered saline solution containing 0.3 μM cadmium: $CdCl^+$, 0.18 μM (61%); Cd^{2+}, 0.04 μM (14%); $CdCl_2$, 0.06 μM (21%). Such conditions may be rare in extracellular or intracellular environments *in vivo*. In the late proximal tubule or more distal segments of the renal tubule, sulfhydryl concentrations in tubular fluid may be sufficiently low to permit free inorganic cadmium forms to predominate. A sulfhydryl-limited environment also may occur in the lumen of the small intestine.

11.4.2 Transport of cadmium–protein complexes in kidney

Binding of cadmium to albumin greatly restricts the uptake of cadmium by the kidney (Foulkes, 1974; Diamond *et al.*, 1986), which would suggest that the dominant uptake mechanisms for cadmium are on the luminal side of the renal tubule. A peritubular component to cadmium uptake can be observed in kidneys perfused *in vivo* or *in vitro* with protein-free perfusate (Diamond *et al.*, 1986). Although Cd–albumin can be expected to predominate in plasma under normal conditions, after prolonged *in vivo* exposure to cadmium, the concentration of Cd–MT in plasma increases and becomes a toxicologically, if not quantitatively, significant form of cadmium in the glomerular filtrate (Shaikh and Hirayama, 1979). Cd–MT is filtered at the glomerulus and undergoes reabsorption, presumably in the proximal tubule where other low-molecular-weight proteins are known to be reabsorbed (Pesce and First, 1979; Maack, 1992). The evidence for this is as follows (Foulkes, 1978; Felley-Bosco and Diezi, 1987):

1 In the rabbit, after injection of ^{109}Cd–MT and inulin into the renal artery, fractional recovery of cadmium in the renal venous blood was virtually identical to that of inulin, suggesting that Cd–MT does not undergo significant extraction from the peritubular blood.
2 Ureteral ligation preceding an arterial injection of ^{109}Cd–MT and inulin decreased the cadmium-to-inulin uptake ratio in kidney by approximately 98% relative to the uptake observed in filtering kidneys, suggesting that all of the cadmium uptake from injected Cd–MT results from uptake from the luminal side of the tubule.
3 In rabbits, first-pass urinary recovery of cadmium after an injection of 50 μg

^{109}Cd–MT and inulin into the thoracic aorta was approximately 50% that of inulin, suggesting that approximately 50% of filtered cadmium had been absorbed. All of the cadmium recovered in urine was identified by its chromatographic characteristics as Cd–MT.

4 Under similar conditions, recovery of cadmium after an injection of 1 mg ^{109}Cd–MT was approximately 90% of that of injected inulin, suggesting that at greater filtered loads, Cd–MT absorption from the tubule lumen is depressed. The same doses of Cd–MT had no effect on absorption of filtered glucose or on secretion of PAH.

5 Fractional urinary recovery (Cd–MT/inulin) was depressed by concurrent injection of myoglobin, a LMW protein known to be reabsorbed by endocytosis in the proximal tubule, suggesting possible competition for a common reabsorptive mechanism.

6 Urinary recovery of cadmium after microinjection of ^{109}Cd–MT and inulin into the lumen of superficial proximal convoluted tubules was 80% that of inulin, suggesting absorption of approximately 20% of the injected dose of Cd–MT (Felley-Bosco and Diezi, 1987). Recovery of injected cadmium increased to approximately 90% when the injected dose of Cd–MT was increased eightfold. Recoveries of cadmium and inulin injected into superficial distal convoluted tubules were nearly identical.

The mechanism of uptake of Cd–MT from the apical side of the renal proximal tubule appears to be endocytosis. In rats, shortly after an injection of ^{109}Cd–MT, ^{109}Cd is found initially associated with lysosomes in the renal proximal tubules and subsequently in the cytosolic fraction (Squibb *et al.*, 1979, 1984). The protein itself is degraded and cadmium is released from metallothionein during the degradation process (Cherian and Shaikh, 1975; Cain and Holt, 1983). This results in a transient increase in the amount of cadmium bound to various intracellular HMW proteins within 4 hours after injection of Cd–MT and prior to the induced synthesis of additional Cd–MT (Cain and Holt, 1983). Net transepithelial transport of Cd–MT has not been demonstrated definitively in kidney; however, Cd–MT is absorbed, apparently intact, across the rat intestine *in vivo* (Cherian *et al.*, 1978; Cherian, 1979; Sugawara and Sugawara, 1991) and *in vitro* across the everted intestinal sac (Ohta *et al.*, 1989). The mechanism of absorption of Cd–MT in small intestine is not understood and, as in kidney, may involve endocytosis. This would be consistent with the slow kinetics of Cd–MT uptake relative to $CdCl_2$ or Cd–cys in the everted rat intestinal sac. It also is consistent with the observation that there is a relatively small amount of binding of cadmium to brush-border membrane vesicles isolated from rat small intestine when incubated with Cd–MT, relative to that observed when the vesicles are incubated with $CdCl_2$ (Sugawara *et al.*, 1988). Cd–MT appears to be more resistant to degradation in lysosomes of small intestine than in those of renal proximal tubules (Cherian, 1979).

11.4.3 Transport of cadmium as NPSH complexes

Specific mechanisms for membrane transport of cadmium or LMW cadmium complexes have not been characterized. Given the similarities in the affinity constants of cadmium and mercury for NPSHs (Andersen, 1984), it is reasonable to speculate that complexes of cadmium and NPSHs may be important transport substrates; however, this has not been examined closely. Distinct mechanisms of uptake of Cd–MT and Cd–cys in kidney are implicated from studies of the acute nephropathy induced by Cd–MT and combinations of $CdCl_2$ and cysteine. The nephropathy induced by cadmium chloride or Cd–MT is characteristically more severe in the proximal convoluted tubule than in the proximal straight tubule; however, an injected dose of cadmium chloride together with cysteine induces an acute injury predominantly in the proximal straight tubule (Murakami and Webb, 1981). Felley-Bosco and Diezi (1987) measured the urinary excretion of cadmium after microinjection of mixtures of cadmium and L-cysteine into the lumen of superficial proximal convoluted tubule of the *in vivo* rat kidney. Injection of a 1 : 1 molar ratio of cadmium and cysteine increased tubular absorption of cadmium relative to that of cadmium chloride alone, whereas a 5 : 1 or 10 : 1 cysteine:cadmium molar ratio decreased tubular absorption of cadmium. Further elucidation of the mechanism of uptake of cadmium–NPSH complexes in kidney has not be achieved. In the *in vivo* perfused rat jejunum, inclusion of glutathione (GSH) in the lumen perfusate had no effect on cadmium uptake or transepithelial transport of cadmium (Foulkes, 1993). Similarly, in the everted rat intestinal sac, uptake and transepithelial transport of cadmium presented to the lumen as Cd–cys were not different from that observed when $CdCl_2$ rather than Cd–cys was initially in the lumen.

11.4.4 Mechanisms of apical membrane transport of cadmium

Foulkes (1985, 1988; Foulkes and McMullen, 1987) proposed a two-step model of apical transport of cadmium in the small intestine in which binding of cadmium to the external surface of the membrane is followed by transfer of the cadmium to the cell and intracellular binding. In the *in vitro* everted jejunum sac, the first step is discerned as a rapid uptake of the metal from the sac lumen into a chelator-accessible pool which is insensitive to incubation temperature or metabolic inhibitors. The second step is discerned as a slower (requiring approximately 40 seconds to complete at 37 °C), temperature-sensitive transfer of cadmium to a chelator-inaccessible pool which is similarly insensitive to metabolic inhibitors. Foulkes (1988) estimated that approximately one-third of the cadmium initially bound to the apical membrane is available for transfer into the cell. This is consistent with the observation that after a 30-minute incubation of everted intestinal sacs with $^{109}CdCl_2$ (0.1 mM) in the lumen, approximately 20% of tissue ^{109}Cd was associated with the cytosolic cell fraction and the remaining portion was associated with the particulate fraction (Ohta *et al.*, 1989). The cytosolic fraction increases to approximately 40% in sacs from animals pretreated with zinc, which induces metallothionein. The same zinc pretreatment increases lumen-to-basolateral transport of cadmium and the fraction of cytosolic cadmium that is bound to metallothionein.

Foulkes (1985) attributed the first step in the apical mechanism to nonspecific binding of cadmium to anionic charged sites on the external surface of the apical membrane; such a mechanism would permit a wide variety of polycations, including polylysine, to compete for binding, as has been observed in preparations of *in situ* perfused rat jejunum (Foulkes, 1985; Foulkes and McMullen, 1986, 1987), in the *in vitro* lumen and vascular perfused rat intestine (Jaeger, 1990) and in everted rat jejunum sacs (Foulkes and McMullen, 1987). Attempts to explore mechanisms of apical-membrane cadmium transfer in membrane vesicles have also revealed extensive binding of cadmium to the membranes. Endo *et al.* (1997) examined the accumulation of cadmium by brush-border membrane vesicles isolated from rat renal cortex. Vesicles were washed with the chelator, EGTA, after incubations with $CdCl_2$ to remove cadmium bound to the outside of the vesicles; the remaining cadmium associated with the vesicles, the 'chelator-inaccessible' fraction, was assumed to represent cadmium that had been transferred into the vesicle interior. A rapid phase of accumulation of cadmium into the intravesicular (chelator-inaccessible) space occurred during the first minute of incubation with 1 μM $CdCl_2$, which was followed by a slower phase that reached an apparent equilibrium after approximately 50 minutes. During the rapid initial phase, a relatively small, osmotically active, chelator-inaccessible cadmium space was detected, representing approximately 25% of total accumulation, which was abolished by the addition of 30 μM $ZnCl_2$ or $CuCl_2$. Intravesicular accumulation of cadmium appeared to be capacity limited (apparent K_m = 14 μM) and competitively inhibited by $ZnCl_2$ (apparent K_i = 14 μM) and $CuCl_2$ (apparent K_i = 12 μM), but was not inhibited by 2 mM $CaCl_2$. These observations suggest that cadmium, zinc and copper may share a common mechanism for binding to, and possibly transfer across, the apical membrane of the renal proximal tubule.

Studies of zinc accumulation by brush-border membrane vesicles from rat and monkey kidney cortex and pig small intestine provide further support for a common transport or binding mechanism of zinc and cadmium in these tissues (Tacnet *et al.*, 1990, 1991; Prasad and Nath, 1993; Prasad *et al.*, 1996). In general, these studies provide evidence for intravesicular accumulation and binding of zinc to a single kinetically distinct class of sites on the membrane which can be inhibited by cadmium, but not by calcium. Accumulation of zinc by the vesicles in most studies has been shown to be nearly completely insensitive to intravesicular volume and, thus, entirely associated with binding. The exception is the observation of a significant osmotic sensitivity of zinc accumulation in membrane vesicles prepared from monkey renal cortex (Prasad and Nath, 1993). Two pools of zinc binding have been detected in brush-border membrane vesicles, a chelator (e.g. EGTA)-accessible fraction and a chelator-inaccessible fraction. Evidence that the two fractions reflect distinct extravesicular and intravesicular zinc pools is supported by the observation that efflux of zinc from the chelator-inaccessible pool is greatly accelerated by the presence of cation ionophores (Tacnet *et al.*, 1990, 1991; Prasad *et al.*, 1996). Extravesicular binding is characterized by two classes of saturable binding sites (differing in affinity and site density), whereas intravesicular binding consisted of a single class of high-affinity binding sites ($K_d = 3 \times 10^{-5}$ M in pig intestine, 1×10^{-4} M rat kidney) (Tacnet *et al.*, 1990; Prasad *et al.*, 1996). Intravesicular binding is inhibited competitively by cadmium, but not calcium, with zinc and cadmium having similar affinities (Tacnet *et al.*, 1991; Prasad and Nath, 1993). Intravesicular binding in vesicles prepared from pig small intestine has been shown to be inhibited by an inward-negative electrical potential (i.e. outward directed K^+ gradient in the presence of the ionophore valinomycin), suggesting that the transfer of zinc into the vesicle may involve the movement of a negatively charged form of zinc or may be coupled to the transfer of an anion across the membrane (Tacnet *et al.*, 1991). Intravesicular binding was stimulated by pre-equilibration of the vesicles with potassium thiocyanate, but not potassium sulfate, in the absence of an electrical potential (i.e. presence of valinomycin), further suggesting that zinc transfer or binding is sensitive to the anion environment (Tacnet *et al.*, 1991).

The kinetics of uptake and efflux of cadmium has been studied in cultured Caco-2 cells (Rossi *et al.*, 1996; Jumarie *et al.*, 1997; Pigman *et al.*, 1997). The Caco-2 cell line originated as an isolate from a human colon adenocarcinoma (Fogh *et al.*, 1977). Caco-2 cells spontaneously differentiate in culture and exhibit structural and functional differentiation patterns that are characteristic of mature enterocytes (Pinto *et al.*, 1983; Chung *et al.*, 1985). The cells grow in culture to confluency and form monolayers that have microvilli, tight junctions and form domes, characteristic of functionally polarized transporting epithelia (Pinto *et al.*, 1983; Gum *et al.*, 1987). The cells express brush-border hydrolases typical of enterocytes (Pinto *et al.*, 1983; Zweibaum *et al.*, 1984; Hauri *et al.*, 1985; Mohrmann *et al.*, 1986). If grown on permeable supports, they exhibit directional ion transport (Grasset *et al.*, 1984; Ramond *et al.*, 1985; Mohrmann *et al.*, 1986) and glucose transport (Blais *et al.*, 1987). Vitamin uptake (Vincent *et al.*, 1985) and gut-neuropeptide hormone response (Chung *et al.*, 1985) have been observed in Caco-2 cells. The Caco-2 cell line exhibits rapid and slow cadmium uptake steps, consistent with observations made in intestinal segment preparations and isolated brush-border membranes (Jumarie *et al.*, 1997). Uptake was shown to be capacity limited, competitively inhibited by zinc, but not by calcium, and insensitive to oubain. A model was proposed in which cadmium undergoes rapid initial specific binding followed by slower transport into the cell and subsequent rapid binding to a single class of high-affinity intracellular sites (apparent $K_m = 7\ \mu M$). The apparent K_m for the apical-membrane transport process was estimated to be approximately 4 μM and the apparent K_m for zinc inhibition was estimated to be approximately 17 μM.

11.4.5 Mechanisms of basolateral-membrane transport of cadmium

The mechanism responsible for basolateral transfer of cadmium has not been as thoroughly examined and very little information is available. In the *in situ* lumen-perfused rat jejunum, mucosal-to-serosal transfer of cadmium occurs at rates that are considerably slower than those of zinc and nickel (Foulkes and McMullen, 1986, 1987). Jaeger (1990) also reported relatively slow rates of lumen-to-vascular transfer of cadmium, relative to zinc, across the *in vitro* lumen- and vascular-perfused rat intestine. Ohta *et al.* (1989) reported that zinc pretreatment increases mucosal-to-serosal transfer of cadmium in the everted rat intestinal sac; however, Foulkes (1993) reported that a similar treatment abolishes lumen-to-vascular transfer in the *in situ* lumen-perfused rat jejunum. Apical-to-basolateral transfer of cadmium has been demonstrated in Caco-2 cell monolayers grown on permeable supports (Rossi *et al.*, 1996; Pigman *et al.*, 1997). Cadmium transfer across the intact monolayer was substantially less than that of zinc (Rossi *et al.*, 1996). A notable observation made in both studies was that at concentrations of cadmium exceeding 60 μM (as $CdCl_2$) in the apical bathing solution, apical-to-basolateral transfer of cadmium appeared to increase substantially. Cadmium concentrations exceeding 50 μM increased the permeability of the Caco-2 monolayer to mannitol and depressed transepithelial electrical resistance (Rossi *et al.*, 1996; Pigman *et al.*, 1997). This may reflect the opening of the paracellular pathway in the monolayer as a toxic manifestation of the high cadmium concentrations.

11.4.6 Membrane cadmium transport mediated by calcium channels

Cadmium appears to be a substrate for at least some types of Ca^{2+} channels. In primary cultures of rat hepatocytes, cadmium uptake is depressed by the channel antagonists, diltiazem and verapamil, and increased by the channel agonist, vasopressin, suggesting that uptake occurs through voltage-gated L-type Ca^{2+} channels (Blazka and Shaikh, 1991). Channel antagonists decreased cadmium uptake by approximately 30% in this preparation, suggesting that movement of Cd^{2+} through Ca^{2+} channels may be an important pathway for membrane transport of cadmium in hepatocytes. In pituitary GH_3 cells grown in culture, cadmium uptake is depressed by L-type Ca^{2+}-channel antagonists and increased by channel agonists (Hinkle *et al.*, 1987). Furthermore, the Ca^{2+}-channel antagonists, nifedipine, verapamil and diltiazem, showed a protective effect against cadmium-induced toxicity in GH_3 cells. Hinkle *et al.* (1992) used the fluorescent dyes fura-2 and quin-2 to monitor intracellular Cd^{2+} in GH_3 cells. The rate of cadmium uptake increased in response to depolarization (elevated extracellular K^+) or the channel agonist BAY K 8644 and decreased in response to the channel antagonist nimodipine, suggesting that cadmium uptake was occurring through voltage-gated L-type Ca^{2+} channels. Depolarization, or BAY K 8644, had no effect on cadmium uptake rate in glial C6 cells which do not have voltage-gated L-type Ca^{2+} channels. These studies suggest that cadmium may enter certain types of cells through calcium channels. Evidence that such mechanisms play an important role in the transport of cadmium in transporting epithelia is lacking. It should be noted that these observations were made in cells incubated with $CdCl_2$ in a protein-free, physiologic salt medium. The extracellular concentration of Cd^{2+} is likely to be considerably higher under these conditions than would be expected in most extracellular fluids *in vivo* in which protein and nonprotein sulfhydryls are present.

11.4.7 Model of epithelial cadmium transport

In summary, our understanding of the mechanisms of the epithelial transport of cadmium is limited (Fig. 11.7). Cd–MT is absorbed from the lumen of the renal proximal tubule and small intestine, probably by an endocytotic mechanism. Evidence is in support of the existence of saturable mechanisms of binding and uptake of cadmium at the apical membrane, however the nature of these mechanisms is poorly understood. Although analyses of cadmium uptake kinetics suggest the possibility of carrier-mediated or facilitated transport mechanisms for cadmium in kidney and intestine (Endo *et al.*, 1997; Jumarie *et al.*, 1997), direct evidence for such mechanisms is lacking. This relates, in part, to the complexity of the ligand chemistry of cadmium and lack of methods for measuring the electrochemical gradients for cadmium across cell membranes. Cadmium appears to be a substrate for at least some types of Ca^{2+} channels, although evidence that such mechanisms play an important role in the transport of cadmium in transporting epithelia is lacking. Once in the cell, cadmium binds to various proteins, most notably metallothionein, which it also induces. Transfer to the basolateral side of the epithelium appears to be a slow process, relative to other divalent cations such as zinc and nickel, and there is no evidence for a facilitated mechanism.

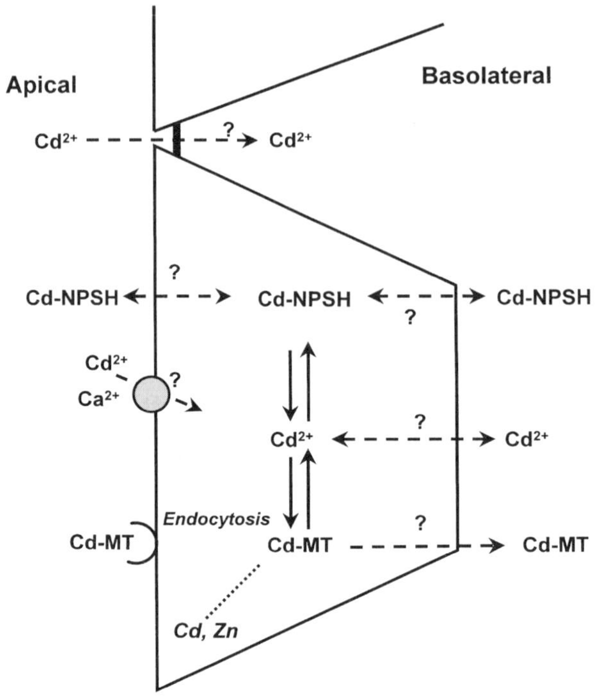

Figure 11.7 Hypothetical model of cadmium transport in epithelia based on our current understanding of transport in the kidney. Cadmium–metallothionein (Cd–MT) crosses the apical membrane of the proximal tubule by endocytosis. In the cell, cadmium released from metallothionein during lysosomal catabolism of the protein induces the synthesis of Cd–MT. Cadmium may also form complexes with other proteins in the cell and nonprotein sulfhydryls (Cd–NPSH). Cd^{2+} may also enter the cell by crossing the apical membrane though Ca^{2+} channels. Other mechanisms of membrane transport of cadmium have not been elucidated, although facilitated mechanisms are suggested from kinetic analyses of radioisotope tracer studies. A paracellular pathway is not evident, although it may exist. Electrochemical gradients for cadmium have not been quantified.

11.5 Mercury transport

The physiologic disposition of mercuric mercury (Hg^{2+}) is dominated by its high affinity for the thiolate anion (Hughes, 1957). In plasma, the predominant thiol ligands include albumin (Cember *et al.*, 1968), the most abundant sulfhydryl (approximately 1 mM) in plasma (Jocelyn, 1972; Brown and Shockley, 1982), and low-molecular-weight (LMW) thiols (10 μM) glutathione (GSH) and cysteine (Lash and Jones, 1985). Within cells, Hg^{2+} forms complexes with the major intracellular thiols, including GSH, cysteine, glycinyl-cysteine and metallothionein (Kagi and Vallee, 1961; Kagi and Nordberg, 1979). Given the abundance and variety of thiols in and outside of cells (1 mM) and the high stability of S–Hg complexes, it is reasonable to expect that the epithelial transport of mercuric mercury may occur, at least in part, by the transport of S–Hg complexes. This concept is supported by observations of the apparent coupling of the movement of mercuric mercury and the transport of nonprotein sulfhydryls (NPHS) across epithelial membranes in kidney (Diamond *et al.*, 1988; Klotzbach and Diamond, 1988), liver (Ballatori and Clarkson, 1985), small intestine (Foulkes, 1993), and evidence that a methylmercury (MeHg)–cysteine complex is a substrate for the amino-acid transport system L in capillary endothelial cells (Mokrzan *et al.*, 1995).

11.5.1 Mercury transport in kidney – evidence for basolateral and apical transport mechanisms

Audioradiographic, histochemical and microdissection techniques have localized the sites of inorganic mercury accumulation in the kidney to the proximal tubule (Taugner *et al.*, 1966; Rodier *et al.*, 1988; Zalups, 1991a,b). In the BALB/c mouse, after administration of a nephrotoxic dose of mercuric chloride, and in the Sprague–Dawley rat, after administration of either a nontoxic or nephrotoxic dose of mercuric chloride, mercury was localized with autometallographic techniques to the S1, S2 and S3 segments with more abundant deposits in the proximal straight tubule, S2 and S3 segments (Hultman *et al.*, 1985; Hultman and Enestrom, 1986; Zalups, 1991a,b). The mechanisms by which Hg^{2+} enters cells of the renal proximal tubule are not completely understood, however, there appears to be uptake mechanisms located at both the luminal and basolateral sides of the tubule.

Glomerular filtration can be prevented or reduced to negligible values by ureteral ligation during a mannitol diuresis (Malvin and Wilde, 1973). In rats, this 'stop-flow' procedure decreases uptake of an injected dose of mercuric chloride into the renal cortex and outer stripe of the outer medulla by approximately 40–50%, suggesting that approximately half of the mercury uptake occurs from the glomerular filtrate and, therefore, from the luminal side of the tubule, and half of the uptake occurs at the basolateral side of the tubule (Zalups and Minor, 1995). Most of this 'basolateral component' of mercury uptake, approximately 75–80%, can be inhibited by an intravenous injection of the organic anion *p*-aminohippurate (PAH) given prior to mercuric chloride during stop-flow conditions, suggesting the involvement of an organic anion transport mechanism in the basolateral uptake of mercury (Zalups and Minor, 1995).

In rats, uptake of mercury into the renal cortex and outer stripe of the outer medulla is also inhibited by prior treatment with the dicarboxylic acids, succinic, glutaric or adipic acid (Zalups and Barfuss, 1998). The inhibition appears to be, at least in part, an inhibition of basolateral uptake of mercury. A 1 mmol/kg intravenous dose of glutaric acid administered 1 hour prior to a 0.5 μmol/kg dose of mercuric chloride decreases mercury uptake in the kidney by approximately 37%. Under stop-flow conditions, the same dose of glutaric acid decreases renal mercury uptake by approximately 48%; thus, at least one-half of the basolateral mercury uptake that occurs under stop-flow conditions appears to be sensitive to inhibition by glutaric acid. A PAH- and probenecid-sensitive organic anion/dicarboxylic acid (OA/DCA) exchange mechanism on the basolateral membrane of the renal proximal tubule has been postulated (Pritchard and Miller, 1993; Dantzler, 1996). The observation that basolateral uptake of Hg^{2+} is inhibited by PAH and dicarboxylic acids, both of which are substrates for the OA/DCA-exchanger, suggests the possibility that Hg^{2+} may be transported across the basolateral membrane as a complex with an anionic substrate for the OA/DCA-exchanger.

In addition to an organic anion-sensitive mechanism previously eluded to, an acivicin-sensitive uptake mechanism has been shown to exist in the mouse kidney (Tanaka *et al.*, 1990; Tanaka-Kagawa *et al.*, 1993) and rat kidney (Berndt *et al.*, 1985; de Ceaurriz *et al.*, 1994; Zalups, 1995). Acivicin is an inhibitor of the enzyme γ-glutamyl transpeptidase (γGT) which cleaves the γ-glutamyl moiety from GSH. The enzyme is located in the kidney, predominantly in the proximal straight tubule (Heinle and Wendel, 1977), and predominantly in the luminal membrane, although the enzyme has also been associated with the basolateral membrane and renal microvasculature (Dass *et al.*, 1981; Spater *et al.*, 1982). In rats, after an injection of a nontoxic dose of mercuric chloride (0.5 mmol/kg), near complete inhibition of renal γGT activity by acivicin decreases uptake of mercury in the renal cortex by approximately 65%. The acivicin-induced decrease in mercury uptake occurs almost exclusively in the renal cortex, whereas PAH inhibits uptake of mercury in both the cortex and outer medulla (Zalups, 1995). Treatment of rats with acivicin and PAH in combination inhibits mercury uptake in the renal cortex by approximately 90% and in the outer medulla by approximately 50% (Zalups, 1995). Furthermore, ureteral ligation abolishes the effect of acivicin on mercury uptake, suggesting that the acivicin-sensitive mechanisms reside predominantly, if not exclusively, on the luminal side of the proximal tubule. Thus, the acivicin-sensitive and PAH-sensitive mechanisms appear to be distinct mechanisms and, together, account for most (possibly as much as 80%) of the mercury uptake in the proximal tubule.

It is possible that the PAH-sensitive and acivicin-sensitive mechanisms involve some common transport substrates. Hg^{2+} readily forms highly stable S–Hg complexes with thiols such as cysteine and GSH (Fuhr and Rabenstein, 1973; Rabenstein, 1989) and such complexes would be expected to form in plasma given that the concentrations of reduced cysteine and glutathione in plasma are approximately 10 mM (Lash and Jones, 1985). Furthermore, the catabolism of GSH to its constituent amino acids by γGT and dipeptidases in the brush-border membrane of the proximal tubule would be expected to give rise to a mercury–cysteine complex (Hg–cys) in the tubular lumen. Studies of basolateral-membrane vesicles isolated from rat kidney cortex and primary cultures of rat proximal tubular cells have revealed PAH- or probenecid-sensitive transport mechanisms for GSH, S–GSH conjugates and S–cysteine conjugates (Lash and Jones, 1983, 1984; Lash and Anders, 1989). Thus, it is conceivable that Hg–cys and Hg–GSH complexes may be substrates for a PAH-sensitive transport mechanism in the proximal tubule, although evidence for this is currently lacking. Both GSH and cysteine, when administered with mercuric chloride, increase the uptake of mercury in the renal cortex and outer medulla (Zalups and Barfuss, 1995a,b). Co-administration of cysteine increases the severity of nephrotoxicity of mercuric chloride (Zalups and Barfuss, 1996). Studies of the transport of Hg–cys and Hg–GSH have been attempted with preparations of luminal- and basolateral-membrane vesicles isolated from rat kidney; however, rapid and extensive binding of mercury to the membranes severely limits the quantification of mercury transport fluxes in this preparation. Therefore, an effect of cysteine or GSH on the amount of mercury taken up by the membranes could represent an effect on binding or transport since binding could occur before, during or after transport has occurred. Cysteine in a threefold or tenfold excess with Hg^{2+} increased mercury uptake by the luminal-membrane vesicles, relative to that which occurred with Hg^{2+} alone (Zalups and Lash, 1997a,b). GSH, on the other hand, decreased mercury uptake in both the luminal- and basolateral-membrane vesicles. The effect of cysteine on mercury uptake may reflect a direct or indirect coupling of Hg^{2+} and cysteine fluxes into, or across, the luminal membrane vesicle or, possibly, a ligand exchange between cysteine and sulfhydryls at the membrane surface. It is also conceivable that a dipeptide carrier on the luminal membrane (Barfuss *et al.*, 1988) may transport a dipeptide complex with Hg^{2+} (GlyCys–Hg–CysGly). Further study is needed to sort out these possible mechanisms.

Additional evidence for luminal uptake of mercuric mercury comes from studies of isolated perfused rabbit proximal tubules. Uptake or binding of mercury by the tubule occurs in preparations of isolated S1, S2 and S3 segments perfused with mercuric chloride in the lumen (Barfuss *et al.*, 1990). In S3 segments, lumen-to-basolateral transfer rates exceeded the transepithelial leak rate predicted from measurements of the lumen-to-basolateral transfer of L-glucose, suggesting net absorptive transfer of mercury. When S1, S2 or S3 segments are perfused with mercuric chloride and GSH in the lumen, net lumen-to-basolateral transfer exceeding leak rates occurred in the S1 and S2 segments, but not in the S3 segments (Zalups *et al.*, 1991). Transepithelial lumen-to-basolateral transfer of mercury is even more pronounced when tubule segments are perfused with methylmercury; lumen-to-basolateral transfer rates (exceeding leak rates) in S2 and S3 segments were nearly equivalent to the rate of extraction of mercury from the perfusate, suggesting that nearly all of the mercury taken up from the lumen perfusate was transferred across the tubule epithelium (Zalups and Barfuss, 1993). Transepithelial transport of mercury persisted in tubules perfused with methylmercury together with GSH, although rates of transfer were greatly reduced in the presence of GSH. The above studies suggest that mercuric mercury (inorganic or methylmercury) can be extracted from the tubule lumen and transferred across the epithelium. Although the studies do not identify transport mechanisms, they provoke interesting questions about possible mechanisms that could be explored in this preparation. For example, net transepithelial transfer of Hg^{2+} or $MeHg^{+}$ could occur through either a paracellular or transcellular pathway. However, it is unlikely that a paracellular pathway could produce rates of lumen-to-basolateral mercury transport that would exceed L-glucose transfer rates if the transported substrate were a relatively large anionic complex of Hg^{2+} or $MeHg^{+}$ with GSH or the peptidase products of GSH, cysteine or glycinyl-cysteine. Therefore, if a GSH or GSH-derived complex of mercury is being transported out of the tubule lumen across the epithelium in this preparation, it is likely that it is occurring through a transcellular rather than a paracellular pathway. It is also possible that a gradient of some ligand for mercury (e.g. GSH) might drive paracellular transfer of Hg^{2+} or $MeHg^{+}$ across the tubule. Measurements of electrochemical gradients for the various mercury species across the tubule epithelium might resolve this issue.

11.5.2 Transport of mercury–protein complexes

In addition to the PAH-sensitive and acivicin-sensitive mechanisms described above, a third possible mechanism of uptake mercury in the proximal tubule is the endocytosis of Hg^{2+} bound to filtered serum albumin. Mercuric mercury in plasma is bound extensively to albumin (Cember *et al.*, 1968), the most abundant (approximately 1 mM) sulfhydryl in plasma (Jocelyn, 1972; Brown and Shockley, 1982). The filtration fraction of albumin in the rat is approximately 0.0004 (Oken *et al.*, 1972; Pesce and First, 1979); therefore, a small fraction of the Hg–albumin in plasma would be expected to enter the lumen of the early proximal tubule in the glomerular filtrate. Crude estimates of the relative magnitudes of filtered Hg–albumin and filtered low-molecular-weight mercury complexes (Hg–LMW) can be made by assuming the following: 50% of the mercury in blood is in the plasma fraction (Zalups, 1993) of which 98% is bound to albumin (Cember *et al.*, 1968) and the remaining 2% consists of LMW complexes of mercury (e.g. Hg–GSH); a glomerular filtration rate of 1 ml/min in the rat; and a filtration fraction of 0.0004 for Hg–albumin complex and 1.0 for LMW complexes of mercury. The above assumptions yield a plasma clearance of Hg–albumin due to glomerular filtration that is approximately 2% that of filtered LMW complexes. Filtered albumin is extracted from the tubular fluid in the proximal tubule through the process of endocytosis, and is transferred to lysosomes for catabolism (Maack, 1992). A similar mechanism could result in the endocytosis of filtered Hg–albumin. Proteinuric rats given an injection of a nontoxic dose of mercuric chloride excrete a substantially larger fraction of urinary mercury as Hg–albumin compared to normal, nonproteinuric rats and have a larger fraction of total cellular mercury associated with the lysosomal fraction (Madsen, 1980; Madsen and Hansen, 1980). These observations suggest that Hg–albumin may be taken up from the lumen of the proximal tubule, possibly by endocytosis. Mercury can induce proteinuria in animals and, in rats, mercuric chloride-induced proteinuria is associated with increased excretion of Hg–albumin in urine (Madsen, 1980). The albuminuria may reflect impaired reabsorption of protein, as it occurs in association with cellular and tubular necrosis (Zalups *et al.*, 1988); however, increased filtration of albumin may also contribute to the albuminuria. If the endocytosis of filtered Hg–albumin is a significant pathway for entry of mercury into epithelial cells of the proximal tubule, then glomerular proteinuria induced by mercury may further exacerbate toxicity by increasing the delivery of Hg–albumin to the lumen of the proximal tubule.

Complexes between Hg^{2+} and filtered proteins other than albumin also may be taken up by the proximal tubule epithelium. Rats administered Hg–metallothionein complex (Hg–MT) accumulate more mercury in the renal cortex and outer medulla and excrete more mercury in urine than rats administered mercuric chloride, suggesting that the complex is more available for uptake in the proximal tubule than Hg^{2+} bound to other endogenous ligands (Zalups *et al.*, 1995a). Metallothionein is a 6000–7000 Da, sulfhydryl-rich protein that forms highly stable complexes (Zalups and Minor, 1995) with Hg^{2+}, Cd^{2+} and other divalent, thiolphilic metals (Kagi and Vallee, 1961; Kagi and Nordberg, 1979). The mechanism for the effect of metallothionein on mercury uptake has not been determined. As noted previously, Cd–MT is filtered at the glomerulus and undergoes reabsorption (Foulkes, 1978; Felley-Bosco and Diezi, 1987), presumably in the proximal tubule where other low-molecular-weight proteins are known to be reabsorbed (Pesce and First, 1979; Maack, 1992). A similar mechanism may explain the increased uptake of mercury when Hg–MT is administered to rats. Furthermore, when mercury is administered as Hg–MT, the segmental distribution of injury is shifted along the proximal tubule. The proximal convoluted and early proximal straight (S1 and S2) tubule segments appear to be more vulnerable to Hg–MT than is the late proximal straight tubule (S3) segment (Chan *et al.*, 1992); whereas the opposite pattern of vulnerability is characteristic of mercuric chloride (Zalme *et al.*, 1976; Murakami and Webb, 1981; Zalups and Lash, 1994). It is tempting to speculate that this change in the location of the injury may reflect endocytosis of Hg–MT in the proximal convoluted tubule. In support of this, the nephropathy induced by cadmium chloride or Cd–MT is characteristically more severe in the proximal convoluted than in the proximal straight segments of the proximal tubule; however, an injected dose of cadmium chloride together with cysteine induces an acute injury predominantly in the proximal straight segment (Murakami and Webb, 1981). As previously noted, co-administration of cysteine with mercuric chloride increases the uptake of mercury in the renal cortex and outer stripe of the outer medulla, where the straight segments of the proximal tubule are located (Zalups and Barfuss, 1996). Isolated perfused rabbit proximal tubule segments do not take up mercury when presented with Hg–MT complex in the tubule lumen or on the basolateral side of the tubule, which is unfortunate since this preparation would be otherwise ideal for examining mechanisms of uptake of Hg–MT (Zalups *et al.*, 1995b). Possibly, the kinetics of endocytosis are too slow to be accommodated in this preparation, or the preparation lacks some necessary property for endocytosis of low-molecular-weight proteins. It also is possible that mercury impairs or inhibits endocytosis. In the rat, mercuric chloride has been shown to inhibit lysosomal catabolism of the low-molecular-weight protein, lysozyme (Madsen and Christensen, 1978). Examination of the transport and metabolism of other low-molecular-weight proteins (e.g. lysozyme, $\alpha_2\mu$-globulin, retinol-binding protein) in the isolated perfused tubule preparation would provide greater insight into these problems.

11.5.3 Mercury transport in the small intestine

Foulkes (1993) examined inorganic mercuric mercury transport across everted intestinal sacs from rats and in the *in situ* perfused rat jejunum. Net transepithelial transport of mercury was observed in the *in situ* perfused jejunum and increased when the lumen perfusate included GSH; GSH had no effect on transepithelial transport of cadmium in the same preparation. Pretreatment of rats with zinc to induce the synthesis of metallothionein on the intestinal epithelium had no effect on net rates of transepithelial transport of mercury, but substantially decreased net transport of cadmium.

In sacs incubated with mercuric chloride in the lumen, mercury accumulated in two pools, distinguished by the rates at which mercury could be extracted from the tissue by luminal incubations with the NPSH complexing agent, dimercaptosuccinic acid (DMSA) at 0 °C (Foulkes and Bergman, 1993). Uptake into 'fast DMSA pool' was faster than the rate of uptake into a 'slow DMSA pool'. These observations may be analogous mechanistically to the chelator-accessible and chelator-inaccessible cadmium pools observed in rat small intestine (Foulkes, 1988). Further similarities to cadmium included the observation that the slow-DMSA pool showed a greater temperature dependence than uptake into the fast-DMSA pool; however, neither pool was sensitive to metabolic inhibitors. Thus, the slow-DMSA pool may represent intracellular mercury and the fast pool may represent mercury bound to the apical membrane (Foulkes and Bergman, 1993). Distinctly different from the behavior of cadmium, however, was the observation that mercury uptake from the lumen was significantly increased by the polycations La^{3+} ($LnCl_3$) and polylysine. Uptake was greater when chloride in the lumen incubation solution was replaced with nitrate, and the enhancement of uptake produced by La^{3+} was no longer evident under the chloride-free conditions. These observations suggest the possibility that mercury may interact with the apical membrane as an anion in this preparation, possibly as $HgCl_3^-$.

11.5.4 Model of epithelial mercury transport

The picture that is emerging from these studies is that mercury enters cells of the renal proximal tubule by pathways from both the luminal and basolateral sides of the tubule (Fig. 11.8). One pathway appears to be dependent on activity of the enzyme γGT; a second pathway appears to involve an organic anion transport mechanism. Both mechanisms may involve the participation of an NPSH complex with mercury. A

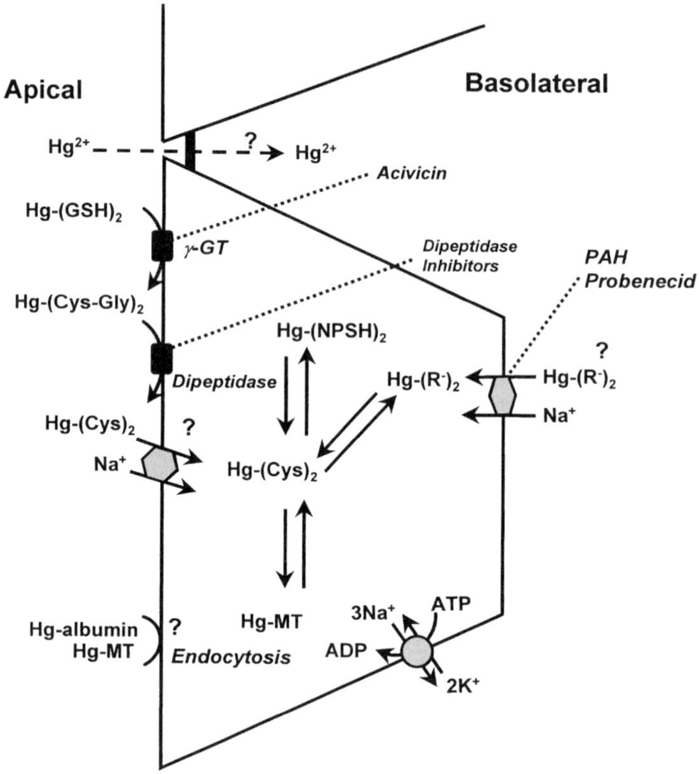

Figure 11.8 Hypothetical model of inorganic mercuric mercury transport in epithelia based on our current understanding of transport in the kidney. Acivicin-sensitive transport of mercury occurs at the apical membrane of the renal proximal tubule. A complex between Hg^{2+} and GSH ($Hg–GSH_2$) is filtered at the glomerulus and catabolized by the enzymes γ-glutamyl transpeptidase (γ-GT) and dipeptidases located on the luminal membrane. The resulting mercury–cysteine complex ($Hg–Cys_2$) is transported across the luminal membrane via a carrier. The nature of a $Hg–Cys_2$ carrier has not be elucidated, but is depicted as a sodium (Na^+)-coupled mechanism. A dipeptide carrier also may transport a dipeptide complex with cysteinyl-glycine, ($Hg-(Cys-Gly)_2$), or the $Hg-GSH_2$ complex could be a transport substrate. PAH-sensitive transport of mercury occurs at the basolateral membrane. An Hg^{2+} complex ($Hg–R^-_2$), possibly $Hg–Cys_2$ or $Hg–GSH_2$, may be transported across the basolateral membrane by a sodium-coupled mechanism that is inhibited by PAH and probenecid. Alternatively, PAH- and probenecid-sensitivity may also result from coupling of the transport of an Hg^{2+} complex via an anion-exchange carrier on the basolateral membrane that is inhibited by PAH or probenecid. Mercury may also cross the apical membrane by endocytosis as a complex with metallothionein (Hg–MT) or other proteins. Electrochemical gradients for mercury have not been quantified.

mercury–cysteine complex is a suspected anionic transport substrate in the kidney. Complexes between mercury and proteins, including metallothionein, may also be taken up at the apical membrane by a mechanism involving endocytosis.

11.6 Unifying concepts

Although our understanding of mechanisms of transport of metals is incomplete, certain general conclusions can be made, based on the existing information gleaned from studies such as those summarized in the preceding sections. Unlike the case for calcium, there is no overwhelming evidence for or against a paracellular pathway for the epithelial transport of lead, cadmium or mercury. Indeed, Foulkes and McMullen (1987) suggest that a *de minis* model capable of describing the uptake and transport of cadmium, need not include a paracellular pathway. There is no direct evidence for the transport of lead, cadmium or mercury against an electrochemical gradient, and therefore, no direct evidence that transport of any of these metals is active. Studies of the effects of anoxia and metabolic inhibitors on lead transport in various preparations suggest there is an energy-requiring step in the transport of lead; however, the nature of the energy-requiring step has not been ascertained. Isotope tracer studies have revealed kinetic characteristics that are consistent with facilitated mechanisms being involved in the transport of lead, cadmium and mercury. These characteristics include evidence for affinity and capacity limitation (saturability), specificity and physiologic regulation. Although definitive identification of mechanisms has not been made, the evidence is strongest for participation of metals or their complexes as substrates for mechanisms that are normally thought to mediate the transport of physiologic substrates ('molecular mimicry'). Examples include the transport of Pb^{2+} and Cd^{2+} through Ca^{2+} channels, the transport of Cd–MT by endocytotic mechanisms and the possible transport of Hg^{2+} and MeHg as complexes with cysteine or glutathione.

References

Akke, M., Forsén, S. and Chazin, W.J. 1995. Solution structure of $(Cd^{2+})_1$-calbindin D_{9k} reveals details of the stepwise structural changes along the Apo → $(Ca^{2+})_1^{II}$ → $(Ca^{2+})_2^{I,II}$ binding pathway. *Journal of Molecular Biology* **252**, 102–21.

Al-Awqati, Q., Chase, H.S. Jr and Kleyman, T.R. 1986. Cellular mechanisms of renal transport and metabolism, in *The Kidney*, Vol. 1, (eds B.M. Brenner and F.C. Rector Jr.) W.B. Saunders, Philadelphia, pp. 61–92.

Allen, G.G. and Barratt, L.J. 1985. Origin of positive transepithelial potential difference in early distal segments of rat kidney. *Kidney International* **272**, 622–29.

Andersen, O. 1984. Chelation of cadmium. *Environmental Health Perspectives* **54**, 249–66.

Armstrong, W.M. 1987. Cellular mechanisms of ion transport in the small intestine. In *Physiology of the Gastrointestinal Tract*, Vol. 2, (ed. L.R. Johnson), Raven Press, New York, pp. 1251–65.

Aungst, B.J. and Fung, H.L. 1981. Kinetic characterization of *in vitro* lead transport across the rat small intestine. *Toxicology and Applied Pharmacology* **61**, 38–47.

Bacskai, B.J. and Friedman, P.A. 1990. Activation of latent Ca^{2+} channels in renal epithelial cells by parathyroid hormone. *Nature* **347**, 388–91.

Ballatori, N. and Clarkson, T.W. 1985. Biliary secretion of glutathione and of glutathione–metal complexes. *Fundamental and Applied Toxicology* **5**, 816–31.

Barfuss, D.W., Ganapathy, V. and Leibach, F.H. 1988. Evidence for active dipeptide transport in isolated proximal straight tubules. *American Journal of Physiology* **255**, F177–81.

Barfuss, D.W., Robinson, M.K. and Zalups, R.K. 1990. Inorganic mercury transport in the proximal tubule of the rabbit. *Journal of the American Society of Nephrology* **1**, 910–17.

Barton, J.C. 1984. Active transport of lead-210 by everted segments of rat duodenum. *American Journal of Physiology* **247**, G193–8.

Berndt, W.O., Baggett, J.McC., Blacker, A. and Houser, M. 1985. Renal glutathione and mercury uptake by kidney. *Fundamental and Applied Toxicology* **5**, 832–9.

Bikle, D.D., Zorlock, D.T. and Munson, S. 1984. Differential response of duodenal epithelial cells to 1,25-dihydroxyvitamin D_3 according to position on the villus: a comparison of calcium uptake, calcium-binding protein, and alkaline phosphatase activity. *Endocrinology* **115**, 2077–84.

Bindels, R.J.M., Hartog, A., Timmermans, J.A.H. and Vanos, C.H. 1991a. Immunocytochemical localization of calbindin D28k, calbindin-D9k and parvalbumin in rat kidney. *Contributions to Nephrology* **91**, 7–13.

Bindels, R.J.M., Timmermans, J.A.H., Hartog, A., Coers, W. and Van OS, C.H. 1991b. Calbindin-D9k and parvalbumin are exclusively located along basolateral membranes in rat distal nephron. *Journal of the American Society of Nephrology* **2**, 1122–9.

Bishop, C.W., Kendrick, N.C. and DeLuca, H.F. 1983. Induction of calcium-binding protein before 1,25-dihydroxyvitamin D_3 stimulation of duodenal calcium uptake. *Journal of Biological Chemistry* **258**, 1305–10.

Bishop, C.W., Kendrick, N.C. and DeLuca, H.F. 1984. The early time course of calcium-binding protein induction by 1,25-dihydroxyvitamin D_3 as determined by computer analysis of two dimensional electrophoresis gels. *Journal of Biological Chemistry* **259**, 3355–60.

Blais, A., Bissonnette, P. and Berteloot, A. 1987. Common characteristic for Na^+-dependent sugar transport in Caco-2 cells and human fetal colon. *Journal of Membrane Biology* **99**, 113–25.

Blazka, M.E. and Shaikh, Z.A. 1991. Differences in cadmium and mercury uptakes by hepatocytes: role of calcium channels. *Toxicology and Applied Pharmacology* **110**, 355–63.

Bomsztyk, K., George, J.P. and Wright F.S 1984. Effects of luminal fluid anions on calcium transport by proximal tubule. *American Journal of Physiology* **250**, F680–9.

Borke, J.L., Minami, J., Verma, A.K., Penniston, J.T. and Kumar, R. 1987. Monoclonal antibodies to human erythrocyte membrane Ca^{2+}–Mg^{2+} adenosine triphosphatase pump recognize an epitope in the basolateral membrane of human kidney distal tubule cells. *Journal of Clinical Investigation* **80**, 1225–31.

Bouhtiauy, I., Lajeunesse, D. and Brunett, M.G. 1991. The mechanism of parathyroid hormone action on calcium reabsorption by the distal tubule. *Endocrinology* **128**, 251–8.

Bouhtiauy, I., Lajeunesse, D., Christakos, S. and Brunette, M.G. 1994a. Two vitamin D_3-dependent calcium binding proteins increase calcium reabsorption by different mechanisms. I. Effect of CaBP 28K. *Kidney International* **45**, 461–8.

Bouhtiauy, I., Lajeunesse, D., Christakos, S. and Brunette, M.G. 1994b. Two vitamin D_3-dependent calcium binding proteins increase calcium reabsorption by different mechanisms. II. Effect of CaBP 9K. *Kidney International* **45**, 469–74.

Boulpaep, E.L., and Seely, J.F. 1971. Electrophysiology of proximal and distal tubules in the autoperfused dog kidney. *American Journal of Physiology* **221**, 1084–96.

Bourdeau, J.E. and Burg, M.B. 1979. Voltage dependence of calcium transport in the thick ascending limb of Henle's loop. *American Journal of Physiology* **236**, F357–64.

Bourdeau, J.E. and Eby, B.K. 1990. cAMP-stimulated rise of $[Ca^{2+}]$ in rabbit connecting tubules: role of peritubular Ca. *American Journal of Physiology* **258**, F751–55.

Bourdeau, J.E. and Lau, K. 1989. Effects of parathyroid hormone on cytosolic free calcium concentration in individual rabbit connecting tubules. *Journal of Clinical Investigation* **83**, 373–9.

Bourdeau, J.E. and Lau, K. 1992. Regulation of cytosolic free calcium concentration in rabbit connecting tubule: a calcium-absorbing renal epithelium. *Journal of Laboratory and Clinical Medicine* **119**, 650–62.

Bourdeau, J.E., Taylor, A.N. and Iacopino, A.M. 1993. Immunocytochemical localization of sodium–calcium exchanger in canine nephron. *Journal of the American Society of Nephrology* **4**, 105–10.

Bronner, F., Pansu, S. and Stein, W.D. 1986. An analysis of intestinal calcium transport across the rat intestine. *American Journal of Physiology* **250**, G561–9.

Brown, J.R. and Shockley, P. 1982. Serum albumin: Structure and characterization of its ligand binding sites, in *Lipid–Protein Interactions*, Vol. 1, (eds P.C. Jost and O.H. Griffith), Wiley, New York, pp. 25–68.

Brunette, M.G., Blouin, S. and Chan, M. 1987. High affinity Ca^{2} Mg^{2+} ATPase in the distal tubule of the mouse kidney. *Canadian Journal of Physiology and Pharmacology* **65**, 2093–8.

Brunette, M.G., Mailloux, J., Chan, M. and Ramachandran, C. 1990. Characterization of the high and low affinity components of the renal Ca^{2+}-Mg^{2+} ATPase. *Canadian Journal of Physiology and Pharmacology* **68**, 718–26.

Brunette, M.G., Mailloux, J. and Lajeunesse, D. 1992. Calcium transport through the luminal membrane of distal tubule. II. Effect of pH, electrical potential and calcium channel inhibitors. *Kidney International* **41**, 289–96.

Cain, K. and Holt, D.E. 1983. Studies of cadmium–thionein induced nephropathy: Time course of cadmium–thionein uptake and degradation. *Chemical–Biological Interactions* **43**, 223–37.

Cember, H., Gallagher, P. and Faulkner, A. 1968. Distribution of mercury among blood fractions and serum proteins. *American Industrial Hygiene Association Journal* **29**, 233–7.

Chan, H.M., Satoh, M., Zalups, R.K. and Cherian, M.G. 1992. Exogenous metallothionein and renal toxicity of cadmium and mercury in rats. *Toxicology* **76**, 15–26.

Cherian, M.G. 1979. Metabolism of orally administered cadmium–metallothionein in mice. *Environmental Health Perspectives* **28**, 127–30.

Cherian, M.G. and Shaikh, Z.A. 1975. Metabolism of intravenously injected cadmium binding protein. *Biochemical and Biophysical Research Communications* **65**, 863–9.

Cherian, M.G., Goyer, R.A. and Valberg, L.S. 1978. Gastrointestinal absorption and organ distribution of oral cadmium chloride and cadmium–metallothionein in mice. *Journal of Toxicology and Environmental Health* **4**, 861–8.

Chung, Y.S., Song, I.S., Erikson, R.H., Sleisenger, H.H. and Kim, Y.S. 1985. Effect of growth and sodium butyrate on brush-border membrane associated hydrolases in human colorectal cancer cell lines. *Cancer Research* **45**, 2976–82.

Cikrt, M. 1970. The uptake of Hg^{203}, Cu^{64}, Mn^{52}, and Pb^{212} by the intestinal wall of the duodenal and ileal segment *in vitro*. *International Journal of Clinical Pharmacology* **3**, 351–7.

Corradino, R.A. and Taylor, A.N. 1985. 1,25-dihydroxyvitamin D_3-induced calcium-binding protein: localization in organ-cultured embryonic chick duodenum. *Journal of Histochemistry and Cytochemistry* **33**, 477–9.

Corradino, R.S. and Wasserman, R.H. 1971. Vitamin D_3: induction of calcium-binding protein in embryonic chick intestine *in vitro*. *Science* **171**, 731–3.

Costanzo, L.S. and Windhager, E.E. 1978. Calcium and sodium transport by the distal convoluted tubule of the rat. *American Journal of Physiology* **235**, F492–506.

Costanzo, L.S. and Windhager, E.E. 1980. Effects of PTH, ADH, and cyclic AMP on distal tubular Ca and Na reabsorption. *American Journal of Physiology* **239**, F478–85.

Dantzler, W.H. 1996. Comparative aspects of renal organic anion transport. *Cellular Physiology and Biochemistry* **6**, 28–38.

Dass, P.D., Misra, R.P. and Welbourne, T.C. 1981. Presence of γ-glutamyltransferase in the renal microvascular compartment. *Canadian Journal of Biochemistry* **59**, 383–6.

de Ceaurriz, J., Pyan, J.P., Morel, G. and Brondeau, M.T. 1994. Role of extracellular glutathione and γ-glutamyltranspeptidase in the disposition and kidney toxicity of mercury. *Journal of Applied Toxicology* **14**, 201–6.

Diamond, G.L., Cohen, J.J. and Weinstein, S.L. 1986. Renal handling of cadmium in perfused rat kidney and effects on renal function and tissue composition. *American Journal of Physiology* **251**, F784–94.

Diamond, G.L., Klotzbach, J.M. and Stewart, J.R. 1988. Complexing activity of 2,3-dimercapto-1-propanesulfonate and its disulfide auto-oxidation product in rat kidney. *Journal of Pharmacology and Experimental Therapeutics* **246**, 270–4.

Diaz, R.S. and Monreal, J. 1995. Protein-independent lead permeation through myelin lipid liposomes. *Molecular Pharmacology* **47**, 766–71.

Dominguez, J.H., Rothrock, J.K., Macias, W.L. and Price, J. 1989. Na^+ electrochemical gradient and Na^+–Ca^{2+} exchange in rat proximal tubule. *American Journal of Physiology* **257**, F531–8.

Dominguez, J.H., Mann, C., Rothrock, J.K. and Bhati, V. 1991. Na^+-Ca^{2+} exchange and depletion in rat proximal tubules. *American Journal of Physiology* **261**, F328–35.

Dominguez, J.H., Juhaszova, M., Kleiboeker, S.B., Hale, C.C. and Feister, H.A. 1992. Na^+–Ca^{2+} exchanger of rat proximal tubule: gene expression and subcellular localization. *American Journal of Physiology* **263**, F945–50.

Doucet, A. and Katz, A.I. 1982. High-affinity Ca-Mg-ATPase along the rabbit nephron. *American Journal of Physiology* **242**, F346–52.

Dragsten, P.R., Blumenthal, R. and Handler, J.S. 1981. Membrane asymmetry in epithelia: Is the tight junction a barrier to diffusion in the plasma membrane? *Nature* **294**, 718–22.

Durante, C.G. and Watson, J.F. 1967. Calcium reabsorption in proximal tubule of the dog nephron. *American Journal of Physiology* **212**, 1355–60.

Endo, T., Kimura, O., Hatakeyama, M., Takada, M. and Sakata M. 1997. Effects of zinc and copper on cadmium uptake by brush border membrane vesicles. *Toxicology Letters* **91**, 111–20.

Evans, H.T. 1994. Ionic radii in crystals. *Handbook of Chemistry and Physics*, CRC Press, Boca Raton, pp. 12-8–9.

Feher, J.J. 1983. Facilitated calcium diffusion by intestinal calcium-binding protein. *American Journal of Physiology* **244**, C303–7.

Feher, J.J. 1984. Measurement of calcium diffusion by intestinal calcium-binding protein. *Biochimica et Biophysica Acta* **773**, 91–8.

Felley-Bosco, E. and Diezi, J. 1987. Fate of cadmium in rat renal tubules: A microinjection study. *Toxicology and Applied Pharmacology* **91**, 204–11.

Flanagan, P.R., Hamilton, D.L., Haist J. and Valberg, L.S. 1979. Inter-relationships between iron and lead absorption in iron-deficient mice. *Gastroenterology* 77, 1074–81.

Fogh, J., Fogh, J.M. and Orfeo, T. 1977. One hundred and twenty-seven cultured human tumor cell lines producing tumors in nude mice. *Journal of the National Cancer Institute* 59, 221–6.

Foulkes, E.C. 1974. Excretion and retention of cadmium, zinc, and mercury by rabbit kidney. *American Journal of Physiology* **227**, 1356–60.

Foulkes, E.C. 1978. Renal tubular transport of cadmium–metallothionein. *Toxicology and Applied Pharmacology* **45**, 505–12.

Foulkes, E.C. 1985. Interactions between metals in rat jejunum: implications on the nature of cadmium uptake. *Toxicology* **37**, 117–25.

Foulkes, E.C. 1988. On the mechanism of transfer of heavy metals across cell membranes. *Toxicology* **52**, 263–72.

Foulkes, E.C. 1993. Metallothionein and glutathione as determinants of cellular retention and extrusion of cadmium and mercury. *Life Sciences* **52**, 1617–20.

Foulkes, E.C. and Bergman, D. 1993. Inorganic mercury absorption in mature and immature rat jejunum: Transcellular and intercellular pathways *in vivo* and in everted sacs. *Toxicology and Applied Pharmacology* **120**, 89–95.

Foulkes, E.C. and McMullen, D.M. 1986. On the mechanisms of nickel absorption in the rat jejunum. *Toxicology* **38**, 35–42.
Foulkes, E.C. and McMullen, D.M. 1987. Kinetics of transepithelial movement of heavy metals in rat jejunum. *American Journal of Physiology* **253**, G134–8.
Fowler, B.A. and DuVal, G. 1991. Effects of lead on the kidney: roles of high-affinity lead-binding proteins. *Environmental Health Perspectives* **91**, 77–80.
Fowler, B.A., Kahng, M.W. and Smith, D.R. 1994. Role of lead-binding proteins in renal cancer. *Environmental Health Perspectives* **102**, 115–16.
Frick, A.G., Rumrich, G., Ullrich, K.J. and Lassiter, W.E. 1965. Microperfusion study of calcium transport in the proximal tubule of the rat kidney. *Pfluegers Arch.* **286**, 109–17.
Friedman, P.A. and Gesek, F.A. 1995. Cellular calcium transport in renal epithelia: Measurement, mechanisms and regulation. *Physiological Reviews* **75**, 429–71.
Friedman, P.A., Figueiredo, J.F., Maack, T. and Windhager, E.E. 1981. Sodium–calcium interactions in the renal proximal convoluted tubule of the rabbit. *American Journal of Physiology* **240**, F558–68.
Fromter, E., Rumrich, G. and Ullrich, K.J. 1973. Phenomenologic description of Na, Cl and bicarbonate reabsorption from proximal tubules of the rat kidney. *Pfluegers Archive* **343**, 189–220.
Fuhr, B.J. and Rabenstein, D.L. 1973. Nuclear magnetic-resonance studies of solution chemistry of metal-complexes. 9. Binding of cadmium, zinc, lead and mercury by glutathione. *Journal of the American Chemical Society* **95**, 6944–50.
Fullmer, C.S. and Rosen, J.F. 1990. Effect of dietary calcium and lead status on intestinal calcium absorption. *Environmental Research* **51**, 91–9.
Fullmer, C.S. and Wasserman, R.H. 1981. The amino acid sequence of bovine intestinal calcium-binding protein. *Journal of Biological Chemistry* **256**, 5669–74.
Fullmer, C.S., Edelstin, S. and Wasserman, R.H. 1985. Lead-binding properties of intestinal calcium-binding proteins. *Journal of Biological Chemistry* **260**, 6816–19.
Gesek, F.A. and Friedman, P.A. 1992a. Mechanism of calcium transport stimulated by chlorothiazide in mouse distal convoluted tubule cells. *Journal of Clinical Investigation* **90**, 279–88.
Gesek, F.A. and Friedman, P.A. 1992b. On the mechanism of parathyroid hormone stimulation of calcium uptake by mouse distal convoluted tubule cells. *Journal of Clinical Investigation* **90**, 749–58.
Gesek, F.A. and Friedman, P.A. 1995. Sodium entry mechanisms in distal convoluted tubule cells. *American Journal of Physiology* **268**, F89–98.
Ghijsen, W., Van Os, C.H., Herzmann, C.W. and Murer, H. 1986. Regulation of duodenal Ca^{2+} pump by calmodulin and vitamin D-dependent Ca^{2+}-binding protein. *American Journal of Physiology* **251**, G223–9.
Gmaj, P., Murer, H. and Kinne, R. 1979. Calcium ion transport across plasma membranes isolated from rat kidney cortex. *Biochemical Journal* **178**, 549–57.
Gmaj, P., Zurini, M., Murer, H. and Carafoli, E. 1983. Localization and properties of a high-affinity (Ca^{2+} + Mg^{2+})-ATPase in isolated kidney cortex plasma membranes. *European Journal of Biochemistry* **136**, 71–6.
Gmaj, P., Malmstrom, K., Biber, J., Amstutz, M., Ghijsen, W. and Murer, H. 1985. Renal proximal tubular transport of calcium and inorganic phosphate studied with isolated vesicles. *Molecular Physiology* **8**, 59–76.
Goering, P.L. and Fowler, B.A. 1984. Regulation of lead inhibition of δ-aminolevulinic acid dehydratase by a low molecular weight, high affinity renal lead-binding protein. *Journal of Pharmacology and Experimental Therapeutics* **231**, 66–71.
Grasset, E., Pinto, M., Dusseaulx, E., Zweibaum, A. and Desjeaux, J.F. 1984. Epithelial properties of human colonic carcinoma cell line Caco-2: electrical parameters. *American Journal of Physiology* **247**, C260–70.

Gross, M. and Kumar, R. 1990. Physiology and biochemistry of vitamin D-dependent calcium binding proteins. *American Journal of Physiology* **259**, F195–209.

Gross, M.D., Kumar, R. and Hunziker, W. 1988. Expression in *E. coli* of full-length and mutant rat brain calbindin D28: comparison with purified native protein. *Journal of Biological Chemistry* **253**, 11426–32.

Gum, J.R., Kam, W.K., Byrd, J.C., Hicks, J.W., Sleisenger, M.H. and Kim, Y.S. 1987. Effects of sodium butyrate on human colonic adenocarcinoma cells: Induction of placental like alkaline phosphate. *Journal of Biological Chemistry* **262**, 1092–7.

Hanai, H., Ishida, M., Liang, C.T. and Sacktor, B. 1986. Parathyroid hormone increases sodium/calcium exchange activity in renal cells and the blunting of the response in aging. *Journal of Biological Chemistry* **261**, 5419–25.

Hanaoka, K., Sakai, O., Imai, M. and Yoshitomi, K. 1993. Mechanisms of calcium transport across the basolateral membrane of the rabbit cortical thick ascending limb of Henle's loop. *Pflügers Archiv* **422**, 339–46.

Harris, C.A., Baer, P.G., Chirito, E. and Dirks, J.H. 1974. Composition of mammalian glomerular filtrate. *American Journal of Physiology* **227**, 972–5.

Hauri, H.P., Sterchi, E.E., Bienz, D., Fransen, J.A. and Marxer, A. 1985. Expression and intracellular transport of microvillus membrane hydrolases in human intestinal epithelial cells. *Journal of Cell Biology* **101**, 838–51.

Heinle, H. and Wendel, A. 1977. The activities of the key enzymes of the γ-glutamyl cycle in microdissected segments of the rat nephron. *FEBS Letters* **73**, 220–4.

Hinkle, P.M., Kinsella, P.A. and Osterhoudt, K.C. 1987. Cadmium uptake and toxicity via voltage-sensitive calcium channels. *Journal of Biological Chemistry* **262**, 333–7.

Hinkle, P.M., Shanshala, E.D., II and Nelson, E.J. 1992. Measurement of intracellular cadmium with fluorescent dyes. Further evidence for the role of calcium channels in cadmium uptake. *Journal of Biological Chemistry* **267**, 25553–9.

Hughes, W.L. 1957. A physicochemical rationale for the biological activity of mercury and its compounds. *Annals of the New York Academy of Sciences* **65**, 454–60.

Hultman, P. and Enestrom, S. 1986. Localization of mercury in the kidney during experimental acute renal tubular necrosis studied by the cytochemical silver amplification method. *British Journal of Experimental Pathology* **67**, 493–503.

Hultman, P., Enestrom, S. and von Schenck, H. 1985. Renal handling of inorganic mercury in mice. The early excretion phase following a single intravenous injection of mercuric chloride studies by the silver amplification method. *Virchows Archiv, B: Cell Pathology Including Molecular Pathology* **49**, 209–24.

Imai, M. 1981. Effects of parathyroid hormone and N^6, O^2-dibutyryl cyclic AMP on calcium transport across the rabbit distal nephron segments perfused *in vitro*. *Pflügers Archiv* **390**, 145–51.

Jaeger, D.E. 1990. Absorption interactions of zinc and cadmium in the isolated perfused rat intestine. *Journal of Trace Elements and Electrolytes in Health and Disease* **4**, 101–5.

James, P., Vorherr, T., Thulin, E., Forsen, S. and Carafoli, E. 1991. Identification and primary structure of a calbindin 9K binding domain in the plasma membrane Ca^{2+} pump. *FEBS Letters* **278**, 155–9.

Jayakumar, A., Cheng, L., Liang, C.T. and Sacktor, B. 1984. Sodium gradient-dependent calcium uptake in renal basolateral membrane vesicles: effect of parathyroid hormone. *Journal of Biological Chemistry* **259**, 10827–33.

Jocelyn, P.C. 1972. *Biochemistry of the SH Group*, Academic Press, London.

Jumarie, C., Campbell, P.G.C., Berteloot, A., Houde, M. and Denizeau, F. 1997. Caco-2 cell line used as an *in vitro* model to study cadmium accumulation in intestinal epithelial cells. *Journal of Membrane Biology* **158**, 31–48.

Kagi, J.H.R. and Nordberg, G.F. 1979. *Metallothionein*, Beerkhouser Verlag, Basel.

Kagi, J.H.R. and Vallee, B.L. 1961. Metallothionein: A cadmium and zinc-containing protein from equine cortex. II. Physiochemical properties. *Journal of Biological Chemistry* **236**, 2435–42.

Kerper, L.E. and Hinkle, P.M. 1997a. Cellular uptake of lead is activated by depletion of intracellular calcium stores. *Journal of Biological Chemistry* **272**, 8346–52.

Kerper, L.E. and Hinkle, P.M. 1997b. Lead uptake in brain capillary endothelial cells: Activation by calcium store depletion. *Toxicology and Applied Pharmacology* **146**, 127–33.

Klotzbach, J.M. and Diamond, G.L. 1988. Complexing activity and excretion of 2,3-dimercapto-1-propane sulfonate in rat kidney. *American Journal of Physiology* **254**, F871–78.

Koeppen, B.M. and Giebisch, G. 1983. Electrophysiology of mammalian renal tubules: Inferences from intracellular microelectrode studies. *Annual Review of Physiology* **45**, 487–517.

Kretsinger, R.H., Mann, J.E. and Simmonds, J.G. 1982. Model of facilitated diffusion of calcium by the intestinal calcium binding protein, in *Vitamin D: Chemical, Biochemical and Clinical Endocrinology of Calcium Metabolism*, (eds A.W. Norman, K. Schaefer, D.V. Herrath and H.-G. Grigoleit), de Druyter, New York, pp. 233–48.

Kubota, T., Hagiwara, N. and Fujimoto, M. 1990. Intracellular calcium measurements with PVC-resin Ca-selective microelectrodes in frog proximal tubules and sartorius muscle fibers. *Japanese Journal of Physiology* **40**, 79–95.

Lajeunesse, D., Bouhtiauy, I. and Brunette, M.G. 1994. Parathyroid hormone and hydrochlorothiazide increase calcium transport by the luminal membrane of rabbit distal nephron segments through different pathways. *Endocrinology* **134**, 35–41.

Lash, L.H. and Anders, M.W. 1989. Uptake of nephrotoxic S-conjugates by isolated proximal tubule cells. *Journal of Pharmacology and Experimental Therapeutics* **248**, 531–7.

Lash, L.H. and Jones, D.P. 1983. Transport of glutathione by renal basal–lateral membrane vesicles. *Biochemical and Biophysical Research Communications* **112**, 55–60.

Lash, L.H. and Jones, D.P. 1984. Renal glutathione transport: Characteristics of the sodium-dependent system in the basal–lateral membrane. *Journal of Biological Chemistry* **259**, 14508–14.

Lash, L.H. and Jones, D.P. 1985. Distribution of oxidized and reduced forms of glutathione and cysteine in rat plasma. *Archives of Biochemistry and Biophysics* **240**, 583–92.

Lassiter, W.E., Gottschalk, C.W. and Mylle, M. 1963. Micropuncture study of renal tubular reabsorption of calcium in normal rodents. *American Journal of Physiology* **204**, 771–5.

Le Grimellec, C., Roinel, N. and Morel, F. 1973. Simultaneous Mg, Ca, P, K, Na and Cl analysis in rat tubular fluid. I. During perfusion of either inulin or ferrocyanide. *Pflügers Arch.* **340**, 181–96.

Le Grimellec, C.P., Pooujeol, P. and De Rouffingnac, C. 1975. ^{3}H-inulin and electrolyte concentrations in Bowman's capsule in rat kidney. Comparison with artificial ultrafiltration. *Pflügers Arch.* **354**, 117–31.

Levy, J., Gavin, J.R. III, Morimoto, S., Hammerman, M.R. and Avioli, L.V. 1986. Hormonal regulation of (Ca^{2+} + Mg^{2+})ATPase activity in canine renal basolateral membrane. *Endocrinology* **119**, 2405–11.

Linse, S., Brodin, P., Drakenberg, T., Thulin, E., Sellers, P., Elmdén, K., Grundström, T. and Forsén, S. 1987. Structure–function relationships in EF-hand Ca^{2+}-binding proteins. Protein engineering and biophysical studies of calbindin D_{9k}. *Biochemistry* **26**, 6723–35.

Linse, S., Johansson, E., Brodin, P., Grundström, T., Drakenberg, T. and Forsén, S. 1991. Electrostatic contributions to the binding of Ca^{2+} in calbindin D_{9k}. *Biochemistry* **30**, 154–62.

Lorenzen, M., Lee, C.O. and Windhager, E.E. 1984. Cytosolic Ca^{2+} and Na^{+} activities in perfused proximal tubules of *Necturus* kidney. *American Journal of Physiology* **247**, F93–102.

Maack, T. 1992. Renal handling of proteins and polypeptides, in *Handbook of Physiology, Section 8, Renal Physiology*, Vol. 2, (ed. E.E. Windhager), Oxford University Press, New York, pp. 2039–83.

Madsen, K.M. 1980. Mercury accumulation in kidney lysosomes of proteinuric rats. *Kidney International* **18**, 445–53.

Madsen, K.M. and Christensen, E.L. 1978. Effects of mercury on lysosomal protein digestion in the kidney proximal tubule. *Laboratory Investigation* **38**, 165–74.

Madsen, K.M. and Hansen, J.C. 1980. Subcellular distribution of mercury in the rat kidney cortex after exposure to mercuric chloride. *Toxicology and Applied Pharmacology* **54**, 443–53.

Malvin, R.L. and Wilde, W.S. 1973. Stop-flow technique, in *Handbook of Physiology, Renal Physiology*, (eds J. Orloff and R.W. Berliner), American Physiological Society, Washington, DC, pp. 119–28.

Mandel, L.J. and Murphy, E. 1984. Regulation of cytosolic free calcium in rabbit proximal renal tubules. *Journal of Biological Chemistry* **259**, 11188–96.

Matsunaga, H., Stanton, B.A. and Friedman, P.A. 1994. Epithelial Ca^{2+} channels sensitive to dihydropyridines and activated by hyperpolarizing voltages. *American Journal of Physiology* **267**, C157–65.

Merrill, A.R., Proulx, P. and Szabo, A.G. 1986. Studies on calcium binding to brush-boarder membranes from rabbit small intestine. *Biochimica et Biophysica Acta* **859**, 237–45.

Mohrmann, I., Mohrmann, M., Biber, J. and Murer, H. 1986. Sodium-dependent transport of Pi by an established intestinal epithelial cell line (CaCo-2). *American Journal of Physiology* **250**, G323–30.

Mokrzan, E.M., Kerper, L.E., Ballatori, N. and Clarkson,T.W. 1995. Methylmercury–thiol uptake into cultured brain capillary endothelial cells on amino acid system L. *Journal of Pharmacology and Experimental Therapeutics* **272**, 1277–84.

Moriuchi, S., Yamamouchi, T. and Hasoya, N. 1975. Demonstration of two different vitamin D-dependent calcium-binding proteins in rat intestinal mucosa. *Journal of Nutritional Science and Vitaminology* **21**, 251–9.

Morton, A.P., Partridge, S. and Blair, J.A. 1985. The intestinal uptake of lead. *Chemistry in Britain* **21**, 923–7.

Murakami, M. and Webb, M. 1981. A morphological and biochemical study of the effects of L-cysteine on the renal uptake and nephrotoxicity of cadmium. *British Journal of Experimental Pathology* **62**, 115–30.

Murayama, Y., Moral, F. and LeGrimellec, C. 1972. Phosphate, calcium and magnesium transfers in proximal tubules and loops of Henle, as measured by single nephron microperfusion experiments in the rat. *Pfluegers Arch.* **333**, 1–16.

Mykkänen, H.M. and Wasserman, R.H. 1981. Gastro-intestinal absorption of lead (^{203}Pb) in chicks: influence of lead, calcium and age. *Journal of Nutrition* **111**, 1757–65.

Mykkänen, H.M. and Wasserman, R.H. 1982. Effect of vitamin D on the intestinal absorption of ^{203}Pb and ^{47}Ca in chicks. *Journal of Nutrition* **112**, 520–7.

Naftalin, R. and Tripathi, S. 1985. Passive water flows driven across the isolated rabbit ileum by osmotic, hydrostatic and electrical gradients. *Journal of Physiology (London)* **360**, 27–50.

Ng, R., Rouse, K.D. and Suki, W.N. 1984. Calcium transport in the rabbit superficial proximal convoluted tubule. *Journal of Clinical Investigation* **74**, 834–42.

Ohta, H., DeAngelis, M.V. and Cherian, M.G. 1989. Uptake of cadmium and metallothionein by rat everted intestinal sacs. *Toxicology and Applied Pharmacology* **101**, 62–9.

Oken, D.E., Cotes, S.C. and Mende, C.W. 1972. Micropuncture study of tubular transport of albumin in rats with aminonucleoside nephrosis. *Kidney International* **1**, 3–11.

Pansu, D., Bellaton, C. and Bronner, F. 1981. The effect of calcium intake on saturable and nonsaturable components of duodenal calcium transport. *American Journal of Physiology* **240**, G32–7.

Pansu, D., Bellaton, C. and Bronner, F. 1983a. Developmental changes in the mechanisms of duodenal calcium transport in the rat. *American Journal of Physiology* **244**, G20–6.

Pansu, D., Bellaton, C., Roche, C. and Bronner, F. 1983b. Duodenal and ileal calcium absorption in the rat and effects of vitamin-D. *American Journal of Physiology* **244**, G695–700.

Parsegian, A. 1969. Energy of an ion crossing a low dielectric membrane: solutions to four relevant electrostatic problems. *Nature* **221**, 834–46.

Perret, C., Delplan, C. and Thomasset, M. 1985. Cholecalcin (a 9-kDA cholecalciferol-induced calcium binding protein) messenger RNA: distribution and induction by calcitriol in the rat digestive tract. *European Journal of Biochemistry* **150**, 211–17.

Pesce, A.J. and First, M.R. 1979. *Proteinuria: An Integrated Review*, Marcel Dekker, New York, pp. 5–31.

Pigman, E.A., Blanchard, J. and Laird, H.E. 1997. A study of cadmium transport pathways using the Caco-2 cell model. *Toxicology and Applied Pharmacology* **142**, 243–7.

Pinto, M., Robine-Leaon, S., Tappay, M., Kedinger, M., Triadou, N., Dussaulx, M., Lacroix, B., Simon-Assman, P., Haffer, K., Fogh, J. and Zweibaum, A. 1983. Enterocyte-like differentiation and polarization of the human colon carcinoma cell line Caco2 in culture. *Biology of Cells* **47**, 323–30.

Poncet, V., Merot, J. and Poujeol, P. 1992. A calcium-permeable channel in the apical membrane of primary cultures of the rabbit distal bright convoluted tubule. *Pflügers Archiv*. **422**, 112–19.

Powell, D. 1981. Barrier function of epithelia. *American Journal of Physiology* **241**, G275–88.

Powell, D. 1987. Intestinal water and electrolyte transport, in *Physiology of the Gastrointestinal Tract*, (ed. L.R. Johnson), Raven, New York, pp. 1267–305.

Prasad, R. and Nath, R. 1993. Zinc transport in monkey renal brush border membrane vesicles and its interaction with cadmium: A kinetic study. *Journal of Trace Elements in Experimental Medicine* **6**, 95–107.

Prasad, R., Kaur, D. and Kumar, V. 1996. Kinetic characterization of zinc binding to brush border membranes from rat kidney cortex: interaction with cadmium. *Biochimica et Biophysica Acta* **1284**, 69–78.

Pritchard, J.B. and Miller, D.S. 1993. Mechanism mediating renal secretion of organic anions and cations. *Physiological Reviews* **73**, 765–88.

Quintanilla-Vega, B., Smith, D.R., Kahng, M.W., Hernandez, J., Albores, A. and Fowler, B.A. 1995. Lead-binding proteins in brain tissue of environmentally lead-exposed humans. *Chemico-Biological Interactions* **98**, 193–209.

Rabenstein, D.L. 1989. Metal complexes of glutathione and their biological significance, in *Glutathione: Chemical, Biochemical and Medical Aspects. Coenzymes and Cofactors*, Vol. 3, (eds D. Dolphin, O. Auromovbic and R. Poulson), Wiley, New York, pp. 147–86.

Ramachandran, C. and Brunette, M.G. 1989., The renal Na^+/Ca^{2+} exchange system is located exclusively in the distal tubule. *Biochemical Journal* **257**, 259–64.

Ramond, M.J., Martinot-Peignoux, M. and Erlinger, S. 1985. Dome formation in the human colon carcinoma cell line Caco-2 in culture. Influence of ouabain and permeable supports. *Biology of Cells* **54**, 88–92.

Rao, M.S.N. and Lal, H. 1958. Metal protein interactions in buffer solutions. III. Interaction of Cu^{2+}, Zn^{2+}, Cd^{2+}, Co^{2+}, and Ni^{2+} with native and modified serum albumins. *Journal of the American Chemical Society* **80**, 3226–35.

Rector, F.C. 1983. Sodium, bicarbonate, and chloride absorption by the proximal tubule. *American Journal of Physiology* **244**, F461–71.

Reeves, J.P. and Hale, C.C. 1984. The stoichiometry of the cardiac sodium–calcium exchange system. *Journal of Biological Chemistry* **259**, 7733–9.

Reilly, R.F., Shugrue, C.A., Lattanzi, D. and Biemesderfer, D. 1993. Immunolocalization of the Na^+/Ca^{2+} exchanger in rabbit kidney. *American Journal of Physiology* **265**, F327–32.

Rodier, P.M., Kates, B. and Simons, R. 1988. Mercury localization in mouse kidney over time: Audoradiography versus silver staining. *Toxicology and Applied Pharmacology* **257**, 235–45.

Rossi, A., Pverini, R., Di Lullo, G., Modesti, A., Modica, A. and Scarino, M.L. 1996. Heavy metal toxicity following apical and basolateral exposure in the human intestine cell line Caco-2. *Toxicology in Vitro* **10**, 27–36.

Saunders, J.C.J. and Isaacson, L.C. 1990. Patch clamp study of Ca channels in isolated renal tubule segments, in *Calcium Transport and Intracellular Calcium Homeostasis*, (eds D. Pansu and F. Bronner), Springer-Verlag, Berlin, pp. 27–34.

Scoble, J.E., Mills, S. and Hruska, K.A. 1985. Calcium transport in canine renal basolateral membrane vesicles. Effects of parathyroid hormone. *Journal of Clinical Investigation* **75**, 1096–105.

Shaikh, Z.A. and Hirayama, K. 1979. Metallothionein in the extracellular fluids as an index of cadmium toxicity. *Environmental Health Perspectives* **28**, 267–71.

Shareghi, G.R. and Stoner, L.C. 1978. Calcium transport across segments of the rabbit distal nephron *in vitro*. *American Journal of Physiology* **235**, F367–75.

Shimizu, T., Yoshitomi, K., Nakamura, M. and Imai, M. 1990. Effects of PTH, calcitonin, and cAMP on calcium transport in rabbit distal nephron segments. *American Journal of Physiology* **259**, F408–14.

Simons, T.J.B. and Pocock, G. 1987. Lead enters bovine adrenal medullary cells through calcium channels. *Journal of Neurochemistry* **48**, 383–9.

Spater, H.W., Prouchynsky, M.S., Quintana, N., Inoue, M. and Novikoff, A.B. 1982. Immunocytochemical localization of γ-glutamyltransferase in rat kidney with protein A-horseradish peroxidase. *Proceedings of the National Academy of Sciences of the United States of America* **79**, 3547–50.

Squibb, K.S., Riddlington, J.W., Carmichael, N.G. and Folwer, B.A. 1979. Early cellular effects of circulating cadmium–thionein on proximal tubules. *Environmental Health Perspectives* **28**, 287–96.

Squibb, K.S., Pritchard, J.B. and Fowler, B.A. 1984. Cadmium–metallothionein nephropathy: Relationships between ultrastructural/biochemical alterations and intracellular cadmium binding. *Journal of Pharmacology and Experimental Therapeutics* **229**, 311–21.

Sugawara, N. and Sugawara, C. 1991. Gastrointestinal absorption of Cd–metallothionein and cadmium chloride in mice. *Archives of Toxicology* **65**, 689–92.

Sugawara, N., Sugawara, C. and Miyake, H. 1988. Binding of cadmium chloride and Cd–metallothionein to mucosal brush-border membrane of the rat small intestinal tract. *Bulletin of Environmental Contamination and Toxicology* **40**, 418–24.

Sutton, R.A.L. and Dirks, J.H. 1975. The renal excretion of calcium: a review of micropuncture data. *Canadian Journal of Physiology and Pharmacology* **53**, 973–88.

Tacnet, F., Watkins, D.W. and Ripoche, P. 1990. Studies of zinc transport into brush-border membrane vesicles isolated from pig small intestine. *Biochimica et Biophysica Acta* **1024**, 323–30.

Tacnet, F., Watkins, D.W. and Ripoche, P. 1991. Zinc binding in intestinal brush-border membrane isolated from pig. *Biochimica et Biophysica Acta* **1063**, 51–9.

Tan, S. and Lau, K. 1993. Patch-clamp evidence for calcium channels in apical membranes of rabbit kidney connecting tubules. *Journal of Clinical Investigation* **92**, 2731–6.

Tanaka, T., Naganuma, A. and Imura, N. 1990. Role of γ-glutamyltranspeptidase in renal uptake and toxicity of inorganic mercury in mice. *Toxicology* **60**, 187–98.

Tanaka-Kagawa, T., Naganuma, A. and Imura, N. 1993. Tubular secretion and reabsorption of mercury compounds in mouse kidney. *Journal of Pharmacology and Experimental Therapeutics* **264**, 776–82.

Taniguchi, J., Takeda, M., Yoshitomi, K. and Imai, M. 1994. Pressure- and parathyroid-hormone-dependent Ca^{2+} transport in rabbit connecting tubule: role of the stretch-activated nonselective cation channel. *Journal of Membrane Biology* **140**, 123–32.

Taugner, R., Winkel, K. and Iravani, J. 1966. Zur lokalization der sublimatanreicherung in der ratenniere. *Virchows Archiv A: Pathologische Anatomie* **340**, 369–83.

Taylor, A.N. 1983. Intestinal vitamin D-induced caclium-binding protein: time-course of immunocytological localization following 1,25-dihydrixyvitamin D_3. *Journal of Histochemcistry and Cytochemistry* **31**, 426–32.

Taylor, A.N. and Wasserman, R.H. 1967. Vitamin D_3-induced calcium-binding protein: partial purification, electrophoretic visualization, and tissue distribution. *Archives of Biochemistry and Biophysics* **119**, 536–40.

Taylor, A.N., McIntosh, J.E. and Bourdeau, J.E. 1982. Immunocytochemical localization of vitamin D-dependent calcium-binding protein in renal tubules of rabbit, rat, and chick. *Kidney International* **21**, 765–73.

Thomasset, M., Parkes, C.O. and Cuisinier-Gleizes, P. 1982. Rat calcium binding proteins: distribution, development and vitamin D dependence. *American Journal of Physiology* **243**, E483–E488.

Thomasset, M., Desplan, C., Warembourg, M. and Perret, C. 1986. Vitamin D-dependent 9kDa calcium-binding protein gene: cDNA cloning, mRNA distribution and regulation. *Biochemie*, **68**, 935–40.

Tomsig. J.L. and Suszkiw, J.B. 1991. Permeation of Pb^{+2} through calcium channels: Fura-2 measurements of voltage- and dihydropyridine-sensitive Pb^{+2} entry in isolated bovine chromaffin cells. *Biochimical et Biophysica Acta* **1069**, 197–200.

Tsukamoto, Y., Saka, S. and Saitoh, M. 1992. Parathyroid hormone stimulates ATP-dependent calcium pump activity by a different mode in proximal and distal tubules of the rat. *Biochimica et Biophysica Acta* **1103**, 163–71.

Ueng, T.H., Golub, E.E. and Bronner, F. 1979. The effects of age and 1,24-dihydroxyvitamin D_3 treatment on intestinal calcium binding protein of sucking rats. *Archives of Biochemistry and Biophysics* **196**, 624–30.

Ullrich, K.J., Rumrich, G., Kloss, S. 1976. Active Ca^{2+} reabsorption in the proximal and distal tubules of the rat kidney. Dependence on sodium and buffer transport. *Pflügers Archiv.*, **364**, 223–8.

Van Corven, E., Roche, C. and Van Os, C.H. 1985. ATP-dependent Ca^{2+}-transport calmodulin and vitamin D-dependent Ca^{2+}-protein along the villus–crypt axis in rat duodenum. *Biochimica et Biophysica Acta* **870**, 274–84.

Vander, A.J. and Johnson, B. 1981. Accumulation of lead by renal slices in the presence of organic anions. *Proceedings of the Society for Experimental Biology and Medicine* **166**, 583–6.

Vander, A.J., Taylor, D.L., Kalitis, K., Mouw, D.R. and Victery, W. 1977. Renal handling of lead in dogs: clearance studies. *American Journal of Physiology* **233**, F532–8.

Vander, A.J., Mouw, D.R. Cox, J. and Johnson, B. 1979. Lead transport by renal slices and its inhibition by tin. *American Journal of Physiology* **236**, F373–8.

Van Heeswijk, M.P.E., Geertsen, J.A.M. and Van Os, C.H. 1984. Kinetic properties of the ATP-dependent Ca^{2+} pump and Na^+/Ca^{2+} exchange system in basolateral membranes from rat kidney cortex. *Journal of Membrane Biology* **79**, 19–31.

Verma, M.O., Sharma, R.P. and Bourcier, D.R. 1982. Macromolecular interactions with cadmium and the effects of zinc, copper, lead and mercury ions. *Biological Trace Element Research* **3**, 35–42.

Victery, W., Vander, A.J. and Mouw, D.R. 1979a. Effect of acid–base status on renal excretion and accumulation of lead in dogs and rats. *American Journal of Physiology* **237**, F398–407.

Victery, W., Vander, A.J. and Mouw, D.R. 1979b. Renal handling of lead in dogs: stop-flow analysis. *American Journal of Physiology* **237**, F408–14.

Victery, W., Miller, C.R. and Fowler, B.A. 1984. Lead accumulation by rat renal brush-border membrane vesicles. *Journal of Pharmacology and Experimental Therapeutics* **231**, 589–96.

Vincent, M.L., Russell, R.M. and Sasak, V. 1985. Folic acid uptake characteristics of a human colon carcinoma cell line, Caco-2. A newly-described cellular model for small intestinal epithelium. *Human Nutrition and Clinical Nutrition* **39C**, 355–60.

Walters, J.R. 1989. Calbindin-D9k stimulates the calcium pump in rat enterocyte basolateral membranes. *American Journal of Physiology* **256**, G124–8.

Walters, J.R., Howard, A., Charpin, M.B., Gniecko, K.C., Brodin, P., Thulin, E. and Forsen, S. 1990. Stimulation of intestinal basolateral membrane calcium-pump activity by recombinant synthetic calbindin-D9k and specific mutants. *Biochemical and Biophysical Research Communications* **170**, 603–8.

Wasserman. R.H., Brindak, M.E., Meyer, S.A. and Fullmer, C.S. 1982. Evidence for multiple effects of vitmamin D_3 on calcium absorption: response of rachitic chicks, with or without partial vitamin D_3 repletion, to 1,25-dihydroxyvitamin D_3. *Proceedings of the National Academy of Science USA* **79**, 7939–43.

Wong, N.L.M., Quamme, G.A., Sutton, R.A.L. and Dirks, J.H. 1979. The effects of mannitol on divalent ion transport in the kidney. *Journal of Laboratory and Clinical Medicine* **94**, 683–92.

Wright, F. 1971. Increasing magnitude of electrical potential along the renal distal tubule. *American Journal of Physiology* **229**, 624–38.

Yang, J.M., Lee, C.O. and Windhager, E.E. 1988. Regulation of cytosolic free calcium in isolate perfused proximal tubules of *Necturus*. *American Journal of Physiology* **255**, F787–99.

Yoshitomi, K., Shimizu, T., Taniguchi, J. and Imai, M. 1989. Electro-physiological characterization of rabbit distal convoluted tubule cell. *Pflügers Archiv*. **414**, 457–63.

Yu, A.S.L., Hebert, S.C., Lee, S.-L., Brenner, B.M. and Lytton, J. 1992. Identification and localization of renal Na^+-Ca^{2+} exchanger by polymerase chain reaction. *American Journal of Physiology* **263**, F680–5.

Zalme, R.C., McDowell, E.M., Nagle, R.B., McNeil, J.S., Flemenbaum, W. and Trump, B.F. 1976. Studies of the pathophysiology of acute renal failure. II. A histochemical study of the proximal tubule of the rat following administration of mercuric chloride. *Virchows Archiv, B: Cell Pathology Including Molecular Pathology* **22**, 197–216.

Zalups, R.K. 1991a. Autometallographic localization of inorganic mercury in the kidneys of rats: Effect of unilateral nephrectomy and compensatory renal growth. *Experimental and Molecular Pathology* **54**, 10–21.

Zalups, R.K. 1991b. Method for studying the *in vivo* accumulation of inorganic mercury in segments of the nephron in the kidneys of rats treated with mercuric chloride. *Journal of Pharmacological Methods* **26**, 89–104.

Zalups, R.K. 1993. Early aspects of the intrarenal distribution of mercury after the intravenous administration of mercuric chloride. *Toxicology* **79**, 215–28.

Zalups, R.K. 1995. Organic anion transport and action of γ-glutamyl transpeptidase in kidney linked mechanistically to renal tubular uptake of inorganic mercury. *Toxicology and Applied Pharmacology* **132**, 289–98.

Zalups, R.K. and Barfuss, D. 1993. Transport and toxicity of methylmercury along the proximal tubule of the rabbit. *Toxicology and Applied Pharmacology* **121**, 176–85.

Zalups, R.K. and Barfuss, D. 1995a. Accumulation and handling of inorganic mercury in the kidney after co-administration with glutathione. *Journal of Toxicology and Environmental Health* **44**, 385–99.

Zalups, R.K. and Barfuss, D. 1995b. Renal disposition of mercury in rats after intravenous injection of inorganic mercury and cysteine. *Journal of Toxicology and Environmental Health* **44**, 401–13.

Zalups, R.K. and Barfuss, D. 1996. Nephrotoxicity of inorganic mercury co-administered with L-cysteine. *Toxicology* **109**, 15–29.

Zalups, R.K. and Barfuss, D.W. 1998. Small aliphatic dicarboxylic acids inhibit renal uptake of administered mercury. *Toxicology and Applied Pharmacology* **148**, 183–93.

Zalups, R.K. and Lash, L.H. 1994. Advances in understanding the renal transport and toxicity of mercury. *Journal of Toxicology and Environmental Health* **42**, 1–44.

Zalups, R.K. and Lash, L.H. 1997a. Depletion of glutathione in the kidney and the renal disposition of administered inorganic mercury. *Drug Metabolism and Disposition* **25**, 516–23.

Zalups, R.K. and Lash, L.H. 1997b. Binding of mercury in renal brush-border and basolateral membrane-vesicles. Implication of a cysteine conjugate of mercury involved in the luminal uptake of inorganic mercury in the kidney. *Biochemical Pharmacology* **53**, 1889–900.

Zalups, R.K. and Minor, K.H. 1995. Luminal and basolateral mechanisms involved in the renal tubular uptake of inorganic mercury. *Journal of Toxicology and Environmental Health* **46**, 73–100.

Zalups, R.K., Cox, C. and Diamond, G.L. 1988. Histological and urinalysis assessment of nephrotoxicity induced by mercuric chloride in normal and uninephrectomized rats, in *Biological Monitoring of Toxic Metals*, (eds T.W. Clarkson, L. Friberg, G.F. Nordberg and P.R. Sager), Plenum, New York, pp. 531–45.

Zalups, R.K., Robinson, M.K. and Barfuss, D.E. 1991. Factors affecting inorganic mercury transport and toxicity in the isolated perfused tubule. *Journal of the American Society of Nephrology* **2**, 866–78.

Zalups, R.K., Cherian, M.G. and Barfuss, D.W. 1995a. Mercury–metallothionein and the renal accumulation and handling of mercury. *Toxicology* **83**, 61–78.

Zalups, R.K., Cherian, M.G. and Barfuss, D.W. 1995b. Lack of luminal or basolateral uptake and transepithelial transport of mercury in isolated perfused proximal tubules exposed to mercury–metallothionein. *Journal of Toxicology and Environmental Health* **44**, 101–13.

Zornitzer, A.E. and Bronner, F. 1971. *In situ* studies of calcium absorption in rats. *American Journal of Physiology* **220**, 1261–5.

Zweibaum, A., Hauri, H.P., Sterchi, E., Chantret, I., Haffen, K., Bamat, J. and Sordat, B. 1984. Immunohistological evidence, obtained with monoclonal antibodies of small intestinal brush border hydrolases in human colon cancers and foetal colons. *International Journal of Cancer* **34**, 591–8.

12

MOLECULAR MECHANISMS OF HEPATIC METAL TRANSPORT

Nazzareno Ballatori

12.1 Introduction

The liver plays a central role in the whole-body homeostasis of essential trace metals and in the disposition of toxic heavy metals. A major hepatic function involves uptake of metals from portal blood, their intracellular metabolism, storage, incorporation into apometalloproteins, and distribution of the resulting metal complexes either into the bloodstream for delivery to other tissues, or excretion into bile. Biliary excretion is a principal pathway for the elimination of a number of metals from the body, including copper, manganese, mercury, silver and arsenic.

Significant progress has been made recently in the identification of the molecular mechanisms by which metals traverse the liver cell sinusoidal (basolateral) and canalicular (apical) plasma membranes. For the sinusoidal membrane, several candidate metal transporters have been described, including the human equivalent of the yeast high-affinity copper transporter (hCTR1), and members of the Nramp family of membrane proteins (e.g. DCT1); however, there is as yet no direct evidence that these proteins are localized to the sinusoidal membrane, and that they are functionally active in the liver. DCT1, a divalent cation transporter with an unusually broad metal substrate specificity, is expressed in rat liver in moderate amounts, although expression is low when compared to that of intestine or kidney. In contrast, *hCTR1*, the human gene for the putative high-affinity copper-uptake transporter, is expressed at its highest levels in the liver. Metal oxyanions such as arsenate, vanadate, chromate and selenate are substrates for phosphate or sulfate transporters, whereas transport of organometallic complexes may occur on the recently cloned sinusoidal organic solute carriers. A key sinusoidal metal-uptake system is the transferrin-receptor-mediated endocytosis of iron and possibly of other metals that bind to transferrin. Hepatic metal uptake also occurs by other receptor-mediated pathways (e.g. ferritin and asialoglycoprotein), and by adsorptive and fluid-phase endocytosis.

Release of metals from liver cells into either blood plasma or bile is also mediated by both metal-selective membrane transport proteins and membrane recycling. Important insight into mechanisms for cellular export of metals was provided by the recent cloning of several putative metal transporters, including the Wilson disease and Menkes syndrome genes, some members of the cation diffusion facilitator (CDF) family of transporters (e.g. ZNT1, ZNT2, ZNT3 and ZNT4), and of some of the canalicular organic solute efflux pumps (of the MDR and MRP families). The Wilson and Menkes diseases genes encode transmembrane P-type ATPases that are required for copper export, but these proteins are localized largely to intracellular vesicles, indicating that copper (and possibly other metals) are first compartmentalized within intracellular vesicles and that export is accomplished by vesicle fusion with the plasma membrane, although this issue remains controversial. In contrast, ZNT1, a member of the CDF family, appears to localize to the plasma membrane, where it may function to release zinc into the extracellular space. However, neither the hepatocellular distribution nor driving force for ZNT1 or of other members of the CDF family have been determined. ATP is not required for ZNT1-mediated metal efflux. Additional mechanisms for the export of metals from liver cells into bile involve the canalicular ATP-dependent, multispecific organic solute transporters, MDR1 and MRP2. These multispecific transport proteins are members of the ATP-binding cassette (ABC) superfamily of proteins, and are expressed in high levels on the liver cell canalicular membrane. Although their preferred substrates are amphiphatic organic molecules, they also transport organometallic complexes. MRP2 mediates biliary transport of metals complexed with glutathione (GSH), including arsenic and cisplatin, and possibly copper, cadmium and mercury. Cells overexpressing MDR1 are more resistant to cationic lipophilic metal complexes, indicating transport of the metal complexes on MDR1.

This chapter will provide an overview of the mechanisms that the liver cell utilizes to transport metals to and from the bloodstream, and into bile, with an emphasis on recent work on the molecular characterization of putative hepatic metal transporters. Previous studies of hepatic uptake and biliary excretion of specific metals are described in several comprehensive reviews (Klaassen 1976; Klaassen and Watkins, 1984; Ballatori 1991a, b, 1994), and are only discussed briefly here.

12.2 Models used to study hepatic transport

The liver's complex microanatomy and reticuloendothelial system create a specialized epithelial barrier separating blood from bile. This complexity poses a special challenge in the design of experiments to investigate mechanisms of hepatobiliary transport of solutes, including metals. The intact animal and the isolated perfused liver model are the only systems that permit the simultaneous measurement of sinusoidal transport and biliary excretion; however, it is often difficult to obtain detailed kinetic data using these model systems. Other *in vitro* models are most often chosen; however, all *in vitro* models share a major limitation, namely the inability to reproduce the ionic and ligand composition of blood and blood plasma, a critical factor in regulating metal uptake. Metals differ from most other transported solutes in that they generally do not exist in their 'free' form in biological fluids and tissues, but are normally complexed with amino acids, peptides, proteins, phospholipids and other tissue constituents. As a consequence, their disposition, including their membrane transport, is regulated by the availability and relative concentrations of competing ligands (Ballatori 1991a, 1994).

Despite the limitations of *in vitro* systems, isolated hepatocytes in culture have provided invaluable information on the characteristics of uptake of a variety of metals. Isolated hepatocytes have been used to determine the apparent K_m and V_{max} and the driving forces for metal uptake, as well as the effects of individual ligands, metals and other drugs on uptake. In contrast, isolated hepatocytes have not been useful for studying export of metals and other solutes from the cell. Mammalian hepatocytes rapidly lose their transport polarity upon isolation, which makes it difficult to distinguish excretion via the sinusoidal versus the canalicular membranes. The hepatocyte couplet system obviates some of these difficulties in that it retains a bile canalicular space (Graf *et al.*, 1984), but has additional limitations, including the relative inaccessibility of the small bile canalicular space.

In the early 1980s another useful model system was introduced: a technique for the separation of plasma membrane vesicles derived from the liver cell sinusoidal and canalicular membrane domains (Meier *et al.*, 1984). This made it possible to examine transport at both poles of the cell separately, without interference from cell metabolic functions. With this system, the electrogenic features, energy dependence, and ion requirements could be determined with relative ease. The limitations of this system include cross-contamination with other intracellular or plasma membranes, orientation of the vesicles, functional integrity of the transport proteins or the vesicles themselves, and the relatively high degree of nonspecific binding that is often observed with metals (Ballatori, 1994).

The most significant progress in the characterization of hepatic transport systems, including metal transporters, has come from recent studies that utilized molecular biological approaches. In particular, the use of complementation strategies in bacteria and yeast, and the use of the *Xenopus laevis* oocyte expression system have allowed for the identification of several candidate hepatic metal-transport proteins.

It is becoming increasingly apparent that most cell functions, including membrane transport functions are highly conserved throughout evolution. Several mammalian transporters have recently been identified after structurally homologous proteins were first characterized and identified in either bacteria or yeast, as described in more detail below. For example, the recently cloned human gene for a putative high-affinity copper uptake transporter, hCTR1, was identified by functional complementation in yeast (Zhou and Gitschier, 1997). A second strategy involves use of *Xenopus* oocytes, which have become a standard tool for the cloning and functional expression of integral membrane proteins (Sigel, 1990; Soreq and Seidman, 1992). Gunshin *et al.* (1997) used this strategy to clone the cDNA for DCT1, a multispecific divalent cation transporter of the Nramp family. *Xenopus* oocytes allow for the study of biogenesis, functional structure, modulation and expression of plasma membrane proteins (Sigel, 1990; Soreq and Seidman, 1992). Oocytes translate exogenous mRNA injected into their cytoplasm efficiently, and are generally able to post-translationally modify and target heterologously expressed proteins correctly to the appropriate cellular compartment. This system has been used to clone and express membrane proteins of several classes, including transporters, channels, receptors and enzymes (Sigel, 1990; Soreq and Seidman, 1992).

12.3 Determinants of hepatic metal transport

12.3.1 Hepatic microanatomy

Mammalian hepatocytes are arranged in plates or sheets that are only one cell thick, and are surrounded by capillaries (sinusoids), thus maximizing surface area for exchange. The sinusoidal endothelium is highly fenestrated and has little, if any, basement membrane (Disse space). The sieve-like plates in the endothelial cells permit solutes up to the size of chylomicrons to pass from blood into the Disse space. As a result, metals associated with particles of this size or smaller in blood plasma have direct access to the basolateral surface of hepatocytes.

Three distinct plasma-membrane domains are present on hepatocytes; the sinusoidal, lateral and canalicular membranes. The sinusoidal or 'basal' surface is covered with numerous irregular microvilli that increase the surface membrane area, thereby facilitating hepatic uptake. The lateral domain is a relatively smooth and straight extension of the sinusoidal membrane that projects between neighboring hepatocytes and lines the intercellular cleft. It contains gap junctions, desmosomes and tight junctions, but no microvilli. In analogy with other epithelial cells, the sinusoidal and lateral domains together form the basolateral membrane. The apical or canalicular membrane differs histochemically and biochemically from the basolateral surface. Although the lumen of the canalicular space is only about 1 μm in diameter, the surface area is enlarged by an extensive network of microvilli which often fill the lumen. The canalicular membrane constitutes approximately 13% of the plasma-membrane surface.

Separation of blood spaces from the lumen of biliary spaces is accomplished largely by two belt-like junctional complexes that encircle each hepatocyte, enclosing the canaliculi and linking adjacent hepatocytes within the hepatic plate. The canalicular wall is therefore a specialized part of the hepatocellular plasma membrane. The tight junctions that separate bile from plasma were initially considered to be impermeable; however, it is now accepted that the paracellular pathway is the major route for the movement of certain solutes and water between plasma and bile. Studies in perfused rat liver demonstrate that greater than 75% of biliary Na^+, K^+, Cl^- and water come directly from plasma via the paracellular pathway (Graf *et al.*, 1984, 1987). Hepatocyte junctional complexes are of low electrical resistance and somewhat cation selective, suggesting the presence of negative charges. Studies by Graf *et al.* (1987) in isolated rat hepatocyte couplets, indicate that the cation selectivity of the paracellular pathway is relatively weak. Paracellular permeability appears to be regulated by hormones, Ca^{2+} and other endogenous factors, and is altered by some drugs, chemicals or during hepatic injury. The role of this pathway in the hepatobiliary transport of metals remains obscure.

12.3.2 Functional polarity

Differences in lipid and protein composition among the three plasma-membrane domains generate the hepatocyte's functional polarity (Meier, 1988). The sinusoidal membrane is rich in receptors, enzymes and transport proteins. Receptors for hormones and transmitters such as insulin, glucagon, vasopressin and α-adrenergic agonists, and for metabolites such as lipoproteins, asialoglycoproteins, transferrin and IgA, are concentrated on this membrane domain. Hormone-sensitive adenylate cyclase and Na^+,K^+-ATPase enzyme activity are also localized to the sinusoidal membrane. Its high phagocytic and endocytic properties are revealed by the presence of clathrin-coated and noncoated vesicles.

Transport proteins are also abundant on the sinusoidal membrane. These proteins facilitate uptake of ions, nutrients, bile acids, and organic anions and cations, including certain drugs and xenobiotics from blood plasma (Fig. 12.1). Hepatic uptake of organic

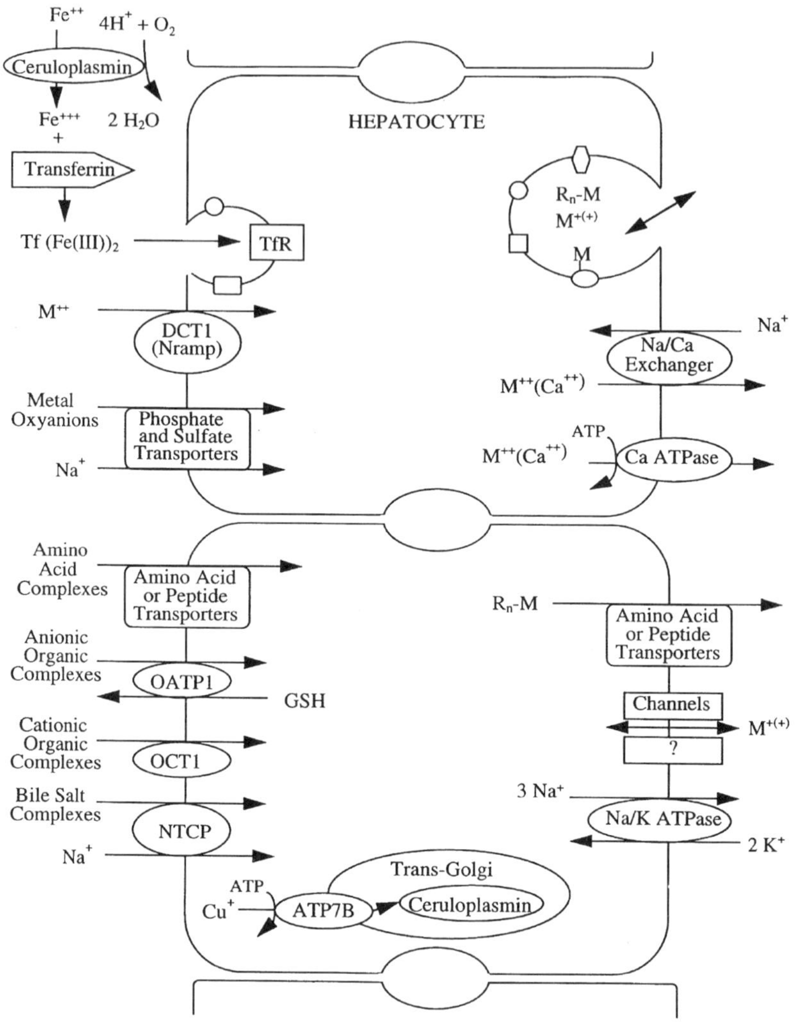

Figure 12.1 Putative hepatic sinusoidal metal uptake and efflux pathways.

solutes across the sinusoidal membrane is mediated by multiple transport systems, and four transporters have now been cloned: NTCP mediates Na^+-coupled transport of bile acids (Hagenbuch *et al.*, 1991); OCT1 mediates H^+-coupled uptake of organic cations (Grundemann *et al.*, 1994); and OATP1 and OATP2 are related multispecific transporters (Jacquemin *et al.*, 1994; Kullak-Ublick *et al.*, 1994; Noé *et al.*, 1997). The driving force for uptake on OATP1 was recently demonstrated to be the GSH electrochemical gradient (Li *et al.*, 1998). OATP1 thereby provides a mechanism for GSH efflux from hepatocytes into blood plasma (Fig. 12.1). The sinusoidal membrane also has transport systems that facilitate efflux of H^+, Ca^{2+}, amino acids and possibly other solutes from liver cells back into blood plasma, and some of these are illustrated in Fig. 12.1.

The lateral membrane participates in cell–cell interaction, but has not yet been shown to contain specific transport processes. Desmosomes and tight junctions within lateral membranes link adjacent hepatocytes and create the canalicular space, whereas its gap junctions are specialized for cell–cell communication.

The canalicular membrane is also replete with membrane transport proteins. Some of these are illustrated in Fig. 12.2. Transport systems for organic anions (MRP2), organic cations (MDR1), phospholipids (MDR2), and bile acids (cBAT or spgp), along with high specific activities of the enzymes γ-glutamyl transpeptidase, leucylnaphthylamidase, Mg^{2+}-ATPase and alkaline phosphodiesterase, are localized to the canalicular membrane. MDR1 and MRP2 have been implicated in biliary excretion of metal complexes, as discussed further below. A Cl^-/HCO_3^- exchanger and a Cl^- selective channel have also been identified (Meier, 1988). Lysosomes and endocytic vesicles fuse with the canalicular membrane and discharge their contents (including metals) into bile (Renston *et al.*, 1980).

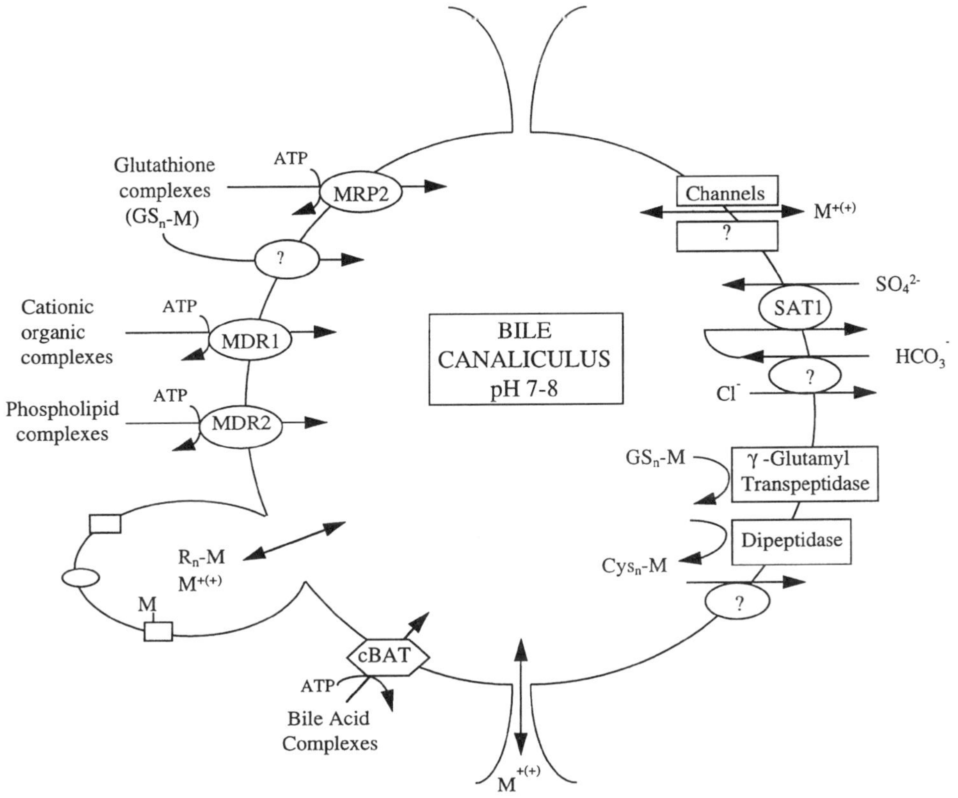

Figure 12.2 Hepatocyte bile canalicular membrane metal transport pathways.

12.3.3 Bile secretion and biliary–hepatic cycle

Hepatic bile is an isosmotic electrolyte solution containing mixed lipid micelles of bile acids, cholesterol and phospholipids, along with porphyrins, proteins, peptides, amino acids, metals and a variety of other endogenous and exogenous compounds (Klaassen and Watkins, 1984; Boyer, 1986). Bile is secreted largely in response to the osmotic gradient between bile, liver tissue and plasma, generated by active secretion of impermeant solutes into the canalicular spaces. Water, electrolytes and other solutes move across the canalicular membrane as well as through the tight junctions to dissipate the osmotic gradient created by active secretion of bile acids (Sperber, 1959), glutathione (Ballatori and Truong, 1992) and possibly other solutes.

After its secretion, canalicular bile winds its way through the labyrinth of canaliculi towards the periphery of the hepatic lobule, and drains into the hepatic ductules and ducts, and finally into the common duct and gallbladder. At each of these sites canalicular bile may be modified by both absorptive and secretory processes, and some of its components metabolized by membrane-bound ectoproteins (Ballatori *et al.*, 1986b, 1988; Ballatori and Truong, 1989), demonstrating that biliary epithelia are not inert conduits for bile, but are actively involved in its postsecretory processing. In analogy with the enterohepatic circulation, there is a biliary–hepatic cycle for amino acids (Ballatori *et al.* 1986a, 1988; Moseley *et al.*, 1988) and an organometallic compound, methylmercury (Dutczak *et al.*, 1991; Dutczack and Ballatori, 1992; Ballatori and Dutczack, 1994). Thus, metals that are secreted by hepatocytes into bile canaliculi may be reabsorbed at more distal sites within the biliary tree (Ballatori, 1994).

12.3.4 Paracellular pathway for biliary excretion

Rather than traversing the hepatocyte basolateral and canalicular membranes, metals may also enter bile via either the paracellular pathway (Fig. 12.2), or the 'trans-hepatocellular' pathway (i.e. a transepithelial route across bile ductular epithelia). However, because bile cannot be sampled simultaneously at its source (canaliculus) and within bile ductules, it is difficult to quantitate the role of the ductular epithelium in solute transport into or reabsorption from bile.

Movement of substances between blood plasma and bile is restricted and regulated primarily by the hepatocellular junctional complex. This junction is normally impermeable to solutes as large as proteins but, as already indicated, appears to be the predominant route for entry of small ions and water into bile (Graf *et al.*, 1984, 1987). Hepatic junctional complexes are considered relatively 'leaky' since the shunt conductance accounts for a large fraction of total tissue conductance. In freeze–fracture replicas, tight junctions normally contain 3–5 parallel strands that line the perimeters of the canaliculi, but the strand number may vary between 1 and 8. The complexes are also relatively cation selective, suggesting a negatively charged barrier. Thus, movement of negatively charged compounds is impeded relative to cationic compounds of similar size; however, both readily cross the junction. Studies in rat hepatocyte couplets indicate a relatively low ion selectivity of the paracellular pathway (Graf *et al.*, 1987).

Of importance to the present discussion, the Ca^{2+} and Mg^{2+} concentrations in bile are similar to those in blood plasma, these cations apparently entering bile via the paracellular pathway (Boyer, 1986). It is reasonable to expect that other free divalent metal cations in plasma could also use the shunt pathway, but there is no direct evidence for this hypothesis. For most heavy metals the role of the paracellular pathway appears to be minimal, probably because these metals are bound to plasma constituents that are normally excluded from the paracellular pathway.

12.3.5 Plasma and intracellular protein binding

Because formation of metal complexes is highly favored thermodynamically, most heavy metals are present in biological tissues and fluids as complexes, rather than as the free cations. Although the thermodynamic stability of coordinate-covalent bonds is typically quite high, they are kinetically labile, so that a given metal may rapidly exchange from one ligand to another. This kinetic lability has proven to be the greatest stumbling block to the isolation and identification of metal complexes in biological fluids and tissues. For example, during tissue homogenization or chromatographic separation additional binding sites may be exposed (or some eliminated), thus altering the distribution of the metal. Reactivity varies between metals, and is influenced by the nature of the ligand, whether mono- or multidentate, and the pH and ionic strength of the media. Copper, for example, forms relatively low-affinity complexes with albumin or amino acids, but is tightly bound to ceruloplasmin. Similarly, mercury and cadmium form kinetically labile complexes with amino acids, glutathione or albumin, but more stable complexes with metallothionein. Chelating agents normally display low specificity for metals, and bind a wide range of metals.

For the heavy metals, detoxification or protection from toxicity usually involves binding to specific proteins, including: metallothioneins, which form complexes with copper, zinc, cadmium, mercury and other metals; ferritin, transferrin and hemosiderin, which are predominantly iron-binding proteins, but also have some affinity for other metals; and ceruloplasmin, which chelates copper and possibly other metals. Similarly, the biological activity of the essential trace metals is due to their ability to attach to specific prosthetic groups on proteins. Manganese may be an exception to this generalization, since at least some of its biological functions are related to the free divalent metal.

Conversely, metal-induced toxicity is attributed to the reactivity of the 'free' metal, and is most often observed in tissues involved in their transport, such as intestine, liver and kidney. Toxic metals, or an excess of trace metals, cause injury by binding to and perturbing the functions of cellular ligands. Mercury binds avidly to reduced sulfhydryl groups on cysteine moieties, and interferes in the function of all sulfhydryl-dependent proteins that have been examined (Clarkson, 1972). Because of the shared physicochemical properties between Ca^{2+}, Pb^{2+} and Cd^{2+}, lead and cadmium probably exert some of their toxic effects by replacing Ca^{2+} on essential proteins or lipids. Similarly, vanadate and arsenate are structurally similar to phosphate and are able to displace phosphate from critical binding sites.

12.4 Sinusoidal uptake of metals from blood plasma and secretion back into the bloodstream

As with other cell types, the major determinants of the chemical interaction that will ensue following exposure of a liver cell to a given metal are the type and speciation of the metal in the extracellular medium (blood plasma), and the presence and activation state of a given metal-transport pathway. Metal binding to the membrane and penetration into the cells is regulated by the affinity and accessibility of a specific chemical form to various membrane ligands, including metal-selective transport systems.

It is becoming increasingly apparent that hepatocytes and other cells possess efficient uptake systems for procuring essential metals, and that these uptake systems are under tight regulatory control (Eide, 1997). Recent evidence indicates the presence of several distinct families of metal-transport proteins. The nonessential metals frequently mimic the essential metals and gain access via the same pathways, or enter via less selective pathways (Ballatori, 1994; Dawson and Ballatori, 1995), as described below.

12.4.1 Transport during membrane recycling: roles of transferrin and ceruloplasmin

Endocytotic and exocytotic mechanisms play a critical role in hepatic metal homeostasis. The importance of receptor-mediated endocytosis in the transport of iron and copper has long been recognized, and many of the individual steps in this process have been characterized at the molecular level (Aisen, 1998). However, the role of receptor-mediated endocytosis in the transport of other metals, and the general roles of fluid-phase and adsorptive endocytosis in metal transport have not been examined to a significant extent.

It is likely that the overall contribution of membrane recycling to metal transport has been underestimated. In rat hepatocytes, the volume of fluid taken up (and released) via the fluid-phase pathway(s) is quite large, and far greater than by receptor-mediated pathways (Oka *et al.*, 1989). The total rate of fluid internalization in rat liver cells is between 1 and 6 pl/cell/hour (Scharschmidt *et al.*, 1986; Oka *et al.*, 1989), or greater than 20% of their volume each hour. From this rate it can be estimated that between 5 and 50 times the hepatocytes' plasma membrane is endocytosed each hour (Scharschmidt *et al.*, 1986; Blomhoff *et al.*, 1989; Oka *et al.*, 1989). This phenomenal rate of plasma membrane and fluid internalization has obvious implications for the transport of substances across the membrane, yet little information is available on its role in hepatic metal transport. The rapid rate of membrane and fluid transport may be particularly important in facilitating movement of metals that either have a high affinity for plasma membrane binding sites, or are bound to ligands that are selectively cleared via their hepatic receptors. Because the process is bidirectional, membrane recycling functions to transport metals to and from blood plasma (Fig. 12.1), and to and from bile (Fig. 12.2).

Internalized endocytotic vesicles move rapidly (within minutes) to lysosomes and the pericanalicular Golgi area. The entrapped metal complexes may be metabolized, released to other intracellular compartments, or transferred in coated and noncoated vesicles to bile canaliculi. As in other epithelia, canalicular membrane biogenesis first requires insertion of integral membrane proteins into the basolateral plasma membrane, followed by their selective endocytosis and delivery to the canalicular domain (Bartles *et al.*, 1987). A tubulo-vesicular transport process has been suggested to function in the transcellular movement of certain proteins into bile (Coleman, 1987; Sakisaka *et al.*, 1988). Although vesicular exocytosis into bile makes only a minimal contribution to the total volume of fluid secreted into bile (<4%), it is the predominant mechanism for biliary excretion of proteins and certain other macromolecules. Similar endocytotic/exocytotic mechanisms are probably quite important in hepatic metal transport, particularly for metals that are complexed with high-molecular-weight plasma constituents that are cleared by the liver. Even if only a small fraction of the metal in plasma were bound to such ligands, the efficiency of the process could make it the predominant pathway. Additional studies are needed to assess the relative contributions of these ubiquitous transport pathways.

The only relatively well-characterized endocytotic process is that for transferrin. Most investigators agree that the predominant mechanism of iron transport from blood plasma into hepatocytes is via the transferrin receptor (Morgan and Baker, 1988), as illustrated in Fig. 12.1. Nearly all of the iron in plasma (~99%) is normally associated with transferrin, a protein that binds iron in the Fe(III) oxidation state. Oxidation of Fe(II) to Fe(III) is catalyzed by the copper-containing enzyme ceruloplasmin (Fig. 12.1).

The interaction of transferrin with its receptor promotes iron uptake by two mechanisms. First, receptor-mediated endocytosis leads to the internalization of diferric transferrin, followed by release of iron within acidic vesicles, and extrusion of iron-depleted transferrin (apotransferrin). The mechanism by which the released iron is subsequently transferred from the endosome/lysosome to the cytosol is not clear. Some possibilities are discussed in a subsequent section (see Fig. 12.5). A second and more speculative uptake mechanism involves the possibility that iron is released at the plasma membrane without internalization of the transferrin–receptor complex (Thorstensen and Romslo, 1988; Jordan and Kaplan, 1994; Thorstensen *et al.*, 1995). In this model, transferrin-bound ferric iron is reduced to ferrous iron extracellularly, removed from the transferrin molecule, and transported into the cell, possibly by the recently identified DCT1 protein (Fig. 12.1).

In addition to transferrin receptors, ferritin receptors play a major role in hepatic iron uptake (Mack *et al.*, 1983; Osterloh and Aisen, 1989). Kupffer cells release a substantial fraction of the iron acquired by erythrophagocytosis in the form of ferritin, which is efficiently internalized by hepatocytes, via their ferritin receptors. Aisen (1998) suggests that ferritin may be more important than transferrin in mediating hepatic iron uptake at physiologic concentrations of these ligands. This issue remains to be resolved.

A third receptor-mediated endocytotic mechanism involves the asialoglycoprotein receptors (Stockert *et al.*, 1980; Oka and Weigel, 1983; Young *et al.* 1983; Regoeczi *et al.*, 1984). Galactose-specific asialoglycoprotein receptors are unique to hepatocytes, and have been identified in every mammalian species examined. One substrate for these receptors is asialotransferrin. Sinusoidal endothelial cells appear to be able to desialate transferrin (Tavassoli, 1988); the iron carried by these sialic acid-depleted molecules can be cleared rapidly from the blood by receptor-mediated endocytosis. Hepatic iron uptake is faster from asialotransferrin than from fully sialylated transferrin (Rudolph *et al.*, 1986).

Although the relative contributions of these auxiliary pathways of iron uptake are unknown, it is becoming quite clear that these pathways are important not only in iron-overload states, but work in parallel with the transferrin receptor to mediate normal hepatic iron uptake (Morley and Bezkorovainy, 1985).

Interestingly, these iron-binding proteins may also be involved in the transmembrane movement of other metal cations, including manganese, zinc and vanadium. Most of the manganese in plasma is bound to transferrin, a finding that may explain the rapid hepatic clearance of Mn from plasma. Similarly, a substantial fraction of vanadium in rat plasma is associated with transferrin. Vanadium accumulates preferentially in tissues that are also abundant in iron (liver, spleen and kidney). Chromatographic separation of vanadium in liver homogenates reveals that the metal co-elutes with fractions corresponding to transferrin and ferritin. It has been speculated that ferritin may serve as a general metal detoxicant because of its ability to bind to a variety of metal cations, including Cd, Zn, Be and Al (Joshi *et al.*, 1989). Metals also bind to albumin, and this complex may also be transported by vesicular mechanisms (Tibaduiza and Bobilya, 1996).

The role of membrane turnover in biliary excretion of metals is discussed in Section 12.6.1.

12.4.2 Role of sinusoidal membrane transporters

12.4.2.1 Uptake on phosphate or sulfate transporters

One of the earliest metal-transport mechanisms to be described was the transport of metal oxyanions on phosphate and sulfate carriers (Wetterhahn-Jennette, 1981; Clarkson, 1993; van Veen *et al.*, 1994). Vanadate and arsenate are structurally similar to phosphate and can compete with phosphate for transport, as well as intracellular binding sites. Indeed, their toxicity is thought to be directly related to this competition. Similarly, chromate, selenate and molybdate are structurally similar to sulfate, and are substrates for sulfate transporters.

Hepatic phosphate or sulfate transporters have been characterized functionally, and recent studies have identified some of the proteins at the molecular level. Hepatic phosphate uptake is mediated in part by a Na^+-dependent mechanism (Ghishan *et al.*, 1993), and genes have been identified that mediate Na^+-dependent phosphate uptake when expressed in *Xenopus* oocytes (Chong *et al.*, 1993; Miyamoto *et al.*, 1995; Kavanaugh and Kabat, 1996; Norbis *et al.*, 1997). These proteins have been termed NaP(i)1, NPT1, PiT1 and PiT2. All four transporters appear to be expressed in the liver, and are presumably localized to the sinusoidal membrane, but this has not been established. In addition to these plasma membrane transporters, phosphate transporters are also present in intracellular organelles, including mitochondria (Pratt *et al.*, 1991). Arsenate and vanadate most likely enter hepatocytes and their organelles on these phosphate transporters (Wetterhahn-Jennette, 1981).

Hepatic sinusoidal sulfate uptake also appears to be mediated by a Na^+-dependent mechanism; however, a liver Na^+–sulfate cotransporter has not been identified at the molecular level. By analogy, this sulfate transporter probably mediates hepatic uptake of chromate, selenate and molybdate. Markovich *et al.* (1993) cloned a cDNA encoding a rat renal Na^+–sulfate cotransporter (NaS(i)1), but were unable to detect its mRNA in rat liver.

12.4.2.2 hCTR1 and hCTR2, putative copper-uptake transporters

Although it has long been recognized that the liver is the central organ of copper homeostasis, the molecular basis for hepatic copper uptake has remained a mystery until recently (Vulpe *et al.*, 1993; Bull and Cox, 1994; Vulpe and Packman, 1995; Yamaguchi *et al.*, 1996). Important insight into this process was provided by recent studies in several laboratories, many of which took advantage of the remarkable similarities between yeast and mammalian cells in terms of copper and iron metabolism, and the relative ease with which genetic manipulations can be carried out in yeast. A great deal is now known about how lower eucaryotes, such as yeast, and procaryotes defend themselves against toxic metals, whereas relatively little is known about defense strategies utilized by mammalian cells. In yeast, resistance to metals is usually due to genes whose products function to sequester metals in intracellular compartments, whereas bacterial defenses are mainly attributed to plasmid-encoded genes whose products function to export metals out of the cell (Solioz *et al.*, 1994; Silver, 1996; Tsai *et al.*, 1997).

A model illustrating the normal pathways of copper metabolism in hepatocytes is provided in Fig. 12.3, and those of the yeast *Saccharomyces cerevisiae* in Fig. 12.4. As depicted in these figures, there is a remarkable evolutionary conservation in the overall mechanisms of metal homeostasis from yeast to humans (Kaplan and O'Halloran, 1996; Perego and Howell, 1997; Solioz *et al.*, 1994; Valentine and Gralla, 1997).

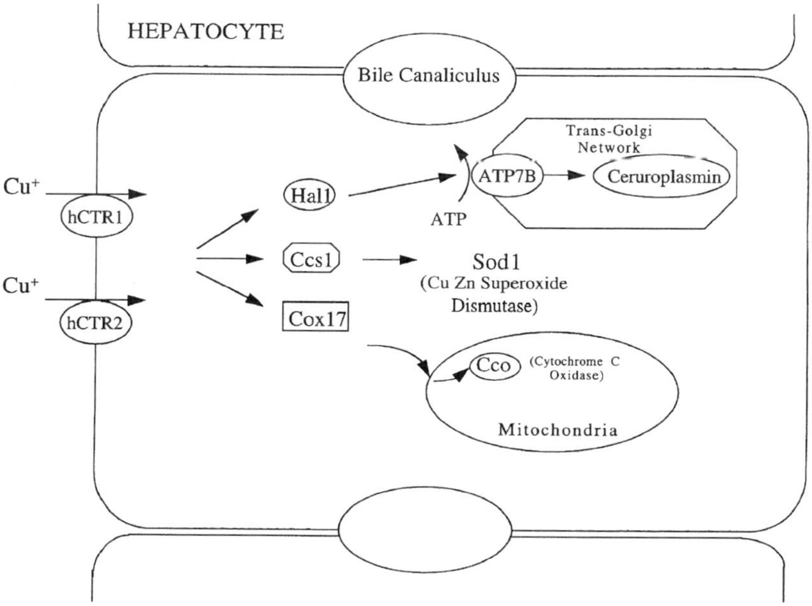

Figure 12.3 Model of hepatocellular copper uptake and utilization.

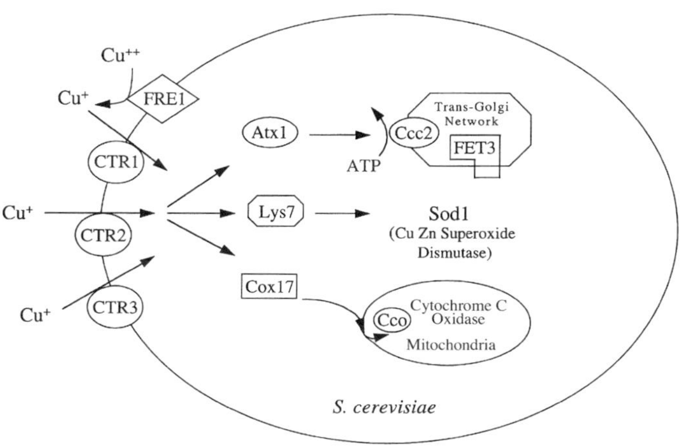

Figure 12.4 Yeast copper uptake mechanisms, and intracellular distribution to various copper acceptors.

Saccharomyces cerevisiae acquires copper as Cu(I) through the action of a plasma-membrane-associated Cu(II)–Fe(III) reductase (FRE1), two high-affinity copper-transport proteins (CTR1 and CTR3), and a low-affinity copper transporter (CTR2) (Fig. 12.4). Of the three transporters, only the CTR1 protein is indispensable for high-affinity copper uptake (Dancis *et al.*, 1994). Once within the yeast cell, Cu(I) is transferred to small cytoplasmic proteins that function as intracellular transporters or chaperones. These intracellular copper-binding proteins deliver the copper to selected intracellular proteins or organelles (Fig. 12.4). The Atx1 protein, for example, delivers copper to Ccc2, a P-type ATPase that pumps copper into endosomal/lysosomal compartments. Ccc2 is structurally and functionally similar to the gene products defective in Wilson disease (ATP7B) and Menkes syndrome (ATP7A), and is localized to the same intracellular compartment (Figs 12.3 and 12.4). Ccc2 functions to deliver copper into this space for storage, as well as for insertion into FET3, a multicopper oxidase that is similar to ceruloplasmin (Yuan *et al.*, 1995).

The similarity between yeast and mammalian copper homeostasis is quite striking, and extends to the membrane transporters, the cytoplasmic chaperones and the terminal copper acceptors (Figs 12.3 and 12.4). Of significance, Zhou and Gitschier (1997) isolated a human gene involved in copper uptake by complementation of the yeast high-affinity copper uptake mutant, *ctr1*. The human gene product (hCTR1; Fig. 12.3) exhibits 29% amino acid identity with yeast CTR1. A database search by Zhou and Gitschier (1997) revealed an additional human gene that was named *hCTR2*. By Northern blot analysis, *hCTR1* and *hCTR2* were expressed in all tissues examined, but the liver exhibited the highest level of expression. However, the cellular localization of the gene products, and their function and mechanism of action in the liver have not yet been described.

The pathways that control iron metabolism in eucaryotic cells, and the central role of copper in regulating iron uptake are also conserved during evolution (Askwith *et al.*, 1996; Kaplan and O'Halloran, 1996). High-affinity iron uptake in yeast and mammalian cells requires the presence of the homologous multicopper-containing enzymes FET3 and ceruloplasmin, respectively (Figs 12.1 and 12.5). Ceruloplasmin catalyzes the oxidation of Fe(II) to Fe(III), the form that can be bound by transferrin (Fig. 12.1), whereas FET3 generates Fe(III) for subsequent uptake on the FTR1 transporter (Fig. 12.5). A major difference between ceruloplasmin and FET3 is that ceruloplasmin is a secreted protein, whereas FET3 is a membrane-bound protein that is cotransported to the cell surface with the iron transporter FTR1 (Fig. 12.5). *Saccharomyces cerevisiae* also has a low-affinity Fe^{2+} uptake system that is encoded by the *FET4* gene (Dix *et al.*, 1997). The low-affinity system is the major route of iron entry in iron-replete environments, whereas the high-affinity system (FTR1) is critical when iron is scarce. FET4 may also mediate uptake of other metals, including cobalt and cadmium (Askwith *et al.*, 1996).

This requirement of a copper-containing protein for iron transport in yeast and mammalian cells provides at least a partial explanation for the observation that animals that have low ceruloplasmin levels due to either copper deficiency or the inability to synthesize the holoprotein are unable to mobilize iron from tissues (Kaplan and O'Halloran, 1996). Delivery of hepatocellular copper to apoceruloplasmin requires the endosomal transporter ATP7B (Fig. 12.3), the protein that is defective in Wilson disease. Other tissues utilize the homologous transporter ATP7A, the protein defective in

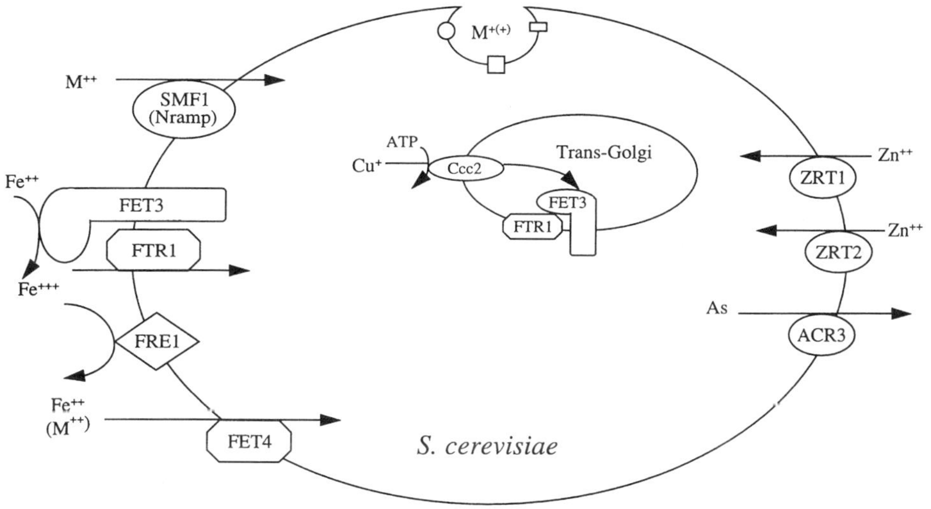

Figure 12.5 Yeast uptake and efflux systems for iron and other metals.

Menkes syndrome, to transport copper into intracellular compartments. Because the liver is a major site of ceruloplasmin synthesis, most Wilson disease patients have diminished ceruloplasmin levels, yet these levels are nevertheless adequate to maintain iron homeostasis.

Ceruloplasmin appears to play a dual role in facilitating iron uptake into hepatocytes. In addition to catalyzing the oxidation of Fe(II) to Fe(III) for incorporation into transferrin, ceruloplasmin stimulates iron uptake by a transferrin-independent mechanism (Mukhopadhyay *et al.*, 1998). Mukhopadhyay *et al.* (1998) reported that although ceruloplasmin had no effect on iron uptake or release in iron-replete HepG2 cells, it stimulated uptake in iron-depleted cells by increasing the apparent affinity of iron for an, as yet undefined, transporter. Ceruloplasmin-stimulated iron uptake was independent of transferrin or of transferrin receptors. In analogy with the pathways controlling iron metabolism in the yeast *S. cerevisiae* (Fig. 12.5), Mukhopadhyay and *et al.* (1998) suggested a model in which iron-deficient hepatocytes increase transcription of both ceruloplasmin and an, as yet undefined, iron transporter. It is tempting to speculate that iron uptake is mediated by a mammalian homolog of yeast FTR1 (Fig. 12.5), but such a transporter has not yet been identified. Barisani *et al.* (1995) provided a functional description of a low K_m transporter for non-transferrin-bound iron in isolated rat hepatocytes that might be the functional equivalent of yeast FTR1.

Saccharomyces cerevisiae also has two zinc-uptake transporters, ZRT1 and ZRT2, and an arsenic efflux system, ACR3 (Fig. 12.5), whose mammalian counterparts have not been described. Zinc uptake by *S. cerevisiae* is mediated by a high-affinity system that is required for growth in zinc-limited medium (ZRT1), and a lower-affinity system that is active in zinc-replete cells (ZRT2) (Zhao and Eide, 1996; Eide, 1997). The driving force for zinc uptake by these proteins has not been defined. ZRT1 and ZRT2 are distinct from the ZNT family of mammalian zinc transporters (see below). However, one feature shared by all zinc transporters is a histidine-rich domain that probably functions as a coordination site for zinc (Eide, 1997).

A putative oxyanion transporter involved in arsenic efflux from *S. cerevisiae* was recently described by Wysocki *et al.* (1997). *ACR3* null mutant cells were found to be hypersensitive to sodium arsenite and sodium arsenate, but not to antimony, tellurite, cadmium or phenylarsine oxide, indicating a relatively narrow substrate selectivity. The predicted amino-acid sequence of ACR3 exhibits 16% identity with the ArsB protein of the procaryotic *ars* operon, providing additional evidence for a role in arsenic efflux. The ArsB protein is the transmembrane component of the procaryotic arsenic resistance pump (Tsai *et al.*, 1997). Movement of the oxyanion through ArsB does not require ATP, but is driven by the protonmotive force (Tsai *et al.*, 1997). Mammalian homologs of these proteins have not been identified.

12.4.2.3 DCT1, a multispecific metal transporter of the Nramp family

An additional mechanism for cellular uptake of free ferrous iron was described recently by Gunshin *et al.* (1997). These investigators cloned and characterized a mammalian iron and divalent cation transporter (DCT1) that is expressed in a number of tissues, including liver. DCT1 is able to transport a variety of divalent metal cations (Fe, Zn, Mn, Co, Cd, Cu, Ni and Pb), by a proton-coupled and membrane-potential-dependent mechanism. In the intestine, this protein probably mediates metal uptake at the lumenal surface by cotransport with protons (with a stoichiometry of $1H^+ : 1M^{2+}$). However, its function in the liver is not known. Although Fig. 12.1 shows DCT1 mediating metal uptake from sinusoidal blood, this is only speculative. The localization, orientation and energy coupling of DCT1 in liver cell membranes has not been established.

Interestingly, DCT1 is highly homologous to the 'natural-resistance-associated macrophage protein' (Nramp) family of membrane proteins, providing strong evidence that this is a family of metal transporters (Cellier *et al.*, 1995; Supek *et al.*, 1996, 1997; Atkinson *et al.*, 1997; Fleming *et al.*, 1997; Gunshin *et al.*, 1997). The Nramp family of membrane-associated proteins displays a high sequence conservation from yeast to humans, with many species expressing at least two discrete gene copies (Cellier *et al.*, 1995). Preliminary studies by Gunshin *et al.* (1997) demonstrated that iron uptake is stimulated in *Xenopus* oocytes expressing mouse Nramp1, providing direct evidence for a role in metal transport. However, its cellular localization and mechanism of action are not defined. Atkinson *et al.* (1997) reported that murine macrophage Nramp1 is localized to a subcellular organelle that displays the structural and functional characteristics of a phagolysosome. Nramp1 is expressed only in reticuloendothelial cells, whereas a closely related protein Nramp2 is expressed in all tissues. Fleming and co-workers (1997) suggest that *Nramp2* is the gene that is defective in mice with microcytic hypochromic anemia, a disease associated with defects in intestinal iron absorption and erythroid iron utilization.

The yeast homolog of Nramp is SMF1 (Fig. 12.5). SMF1 displays 55% sequence similarity with Nramp1, and its deletion in yeast renders cells sensitive to low manganese concentrations, indicating that SMF1 is a high-affinity manganese-uptake transporter (Blackwell 1996; Supek *et al.*, 1997). In contrast to the lysosomal localization of Nramp1, antibodies to SMF1 have localized this protein to the yeast cell membrane (Blackwell, 1996) (Fig. 12.5).

12.4.2.4 *ZNT1, a putative zinc-efflux transporter*

Palmiter and Findley (1995) isolated a cDNA encoding a zinc transporter (ZNT1) from a rat kidney expression library by complementation of a mutated, zinc-sensitive BHK (baby hamster kidney) cell line. The transporter was localized to the plasma membrane of these cells, and was found to mediate both uptake and efflux of zinc, although efflux was considered the physiologic direction of transport. The energetics of transport were not identified: zinc transport was unaffected by metabolic poisons, or by the ionic composition of the culture medium (Palmiter and Findley, 1995).

At present, it is unknown whether the liver expresses ZNT1, and if so, where it is located within the cell, and whether it mediates metal uptake or efflux. The metal ion specificity of ZNT1 is also not clear. Palmiter *et al.* (1996a) have also identified a related transporter that is localized to intracellular endosomal/lysosomal vesicles, ZNT2. This transporter is relatively selective for zinc, but is not expressed in the liver (Palmiter *et al.*, 1996a). This same group has also identified a related protein, ZNT3 whose expression is restricted to brain and testes (Palmiter *et al.*, 1996b). Immunohistochemical analysis of murine brain suggests localization of ZNT3 to synaptic vesicles (Palmiter *et al.*, 1996b). Huang and Gitschier (1997) recently identified another member of this group of transporters, ZNT4, as the protein that is defective in the inherited zinc deficiency in the lethal milk mouse. ZNT4 was shown to confer zinc resistance to a zinc-sensitive yeast strain, and was abundantly expressed in mammary epithelia and brain (Huang and Gitschier, 1997). ZNT proteins belong to the so-called CDF (cation diffusion facilitator) family of metal ion transporters (Silver, 1996; Paulsen and Saier, 1997).

12.4.2.5 *Organic solute carriers: OATP1, OCT1, NTCP, amino acid and peptide transporters*

Sinusoidal uptake and efflux of metals may also occur on the recently cloned organic solute carriers. As already mentioned, metals are normally present in biological fluids as complexes with endogenous molecules such as amino acids, peptides or phospholipids, substances that are themselves transported across cell membranes by carrier proteins. These proteins also appear to be able to transport metal–substrate complexes (Kerper *et al.* 1992; Ballatori, 1994; Dutczak and Ballatori 1994). Organic solute carriers are generally multispecific, that is, they accept substrates that differ considerably in their chemical structures, and most are not able to discriminate between substrates whose only modification is the presence of a metal ion. This mode of transport appears to be particularly important for toxic metals, given the absence of selective transport systems for these metals.

Several organic solute transporters have been identified on the sinusoidal membrane of hepatocytes, including the organic anion transporting polypeptides, OATP1 and OATP2 (Jacquemin *et al.*, 1994; Kullak-Ublick *et al.*, 1994; Noé *et al.*, 1997), the organic cation transporter, OCT1 (Grundemann *et al.*, 1994), and the Na^+-taurocholate (bile acid) cotransporting polypeptide, NTCP (Hagenbuch *et al.*, 1991) (Fig. 12.1). Amino-acid transporters (Malandro and Kilberg, 1996) and peptide transporters (Fei *et al.*, 1994) have also been identified that could function both for hepatic uptake and release of these organic solutes. Theoretically, these sinusoidal carriers may facilitate transport of metals bound to their respective substrates (Fig. 12.1). Indeed, several groups have provided evidence for cellular copper or zinc uptake as histidine complexes (Aiken *et al.*, 1992; Harris, 1993; Horn *et al.*, 1995; Horn and Thomas, 1996), and our group has demonstrated transport of methylmercury as a cysteine complex (Kerper *et al.*, 1992; Ballatori, 1994; Dutczak and Ballatori 1994; Ballatori and Truong, 1995b; Mokrzan *et al.*, 1995). In erythocytes, zinc also appears to be taken up as an anionic complex ($[Zn(HCO_3)_2\,Cl]^-$) through the anion exchanger (Torrubia and Garay, 1989).

12.4.3 Uptake and efflux on sinusoidal membrane channels and pumps

The role of ion channels and primary active pumps in facilitating hepatic sinusoidal metal transport appears to be minimal, although these pathways may be important for metal transport in other cell types. For example, in excitable tissues Cd^{2+} and Pb^{2+} may enter cells via voltage-sensitive channels (Reuter, 1983; Hinkle *et al.*, 1987; Simons and Pocock 1987). Studies by Hinkle *et al.* (1987) in a pituitary cell line demonstrate that one route of cadmium uptake in these cells is via voltage-gated dihydropyridine-sensitive calcium channels. The voltage-gated calcium channels also admit Ba^{2+} and Sr^{2+}, and are inhibited by a number of divalent metal cations.

In hepatocytes, however, voltage-sensitive calcium channels have not been identified (Hughes and Barritt, 1989; Kawanishi *et al.*, 1989). Calcium uptake by hepatocytes occurs largely through receptor-activated calcium channels, which could theoretically allow other divalent cations to enter the cell. Support for this possibility is provided by the observation that hepatocyte receptor-activated calcium channels are inhibited by Zn, Cd, Ni, Co and Mn (Hughes and Barritt, 1989); however, the nature of the inhibition by the metals is unknown. Crofts and Barritt (1990) indicate that Mn^{2+} can move into hepatocytes through the receptor-activated Ca^{2+} inflow system, identifying a potential regulated mechanism for hepatic manganese uptake.

Several other interactions between hepatic uptake of calcium and other metals have been described. Studies *in vivo* (Tang *et al.*, 1987) and with isolated rat hepatocytes (Sorensen, 1988) demonstrate higher cadmium uptake as the concentration of calcium is lowered. However, this effect is probably indirect since cadmium has no effect on ^{45}Ca influx (Sorensen, 1988). This investigator describes several mechanisms whereby changes in Ca^{2+} may alter the plasma membrane's physicochemical properties, and indirectly influence its transport properties. In contrast to these findings in intact hepatocytes, passive Ca^{2+} influx in isolated plasma membrane vesicles is inhibited by Zn^{2+}, Cu^{2+}, Ni^{2+}, Co^{2+}, Hg^{2+} and *p*-chloromercuribenzoate (Bygrave *et al.*, 1989). The involvement of sulfhydryl groups in this interaction has been suggested by Bygrave *et al.* (1989).

Metals also interact with plasma membrane Ca^{2+}-ATPases, the predominant proteins for hepatic Ca^{2+} efflux (Pavoine *et al.*, 1987; Blitzer *et al.*, 1989; Suzuki and Kawakita, 1993), raising the possibility that other metals may utilize these ATPases to exit the liver cell and enter blood plasma (Fig. 12.1). Distinct Ca^{2+}-ATPases may be present on the canalicular and basolateral membranes (Blitzer *et al.*, 1989). However, the existence of a Na^+-dependent hepatic Ca^{2+} efflux system (i.e. Na^+/Ca^{2+}-exchanger), has not been established, although its existence and a hypothetical role in metal transport remains (Fig. 12.1).

12.4.4 Permeation as lipid-soluble forms or by convection

Although often invoked to explain the 'nonsaturable' component of metal transport in *in vitro* hepatocyte systems, it is unlikely that simple diffusion and convection play a role in metal transport across the plasma membrane. Diffusion denotes the nonmediated movement of a solute down its electrochemical gradient, leading to the dissipation of free energy in that gradient. Convection or bulk flow is the vectorial movement of an assembly of molecules under an imposed external potential, such as hydrostatic pressure.

Most metals and metal complexes are polar, hydrophilic and quite often charged, properties that would exclude them from the hydrophobic interior of the plasma membrane. Simple diffusion through a membrane protein matrix or ion channel is also highly unlikely. The reason for the preponderance of 'nonsaturable' components of transport in *in vitro* systems is unclear. It is most likely an artifact of either the cell or tissue isolation procedure, or of the experimental conditions used to examine transport. Kinetics consistent with simple diffusion may be obtained if:

1. the concentration of metal substrate is below the K_m of the transport system (the effective metal substrate concentration in the incubation media is often artificially lowered by extensive binding to nonsubstrate ligands);
2. there is a large component of nonspecific, high-affinity (or essentially irreversible) binding to cellular constituents;
3. transport occurs across a cell membrane damaged either during cell isolation or by increasing concentrations of the metal itself; for example, protease contaminants in the collagenase used for hepatocyte isolation can degrade transferrin receptors, thereby inhibiting the predominant iron-uptake mechanism, and accelerating other nonspecific transport pathways (Morgan and Baker, 1986); and
4. uptake occurs via vesicular endocytosis/exocytosis; in this case, the metal would be transported nonspecifically along with the entrapped fluid and/or membrane vesicle components.

All of these mechanisms need to be considered before attributing a particular component of transport to 'simple diffusion'.

Studies of metal transport in isolated hepatocyte systems are particularly susceptible to these artifacts. First, since a given metal normally binds to many endogenous ligands, some of which may be unidentified, it is difficult to reproduce *in vivo* conditions in a cell culture system. As discussed above, metals are normally complexed with amino acids, peptides, proteins, phospholipids and other tissue constituents, and are distributed among all competing ligands on the basis of concentrations and relative affinities of available ligands. Because it is difficult to reproduce *in vivo* conditions in a culture dish, interpretation of transport studies performed *in vitro* should be performed with care.

Secondly, the propensity of metals to bind to things such as culture dishes, filters, components of the culture media or, more importantly, cell membranes, creates both a practical and theoretical dilemma. This 'nonspecific' attachment is difficult to eliminate and even more difficult to quantitate. When metal salts are added to protein and amino-acid-free culture media, the metal frequently associates rapidly with the cells. This rapid uptake is not due to the presence of an efficient 'transport system' as often suggested, but may simply reflect the absence of competing ligands.

Thirdly, as already indicated, the loss of hepatocyte polarity, the dedifferentiation, and the damage that may occur during cell isolation and culture may alter the hepatocyte's basic transport mechanisms. Many investigators using isolated hepatocytes tacitly assume that the transport that they observe is localized to the basolateral membrane. While this is usually a good approximation, there are exceptions. Loss of cell polarity in isolated hepatocytes implies that all plasma membrane domains contribute to the transport processes.

12.5 Intracellular binding, sequestration and metabolism

12.5.1 Intracellular ligands

Intracellular sequestration of essential metals is a regulated process. As discussed above for copper (Fig. 12.3), there are specific metal-binding ligands that direct essential intracellular metals to their appropriate destination within the cell, or that control the reactivity (toxicity) of both essential and nonessential metals. However, only a few of these intracellular metal chaperones or transporters have been characterized at the molecular level (Cousins, 1985; Aballay *et al.*, 1996; Koch *et al.*, 1997). In addition to these relatively selective ligands, there are a number of other compounds that play a more general role in metal disposition, including metallothionein, ferritin, citrate, ascorbate, GSH and amino acids. The relative abundance and affinity of these ligands is an important determinant of the transport of metals to various intracellular organelles, or back into the extracellular space (bile or blood plasma).

12.5.2 Metal uptake and release from endosomal/lysosomal compartments: ATP7A, ATP7B and ZNT2

As already indicated, endosomal and lysosomal vesicles are key intracellular organelles for the storage and metabolism of copper, iron and possibly other metals. For example, normal hepatocytes transport copper into the *trans*-Golgi network using the ATP-dependent transporter ATP7B, the transporter that is defective in Wilson disease (Fig. 12.6). The valence state of the copper that is transported by ATP7B is unknown, although Bingham *et al.* (1996) suggested that it is Cu(II) rather than Cu(I). Within this intracellular compartment, copper is then used for the synthesis of copper-containing proteins such as ceruloplasmin, or it is stored for subsequent excretion (Davis *et al.*, 1996). Some of the intravesicular copper is presumably sorted into vesicles destined for the lysosomal-biliary excretory pathway. According to this model, fusion of exocytic vesicles with the canalicular membrane delivers copper into bile (Fig. 12.2); however, neither the site nor the mechanism by which this vesicular sorting occurs is known.

In tissues other than the liver, a comparable P-type ATPase (ATP7A) pumps copper into endosomal/lysosomal compartments. The Wilson and Menkes diseases genes have 56% overall identity. The Menkes gene (*ATP7A*) is ubiquitously expressed in adult tissues, with little or no expression in the liver (Kuo *et al.*, 1997), whereas the Wilson disease gene is expressed in only a few cell types, notably liver and certain neuronal cells. Mutations in *ATP7A* lead to significant copper accumulation in intestinal mucosa, kidney and selected other tissues. This inability to deliver copper from sites of its absorption and storage results in a systemic copper insufficiency in Menkes patients. In

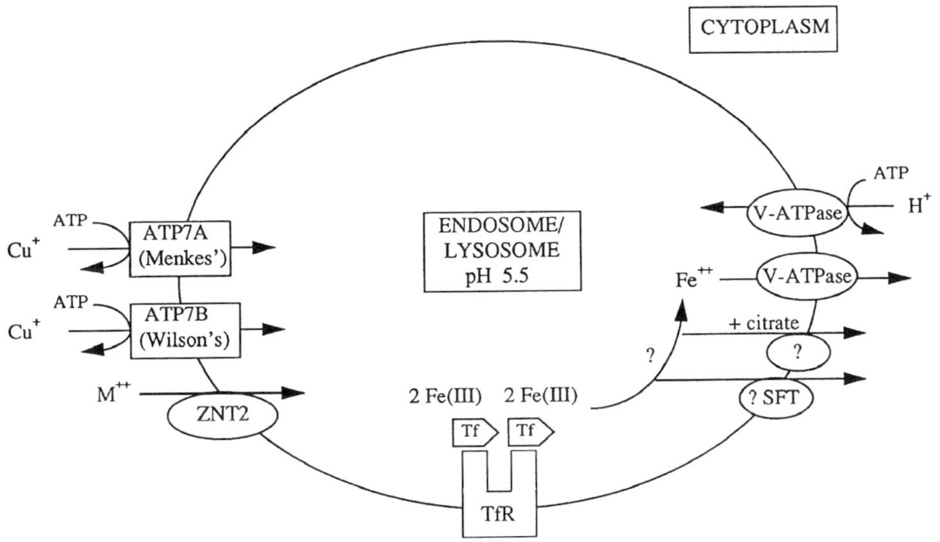

Figure 12.6 Possible mechanisms for metal sequestration and release from endosomal/lysosomal compartments.

contrast with the copper insufficiency of Menkes patients, Wilson disease patients accumulate excess copper in many tissues (e.g. liver, brain, kidney, cornea) due to the inability to excrete copper into bile, the main route of its elimination. At the cellular level, copper accumulates in the cytosol and the cell eventually succumbs to copper toxicity.

Cells utilize different mechanism to transfer iron to and from intracellular vesicles (Fig. 12.6). Transferrin receptor-mediated endocytosis leads to the internalization of ferric iron (Fig. 12.6). Acidification of the organelle by V-ATPases promotes the dissociation of ferric iron from transferrin, and of transferrin from its receptor (Fig. 12.6). However, the mechanism utilized to transport iron from these compartments back into the cytosol are not well defined. Li *et al.* (1994) suggested that the V-ATPase may itself function to transport iron back into the cytosol (Fig. 12.6). These investigators suggested that iron uses the proton pore of the V-ATPase to gain access to the cytosol. Transport occurs preferentially for Fe(II), is not associated with ATP hydrolysis and is inhibited by dicyclohexylcarbodiimide (Li *et al.*, 1994). If correct, this mechanism would necessitate the conversion of Fe(III) to Fe(II) by some unknown reductant or enzyme (Fig. 12.6). Other speculative possibilities include transport as the citrate complex, or possibly by the recently described endosomal protein STF (stimulator of iron transporter; Gutierrez *et al.*, 1997) (Fig. 12.6).

An additional metal transporter that is localized to these intracellular vesicles is ZNT2 (Palmiter *et al.*, 1996a), a member of the CDF family of metal transporters (Fig. 12.6). Palmiter *et al.* (1996a) reported that BHK (baby hamster kidney) cells that express ZNT2 are able to accumulate zinc to levels that are much higher than are tolerated by cells that lack this gene product. ZNT2 was able to confer resistance to zinc, but not to cadmium, cobalt or copper, indicating that it is relatively specific for zinc. Zinc accumulation into endosomal/lysosomal vesicles was unaffected by bafilomycin, an inhibitor of the vacuolar-type H^+-ATPase, indicating that zinc uptake does not depend on the proton gradient. As with ZNT1, ZNT2 does not have recognizable nucleotide-binding domains, indicating that ATP is not directly required for transport function. ZNT2 is expressed in many tissues, including kidney, intestine, seminal vesicles and testes, but not liver (Palmiter *et al.*, 1996a).

12.5.3 Nuclear uptake of metals

Nuclear metal homeostasis is probably also regulated by transport pathways, but few studies have examined this issue. Gurgueira and Meneghini (1996) recently demonstrated the presence of an ATP-dependent iron-uptake system in isolated rat liver nuclei. Nuclear iron accumulation was mediated by an ATP-dependent mechanism when nuclei were incubated with either Fe(III)–citrate or Fe(III)–ATP, two physiologically important iron complexes. The characteristics of uptake indicate that a P-type ATPase mediates iron uptake, and that this ATPase is distinct from the nuclear membrane Ca^{2+}-ATPase (Gurgueira and Meneghini, 1996).

12.5.4 Mitochondrial metal uptake

Mitochondria have regulated calcium transporters that are important for Ca^{2+} homeostasis; however, the molecular mechanisms for transport of other metals in mitochondria have not been elucidated. One possibility is that the calcium transporters participate in mitochondrial uptake of heavy metals, but this has not been tested directly.

A cobalt transporter (COT1) was identified in mitochondria from the yeast *S. cerevisiae* (Conklin *et al.*, 1994). COT1 confers resistance to cobalt, and to a lesser extent to zinc, presumably by sequestering the metal in the mitochondria. COT1 is homologous to yeast ZRC1 and mammalian ZNT transporters, proteins that confer resistance to zinc. Each of these proteins has six membrane-spanning domains, a large histidine-rich intracellular loop, and long C-terminal tails; however, the energetics of metal-ion transport have not been established for any of these proteins (Conklin *et al.*, 1994; Palmiter *et al.*, 1996a, b).

12.6 Transport across the canalicular membrane into bile

A model illustrating the mechanisms of metal transport at the liver cell bile canaliculus is provided in Fig. 12.2, and the individual transport systems are described below. Interestingly, many of the transporters and enzymes present on the canalicular membrane are homologous to those of vacuolar membranes of the yeast *S. cerevisiae* (Fig. 12.7), providing important insight into function and regulation of metal excretion.

12.6.1 Transport during membrane recycling: role of the gene products defective in Wilson disease (ATP7B) and Menkes syndrome (ATP7A)

Biliary excretion is the most important route for the elimination of copper, and two mechanisms for copper transport across the canalicular membrane have been identified: lysosomal exocytosis and transport as a GSH complex. In copper overload conditions, the major mechanism for removing the excess copper is via lysosomal exocytosis into

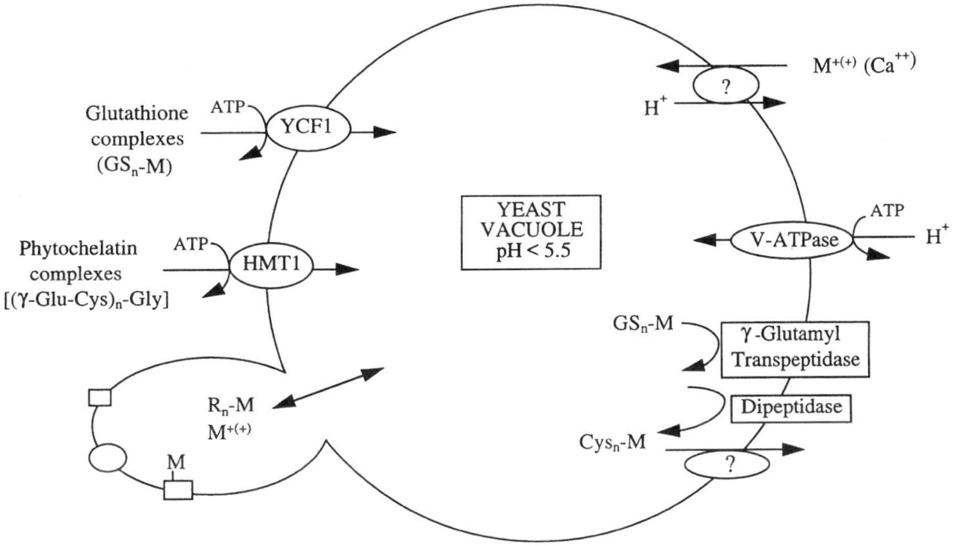

Figure 12.7 Metal transport into yeast intracellular vacuoles.

bile. Sternlieb *et al.* (1973) found that the specific activity of copper in hepatic lysosomal fractions from a patient with Wilson disease was similar to that in bile, suggesting that lysosomes might be the source of biliary copper. However, Gross *et al.* (1989) suggested that this process may not be as important when hepatocytes contain a normal amount of copper. Nevertheless, copper is normally present in relatively high concentrations in lysosomes and this mechanism could contribute to its basal excretion.

Compelling evidence in support of this possibility was provided by the recent demonstration that the gene products defective in Wilson disease (ATP7B) and Menkes syndrome (ATP7A) are localized largely to intracellular vesicles (Dierick *et al.*, 1997; Yang *et al.*, 1997; Nagano *et al.*, 1998). ATP7A, ATP7B and other ATP-driven heavy-metal pumps appear to form a separate branch of the ion-transporting P-type ATPases, the CPx-type ATPases, based on a common conserved intramembranous cysteine–proline–cysteine or cysteine–proline–histidine motif (Solioz and Vulpe, 1996). The major defect in Wilson disease is the inability to excrete copper into bile, which causes an accumulation of copper in various tissues, including liver and brain. Most Wilson disease patients also have a reduction in serum ceruloplasmin activity. These clinical findings are most easily explained by the inability to transport copper into endosome/lysosomes. The Wilson disease gene is expressed predominantly in the liver, whereas the Menkes syndrome gene is expressed in low levels in the liver and higher levels in many other tissues. Menkes syndrome patients accumulate copper at high levels in many tissues, but liver copper concentrations are normal.

The second mechanism of copper export involves transport as a GSH complex on the ATP-dependent canalicular membrane pump MRP2, or possibly on a related transporter (Fig. 12.2; Ballatori, 1994), as discussed further below. Interestingly, patients with Wilson disease have markedly lower hepatic concentrations of GSH (Summer and Eisenburg, 1985). Whether this is a consequence or a cause of hepatic copper accumulation is unknown.

Vesicular exocytosis may also be involved in biliary excretion of iron and possibly other metals (LeSage *et al.*, 1986; Ramm *et al.*, 1994; Regoeczi and Chindemi, 1995). Regoeczi and Chindemi (1995) measured the translocation of different forms of transferrin from blood to bile in the rat, and found that only a small fraction of the metal-containing protein is excreted in bile.

12.6.2 Canalicular membrane transporters and pumps

12.6.2.1 MRP2-mediated excretion of GSH complexes

Among the metal-binding ligands, GSH is one of the most versatile and pervasive. GSH is the most abundant nonprotein sulfhydryl-containing compound within cells, at concentrations of 1–10 mM. The sulfhydryl group on the cysteine moiety has a high affinity for metals, forming thermodynamically stable but kinetically labile mercaptides with a number of metals including mercury, silver, cadmium, arsenic, lead, gold, zinc and copper. Recent studies demonstrate that the abundance of GSH, along with the high turnover rate for this tripeptide and its adducts via both intra- and inter-organ cycles, play key roles in metal transport and metabolism (Ballatori and Clarkson, 1982, 1983; Ballatori, 1994). Of significance, membrane transport of several metals is accomplished by efflux of the corresponding GSH mercaptides on membrane proteins that normally transport either glutathione *S*-conjugates or GSH itself (Ballatori, 1994; Dutczak and Ballatori, 1994; Ballatori and Truong, 1995a; Dijkstra *et al.*, 1995). The transporters for the glutathione *S*-conjugates have recently been identified at the molecular level (Cole *et al.*, 1994; Kool *et al.*, 1997; Paulusma *et al.*, 1996, 1997; Deeley and Cole, 1997; Ito *et al.*, 1997; Keppler *et al.*, 1997), whereas GSH transporters have not yet been identified (Lee *et al.*, 1997; L. Li *et al.*, 1997).

Efflux of glutathione *S*-conjugates from hepatocytes and other mammalian cells is mediated by the multidrug resistance-associated proteins (MRP), of which several have now been identified in the human genome (MRP1, MRP2, MRP3, MRP4, MRP5 and MRP6) (Cole *et al.*, 1994; Buchler *et al.*, 1996; Paulusma *et al.*, 1996, 1997; Borst *et al.*, 1997; Deeley and Cole, 1997; Ito *et al.*, 1997; Keppler *et al.*, 1997; Kool *et al.*, 1997; Wijnholds *et al.*, 1997). However, to date only MRP1 and MRP2 have been characterized at the functional level. MRP1 is found on the plasma membrane of many cell types, but not the liver (Kool *et al.*, 1997), whereas MRP2, which is also referred to as cMOAT (canalicular multispecific organic anion transporter) or cMRP (canalicular MRP), is selectively localized to the apical plasma membrane of specific transporting epithelia, including the hepatocyte canalicular membrane (Fig. 12.2), and the apical membrane of kidney proximal tubules. Both MRP1 and MRP2 have a broad specificity for conjugates formed from GSH, glucuronides or sulfate.

There is now considerable evidence that MRP1, MRP2 and related membrane proteins also mediate cellular efflux of glutathione–metal complexes. First, cells overexpressing MRP1 are more resistant to arsenite, antimonite and cisplatin, consistent with accelerated metal efflux, although resistance to the metals is relatively low (Borst *et al.*, 1997; Chen *et al.*, 1997; Deeley and Cole, 1997). These metals can form divalent and trivalent complexes with GSH *in vitro*, and presumably are forming these GSH complexes *in vivo* as well. Zaman *et al.* (1995) demonstrated MRP-mediated extrusion of [^{35}S]cysteine-containing compounds after addition of arsenite, suggesting coupled efflux. Secondly, Ishikawa *et al.* (1996, 1997) demonstrated enhanced transport of a glutathione–platinum complex in membrane vesicles prepared from a tumor cell line overexpressing MRP. Thirdly, mutant rats that lack mrp2 activity (TR^-, GY or EHBR rats), exhibit impaired ability to transport several metals into bile, including zinc, excess copper, silver, cadmium and methylmercury (Houwen *et al.* 1990; Ballatori *et al.*, 1995; Dijkstra *et al.*, 1996, 1997; Sugawara *et al.*, 1997). In contrast, basal excretion of endogenous copper is unaffected in these mutant rats (Houwen *et al.*, 1990). Fourthly, the predicted MRP1 protein shows 43% amino acid identity with the YCF1 protein of *S. cerevisiae*, a protein that confers cadmium resistance by transporting and sequestering the glutathione–cadmium complex ($Cd(GS)_2$) in the yeast vacuole (Szczypka *et al.*, 1994; Wemmie *et al.*, 1994; Z.S. Li *et al.*, 1996, 1997). Of significance, human *MRP1* cDNA can complement the loss of cadmium resistance in *YCF1*-disrupted yeast cells (Tommasini *et al.*, 1996). The predicted MRP1 protein sequence also shows 34% amino acid identity with the *Leishmania* PgpA, ltpgpA (Cole *et al.*, 1994), a transporter involved in resistance to arsenite and antimonite (Callahan and Beverley, 1991; Dey *et al.*, 1996). LtpgpA confers resistance to arsenite and trivalent antimonials, but not to pentavalent antimonials, zinc, cadmium or the typical MDR substrates vinblastine and puromycin (Callahan and Beverley, 1991). Dey *et al.* (1996) demonstrated ATP-dependent transport of the As(III)–glutathione complex ($As(GS)_3$) in membrane vesicles of *Leishmania tarentolae*. In addition, a family of four MRP-related proteins has recently been identified in the nematode *Caenorhabditis elegans*, one of which (ceMRP1) confers resistance to cadmium and arsenite (Broeks *et al.*, 1996). ceMRP1 is 47% identical to human MRP1 at the predicted amino-acid level (Broeks *et al.*, 1996).

12.6.2.1.1 INTRABILIARY DEGRADATION OF GSH–METAL COMPLEXES

Once secreted across the canalicular membrane into bile, GSH and glutathione *S*-conjugates, as well as glutathione–complexes, are degraded by γ-glutamyl transpeptidase and dipeptidase activities on the canalicular membrane of hepatocytes and the luminal surface of bile ductular epithelia (Fig. 12.2; Ballatori *et al.*, 1986b; Hinchman *et al.*, 1991; Simmons *et al.*, 1991; Hinchman and Ballatori, 1994), and the resulting products are partially reabsorbed from bile back to liver (Ballatori *et al.*, 1986a, 1988; Moseley *et al.*, 1988; Hinchman *et al.*, 1991; Simmons *et al.*, 1992). Species differences in biliary glutathione degradation result from differences among species in γ-glutamyl transpeptidase (γ-GT) activity and localization within the liver (Ballatori *et al.*, 1988; Hinchman and Ballatori 1990).

Our recent studies in γ-glutamyl transpeptidase-deficient mice provide direct evidence for a major role of this enzyme in regulating the metabolism, tissue distribution and elimination of methylmercury and inorganic mercury (Ballatori *et al.*, 1998a). To evaluate the role of γ-GT in the whole-body disposition of methylmercury, we compared the elimination of [^{203}Hg]methylmercury in γ-GT-deficient mice to that in wild-type mice and mice heterozygous for this deficiency. The effects of *N*-acetyl cysteine (NAC), a drug used to maintain the cysteine and GSH levels of γ-GT-deficient mice, were also examined. There were no differences in methylmercury excretion between the wild-type and heterozygous mice; however, the γ-GT-deficient mice excreted methylmercury more rapidly at both dose levels (Ballatori *et al.*, 1998a). Wild-type and heterozygous mice excreted from 11 to 24% of the dose in the first 48 h, whereas the γ-GT-deficient mice excreted 55–66% of the dose, with most of the methylmercury being excreted in urine. Urinary methylmercury excretion was further accelerated in mice that received NAC. In contrast to methylmercury, the whole-body elimination of inorganic mercury was not affected by γ-GT deficiency, although the tissue distribution of inorganic mercury was markedly different in γ-GT-deficient male mice, with only 13% of the ^{203}Hg body burden in the kidneys of γ-GT-deficient mice, versus ~50% in kidneys of wild-type male mice. Our findings provide direct evidence for a major role of γ-GT in regulating the tissue distribution and elimination of methylmercury and inorganic mercury (Ballatori *et al.*, 1998a), and provide further support for our previous finding that NAC may be an excellent antidote in methylmercury poisoning (Ballatori *et al.*, 1998b).

12.6.2.1.2 REABSORPTION OF METALS FROM BILE

The methylmercury that is secreted into bile is extensively reabsorbed from the gallbladder (Dutczak *et al.* 1991) and from intrahepatic biliary spaces (Dutczak and Ballatori, 1992). Reabsorption diminishes the amount excreted in bile, the main route of methylmercury elimination, and may account for the relatively long biological half-life and toxicity of this metal. Methylmercury reabsorption from guinea-pig biliary spaces back into the liver is facilitated by the γ-glutamyl transpeptidase-mediated conversion of the GSH complex to the cysteine complex (Fig. 12.2). The guinea-pig liver has relatively high γ-glutamyl transpeptidase activity (Hinchman and Ballatori 1990), such that nearly all of the GSH (Ballatori *et al.*, 1988) and GSH conjugates (Hinchman *et al.*, 1991) secreted into bile are catabolized within the biliary tree. Dutczak and Ballatori (1992) demonstrated that γ-glutamyl transpeptidase also catalyzes the initial step in the conversion of the biliary methylmercury–glutathione complex to the cysteine complex and that in the latter form methylmercury is readily reabsorbed from the biliary tree.

Similarly, methylmercury reabsorption from guinea-pig gallbladder was more extensive when the metal was given as a cysteine or GSH complex, as compared with an albumin complex (Dutczak *et al.*, 1991). Because the gallbladder also has γ-glutamyl transpeptidase activity (Hinchman and Ballatori, 1990), a similar mechanism may be involved; however, there is as yet no evidence for this hypothesis.

Interestingly, similar degradative and reabsorptive processes occur within the yeast vacuole (Fig. 12.7). Glutathione complexes are transported into the vacuole by the ATP-dependent transporter YCF1 (yeast cadmium factor), the yeast homolog of mammalian MRP. YCF1 was initially identified by its ability to confer resistance to cadmium toxicity (Wemmie *et al.*, 1994). Yeast vacuoles also possess γ-glutamyl transpeptidase and dipeptidase activities that function to degrade GSH and GSH-containing compounds (Penninckx and Elskens, 1993). The amino acids liberated from GSH hydrolysis are transported back into the cytoplasm of the yeast cell.

Yeast vacuoles have an additional ABC transporter, HMT1 (Fig. 12.7), whose principal substrates are peptides and peptide complexes, and in particular the GSH-derived phytochelatins and phytochelatin–metal complexes (Ortiz *et al.*, 1995). However, phytochelatins are not found in mammalian cells. These peptides bind metals in the cytosol and are then transported into the vacuole by HMT1. High-molecular-weight phytochelatin complexes also contain sulfide (S^{2-}) which increases their capacity for binding cadmium and other metals. Binding of metals to these peptides within the yeast vacuole is an effective mechanism for sequestering the metals and diminishing their toxicity.

12.6.3 *Sulfate exchanger*

As noted earlier, vanadate and arsenate are structurally similar to phosphate and can compete with phosphate for transport, whereas chromate, selenate and molybdate are structurally similar to sulfate, and are substrates for sulfate transporters. A sulfate transporter has been identified on the liver cell canalicular membrane (sat1; Bissig *et al.*, 1994), providing a mechanism by which these oxyanions may be secreted into bile, although this has not been tested directly. Canalicular sulfate efflux on sat1 is driven by exchange with bicarbonate; the latter is secreted into bile by a chloride–bicarbonate exchanger (Fig. 12.2).

12.6.4 *Additional organic solute carriers: MDR1, MDR2 and cBAT*

P-Glycoproteins are involved in multidrug resistance (MDR) in tumor cell lines, but are also constitutively expressed in several epithelial and endothelial tissues, including the liver cell bile canalicular membrane. Bile canalicular membranes contain MDR1, which transports cationic drugs into bile, and MDR2 which transports phospholipids into bile (Smit *et al.*, 1993) (Fig. 12.2). Sharma *et al.* (1996) demonstrated that cationic organometallic complexes are also substrates for MDR1. These investigators demonstrated that human epidermal carcinoma KB cells overexpressing MDR1 P-glycoprotein are more resistant to the cytotoxic activities of Al(III)-, Fe(III)- and Ga(III)-complexes of (ethylenediamine)-*N*,*N'*-*bis*{propyl[(2-hydroxy-4,6-dimethoxybenzyl)-imino]}. Broeks *et al.* (1996) reported that the *C. elegans* equivalent MDR protein (cePGP) also contributes to heavy-metal (cadmium and arsenite) resistance in this organism, but the mechanism involved is unknown. These observations raise the possibility that MDR proteins on the canalicular membrane may be involved in transport of metal complexes into bile.

An additional canalicular transporter that might be involved in biliary metal excretion is the ATP-dependent bile acid transporter, cBAT (Fig. 12.2) (Ortiz *et al.*, 1997), although its role is probably minimal. Bile acid–metal complexes may form under certain conditions, and these would most likely be substrates for the multispecific cBAT protein.

12.7 Summary

Several metal-selective transport mechanisms on hepatocyte cell membranes allow the liver to extract metals efficiently from plasma, metabolize, store and redistribute them in various forms, either back into the bloodstream or into bile. Bidirectional transport across the sinusoidal plasma membrane allows the liver to regulate plasma concentrations, and therefore availability to other tissues. Transport across the canalicular membrane is largely, but not exclusively, unidirectional, and is a major excretory pathway for potentially toxic metals.

Some of the key transport mechanisms have recently been identified at the molecular level, providing an unprecedented opportunity to examine their structure, function and regulation. However, many of these transporters have only been identified at the cDNA level, and there is little or no information as to their cellular and subcellular localization in the liver, their functional orientation in the membrane (uptake or efflux), driving force, substrate selectivity or regulation under physiologic and pathophysiologic conditions. It is likely that additional families of metal-specific transporters will be identified in the near future, and that many new members of existing families will be described. This more complete picture of metal-transport systems will provide a solid framework for understanding metal homeostasis in both health and disease.

Acknowledgments

Preparation of this overview was supported in part by National Institutes of Health Grants DK48823 and ES06484, and NIEHS Center Grants ES01247 and ES03828.

References

Aballay A, Arenas GN and Mayorga LS 1996. Calcium- and zinc-binding proteins in intracellular transport. *Biocell* **20**, 339–42.

Aiken SP, Horn NM and Saunders NR 1992. Effects of amino acids on zinc transport in rat erythrocytes. *Journal of Physiology* **445**, 69–80.

Aisen P 1998. Transferrin, the transferrin receptor, and the uptake of iron by cells. *Metal Ions in Biological Systems* **35**, 585–631.

Askwith CC, de Silva D and Kaplan J 1996. Molecular biology of iron acquisition in *Saccharomyces cerevisiae. Molecular Microbiology* **20**, 27–34.

Atkinson PG, Blackwell JM and Barton CH 1997. Nramp1 locus encodes a 65 kDa interferon-gamma-inducible protein in murine macrophages. *Biochemical Journal* **325**, 779–86.

Ballatori N 1991a. Mechanisms of metal transport across liver cell plasma membranes. *Drug Metabolism Reviews* **23**, 83–132.

Ballatori N 1991b. Biliary excretion of metals, in *Biliary Excretion of Drugs and Other Chemicals*, (eds C-P Siegers and JB Watkins III), Fischer Verlag, Stuttgart/New York, Chapter 3.7, pp. 283–306.

Ballatori N 1994. Glutathione mercaptides as transport forms of metals. *Advances in Pharmacology* **27**, 271–98.

Ballatori N and Clarkson TW, 1982, Developmental changes in the biliary excretion of methylmercury and glutathione. *Science*, 216, 61–63.

Ballatori N and Clarkson TW 1983. Biliary transport of glutathione and methylmercury. *American Journal of Physiology* **244**, G435–G441.

Ballatori N and Dutczak WJ 1994. Identification and characterization of high and low-affinity transport systems for reduced glutathione in liver cell canalicular membranes. *Journal of Biological Chemistry* **269**, 19731–7.

Ballatori, N and Truong AT 1989. Relation between biliary glutathione excretion and bile acid-independent bile flow. *American Journal of Physiology* **256**, G22–G30.

Ballatori N and Truong AT 1992. Glutathione as a primary osmotic driving force in hepatic bile formation. *American Journal of Physiology* **263**, G617–G624.

Ballatori N and Truong AT 1995a. Multiple canalicular transport mechanisms for glutathione *S*-conjugates. Transport on both ATP- and voltage-dependent carriers. *Journal of Biological Chemistry* **270**, 3594–601.

Ballatori N and Truong AT 1995b. Mechanisms of hepatic methylmercury uptake. *Journal of Toxicology and Environmental Health* **46**, 343–353.

Ballatori N, Moseley RH and Boyer JL 1986a. Sodium gradient-dependent L-glutamate transport is localized to the canalicular domain of liver plasma membranes. *Journal of Biological Chemistry* **261**, 6216–21.

Ballatori N, Jacob R and Boyer JL 1986b. Intrabiliary glutathione hydrolysis – a source of glutamate in bile. *Journal of Biological Chemistry* **261**, 7860–5.

Ballatori N Jacob R Barrett C and Boyer JL 1988. Biliary catabolism of glutathione and differential reabsorption of its amino acid constituents. *American Journal of Physiology* **254**, G1–G7.

Ballatori N, Gatmaitan Z and Truong AT 1995. Impaired biliary excretion and whole body elimination of methylmercury in rats with congenital defect in biliary glutathione excretion. *Hepatology* **22**, 1469–73.

Ballatori N, Wang W and Lieberman MW 1998a. Accelerated methylmercury elimination in γ-glutamyl transpeptidase-deficient mice. *American Journal of Pathology* **152**, 1049–55.

Ballatori N, Lieberman MW and Wang W 1998b. *N*-Acetylcysteine as an antidote in methylmercury poisoning. *Environmental Health Perspectives* **106**, 267–71.

Barisani D, Berg CL, Wessling-Resnick M and Gollan JL 1995. Evidence for a low Km transporter for non-transferrin-bound iron in isolated rat hepatocytes. *American Journal of Physiology* **269**, G570–G576.

Bartles JR, Feracci HM, Stieger B and Hubbard AL 1987. Biogenesis of the rat hepatocyte plasma membrane *in vivo*: comparison of the pathways taken by apical and basolateral proteins using subcellular fractionation. *Journal of Cell Biology* **105**, 1241–51.

Bingham MJ, Ong TJ, Ingledew WJ and McArdle HJ 1996. ATP-dependent copper transporter in the Golgi apparatus of rat hepatocytes transports Cu(II) not Cu(I). *American Journal of Physiology* **271**, G741–G746.

Bissig M, Hagenbuch B, Stieger B, Koller T and Meier PJ 1994. Functional expression cloning of the canalicular sulfate transport system of rat hepatocytes. *Journal of Biological Chemistry* **269**, 3017–21.

Blackwell JM 1996. Structure and function of the natural-resistance-associated macrophage protein (Nramp1), a candidate protein for infectious and autoimmune disease susceptibility. *Molecular Medicine Today* **2**, 205–11.

Blitzer BL, Hostetler BR and Scott KA 1989. Hepatic adenosine triphosphate-dependent Ca^{2+} transport is mediated by distinct carriers on rat basolateral and canalicular membranes. *Journal of Clinical Investigation* **83**, 1319–25.

Blomhoff R, Nenseter MS, Green MH and Berg T 1989. A multicompartmental model of fluid-phase endocytosis in rabbit liver parenchymal cells. *Biochemical Journal* **262**, 605–10.

Borst P, Kool M. and Evers R, 1997. Do CMOAT (MRP2), other MRP homologues, and LRP play a role in MDR? *Seminars in Cancer Biology* **8**, 205–13.

Boyer JL 1986. Mechanisms of bile secretion and cholestasis, in *Physiology of Membrane Disorders*, (eds TE Andreoli, JF Hoffman, DD Fanestil and SG Schultz), Plenum Publishing, New York, pp. 609–36.

Broeks A, Gerrard B, Allikmets R, Dean M. and Plasterk RH 1996. Homologues of the human multidrug resistance genes MRP and MDR contribute to heavy metal resistance in the soil nematode *Caenorhabditis elegans*. *EMBO Journal* **15**, 6132–43.

Buchler M, Konig J, Brom M, Kartenbeck J, Spring H, Horie T and Keppler D 1996. cDNA cloning of the hepatocyte canalicular isoform of the multidrug resistance protein, cMrp, reveals a novel conjugate export pump deficient in hyperbilirubinemic mutant rats. *Journal of Biological Chemistry* **271**, 15091–8.

Bull PC and Cox DW 1994. Wilson disease and Menkes disease: new handles on heavy-metal transport. *Trends in Genetics* **10**, 246–52.

Bygrave FL, Karjalainen A and Altin JG 1989. Passive calcium influx by plasma membrane vesicles isolated from rat liver. *Cell Calcium* **10**, 235–40.

Callahan HL and Beverley SM 1991. Heavy metal resistance: a new role for P-glycoprotein in *Leishmania*. *Journal of Biological Chemistry* **266**, 18427–30.

Cellier M. Prive G. Belouchi A. Kwan T. Rodrigues V. Chia W and Gros P 1995. Nramp defines a family of membrane proteins. *Proceedings of the National Academy of Sciences of the United States of America* **92**, 10089–93.

Chen Z-S, Mutoh M, Sumizawa T, Furukawa T, Haraguchi M, Tani A and Akiyama S 1997. Reversal of heavy metal resistance in multidrug-resistant human KB carcinoma cells. *Biochemical and Biophysical Research Communications* **236**, 586–90.

Chong SS, Kristjansson K, Zoghbi HY and Hughes MR, 1993. Molecular cloning of the cDNA encoding a human renal sodium phosphate transport protein and its assignment to chromosome 6p21.3–p23. *Genomics* **18**, 355–9.

Clarkson TW 1972. The pharmacology of mercury compounds. *Annual Review of Pharmacology* **12**, 375–406.

Clarkson TW 1993. Molecular and ionic mimicry of toxic metals. *Annual Review of Pharmacology and Toxicology* **33**, 545–71.

Cole SP, Sparks KE, Fraser K, Loe DW, Grant CE, Wilson GM and Deeley RG 1994. Pharmacological characterization of multidrug resistant MRP-transfected human tumor cells. *Cancer Research* **54**, 5902–10.

Coleman R 1987. Biochemistry of bile secretion. *Biochemical Journal* **244**, 249–61.

Conklin DS, Culbertson MR and Kung C 1994. Interactions between gene products involved in divalent cation transport in *Saccharomyces cerevisiae. Molecular Genetics* **244**, 303–11.

Cousins RJ 1985. Absorption, transport, and hepatic metabolism of copper and zinc: Special reference to metallothionein and ceruloplasmin. *Physiological Reviews* **65**, 238–309.

Crofts, JN and Barritt GJ 1990. The liver cell plasma membrane Ca^{++} inflow systems exhibit a broad specificity for divalent metal ions. *Biochemical Journal* **269**, 579–87.

Dancis A, Haile D, Yuan DS and Klausner RD 1994. The *Saccharomyces cerevisiae* copper transport protein Ctr1p – biochemical characterization, regulation by copper, and physiologic role in copper uptake. *Journal of Biological Chemistry* **269**, 25660–7.

Davis W, Chowrimootoo GF and Seymour CA 1996. Defective biliary copper excretion in Wilson's disease: the role of caeruloplasmin. *European Journal of Clinical Investigation* **26**, 893–901.

Dawson DC and Ballatori N 1995. Membrane transporters as sites of action and routes of entry for toxic metals. *Handbook Experimental Pharmacology* **115**, 53–76.

Deeley RG and Cole SPC 1997. Function, evolution and structure of multidrug resistance protein (MRP). *Seminars in Cancer Biology* **8**, 193–204.

Dey S, Ouellette M, Lightbody J, Papadopoulou B and Rosen BP 1996. An ATP-dependent As(III)-glutathione transport system in membrane vesicles of *Leishmania tarentolae. Proceedings of the National Academy of Sciences of the United States of America* **93**, 2192–7.

Dierick HA, Adam AN, Escara-Wilke JF and Glover TW 1997. Immunocytochemical localization of the Menkes copper transport protein (ATP7A) to the trans-Golgi network. *Human Molecular Genetics* **6**, 409–16.

Dijkstra M, In 't Veld G, van den Berg GJ, Muller M, Kuipers F and Vonk RJ 1995. Adenosine triphosphate-dependent copper transport in isolated rat liver plasma membranes. *Journal of Clinical Investigation* **95**, 412–16.

Dijkstra M, Havinga R, Vonk RJ and Kuipers F 1996. Bile secretion of cadmium, silver, zinc and copper in the rat. Involvement of various tranport systems. *Life Sciences* **59**, 1237–46.

Dijkstra M, Kuipers F, van den Berg GJ, Havinga R and Vonk RJ 1997. Differences in hepatic processing of dietary and intravenously administered copper in rats. *Hepatology* **26**, 962–6.

Dix D, Bridgham J, Broderius M and Eide D 1997. Characterization of the FET4 protein of yeast. Evidence for a direct role in the transport of iron. *Journal of Biological Chemistry* **272**, 11770–7.

Dutczak WJ and Ballatori N 1992. γ-Glutamyltransferase-dependent biliary-hepatic recycling of methyl mercury in the guinea pig. *Journal of Pharmacology and Experimental Therapeutics* **262**, 619–23.

Dutczak WJ and Ballatori N 1994. Transport of the glutathione-methylmercury complex across liver canalicular membranes on reduced glutathione carriers. *Journal of Biological Chemistry* **269**, 9746–51.

Dutczak WJ, Clarkson TW and Ballatori N 1991. Biliary-hepatic recycling of a xenobiotic : gallbladder absorption of methylmercury. *American Journal of Physiology* **260**, G873–G880.

Eide D 1997. Molecular biology of iron and zinc uptake in eucaryotes. *Current Opinion in Cell Biology* **9**, 573–7.

Fei YJ, Kanai Y, Nussberger S, Ganapathy V, Leibach FH, Romero MF, Singh SK, Boron WF and Hediger MA 1994. Expression cloning of a mammalian proton-coupled oligopeptide transporter. *Nature* **368**, 563–6.

Fleming MD, Trenor CC 3rd, Su MA, Foernzler D, Beier DR, Dietrich WF and Andrews NC 1997. Microcytic anaemia mice have a mutation in Nramp2, a candidate iron transporter gene. *Nature Genetics* **16**, 383–6.

Ghishan FK, Rebeiz R, Honda T and Nakagawa N 1993. Characterization and expression of a novel Na(+)-inorganic phosphate transporter at the liver plasma membrane of the rat. *Gastroenterology* **105**, 519–26.

Graf J, Gautam A and Boyer JL 1984. Isolated rat hepatocyte couplets: a primary secretory unit for electrophysiologic studies of bile secretory function. *Proceedings of the National Academy of Sciences of the United States of America* **81**, 6516–20.

Graf J, Henderson RM, Krumpholz B and Boyer JL 1987. Cell membrane and transepithelial voltages and resistances in isolated rat hepatocyte couplets. *Journal of Membrane Biology* **95**, 241–54.

Gross JB Jr, Myers BM, Kost LJ, Kuntz SM and LaRusso NF 1989. Biliary copper excretion by hepatocyte lysosomes in the rat. Major excretory pathway in experimental copper overload. *Journal of Clinical Investigation* **83**, 30–9.

Grundemann D, Gorboulev V, Gambaryan S, Veyhl M and Koepsell H 1994. Drug excretion mediated by a new prototype of polyspecific transporter. *Nature* **372**, 549–52.

Gunshin H, Mackenzie B, Berger UV, Gunshin Y, Romero MF, Boron WF, Nussberger S, Gollan JL and Hediger MA 1997. Cloning and characterization of a mammalian proton-coupled metal-ion transporter. *Nature* **388**, 482–8.

Gurgueira SA and Meneghini R 1996. An ATP-dependent iron transport system in isolated rat liver nuclei. *Journal of Biological Chemistry* **271**, 13616–20.

Gutierrez JA, Yu J, Rivera S and Wessling-Resnick M 1997. Functional expression cloning and characterization of SFT, a stimulator of Fe transport. *Journal of Cell Biology* **139**, 895–905.

Hagenbuch B, Stieger B, Foguet M, Lubbert H and Meier PJ 1991. Functional expression cloning and characterization of the hepatocyte Na^+/bile acid cotransport system. *Proceedings of the National Academy of Sciences of the United States of America* **88**, 10629–33.

Harris ED 1993. The transport of copper. *Progress in Clinical and Biological Research* **380** 163–79.

Hinchman CA and Ballatori N 1990. Glutathione-degrading capacities of liver and kidney in different species. *Biochemical Pharmacology* **40**, 1131–5.

Hinchman CA and Ballatori N 1994. Glutathione conjugation and conversion to mercapturic acids can occur as an intrahepatic process. *Journal of Toxicology and Environmental Health* **41**, 387–409.

Hinchman C, Matsumoto H, Simmons TW and Ballatori N 1991. Intrahepatic conversion of a glutathione conjugate to its mercapturic acid. *Journal of Biological Chemistry* **266**, 22179–85.

Hinkle PM, Kinsella PA and Osterhoudt KC 1987 Cadmium uptake and toxicity via voltage-sensitive calcium channels. *Journal of Biological Chemistry* **262**, 16333–7.

Horn NM and Thomas AL 1996. Interactions between the histidine stimulation of cadmium and zinc influx into human erythrocytes. *Journal of Physiology* **496**, 711–18.

Horn NM, Thomas AL and Tompkins JD 1995. The effect of histidine and cysteine on zinc influx into rat and human erythrocytes. *Journal of Physiology* **489**, 73–80.

Houwen R, Dijkstra M, Kuipers F, Smit EP, Havinga R and Vonk RJ 1990. Two pathways for biliary copper excretion in the rat. The role of glutathione. *Biochemical Pharmacology* **39**, 1039–44.

Huang LP, and Gitschier J 1997. A novel gene involved in zinc transport is deficient in the lethal milk mouse. *Nature Genetics* **17** 292–7.

Hughes BP, and Barritt GJ 1989. Inhibition of the liver cell receptor-activated Ca^{2+} inflow system by metal ion inhibitors of voltage-operated Ca^{2+} channels but not by other inhibitors of Ca^{2+} inflow. *Biochimica et Biophysica Acta* **1013**, 197–205.

Ishikawa T, Bao JJ, Yamane Y, Akimaru K, Frindrich K, Wright CD and Kuo MT 1996. Coordinated induction of MRP/GS-X pump and gamma-glutamylcysteine synthetase by heavy metals in human leukemia cells. *Journal of Biological Chemistry* **271**, 14981–8.

Ishikawa T, Li ZS, Lu YP and Rea PA 1997. The GS-X pump in plant, yeast, and animal cells: structure, function, and gene expression. *Bioscience Reports* **17**, 189–207.

Ito K, Suzuki H, Hirohashi T, Kume K, Shimizu T and Sugiyama Y. 1997. Molecular cloning of canalicular multispecific organic anion transporter defective in EHBR. *American Journal of Physiology*, **272**, G16–G22.

Jacquemin E, Hagenbuch B, Stieger B, Wolkoff AW and Meier PJ 1994. Expression cloning of a rat liver Na(+)-independent organic anion transporter. *Proceedings of the National Academy of Sciences of the United States of America* **91**, 133–7.

Jordan I and Kaplan J 1994. The mammalian transferrin-independent iron transport system may involve a surface ferrireductase activity. *Biochemical Journal* **302**, 875–9.

Joshi JG, Sczekan SR and Fleming JT 1989. Ferritin – a general metal detoxicant. *Biological Trace Element Research* **21**, 105–10.

Kaplan J and O'Halloran TV 1996. Iron metabolism in eucaryotes: Mars and Venus at it again. *Science* **271**, 1510–12.

Kavanaugh MP and Kabat D 1996. Identification and characterization of a widely expressed phosphate transporter/retrovirus receptor family. *Kidney International* **49**, 959–63.

Kawanishi T, Blank LM, Harootunian AT, Smith MT and Tsien RY 1989. Ca^{2+} oscillations induced by hormonal stimulation of individual fura-2-loaded hepatocytes. *Journal of Biological Chemistry* **264**, 12859–66.

Keppler D, Leier I and Jedlitschky G 1997. Transport of glutathione conjugates and glucuronides by the multidrug resistance proteins MRP1 and MRP2. *Biological Chemistry* **378**, 787–91.

Kerper LE, Ballatori N and Clarkson TW 1992. Methylmercury transport across the blood–brain barrier by an amino acid carrier. *American Journal of Physiology* **262**, R761–R765.

Klaassen CD 1976. Biliary excretion of metals. *Drug Metabolism Review* **5**, 165–96.

Klaassen CD and Watkins JB III 1984. Mechanisms of bile formation, hepatic uptake, and biliary excretion. *Pharmacological Reviews* **36**, 1–67.

Koch KA, Pena MM and Thiele DJ 1997. Copper-binding motifs in catalysis, transport, detoxification and signaling. *Chemistry and Biology* **4**, 549–60.

Kool M, Dehaas M, Scheffer GL, Scheper RJ, Vaneijk MJT, Juijn JA, Baas F and Borst P 1997. Analysis of expression of cMOAT (MRP2), MRP 3, MRP 4, and MRP 5, homologues of the multidrug resistance-associated protein gene (MRP 1), in human cancer cell lines. *Cancer Research* **57**, 3537–47.

Kullak-Ublick GA, Hagenbuch B, Stieger B, Wolkoff AW and Meier PJ 1994. Functional characterization of the basolateral rat liver organic anion transporting polypeptide. *Hepatology* **20**, 411–16.

Kuo YM, Gitschier J and Packman S 1997. Developmental expression of the mouse mottled and toxic milk genes suggests distinct functions for the Menkes and Wilson disease copper transporters. *Human Molecular Genetics* **6**, 1043–9.

Lee TK, Li L and Ballatori N 1997. Hepatic glutathione and glutathione S-conjugate transport mechanisms. *Yale Journal of Biology and Medicine* **70**, 287–300.

LeSage GD, Kost LJ, Barham SS and LaRusso NF 1986. Biliary excretion of iron from hepatocyte lysosomes in the rat. A major excretory pathway in experimental iron overload. *Journal of Clinical Investigation* 77, 90–7.

Li CY, Watkins JA and Glass J 1994. The H(+)-ATPase from reticulocyte endosomes reconstituted into liposomes acts as an iron transporter. *Journal of Biological Chemistry* **269**, 10242–6.

Li L, Lee TK and Ballatori N, 1997. Functional re-evaluation of the putative glutathione transporters, RcGshT and RsGshT. *Yale Journal of Biology and Medicine* **70**, 301–10.

Li L, Lee TK, Meier PJ and Ballatori N 1998. Identification of glutathione as a driving force and leukotriene C_4 as a substrate for oatp1, the hepatic sinusoidal organic solute transporter. *Journal of Biological Chemistry* **273**, 16184–91.

Li ZS, Szczypka M, Lu YP, Thiele DJ and Rea PA 1996. The yeast cadmium factor protein (YCF1) is a vacuolar glutathione S-conjugate pump. *Journal of Biological Chemistry* **271**, 6509–17.

Li ZS, Lu YP, Zhen RG, Szczypka M, Thiele DJ and Rea PA 1997. A new pathway for vacuolar cadmium sequestration in *Saccharomyces cerevisiae*: YCF1-catalyzed transport of bis-(glutathionato)cadmium. *Proceedings of the National Academy of Sciences of the United States of America* **94**, 42–7.

Mack U, Powell LW and Halliday JW 1983., Detection and isolation of a hepatic membrane receptor for ferritin. *Journal of Biological Chemistry* **258**, 4672–5.

Malandro MS and Kilberg MS 1996. Molecular biology of mammalian amino acid transporters. *Annual Review of Biochemistry* **65**, 305–36.

Markovich D, Forgo J, Stange G, Biber J and Murer H 1993. Expression cloning of rat renal $Na^+/SO_4(^{2-})$ cotransport. *Proceedings of the National Academy of Sciences of the United States of America* **90**, 8073–7.

Meier PJ 1988. Transport polarity of hepatocytes. *Seminars Liver Diseases* **8**, 293–307.

Meier PJ, Sztul ES, Reuben A and Boyer JL 1984. Structural and functional polarity of canalicular and basolateral plasma membrane vesicles isolated in high yield from rat liver. *Journal of Cell Biology* **98**, 991–1000.

Miyamoto K, Tatsumi S, Sonoda T, Yamamoto H, Minami H, Taketani Y and Takeda E 1995. Cloning and functional expression of a Na^+-dependent phosphate co-transporter from human kidney: cDNA cloning and functional expression. *Biochemical Journal* **305**, 81–5.

Mokrzan EM, Kerper LE, Ballatori N and Clarkson TW 1995. Methylmercury–thiol uptake into cultured brain capillary endothelial cells on amino acid system L. *Journal of Pharmacology and Experimental Therapeutics* **272**, 1277–84.

Morgan EH and Baker E 1986. Iron uptake and metabolism by hepatocytes. *Federation Proceedings* **45**, 2810–16.

Morgan EH and Baker E 1988. Role of transferrin receptors and endocytosis in iron uptake by hepatic and erythroid cells. *Annals New York Academy of Sciences* **526**, 65–82.

Morley CG and Bezkorovainy A 1985. Cellular iron uptake from transferrin: is endocytosis the only mechanism? *International Journal of Biochemistry* **17**, 553–64.

Moseley RH, Ballatori N and Murphy, SM 1988. Na^+-Glycine cotransport in canalicular liver plasma membrane vesicles. *American Journal of Physiology* **255**, G253–G259.

Mukhopadhyay CK, Attieh ZK and Fox PL 1998. Role of ceruloplasmin in cellular iron uptake. *Science* **279**, 714–17.

Nagano K, Nakamura K, Urakami K-I, Umeyama K, Uchiyama H, Koiwai K, Hattori S, Yamamoto T, Matsuda I and Endo F 1998. Intracellular distribution of the Wilson's disease gene product (ATPase7B) after *in vitro* and *in vivo* exogenous expression in hepatocytes from the LEC rat, an animal model of Wilson's disease. *Hepatology* **27**, 799–807.

Noé B, Hagenbuch B, Stieger B and Meier PJ 1997. Isolation of a multispecific organic anion and cardiac glycoside transporter from rat brain. *Proceedings of the National Academy of Sciences of the United States of America* **94**, 10346–50.

Norbis F, Boll M, Stange G, Markovich D, Verrey F, Biber J and Murer H 1997. Identification of a cDNA/protein leading to an increased Pi-uptake in *Xenopus laevis* oocytes. *Journal of Membrane Biology* **156**, 19–24.

Oka JA and Weigel PH 1983. Recycling of the asialoglycoprotein receptor in isolated rat hepatocytes. Dissociation of internalized ligand from receptor occurs in two kinetically and thermally distinguishable compartments. *Journal of Biological Chemistry* **258**, 10253–62.

Oka JA, Christensen MD and Weigel PH, 1989. Hyperosmolarity inhibits galactosyl receptor-mediated but not fluid phase endocytosis in isolated rat hepatocytes. *Journal of Biological Chemistry* **264**, 12016–24.

Ortiz DF, Ruscitti T, McCue KF and Ow DW 1995. Transport of metal-binding peptides by HMT1, a fission yeast ABC-type vacuolar membrane protein. *Journal of Biological Chemistry* **270**, 4721–8.

Ortiz DF, St. Pierre MV, Abdulmessih A and Arias IM 1997. A yeast ATP-binding cassette-type protein mediating ATP-dependent bile acid transport. *Journal of Biological Chemistry* **272**, 15358–65.

Osterloh K and Aisen P 1989. Pathways in the binding and uptake of ferritin by hepatocytes. *Biochimica et Biophysica Acta* **1011**, 40–5.

Palmiter RD and Findley SD 1995. Cloning and functional characterization of a mammalian zinc transporter that confers resistance to zinc. *EMBO Journal* **14**, 639–49.

Palmiter RD, Cole TB and Findley SD 1996a. ZnT-2, a mammalian protein that confers resistance to zinc by facilitating vesicular sequestration. *EMBO Journal* **15**, 1784–91.

Palmiter RD, Cole TB, Quaife CJ and Findley SD 1996b. ZnT-3, a putative transporter of zinc into synaptic vesicles. *Proceedings of the National Academy of Sciences of the United States of America* **93**, 14934–9.

Paulsen IT and Saier MH Jr, 1997. A novel family of ubiquitous heavy metal ion transport proteins. *Journal of Membrane Biology* **156**, 99–103.

Paulusma C, Bosma PJ, Zaman GJR, Bakker CTM, Otter M, Scheffer GL, Scheper RJ, Borst P and Oude Elferink RPJ 1996. Congenital jaundice in rats with a mutation in a multidrug resistance-associated protein gene. *Science* **271**, 1126–8.

Paulusma CC and Oude Elferink RPJ 1997. The canalicular multispecific organic anion transporter and conjugated hyperbilirubinemia in rat and man. *Journal of Molecular Medicine* **75**, 420–8.

Pavoine C, Lotersztajn S, Mallat A and Pecker F 1987. The high affinity (Ca^{2+}-Mg^{2+})-ATPase in liver plasma membranes is a Ca^{2+} pump. Reconstitution of the purified enzyme into phospholipid vesicles. *Journal of Biological Chemistry* **262**, 5113–17.

Penninckx MJ and Elskens MT 1993. Metabolism and functions of glutathione in micro-organisms. *Advances in Microbial Physiology* **34**, 239–301.

Perego P and Howell SB 1997. Molecular mechanisms controlling sensitivity to toxic metal ions in yeast. *Toxicology and Applied Pharmacology* **147**, 312–18.

Pratt RD, Ferreira GC and Pedersen PL 1991. Mitochondrial phosphate transport. Import of the H+/Pi symporter and role of the presequence. *Journal of Biological Chemistry* **266**, 1276–80.

Ramm GA, Powell LW and Halliday JW 1994. Pathways of intracellular trafficking and release of ferritin by the liver *in vivo*: the effect of chloroquine and cytochalasin D. *Hepatology* **19**, 504–13.

Regoeczi E and Chindemi PA 1995. Translocation of different forms of transferrin from blood to bile in the rat. *Hepatology* **21**, 1063–9.

Regoeczi E, Chindemi PA and Debanne MT 1984. Partial resialylation of human asialotransferrin types 1 and 2 in the rat. *Canadian Journal of Biochemistry and Cell Biology* **62**, 853–8.

Renston RH, Jones AL, Christiansen WD and Hradek GT 1980. Evidence for a vesicular transport mechanism in hepatocytes for biliary secretion of immunoglobulin A. *Science* **208**, 1276–8.

Reuter H 1983. Calcium channel modulation by neurotransmitters, enzymes and drugs. *Nature* **301**, 569–74.

Rudolph JR, Regoeczi E, Chindemi PA and Debanne MT 1986. Preferential hepatic uptake of iron from rat asialotransferrin: possible engagement of two receptors. *American Journal of Physiology* **251**, G398–G404.

Sakisaka S, Ng OC and Boyer JL 1988. Tubulovesicular transcytotic pathway in isolated rat hepatocyte couplets in culture. Effect of colchicine and taurocholate. *Gastroenterology* **95**, 793–804.

Scharschmidt BF, Lake JR, Renner EL, Licko V and Van Dyke RW 1986. Fluid phase endocytosis by cultured rat hepatocytes and perfused rat liver: implications for plasma membrane turnover and vesicular trafficking of fluid phase markers. *Proceedings of the National Academy of Sciences of the United States of America* **83**, 9488–92.

Sharma V, Crankshaw CL and Piwnica-Worms D 1996. Effects of multidrug resistance (MDR1)P-glycoprotein expression levels and coordination metal on the cytotoxic potency of multidentate (N4O2) (ethylenediamine)bis[propyl(R-benzylimino)]metal(III) cations. *Journal of Medicinal Chemistry* **39**, 3483–90.

Sigel E 1990. Use of *Xenopus* oocytes for the functional expression of plasma membrane proteins. *Journal of Membrane Biology* **117**, 201–21.

Silver S. 1996. Bacterial resistances to toxic metal ions – a review. *Gene* **179**, 9–19.

Simmons TW, Hinchman CA and Ballatori N 1991. Polarity of hepatic glutathione and glutathione *S*-conjugate efflux, and intraorgan mercapturic acid formation in the skate. *Biochemical Pharmacology* **42**, 2221–8.

Simmons TW, Anders MW and Ballatori N 1992. Cysteine and *S*-(1,2-dichlorovinyl)-L-cysteine transport in rat liver canalicular membrane vesicles: Potential reabsorption mechanisms for biliary metabolites of glutathione and its *S*-conjugates. *Journal of Pharmacology and Experimental Therapeutics* **262**, 1182–8.

Simons TJ and Pocock G 1987. Lead enters bovine adrenal medullary cells through calcium channels. *Journal of Neurochemistry* **48**, 383–9.

Smit JJM, Schinkel AH, Oude Elferink RPJ, Groen AK, Wagenaar E, van Deemter L, Mol CAAM, Ottenhoff R, van der Lugt NMT, van Roon MA, van der Valk MA, Berns AJM and Borst P 1993. Homogeneous disruption of the murine mdr2 P-glycoprotein gene leads to a complete absence of phospholipid from bile and to liver disease. *Cell* **75**, 451–62.

Solioz M and Vulpe C 1996. CPx-type ATPases: a class of P-type ATPases that pump heavy metals. *Trends in Biochemical Sciences* **21**, 237–41.

Solioz M, Odermatt A and Krapf R 1994. Copper pumping ATPases: common concepts in bacteria and man. *FEBS Letters* **346**, 44–7.

Sorensen EM 1988. Modulatory effect of calcium on the influx and binding of cadmium in primary cultures of neonatal rat hepatocytes. *Toxicology Letters* **41**, 39–48.

Soreq H and Seidman S 1992. *Xenopus* oocyte microinjection: from gene to protein. *Methods Enzymology* **207**, 225–65.

Sperber I 1959. Secretion of organic anions in the formation of urine and bile. *Pharmacological Reviews* **11**, 109–34.

Sternlieb I, van den Hamer CJA, Morell AG, Alpert S, Gregoriadis G and Scheinberg IH 1973. Lysosomal defect of hepatic copper excretion in Wilson's disease (hepatolenticular degeneration). *Gastroenterology* **64**, 99–105.

Stockert, RJ, Haimes HB, Morell AG, Novikoff PM, Novikoff AB, Quintana N and Sternlieb I 1980. Endocytosis of asialoglycoprotein-enzyme conjugates by hepatocytes. *Laboratory Investigation* **43**, 556–63.

Sugawara N, Lai YR, Arizono K, Kitajima T and Inoue H 1997. Lack of biliary excretion of Cd linked to an inherent defect of the canalicular isoform of multidrug resistance protein (cMrp) does not abnormally stimulate accumulation of Cd in the Eisai hyperbilirubinemic (EHB) rat liver. *Archives of Toxicology* **71**, 336–9.

Summer KH and Eisenburg J 1985. Low content of hepatic reduced glutathione in patients with Wilson's disease. *Biochemistry and Medicine* **34**, 107–11.

Supek F, Supekova L, Nelson H and Nelson N 1996. A yeast manganese transporter related to the macrophage protein involved in conferring resistance to mycobacteria. *Proceedings of the National Academy of Sciences of the United States of America* **93**, 5105–10.

Supek F, Supekova L, Nelson H and Nelson N 1997. Function of metal-ion homeostasis in the cell division cycle, mitochondrial protein processing, sensitivity to mycobacterial infection and brain function. *Journal of Experimental Biology* **200**, 321–30.

Suzuki T and Kawakita M 1993. Uncoupling of ATP splitting from Ca(2+)-transport in Ca(2+)-transporting ATPase of the sarcoplasmic reticulum as a result of modification by *N*-(3-pyrene)maleimide: activation of a channel with a specificity for alkaline earth metal ions. *Journal of Biochemistry* **114**, 203–9.

Szczypka MS, Wemmie JA, Moye-Rowley WS and Thiele DJ 1994. A yeast metal resistance protein similar to human cystic fibrosis transmembrane conductance regulator (CFTR) and multidrug resistance-associated protein. *Journal of Biological Chemistry* **269**, 22853–7.

Tang DT, Zhang S and Zhong X 1987. Studies on the influence of calcium on cadmium absorbed by rat liver and the resistivity against the toxicity of cadmium in rat. *Bulletin of Environmental Contamination and Toxicology* **39**, 453–9.

Tavassoli M 1988. The role of liver endothelium in the transfer of iron from transferrin to the hepatocyte. *Annals of the New York Academy of Sciences* **526**, 83–92.

Thorstensen K and Romslo I 1988. Uptake of iron from transferrin by isolated rat hepatocytes. A redox-mediated plasma membrane process? *Journal of Biological Chemistry* **263**, 8844–50.

Thorstensen K, Trinder D, Zak O and Aisen P 1995. Uptake of iron from N-terminal half-transferrin by isolated rat hepatocytes. Evidence of transferrin-receptor-independent iron uptake. *European Journal of Biochemistry* **232**, 129–33.

Tibaduiza EC and Bobilya DJ 1996. Zinc transport across an endothelium includes vesicular cotransport with albumin. *Journal of Cellular Physiology* **167**, 539–47.

Tommasini R, Evers R, Vogt E, Mornet C, Zaman GJ, Schinkel AH, Borst P and Martinoia E 1996. The human multidrug resistance-associated protein functionally complements the yeast cadmium resistance factor 1. *Proceedings of the National Academy of Sciences of the United States of America* **93**, 6743–8.

Torrubia JOA and Garay R 1989. Evidence for a major route for zinc uptake in human red blood cells: $[Zn(HCO_3)_2 Cl]^-$ influx through the $[Cl^-/HCO_3^-]$ anion exchanger. *Journal of Cellular Physiology* **138**, 316–22.

Tsai K-J, Hsu C-M and Rosen BP 1997. Efflux mechanisms of resistance to cadmium, arsenic and antimony in procaryotes and eucaryotes. *Zoological Studies* **36**, 1–16.

van Veen HW, Abee T, Kortstee GJ, Konings WN and Zehnder AJ 1994. Translocation of metal phosphate via the phosphate inorganic transport system of *Escherichia coli*. *Biochemistry* **33**, 1766–70.

Valentine JS and Gralla EB 1997. Delivering copper inside yeast and human cells. *Science* **278**, 817–18.

Vulpe CD and Packman S 1995. Cellular copper transport. *Annual Review of Nutrition* **15**, 293–322.

Vulpe C, Levinson B, Whitney S, Packman S and Gitschier J 1993. Isolation of a candidate gene for Menkes disease and evidence that it encodes a copper-transporting ATPase. *Nature Genetics* **3**, 7–13.

Wemmie JA, Szczypka MS, Thiele DJ and Moye-Rowley WS 1994. Cadmium tolerance mediated by the yeast AP-1 protein requires the presence of an ATP-binding cassette transporter-encoding gene, YCF1. *Journal of Biological Chemistry* **269**, 32592–7.

Wetterhahn-Jennette K 1981. The role of metals in carcinogenesis: Biochemistry and metabolism. *Environmental Health Perspectives* **40**, 233–52.

Wijnholds J, Evers R, Vanleusden MR, Mol CAAM, Zaman GJR, Mayer U, Beijnen JH, Vandervalk M, Krimpenfort P and Borst P 1997. Increased sensitivity to anticancer drugs and decreased inflammatory response in mice lacking the multidrug resistance-associated protein. *Nature Medicine* **3**, 1275–9.

Wysocki R, Bobrowicz P and Ulaszewski S, 1997. The *Saccharomyces cerevisiae* ACR3 gene encodes a putative membrane protein involved in arsenite transport. *Journal of Biological Chemistry* **272**, 30061–6.

Yamaguchi Y, Heiny ME, Suzuki M and Gitlin JD 1996. Biochemical characterization and intracellular localization of the Menkes disease protein. *Proceedings of the National Academy of Sciences of the United States of America* **93**, 14030–5.

Yang XL, Miura N, Kawarada Y, Terada K, Petrukhin K, Gilliam T and Sugiyama T 1997. Two forms of Wilson disease protein produced by alternative splicing are localized in distinct cellular compartments. *Biochemical Journal* **326**, 897–902.

Young SP, Bomford AB and Williams R 1983. Dual pathways for the uptake of rat asialo-transferrin by rat hepatocytes. *Journal of Biological Chemistry* **258**, 4972–6.

Yuan DS, Stearman R, Dancis A, Dunn T, Beeler T and Klausner RD 1995. The Menkes/Wilson disease gene homologue in yeast provides copper to a ceruloplasmin-like oxidase required for iron uptake. *Proceedings of the National Academy of Sciences of the United States of America* **92**, 2632–6.

Zaman GJR, Lankelma J, van Tellingen O, Beijnen J, Bekker H, Paulusma C, Oude Elferink RPJ, Baas F and Borst P 1995. Role of glutathione in the export of compounds from cells by the multidrug-resistance-associated protein. *Proceedings of the National Academy of Sciences of the United States of America* **92**, 7690–4.

Zhao H and Eide D 1996. The yeast ZRT1 gene encodes the zinc transporter protein of a high-affinity uptake system induced by zinc limitation. *Proceedings of the National Academy of Sciences of the United States of America* **93**, 2454–8.

Zhou B and Gitschier J 1997. hCTR1: a human gene for copper uptake identified by complementation in yeast. *Proceedings of the National Academy of Sciences of the United States of America* **94**, 7481–6.

13

ROLE OF GLUTATHIONE IN THE METABOLISM, TRANSPORT AND TOXICITY OF METALS

Lawrence H. Lash

13.1 Introduction

Low-molecular-weight thiols and protein sulfhydryl groups play major roles in the metabolism and transport of several metals because of the high affinity of the metals for these functional groups. This chapter will focus on the metabolism, transport and accumulation of three heavy metals, mercury (both inorganic mercury (Hg^{2+}) and methylmercury (MeHg)), cadmium (Cd) and arsenic (As), in selected target tissues. Specifically, the roles of glutathione (GSH) and thiols derived from GSH (i.e. L-cysteinylglycine (CySH–Gly) and cysteine (CySH)) in the metabolism, transport and accumulation of these metals will be highlighted. These metals are those that most commonly interact with GSH. Other thiols, such as serum albumin or metallothionein, or synthetic compounds such as 2,3-dimercapto-1-propanesulfonic acid (DMPS) will not be discussed as these are subjects of other chapters.

The primary target organs for these metals include the kidneys, liver, lung and nervous system. Target organ specificity is determined by several factors, including the route of administration or exposure to the metal, the selective tissue localization of membrane carrier proteins and various enzymes, and the availability of cellular molecules in the target organ with which the metal can interact. The manner in which the body handles GSH and GSH *S*-conjugates is illustrative of target-organ specificity and appears to be of relevance to heavy metals because of the importance of GSH as a physiological ligand for Hg^{2+}, MeHg, Cd and As. As summarized by Ballatori (1994), GSH is important for the disposition and toxicity of metals by functioning in four ways:

1 it functions in the mobilization and delivery of metals between ligands;
2 it functions to transport metals across cell membranes;
3 it serves as a source of CySH; and
4 it serves as a cofactor for redox reactions.

Hence, GSH would appear to play a key role in multiple stages in the disposition of metals.

The process of interorgan GSH and GSH S-conjugate metabolism has been reviewed previously (Lash *et al.*, 1988) and is summarized schematically in Fig. 13.1. The term 'interorgan metabolism' indicates that enzymes and membrane transport proteins in various tissues are required for metabolism and that one or more of the metabolites circulates through the plasma and is extracted by various tissues for specific metabolic steps. Although different carrier proteins may be involved in certain transport steps with metals, the handling of GSH and GSH conjugates of organic compounds exhibits essentially the same pattern as that for metals.

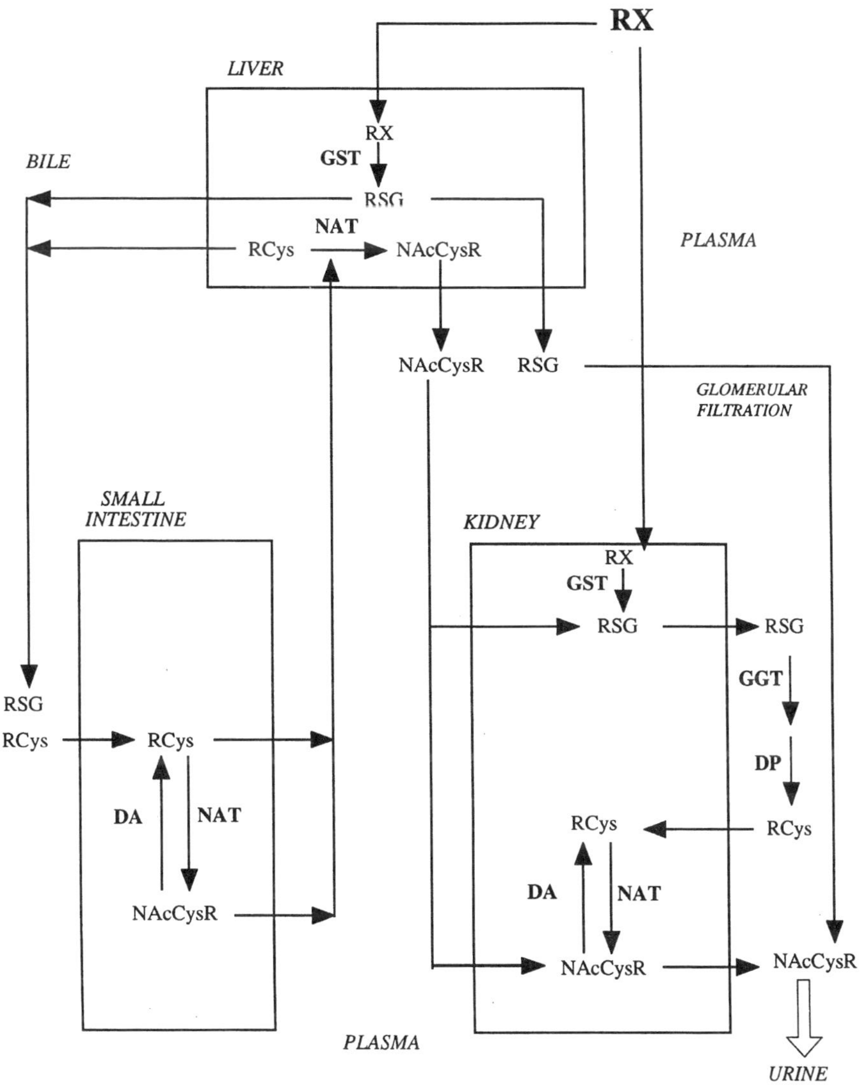

Figure 13.1 Interorgan processing of GSH conjugates. This scheme depicts the metabolism and patterns of translocation between tissues (principally the liver, small intestine and kidneys) of GSH conjugates and associated metabolites. Abbreviations: DA, deacetylase; DP, dipeptidase; GGT, γ-glutamyltransferase; GST, GSH *S*-transferase; NAT, cysteine conjugate *N*-acetyltransferase; NAcCysR, *N*-acetylcysteine conjugate; RX, electrophile with X as a good leaving group; RSG, glutathione conjugate; RCys, cysteine conjugate.

The liver is the primary site of GSH conjugation, although GSH conjugation can occur in most tissues, including the kidneys (Lash *et al.*, 1995). Once formed in the liver, efflux of the GSH conjugate into either the bile or plasma occurs readily. Determinants of the compartment into which efflux occurs is uncertain, but the substrate specificity of carrier proteins on the canalicular and sinusoidal plasma membranes and the molecular properties of the conjugate (i.e. size, charge, lipophilicity) must be important factors. Biliary GSH conjugate is either degraded by γ-glutamyltransferase (GGT) and dipeptidase activities to the corresponding cysteine conjugate, which is then delivered to the small intestines, or is delivered directly to the small intestines. The small-intestinal epithelium degrades the GSH conjugate to the cysteine conjugate and this form, or the corresponding mercapturate, is then translocated back to the liver via enterohepatic circulation. The kidneys, therefore, are primarily exposed to two forms, the GSH conjugate and the mercapturate.

There are two mechanisms by which the kidneys extract these compounds from plasma. The luminal process involves either glomerular filtration and excretion of the mercapturate or degradation of the GSH conjugate by brush-border membrane enzymes to the cysteine conjugate, uptake into the cell and acetylation. The basolateral process involves uptake of the mercapturate and secretion into the lumen or uptake of the GSH conjugate by a sodium-dependent carrier (Lash and Jones, 1984, 1985a), secretion into the lumen, and further processing as described above. Hence, it is the selective localization of transporters on the brush-border and basolateral plasma membranes of renal proximal tubules that makes the kidneys a site of accumulation of GSH conjugates. Additionally, in the case of nephrotoxic GSH conjugates, the presence of these membrane transporters contributes to the target-organ specificity of these chemicals.

Because of the high affinity of many metals for thiol groups and the interorgan processing of GSH and GSH conjugates, the thiol-disulfide composition of plasma should be considered to play an important role in determining the precise chemical form of the metal to which target tissues are exposed. Albumin is the major protein sulfhydryl component of plasma. In the case of Hg^{2+}, most of it is bound to serum albumin (Friedman, 1957; Cember *et al.*, 1968; Foulkes, 1974; Lau and Sarkar, 1979). However, plasma also contains significant amounts of several forms of low-molecular-weight thiols, including GSH, CySH–Gly and CySH (Jones *et al.*, 1997; Lash and Jones, 1985b). Exchange of thiol-ligands can occur and will modify the form of the metal presented to the target tissue.

13.2 Chemistry and metabolism of metals

13.2.1 Inorganic mercury and methylmercury

Three major chemical forms of mercury must be distinguished: mercury vapor (elemental mercury, Hg^0), salts of inorganic mercury (Hg^{2+}) and organic mercurials. Although there are several types of organic mercurials, the best studied is MeHg. Mercury readily forms covalent bonds with sulfur, and it is this property that accounts for most of the biological properties of mercurials. When the sulfur is in the form of a sulfhydryl group, Hg^{2+} or the organic mercurial cation replaces the hydrogen atom to form a mercaptide. The affinity of mercurials for thiols also provides the basis for treatment of mercury poisoning with agents such as dimercaprol and penicillamine.

Hg^{2+} forms 1 : 2 complexes with thiols, including GSH, cysteine and protein sulfhydryl groups. Although the thermodynamic stability of these complexes is very high, the bonds formed between Hg^{2+} and the thiol groups are labile (Rabenstein, 1989; Oram *et al.*, 1996). This lability allows for exchange of thiol ligands, which, as described below, plays a critical role in the distribution and tissue extraction of mercury-containing compounds.

Clarkson (1993) elaborated the concept of molecular and ionic mimicry of metals. The basis of metal toxicity is thus seen as complexes of metals with various compounds functioning as structural analogs of endogenous compounds. As shown in Fig. 13.2, complexes of MeHg with a molecule of cysteine, or that of Hg^{2+} with two molecules of cysteine, can function as structural analogs of methionine and cystine, respectively. Similarly, complexes of Hg^{2+} with two molecules of GSH can function as a structural analog of glutathione disulfide (GSSG). Besides these analogs functioning as replacements for the respective endogenous compounds, the disposition of the mercury–thiol complexes is determined by transport processes that are similar to those of the respective endogenous compounds.

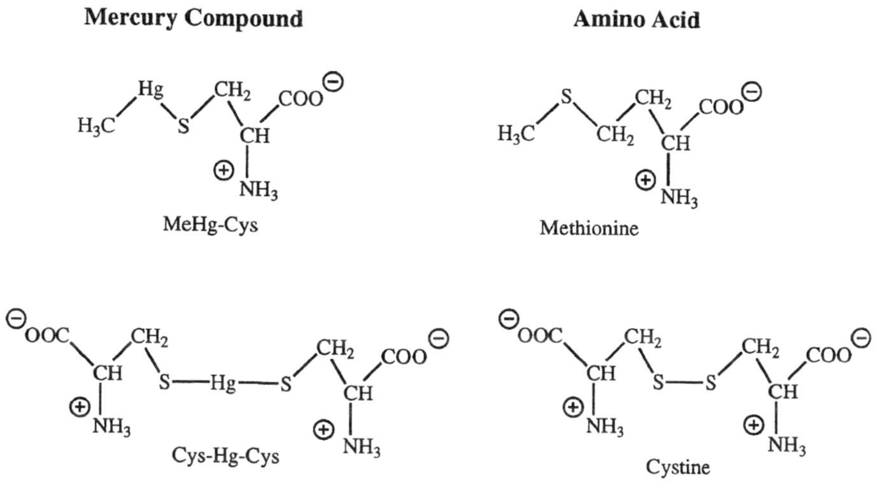

Figure 13.2 Chemical structures of two mercury–thiol complexes and their amino-acid analogs. Shown are structures of the cysteine (Cys) complex with methylmercury and with inorganic mercury (Hg) and their analogous amino acids, methionine and cystine, respectively.

13.2.2 Cadmium

Cadmium exists in only one valency state, 2+, and does not generally form stable alkyl compounds or other organometallic compounds of known toxicological significance. Hence, exposure to cadmium is generally as salts. In spite of this, thiols play a major role in the disposition of cadmium. Cadmium associates with metallothionein (MT) (Viljoen and Tappel, 1988) and with the reduced form of serum albumin (Suzuki *et al.*, 1986). While there are several target organs for cadmium, it accumulates preferentially in the liver (Bose *et al.*, 1994).

13.2.3 Arsenic

Arsenic exists in three forms, elemental arsenic, the trivalent form and the pentavalent form. The trivalent and pentavalent forms can exist as inorganic salts and as organic compounds. Organic arsenicals contain arsenic linked to a carbon atom by a covalent bond. The pentavalent oxidation state is found in arsenates, such as lead arsenate ($PbHAsO_4$), which are salts of arsenic acid (H_3AsO_4). Arsenites (e.g. potassium arsenite, $KAsO_2$) and salts of arsenious acid contain trivalent arsenic. Arsine (AsH_3) is a gaseous hydride of trivalent arsenic.

The oxidation state and form of the As atom is critical to its biological effects, as the two oxidation states and the various forms act very differently. As^{5+} can substitute for phosphate in a form of molecular mimicry (Clarkson, 1993), and this effect is the basis for the potent inhibition of oxidative phosphorylation by arsenates. Pentavalent arsenic uncouples oxidative phosphorylation by replacing phosphate in ATP and forming an unstable arsenate ester that undergoes rapid hydrolysis. While the pentavalent arsenicals have very low affinities for thiol groups, the trivalent compounds are characterized by their high affinities for both low- and high-molecular-weight thiols. For example, binding of As^{3+} species to protein sulfhydryl groups has been shown *in vivo* to be a key mechanism of action in the inhibition of the pyruvate dehydrogenase complex, which presumably occurred by binding of As^{3+} to the lipoic acid residue of the complex (Stevenson *et al.*, 1978).

Selected interactions in biological systems and metabolism of arsenic are illustrated in Fig. 13.3. Besides interactions with cellular phosphate pools, As^{5+} can be reduced by a GSH-dependent mechanism to As^{3+} (Bertolero *et al.*, 1987; Scott *et al.*, 1993). Although the precise nature of the oxidation-reduction reaction is unclear, evidence for formation of a complex in solution with GSH to form triglutathionyl arsenite, $As(SG)_3$, was indicated by NMR spectroscopy (Scott *et al.*, 1993). The reaction was postulated to occur as follows:

$$3H^+ + AsO_3^{3-} + 3GSH \rightarrow As(SG)_3 + 3H_2O.$$

Incubations of arsenate salts (Na_3AsO_4) with GSH showed that GSH was oxidized to GSSG, and the following reaction was suggested:

$$3H^+ + AsO_4^{3-} + 5GSH \rightarrow As(SG)_3 + GSSG + 4H_2O.$$

The other major fate of arsenicals is that they undergo methylation reactions. In fact, arsenate, monomethylarsonic acid (MMA) and dimethylarsonic acid (DMA) are the major arsenic species found in urine (Rosner and Carter, 1987). Each of these species is in the pentavalent form. GSH is required for the methylation reactions, suggesting that As^{5+} is first reduced to As^{3+}, which forms a complex with GSH, and it is the GSH complex that is likely to be the direct substrate for the monomethylation (Scott *et al.*, 1993). A GSH complex with MMA is then thought to be the substrate for the dimethylation. Separate methyltransferases are believed to be involved in the two steps and these reactions can be considered a detoxification mechanism because the pentavalent organoarsenic species have much lower affinities for thiols (Thompson, 1993).

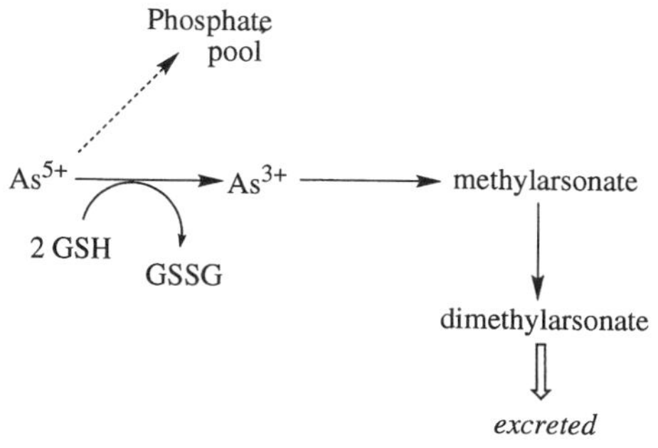

Figure 13.3 Interactions of arsenic with glutathione. Arsenic in the pentavalent form can either interact with the cellular phosphate pool, thereby uncoupling mitochondrial oxidative phosphorylation, or it is reduced by a GSH-dependent reaction to trivalent As^{3+}. As^{3+} is the species most closely associated with arsenic-induced cytotoxicity, but it can also undergo a series of methylation reactions.

13.3 Modulation of tissue accumulation and excretion of metals by GSH

13.3.1 Inorganic mercury and methylmercury

Although there is conversion of Hg^0 and organic mercurials to Hg^{2+}, there are differences in the disposition of each form of mercury. Ingested Hg^0 is not particularly toxic because of poor gastrointestinal absorption and the inability of this form to react with biological molecules. In contrast, inhaled mercury vapor is well absorbed by the pulmonary epithelium and is readily oxidized to Hg^{2+} by catalase in the erythrocytes (Magos *et al.*, 1978). Hence, the tissue distribution of inhaled mercury vapor resembles that after ingestion of mercuric salts, with one critical difference. Because mercury vapor crosses biological membranes more readily than Hg^{2+}, a significant fraction of the Hg^0 crosses the blood–brain barrier before oxidation to Hg^{2+}. CNS toxicity is, therefore, more prominent after exposure to mercury vapor than to divalent forms of mercury.

Gastrointestinal absorption of inorganic mercuric salts is only about 10%. Of the absorbed Hg^{2+}, tissue distribution after absorption is markedly nonuniform. Inorganic mercurials (Hg^{2+}) are primarily concentrated in the kidneys and they do not readily cross the blood–brain barrier. Excretion is in both urine and feces. In contrast, organic mercurials such as MeHg exhibit more complete absorption from the gastrointestinal tract (>90%) because they are more lipophilic and produce less corrosive damage to the intestinal mucosa than the inorganic mercury salts. Organic mercurials cross the blood–brain barrier, thus producing neurological effects, and tissue distribution is more uniform than that of inorganic mercury salts.

The central role of sulfhydryl groups, and in particular, GSH and GSH-derived amino acids, in the tissue distribution of MeHg and Hg^{2+} is illustrated in Fig. 13.4. Following absorption of either form of mercury, mercaptides with either albumin, GSH, CySH–Gly or CySH are formed. These thiol conjugates are translocated through the body, undergoing enterohepatic and renal–hepatic circulation and are ultimately presented to the kidneys. The precise pattern of interorgan translocation will depend on the route of exposure or administration. For example, the first-pass effect is critical with intravenously administered mercury, indicating that hepatic accumulation and biliary efflux of mercury will occur. Ingested mercury will initially undergo gastrointestinal absorption, followed by enterohepatic circulation. The kidneys then extract the conjugates by either glomerular filtration or by basolateral transport mechanisms from

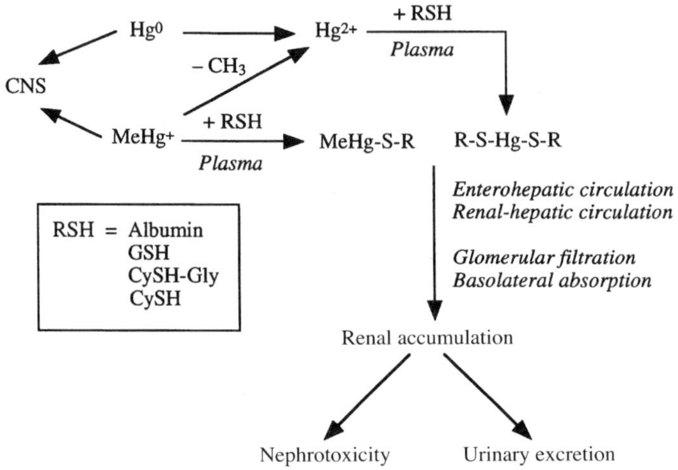

Figure 13.4 Role of sulfhydryl-containing compounds in the tissue distribution of the various forms of mercury. Elemental mercury (Hg^0) can either be accumulated in the central nervous system (CNS) or it can be oxidized to inorganic divalent mercury (Hg^{2+}). Hg^{2+} is bound to a sulfhydryl-containing ligand (R) in plasma and is translocated eventually to the kidneys. Similarly, methylmercury ($MeHg^+$) can accumulate in the CNS, it can be demethylated to form Hg^{2+}, or it can be transported to tissues such as the kidneys bound to a sulfhydryl-containing ligand. Abbreviations: GSH, reduced glutathione; CySH–Gly, L-cysteinylglycine; CySH, L-cysteine.

the renal plasma. Depending on renal cellular thiol status, as described below, the various thiol conjugates of mercury are either secreted into urine or produce nephrotoxicity.

Evidence for the involvement of thiols, and in particular GSH, in the tissue distribution and target-organ accumulation of mercury-containing compounds comes from numerous *in vivo* and *in vitro* studies over the past couple of decades. Tanaka *et al.* (1991) demonstrated a direct correlation between strain- and sex-dependent differences in males and females of five different strains of mice in renal and hepatic GSH content and renal accumulation of MeHg. They similarly found a direct correlation between renal levels of GGT activity and renal accumulation of MeHg. Mulder and Kostyniak (1985) described a strain of mice (CFW Swiss) that exhibits elevated urinary excretion of MeHg as compared with that in other mouse strains. This increased urinary MeHg content correlated with higher urinary GSH content and lower urinary GGT activity in this strain as compared with other mouse strains. These findings imply that both concentrations of GSH and rates of GSH or GSH-conjugate metabolism determine renal capacity to accumulate MeHg.

Alteration of tissue GSH content by administration of various agents that alter GSH synthesis or degradation, or administration of mercury-containing compounds with exogenous GSH, can have profound effects on cellular accumulation of both MeHg and Hg^{2+}. Depletion of renal GSH with diethyl maleate (DEM) or buthionine sulfoximine (BSO), which act by conjugating GSH and inhibiting GSH synthesis, respectively, specifically reduces renal but not hepatic accumulation of both MeHg and Hg^{2+} (Johnson, 1982; Berndt *et al.*, 1985; Baggett and Berndt, 1986; Girardi and Elias,

1993). Similar results have been obtained with exposure of mice to Hg^0: BSO reduced GSH content in liver, kidney and lung tissue, and these decreases were correlated with increases in Hg^{2+} accumulation in liver, lung and plasma but a decrease in Hg^{2+} accumulation in the kidneys (Kim *et al.*, 1995). Zalups and Lash (1997a) found that administration of either BSO or DEM to rats prior to administration of a subtoxic dose of $HgCl_2$ decreased both total renal and hepatic contents of Hg^{2+} when GSH content was expressed as total μmol. When contents of Hg^{2+} were expressed as a percentage of dose administered, only DEM decreased renal Hg^{2+} content and neither compound had any effect on hepatic Hg^{2+} content. These results indicate that the mechanisms for accumulation of MeHg and Hg^{2+} differ in liver and kidney. However, hepatic GSH status does appear to be a determinant of the renal accumulation of mercury-containing compounds. Tanaka *et al.* (1990) specifically depleted liver GSH in mice with 1,2-dichloro-2,4-nitrobenzene prior to injection of $HgCl_2$ and observed a significant reduction in renal Hg^{2+} accumulation.

Further support for the concept that GSH status is critical in target-tissue accumulation of MeHg and Hg^{2+} comes from studies in which cellular contents of GSH are increased. Administration of *N*-acetylcysteine to rats, which increases renal non-protein sulfhydryl content, increased renal accumulation and decreased renal clearance of Hg^{2+} (Girardi and Elias, 1993). Co-administration of GSH with $HgCl_2$ in mice resulted in an increase in renal Hg^{2+} content (Tanaka *et al.*, 1990). Administration of a twofold molar excess of GSH with a subtoxic dose of $HgCl_2$ (0.25 μmol/kg body weight) to rats also resulted in a significant, although transient, increase in the renal content of Hg^{2+}, no change in hepatic content of Hg^{2+}, and a redistribution of Hg^{2+} in the blood out of erythrocytes and into the plasma fraction (Zalups and Barfuss, 1995a). These findings are interpreted as being consistent with transport of Hg^{2+} to the kidneys as an Hg–GSH complex. Co-administration of a twofold molar excess of CySH with a subtoxic dose of $HgCl_2$ (0.25 μmol/kg body weight) to rats produced the same patterns of increase in renal Hg^{2+} content, no effect on hepatic Hg^{2+} content, and a redistribution of Hg^{2+} into the plasma fraction of blood (Zalups and Barfuss, 1995b), suggesting that complexes of Hg^{2+} with CySH may also be transport forms for the renal accumulation of Hg^{2+}. Although exogenous GSH does not appear to affect hepatic accumulation of Hg^{2+}, Foulkes (1993) showed that GSH increased the extrusion of Hg^{2+} from the jejunal mucosa, suggesting that complexes of Hg^{2+} with GSH may be important in the enterohepatic circulation of Hg^{2+}. Naganuma *et al.* (1988) observed that renal accumulation of MeHg in mice after administration of MeHgGSH was significantly greater than that after administration of MeHgCl. This indicates that hepatic GSH also plays an important role in the renal accumulation of MeHg.

Animals that have undergone uninephrectomy and compensatory renal growth have significantly elevated contents of renal GSH, particularly in the outer stripe of the outer medulla (Zalups and Veltman, 1988; Zalups and Lash, 1990). These animals thus represent a unique model in which to study the renal disposition of mercury-containing compounds. Indeed, the remnant kidney of rats that have undergone uninephrectomy and compensatory renal growth accumulates higher amounts of Hg^{2+} on a per gram tissue or a per mg protein basis than the kidneys of control rats (Zalups and Diamond, 1987; Zalups *et al.*, 1987; Houser and Berndt, 1988; Zalups and Lash, 1990).

Assessment of the role of GSH in renal Hg^{2+} or MeHg accumulation from *in vitro* studies is complex because of the variety of experimental models and incubation conditions used and the steep concentration–toxicity relationship that is even more pronounced *in vitro* than it is *in vivo*. Results from studies in which a cytotoxic dose of Hg^{2+} or MeHg is administered are often difficult to interpret because of the complications due to cellular necrosis. It is critical, therefore, that information on the effects of thiols and mechanistic information on transport be obtained with subtoxic concentrations of Hg^{2+} or MeHg.

Zalups *et al.* (1991) characterized the effects of several exogenous thiols on the transport of Hg^{2+} in isolated perfused tubules from the S1, S2, and S3 segments of rabbit proximal tubules. When CySH or GSH (80 μM) or an ultrafiltrate of rabbit plasma were included with 18.4 μM $HgCl_2$ in the luminal perfusate, the disappearance flux of Hg^{2+} from the tubular lumen into the tubular epithelium was decreased. A similar study with MeHgCl (Zalups and Barfuss, 1993) gave the same results with regard to the ability of exogenous GSH to decrease luminal disappearance flux. In another study with rabbit proximal tubular fragments, Zalups *et al.* (1993) showed that a fourfold molar excess of either GSH, CySH or bovine serum albumin decreased the tubular accumulation of 10 μM $HgCl_2$. Burton *et al.* (1995) incubated kidney slices from New Zealand White rabbits with 100 μM $HgCl_2$ alone or in the presence of a one- to threefold molar excess of GSH or with 100 μM diglutathionyl mercury complex $[Hg(SG)_2]$ and observed no difference in the renal accumulation of Hg^{2+} with either $HgCl_2$ or $Hg(SG)_2$ but did show that exogenous GSH markedly decreased renal accumulation of Hg^{2+}. These results are consistent with the ability of GSH to chelate Hg^{2+} and MeHg but are not necessarily consistent with the function of complexes of Hg^{2+} or MeHg with GSH as physiologic transport forms.

A recent study in suspensions of isolated proximal tubular cells from rat kidney, however, does provide evidence for complexes of Hg^{2+} and GSH or CySH as physiologic transport forms (Lash *et al.*, 1998). As shown in Fig. 13.5, proximal tubular cells co-incubated with Hg^{2+} and either bovine serum albumin (BSA) or DMPS exhibited markedly lower amounts of

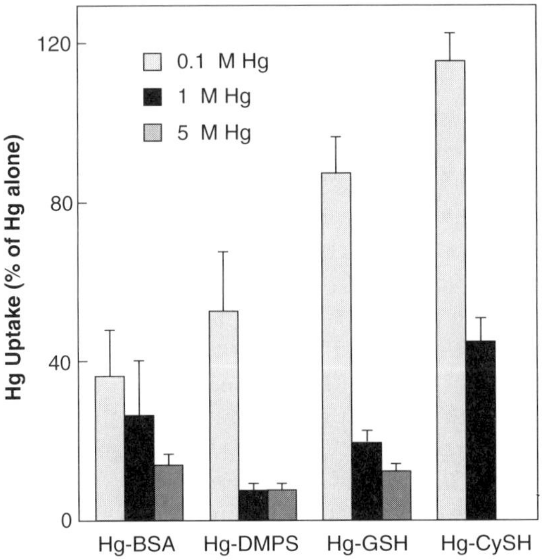

Figure 13.5 Modulation of renal proximal tubular accumulation of inorganic mercury by extracellular thiols. Incubation mixtures (1.0 ml) contained 0.75 ml of appropriate $HgCl_2$ stock solution (0.05 μCi of $^{203}HgCl_2$, cold $HgCl_2$ to give a final Hg^{2+} concentration of either 0.1, 1 or 5 μM, and either buffer or a fourfold molar excess with respect to Hg^{2+} of either bovine serum albumin (BSA), 2,3-dimercapto-1-propanesulfonate (DMPS), glutathione (GSH), or L-cysteine (CySH)) and 0.25 ml of proximal tubular cells (1.2×10^6/ml) from male Sprague–Dawley rats. Mixtures were incubated at 37 °C with shaking in a Dubnoff water bath for 10 min. Hg^{2+} accumulation was stopped by centrifugation for 2 min at 13 000 × *g*. Radioactivity in cell pellets was then determined in a gamma counter. Results are the means ± SE of measurements from four or five separate cell preparations. (Data are derived from Lash *et al.*, 1998.)

cellular Hg^{2+} accumulation than cells incubated with Hg^{2+} alone. Of special note is the observation that the degree of inhibition of Hg^{2+} accumulation by co-incubation with BSA or DMPS exhibited a concentration dependence, even though the ratio of Hg^{2+} : thiol remained constant. Proximal tubular cells co-incubated with 0.1 μM Hg^{2+} and 0.4 μM GSH exhibited significantly higher accumulation of Hg^{2+} than those co-incubated with BSA or DMPS. Proximal tubular cells co-incubated with 0.1 μM Hg^{2+} and 0.4 μM CySH actually exhibited higher accumulation of Hg^{2+} than those incubated with Hg^{2+} alone. The stimulatory effect of GSH or CySH was also concentration dependent, being greater at the lower concentrations of Hg^{2+}. These results are consistent with complexes of Hg^{2+} with CySH or GSH being physiologic transport forms. It is unclear whether Hg–GSH complexes function as a transport form or whether they are degraded to CySH complexes before being transported in the proximal tubular cells. The concentration dependence exhibited in these studies highlights the difficulties in assessment of Hg^{2+} transport and indicates that appropriate conclusions about the role of thiol complexes in Hg^{2+} transport can be missed if a broad range of incubation conditions is not used.

13.3.2 Cadmium

The role of GSH in the accumulation of cadmium in its various target tissues is complex and varies in different tissues. For example, administration of GSH monoisopropyl ester to mice 2 hours prior to administration of cadmium resulted in increased cadmium accumulation in kidney, liver, spleen and testes but decreased cadmium accumulation in the brain (Gale *et al.*, 1990). Cadmium has a much higher affinity for MT than for other thiols and GSH does not stimulate extrusion of cadmium from jejunal mucosa (Foulkes, 1993). Administration of BSO to rats to deplete GSH prior to administration of a nephrotoxic dose of Cd–MT produces a significant reduction in renal cadmium accumulation (Suzuki and Cherian, 1989). Diamide, which depletes GSH by promoting its oxidation to GSSG, decreased cadmium accumulation by human lung carcinoma A549 cells (Kang *et al.*, 1990). Addition of GSH to normal rat kidney fibroblasts (NRK-49F cells) decreased cellular cadmium accumulation; this decrease was presumed to occur by formation of a Cd–GSH complex (Kang, 1992). Results from these studies are all consistent with a direct role for cellular GSH status in the uptake and accumulation of cadmium.

In spite of the studies described above, results from several other studies suggest that the function of GSH in cadmium accumulation is not that clear and is certainly not identical to that described above for GSH in the accumulation of mercury-containing compounds. Rather, other studies suggest a distinct role for GSH in the accumulation of cadmium. Marked depletion of liver GSH content with BSO did not alter accumulation of cadmium by liver slices (Chan and Cherian, 1992). In contrast, addition of exogenous GSH to suspensions of isolated hepatocytes at either the same time or 15 min after administration of cadmium resulted in a decrease in the cellular content of cadmium (Stacey, 1986). Because GSH does not enter hepatocytes and remains extracellular, this effect cannot be due to direct complexation of GSH with cadmium. Studies on the development of resistance to cadmium by human lung carcinoma A549 cells have shown that while cellular GSH content is markedly elevated in these resistant cells, resistance was not associated with decreased cadmium accumulation (Hatcher *et al.*, 1995). Similarly, although GSH depletion sensitizes A549 cells to cadmium, sensitivity is not associated with enhanced cadmium accumulation (Kang *et al.*, 1989). The complex nature of the interaction between GSH and cadmium is illustrated further by a study comparing the tissue distribution of cadmium after administration of three different forms of cadmium (Cd-acetate, Cd–MT, Cd–GSH) to male Wistar rats (Muller *et al.*, 1986). Cadmium accumulation in the small intestine was modestly increased by Cd-acetate, markedly increased by Cd–GSH, but not affected by Cd–MT. In contrast, renal and hepatic cadmium accumulation was lower with Cd–GSH than with the other two forms. Hence, the tissue distribution of cadmium is determined by the form of administered cadmium, and the pattern differs with different tissues.

13.3.3 Arsenic

Similar to the situation with cadmium, the effects of GSH on cellular arsenic status appear to be dependent on the cell type that is being studied and the experimental design. Burton *et al.* (1995) found no difference in the accumulation of arsenic when renal cortical slices from male New Zealand White rabbits were incubated with either $NaAsO_2$ or $As(SG)_3$ and found no effect of exogenous GSH on arsenic accumulation. In contrast to this, Ramos *et al.* (1995) pretreated female rats with *N*-acetylcysteine, which significantly elevated GSH contents in liver, heart and kidney, and observed decreased tissue arsenic contents. They also depleted tissues of GSH with BSO and observed a redistribution of arsenic among several tissues and a decrease in the renal content of arsenic. Hirata *et al.* (1990) treated hamsters with BSO and observed greater than a fivefold decrease in excretion of arsenic in urine and feces. At the same time, methylation of arsenic decreased but renal content of arsenic increased 1.7-fold. Thus, although the studies of Ramos *et al.* (1995) and Hirata *et al.* (1990) are consistent with each other in terms of renal contents of GSH and arsenic being inversely related, it is not clear exactly how GSH interacts with arsenic.

Other *in vitro* studies, however, provide evidence for a different type of relationship between GSH status and arsenic uptake and accumulation. Huang *et al.* (1993) noted that GSH is important in facilitating the metabolism of arsenic and found that GSH pretreatment elevated arsenic content and BSO pretreatment decreased arsenic content in Chinese hamster ovary cells. Similarly, Bertolero *et al.* (1987) showed that depletion of GSH with DEM inhibited the reduction of As^{5+} to As^{3+}. Hence, these two studies suggest a direct correlation, rather than an inverse one, between GSH and arsenic.

13.4 Metal-induced changes in GSH status

While the section above described the effects of GSH status on metal accumulation and transport, the converse situation, in which one looks at the effects of metal exposure on GSH status, is also important in terms of understanding mechanism of action. In considering the effects of metals on cellular GSH status, it is critical to specify the nature of the administered dose. Pretreatment with, or direct exposure to, a subtoxic dose of many metals has been associated with induction of various proteins, analogous to the heat shock response. Exposure to higher doses, however, is often associated with acute tissue necrosis and a decrease in protein expression. Hence, some of the confusion in the literature about whether or not specific metals increase or decrease target-tissue GSH content or selected enzyme activities relates directly to the dose of the metal.

13.4.1 *Inorganic mercury and methylmercury*

Several studies on the effects of Hg^{2+} on cellular thiol status illustrate the point discussed above about toxic versus subtoxic doses of metals. Klonne and Johnson (1988) administered $HgCl_2$ to rats and found a marked depletion of renal cortical GSH and protein sulfhydryl content. Gstraunthaler *et al.* (1983) administered a highly nephrotoxic dose of $HgCl_2$ to rats and observed a marked depletion of GSH content in renal tubular cells and an increase in formation of free radicals and lipid peroxides. Similarly, Bose *et al.* (1994) compared the effects of cadmium, zinc and Hg^{2+} on liver and kidney levels of thiols, finding depletion of GSH due to Hg^{2+}, particularly in the liver. Nielsen *et al.* (1994) also found a concentration dependent decrease in renal GSH content with increasing doses of $HgCl_2$. Decreases in or no effect on GSH-dependent enzymes, such as GSH peroxidase, due to Hg^{2+} exposure have also been described (Klonne and Johnson, 1988; Nielsen *et al.*, 1994). In contrast to these findings, administration of subtoxic doses of Hg^{2+} have been associated with increases in renal content of GSH and increases in activities of selected GSH-dependent enzymes, including GSH peroxidase, GSSG reductase and γ-glutamylcysteine synthetase (Fukino *et al.*, 1986; Zalups and Lash, 1990; Girardi and Elias, 1995; Lash and Zalups, 1996). From these studies, we can conclude that administration of a subtoxic dose of Hg^{2+} leads to an adaptive response that is characterized by increases in antioxidants and antioxidant enzymes.

Similar findings have been reported for MeHg. While Di Simplicio *et al.* (1990) reported a 50% decrease in liver and kidney GSH content and Choi *et al.* (1996) reported marked depletion of brain and liver GSH in C57BL/6J mice after administration of MeHg, several studies from Woods and colleagues (Woods *et al.*, 1992; Woods and Ellis, 1995; Li *et al.*, 1996a,b) showed that chronic exposure to subtoxic doses of MeHg results in increases in renal and brain content of GSH and in the mRNA for γ-glutamylcysteine synthetase, which is the rate-limiting enzyme for the synthesis of GSH. In one of these studies, Li *et al.* (1996b) showed that MeHg increased expression of γ-glutamylcysteine synthetase mRNA specifically in cerebellar and hippocampal regions, which are known to be resistant to MeHg-induced injury. Thus, they concluded that resistance in these brain regions reflects the ability of specific neuronal cell types to upregulate GSH synthesis.

13.4.2 *Cadmium*

With the exception of differences in target-organ specificity (cadmium accumulates mostly in liver and kidney (Sidhu *et al.*, 1997) whereas Hg^{2+} accumulates predominantly in kidney), cadmium appears generally to produce similar effects to Hg^{2+} on tissue GSH content and GSH-dependent enzymes. Thus, high-dose, acute or chronic exposures to cadmium have been associated with depletion of hepatic, renal and erythrocyte GSH content (Bell *et al.*, 1991; Chan and Cherian, 1992; Ossola and Tomaro, 1995; Strubelt *et al.*, 1996; Nehru and Bansal, 1997; Sarkar *et al.*, 1997). In contrast, low-dose, chronic or subtoxic, acute exposures to cadmium produce increases in renal, hepatic and pulmonary contents of GSH in various *in vivo* and *in vitro* models (Gambhir and Nath, 1992; Chin and Templeton, 1993; Bose *et al.*, 1994; Kamiyama *et al.*, 1995). Resistance of lung A549 cells (Hatcher *et al.*, 1995) and Swiss mouse 3T3 cells (Li *et al.*, 1994) to cadmium is associated with elevated levels of GSH. Therefore, cadmium can induce compensatory changes in the GSH status of target organs under the appropriate exposure conditions, similar to those of Hg^{2+} or MeHg described above. The ability of low doses of cadmium to alter gene regulation and cellular signaling processes has been highlighted in a recent review (Beyersmann and Hechtenberg, 1997). This emphasizes the critical importance of understanding subtle, regulatory effects that only occur at doses that are often significantly below those at which overt toxicity occurs.

13.4.3 Arsenic

Although cellular GSH has been clearly shown to be involved in arsenic metabolism (i.e. reduction and methylation reactions), there are few studies on the effects of arsenic exposure on cellular GSH status. As with the other metals, exposure to relatively high levels of arsenic, in the form of $NaAsO_2$, results in marked depletion of GSH in liver, kidney and heart (Ramos *et al.*, 1995). In contrast, exposure of Chinese hamster ovary V79 cells to subtoxic doses of As^{3+} produced a marked increase in GSH content by 8 hours, followed by a gradual decline in GSH content (Ochi, 1997). The increase in cellular GSH content was associated with increased uptake of cystine but not increased γ-glutamylcysteine synthetase activity. DMA, but not MMA, was also found to increase the cellular content of GSH. In this case, however, the increase in GSH content was associated with both an increase in cystine uptake and an increase in γ-glutamylcysteine synthetase activity.

13.5 Hepatobiliary transport of metals

Secretion of several metals from the liver is associated in one manner or another with that of GSH. Because of molecular weight and lipid solubility, secretion is primarily across the canalicular membrane into the bile, suggesting that enterohepatic circulation plays an important role in the ultimate delivery of metals to their target tissues. Separate canalicular transport systems exist for GSH (Fernandez-Checa *et al.*, 1992, 1993; Garcia-Ruiz *et al.*, 1992; Ballatori and Dutczak, 1994) and GSSG/GSH *S*-conjugates (Akerboom *et al.*, 1982, 1991; Inoue *et al.*, 1984; Nicotera *et al.*, 1985; Awashti *et al.*, 1991). Ballatori and Dutczak (1994) have defined the kinetics of GSH transport across the canalicular membrane, identifying both high- and low-affinity processes. The high-affinity, low-capacity process transports some GSH conjugates, γ-glutamyl amino acids and anions whereas the low-affinity, high-capacity process is the primary system that delivers GSH into bile. In contrast, the transporter for GSSG and most GSH *S*-conjugates is an ATPase (Nicotera *et al.*, 1985) and has been identified with the multidrug resistance-associated protein (MRP) or P-glycoprotein (Jedlitschky *et al.*, 1994, 1996; Leier *et al.*, 1996; Keppler *et al.*, 1997).

One can consider that complexes of metals with GSH may be handled by the carriers as either modified GSH molecules or as GSH conjugates and may thus be substrates for one or more of these carriers. This section will summarize studies on the role of GSH in the hepatobiliary transport of selected metals.

13.5.1 Inorganic mercury and methylmercury

The importance of biliary secretion in the renal accumulation of Hg^{2+} is highlighted by the demonstration that biliary duct ligation in rats produces a marked decrease in the renal uptake and accumulation of Hg^{2+} (Zalups and Barfuss, 1996). Hence, enterohepatic circulation is an important step in the process whereby Hg^{2+} gets to its primary target organ (i.e. the kidneys). Studies by Ballatori and Clarkson (1984a,b) demonstrated a correlation between the biliary efflux of Hg^{2+} and GSH and that Hg^{2+} was secreted into bile as a complex with a low-molecular-weight substance that was tentatively identified as GSH.

Biliary secretion of MeHg also occurs as the GSH complex. Several studies showed correlations between GSH status and biliary secretion of MeHg. For example, Ballatori and Clarkson (1982) showed that the ability of rats to secrete both GSH and MeHg exhibits parallel development with age from the neonate to the adult rodent. Depletion of hepatic GSH content also decreases the rate of MeHg efflux into bile (Ballatori and Clarkson, 1983; Gregus and Varga, 1985). Additional evidence comes from study of the Eisai hyperbilirubinemic (EHBR) rat, which is a mutant Sprague–Dawley strain with a defect in the biliary excretion of a variety of organic anions, including GSH and GSH *S*-conjugates; these rats were also deficient in the ability to excrete MeHg into the bile (Ballatori *et al.*, 1995). Dutczak and Ballatori (1994) also characterized the transport of a synthetic MeHg–GSH complex in canalicular membrane vesicles from rat liver. They found that MeHg–GSH is not a substrate for the ATP-dependent, canalicular GSSG or GSH *S*-conjugate transporter, but rather, interacts with GSH, the GSH analog ophthalmic acid, and some *S*-alkyl GSH analogs. Furthermore, MeHg–GSH uptake was described kinetically by two processes, a high-affinity, low-capacity and a low-affinity, high-capacity system, similar to that which they described for GSH (Ballatori and Dutczak, 1994).

While the form of MeHg that is transported from the hepatocyte into the bile is the GSH conjugate, most of the biliary MeHg is in the form of a CySH–Gly conjugate (Urano *et al.*, 1988a). Biliary GSH and GSH conjugates can be readily degraded in the bile by GGT and dipeptidase activities (Lash *et al.*, 1988). In another study, Urano *et al.* (1988b) found that species-dependent differences in the rates of MeHg excretion into bile correlated with species-dependent differences in thiol content and composition of bile. Hence, rabbits and guinea-pigs exhibited lower rates of MeHg excretion into bile than rats, mice or hamsters and had lower total nonprotein sulfhydryl content in bile.

Naganuma *et al.* (1988) studied the effect of bile duct ligation on the renal accumulation of MeHg. No effect was observed, which was interpreted to mean that enterohepatic circulation is not important for renal accumulation of MeHg. Another possible interpretation is that inhibition of biliary secretion leads to activation of other secretory pathways that compensate for the absence of bile flow, thereby keeping the delivery of MeHg to the kidneys unchanged.

13.3.2 Cadmium

As with Hg^{2+} and MeHg, biliary excretion of cadmium is linked to that of GSH and likely occurs as a complex of cadmium and GSH. The mutant EHBR rat, which has almost no biliary GSH excretion, exhibited only 2.5% of the rate of cadmium excretion as normal rats (Sugawara *et al.*, 1996). In another study, Dijkstra *et al.* (1996) showed that mutant GY Wistar rats, which express defective canalicular ATP-dependent GSH conjugate transport, and have virtually no GSH in bile, had no detectable cadmium secretion into bile. Biliary cadmium excretion is also reduced by prior depletion of GSH (Gregus and Varga, 1985). Shaikh *et al.* (1995) studied cadmium uptake by isolated hepatocytes and found that it involved interaction with sulfhydryl groups and was significantly inhibited by Hg^{2+} and zinc.

13.5.3 Arsenic

The relationship between biliary secretion of arsenic and GSH is less clear than that between mercuric compounds and GSH or cadmium and GSH. Gyurasics *et al.* (1991) published the only study on the interaction between arsenicals and biliary GSH excretion. Injection of sodium arsenite or arsenate into rats increased biliary secretion of GSH and nonprotein sulfhydryls up to 24- or 31-fold, respectively, in a dose-dependent manner. Thus, the hepatobiliary transport of arsenic compounds generates transport of several GSH molecules. In contrast to this, the arsenic-induced increase in GSH excretion only increased that of Hg^{2+} twofold and had no effect on the excretion of other metals. The authors of this study thus concluded that arsenic markedly enhanced the biliary transport of GSH with a free thiol group but not that of GSH bound to another metal ion.

The various transport proteins on the liver canalicular membrane are summarized schematically in Fig. 13.6. Five carriers are depicted:

1 the bile acid carrier;
2 MDR2 P-glycoprotein, which transports phospholipids and is linked to ATP hydrolysis;
3 MDR1 P-glycoprotein, which transports drugs such as doxorubicin and is linked to ATP hydrolysis;
4 the multiple organic anion transporter (MOAT), which transports GSH *S*-conjugates, bilirubin, and various organic acids and is linked to ATP hydrolysis; and
5 the GSH carrier, which transports GSH and complexes of GSH with MeHg, Hg^{2+} and cadmium, and is stimulated by As^{3+}.

13.6 Renal transport of metals

Renal uptake of metals can occur by one of three mechanisms:

1 glomerular filtration of protein-bound metal followed by endocytosis at the brush-border membrane;
2 glomerular filtration followed by transport across the brush-border membrane as catalyzed by a carrier protein; or
3 uptake of metal from renal plasma by transport across the basolateral membrane as catalyzed by a carrier protein.

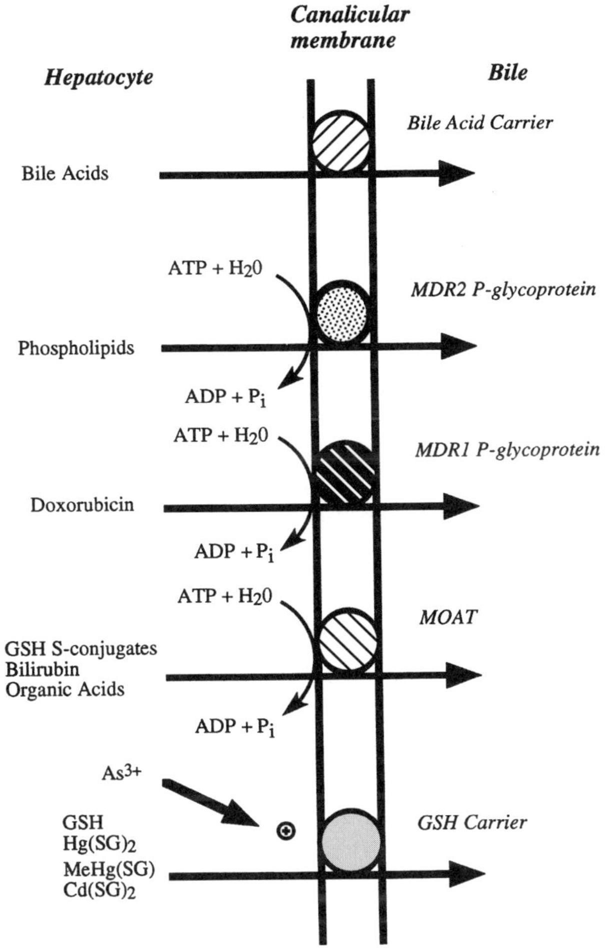

Figure 13.6 Anion transport systems in liver canalicular membranes. This scheme illustrates the function of five different anion transporters in the liver canalicular plasma membrane. Complexes of inorganic mercury (Hg^{2+}), methylmercury (MeHg), or Cd with GSH are believed to be transported into the bile by the GSH carrier, which is distinct from the bile acid carrier, the two multidrug resistance (MDR) P-glycoprotein carriers, and the multiple organic anion transporter (MOAT). As^{3+} is believed to enhance the transport of GSH by the GSH carrier but is not necessarily transported by that carrier.

In this context, current knowledge about the transport of Hg^{2+}, MeHg, cadmium and arsenic into renal epithelial cells will be discussed.

13.6.1 Inorganic mercury and methylmercury

Because most of the Hg^{2+} in plasma is bound to serum albumin (Friedman, 1957; Cember *et al.*, 1968; Foulkes, 1974; Lau and Sarkar, 1979), there are three potential mechanisms by which the mercuric ion in Hg–albumin complexes in plasma can be taken up by the kidneys:

1 exchange of the albumin ligand with plasma GSH or CySH and transport of the Hg–GSH or Hg–CySH complexes into renal cells;
2 exchange of the albumin ligand with other protein sulfhydryl groups, such as those on cellular plasma membranes; or
3 endocytosis of Hg with filtered albumin, as described above.

Besides being delivered to the kidneys as complexes with albumin, Hg^{2+} may be bound directly to GSH, CySH or even CySH–Gly, which are all present at significant levels in plasma (Lash and Jones, 1985b; Jones *et al.*, 1997). As described above, the lability of the Hg–S bond in biological fluids (Rabenstein, 1989; Oram *et al.*, 1996) implies that ligand exchange is likely to be an important component of the handling of Hg^{2+} by the kidneys.

Whatever the form in which the kidneys are presented with Hg^{2+}, there is evidence for at least two specific, carrier-mediated mechanisms of uptake, one on the brush-border or luminal membrane and one on the basolateral membrane. Zalups and Minor (1995) showed that reduction of glomerular filtration in rats to negligible levels by ureteral ligation only partially reduced Hg^{2+} uptake, suggesting that the portion of Hg^{2+} not taken up during ureteral ligation represents uptake by a luminal mechanism. *P*-Aminohippurate (PAH) also causes a significant reduction in Hg^{2+} uptake (Zalups and Barfuss, 1995c; Zalups and Minor, 1995; Zalups, 1997) and a combination of ureteral ligation and PAH treatment produces additive inhibition of Hg^{2+} uptake (Zalups and Minor, 1995), indicating that the effect of PAH must occur at the basolateral membrane.

Several studies have also shown that inhibition of renal GGT activity with acivicin significantly reduces the renal accumulation of Hg^{2+} and increases the urinary excretion of GSH and Hg^{2+} (Berndt *et al.*, 1985; Tanaka *et al.*, 1990; Tanaka-Kagawa *et al.*, 1993; de Ceaurriz *et al.*, 1994; Zalups, 1995). These results imply that GGT activity is involved in the luminal uptake of Hg^{2+}. Recent studies in isolated brush-border membrane vesicles from rat renal cortex (Zalups and Lash, 1997b) support a CySH conjugate of Hg^{2+} as a transport form of Hg^{2+}. Using the molecular mimicry concept discussed above (see Fig. 13.2), Hg-$(CySH)_2$ may function analogously to cystine. Indeed, there is evidence that Hg^{2+} interacts with rBAT-induced amino-acid transport (Waldegger *et al.*, 1995). rBAT is a protein that is present in renal brush-border membranes and catalyzes the Na^+-independent uptake of dibasic and neutral amino acids, including cystine. Hence, the inhibitory effect of acivicin may be due to decreased generation of Hg–CySH complexes. A schematic summary of renal brush-border membrane transport of Hg^{2+} is presented in Fig. 13.7. Hg–$(CySH)_2$ is the likely transport form, with metabolism of Hg–$(SG)_2$ by GGT and dipeptidase playing a critical role in generation of the ultimate transport form. Endocytosis of an Hg–albumin complex can also occur.

Inhibition of renal uptake of Hg^{2+} by PAH (Zalups, 1995, 1997; Zalups and Minor, 1995) and small, aliphatic dicarboxylic acids (Zalups and Barfuss, 1998) suggests that both the organic anion transporter and the dicarboxylic acid transporter are involved in the basolateral transport of various Hg^{2+}–thiol complexes. As illustrated in Fig. 13.8, either the GSH or the CySH complex with Hg^{2+} can be transport forms. Because of the anionic nature of the various Hg–thiol conjugates, both the Hg–GSH or the Hg–CySH complexes can be transported by either the organic anion or dicarboxylic acid transporters. GSH and various GSH conjugates are transported across the renal basolateral

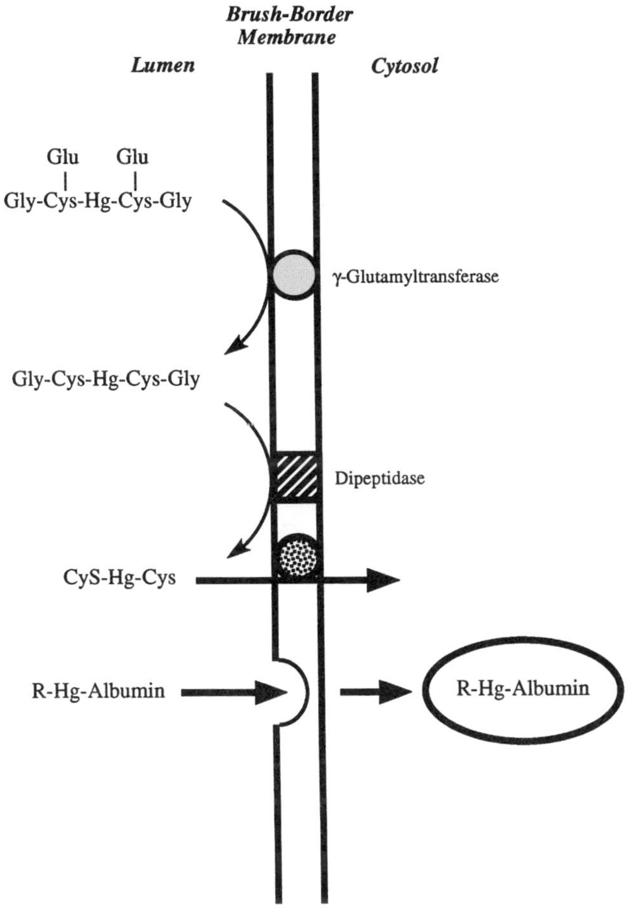

Figure 13.7 Transport of mercury–thiol complexes across the renal brush-border plasma membrane. Complexes of Hg^{2+} with albumin are transported into renal proximal tubular cells by endocytosis. Complexes of Hg^{2+} with glutathione are sequentially degraded by γ-glutamyltransferase and dipeptidase activities to complexes of Hg^{2+} with L-cysteine, which are then transported into renal proximal tubular cells, possibly by the dibasic amino-acid carrier.

membrane by Na^+-dependent and PAH- and probenecid-sensitive mechanisms (Lash and Jones, 1984, 1985a). Cysteine conjugates are also transported by a PAH- and probenecid-sensitive process (Lash and Anders, 1989). Thus, Hg–GSH or Hg–CySH complexes may be handled analogously to other GSH and CySH conjugates. The possible identity of the system(s) that transport(s) GSH and CySH conjugates of aliphatic hydrocarbons with the organic anion and/or the dicarboxylic acid carriers remains to be determined.

Less work has been done on the study of MeHg transport into the kidneys. However, it is likely that the same mechanisms apply for MeHg as for Hg^{2+}. For example, inhibition of GGT activity with either acivicin (Naganuma *et al.*, 1988; Tanaka-Kagawa *et al.*, 1993) or serine–borate (Yasutake *et al.*, 1989) markedly decreases the renal accumulation of MeHg. In the absence of acivicin treatment, most of the urinary MeHg

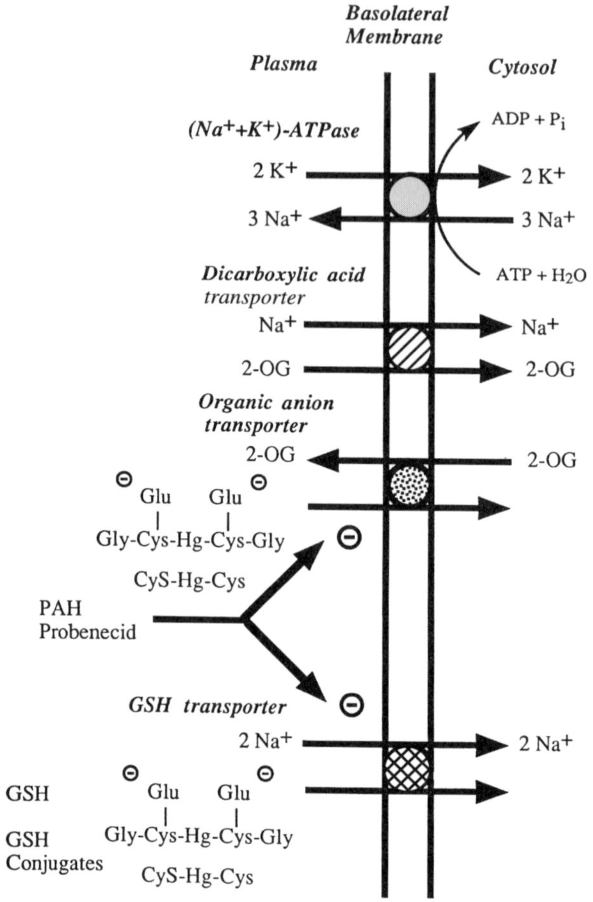

Figure 13.8 Transport of mercury–thiol complexes across the renal basolateral plasma membrane. Complexes of Hg^{2+} with glutathione or L-cysteine are transported into renal proximal tubular cells by either the GSH transporter, which is cotransported with two sodium ions, or by the organic anion transporter. Uptake by the organic anion transporter occurs in exchange for 2-oxoglutarate (2-OG), which is taken up by cotransport with sodium by the dicarboxylate transporter. Also shown is the (Na^+ + K^+)-stimulated ATPase, which is the ultimate active transport system that generates the transmembrane sodium gradient.

is in the form of a complex with CySH. With acivicin pretreatment, however, a marked increase in the MeHg complex with GSH is observed. These results suggest that MeHg–thiol complexes are processed in the same manner as Hg^{2+}–thiol complexes. Although Urano *et al.* (1990) studied MeHg absorption in the small intestine, it is likely that many of the same transport mechanisms are operative in the renal proximal tubules. In their study of MeHg absorption, Urano *et al.* (1990) found that the extent of absorption of the CySH–Gly conjugate was similar to that of the CySH conjugate and that both were 1.5-fold higher than that of the GSH conjugate. Additionally, pretreatment of intestinal tissue with probenecid inhibited the absorption of MeHg–GSH, suggesting that the organic anion transporter is also involved in the intestinal uptake of MeHg–thiol complexes.

13.6.2 *Cadmium*

The renal transport of cadmium has been less thoroughly studied than that of Hg^{2+}. As with Hg^{2+}-containing compounds, the high affinity of cadmium for sulfhydryl groups should play an important role in its transport. Indeed, complexes of cadmium with MT are believed to be a major form in which cadmium is delivered to the kidneys. As with Hg^{2+}–albumin complexes, Cd–MT complexes are translocated into renal proximal tubular cells by endocytosis. Any specific role for GSH in the renal transport of cadmium is unknown at present. Shaikh *et al.* (1995) studied some basic characteristics of cadmium transport in renal cortical epithelial cells from the rat and found that only 30% of total cadmium transport through the basolateral membrane, and none of that through the brush-border membrane, was temperature-sensitive, suggesting that the majority of cadmium transport into renal cells occurs by diffusion.

Robinson *et al.* (1993) studied transport of $CdCl_2$ in isolated perfused S1, S2 and S3 segments of rabbit proximal tubules and found that cadmium is taken up avidly at the luminal membrane in all three segments, but most rapidly in S1 segments. They also found that less than 10% of the cadmium that was taken up at the luminal membrane was subsequently transported across the basolateral membrane, indicating that most of the transported cadmium was sequestered within the epithelial cells.

Another potential route of cadmium transport into renal cells is through calcium channels. Although some evidence has been obtained for the function of calcium channels in the transport of cadmium into hepatocytes (Blazka and Shaikh, 1991), other data obtained in renal cells (Templeton, 1990; Robinson *et al.*, 1993) argues against a significant function of calcium channels in the renal transport and accumulation of cadmium.

13.6.3 *Arsenic*

Even less information about renal transport is available for arsenic. Burton *et al.* (1995) studied the influence of GSH on the transport and toxicity of $NaAsO_2$ in rabbit and renal cortical slices. No differences were observed in the accumulation of arsenic when slices were incubated with equal concentrations of either $NaAsO_2$ or $As(SG)_3$. The authors concluded that while GSH is necessary in the methylation of As^{3+} once it is accumulated inside the renal cell, the role of GSH in the uptake of As^{3+} is still unclear.

13.7 Role of GSH in target-organ toxicity of metals

Inasmuch as the goal of this chapter is to discuss the role of GSH in metal transport, metabolism and toxicity (with a focus on mercury, cadmium, and arsenic), a detailed discussion of mechanisms of toxicity is beyond the present scope. Previous sections have presented evidence that GSH is involved in the intracellular transport and accumulation of certain metals under specified conditions in target organs. A general pharmacological principle is that dose correlates with toxic response. Because of the ability of thiols, on the one hand, to chelate many metals and, on the other hand, to facilitate their transport across biological membranes, a precise description of the role of thiols in toxicity is complex. Thus, evidence is available that indicates that the presence of thiols protects target tissues from the toxic effects of metals. Evidence is also available that the presence of thiols can also exacerbate toxicity in target organs. This section will, therefore, summarize some of the data on the role of GSH in determining the extent of toxicity of Hg^{2+}, MeHg, cadmium and arsenic in their primary target organs.

13.7.1 Inorganic mercury and methylmercury

GSH has at least two distinct roles in the nephrotoxicity of Hg^{2+}. On the one hand, formation of a complex of Hg^{2+} with GSH can function as a means to promote delivery of Hg^{2+} to the target organ, as described above. Experimentally, this function has been studied by either co-administration of Hg^{2+} with GSH or by direct administration of a Hg^{2+}–GSH complex. Alternatively, one of the major biochemical effects of Hg^{2+} in renal tissue is to cause oxidative stress, both at the level of the whole cell (e.g. Nath *et al.*, 1996) and at the level of mitochondria (Weinberg *et al.*, 1982a,b; Lund *et al.*, 1993). GSH plays a critical role in maintenance of cellular redox status and protects against oxidative injury. In this manner, increased cellular levels of GSH should protect against Hg^{2+}-induced oxidative injury.

Indeed, addition of exogenous GSH to various *in vitro* preparations of renal tissue protects against Hg^{2+}-induced toxicity. For example, exogenous GSH protected rabbit renal cortical slices from Hg^{2+}-induced toxicity and this was correlated with decreased accumulation of Hg^{2+} (Burton *et al.*, 1995). Isolated proximal tubule fragments from rabbit kidneys were also protected from Hg^{2+}-induced cytotoxicity by exogenous GSH, and protection was correlated with decreased Hg^{2+} accumulation (Zalups *et al.*, 1993). Hg^{2+}-induced cytotoxicity in isolated proximal tubular cells from rats is also diminished by either pre-incubation or co-incubation with GSH or other thiols (Lash and Zalups, 1992). Other means of increasing intracellular content of GSH have also demonstrated protection from Hg^{2+}-induced toxicity. Administration of GSH monoethyl ester, which enhances intracellular GSH content, protected against Hg^{2+}-induced acute renal failure (Houser *et al.*, 1992), presumably by providing an increased thiol pool to reduce the reactivity of Hg^{2+} towards essential protein sulfhydryls in the cell. Fukino *et al.* (1986) determined that the ability of zinc to protect rats from Hg^{2+}-induced nephrotoxicity was due to its ability to increase GSH synthesis and renal GSH content. GSH glycoside increases brain and liver GSH content in C57BL/6J mice and prevents neurotoxicity due to MeHg (Choi *et al.*, 1996). The converse situation, where renal GSH content is decreased, such as by administration of DEM or BSO (Baggett and Berndt, 1986; Fukino *et al.*, 1986), has been associated with increased nephrotoxicity due to Hg^{2+}, presumably because more Hg^{2+} is available to interact with critical thiols in the renal cell.

Other studies, however, point to a different role for GSH in Hg^{2+}-induced nephrotoxicity. Zalups and Barfuss (1993) found that coexposure of isolated, perfused proximal tubules from rabbit kidney to MeHg and a fourfold molar excess of GSH greatly increased cytotoxicity. Inhibition of renal GSH metabolism with acivicin decreased renal Hg^{2+} accumulation and decreased nephrotoxicity (Tanaka *et al.*, 1990; de Ceaurriz *et al.*, 1994). Under these circumstances, GSH is seen to function as a means to enhance the delivery of Hg^{2+} to the renal cell, in a manner that is dependent on its metabolism by GGT and dipeptidase. Tanaka-Kagawa *et al.* (1998) have also shown that strain-dependent differences in the susceptibility of mice to the renal toxicity of Hg^{2+} correlate with GGT activity in each strain. Hence, those mice with higher renal activities of GGT exhibited greater toxicity. Another circumstance in which increased cellular GSH content is associated with increased renal cellular injury from Hg^{2+} is the model of uninephrectomy and compensatory renal growth. Renal proximal tubular cells from remnant kidneys that have undergone compensatory hypertrophy have elevated levels of GSH (Zalups and Veltman, 1988; Zalups and Lash, 1990) and kidneys from rats that have undergone uninephrectomy and compensatory renal hypertrophy exhibit enhanced susceptibility to Hg^{2+}-induced toxicity both *in vivo* (Zalups and Diamond, 1987; Zalups and Lash, 1990) and *in vitro* (Lash and Zalups, 1992).

Besides the proximal tubules, other renal cell types may be susceptible to Hg^{2+}-induced injury. Wilks *et al.* (1990) found that isolated glomeruli and proximal tubules from rats were both susceptible to injury from exposure to Hg^{2+}. However, preincubation with GSH provided protection only to the proximal tubules and not to the isolated glomeruli. This would be consistent with the differences in handling of GSH by the various cell populations that comprise the kidneys (Lash *et al.*, 1988). Hence, cell type differences must also be taken into account when assessing the role of GSH in metal-induced nephrotoxicity.

The potential roles of GSH in modulating Hg^{2+}- or MeHg-induced toxicity are summarized schematically in Fig. 13.9. In the first role (a), exogenous GSH inhibits

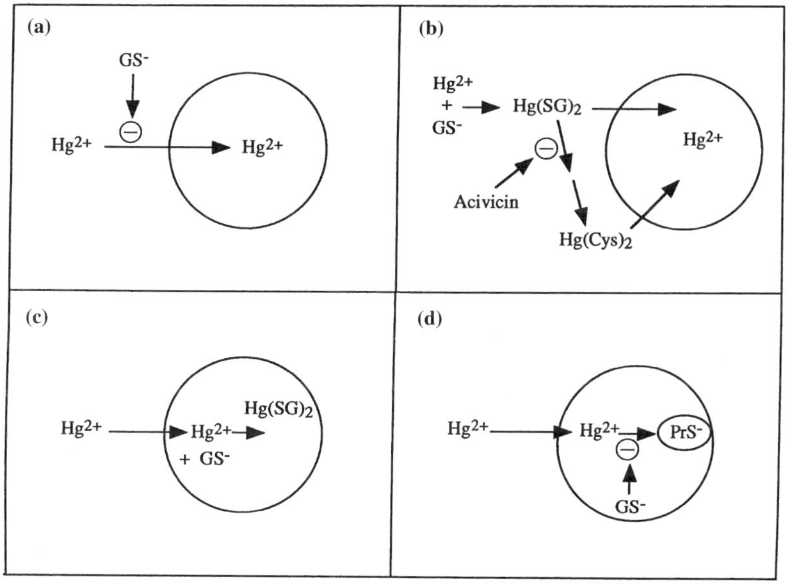

Figure 13.9 Potential roles of glutathione in modulation of Hg^{2+}-induced nephrotoxicity. Four potential scenarios for the interaction of glutathione (GSH) with Hg^{2+} in modulating Hg^{2+}-induced cytotoxicity in renal cells are shown in an attempt to explain data on the role of GSH in Hg^{2+}-induced nephrotoxicity. (a) GSH protects cells from Hg^{2+} by inhibiting cellular accumulation. (b) GSH forms a complex with Hg^{2+} and either the Hg^{2+}–GSH ($Hg(SG)_2$) complex or the Hg^{2+}–cysteine ($Hg(Cys)_2$) complex is transported into the renal cell. The $Hg(Cys)_2$ complex is generated by the successive action of γ-glutamyltransferase and dipeptidase. Acivicin, which irreversibly inhibits γ-glutamyltransferase, protects renal cells by decreasing generation of the major transport form of Hg^{2+}. (c) Intracellular GSH forms a complex with transported Hg^{2+}, thereby enhancing cellular accumulation of Hg^{2+} and increasing Hg^{2+}-induced cellular injury. (d) Intracellular GSH protects renal cells from Hg^{2+}-induced cellular injury by binding to Hg^{2+}, thereby preventing it from interacting with essential proteins or enzymes in the cell.

cellular accumulation of Hg^{2+} outside of the cell, thereby protecting against Hg^{2+}-induced toxicity. In the second role (b), protection of renal cells from Hg^{2+}-induced toxicity due to acivicin is associated with inhibition of $Hg(SG)_2$ complex metabolism and transport. In the third role (c), elevations in intracellular GSH exacerbate injury because it increases cellular retention of Hg^{2+}. Finally, in the fourth role (d), elevations in intracellular GSH protect cells from Hg^{2+} by binding to it and preventing interaction of Hg^{2+} with cellular proteins or enzymes. More research is needed to define precisely the conditions under which each of these hypothesized roles of GSH predominates.

13.7.2 Cadmium

In contrast to Hg^{2+}, which has the kidneys as its predominant target organ, cadmium produces toxicity in several tissues, including the lungs, liver and kidney. Most of the experimental data indicate a protective role for GSH in each of these target tissues. Cadmium produces marked depletion of cellular GSH in several tissues, including liver, erythrocytes and kidney, and this is often associated with increases in levels of reactive oxygen species and lipid peroxidation (Stohs and Bagchi, 1995; Strubelt *et al.*, 1996; Sarkar *et al.*, 1997). Cytotoxicity of cadmium in several cell types is increased with prior depletion of GSH, such as by pretreatment with BSO or phorone (Siegers *et al.*, 1986; Singhal *et al.*, 1987; Ochi *et al.*, 1988; Suzuki and Cherian, 1989; Chan and Cherian, 1992; Sengupta and Bhattacharyya, 1996). Hence, there is a clear association between depletion of cellular GSH and the severity of cadmium cytotoxicity. Kang and colleagues (Kang and Enger, 1987, 1988, 1990; Kang *et al.*, 1989; Hatcher *et al.*, 1995) have shown that cadmium-resistant lung A549 cells have elevated intracellular levels of GSH. Similarly, Li *et al.* (1994) found that cadmium-resistant Swiss mouse 3T3 cells have higher intracellular levels of both GSH and protein sulfhydryls. Rana and Verma (1996) found that GSH substantially prevented liver and kidney toxicity in rats chronically exposed to cadmium in their diet.

An interesting dissociation between some of the cytotoxic effects of cadmium and GSH status was reported by Prozialeck and Lamar (1995). These investigators found that one of the earliest effects of exposure of monolayers of LLC-PK1 cells to cadmium was a disruption of junctions between cells. Pretreatment of these cells with GSH had no effect on the disruption of cytoskeletal proteins. However, more pronounced cytotoxicity that occurred with prolonged exposure to cadmium was reversed by GSH, indicating the presence of both GSH-dependent and GSH-independent mechanisms of cadmium-induced cytotoxicity. A temporal relationship in the ability of exogenous GSH to protect isolated rat hepatocytes from cadmium-induced cytotoxicity was revealed by Stacey (1986): addition of GSH just prior to cadmium produced almost complete protection whereas addition of GSH as little as 15 min after cadmium addition produced only partial protection.

One of the major forms of cadmium to which kidneys are believed to be exposed is the complex of cadmium with MT. The Cd–MT complex is taken up at the luminal membrane by endocytosis, similar to Hg^{2+}–albumin complexes, and is digested in lysosomes to release ionic cadmium inside the renal proximal tubular cell. Although Suzuki and Cherian (1989) found that depletion of renal GSH with BSO increased nephrotoxicity due to Cd–MT, renal cadmium accumulation was decreased, thus showing an apparent dissociation between target organ dose and toxicity.

13.7.3 *Arsenic*

Arsenic-containing compounds exhibit a variety of target organs, including the liver and kidneys. Although GSH plays a well-established role in As^{5+} reduction and arsenic methylation, its role in arsenic transport is less clear. The role of GSH in arsenic-induced cytotoxicity is also somewhat uncertain. Several studies have shown an inverse correlation between cellular GSH levels and arsenic-induced toxicity. For example, Ramos *et al.* (1995) pretreated female rats with either *N*-acetylcysteine to increase GSH or BSO to deplete GSH, and then treated the rats with $NaAsO_2$. As^{3+} produced GSH depletion and increases in lipid peroxidation in liver, kidney and heart. With *N*-acetylcysteine pretreatment, GSH depletion and As^{3+}-induced lipid peroxidation were prevented in liver but not in kidney or heart. In contrast, BSO produced additive effects with As^{3+} on GSH depletion and lipid peroxidation in all three tissues. Pretreatment of Chinese hamster ovary cells (Huang *et al.*, 1993) or cultured endothelial cells from bovine carotid artery (Chang *et al.*, 1991) with GSH protected against As^{3+}-induced cytotoxicity. Similarly, depletion of GSH with BSO in hamsters not only impaired arsenic metabolism and excretion, but produced renal tubular necrosis (Hirata *et al.*, 1990).

All of the studies described above clearly showed a correlation between cellular GSH content and As^{3+}-induced toxicity, with GSH depletion associated with toxicity and GSH elevation associated with protection. There are other studies, however, that found no association between cytotoxicity and GSH or that showed differential responses, depending on the species of arsenic tested. Thus, Burton *et al.* (1995) found that exogenous GSH applied to rabbit renal cortical slices readily protected the tissue from $HgCl_2$-induced toxicity but had no effect on $NaAsO_2$-induced toxicity. Similarly, Mückter *et al.* (1993) used freshly isolated renal tubules from rats to screen a large series of thiol-containing compounds as antidotes against arsenic-induced toxicity. 2,3-Dimercapto-1-propanol, DMPS, and meso-2,3-dimercaptosuccinic acid (DMSA) were clearly protective against organic arsenicals, while those three thiols and both dithiothreitol and dithioerythritol were protective against inorganic arsenite. In no case was GSH or any other monothiol protective.

Ochi *et al.* (1996) compared the cytotoxicity of As^{3+}, As^{5+}, MMA, and DMA in cultured human HL-60 cells and found a differential response with regard to GSH status. All four arsenicals, but DMA most potently, produced morphological changes that are characteristic of apoptosis. Depletion of GSH with BSO pretreatment increased the cytotoxicity of As^{3+}, As^{5+} and MMA, but decreased the cytotoxicity of DMA. These authors suggested that GSH is involved causally in DMA-induced apoptosis.

The various studies on arsenic described above demonstrate heterogeneous responses, depending on target tissue and chemical form of arsenic. Although redox processes are clearly involved in the mechanism of arsenic-induced toxicity, GSH is not always sufficient to counteract these effects.

13.8 Summary and conclusions

This chapter has focused on the role of GSH in the metabolism, tissue disposition, excretion, transport and target organ toxicity of Hg^{2+}, MeHg, cadmium and arsenic-containing compounds. A key property of all these metals is their relatively high affinity towards thiols. By virtue of this fact alone, GSH, being the predominant intracellular thiol, should be important in the handling of these compounds. Indeed, complexes of GSH with several of these metals are physiologic transport forms. In some cases, the ability of GSH to chelate the metal acts to protect target tissues from metal-induced toxicity. In other cases, the situation is more complex and a clear association between GSH status, metal accumulation and toxicity is unclear. Further research in delineation of cellular transport mechanisms should help clarify the apparent discrepancies about the role of GSH status and metal toxicity.

References

Akerboom, T.P.M., Bilzer, M. and Sies, H. 1982. Competition between transport of glutathione disulfide (GSSG) and glutathione S-conjugates from perfused rat liver into bile. *FEBS Letters* **140**, 73–6.

Akerboom, T.P.M., Narayamaswami, V., Kunst, M. and Sies, H. 1991. ATP-dependent *S*-(2,4-dinitrophenyl)glutathione transport in canalicular plasma membrane vesicles from rat liver. *Journal of Biological Chemistry* **266**, 13147–52.

Awashti, Y.C., Singhal, S.S., Gupta, S., Ahmad, H., Zimniak, P., Radominska, A., Lester, R. and Sharma, R. 1991. Purification and characterization of an ATPase from human liver which catalyzes ATP hydrolysis in the presence of conjugates of bilirubin, bile acids and glutathione. *Biochemical Biophysical Research Communications 175*,1090–6.

Baggett, J.M. and Berndt, W.O. 1986. The effect of depletion of nonprotein sulfhydryls by diethyl maleate plus buthionine sulfoximine on renal uptake of mercury in the rat. *Toxicology and Applied Pharmacology* **83**, 556–62.

Ballatori, N. 1994. Glutathione mercaptides as transport forms of metals. *Advances in Pharmacology* **27**, 271–98.

Ballatori, N. and Clarkson, T.W. 1982. Developmental changes in the biliary excretion of methylmercury and glutathione. *Science* **216**, 61–3.

Ballatori, N. and Clarkson, T.W. 1983. Biliary transport of glutathione and meethylmercury. *American Journal of Physiology* **244**, G435–G441.

Ballatori, N. and Clarkson, T.W. 1984a. Inorganic mercury secretion into bile as a low molecular weight complex. *Biochemical Pharmacology* **33**, 1087–92.

Ballatori, N. and Clarkson, T.W. 1984b. Dependence of biliary secretion of inorganic mercury on the biliary transport of glutathione. *Biochemical Pharmacology* **33**, 1093–98.

Ballatori, N. and Clarkson, T.W. 1985. Biliary secretion of glutathione and of glutathione–metal complexes. *Fundamental and Applied Toxicology* **5**, 816–31.

Ballatori, N. and Dutczak, W.J. 1994. Identification and characterization of high and low affinity transport systems for reduced glutathione in liver cell canalicular membranes. *Journal of Biological Chemistry* **269**, 19731–7.

Ballatori, N., Gatmaitan, Z. and Truong, A.T. 1995. Impaired biliary excretion and whole body elimination of methylmercury in rats with a congenital defect in biliary glutathione excretion. *Hepatology* **22**, 1469–73.

Bell, R.R., Nonavinakere, V.K., Soliman, M.R. and Early J.L. 2nd 1991. Effect of *in vitro* treatment of rat hepatocytes with selenium, and/or cadmium on cell viability, glucose output, and cellular glutathione. *Toxicology* **69**, 111–19.

Berndt, W.O., Baggett, J.M., Blacker, A. and Houser, M. 1985. Renal glutathione and mercury uptake by kidney. *Fundamental and Applied Toxicology* **5**, 832–9.

Bertolero, F., Pozzi, G., Sabbioni, E. and Saffiotti, U. 1987. Cellular uptake and metabolic reduction of pentavalent to trivalent arsenic as determinants of cytotoxicity and morphological transformation. *Carcinogenesis* **8**, 803–8.

Beyersmann, D. and Hechtenberg, S. 1997. Cadmium, gene regulation, and cellular signaling in mammalian cells. *Toxicology and Applied Pharmacology* **144**, 247–61.

Blazka, M.E. and Shaikh, Z.A. 1991. Differences in cadmium and mercury uptake by hepatocytes: Role of calcium channels. *Toxicology and Applied Pharmacology* **110**, 355–63.

Bose, S., Mukhopadhyay, S., Chaudhury, S. and Bhattacharya, S. 1994. Correlation of metal distribution, reduced glutathione and metallothionein levels in liver and kidney of rat. *Indian Journal of Experimental Biology* **32**, 679–81.

Burton, C.A., Hatlelid, K., Devine, K., Carter, D.E., Fernando, Q., Brendel, K. and Gandolfi, A.J. 1995. Glutathione effects on toxicity and uptake of mercuric chloride and sodium arsenite in rabbit renal cortical slices. *Environmental Health Perspectives* **103** (Suppl. 1), 81–4.

Cember, H., Gallagher, P. and Faulkner, A. 1968. Distribution of mercury among blood fractions and serum proteins. *American Industrial Hygiene Association Journal* **29**, 233–7.

Chan, H.M. and Cherian, M.G. 1992. Protective role of metallothionein and glutathione in hepatotoxicity of cadmium. *Toxicology* **72**, 281–90.

Chang, W.C., Chen, S.H., Wu, H.L., Shi, G.Y., Murota, S. and Morita, I. 1991. Cytoprotective effect of reduced glutathione in arsenical-induced endothelial cell injury. *Toxicology* **69**, 101–10.

Chin, T.A. and Templeton, D.M. 1993. Protective elevations of glutathione and metallothionein in cadmium-exposed mesangial cells. *Toxicology* **77**, 145–56.

Choi, B.H., Yee, S. and Robles, M. 1996. The effects of glutathione glycoside in methyl mercury poisoning. *Toxicology and Applied Pharmacology* **141**, 357–64.

Clarkson, T.W. 1993. Molecular and ionic mimicry of toxic metals. *Annual Review of Pharmacology and Toxicology* **32**, 545–71.

de Ceaurriz, J., Payan, J.P., Morel, G. and Brondeau, M.T. 1994. Role of extracellular glutathione and γ-glutamyltranspeptidase in the disposition and kidney toxicity of inorganic mercury in rats. *Journal of Applied Toxicology* **14**, 201–6.

Dijkstra, M., Havinga, R., Vonk, R.J. and Kuipers, F. 1996. Bile secretion of cadmium, silver, zinc and copper in the rat. Involvement of various transport systems. *Life Sciences* **59**, 1237–46.

Di Simplicio, P., Gorelli, M., Cluffreda, P. and Leonzio, C. 1990. The relationship between γ-glutamyl transpeptidase and Hg levels in Se/Hg antagonism in mouse liver and kidney. *Pharmacology Research* **22**, 515–26.

Dutczak, W.J. and Ballatori, N. 1994. Transport of the glutathione–methylmercury complex across liver canalicular membranes on reduced glutathione carriers. *Journal of Biological Chemistry* **269**, 9746–51.

Fernandez-Checa, J.C., Takikawa, H., Horie, T., Ookhtens, M. and Kaplowitz, N. 1992. Canalicular transport of reduced glutathione in normal and mutant Eisai hyperbilirubinemic rats. *Journal of Biological Chemistry* **267**, 1667–73.

Fernandez-Checa, J.C., Ookhtens, M. and Kaplowitz, N. 1993. Selective induction by phenobarbital of the electrogenic transport of glutathione and organic anions in rat liver canalicular membrane vesicles. *Journal of Biological Chemistry* **268**, 10836–41.

Foulkes, E.C. 1974. Excretion and retention of cadmium, zinc, and mercury by rabbit kidney. *American Journal of Physiology* **227**,1356–60.

Foulkes, E.C. 1993. Metallothionein and glutathione as determinants of cellular retention and extrusion of cadmium and mercury. *Life Sciences* **52**, 1617–20.

Friedman, H.L. 1957. Relationship between chemical structure and biological activity in mercurial compounds. *Annals of the New York Academy of Sciences* **65**, 461–70.

Fukino, H., Hirai, M., Hsueh, Y.M., Moriyasu, S. and Yamane, Y. 1986. Mechanism of protection by zinc against mercuric chloride toxicity in rats: Effects of zinc and mercury on glutathione metabolism. *Journal of Toxicology and Environmental Health* **19**, 75–89.

Gale, G.R., Smith, A.B., Atkins, L.M., Jones, M.M., Singh, P.K. and Meredith, M.J. 1990.

Glutathione status, glutathione monoisopropyl ester, and cadmium metabolism in mice. *Research Communications in Chemistry Pathology and Pharmacology* **70**, 71–91.

Gambhir, J. and Nath, R. 1992. Effect of cadmium on tissue glutathione and glutathione peroxidase in rats: influence of selenium supplementation. *Indian Journal of Experimental Biology* **30**, 597–601.

Garcia-Ruiz, C., Fernandez-Checa, J.C. and Kaplowitz, N. 1992. Bidirectional mechanism of plasma membrane transport of reduced glutathione in intact rat hepatocytes and membrane vesicles. *Journal of Biological Chemistry* **267**, 22256–64.

Girardi, G. and Elias, M.M. 1993. Effect of different renal glutathione levels on renal mercury disposition and excretion in the rat. *Toxicology* **81**, 57–67.

Girardi, G. and Elias, M.M. 1995. Mercuric chloride effects on rat renal redox enzymes activities: SOD protection. *Free Radical Biology and Medicine* **18**, 61–6.

Gregus, Z. and Varga, F. 1985. Role of glutathione and hepatic glutathione S-transferase in the biliary excretion of methyl mercury, cadmium and zinc: a study with enzyme inducers and glutathione depletors. *Acta Pharmacology and Toxicology* **56**, 398–403.

Gstraunthaler, G., Pfaller, W. and Kotanki, P. 1983. Glutathione depletion and *in vitro* lipid peroxidation in mercury or maleate induced acute renal failure. *Biochemical Pharmacology* **32**, 2969–72.

Gyurasics, A., Varga, F. and Gregus, Z. 1991. Effects of arsenicals on biliary excretion of endogenous glutathione and xenobiotics with glutathione-dependent hepatobiliary transport. *Biochemical Pharmacology* **41**, 937–44.

Hatcher, E.L., Chen, Y. and Kang, Y.J. 1995. Cadmium resistance in A549 cells correlates with elevated glutathione content but not antioxidant enzymatic activities. *Free Radical Biology and Medicine* **19**, 805–12.

Hirata, M., Tanaka, A., Hisanaga, A. and Ishinishi, N. 1990. Effects of glutathione depletion on the acute nephrotoxic potential of arsenite and on arsenic metabolism in hamsters. *Toxicology and Applied Pharmacology* **106**, 469–81.

Houser, M.T. and Berndt, W.O. 1988. Unilateral nephrectomy in the rat: Effects on mercury handling and renal cortical subcellular distribution. *Toxicology and Applied Pharmacology* **93**, 187–94.

Houser, M.T., Milner, L.S., Kolbeck, P.C., Wei, S.H. and Stohs, S.J. 1992. Glutathione monoethyl ester moderates mercuric chloride-induced acute renal failure. *Nephron* **61**, 449–55.

Huang, H., Huang, C.F., Wu, D.R., Jinn, C.M. and Jan, K.Y. 1993. Glutathione as a cellular defense against arsenite toxicity in cultured Chinese hamster ovary cells. *Toxicology* **79**, 195–204.

Inoue, M., Akerboom, T.P.M., Sies, H., Kinne, R., Thao, T. and Arias, I.M. 1984. Biliary transport of glutathione S-conjugates by rat liver canalicular membrane vesicles. *Journal of Biological Chemistry* **259**, 4998–5002.

Jedlitschky, G., Leier, I., Buchholz, U., Center, M. and Keppler, D. 1994. ATP-dependent transport of glutathione S-conjugates by the multidrug resistance-associated protein. *Cancer Research* **54**, 4833–6.

Jedlitschky, G., Leier, I., Buchholz, U., Barnouin, K., Kurz, G. and Keppler, D. 1996. Transport of glutathione, glucuronate, and sulfate conjugates by the MRP gene-encoded conjugate export pump. *Cancer Research* **56**, 988–94.

Johnson, D.R. 1982. Role of renal cortical sulfhydryl groups in development of mercury-induced renal toxicity. *Journal of Toxicology and Environmental Health* **9**, 119–26.

Jones, D.P., Kurtz, J.C. and Sternberg, P. Jr 1997. Thiol and disulfide forms of glutathione (GSH), cysteine (CYS), cysteinylglycine (CG) and homocysteine (HC) in human plasma. *FASEB Journal* **11**, A150.

Kamiyama, T., Miyakawa, H., Li, J.P., Akiba, T., Liu, J.H., Liu, J., Marumo, F. and Sato, C. 1995. Effects of one-year cadmium exposure on livers and kidneys and their relation to glutathione levels. *Research Communications in Chemistry Pathology and Pharmacology* **88**, 177–86.

Kang, Y.J. 1992. Exogenous glutathione decreases cellular cadmium uptake and toxicity. *Drug Metabolism and Disposition* **20**, 714–18.

Kang, Y.J. and Enger, M.D. 1987. Effect of cellular glutathione depletion on cadmium-induced cytotoxicity in human lung carcinoma cells. *Cell Biology and Toxicology* **3**, 347–60.

Kang, Y.J. and Enger, M.D. 1988. Glutathione is involved in the early cadmium cytotoxic response in human lung carcinoma cells. *Toxicology* **48**, 93–101.

Kang, Y.J. and Enger, M.D. 1990. Cadmium cytotoxicity correlates with the changes in glutathione content that occur during the logarithmic growth phase of A549-T27 cells. *Toxicology Letters* **51**, 23–8.

Kang, Y.J., Clapper, J.A. and Enger, M.D. 1989. Enhanced cadmium cytotoxicity in A549 cells with reduced glutathione levels is due to neither enhanced cadmium accumulation nor reduced metallothionein synthesis. *Cell Biology and Toxicology* **5**, 249–59.

Kang, Y.J., Liu, M.S. and Enger, M.D. 1990. Diamide reduces cadmium accumulation by human lung carcinoma A549 cells. *Toxicology* **62**, 53–8.

Keppler, D., Leier, I. and Jedlitschky, G. 1997. Transport of glutathione conjugates and glucuronides by the multidrug resistance proteins MRP1 and MRP2. *Biological Chemistry* **378**, 787–91.

Kim, C.Y., Watanabe, C. and Satoh, H. 1995. Effects of buthionine sulfoximine (BSO) on mercury distribution after Hg(0) exposure. *Toxicology* **98**, 67–72.

Klonne, D.R. and Johnson, D.R. 1988. Enzyme activity and sulfhydryl status in rat renal cortex following mercuric chloride and dithiothreitol administration. *Toxicology Letters* **42**, 199–205.

Lash, L.H. and Anders, M.W. 1989. Uptake of nephrotoxic S-conjugates by isolated rat renal proximal tubular cells. *Journal of Pharmacology and Experimental Therapeutics* **248**, 531–7.

Lash, L.H. and Jones, D.P. 1984. Renal glutathione transport: Characteristics of the sodium-dependent system in the basal-lateral membrane. *Journal of Biological Chemistry* **259**, 14508–14.

Lash, L.H. and Jones, D.P. 1985a. Uptake of the glutathione conjugate *S*-(1,2-dichloro-vinyl)glutathione by renal basal-lateral membrane vesicles and isolated kidney cells. *Molecular Pharmacology* **28**, 278–82.

Lash, L.H. and Jones, D.P. 1985b. Distribution of oxidized and reduced forms of glutathione and cyst(e)ine in rat plasma. *Archives of Biochemistry and Biophysics* **240**, 583–92.

Lash, L.H. and Zalups, R.K. 1992. Mercuric chloride-induced cytotoxicity and compensatory hypertrophy in rat kidney proximal tubular cells. *Journal of Pharmacology and Experimental Therapeutics* **261**, 819–29.

Lash, L.H. and Zalups, R.K. 1996. Alterations in renal cellular glutathione metabolism after *in vivo* administration of a subtoxic dose of mercuric chloride. *Journal of Biochemical Toxicology* **11**, 1–9.

Lash, L.H., Jones, D.P. and Anders, M.W. 1988. Glutathione homeostasis and glutathione S-conjugate toxicity in kidney. *Reviews in Biochemical Toxicology* **9**, 29–67.

Lash, L.H., Xu, Y., Elfarra, A.A., Duescher, R.J. and Parker, J.C. 1995. Glutathione-dependent metabolism of trichloroethylene in isolated liver and kidney cells of rats and its role in mitochondrial and cellular toxicity. *Drug Metabolism and Disposition* **23**, 846–53.

Lash, L.H., Putt, D.A. and Zalups, R.K. 1998. Role of extracellular thiols in accumulation and distribution of inorganic mercury in rat renal proximal and distal tubular cells. *Journal of Pharmacology and Experimental Therapeutics* **285**, 1039–50.

Lau, S.-J. and Sarkar, B. 1979. Inorganic mercury(II)-binding components in normal human blood stream. *Journal of Toxicology and Environmental Health* **5**, 907–16.

Leier, I., Jedlitschky, G., Buchholz, U., Center, M., Cole, S.P.C., Deeley, R.G. and Keppler, D.

1996. ATP-dependent glutathione disulphide transport mediated by MRP gene-encoded conjugate export pump. *Biochemical Journal* **314**, 433–7.

Li, S., Thompson, S.A., Kavanagh, T.J. and Woods, J.S. 1996a. Localization by *in situ* hybridization of γ-glutamylcysteine synthetase mRNA expression in rat kidney following acute methylmercury treatment. *Toxicology and Applied Pharmacology* **141**, 59–67.

Li, S., Thompson, S.A. and Woods, J.S. 1996b. Localization of γ-glutamylcysteine synthetase mRNA expression in mouse brain following methylmercury treatment using reverse transcription *in situ* PCR amplification. *Toxicology and Applied Pharmacology* **140**, 180–7.

Li, W., Kagan, H.M. and Chou, I.N. 1994. Alterations in cytoskeletal organization and homeostasis of cellular thiols in cadmium-resistant cells. *Toxicology and Applied Pharmacology* **126**, 114–23.

Lund, B.O., Miller, D.M. and Woods, J.S. 1993. Studies on Hg(II)-induced H_2O_2 formation and oxidative stress *in vivo* and *in vitro* in rat kidney mitochondria. *Biochemical Pharmacology* **45**, 2017–24.

Magos, L., Clarkson, T.W. and Allen, J. 1978. The interrelationship between nonprotein bound thiols and the biliary excretion of methylmercury. *Biochemical Pharmacology* **27**, 2203–8.

Mückter, H., Islambouli, S., Doklea, E., Hopfer, C., Szinicz, L., Fichtl, B. and Forth, W. 1993. Isolated rat kidney tubules as a screening system for arsenic antidotes. *Toxicology and Applied Pharmacology* **121**, 118–28.

Mulder, K.M. and Kostyniak, P.J. 1985. Involvement of glutathione in the enhanced renal excretion of methyl mercury in CFW Swiss mice. *Toxicology and Applied Pharmacology* **78**, 451–7.

Muller, L., Abel, J. and Ohnesorge, F.K. 1986. Absorption and distribution of cadmium (Cd), copper and zinc following oral subchronic low level administration to rats of different binding forms of cadmium (Cd-acetate, Cd-metallothionein, Cd-glutathione). *Toxicology* **39**, 187–95.

Naganuma, A., Oda-Urano, N., Tanaka, T. and Imura, N. 1988. Possible role of hepatic glutathione in transport of methymercury into mouse kidney. *Biochemical Pharmacology* **37**, 291–6.

Nath, K.A., Croatt, A.J., Likely, S., Behrens, T.W. and Warden, D. 1996. Renal oxidant injury and oxidant response induced by mercury. *Kidney International* **50**, 1032–43.

Nehru, L.B. and Bansal, M.P. 1997. Effect of selenium supplementation on the glutathione redox system in the kidney of mice after chronic cadmium exposures. *Journal of Applied Toxicology* **17**, 81–4.

Nicotera, P., Moore, M., Bellomo, G., Mirabelli, F. and Orrenius, S. 1985. Demonstration and partial characterization of glutathione disulfide-stimulated ATPase activity in the plasma membrane fraction from rat hepatocytes. *Journal of Biological Chemistry* **260**, 1999–2002.

Nielsen, J.B., Andersen, H.R., Anderson, O. and Starklint, H. 1994. Dose and time relations in $Hg^{(++)}$-induced tubular necrosis and regeneration. *Environmental Health Perspectives* **102** (Suppl. 3), 317–20.

Ochi, T. 1997. Arsenic compound-induced increases in glutathione levels in cultured Chinese hamster V79 cells and mechanisms associated with changes in γ-glutamylcysteine synthetase activity, cystine uptake and utilization of cysteine. *Archives of Toxicology* **71**, 730–40.

Ochi, T., Otsuka, F., Takahashi, K. and Ohsawa, M. 1988. Glutathione and metallothioneins as cellular defense against cadmium toxicity in cultured Chinese hamster cells. *Chemico-Biological Interactions* **65**, 1–14.

Ochi, T., Nakajima, F., Sakurai, T., Kaise, T. and Oya-Ohta, Y. 1996. Dimethylarsinic acid causes apoptosis in HL-60 cells via interaction with glutathione. *Archives of Toxicology* **70**, 815–21.

Oram, P.D., Fang, X., Fernando, Q., Letkeman, P. and Letkeman, D. 1996. The formation constants of mercury(II)–glutathione complexes. *Chemical Research in Toxicology* **9**, 709–12.

Ossola, J.O. and Tomaro, M.L. 1995. Heme oxygenase induction by cadmium chloride: evidence for oxidative stress involvement. *Toxicology* **104**, 141–7.

Prozialeck, W.C. and Lamar, P.C. 1995. Effects of glutathione depletion on the cytotoxic actions of cadmium in LLC-PK1 cells. *Toxicology and Applied Pharmacology* **134**, 285–95.

Rabenstein, D.L. 1989. Metal complexes of glutathione and their biological significance, in *Glutathione: Chemical, Biochemical and Medical Aspects*, (eds D. Dolphin, O. Auramovic and R. Poulson), Wiley, New York, pp. 147–86.

Ramos, O., Carrizales, L., Yanez, L., Mejia, J., Batres, L., Ortiz, D. and Diaz-Barriga, F. 1995. Arsenic increased lipid peroxidation in rat tissues by a mechanism independent of glutathione levels. *Environmental Health Perspectives* **103** (Suppl. 1), 85–8.

Rana, S.V. and Verma, S. 1996. Protective effects of GSH, vitamin E, and selenium on lipid peroxidation in cadmium-fed rats. *Biological Trace Elements Research* **51**, 161–8.

Robinson, M.K., Barfuss, D.W. and Zalups, R.K. 1993. Cadmium transport and toxicity in isolated perfused segments of the renal proximal tubule. *Toxicology and Applied Pharmacology* **121**, 103–11.

Rosner, M.H. and Carter, D.E. 1987. Metabolism and excretion of gallium arsenide and arsenic oxides by hamsters following intratracheal instillation. *Fundamental and Applied Toxicology* **9**, 730–7.

Sarkar, S., Yadav, P. and Bhatnagar, D. 1997. Cadmium-induced lipid peroxidation and the antioxidant system in rat erythrocytes: the role of antioxidants. *Journal of Trace Element and Medical Biology* **11**, 8–13.

Scott, N., Hatlelid, K.M., MacKenzie, N.E. and Carter, D.E. 1993. Reactions of arsenic(III) and arsenic(V) species with glutathione. *Chemical Research in Toxicology* **6**, 102–6.

Sengupta, S. and Bhattacharyya, N.P. 1996. Possible role of glutathione in resistance to heavy metals and hydrogen peroxide in a radioresistant Chinese hamster V79 cell strain. *Indian Journal of Experimental Biology* **34**, 905–8.

Shaikh, Z.A., Blazka, M.E. and Endo, T. 1995. Metal transport in cells: Cadmium uptake by rat hepatocytes and renal cortical epithelial cells. *Environmental Health Perspectives* **103** (Suppl. 1), 73–5.

Sidhu, M., Prasad, R., Gill, K.D. and Nath, R. 1997. Alterations in isoforms of glutathione *S*-transferase in liver and kidney of cadmium exposed rhesus monkeys: purification and kinetic characterization. *Molecular and Cellular Biochemistry* **166**, 55–63.

Siegers, C.P., Sharma, S.C. and Younes, M. 1986. Hepatotoxicity of metals in glutathione-depleted mice. *Toxicology Letters* **34**, 185–91.

Singhal, R.K., Andersen, M.E. and Meister, A. 1987. Glutathione, a first line of defense against cadmium toxicity. *FASEB Journal* **1**, 220–3.

Stacey, N.H. 1986. The amelioration of cadmium-induced injury in isolated hepatocytes by reduced glutathione. *Toxicology* **42**, 85–93.

Stevenson, K.J., Hale, G. and Perham, R.N. 1978. Inhibition of pyruvate dehydrogenase multienzyme complex from *Escherichia coli* with mono- and bifunctional arsenoxides. *Biochemistry* **17**, 2189–92.

Stohs, S.J. and Bagchi, D. 1995. Oxidative mechanisms in the toxicity of metal ions. *Free Radical Biology and Medicine* **18**, 321–36.

Strubelt, O., Kremer, J., Tilse, A., Keogh, J., Pentz, R. and Younes, M. 1996. Comparative studies on the toxicity of mercury, cadmium, and copper toward the isolated perfused rat liver. *Journal of Toxicology and Environmental Health* **47**, 267–83.

Sugawara, N., Lai, Y.R., Arizono, K. and Ariyoshi, T. 1996. Biliary excretion of exogenous cadmium, and endogenous copper and zinc in the Eisai hyperbilrubinuric (EHB) rat with a near absence of biliary glutathione. *Toxicology* **112**, 87–94.

Suzuki, C.A. and Cherian, M.G. 1989. Renal glutathione depletion and nephrotoxicity of cadmium–metallothionein in rats. *Toxicology and Applied Pharmacology* **98**, 544–52.

Suzuki, K.T., Sunaga, H., Kobayashi, E. and Shimojo, N. 1986. Mercaptalbumin as a selective cadmium-binding protein in rat serum. *Toxicology and Applied Pharmacology* **86**, 466–73.

Tanaka, T., Naganuma, A. and Imura, N. 1990. Role of γ-glutamyltranspeptidase in renal uptake and toxicity of inorganic mercury in mice. *Toxicology* **60**, 187–98.

Tanaka, T., Naganuma, A., Kobayashi, K. and Imura, N. 1991. An explanation for strain and sex differences in renal uptake of methylmercury in mice. *Toxicology* **69**, 317–29.

Tanaka-Kagawa, T., Naganuma, A. and Imura, N. 1993. Tubular secretion and reabsorption of mercury compounds in mouse kidney. *Journal of Pharmacology and Experimental Therapeutics* **264**, 776–82.

Tanaka-Kagawa, T., Suzuki, M., Naganuma, A., Yamanaka, N. and Imura, N. 1998. Strain difference in sensitivity of mice to renal toxicity of inorganic mercury. *Journal of Pharmacology and Experimental Therapeutics* **285**, 335–41.

Templeton, D.M. 1990. Cadmium uptake by cells of renal origin. *Journal of Biological Chemistry* **265**, 21764–70.

Thompson, D.J. 1993. A chemical hypothesis for arsenic methylation in mammals. *Chemico-Biological Interactions* **88**, 89–114.

Urano, T., Naganuma, A. and Imura, N. 1988a. Methylmercury-cysteinylglycine constitutes the main form of methylmercury in rat bile. *Research Communications in Chemistry Pathology and Pharmacology* **60**, 197–210.

Urano, T., Naganuma, A. and Imura, N. 1988b. Species differences in biliary excretion of methylmercury: role of non-protein sulfhydryls in bile. *Research Communications in Chemistry Pathology and Pharmacology* **62**, 339–51.

Urano, T., Iwasaki, A., Himeno, S., Naganuma, A. and Imura, N. 1990. Absorption of methylmercury compounds from rat intestine. *Toxicology Letters* **50**, 159–64.

Viljoen, A.J. and Tappel, A.L. 1988. Interactions of selenium and cadmium with metallothionein-like and other cytosolic proteins of rat kidney and liver. *Journal of Inorganic Biochemistry* **34**, 277–90.

Waldegger, S., Schmidt, F., Herzer, T., Gulbins, E., Schuster, A., Biber, J., Markovich, D., Murer, H., Busch, A.E. and Lang, F. 1995. Heavy metal mediated inhibition of rBAT-induced amino acid transport. *Kidney International* **47**, 1677–81.

Weinberg, J.M., Harding, P.G. and Humes, H.D. 1982a. Mitochondrial bioenergetics during the initiation of mercuric chloride-induced renal injury. I. Direct effects of *in vitro* mercuric chloride on renal cortical mitochondrial function. *Journal of Biological Chemistry* **257**, 60–7.

Weinberg, J.M., Harding, P.G. and Humes, H.D. 1982b. Mitochondrial bioenergetics during the initiation of mercuric chloride-induced renal injury. II. Functional alterations of renal cortical mitochondria isolated after mercuric chloride treatment. *Journal of Biological Chemistry* **257**, 68–74.

Wilks, M.F., Kwizera, E.N. and Bach, P.H. 1990. Assessment of heavy metal nephrotoxicity *in vitro* using isolated rat glomeruli and proximal tubular fragments. *Renal Physiology and Biochemistry* **13**, 275–84.

Woods, J.S. and Ellis, M.E. 1995. Up-regulation of glutathione synthesis in rat kidney by methyl mercury. Relationship to mercury-induced oxidative stress. *Biochemical Pharmacology* **50**, 1719–24.

Woods, J.S., Davis, H.A. and Baer, R.P. 1992. Enhancement of γ-glutamylcysteine synthetase mRNA in rat kidney by methyl mercury. *Archives of Biochemistry and Biophysics* **296**, 350–3.

Yasutake, A., Hirayama, K. and Inoue, M. 1989. Mechanism of urinary excretion of methylmercury in mice. *Archives of Toxicology* **63**, 479–83.

Zalups, R.K. 1995. Organic anion transport and action of γ-glutamyl transpeptidase in kidney mechanistically linked to renal tubular uptake of inorganic mercury. *Toxicology and Applied Pharmacology* **132**, 289–98.

Zalups, R.K. 1997. Enhanced renal outer medullary uptake of mercury associated with uninephrectomy: Implication of a luminal mechanism. *Journal of Toxicology and Environmental Health* **50**, 173–94.

Zalups, R.K. and Barfuss, D.W. 1993. Transport and toxicity of methylmercury along the proximal tubule of the rabbit. *Toxicology and Applied Pharmacology* **121**, 176–85.

Zalups, R.K. and Barfuss, D.W. 1995a. Accumulation and handling of inorganic mercury in the kidney after coadministration with glutathione. *Journal of Toxicology and Environmental Health* **44**, 385–99.

Zalups, R.K. and Barfuss, D.W. 1995b. Renal disposition of mercury in rats after intravenous injection of inorganic mercury and cysteine. *Journal of Toxicology and Environmental Health* **44**, 401–13.

Zalups, R.K. and Barfuss, D.W. 1995c. Pretreatment with *p*-aminohippurate inhibits the renal uptake and accumulation of injected inorganic mercury in the rat. *Toxicology* **103**, 23–35.

Zalups, R.K. and Barfuss, D.W. 1996. Diversion or prevention of biliary outflow from the liver diminishes the renal uptake of injected inorganic mercury. *Drug Metabolism and Disposition* **24**, 480–6.

Zalups, R.K. and Barfuss, D.W. 1998. Small aliphatic dicarboxylic acids inhibit renal uptake of administered mercury. *Toxicology and Applied Pharmacology* **148**, 183–93.

Zalups, R.K. and Diamond, G.L. 1987. Mercuric chloride-induced nephrotoxicity in the rat following unilateral nephrectomy and compensatory renal growth. *Virchows Archives B* **53**, 336–46.

Zalups, R.K. and Lash, L.H. 1990. Effects of uninephrectomy and mercuric chloride on renal glutathione homeostasis. *Journal of Pharmacology and Experimental Therapeutics* **254**, 962–70.

Zalups, R.K. and Lash, L.H. 1997a. Depletion of glutathione in the kidney and the renal disposition of administered inorganic mercury. *Drug Metabolism and Disposition* **25**, 516–23.

Zalups, R.K. and Lash, L.H. 1997b. Binding of mercury in renal brush-border and basolateral membrane-vesicles: Implication of a cysteine conjugate of mercury involved in the luminal uptake of inorganic mercury in the kidney. *Biochemical Pharmacology* **53**, 1889–900.

Zalups, R.K. and Minor, K.H. 1995. Luminal and basolateral mechanisms involved in the renal tubular uptake of inorganic mercury. *Journal of Toxicology and Environmental Health* **46**, 73–100.

Zalups, R.K. and Veltman, J.C. 1988. Renal glutathione homeostasis in compensatory renal growth. *Life Sciences* **42**, 2171–6.

Zalups, R.K., Klotzbach, J.M. and Diamond, G.L. 1987. Enhanced accumulation of injected inorganic mercury in renal outer medulla after unilateral nephrectomy. *Toxicology and Applied Pharmacology* **94**, 226–36.

Zalups, R.K., Robinson, M.K. and Barfuss, D.W. 1991. Factors affecting inorganic mercury transport and toxicity in the isolated perfused proximal tubule. *Journal of the American Society of Nephrology* **2**, 866–78.

Zalups, R.K., Knutson, K.L. and Schnellmann, R.G. 1993. *In vitro* analysis of the accumulation and toxicity of inorganic mercury in segments of the proximal tubule isolated from the rabbit kidney. *Toxicology and Applied Pharmacology* **119**, 221–7.

14

ROLE OF METALLOTHIONEIN IN THE METABOLISM, TRANSPORT AND TOXICITY OF METALS

Michael P. Waalkes and Raül Pérez-Ollé

14.1 Introduction

Metals are an indispensable part of organic life. Many are, in fact, essential for survival. They participate in a vast array of critical physiologic functions. Essential metals function as components in the active site of many enzymes, as well as in neuro-transmission, muscle contraction, oxygen transport and control of gene expression. For instance, zinc is essential to over 300 different enzymes of 50 different types (Vallee, 1995; Vallee and Auld, 1995). Because of the critical nature of metals to so many cellular functions, intentional accumulation is a common biologic strategy. However, all metals and metal compounds are potentially toxic. This makes the safe transport, storage and usage of metals a physiologic necessity.

In addition to naturally occurring metals in the environment, humans have been intentionally concentrating metals, as a by-product of industrialization, within the biosphere for thousands of years. The intensive and deliberate use of metals is inalterably linked with the success of modern cultures. Therefore human exposure to metals is inevitable. As a result, metals are an important class of toxic agents and, in fact, some of the most toxic agents known are metals. Metal exposure in humans can result in a wide variety of adverse effects, ranging from debilitating neurotoxicity to highly aggressive malignancies.

Metals, as toxic agents, have several key characteristics. First, they often accumulate in biologic material. This is, in part, because many nonessential metals will mimic essential elements and gain access through systems intended for the transport and storage of the essential element. Secondly, although the precise form of a particular metal many change, the basic unit is, to all intents and purposes, biologically inde-structible. Thus, unlike organic toxicants which can often be broken down metabolically to less toxic subunits, metals remain intact in biological systems.

Metallothionein (MT) was first isolated from the equine renal cortex over 40 years ago (Margoshes and Vallee, 1957). Since that time a large amount of research has focused on this low-molecular-weight, metal-binding protein. Despite all these efforts, the precise role of MT remains elusive. The binding of metals by MT is clearly a chemical function of this protein but does not provide direct evidence of a physiologic function.

There are indications that MT participates as a factor in the homeostatic control of the essential metals zinc and copper. The reduction of the adverse effects of toxic metals, such as cadmium or mercury, by MT is also well established. However, functionally speaking, reduced toxicity of these nonessential metals may be more fortuitous than intentional. It is difficult to envision the evolution of a biologic system designed to protect against toxic metals, such as cadmium or mercury, when significant exposure to these metals has only occurred very recently in human history due to anthropogenic emissions. This does not mean that protection against toxic metals is not a real function of MT but rather serves to create the distinction between physiologic and toxicologic function.

This chapter will focus on the role of the MT protein in the metabolism and toxicity of metals. Other potential roles of MT that would be distinct from its metal-binding functions, such as in the prevention of toxicity from organic compounds, will be left largely to other reviews (Vallee, 1995; Lazo and Pitt, 1995; Penketh *et al.*, 1996). Additionally, although MTs are widely expressed in many, diverse species, this review will focus on mammalian MTs.

14.2 Physiology and biochemistry of metallothionein

14.2.1 Introduction

Metallothioneins were initially discovered over four decades ago as cadmium-binding proteins in equine renal cortex (Margoshes and Vallee, 1957). Since that time this family of proteins has proven to be of great interest in a variety of scientific disciplines, including toxicology, biological and physical chemistry, molecular biology and cancer research. As a class of proteins MTs have several unusual characteristics. They are small proteins, of the order of 6000 Da, that are perhaps most distinguished by their high metal content. These metal-binding proteins are found in a very wide variety of organisms, ranging from invertebrates to humans (Henry *et al.*, 1994; Dallinger, 1996). As the name implies, MTs contain numerous thiol groups, due to their very high cysteine content, which provide the basis for high-affinity binding of many metals, doubtless a major purpose of these proteins. MTs can, in fact, bind a variety of transition metals including cadmium, copper, mercury, silver and zinc. In most cases MTs are highly inducible and protein levels can be accumulated in response to a wide variety of stimulants. MTs are thought to have several diverse, potential functions, including essential element homeostasis and toxic metal detoxification.

14.2.2 Structure and biological characteristics of MT

The distinguishing biochemical characteristics of MTs are listed in Table 14.1 and have been the subject of many reviews over the years (e.g. Kägi and Schäffer, 1988; Waalkes and Goering, 1990; Kägi, 1991; Cherian and Chan, 1993; Kägi, 1993; Vallee and Maret, 1993; Vallee, 1995; Aschner, 1997; Moffatt and Denizeau, 1997). The structure of MT is typified by a preponderance of cysteine, which accounts for up to a third of all the amino acids in this protein. MTs also have a relatively high content of serine and glycine. The MTs lack the aromatic amino acids and histidine. The cysteinyl residues occur in a highly conserved pattern and the 60+ residue chain is interspersed with a

Table 14.1 Characteristics of metallothioneins (modified from Waalkes and Goering, 1990)

Molecular weight 6000–7000 Da
Metal clusters with bridging thiolate ligands
High cysteine content (~30%)
High metal content
High metal affinity
Deficient in aromatic amino acids and histidine
Optical properties characteristic of metal–thiolate bonds
Highly ordered and conserved cysteine arrangement
Often, but not always, inducible by various chemicals and physical treatments
Polymorphic

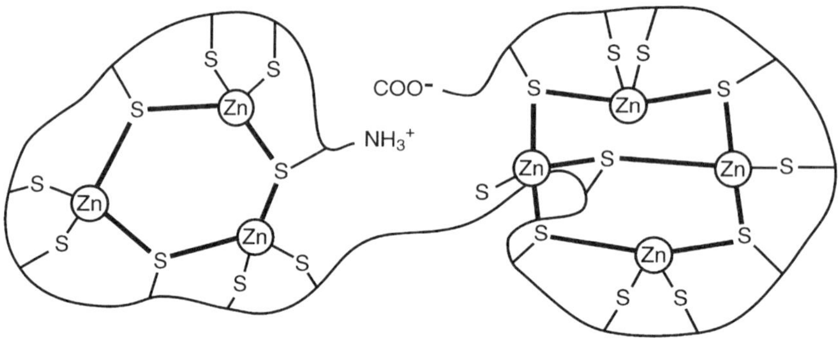

Figure 14.1 Schematic diagram of MT, showing seven zinc atoms bound into two clusters, one of three metals atoms and one of four, with each metal coordinated tetrahedrally to the internally directed cysteines. (Adapted from Nielson and Winge, 1983.)

series of cysteine-X-cysteine, cysteine–X–X–cysteine, and cysteine–X–X–X–cysteine units (where X = noncysteine amino acid). This gives MT the ability to bind a remarkable number of metal atoms (up to seven in the case of cadmium, mercury and zinc, 12 in the case of copper, and 18 in the case of silver). When there are seven metal atoms bound to MT they are chelated into two clusters, one of three metal atoms and one of four, with each metal coordinated tetrahedrally to cysteines (Fig. 14.1). The binding of metal ions to the clusters is selective and specific in nature and is a function of the known differences in binding affinity, which may facilitate intermolecular metal ion transfer between MTs and metal-requiring proteins such as enzymes (Vallee and Maret, 1993). Disulfide bridges do not normally occur in MT and all cysteinyl sulfurs appear to participate in the complexation of metals. MT is an elongated, single-chain molecule with a high degree of random structure and has no known enzymatic activity. A very heat-stable protein, MT is also resistant to precipitation even at low pH and these characteristics are often used in purification or quantification of MT (Waalkes, 1996). MTs have been characterized from a wide variety of both mammalian and nonmammalian sources (Kägi and Kojima, 1987a, b; Waalkes and Goering, 1990; Kägi, 1991, 1993; Henry *et al.*, 1994; Dallinger, 1996).

The dynamic nature of metal binding to MT includes not only the selectivity of the MT clusters for metal atoms but also a cooperativity of metal binding and release from MT (Otvos *et al.*, 1993; Vallee and Maret, 1993). Together with the known reactivity of the binding sites with electrophiles (Otvos *et al.*, 1993) these characteristics indicate that the release of metal from MT could be specifically controlled and potentially limited when necessary (Vallee and Maret, 1993). In this regard, during zinc deficiency, the release of zinc from MT is more rapid than can be accounted for by simple proteolysis (Krezoski *et al.*, 1988). This supports a control mechanism for metal release beyond simple degradation (Vallee and Maret, 1993). It is interesting to speculate that bioeffector molecules may displace MT-bound metal as part of a control system, and, in fact, nitric oxide can displace metals bound to MT within cells (Misra *et al.*, 1996).

Four major isoforms of MT have been discovered, namely MT-I, MT-II, MT-III and MT-IV (Uchida *et al.*, 1991; Tsuji *et al.*, 1992; Palmiter *et al.*, 1992; Kägi, 1993; Quaife *et al.*, 1994). The most common forms, and those that were intially isolated, are MT-I and MT-II. MT-I and MT-II are coordinately expressed in most tissues. Although MT-I and MT-II can be separated by anion exchange chromatography, these two isoforms actually show very little difference, and there are only slight differences in the amino-acid sequence and in the net charge.

More recently a third MT, MT-III, was added to the MT family. This MT isoform is expressed only in the brain (Uchida *et al.*, 1991; Palmiter *et al.*, 1992; Tsuji *et al.*, 1992). MT-III differs from MT-I or MT-II because it contains 68 amino acids instead of 61 or 62. In the human MT-III protein a total of 38 of the amino acids, including all the cysteines, show identical alignment with the MT-I and MT-II isoforms. MT-III differs from the other isoforms in that it contains two additional amino-acid insertions, including a single threonine in the N-terminal region, and a six-residue insert in the C-terminal region. Unlike other isoforms, MT-III expression does not appear to be inducible.

MT-IV is the most recent addition to the MT family and is specifically expressed in specialized epithelia (Quaife *et al.*, 1994). Although 30 of the amino acid residues in MT-IV, including the cysteines, are identical to those in MT-I and MT-II, MT-IV differs in approximately 24–27 amino-acid residues from the other isoforms (Palmiter *et al.*, 1993a; Quaife *et al.*, 1994). Thus most of the amino acids in MT-IV that are nonessential to the metal-binding function are different from other isoforms.

All MT isoforms will complex metals, including zinc and copper, and do not contain aromatic amino acids. It has been hypothesized that MTs may be associated with other macromolecules during the process of metal donation and that this complexation may be related to the surface properties of the particular MT (Quaife *et al.*, 1994). In this way the individual MT isoform could dictate a tissue-specific function in metal donation, although there is no direct evidence for this.

For two major isoforms of MT, MT-I and MT-II, synthesis is highly inducible by a variety of metals and other agents (Table 14.2; Onosaka and Cherian, 1981, 1982; Waalkes and Klaassen, 1984; Waalkes and Goering, 1990; Kägi, 1991, 1993; Cherian and Chan, 1993; Vallee, 1995; Moffatt and Denizeau, 1997). Cadmium and zinc are among the best metal inducers of MT synthesis. Other inducer metals include mercury, bismuth, copper and certain organometallic compounds such as cisplatin. This remarkable inducibility of MT by metals such as cadmium or zinc has justifiably influenced

Table 14.2 Examples of treatments that induce metallothionein synthesis (modified from Waalkes and Goering, 1990)

Metals:	Glucocorticoids
Cd, Zn, Hg, Cu, Bi, As, etc.	Indomethacin
	Insulin
Chemical agents:	Interferon-α, -γ
Acetaminophen	Interleukin-1, -6
Adriamycin	Paraquat
Angiotensin II	Phorbol esters
Ascorbic acid	2-Propanol
Azacytidine	Progesterone
Bromobenzene	Retinoic acid
Butyric acid	Streptozotocin
Calmodulin inhibitors	Tert-butylhydroquinone
Carbon tetrachloride	Tumor necrosis factor
Carrageenan	Urethane
Catecholamines	
Chloroform	**Stress-producing conditions:**
Chlorpromazine	Food deprivation
Cisplatin	Hypothermia
Cyproterone acetate	Infection
Dextran	Inflammation
Di(ethylhexyl)phthalate	Oxidative stress
Endotoxin	UV irradiation
Estradiol	X irradiation
Ethanol	
Ethionine	**Surgical procedures:**
Formaldehyde	Partial hepatectomy
Glucagon	Unilateral nephrectomy

In many cases the precise form of MT was not determined in these studies but probably represents MT-I and MT-II.

thinking about the function MT in playing important roles in metal homeostasis or toxicology.

MT synthesis is also induced by a multitude of nonmetallic agents, including a very wide variety of hormones, pharmaceuticals, vitamins, alcohols, cytokines, antibiotics and other diverse chemical and physical treatments (for examples, see Table 14.2). Some nonmetallic inducers of MT include acetaminophen, azacytidine, butyrate, endotoxin, ethanol, ethionine, estradiol, dexamethasone, formaldehyde, paraquat, penicillamine, retinoic acid, and urethane among many others. Physical stresses that are capable of stimulating MT production include both UV- and X-irradiation as well as oxidative stress, hypothermia, infection and inflammation. This inducibility is regulated at the transcriptional level by various *cis*-acting DNA sequences located in the promoter region of MT genes (see Section 14.2.3). The inducibility of MT is very likely an important functional factor of the protein, at least when it comes to protection of cells from excess metal.

14.2.3 Molecular biology of MT

One of the most striking characteristics of the molecular biology of MT is the existence of multiple genes and the regulation of their expression by multiple factors (Palmiter,

1987; Hamer and Winge, 1989; Palmiter *et al.*, 1993a,b; Sanders *et al.*, 1996; Zhu and Thiele, 1996; Moffatt and Denizeau, 1997; Samson and Gedamu, 1998). MTs are encoded by a multigene family consisting of at least 12–15 members (Karin and Richards, 1984; Karin *et al.*, 1984; Varshney *et al.*, 1984). All known functional human *MT* genes are located in a locus on chromosome 16 (Karin *et al.*, 1984; Richards *et al.*, 1984; Varshney *et al.* 1986). It appears that about half of these members are nonfunctional (Karin *et al.*, 1984; Richards *et al.*, 1984; Varshney and Gedamu, 1984). Earlier reports located the MT locus by *in situ* hybridization at 16q22 (Le Beau *et al.*, 1985), a region involved in a translocation associated with myelomonocytic leukemia. However, later experiments suggested that the locus is located at 16q13 (Simmers *et al.*, 1987; Sutherland *et al.*, 1990). The structure of the functional human MT locus on chromosome 16 (locus 16q13) which contains 14 linked MT genes in a region spanning 82.1 kb has been reported (West *et al.*, 1990). It has been suggested that chromosomal loci external to this site contain the inactive pseudogenes (Karin and Richards, 1982). In the mouse the genes encoding the MT-I and MT-II isoforms are closely linked and located on chromosome 8 (Durnam *et al.*, 1980; Glanville *et al.*, 1981; Searle *et al.*, 1984).

Stennard *et al.* (1994) suggested that more MT isoforms will be discovered, based on the fact that fewer MT proteins have been identified than *MT* genes. It is suggested that the various *MT* genes exist in humans to provide fine control of *MT* gene expression, but there may be functional implications as well. The main isoforms expressed in human tissues are hMT-I_A and hMT-II_A.

In the rat the sequences of the structural *MT* gene, three pseudogenes and a single transcript of approximately 550 bp are known (Andersen *et al.*, 1983a, 1986). A comparison of rat and mouse MT-I sequences reveals extensive homology in both the translated (95% residue homology) and untranslated regions. The differences between the amino-acid sequences of rat MT-I and mouse MT-I are due to single-base differences. These changes are conservative and do not affect the charge properties of the molecule or cysteine residues (Andersen *et al.*, 1983a). Greater divergence exists between rat MT-I and human MT-II coding sequences (85%; Karin and Richards, 1982). The human MT-II sequence shows nine amino-acid changes from rat MT-I, involving single-, double- and triple-base changes, resulting in two amino-acid substitutions which are not conservative in respect to charge preservation.

All mammalian *MT* genes share the same structure, with three exons (spanning about 1–2 kb) and two introns. There are greater differences in the control regions of the different *MT* genes than there are in the proteins that they encode. Vertebrate *MT* genes have a 5′ flanking region, a 5′ untranslated region and a 3′ flanking end. A number of DNA sequences in the promoter of *MT* genes act as *cis*-acting regulatory elements to induce transcription of *MT* genes (Hamer, 1986; Palmiter, 1987). Different types of regulatory sequences have been identified: metal-regulatory elements (MREs), glucocorticoid-regulatory elements (GREs) and interferon-regulatory elements (IFN). The MREs, GREs and IREs regulate the expression of *MT* genes by metals, glucocorticoids and interferon, respectively.

The MREs are located in the 5′ untranslated sequence of the gene (Andersen *et al.*, 1987). Studies by Andersen *et al.* (1987) suggest that a transcription factor (TF) binds to the MRE in the *MT* gene in the presence of metal ions and subsequently dissociates when the metal ions are removed, thus controlling induction of *MT* gene expression.

Also, *trans*-acting DNA-binding proteins bind MT-induced metals, facilitating the interaction of the metal with the corresponding MRE to induce the transcription of the gene (Hamer, 1986; Palmiter, 1987). Brugnera *et al.* (1994) and Palmiter (1994) have shown that the transcription-activating protein factor (MTF-1) binds to the MRE.

Differences in the basal activity of MREs seems to be determined by both the sequence of the MRE and its position in the promoter region (Samson and Gedamu, 1998). It is thought that the MRE-binding transcription activator, MTF-1, interacts with MRE sequences to regulate the expression of *MT* genes when induced by metals (Palmiter, 1994). In this model, zinc binding releases MTF-1 from an inhibitor. In addition, MREs may also play a role in the H_2O_2-mediated induction of the *MT-I* gene expression (Dalton *et al.*, 1994). A model for the regulation of *MT* gene expression was recently proposed by Palmiter (1994) after isolating a transcription factor that binds to MREs. According to this model, the uninduced basal state of transcription provides the background levels of MT, while MTF (a zinc-finger-containing transcription factor) is complexed with a protein called MT inhibitor (MTI). The presence of additional zinc provokes the disruption of the MTF–MTI complex freeing MTF to bind to the MRE and stimulate transcription of the *MT* gene. This transcriptional increase results in an increased synthesis of apoMT, which then binds excess zinc (including the zinc atoms bound to MTI) and effectively reduces the concentration of free zinc. This then allows the MTF–MTI complex to reform, withdrawing the stimulation provided by the MTF and the level of transcription returns to uninduced background levels. Metals such as cadmium and mercury presumably stimulate *MT* gene transcription by displacing zinc from MT or from other binding sites within the cell, resulting in an increase in free cellular zinc which is then able to combine with the MTI, releasing the MTF which is then available to bind the MRE and induce *MT* gene transcription. Alternatively, direct binding to the MTI by metals other than zinc might be possible.

A 455 bp region in the *MT* gene containing two GREs that bind the glucocorticoid receptor and prompt the induction of MT expression by dexamethasone has also been identified (Kelly *et al.*, 1997). These two GREs are located about 1 kb and 7 kb upstream of the *MT-II* and *MT-I* genes, respectively, and can regulate induction of *MT* genes by

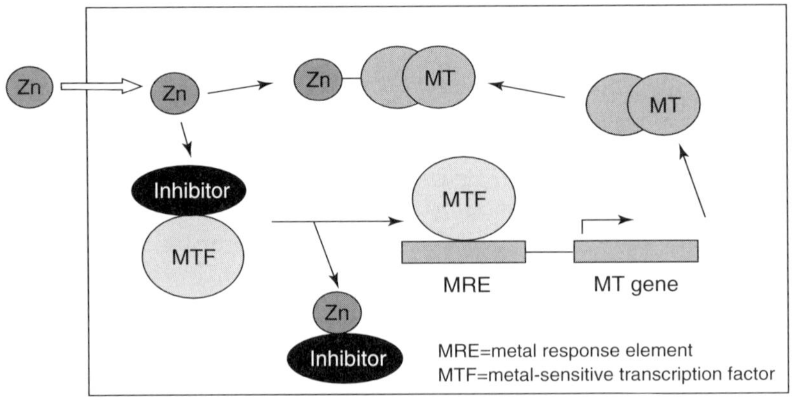

Figure 14.2 Zinc activation of the *MT* gene and subsequent production of MT protein. (Adapted from the scheme proposed by Palmiter, 1994.)

glucocorticoids, and potentially other steroids, independently of other regulatory sequences located in the proximal promoter region of either MT isoforms.

The two introns of *hMT-I*$_A$ gene are highly conserved in the rat *MT-I* gene as well as in the *hMT-II*$_A$ gene (Richards *et al.*, 1984). The protein encoded by the *hMT-II*$_A$ gene only differs from that encoded by the *hMT-I*$_A$ gene by four amino acids (Hamer, 1986). This small difference in amino-acid sequence results in differences in the basal transcription level and responsiveness to induction by cadmium or zinc.

MT gene regulation *in vivo* appears to be a complex function of both exposure to appropriate stimuli and the presence of appropriate signal transduction pathways (Searle *et al.*, 1984; De *et al.*, 1990). For instance, synthesis of mouse MT-I mRNA is regulated transcriptionally both by heavy metals and glucocorticoids (Karin and Herschman, 1979; Beach and Palmiter, 1981; Durnam and Palmiter, 1981; Karin *et al.*, 1981; Mayo and Palmiter 1981). However, when the mouse *MT-I* gene is transfected into human cells, its retains the transcriptional regulation by cadmium but regulation by glucocorticoids is lost (Mayo and Palmiter, 1982; Mayo *et al.*, 1982).

In most rodent cell lines expression of both MT-I and MT-II is coordinate (Griffith, 1985) and mouse *MT-I* and *MT-II* genes are largely expressed to the same extent in the presence of either metals or glucocorticoids (Searle *et al.*, 1984). However, the expression of the different isoforms of MT can be regulated differentially by zinc, cadmium and glucocorticoids (Richards *et al.*, 1984), and it appears that there is tissue-specific expression (Schmidt and Hamer, 1986; Varshney *et al.*, 1986). Thus, although the sequence of the promoter region is highly homologous in different species, the level of transcription of *MT* genes differs widely in different tissues. This may be due to the availability of *trans*-elements and the proportion of the different MT isoforms in different cell types.

DNA methylation is an epigenetic mechanism involved in the regulation of gene expression (Bird, 1992). Generally speaking, DNA hypomethylation will facilitate gene expression while hypermethylation can inhibit expression. There is good evidence that *MT* gene expression and inducibility are inversely correlated with DNA methylation (Compere and Palmiter, 1981; Liebermann *et al.*, 1983; Waalkes *et al.*, 1988). Early studies showed that demethylation of the *MT-I* gene in mouse T-cell lines resulted in enhanced responsiveness to glucocorticoids (Compere and Palmiter, 1981; Liebermann *et al.*, 1983). Moreover, induction of a hypomethylated state of the DNA resulted in activation of human *MT-I* genes (Heguy *et al.*, 1986). In fact, DNA hypomethylation induced by a variety of agents, such as 5-azacytidine, deoxyazacytidine, butyric acid or dimethyl sulfoxide, results in an increased expression and hyperinducibility of MT. Hypomethylation is also associated with tolerance to cadmium toxicity due to the increased synthesis of MT (Waalkes *et al.*, 1985).

One system that has been used to investigate the role of DNA methylation and MT expression is a thymoma-derived cell line, the S49, which is unusual in that neither of the *MT* genes is basally expressed or inducible by metals, apparently because of methylation (Compere and Palmiter, 1981). MacArthur and Lieberman (1987) mapped DNase I-hypersensitive sites in S49 cells and developed cadmium-resistant S49 variants that expressed high levels of MT. DNase I-hypersensitive sites flanking the *MT* genes (about 6 kb 5′ of the *MT-II* gene and 4 kb 3′ of the *MT-I* gene) were present regardless of whether the *MT* genes were expressed, but new hypersensitive sites appeared in the promoter regions of the *MT* genes that were expressed only after selection for resistance

(MacArthur and Lieberman, 1987). Additional studies showed that the induction of cadmium resistance in cadmium-sensitive, MT-negative S49 mouse cells by treatment with direct-acting chemical carcinogens was due, at least in part, to the activation of quiescent *MT* genes (MacArthur *et al.*, 1985, 1986). DNA demethylation had occurred 5′ to each *MT* gene in cells expressing MT (MacArthur *et al.*, 1986).

Gene amplification of the *MT* gene is one of the mechanisms whereby cells become resistant to cadmium toxicity. Crawford *et al.* (1985) isolated cadmium-resistant cells that showed coordinate amplification of *MT-I* and *MT-II* genes. Amplification of the *MT-I* gene has also been shown in cadmium-resistant mouse cells (Beach and Palmiter, 1981).

The remarkable inducibility of the MT promoter by a wide range of agents has been exploited in biomedical research and biotechnology (Palmiter, 1987). This often involves the construction of fusion genes made up of a structural gene regulated by the MT promoter that can, in turn, be induced easily by metals or glucocorticoids. These constructs have been used to investigate gene function.

14.2.4 Phylogeny and classification of MTs

MT occurs in a great variety of different species throughout the animal kingdom in both mammalian and nonmammalian species (Hamer, 1986; Kägi and Kojima, 1987a, b; Kägi and Schäffer, 1988; Waalkes and Goering, 1990). MTs are also found in higher plants (Robinson *et al.*, 1993), eucaryotic microorganisms (Butt and Ecker, 1987) and also in some procaryotic organisms (Olafson *et al.*, 1988).

Kägi and Schäffer (1988) have reviewed the classification of the different MT proteins. After their initial classification (Kägi and Kojima, 1979) and according to their structural relationships, MTs have been classified into three major groups (Fowler *et al.*, 1987). The MT superfamily is described as ‘all polypeptides resembling the equine renal MT in several of their features’. A family of MT proteins is comprised of MTs sharing a set of sequence-specific characters in which amino-acid sequences are alignable with those of the rest of the members of the family. An MT subfamily includes MT proteins sharing a set of more stringent phylogenetic features, and usually they display specific monophyletic relationships among the sequences of proteins or of nucleotide segments of the gene. The MT isoforms or allelic forms are speciable as part of subgroups, subfamilies and families. The MT family 1 comprises vertebrate MT and among its subfamilies special attention should be directed towards subfamilies m1 (mammalian MT-I), m2 (mammalian MT-II), m3 (mammalian MT-III) and m4 (mammalian MT-IV). There is more genetic polymorphism in the MTs belonging to class I than other classes (Kägi and Kojima, 1987a, b). MT families 2, 3, 4, 5, 6, and 7 correspond to mollusc MT, crustacean MT, echinoderm MT, dipterian MT, nematode MT and ciliate MT, respectively. MT families 8–13 include fungal MT, while MT family 14 includes procaryote MT. MT family 15 includes all plant MT, with six subfamilies. Class I MT includes mammalian MT and other polypeptides with related primary structure, while class II MT comprises MT displaying no or only very distant correspondence to the mammalian forms (i.e. from sea urchin, wheat, yeast and certain procaryotes), and class III MT is formed by atypical polypeptides containing γ-glutamylcysteinyl units (Robinson *et al.*, 1993).

Amino-acid composition of class I MTs of arthropods, chickens and mammals indicates that evolutionary divergence occurred in the range of 14 to 16.5 million years ago (Hunziker and Kägi, 1985; McCormick *et al.*, 1988). Most of the amino-acid substitutions are conservative and the majority of them are located nearer to the amino terminus, indicating fewer evolutionary constraints in this portion of the MT molecule. Thus, class I MTs are rather slowly evolving proteins with an evolutionary rate intermediate between that of cytochrome *c* and hemoglobin. The occurrence of closely related MT isoforms in distant lines of descent suggests various independent gene-duplication events. Evolutionary relationships between MTs may potentially be important in elucidating the functional roles of these proteins.

14.2.5 Organ distribution of MT

Generally speaking, MTs are expressed in most tissues, although the relative expression can vary extensively. Tissues with typically high MT expression include liver, kidney, pancreas, intestine, stomach, lung, heart, brain, spleen, placenta, thymus, gonad and muscle (Chen and Ganther, 1975; Onosaka and Cherian, 1981, 1982; Waalkes and Klaassen, 1984; Nishimura *et al.*, 1989a,b). MTs are most abundant in parenchymatous tissues such as liver, kidney, pancreas and intestine. The levels of MT show wide variations between different species and in different organs and tissues of the same species. This variability in MT levels is due to a number of factors, including such things as age, developmental stage, diet and nutritional status, and probably other, as yet, not fully identified factors (Andersen *et al.*, 1983b; Hamer, 1986; Mercer and Grimes, 1986; Kägi and Kojima, 1987a, b; Kägi and Schäffer, 1988; Waalkes and Goering, 1990).

Under normal circumstances some of the highest concentrations of MT occur in the livers of perinatal animals (Wong and Klaassen, 1979; Bakka and Webb, 1981; Waalkes and Klaassen, 1984; Kershaw *et al.*, 1990; Chan and Cherian, 1993), including humans (Cherian, 1993). Other tissues of the perinate are generally higher in MT content than in the adult, including the kidney, pancreas, spleen, heart, lung and testes but the levels here are much less than in the perinatal liver (Waalkes and Klaassen, 1984).

In mature animals the liver, pancreas, intestine and kidney will contain high concentrations of MT following zinc or cadmium exposure (Onosaka and Cherian, 1981, 1982; Waalkes and Klaassen, 1985). Reproductive tissues and sex accessory tissues, such as the rat dorsolateral prostate (Tohyama, *et al.*, 1993a; Coogan *et al.*, 1995, 1997), can also contain substantial amounts of MT under normal conditions. The expression of MT in tissues is often correlated with the tissue distribution of zinc. For example, pancreas and dorsolateral prostate, which have relatively high zinc contents, also express MT in relatively high amounts.

Even in tissues with substantial expression of MT, its distribution can be quite heterogeneous in different cell types. For example, in the liver MT is found in hepatocytes but not in fibroblasts or Kupffer cells (Nartey *et al.*, 1987a). In the brain, pronounced differences in MT expression occur that are cell-specific in nature (Nakajima and Suzuki, 1995).

Organ contents of MT can vary widely with species. For instance when the level of MT constitutively expressed in the liver was measured in a variety of species the following levels were obtained: dog, cat, and goat 500–700 in μg MT/g liver; pig and human, 350–500 μg MT/g liver; cow and sheep, 150–250 μg MT/g liver; hamster, rat, rabbit and guinea-pig, 5–25 μg MT/g liver (Henry *et al.*, 1994). Mouse liver MT levels are comparable to those of other rodents, in the range of 5–10 μg MT/g liver under normal conditions (Waalkes *et al.*, 1984). These data indicate that the commonly used laboratory rodent species (rats, mice, hamsters) may, in fact, be very poor models for defining the functional role of MT in humans because of the dramatic differences in basal MT expression.

Much of the data generated on species comparisons, organ distribution and induction of MT were collected using methods that do not necessarily distinguish between the various isoforms of MT (Onosaka and Cherian, 1981, 1982; Waalkes and Klaassen, 1984, 1985; Henry *et al.*, 1994). However, the MT-III and MT-IV isoforms are very limited in their distribution and studies not specifically looking at brain or gastrointestinal tract probably would involve the major isoforms, MT-I and MT-II.

14.3 Possible functions of MT

14.3.1 Introduction

MTs may play important roles in physiologic, toxicologic and pathologic responses (Waalkes and Goering 1990; Bremner, 1993; Cherian and Chan, 1993; Vallee and Maret, 1993; Waalkes 1993; Vallee 1995). The high metal complement of MT has led to the reasonable conclusion that a biological function of this protein is to bind metals. Since many essential metals bind to MT, this implicates a role in essential metal storage and metabolism. Additionally, the consistent localization of substantial levels of MT in the liver, an organ involved in storage and processing of nutrient substances, indirectly supports a role in metal homeostasis. MT is also involved in aberrant metal metabolism, which includes various storage diseases of essential metals such as copper. Beyond a role in normal metal metabolism, it is clear that MT impacts transition ('heavy') metal toxicity and most toxicologic research in this area has centered around the role of MT in the toxicity of cadmium. The function of MT in the brain via the special form of MT, MT-III, appears to have some unique aspects, potentially involving neurotransmission. Another special form of MT, MT-IV, may play a role in differentiation of squamous epithelia in some locations. MT may also be involved in proliferative diseases, including cancer.

The following section discusses, in a general fashion, the potential roles of MT. In earlier work on MT, the precise isoform was often not determined but probably represents MT-I and MT-II in most studies, excluding those specifically directed at the brain or gastrointestinal tract. Most early studies investigating inducibility of MT again did not specify the isoform but likely involved MT-I and MT-II, particularly since MT-III is a noninducible form (Palmiter *et al.*, 1992) and the MT-IV gene does not appear to respond to the control elements involved in the regulation of *MT-I* and *MT-II* gene expression (Quaife *et al.*, 1994). Specific sections discuss the unique functions of the more specialized forms of MT (MT-III and MT-IV).

14.3.2 Role of MT in normal metal metabolism

The high metal complement of MTs under normal conditions have led to the reasonable conclusion that the function of MTs may be related to metal metabolism and homeostasis (Bremner, 1993). Since all metals and metal compounds have the potential to be toxic, although many are essential, it is also reasonable to assume that systems would evolve for the safe transport, storage and use of these potentially toxic agents. Thus, MT may be a component in systems evolved to deal with zinc and copper. It is clear that these particular metals are critical to survival and their involvement as functional components in so many biomolecules implies the existence of systems that control their availability (Vallee, 1995; Luza and Speisky, 1996).

Under normal conditions, the highest tissue concentrations of MTs occur during periods of high zinc demand, such as during the rapid growth period in perinatal animals (Wong and Klaassen, 1979; Brady, 1982; Waalkes and Klaassen, 1984; Andrews *et al.*, 1993; Datson and Lehman-McKeeman, 1996) or with tissue regeneration such as after partial hepatectomy (Ohtake and Koga, 1979). There is a very stringent nutritional requirement for zinc in dividing normal cells (Petering *et al.*, 1993) and it is suspected that MT provides a supply of zinc within rapidly growing tissues, such as would occur during the perinatal period (Brady, 1982; Datson and Lehman-McKeeman, 1996). The fact that zinc is essential for several critical aspects of cell division, including DNA, RNA and protein synthesis (Brady, 1982), points out the clear need for this trace element during periods of rapid proliferation.

In perinatal mammals, MT levels are very high at or near term and in the initial postpartum period in many tissues, but most notably the liver (Wong and Klaassen, 1979; Brady, 1982; Andrews *et al.*, 1984; Waalkes and Klaassen, 1984; Datson and Lehman-McKeeman, 1996). For instance, in developing mouse liver *MT* gene expression is first detected at gestation day 12 and increases steadily through the end of term, but then decreases rapidly after parturition (Andrews *et al.*, 1984). MT levels start to decrease rapidly during postpartum development (Wong and Klaassen, 1979; Waalkes and Klaassen, 1984; Datson and Lehman-McKeeman, 1996), a stage that includes substantial proliferative growth. Both in humans and other species, MT levels in fetal liver correlate well with zinc levels and gestational age, reinforcing the concept that MT functions as a storage depot of zinc during periods of rapid growth (Wong and Klaassen, 1979; Clough *et al.*, 1986; Datson and Lehman-McKeeman, 1996). Similar results have been observed with copper and MT during early development and it is suspected that MT also serves as a storage protein for copper during this period (Luza and Speisky, 1996).

In adult livers much lower levels of MT exist in comparison to the perinate (Wong and Klaassen, 1979; Cherian, 1993; Datson and Lehman-McKeeman, 1996). Furthermore there appears to be a difference in subcellular localization of MT between adults and perinates. In adult liver, MT is primarily located in the cell cytoplasm where in the fetal/neonatal liver MT is concentrated within the cell nucleus (Cherian, 1993). The concentration of MT within the nucleus also occurs in tumors at the proliferative edge and in tumor cells in the S phase of the cell cycle (Cherian, 1993). Additionally, localization of MT in the nucleus occurs after stimulation of proliferation in liver by partial hepatectomy (Tohyama *et al.*, 1993b; Tsujikawa *et al.*, 1994). MT is also localized in the nucleus of hepatocytes in primary culture after stimulation of cell division (Tsujikawa *et al.*, 1991). It appears that MT in adult hepatocytes is translocated into the nucleus during the S to G_2M phases of the cell cycle following stimulation of proliferation by hepatocyte growth factors (Tsujikawa *et al.*, 1993). The nuclear location of MT is indicative of proliferative activity and may play a role in signal transduction of growth stimulation (Cherian, 1993; Tsujikawa *et al.*, 1993). Recent observations suggest that there is a cell type-specific and energy-dependent mechanism for MT retention in the nulceus (Woo *et al.*, 1996). The exact role of MT in the nucleus is unknown, but may be related to its zinc binding function. The zinc in MT may potentially be available to interact with various nuclear components, including histones, nuclear acidic proteins and transcription factors, and such interactions could modify many cellular processes, including gene expression (Cherian, 1993). There is, in fact, evidence that MT can interact with zinc-finger proteins (Zeng *et al.*, 1991).

It has been hypothesized that MT may play a role in control of intestinal absorption of copper or zinc (Bremner, 1993). For instance, fluctuations in dietary zinc status are reflected in intestinal MT levels (Bremner, 1993). Intestinal MT levels do appear to be inversely correlated with efficiency of zinc absorption, indicating that MT may block excess zinc from passing to systemic circulation (Bremner, 1993). Most studies of this nature have been done using nonphysiological levels of zinc or copper rendering them difficult to interpret (Bremner, 1993).

Induction of hepatic MT synthesis by most inducing agents, including nonmetallic agents, mobilizes zinc to the liver (Waalkes and Goering, 1990; Daston and Lehman-McKeeman, 1996). This observation suggests that MT may play a role in zinc distribution and possibly homeostasis.

Compensatory organ hypertrophy, as a result of heightened cellular proliferation, can also stimulate MT activation. Uninephrectomy in rats results in hypertrophy of the remaining organ due to enhanced cellular proliferation. During this time there is a marked increase in renal MT-I and MT-II expression, particularly in the renal cortex and outer medulla (Zalups *et al.*, 1995). Liver MT expression also increases with uninephrectomy (Zalups *et al.*, 1995). An increased demand for zinc in the proliferating tissue may be responsible for the MT activation in the kidney, whereas in the liver activation may be due to an increase in whole-body demand for zinc.

In mice large, gender-related differences exist in the basal and metal-induced expression of MT in the liver (Iszard *et al.*, 1995). Female mice have hepatic MT levels between two- and fourfold above those observed in male mice (Iszard *et al.*, 1995). This same gender-based pattern is found in the livers of MT-I transgenic mice overexpressing MT-I but does not appear to occur in tissues other than liver, although MT is hyperexpressed in many of the tissues of these mice (Iszard *et al.*, 1995). Although some sex steroids can induce MT synthesis (Shimada *et al.*, 1998) the basis of gender differences in hepatic MT remains unexplained.

On a molecular level, several studies show that MT can act as a donor of copper or zinc to other proteins within the cell (Li *et al.*, 1980; Udom and Brady, 1980; Geller and Winge, 1982; Petering and Fowler, 1986; Goering and Fowler, 1987; Otvos *et al.*, 1993). For instance, several apoenzymes that require zinc or copper as cofactors, such as carbonic anhydrase, alkaline phosphatase and superoxide dismutase, have been reactivated *in vitro* after incubation with zinc or copper MTs (Li *et al.*, 1980; Udom and Brady, 1980; Geller and Winge, 1982). It is also known that zinc MT in kidney activates aminolevulinic acid dehydratase, a zinc-requiring enzyme involved in heme biosynthesis, by donating zinc to the enzyme (Goering and Fowler, 1987). The precise mechanism by which the donation of metal to apoenzymes by MT is controlled or regulated is not yet defined. In fact, very little is known about the fate of the MT protein or of the metals bound to it (Bremner, 1993). It has been hypothesized that MTs may be associated with other macromolecules during metal donation in order to facilitate donation (Quaife *et al.*, 1994). By contrast, in certain cases overexpression of MT is correlated with high levels of the inactive proenzyme form of copper/zinc superoxide dismutase in yeast, suggesting that MT sequesters available metal and thereby inhibits enzyme activation (Galiazzo *et al.*, 1991).

Under normal conditions, MT occurs at very low levels in the circulation (Dudley *et al.*, 1985), indicating that metal transport between tissues is probably not a primary function of the protein. At a critical stage in chronic cadmium intoxication the transport of cadmium that occurs from the liver to kidney is largely in the form of cadmium–MT (Dudley *et al.*, 1985), but this is a pathologic rather than physiologic role (see Section 14.4).

Speculation about a normal role for MT in metal homeostasis should take into account data collected using cells or animals with targeted disruptions of the *MT-I/MT-II* genes. Initial studies with *MT-I/MT-II* knockout mice (Michalska and Choo, 1993; Palmiter *et al.*, 1993a; Masters *et al.*, 1994) suggested that MT is not essential for development or normal cellular function. In these studies *MT-I/MT-II* knockout mice appeared to develop normally (Michalska and Choo, 1993; Masters *et al.*, 1994), at least under the conditions of *ad libitum* access to typical laboratory rodent foods. Frequently, laboratory rodent foods are in great excess of essential nutrient metals, such as zinc, when compared to the minimum levels required for appropriate growth and development (Waalkes, 1986). This fact may have obscured potential effects on homeostasis that may occur at minimally optimal dietary zinc levels. *MT-I/MT-II* knockout mice have about 50% less zinc in the pancreas (Liu *et al.*, 1996) as compared to controls, but this did not perturb development when animals consumed normal rodent foods (Michalska and Choo, 1993; Masters *et al.*, 1994). Studies of MT-deficient mice under conditions of marginal or severe zinc deficiency showed that perinatal *MT-I/MT-II* null mice were much more sensitive to zinc deficiency, showing clear signs of perturbed postpartum renal development (Kelly *et al.*, 1996). Transgenic mice that overexpress *MT-I* are resistant to dietary zinc deficiency (Dalton, *et al.*, 1996a). Additionally, *MT-I/MT-II* null mice are very sensitive to the adverse effects of excess zinc (Kelly *et al.*, 1996). The fact that *MT-I/MT-II* knockout mice show a much lower tissue retention and shorter biological half-life for cadmium (Liu *et al.*, 1996) suggests that other metals bound to MT, like zinc, may also be more poorly retained, which could have a negative impact during periods of high zinc demand. Studies involving animals with simultaneous disruption of all four *MT* genes are not available, but *MT-III* knockout mice also appear to develop normally (Aschner *et al.*, 1997). Cultured cells lacking the ability to express the primary forms of MT can also function normally (Crawford *et al.*, 1985; Palmiter *et al.*, 1993a). From this it is clear that MT may not be the only mechanism whereby metals are donated to apoenzymes or transcription factors, although the data suggest that at some point during diminished zinc status MT is needed to preserve available zinc. It is possible that the roles that MT plays in cellular metal homeostasis are so important that redundancies or back-up systems exist that can compensate for the loss of MT. This has yet to be determined, and the development and senescence of MT knockouts needs to be tested further under conditions of limited zinc or copper before assigning or dismissing a role for MT in metal homeostasis.

14.3.3 *Special functions of brain MT*

Early studies provided evidence of the existence of brain MT as a low-molecular-weight, inducible, zinc-binding protein in the rat brain (Itoh *et al.*, 1983). Other evidence indicated that copper can also induce a MT-like protein in the rat brain (Waalkes *et al.*, 1982). Ebadi (1984) further characterized this protein as a member of the MT family while studying the mechanisms underlying convulsive seizures associated with Pick's disease, a rare neurological disorder in which higher than normal levels of zinc accumulate in the hippocampus and cause inhibition of γ-aminobutyric acid (GABA) transmission.

Brain MTs have been reviewed by several authors (Aschner, 1996; Aschner *et al.*, 1997; Hidalgo *et al.*, 1997). One of the more recently discovered forms of MT, MT-III, is the only brain-specific member of the MT family (Palmiter *et al.*, 1992). Both human and rodent brain contain MT-I, MT-II and MT-III (Ebadi *et al.*, 1995; Choudhuri *et al.*, 1996a,b). In the brain, MT is apparently synthesized at a higher constitutive level than that found in many peripheral tissues. In the human brain, both MT-I and MT-II are expressed in glial cells (mostly in astrocytes), with highest levels found in the cerebral cortex and the basal ganglia. Some studies have identified MT in the cytoplasm and in the axons and dendritic processes of motor neurons in the ventral spinal cord. Although the glial cells of adult brains exhibit strong MT staining, MT is not detected in glial cells of fetal brains. Expression of MT-III is higher in large neuronal bodies in areas with high levels of vesicular zinc, such as the hippocampus, the piriform cortex and the amygdala (Aschner, 1996), and is also abundant in the choroid plexus epithelium.

Brain MTs, like MTs found in other tissues, appear to have several important potential functions, such as the homeostasis of essential metals (Aschner, 1996; Aschner *et al.*, 1997; Hidalgo *et al.*, 1997). Brain MT may also protect nervous tissue from oxidative damage, which could be quite important in the central nervous system (CNS) where there is a continuous production of oxygen free-radicals and an abundance of molecules sensitive to oxidative attack. This function may be partially due to the release of zinc (which has antioxidant properties) from MT. However, the MT protein may itself scavenge free-radicals through its numerous cysteine residues, thereby preventing interaction with other, more sensitive, cellular components. Additionally, the localization and the inducibility of MT-I and MT-II in astrocytes, in the chroroid plexus, and in ependymal cells, indicates that these MT isoforms may play a protective role against damage caused by neurotoxic heavy metal compounds such as methylmercury. Presumably, the neurotoxic effects of these agents are propagated when the binding capacity of MT is exceeded (Aschner, 1996). Therefore the MT expressed in brain may be involved during the metal exposure thought to play a role in neurodegenerative diseases, such as amyotrophic lateral sclerosis, Parkinson's disease and Alzheimer's disease (Ebadi *et al.*, 1995). The induction of MT by increased levels of interleukin-1 could also protect against the oxygen free-radicals and the oxidative stress caused by inflammation and tissue injury in the CNS (Hidalgo *et al.*, 1997).

The vesicular pool of zinc is localized in zinc-containing neurons of the limbic, cerebrocortical, and corticofugal systems (Ebadi *et al.*, 1995). The highest concentration of zinc in the brain is found in the mossy fibers of the hippocampus. These neurons can concentrate zinc in their terminals and release it in a calcium- and impulse-dependent manner. The high-affinity uptake of zinc takes place in the axonal boutons. Glutamatergic, but not cholinergic, pathways are zinc-containing pathways. Under normal physiologic conditions, zinc is thought to affect synaptic transmission in areas of CNS with high levels of zinc, such as the hippocampus, through the inhibition of $GABA_B$ receptor activation after its release to the synaptic cleft. Thus, zinc plays a role as an inhibitory modulator of receptors for both inhibitory (GABA) and excitatory (glutamate/NMDA) neurotransmitters. In this regard, altered zinc metabolism has been observed in a variety of neurodegenerative diseases (epilepsy, Friedrrich's ataxia, Pick's disease, amyotrophic lateral sclerosis, multiple sclerosis, Guillain–Barré syndrome, Parkinson's disease), but to date it is not known whether there are concomitant changes in the levels of brain MT associated with these diseases (Ebadi *et al.*, 1996). Uchida *et al.* (1991) initially discovered MT-III as a protein from brain extracts of people with Alzheimer's disease that inhibited survival of cultured rat neurons, indicating a potential role in neurodegenerative disorders. However further work indicates that MT-III is not deficient in Caucasian patients with Alzheimer's disease (Erickson *et al.*, 1994; Aschner *et al.*, 1997). Additionally *MT-III* knockout since mice do not show any evidence of accelerated neurodegenerative disease of learning disorders. Thus a direct role of MT in neurodegenerative disease seems unlikely.

The specificity of the MT-III isoform distribution in the brain suggests that it has specific functions (Aschner, 1996; Aschner *et al.*, 1997; Hidalgo *et al.*, 1997). The concomitant release of both zinc and glutamate or GABA in the synapses suggests a role for MT-III in the modulation of neurotransmission. The activation of both NMDA and non-NMDA receptors is important for excitatory synaptic transmission, developmental and synaptic plasticity and neurotoxicity. Non-NMDA receptor-mediated neurotoxicity, such as kainic acid (K)-induced neurotoxicity, involves the production of free-radicals. Aging has also been correlated with decreased brain antioxidant capability. However, a parallel disease in MT content with aging has not been observed. In fact, MT content is high in aged brains.

A number of studies have linked zinc metabolism with seizure activity (reviewed by Kuzuya *et al.* 1993; Aschner *et al.*, 1997). Mice can be protected against electrically induced seizures by a zinc-deficient diet, while zinc treatment causes exacerbation of seizure activity. Intrahippocampal injections of zinc sulfate have been observed to induce seizures in rabbits, while others have described higher serum zinc levels in epileptic patients not treated with antiepileptic drugs, as compared with healthy volunteers. Moreover, it has been shown that intracerebroventricular administration of zinc caused seizures that could be prevented by GABA (Ebadi *et al.*, 1995). However, higher doses of zinc administered intravenously or intraperitoneally do not cause convulsive seizures in rats. This could be explained by the fact that when many metals, such as zinc, are administered systematically they do not easily cross the blood–brain barrier and do not readily change the concentration of that metal in the brain.

Kainic acid (KA)-induced seizures are a useful model for the study of the biochemical basis underlying excitotoxic brain injury. KA is a rigid analog of glutamate and a CNS excitotoxin. KA stimulates many neurons, including zinc-ergic neurons, and its systemic administration produces a status epilepticus (Dalton *et al.*, 1995). After KA exposure there are extensive neuropathologic changes, with cellular necrosis in different brain regions, mainly in the piriform cortex, amygdala, hippocampus, entorhinal cortex and lateral septum. Oxygen free-radicals, produced during KA-induced seizures, have been implicated in the KA-associated neuropathology. Ebadi *et al.* (1995) reported that seizures induced by KA in turn induced the expression of MT-I, with little or no alterations in expression of MT-III. Others have found that KA induces modest increases in MT-III in the cortical region (Zheng *et al.*, 1995). The induction of MT-I is dependent on seizures, and higher levels are found in areas of the brain undergoing extensive damage as compared to areas with minimal injury (Ebadi *et al.*, 1995). This has led to the suggestion that MT may be involved in hypoxia-induced protection against the KA-induced neurotoxicity. This process could be mediated by oxidative stress, which could itself induce MT synthesis, or alternatively by zinc released during the seizures that would directly induce MT synthesis (Ebadi *et al.*, 1995). The latter hypothesis is supported by the observation of decreased zinc in axonal boutons during seizures and a subsequent increase in total brain zinc content. The fact that the levels of MT-III remained relatively unchanged during seizures confirms previous reports of unresponsiveness of the MT-III genes to agents known to induce MT-I or MT-II synthesis (Palmiter *et al.*, 1992).

Transgenic mice overexpressing MT-III show no overt changes in the brain (Erickson *et al.*, 1995; Aschner *et al.*, 1997). *MT-III* null mice have decreased levels of zinc in the hippocampus and other brain areas, without any apparent neuropathology or behavioral deficit, in spite of the observation that zinc-ergic neurons containing MT-III may be involved in learning and memory processes (Aschner *et al.*, 1997; Erickson *et al.*, 1997). MT-III-deficient mice are more susceptible to KA-induced seizures. Mice overexpressing MT-III were more resistant to neuronal injury induced by seizures (Aschner *et al.*, 1997). These results support the idea that MT-III may act as an antioxidant in the brain, although this capability is limited and does not offer protection against extensive injuries. However, the normal development of MT-III-deficient mice suggests that MT-III is not the primary agent involved in Alzheimer's disease or other neurodegenerative diseases.

In summary, three major MT isoforms, MT-I, MT-II and MT-III, are expressed in the brain, the latter being brain specific. Several proposed functions for MT in the brain are shared by these various isoforms. However, the brain-specific expression of MT-III also suggests a unique function or functions for this particular isoform. Brain MT is probably involved in the normal cellular homeostasis of essential metals and protection against toxic metals. In addition, brain MT may also have an antioxidant function, either through direct interaction of the cysteine residues with oxygen free-radicals or through modulation of zinc metabolism. Both zinc and MT seem to be involved in neurotransmission, the development of seizures and the protection against seizure-induced oxidative stress and neuronal damage.

14.3.4 MT and heavy metal toxicity

MT was originally discovered as a cadmium-binding protein (Margoshes and Vallee, 1957; Kägi and Vallee, 1961), and this observation provided the basis for the hypothesis that MT is involved in the mitigation of cadmium toxicity. A later study showed that MT levels are increased after exposure to cadmium (Piscator, 1964) leading to the speculation that MT is involved the intracellular sequestration of toxic metal ions. Many additional studies showed that MT plays a central role in reducing the toxicity of cadmium (for example: Webb, 1972; Leber and Miya, 1976; Webb and Verschoyle,

1976; Probst *et al.*, 1977; Goering and Klaassen, 1984) as well as other metals (Yoshikawa, 1970). Cellular tolerance to cadmium is thought to occur through a high-affinity sequestration of the toxic metal ion by MT, and early on it was demonstrated that following cadmium exposure the metal is, in fact, predominantly bound to MT (Piscator, 1964). Subsequently it was shown that pretreatment with metals known to stimulate MT synthesis prevents the toxicity of cadmium (Webb, 1972; Leber and Miya, 1976; Webb and Verschoyle, 1976; Probst *et al.*, 1977; Goering and Klaassen, 1984). Pretreatment with zinc or low, nontoxic doses of cadmium are particularly effective in inducing tolerance to cadmium (Webb, 1972; Yoshikawa, 1973; Leber and Miya, 1976; Probst *et al.*, 1977; Goering and Klaassen, 1984). After the initial pretreatment the synthetic mechanisms for MT are induced, and newly synthesized protein provides the sequestrational capacity necessary to block or substantially reduce the toxic actions of cadmium. Under these conditions free cadmium in the cell, presumably the toxic species, is effectively limited.

The most definitive evidence that MT is involved in protection against cadmium toxicity comes from studies using transgenic mice with selective disruption of the two primary isoforms, MT-I and MT-II (Michalska and Choo, 1993; Masters *et al.*, 1994). These studies clearly show that *MT-I/MT-II* knockout mice are much more sensitive to cadmium toxicity than control mice (Michalska and Choo, 1993; Masters *et al.*, 1994). The *MT-I/MT-II* knockout mice are not only more sensitive to lethality but also to the more subtle effects of cadmium, such as aberrant gene activations (Liu *et al.*, 1995; Zheng *et al.*, 1996). The *MT-I/MT-II* knockout mice are also much more sensitive to the toxicity of cisplatin (Liu *et al.*, 1998). Conversely, transgenic mice overexpressing MT-I are clearly insensitive to cadmium lethality and hepatotoxicity in comparison to MT normal controls (Liu *et al.*, 1995).

Not all the adverse effects of cadmium can be blocked by MT. For instance, overexpression of MT-I does not offer any significant protection against cadmium-induced testicular necrosis in mice (Dalton *et al.*, 1996b), indicating that levels of testicular MT do not play a major role in the sensitivity of the testes to cadmium. Similarly, in cultured rat liver cells, progesterone treatment will activate the *MT* gene, resulting in accumulation of MT protein, but actually increases cadmium cytotoxicity (Shimada *et al.*, 1997). Thus, increasing the MT level is not always associated with induced tolerance to cadmium.

Other metals can be detoxicated by MT. This includes compounds of mercury (Zalups and Cherian, 1992; Aschner *et al.*, 1997) or platinum (Naganuma *et al.*, 1987; Kondo *et al.*, 1995; Lazo and Pitt, 1995; Liu *et al.*, 1998). Presumably many potentially toxic metals would be able to displace zinc from MT, based on relative affinities for MT ($Hg^{2+} > Ag^{+} > Cu^{+} > Cd^{2+} > Zn^{2+}$; Sanders *et al.*, 1996). The displaced zinc would provide further stimulation for MT synthesis and, in turn, further reduce the toxic impact of the displacing metal. Cadmium, for instance, functions as a primary inducer of MT synthesis and displaces zinc from MT (Waalkes and Goering, 1990; Sanders *et al.*, 1996). Other examples of metals that will increase MT tissue levels include chromium, iron, lead, manganese, mercury and nickel (Waalkes and Klaassen, 1985).

The initial distribution of cadmium is not modified in most tissues by MT-I and MT-II deficiency (Liu *et al.*, 1996). Initial disposition of cadmium after a single intraperitoneal dose of cadmium to major sites of cadmium storage, such as liver, kidney and pancreas, is not different in *MT-I/MT-II* null mice compared to controls (Liu *et al.*,

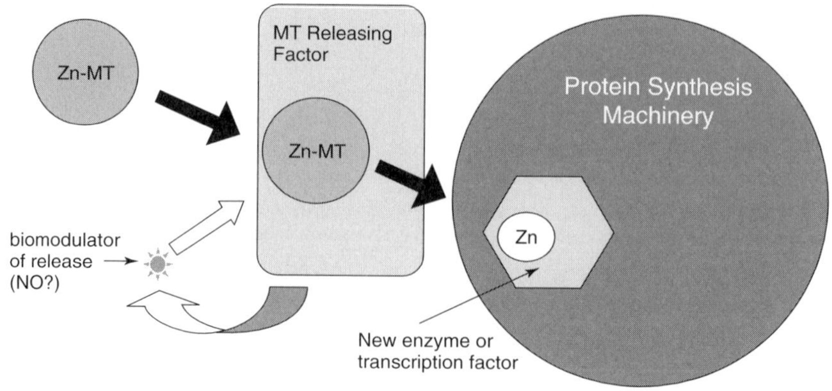

Figure 14.3 Hypothetical mechanism for the controlled donation of zinc from MT to metal-containing proteins such as enzymes or transcription factors. In this scheme MT containing zinc would be bound to an MT-releasing factor in the vicinity of the protein synthesis machinery. Some sort of biomodulator of release (nitric oxide?) could then be produced, displacing zinc from MT and making it readily available for the new enzyme or transcription factor. (Based in part on the proposed scheme of Quaife *et al.*, 1994.)

1996). However, the retention of cadmium is dramatically altered in *MT-I/MT-II* null mice (Liu *et al.*, 1996). These mice show a whole-body retention of cadmium that is approximately half that of MT-sufficient controls (Liu *et al.*, 1996). Retention in liver, kidney and pancreas are similarly reduced by the absence of MT-I and MT-II expression (Liu *et al.*, 1996). Retention in some tissues, such as lung, muscle, bone and reproductive tissues, is not affected by lack of MT-I and MT-II expression (Liu *et al.*, 1996) but these tissues normally only get a very small amount of cadmium. Overall it is clear that expression of MT is responsible for the long residence time of cadmium in the body under normal circumstances of exposure. MT also protects against mercury toxicity but, again, the protein appears to be responsible for a longer residence time of the metal in target organs (Satoh *et al.*, 1997).

Cadmium is a highly toxic, nonessential metal with no known function. Additionally, cadmium, like many other nonessential, toxic metals, is present at measurable, and often high, levels in biologic material of all sorts. Under these circumstances, the existence of a biologic system that can deal with the toxic potential of such an agent is perhaps a logical conclusion. Unlike the case with many organic chemical toxicants, metals such as cadmium cannot be metabolized into subunits of lesser toxicity, and sequestration in a toxicologically inert form would seem to be a reasonable alternative. However, detoxication of cadmium as a primary, intended function for MT seems unlikely. For a protein to evolve specifically for protection from a metal that has only relatively recently been concentrated within the biosphere through anthropogenic emissions seems improbable. Rather, the ability of MT to detoxicate cadmium would appear to be a fortuitous, but evolutionary unintentional, occurrence. The ability to detoxicate cadmium is probably related to the chemical mimicry of cadmium for zinc, and may apply largely to acute toxic effects. The fact that MT contributes to the persistence of cadmium in the body (Liu *et al.*, 1996) may actually increase the potential for chronic toxic effects.

Although MT induction usually leads to a reduction in toxicity, an interesting aspect of MT induction can actually lead to adverse effects in the maternal–fetal system. Maternal induction of MT in the liver will result in systemic changes in zinc distribution, reducing available zinc to the fetus (Datson and Lehman-McKeeman, 1996). A variety of MT-inducing agents, including ethanol and cadmium, has been shown to affect embryonic zinc status adversely in rodents (Waalkes and Bell, 1980; Datson and Lehman-McKeeman, 1996). Systemic exposure of the maternal animal to MT-inducing agents results in the induction of maternal hepatic MT which contains substantial levels of zinc (Waalkes and Bell, 1980; Datson and Lehman-McKeeman, 1996). The mobilized zinc, in turn, reduces distribution of this limited resource to the fetus resulting in a reduced fetal zinc status (Waalkes and Bell, 1980; Datson and Lehman-McKeeman, 1996). A decrease in embryonic zinc is associated with a variety of adverse effects, including abnormal development and, in extreme cases, teratogenesis (Datson and Lehman-McKeeman, 1996).

14.3.5 Possible functions of the special forms of MT (MT-III and MT-IV)

14.3.5.1 MT-III

MT-III is a relatively new member of the MT family. It was initially called growth inhibitory factor (GIF; Uchida *et al.*, 1991; Palmiter *et al.*, 1992; Tsuji *et al.*, 1992) and was isolated as a protein that inhibited survival of cultured rat neurons. These initial studies suggested that the downregulation of the *MT-III* gene may potentially play a role in neurodegenerative disorders such as Alzheimer's disease (Uchida *et al.*, 1991; Tsuji *et al.*, 1992). Although a role in Alzheimer's disease now seems less likely (Erickson *et al.*, 1994) MT-III is certainly unique in function and location. Neither MT-I or MT-II inhibit neuronal growth (Uchida *et al.*, 1991; Erickson *et al.*, 1994; Sewell *et al.*, 1995). Furthermore, while MT-I and MT-II are widely expressed, MT-III is expressed only in the CNS (Palmiter *et al.*, 1992). MT-III differs from MT-I or MT-II because it contains 68 amino acids instead of 61 or 62. A total of 38 amino acids show identical alignment with human MT-I and MT-II isoforms. Moreover, the position of all 20 cysteine residues are conserved (Aschner, 1996; Aschner *et al.*, 1997). MT-III differs from the other MT isomers in that it contains two additional amino-acid insertions: a single threonine residue at position 5 in the N-terminal region, and a six-residue insert at residue 55 in the C-terminal region, the six additional amino acids consisting of glutamic acid and alanine. These additional amino acids in MT-III, its acidic properties and charged surface are likely to be responsible for the specificity of the functions of this particular MT isoform. Thus, the unique activity of MT-III may correlate with its novel N-terminal β domain rather than with its metal-binding properties, which are similar to those of MT-I (Sewell *et al.*, 1995).

Studies in the early 1990s led to the hypothesis that downregulation of a growth inhibitory factor in the brain of Alzheimer's disease patients may play a role in the development of the neurofibrillary tangles characteristic of this disease (Uchida *et al.*, 1991; Tsuji *et al.*, 1992). Initial studies indicated that MT-III was reduced in the brains of Alzheimer's disease patients (Uchida *et al.*, 1991; Tsuji *et al.*, 1992). Later studies contradicted these early reports. The expression of MT-III was unchanged in a sample of Caucasian Alzheimer's disease patients (Erickson *et al.*, 1994). The finding that senescent *MT-III* knockout mice do not show signs of accelerated neurodegeneration or loss of memory function (Aschner *et al.*, 1997) is strong evidence against a role for MT-III in neurodegererative diseases. Studies have, however, consistently shown that MT-III has a clear inhibitory effect on neuronal survival (Uchida *et al.*, 1991; Tsuji *et al.*, 1992; Erickson *et al.*, 1994).

Even though MT-III is brain specific, the content of MT-I mRNA in the brain is higher than that of MT-III mRNA (Aschner, 1996; Choudhuri *et al.*, 1996a,b; Aschner *et al.*, 1997). MT-III is expressed constitutively in the adult mouse brain, although it is almost absent from the white matter which is rich in glial cells (where the expression of MT-I and MT-II is prevalent). Also the expression of MT-III in the cerebellum is low as compared to that of MT-I and MT-II (Aschner *et al.*, 1997).

Nakajima and Suzuki (1995), in a study of the expression of brain MT, found that in normal human aged brains, MT-III was well expressed in layers 2–6 of the gray matter, but weakly expressed in the subcortical white matter. Astrocytes, but not neurons, were stained by antibodies against MT-III in all areas of the brain, except oligodendroglia, microglia or neurons (Nakajima and Suzuki, 1995). The distribution of MT-III in the gray matter was similar to that of MT-I and MT-II. These authors postulated that MT-III functions in growth inhibition while the MT-I and MT-II function in growth promotion or proliferation. Anezaki *et al.* (1995) studied the expression of MT-III in the brains of normal and excitotoxically injured rats and found that MT-III was expressed in neurons and astrocytes in gray matter. MT-III was also expressed in the glial cell layer on the surface of the cerebral cortex and the cells surrounding Purkinje cells. Expression was weaker in neurons than in astrocytes (Anezaki *et al.*, 1995). There was a strong expression of MT-III in the reactive astrocytes around a site of excitotoxic injury caused by intraventricular injection of kainic acid (Anezaki *et al.*, 1995). Based on these findings, a role for MT-III in the control of neurite outgrowth after brain damage has been suggested (Hidalgo *et al.*, 1997).

The MT-III isoform is more abundant in brain regions rich in vesicular zinc, which is sequestered in the presynaptic vesicles of a special class of neurons (Erickson *et al.*, 1995). When this zinc is released, it may serve to modulate synaptic transmission (Erickson *et al.*, 1995). The expression of MT-III in neurons containing zinc suggests that MT-III could be involved in protection from zinc or in recycling of zinc into synaptic vesicles (Aschner *et al.*, 1997). However, experiments involving transgenic mice that overexpress MT-III (Erickson *et al.*, 1995) showed that although this overexpression affects the total zinc content in the brain, it does not change the synaptic pool of zinc. In fact, transgenic mice overexpressing the human MT-III *(hMT-III)* gene mature without any deleterious effect in either the development or mature functioning of the CNS (Erickson *et al.*, 1995). The increase in brain zinc in mice overexpressing *hMT-III* can be accounted for by zinc bound to the additional MT-III. The observation that MT-III does not show induction after zinc exposure (Palmiter *et al.*, 1992), also argues against a role in homeostatic control. Not only does MT-III fail to respond to zinc, but cadmium, dexamethasone or bacterial endotoxin are similarly ineffective in altering levels of MT-III *in vivo* (Palmiter *et al.*, 1992). This is a clear distinction from other MT isoforms (Palmiter *et al.*, 1992). The failure to be activated by these various treatments may be explained by the differences in the promoter regions of MT-III and the MT-I/MT-II isoforms (Palmiter *et al.*, 1992). Interestingly, agents that induce MT-I mRNA actually downregulated the expression of MT-III mRNA in some areas (Aschner *et al.*, 1997). Changes in the expression of both MT-I and MT-III observed in mouse brain can be reproduced *in vitro* using cultures of astrocytes and neurons (Aschner *et al.*, 1997).

Experiments with BHK cells transfected with either the mouse *MT-I* or *MT-III* genes have shown that constitutive MT-III expression inhibits cell proliferation when cells are zinc deficient, while constitutive MT-I expression does not (Palmiter, 1995). MT-III appears to compete with other zinc-requiring metalloproteins and actually exacerbates zinc deficiency (Palmiter, 1995). Several *in vitro* studies using astrocytes obtained from different transgenic mice expressing different levels of MT-III have suggested that this isoform is not essential for the protection against exogenous heavy metals such as cadmium or zinc (Aschner *et al.*, 1997).

In summary, it is clear that the function of MT-III is probably distinct from the function of MT-I and MT-II (Palmiter, 1995). A function in metal protection or homeostasis seems less likely for MT-III than the other major isoforms, and a more specialized role in neuromodulation is possible for this unique member of the MT family (Palmiter, 1995).

14.3.5.2 MT-IV

A new isoform of MT, MT-IV, was recently discovered by Quaife *et al.* (1994). This form is unique among the MT family because it is only expressed in cornified, stratified, squamous epithelium of the oral cavity, esophagus, upper stomach, skin and vagina. Although 30 residues, including the cysteines, are identical in MT-IV and other MTs, MT-IV differs from the other murine MT isoforms in approximately 24–27 amino-acid residues (Palmiter *et al.*, 1993a; Quaife *et al.*, 1994). Thus most of the amino acids in MT-IV that are nonessential to metal-binding function are different from other MTs. The murine and human MT-IV proteins differ only in four amino acids, and their genes are also conserved in the coding and promoter region at the nucleic acid level. However, this sequence conservation between the mouse and human genes does not apply to the introns or the 5′ or 3′ untranslated regions of the MT-IV (Quaife *et al.*, 1994).

The observation that MT-IV is only expressed in differentiating cells of squamous epithelia (Quaife *et al.*, 1994) could be critical to defining its physiologic function. As with the MT-III gene, the MT-IV gene does not appear to respond to control elements involved in the regulation of both MT-I and MT-II gene expression (Quaife *et al.*, 1994). In cell culture the expression of MT-IV protected against the cytotoxicity of cadmium, and MT-IV expression was induced in the stomach of mice by oral exposure to zinc (Quaife *et al.*, 1994). These observations point out a potential, perhaps fortuitous, role for MT-IV in the protection against the toxicity of heavy metals that is shared by all MTs. In addition, the location of MT-IV would perhaps provide a barrier function to sequester toxic metals prior to absorption into the systemic circulation.

MT-IV in normal tongue contains both zinc and copper (Quaife *et al.*, 1994) suggesting a possible role in storage and provision of these essential elements. Additionally, there is an inverse relationship between sensitivity of given tissue to zinc deprivation and the expression of the MT-IV gene (Quaife *et al.*, 1994). This may have implications in the hyperkeratosis frequently observed in zinc deficiencies because this disorder may be based in the disruption of the differentiation processes of keratinocytes.

Quaife *et al.* (1994) observed an apparent switching of MT forms in the cornified epithelium, with MT-I expressed mainly in the basal layer that has a high proliferative activity, and MT-IV being expressed in the spinous layer, that is characterized by a high differentiation activity. Thus, MT-IV may have evolved to interact with the process of cytoskeletal differentiation in keratinocytes (Quaife *et al.*, 1994). In this regard is MT-IV could perhaps regulate the availability of zinc for the assembly of keratins or influence the proteolytic processing of these proteins, or it could be involved more directly in the assembly of keratins.

Liang *et al.* (1996) studied MT-IV expression in the maternal reproductive tract, developing embryo, and fetus. MT-IV expression was not observed in a number of tissues (visceral yolk sac, placenta and fetal liver) in which high levels of MT-I and MT-II expression occur. However, MT-IV was expressed at high levels in the maternal deciduum, where the mRNA of both MT-I and MT-II are expressed. Neither zinc nor lipopolysaccharide (LPS) induced MT-IV mRNA in the decidua *in vivo* or in primary cultures *in vitro*. These authors concluded that the expression of MT genes in the mouse decidua seems to be regulated developmentally, being refractory to induction by zinc or inflammation.

14.3.6 MT and metal storage diseases

Excessive accumulation of even essential metals can have adverse effects on an organism. For instance toxicosis arising from excessive accumulation of copper has been reported in many different species (Harris and Gitlin, 1996). Due to the potential toxicity of copper, its metabolism and distribution is generally highly regulated. It is suspected that ionic copper, or copper that is easily dissociable from ligands such as proteins, is the actual toxic species. There are two diseases caused by abnormal transport of copper, specifically Wilson disease and Menkes disease. Due to tissue-specific pathology and the relationship of these disease states with normal copper homeostasis, Wilson disease is considered a hypercupremic state, while Menkes disease is considered as a hypocupremic state. Metallothionein is involved in both of these disease states.

14.3.6.1 Wilson disease

Wilson disease was first described over 80 years ago as progressive lenticular degeneration (Wilson, 1912). It later became known as hepatolenticular degeneration. The pathophysiology underlying Wilson disease is the failure to excrete excess copper into the bile and its subsequent incorporation into ceruloplasmin in the liver (Brewer and Yuzbasiyan-Gurkan, 1992; Scheinberg and Sternlieb, 1996). The accumulation of copper occurs mainly in the liver, brain, and cornea, where it produces characteristic Kayser–Fleischer rings. Wilson disease patients show excess urinary copper and excessive tissue accumulation of copper but low plasma levels of ceruloplasmin. In the liver of Wilson disease patients there is fatty deposition in early childhood, leading to further degeneration of the hepatocytes, inflammation, fibrosis, chronic hepatitis and cirrhosis. At any given time, copper may be released from the liver into the blood, being transported to virtually any tissue. However, the excessive accumulation of copper in the brain is important etiologically and causes movement disorders (such as tremors, rigidity, dysarthria and dysphagia) and psychiatric disorders in up to 15% of patients.

The localization of ceruloplasmin gene on chromosome 3 initially suggested that it could be the product of the Wilson disease gene. However, Frydman *et al.* (1985) used linkage analysis to locate the Wilson disease gene in the long arm of chromosome 13. The gene was more recently cloned and characterized (Ferenci *et al.*, 1996). The gene (*ATP7B*) is located on 13q14.3 and encodes for a P-type ATPase involved in cation transport (Bull *et al.*, 1993). The Wilson disease gene product is thought to be localized within the late secretory pathway of cells and involved in transporting copper from the cytosol towards its site of incorporation into ceruloplasmin prior to being sequestered by the endosomes for transport into the bile (Harris and Gitlin, 1996). Thomas *et al.* (1995) characterized multiple mutations in the Wilson disease gene. The molecular heterogeneity of the alterations in the Wilson disease gene is important, and results in the heterogeneous phenotype of Wilson disease patients. However, it has also been suggested that other factors, such as oxidative stress induced by excessive copper disposition, may be involved in the onset of certain forms of the hepatic pathology observed. Morphological changes in mitochondria and lipid peroxidation in rats with dietary copper overload, in rats genetically predisposed to copper accumulation (Long–Evans Cinnamon [LEC] rats), in Bedlington terriers with hepatic toxicosis, and in explanted livers from Wilson disease patients, suggest the involvement of copper-induced oxidative stress in Wilson disease (Ferenci *et al.*, 1996). Yang *et al.* (1997) recently observed that the Wilson disease protein localizes mainly in the Golgi apparatus, while a shorter form that is produced by alternative splicing in the human brain is present in the cytosol. The authors suggested that the alternative Wilson disease proteins act as regulators of copper metabolism, and that the short Wilson disease protein could be involved in the binding of copper in the cytosol, acting as a copper-storage depot in the brain.

Analysis of Wilson disease patients suggests that MT is the major copper-binding protein in the liver, and is saturated with copper (Nartey *et al.*, 1987b). Prior to the onset of hepatocellular necrosis, copper–MT is localized in the cytosol but after necrosis it is found in both the nucleus and cytosol (Nartey *et al.*, 1987b). Thus, the increased liver MT levels would presumably assist in the detoxication of the accumulated copper (Nartey *et al.*, 1987b; Mulder *et al.*, 1992) until such time that the capacity of this mechanism is saturated. The successful use of agents that can induce hepatic MT, such as zinc (Brewer, 1995a,b), as therapies to reduce the progression of Wilson disease also points towards a role for hepatic MT in the reduction of copper toxicity. Additionally, oral zinc exposure is thought to block intestinal absorption of copper by inducing MT within intestinal cells and thereby sequestering the copper prior to its reaching the systemic circulation (Brewer, 1995a,b).

The best animal model of Wilson disease is the LEC rat, a mutant strain derived from the Long–Evans Agouti (LEA) rat (Suzuki, 1995). LEC rats are deficient in oxidase-active ceruloplasmin in serum, have a marked increased in the hepatic copper concentration, and show a reduced biliary excretion of copper. LEC rats suffer an autosomal recessive form of hepatitis (Li *et al.*, 1991). Initial studies indicated a defect in the incorporation of copper into the ATP7B apoprotein due to a mutation resulting in a deletion of a major portion of the encoding region of the Wilson disease gene. Further studies demonstrated the full absence of Wilson disease gene transcripts in the liver of LEC rats, together with a deletion of the rat homolog of the Wilson disease gene (Wu *et al.*, 1994). LEC rats are characterized by a very high incidence of hepatocellular carcinoma (Harris and Gitlin, 1996), which is not observed in the parent LEA strain and can be prevented by early chelation of copper. Wilson disease patients do not show an elevated incidence of liver tumors. These liver cancers in LEC rats are likely due to the oxidative damage induced by excess copper and should be seen as an example of metal carcinogenesis. Transplantation of normal LEA hepatocytes into LEC rats prevents copper toxicosis in the latter (Yoshida *et al.*, 1996).

Expression of MT is selectively enhanced in LEC rats by copper in spite of the fact that copper uptake is comparable in LEC cells with that of normal cells (Kanno *et al.*, 1994a). This high inducibility of the MT gene in LEC rats is probably responsible for the accumulation of high levels of hepatic copper (Kanno *et al.*, 1994b). More recently, Klein *et al.* (1997) demonstrated that MT levels in the liver and kidney of LEC rats correlated with cytosolic levels of copper. Non-MT-bound copper, which correlated well with the total level of cytosolic copper, probably plays an important role in the development of copper toxicosis.

The toxic milk mouse is another model of inherited copper toxicosis (Rauch, 1983; Deng *et al.*, 1998). The homozygotes for the mutant gene are born with copper deficiency, and if rescued with copper supplement early in their lives, the pups accumulate an excessive amount of copper, causing toxicity at both the morphologic and functional levels. These homozygotes have a 100-fold increase in hepatic MT levels when compared with the wild-type mice. It has been suggested the increased stability of the MT protein and a defective transport of maternal copper to the embryos are important etiological factors in this disorder (Ferenci *et al.*, 1996). The presence of high levels of copper–MT and copper in the nucleus of cells of toxic milk mutant mice appears to be genotoxic and enhances apoptosis (Deng *et al.*, 1998).

14.3.6.2 Menkes disease

Trichopoliodystrophy, or Menkes disease, was first described by Menkes *et al.* (1962) as an X-linked disorder of copper metabolism, leading to pronounced mental retardation and motor dysfunction, with most patients dying before the age of 3 years (Schmidt *et al.*, 1984). Menkes disease is also known as the 'kinky-hair syndrome' but includes various other symptoms, such as brittle and depigmented hair, skeletal abnormalities, severe mental retardation and a number of other neurological defects (Leone *et al.*, 1985). Studies by Schmidt *et al.* (1984) suggested that because the X chromosome did not contain any MT genes, Menkes disease could not be due to alteration in the expression of MT genes or of their *cis*-acting regulatory sequences. Thus, Menkes disease only affects MT gene expression through a *trans*-acting alteration, or indirectly through alteration of copper homeostasis (Schmidt *et al.*, 1984).

Menkes disease is characterized by normal copper uptake with deficient copper-dependent enzyme activity in the face of excessive intracellular accumulation of copper, resulting in altered placental transport of copper prior to birth and intestinal accumulation of copper after birth, thereby creating a copper-deficient state (Harris and Gitlin, 1996). Copper concentrations are decreased in the brain, liver and serum of affected male infants, with corresponding decreases in the activity of essential copper-dependent enzymes (Leone *et al.*, 1985). However, copper concentrations in different extrahepatic tissues (intestine, kidney, lung, spleen, muscle, pancreas and skin) are higher than normal. In addition, cultured fibroblasts or lymphoblasts from Menkes disease patients also retain more copper, but are deficient in some copper enzymes. A decreased efflux and an increased biological half-life of copper in these mutant lines has also been observed.

A number of studies describe the cloning of the Menkes disease gene (*ATP7A*). ATP7A is a membrane protein, presumably a copper-transporting P-type ATPase (reviewed Harris and Gitlin, 1996). Wilson disease and Menkes disease ATPases are 65% homologous, but the Menkes disease gene is mostly expressed in nonhepatic tissues. The structure of both ATPases reveals an important conservation of sequences specific for the cation-transport proteins. It has also been suggested that variations in the severity of Menkes disease may be due to splicing defects in distinct regions of the Menkes disease gene.

The excessive copper in tissues and cultured cells from Menkes disease patients is accumulated in combination with MT (Riordan and Jolicoeur-Paquet, 1982). Cells from Menkes disease patients appear to be more sensitive to copper-induced toxicity than normal cells, and lower amounts of copper can induce MT synthesis in Menkes disease cells when compared with normal cells (Leone and Hamer, 1987). In cultured muscle cells and fibroblasts of Menkes disease patients, the induction of MT by copper is higher when compared to control cells (Herzberg *et al.*, 1990). The higher inducibility of MT under certain circumstances in Menkes disease cells would be consistent with the observed resistance to cadmium toxicity (Hayashi *et al.*, 1986). However, there appears to be no difference in the induction of MT by zinc or glucocorticoids in Menkes disease cells and normal cells (Waldrop and Ettinger, 1990). The increased MT gene expression in Menkes disease cells appears to be due, at least in part, to an abnormality in a diffusible *trans*-acting factor (Leone *et al.*, 1985; Leone, 1986; Hamer, 1987; Leone and Hamer, 1987). Two additional proteins also correlate with greater copper toxicity in Menkes disease cells (Leone *et al.*, 1985; Leone and Hamer, 1987).

Mice carrying a mottled (blotchy) allele have been used as a model of Menkes disease. It has been suggested that in more severe forms no functional copper-transporting ATPase protein is produced, while in less severe cases a small amount of normal functioning protein may be synthesized (Ferenci *et al.*, 1996). Male mice with the mottled–brindled allele (*Mo-brJ*) accumulate copper in the intestine and fail to export copper to peripheral organs and tissues, thus dying a few weeks after birth (Kelly and Palmiter, 1996). The retention of copper in the intestine induces the synthesis of MT-I and MT-II, resulting in the accumulation of copper–MT complex. The results of Kelly and Palmiter (1996), using a cross between Mo-brj mice and mice with a targeted disruption of MT-I and MT-II, suggest that MT is essential for protection of the embryo against copper toxicity and that MT is also required to protect intestinal cells after birth. Thus, MT may provide protection against copper toxicity as part of its physiologic function. In the macular mutant mice, a model for Menkes disease, high levels of copper–MT are found in the kidney (Suzuki-Kurasaki *et al.*, 1997). These mice are also quite sensitive to acute exposure to copper (Shiraishi *et al.*, 1991).

14.3.6.3 Other forms of copper toxicosis

Indian childhood cirrhosis (ICC) is a disease with a high incidence in India, but a few new cases have been diagnosed in Western countries. Epidemiological data suggest excessive copper intake as the underlying cause of ICC, but the appearance of ICC cases in Western countries has pointed out a possible involvement of the Wilson disease gene. ICC could be a specific form of idiopathic copper toxicosis (ICT) or copper-associated childhood cirrhosis requiring ingestion of high levels of copper for its expression (Pandit and Bhave, 1996). ICT, otherwise called non-Indian childhood cirrhosis, is a rare disorder with only a few proven cases (Scheinberg and Sternlieb, 1996). ICT occurs in the first 2 years of life, with a progressive and severe hepatic insufficiency due to liver cirrhosis. These patients, as opposed to Wilson disease patients, have normal or elevated levels of serum ceruloplasmin, and increased hepatic copper accumulation without significant cholestasis. The precise role of MT in these disease states is unknown but it appears that, in general, copper–MT is the major form of protein-bound copper in patients with hepatic copper overload (Hunziker and Sternlieb, 1991).

14.3.6.4 Iron overload

The relationship between iron homeostasis, overload and oxidative stress and MT has been reviewed recently (Ebadi *et al.*, 1996; Meneghini, 1997). Hemostatic control of iron depends on transport and storage proteins, such as ferritin, while excess tissue iron often accumulates in the form of hemosiderin, and iron overload frequently results in chronic hepatotoxicity. The iron found in ferritin or in hemosiderin may participate in the formation of OHx, a reactive oxygen species that can cause tissue damage due to oxidative interactions with lipids, nucleic acids and proteins. It has been proposed that release of zinc from MT could suppress the damage caused by free-radicals, but it appears that the MT protein itself is the most important factor in reducing iron-induced oxidative damage (Markant and Pallauf, 1996). MT may also sequester iron in a form that is no longer able to contribute to the Fenton chemistry required to generate oxidative radicals (Ebadi *et al.*, 1996). Good and Vasak (1986) demonstrated that under appropriate conditions Fe^{2+} forms the same well-defined complexes with MT as do Zn^{2+}, Cu^{2+} or Co^{2+}, which establishes the potential to detoxicate by binding. In fact, cultured cells poorly expressing MT are more sensitive to DNA damage induced by the hydroxyl radical, strongly suggesting that MT acts as an antioxidant species in the nucleus, protecting DNA from attack by oxidants. This view of MT as a free-radical scavenger is supported by the observation that transgenic mice deficient in MT-I and MT-II show an enhanced sensitivity to oxidative stress. Other studies conducted with hippocampal neurons in primary culture and in Chang liver cells in culture suggest a major role for MT in nullifying the iron-mediated generation of free-radicals and in protecting the brain against oxidative stress (Ebadi *et al.*, 1996). Goncharova and Rossman (1994) observed that Chinese hamster cells transfected with extra copies of the MT-I gene were less sensitive to spontaneous mutagenesis, thought to be primarily caused by oxidant species. Studies conducted with patients with liver cirrhosis due to primary hemochromatosis or Wilson disease showed that although the intracellular excess of iron and copper was bound to ferritin and MT, their excessive accumulation could cause tissue injury due to production of oxyradicals and lipid peroxidation (Kobune *et al.*, 1994). In addition, areas of the brain such as the striatum, containing high levels of iron but low levels of MT, are particularly vulnerable to oxidative stress (Ebadi *et al.*, 1996).

14.3.7 MT and cancer

MT may play several potential roles in cancer (Cherian, 1993; Waalkes, 1993). MT appears to be involved in a complex fashion in tumor cell pathobiology, and it may also be an important resistance factor in cancer chemotherapy. Furthermore, the limited expression of MT may be a predisposing factor in the sensitivity of a given tissue to cadmium carcinogenesis.

Elevated levels of MT expression have been detected in a variety of tumors and tumor cells (Nartey *et al.*, 1987c; Hashiba *et al.*, 1989; Kontozoglou *et al.*, 1989; Bahnson *et al.*, 1991; Cherian, 1993; Maier *et al.*, 1997), although this is by no means a totally consistent finding (Waalkes, 1993; Waalkes *et al.*, 1996; Woo *et al.* 1997). The over-expression of MT in tumors is possibly a consequence of the rapid cellular proliferation associated with malignancy. It is not unusual for proteins that are typically restricted to expression during the fetal or embryonic periods to be re-expressed in tumors, such as α-fetoprotein (Cherian, 1993). Examples of tumors where MT expression appears to be elevated, at least in some cases, include tumors of the thyroid (adenoma and adenocarcinoma), testis (seminoma and carcinoma), urinary bladder (carcinoma), salivary gland (adenoma and carcinoma), skin (papillomas), esophagus (carcinoma), breast and brain (Nartey *et al.*, 1987c; Hashiba *et al.*, 1989; Kontozoglou *et al.*, 1989; Cherian, 1993; Haerslev *et al.*, 1995; Hishikawa *et al.*, 1997; Maier *et al.*, 1997). In the tumors overexpressing MT, it is often found at higher levels on the proliferative edge of the tumor (Cherian, 1993). Poorly vascularized, non-small-cell carcinomas of the lung also appear to overexpress MT (Koomagi *et al.*, 1995).

Some human tumors show very heterogeneous expression of MT, for example ovarian tumors (Germain *et al.*, 1996) and certain gliomas (Maier *et al.*, 1997). In some cases human or rodent tumors express only minimal amounts of MT, such as with tumors of the mouse lung or liver (Waalkes, 1993; Waalkes *et al.*, 1996) or human pulmonary small-cell carcinomas (Matsumoto *et al.*, 1997). Human hepatocellular carcinoma appears to be generally deficient in MT in most cells (Waalkes *et al.*, 1996). Likewise, relatively low MT levels occur in the malignant portions versus the surrounding normal areas of surgically resected human liver tumors (Onosaka *et al.*, 1986). A comprehensive study of the 53 adherent cell lines derived from the National Cancer Institute tumor panel showed a great diversity of MT levels, from minimal to high, and major differences in nuclear or cytoplasmic distribution (Woo *et al.*, 1997).

The expression of MT can be associated with the level of tumor differentiation or prognosis, but this is apparently related to the tissue of origin. In some cases, expression of MT in tumors is increased as the level of tumor differentiation is decreased. For instance, MT is more prominently expressed in poorly differentiated ductal carcinomas of the breast (Douglas-Jones *et al.*, 1997), hepatocellular carcinomas (Tashiro-Itoh *et al.*, 1997) and invasive renal cell carcinomas (Zhang and Takenaka, 1998) than in tumors of earlier or more differentiated stages. In gastric cancers, higher MT expression is associated with a poorer outcome (Monden *et al.*, 1997) and MT expression is increased in tumors of the breast (Goulding *et al.*, 1995; Haerslev *et al.*, 1995; Helal *et al.*, 1997) and esophagus (Hishikawa *et al.*, 1997) with a poor prognosis. In some cases MT expression is not associated with prognostic factors such as stage or grade, as with ovarian carcinomas (Germain *et al.*, 1996) or osteosarcoma (Uozaki *et al.*, 1997). In other cases the expression of MT is reduced in more poorly differentiated tumors with poor prognosis. For example, in germ-cell testicular tumors high MT expression is predictive of a better rate of response to chemotherapy, whereas a low level of MT expression is associated with a poor prognosis (Eid *et al.*, 1998). Likewise, the worst grades of prostatic carcinoma show the least expression of MT (Moussa *et al.*, 1997).

Although little is known about MT levels in metastatic tumors, it appears that in patients with liver metastasis of colorectal cancers the metastatic lesion contains much less MT than the surrounding normal tissue (Mulder *et al.*, 1992). Likewise, in lymph node metastases of breast cancer there is typically minimal evidence of MT expression but the expression of MT in the metastasis is correlated with the expression in the primary tumor (Haerslev *et al.*, 1994). In rats with pulmonary metastasis of distant sarcomas there appears to be less MT in the metastasis than the primary tumor (Waalkes *et al.*, 1998).

In mice, MT concentrations are elevated in liver with tumor promoters that produce cellular proliferation (i.e. acetaminophen and di(2-ethylhexyl)phthalate; Waalkes and Ward, 1989). Liver tumor promoters that induce cellular enlargement (barbiturates) but not cellular hyperplasia, do not elevate hepatic MT, indicating a relationship to stimulation of cell division (Waalkes and Ward, 1989). Likewise, the skin tumor-promoting phorbol ester, 12-O-tetradecanoylphorbol-13-acetate (TPA), which acts, at least in part, by inducing cellular proliferation, can enhance *MT* gene expression (Hashiba *et al.*, 1989).

It is clear that the association of MT with various malignancies depends on a variety of factors, including tumor type, cell of origin, morphological heterogeneity and stage (Cherian, 1993). The frequent development of distinctive subpopulations within tumors creates an additional problem in assigning a general role for MT in tumor-cell pathobiology (Cherian, 1993). The role of MT in tumor-cell pathobiology requires further definition.

It is interesting to speculate that the great variability in tumor expression of MT may be linked to methylation changes in DNA, a common finding in cancers (Laird and Jaenisch, 1994; Counts and Goodman, 1995). MT expression is clearly linked to methylation status of the *MT* gene and hypomethylation facilitates expression while hypermethylation reduces expression (Compere and Palmiter, 1981; Waalkes *et al.*, 1988). If the 5′-flanking regions of the *MT* gene are highly methylated, it will be poorly expressed (Compere and Palmiter, 1981) and minimally reactive to inducing stimuli, making MT inducibility a strong indicator of DNA hypomethylation (Levine *et al.*, 1991; Salehi-Ashitiani *et al.*, 1993). For example, in cellular malignant transformation induced by arsenic, which in this case is linked to genomic DNA hypomethylation, a remarkable hyperinducibility of MT occurs which is indicative of gene hypomethylation (Zhao *et al.*, 1997). Treatment with hypomethylating agents, such as azacytidine, will also cause an exaggerated response to stimulation of MT with inducing agents (Compere and Palmiter, 1981; Waalkes *et al.*, 1985, 1988). However, any direct association between DNA methylation changes in tumors and MT gene expression has yet to be investigated.

Another area in which MTs may have an important role is in chemotherapy of certain cancers, both in the development of tolerance to chemotherapeutic agents and as a potential therapeutic adjunct to reduce toxic side-effects (Kondo *et al.*, 1993; Lazo *et al.*, 1993; Naganuma *et al.*, 1993; Saijo *et al.*, 1993; Waalkes, 1993). For instance, induction of MT in the kidney reduces the renal toxicity of cisplatin (CDDP) (Saijo *et al.*, 1993). Similarly *MT-I/MT-II* knockout mice are much more sensitive to the nephrotoxic effects of CDDP (Liu *et al.*, 1998). The nephrotoxicity of CDDP is a primary factor limiting its chemotherapeutic usefulness.

It is also clear that the expression of MT is often a primary cause of tumor-cell resistance to electrophilic antineoplastic agents, including CDDP (Lazo *et al.*, 1993). It is suspected that overexpression of MT may be a factor in poor response to chemotherapeutic intervention with a variety of human malignancies including, for example, tumors of the brain (Maier *et al.*, 1997), lung (Matsumoto *et al.*, 1997) and esophagus (Hishikawa *et al.*, 1997). In fact, as a generalized finding in tumor cells, MT expression inversely correlates with responsiveness to metallochemotherapeutics (Woo *et al.*, 1997). However, some tumor types do not fall into this pattern and, for example, higher expression of MT correlates with a better rate of response to chemotherapy in tumors of the testes (Eid *et al.*, 1998). Thus the correlation between tumor chemoresistance and MT expression has to be determined for each individual type of cancer.

In some rodent models of chemically induced cancers, treatment with cadmium can markedly reduce tumor incidence and multiplicity (Waalkes *et al.*, 1991, 1993, 1996). This applies to tumors of the liver and lung induced in mice with nitrosamines (Waalkes *et al.*, 1991, 1993, 1996). The reduced incidence of tumors induced by cadmium is associated with tumor-specific necrosis and correlated with poor expression of MT in these particular tumors (Waalkes *et al.*, 1993, 1996). Thus the poor expression of MT renders these tumors sensitive to the toxic effects of cadmium, while preserving the normal surrounding tissue. The chemotherapeutic utility of cadmium may be questionable because of its severe toxic and carcinogenic potential.

There is some evidence that the poor expression of MT may be a factor in the tissue specificity of cadmium carcinogenesis. Cadmium has been associated with a variety of tumors in humans and animals including tumors of the lung and prostate (IARC, 1993; Waalkes, 1995). Rat tumors of the prostate are largely induced in the ventral lobe (Waalkes and Rehm, 1994; Waalkes *et al.*, 1997), a portion of the prostate that only poorly expresses MT relative to other areas (Tohyama *et al.*, 1993a; Coogan *et al.*, 1995, 1997). The *MT* gene in the rat ventral prostate is at best minimally responsive to metal stimulation (Tohyama *et al.*, 1993a; Coogan *et al.*, 1995, 1997). The dorsolateral prostate in the rat shows a high expression of MT (Tohyama *et al.*, 1993a; Coogan *et al.*, 1995; 1997) and is generally not a target for cadmium carcinogenesis (Waalkes and Rehm, 1994; Waalkes *et al.*, 1997). In comparative studies of species sensitivity to cadmium carcinogenesis, mice appear to be resistant to pulmonary carcinogenesis after inhalation of cadmium, whereas rats are sensitive (IARC, 1993). Generally speaking, mice have higher pulmonary MT levels than rats in both control settings and after cadmium exposure (Oberdörster *et al.*, 1994a,b; Kenaga *et al.*, 1996; McKenna *et al.*, 1998) and this correlates with the severity of early reactions to inhaled cadmium (McKenna *et al.*, 1997). The role of MT in cadmium carcinogenesis in humans is not known.

14.4 Toxicity of MT

There is clear evidence that, under certain conditions, MT can be quite toxic. This is dependent on the metal complement of the MT and occurs only under pathologic conditions. Although the cadmium–MT complex is essentially nontoxic when localized within the cell, it becomes a potent nephrotoxin if it reaches the systemic circulation (Nordberg *et al.*, 1972, 1982; Nordberg and Piscator, 1975; Cherian *et al.*, 1976; Foulkes, 1978; Suzuki and Cherian, 1987; Klaassen and Liu, 1997).

The liver accumulates large amounts of cadmium, bound to MT, during chronic cadmium exposure. At some point in chronic exposure situations cadmium-containing MT is probably extruded from the liver, possibly through release after toxic insult, and reaches the systemic circulation (Dudley *et al.*, 1982). Once in the systemic circulation the metal–protein complex is transported to the kidney where it is filtered by the glomerulus, and subsequently reabsorbed by the proximal tubule cell in a manner similar to other low-molecular-weight proteins (Foulkes, 1978; Squibb *et al.*, 1979, 1984; Squibb and Fowler, 1984). The cadmium–MT complex shows preferential uptake into the S1 and S2 segments of the proximal tubules, the site of cadmium nephrotoxicity (Dorian *et al.*, 1992). The cadmium–MT protein complex is highly toxic to the kidney after renal tubular reabsorption (Nordberg *et al.*, 1972; Cherian *et al.*, 1976; Foulkes, 1978; Suzuki and Cherian, 1987; Klaassen and Lui, 1997). The reabsorbed complex is rapidly degraded, which probably releases free cadmium at high concentrations (Squibb *et al.*, 1979, 1984; Squibb and Fowler, 1984).

Typically, cadmium-induced chronic renal injury causes a low-molecular-weight proteinuria (Nordberg *et al.*, 1972; Nordberg and Piscator, 1975), a finding that in the past was common in occupationally exposed individuals (Friberg, 1950). The proteinuria is secondary to proximal tubular necrosis (Nordberg *et al.*, 1975; Cherian *et al.*, 1976; Foulkes, 1978; Suzuki and Cherian, 1987; Klaassen and Lui, 1997). The proximal tubular lesions are an early and specific effect, resulting from exposure of the kidney to cadmium–MT and generally do not occur after ionic cadmium exposure (Squibb *et al.*, 1979, 1984; Squibb and Fowler, 1984). It appears that the extent of renal injury is correlated with the levels of plasma MT (Dudley *et al.*, 1982). Repeated, short-interval exposures of the kidney to cadmium–MT, as opposed to large, infrequent exposures, appear to account for the chronic nephrotoxic effects (Leffler *et al.*, 1996).

Recent evidence clearly indicates that MT is not obligatory in cadmium-induced renal injury (Klaassen *et al.*, 1998). When *MT-I/MT-II* null mice, which cannot form cadmium–MT complexes in the liver or kidney, are given chronic treatments of cadmium salts, chronic nephrotoxicity related to the cadmium still results (Klaassen *et al.*, 1998). This is likely to be due to the availability of other low-molecular-weight proteins to carry cadmium to its site of toxic action in the kidney. The *MT-I/MT-II* null mice in fact show more severe nephropathic changes at lower doses of cadmium than the control mice (Klaassen *et al.*, 1998).

In cases of renal dysfunction caused by chronic cadmium exposure in humans, metallothioneinuria is a consistent finding (Shaikh *et al.*, 1990a,b). Metallothioneinuria occurs in populations exposed to cadmium through the environment or through occupational exposure (Shaikh *et al.*, 1990a,b). The presence of MT in the urine is clearly diagnostic of the etiological role for cadmium in these cases. The levels of MT in the urine can be used as a sensitive biomarker of cadmium exposure even before the onset of severe renal dysfunction (Shaikh *et al.*, 1990a,b).

There is also good evidence that the cadmium–MT complex is absorbed at least partially intact from the gastrointestinal tract (Maitani *et al.*, 1984; Klein *et al.*, 1986; Ohta and Cherian, 1991). Cadmium from this source is more readily deposited in the kidney than ionic cadmium (Maitani *et al.*, 1984). Thus cadmium–MT in the diet has the potential to contribute to cadmium nephrotoxicity.

Several factors can modify the sensitivity to cadmium–MT–induced nephrotoxicity. Genetically based diabetes increases susceptibility to cadmium–MT-induced nephropathy in hamsters (Jin and Frankel, 1996). The later stages of pregnancy in rats appears to be a period of hypersensitivity to cadmium–MT-induced nephropathy, for unknown reasons (Payan *et al.*, 1990). Treatments with zinc, either as zinc salts or as zinc–MT, will reduce nephropathy resulting from acute treatments with cadmium–MT (Dorian and Klaassen, 1995; Klaassen and Liu, 1997) but this does not appear to be due to prior induction of MT by zinc in the kidney (Liu *et al.*, 1996). The metal content of the MT will also modify its nephrotoxic effects and MTs containing higher cadmium/zinc ratios are more toxic (Kojima *et al.*, 1991). Unilateral nephrectomy increases the renal toxicity of cadmium–MT by a mechanism apparently related to a greater renal accumulation of cadmium (Zalups *et al.*, 1992). Although strain-related differences to acute treatment with cadmium salts can be quite dramatic in mice, this genetic variability does not appear to cause sensitivity to cadmium–MT nephrotoxicity (Sendelbach *et al.*, 1992).

14.5 Summary

In summary, metallothionein is an important protein both in the normal and toxic metabolism of metals. It may well have functions beyond metal metabolism. Although there has been a great deal of research on this remarkable protein, much still remains to be elucidated.

References

Andersen, R.D., Birren, B.W., Ganz, T., Piletz, J.E. and Herschman, H.R. 1983a. Molecular cloning of the rat metallothionein 1 (MT-1) mRNA sequence. *DNA* **2**, 15–22.

Andersen, R.D., Piletz, J.E., Birren, B.W. and Herschman, H.R. 1983b. Levels of metallothionein messenger RNA in foetal, neonatal and maternal rat liver. *European Journal of Biochemistry* **131**, 497–500.

Andersen, R.D., Birren, B.W., Taplitz, S.J. and Herschman, H.R. 1986. Rat metallothionein-1 structural gene and three pseudogenes, one of which contains 5′-regulatory sequences. *Molecular and Cellular Biology* **6**, 302–14.

Andersen, R.D., Taplitz, S.J., Wong, S., Bristol, G., Larkin, B. and Herschman, H.R. 1987. Metal-dependent binding of a factor *in vivo* to the metal-responsive elements of the metallothionein 1 gene promoter. *Molecular and Cellular Biology* **7**, 3574–81.

Andrews, G.K., Adamson, E.D. and Gedamu, L. 1984. The ontogeny of expression of murine metallothionein: Comparison with α-fetoprotein gene. *Developmental Biology* **103**, 294–303.

Andrews, G.K., McMaster, M.T., De, S.K., Paria, B.C. and Dey, S.K. 1993. Cell-specific expression and regulation of the mouse metallothionein-I and -II genes in the reproductive tract and preimplanation embryo, in *Metallothionein III*, (eds K.T. Suzuki, N. Imura and M. Kimura), Birkhäuser Verlag, Basel, pp. 351–62.

Anezaki, T., Ishiguro, H., Hozumi, I., Inuzuka, T., Hiraiwa, M., Kobayashi, H., Yoguchi, T., Wanaka, A., Uda, Y., Miyatake, T., Yamada, K., Tohayama, M. and Tsuji, S. 1995. Expression of growth inhibitory factor (GIF) in normal and injured rat brains. *Neurochemistry International* **27**, 89–94.

Aschner, M. 1996. The functional significance of brain metallothioneins. *FASEB Journal* **10**, 1129–36.

Aschner, M. 1997. Astrocyte metallothioneins (MTs) and their neuroprotective role. *Annals of the New York Academy of Science* **825**, 334–47.

Aschner, M., Cherian, M.G., Klaassen, C.D., Palmiter, R.D., Erickson, J.C. and Bush, A.I. 1997.

Metallothioneins in brain – The role in physiology and pathology. *Toxicology and Applied Pharmacology* **142**, 229–42.

Bahnson, R.R., Banner, B.F., Ernstoff, M.S., Lazo, J.S., Cherian, M.G., Banerjee, D. and Chin, J.L. 1991. Immunohistochemical localization of metallothionein in transitional cell carcinoma of the bladder. *Journal of Urology* **146**, 1518–20.

Bakka, A. and Webb, M. 1981. Metabolism of zinc and copper in the neonate: changes in the concentrations and contents of thionein-bound Zn and Cu with age in the livers of the newborn of various mammalian species. *Biochemical Pharmacology* **30**, 721–5.

Beach, L.R. and Palmiter, R.D. 1981. Amplification of the metallothionein-I gene in cadmium-resistant mouse cells. *Proceedings of the National Academy of Sciences, USA* **78**, 2110–14.

Bird, A.P. 1992. The essentials of DNA methylation. *Cell* **70**, 5–8.

Brady, F.O. 1982. The physiological function of metallothionein. *Trends in Biochemical Sciences* **7**, 143–5.

Bremner, I. 1993. Involvement of metallothionein in the regulation of mineral metabolism, in *Metallothionein III*, (eds K.T. Suzuki, N. Imura and M. Kimura), Birkhäuser Verlag, Basel, pp. 111–24.

Brewer, G.J. 1995a. Interactions of zinc and molybdenum with copper in therapy of Wilson disease. *Nutrition* **11**, 114–16.

Brewer, G.J. 1995b. Practical recommendations and new therapies for Wilson disease. *Drugs* **50**, 240–9.

Brewer, G.J. and Yuzbasiyan-Gurkan, V. 1992. Wilson disease. *Medicine* **71**, 139–64.

Brugnera, E., Georgiev, O., Radtke, F., Heuchel, R., Baker, E., Sutherland, G.R. and Schaffner, W. 1994. Cloning, chromosomal mapping and characterization of the human metal-regulatory transcription factor MTF-1. *Nucleic Acids Research* **22**, 3167–73.

Bull, P.C., Thomas, G.R., Rommens, J.M., Forbes, J.R. and Cox, D.W. 1993. The Wilson disease gene is a putative copper transporting P-type ATPase similar to the Menkes gene. *Nature Genetics* **5**, 327–37.

Butt, T.R., and Ecker, D.J. 1987. Yeast metallothionein and applications in biotechnology. *Microbiological Reviews* **51**, 351–64.

Chan, H.M. and Cherian, M.G. 1993. Ontogenic changes in hepatic metallothionein isoforms in prenatal and newborn rats. *Biochemistry and Cell Biology* **71**, 133–40.

Chen, R.W. and Ganther, H.E. 1975. Relative-cadmium-binding capacity of metallothionein and other cytosolic fractions in various tissues of the rat. *Environmental Physiology and Biochemistry* **5**, 378–88.

Cherian, M.G. 1993. Nuclear and cytoplasmic localization of metallothionein in human liver during development and in tumor cells, in *Metallothionein III*, (eds K.T. Suzuki, N. Imura and M. Kimur), Birkhäuser Verlag, Basel, pp. 175–87.

Cherian, M.G. and Chan, H.M. 1993. Biological functions of metallothionein – a review, in *Metallothionein III*, (eds K.T. Suzuki, N. Imura and M. Kimura), Birkhäuser Verlag, Basel, pp. 87–109.

Cherian, M.G., Goyer, R.A. and Delaquerriere-Richardson, L. 1976. Cadmium-metallothionein-induced nephrotoxicity. *Toxicology and Applied Pharmacology* **38**, 399–408.

Choudhuri, S., Liu, W.L., Berman, N.E. and Klaassen, C.D. 1996a. Cadmium accumulation and metallothionein expression in brain of mice at different stages of development. *Toxicology Letter* **84**, 127–33.

Choudhuri, S., Kramer, K.K., Berman, N.E., Dalton, T.P., Andrews, G.K. and Klaassen, C.D. 1996b. Constitutive expression of metallothionein genes in mouse brain. *Toxicology Applied Pharmacology* **131**, 144–54.

Clough, S.R., Mitra, R.S. and Kulkami, A.P. 1986. Qualitative and quantitative aspects of human-fetal liver metallothioneins. *Biology of the Neonate* **49**, 241–54.

Compere, S.J. and Palmiter, R.D. 1981. DNA methylation controls the inducibility of the mouse metallothionein-I gene in lymphoid cells. *Cell* **25**, 23340.

Coogan, T.P., Shiraishi, N. and Waalkes, M.P. 1995. Minimal basal activity and lack of metal-induced activation of the metallothionein gene correlates with lobe-specific sensitivity to the carcinogenic effects of cadmium in the rat prostate. *Toxicology and Applied Pharmacology* **132**, 164–73.

Coogan, T.P., Shiraishi, N. and Waalkes, M.P. 1997. Metallothionein gene expression in the reproductive tissues of the male Wistar rat: Effects of treatment with metals or glucocorticoids. *Toxic Substance Mechanisms* **16**, 357–70.

Counts, J.L. and Goodman, J.I. 1995. Alterations in DNA methylation may play a variety of roles in carcinogenesis. *Cell* **83**, 13–15.

Crawford, B.D., Enger, M.D., Griffith, B.B., Griffith, J.K., Hanners, J.L., Longmire, J.L., Munk, A.C., Stallings, R.L., Tesmer, J.G., Walters, R.A., and Hildebrand, C.E. 1985. Coordinate amplification of metallothionein I and II genes in cadmium-resistant Chinese hamster cells: implications for mechanisms regulating metallothionein gene expression. *Molecular and Cellular Biology* **5**, 320–9.

Dallinger, R. 1996. Metallothionein research in terrestrial invertebrates: synopsis and perspectives. *Comparative Biochemistry and Physiology* **113C**, 125–33.

Dalton,T., Palmiter, R.D. and Andrews, G.K. 1994. Transcriptional induction of the mouse metallothionein-I gene in hydrogen peroxide-treated Hepa cells involves a composite major late transcription factor/antioxidant response element and metal response promoter elements. *Nucleic Acids Research* **22**, 5016–23.

Dalton, T.P., Pazdernik, T.L., Wagner, J., Samson, F. and Andrews, G.K. 1995. Temporospatial patterns of expression of metallothionein-I and -III and other stress related genes in rat brain after kainic acid induced seizures. *Neurochemistry International* **27**, 59–71.

Dalton, T., Fu, K., Palmiter, R.D. and Andrews, G.K. 1996a. Transgenic mice that overexpress metallothionein-I resist dietary zinc deficiency. *Journal of Nutrition* **126**, 825–33.

Dalton, T., Fu, K., Enders, G.C., Palmiter, R.D. and Andrews, G.K. 1996b. Analysis of the effects of overexpression of metallothionein-I in transgenic mice on the reproductive toxicology of cadmium, *Environmental Health Perspectives* **104**, 68–76.

Datson, G.P. and Lehman-McKeeman, L.D. 1996. Constitutive and induced metallothionein expression in development, in *Toxicology of metals*, (eds L.W. Chang, L. Magos and K.T. Suzuki), CRC Lewis Publishing, Boca Raton, pp. 1139–67.

De, S.K., McMaster, M. and Andrews, G.K. 1990. Endotoxin induction of murine metallothionein genes. *The Journal of Biological Chemistry* **265**, 15267–74.

Deng, D.X., Ono, S., Koropatnick J. and Cherian, M.G. 1998. Metallothionein and apoptosis in the toxic milk mutant mouse. *Laboratory Investigations* **78**, 175–83.

Dorian, C. and Klaassen, C. D. 1995. Protection by zinc-metallothionein (ZnMT) against cadmium-metallothionein-induced nephrotoxicity. *Fundamental Applied Toxicology* **26**, 99–106.

Dorian C., Gattone, V.H. 2d and Klaassen, C. D. 1992. Renal cadmium deposition and injury as a result of accumulation of cadmium-metallothionein (CdMT) by the proximal convoluted tubules – A light microscopic autoradiography study with 109CdMT. *Toxicology and Applied Pharmacology* **114**, 173–81.

Douglas-Jones, A.G., Navabi, H., Morgan, J.M. and Jasani, B. 1997. Immunoreactive p53 and metallothionein expression in duct carcinoma *in situ* of the breast; No correlation. *Virchows Archives* **430**, 373–9.

Dudley, R.E., Gammal, L.M. and Klaassen, C.D. 1982. Cadmium-induced hepatic and renal injury in chronically exposed rats: Likely role of hepatic cadmium–metallothionein in nephrotoxicity. *Toxicology and Applied Pharmacology* **77**, 414–26.

Durnam, D.M. and Palmiter, R.D. 1981. Transcriptional regulation of the mouse metallothionein-I gene by heavy metals. *The Journal of Biological Chemistry* **256**, 5712–16.

Durnam, D.M., Perrin, F., Gannon, F. and Palmiter, R.D. 1980. Isolation and characterization of the mouse metallothionein-I gene. *Proceedings of the National Academy of Sciences of the USA* 77, 6511–15.

Ebadi, M. 1984. The presence of metallothionein-like protein (MLT) in rat brain. *Federation Proceedings* **43**, 3317.

Ebadi, M., Iversen, P.L., Hao, R., Cerutis, D.R., Rojas, P., Happe, H.K., Murrin, L.C. and Pfeiffer, R.F. 1995. Expression and regulation of brain metallothionein. *Neurochemistry International* **27**, 1–22.

Ebadi, M., Leuschen, M.P., el-Refaey, H., Hamada, F.M. and Rojas, P. 1996. The antioxidant properties of zinc and metallothionein. *Neurochemistry International* **29**, 159–66.

Eid, H., Geczi, L., Bodrogi, I., Institoris, E. and Bak, M. 1998. Do metallothioneins affect the response to treatment in testis cancers. *Journal of Cancer Research Clinical Oncology* **124**, 31–6.

Erickson, J.C., Sewell, A.K., Jensen, L.T., Winge, D.R. and Palmiter, R.D. 1994. Enhanced neurotrophic activity in Alzheimer's disease cortex is not associated with down-regulation of metallothionein-III (GIF). *Brain Research* **649**, 297–304.

Erickson, J.C., Masters, B.A., Kelly, E.J., Brinster, R.L. and Palmiter, R.D. 1995. Expression of metallothionein-III in transgenic mice. *Neurochemistry International* **27**, 35–41.

Erickson, J.C., Hollopeter, G., Thomas, S.A., Froelick, G.J. and Palmiter, R.D. 1997. Disruption of the metallothionein-III gene in mice: analysis of brain zinc, behavior, and neuron vulnerability to metals, aging, and seizures. *Journal of Neuroscience* **17**(4), 1271–81.

Ferenci, P., Gilliam, T.C., Gitlin, J.D., Packman, S., Schilsky, M.L., Sokol, R.J. and Sternlieb, I. 1996. An international symposium on Wilson's and Menkes diseases. *Hepatology* **24**, 952–8.

Foulkes, E.C. 1978. Renal tubular transport of cadmium metallothionein. *Toxicology and Applied Pharmacology* **45**, 505–12.

Fowler, B.A., Hildebrand, C.E., Kojima, Y. and Webb, M. 1987. Nomenclature of metallothionein. *EXS* **52**, 19–22.

Friberg, L. 1950. Health hazards in the manufacture of alkaline accumulators with special reference to chronic cadmium poisoning. *Acta Medica Scandinavica* **138**, 1–124.

Frydman, M., Bonne-Tamir, B., Farrer, L.A., Conneally, P.M., Magazanik, A., Ashbel, S. and Goldwitch, Z. 1985. Assignment of the gene for Wilson disease to chromosome 13: linkage to the esterase D locus. *Proceedings of the National Academy of Sciences USA* **82**(6), 1819–21.

Galiazzo, F., Ciriolo, M.R., Carri, M.T., Civitareale, P., Marcocci, L., Marmocchi, F. and Rotilio, G. 1991. Activation and induction by copper of Cu/Zn superoxide dismutase in *Saccharomyces cerevisiae*. Presence of an inactive proenzyme in anaerobic yeast. *European Journal of Biochemistry* **196**, 545–9.

Geller, B.L. and Winge, D.R. 1982. Metal binding sites of rat liver Cu-thionein. *Archives of Biochemistry and Biophysics* **219**, 109–17.

Germain, I., Tetu, B., Brisson, J., Mondor, M. and Cherian, M.G. 1996. Markers of chemoresistance in ovarian carcinomas: an immunohistochemical study of 86 cases. *International Journal of Gynecologic Pathology* **15**, 54–62.

Glanville, N., Durnam, D.M. and Palmiter, R.D. 1981. Structure of mouse metallothionein-I gene and its mRNA. *Nature* **292**, 267–9.

Goering, P.L. and Fowler, B.A. 1987. Kidney zinc–thionein regulation of aminolevulinic acid dehydratase inhibition by lead. *Archives of Biochemistry and Biophysics* **253** 48–55.

Goering, P.L. and Klaassen, C.D. 1984. Zinc-induced tolerance to cadmium hepatotoxicity. *Toxicology and Applied Pharmacology* **74**, 299–307.

Goncharova, E.I. and Rossman, T.G. 1994. A role for metallothionein and zinc in spontaneous mutagenesis. *Cancer Research* **54**, 5318–23.

Good, M. and Vasak, M. 1986. Iron(II)-substituted metallothionein: evidence for the existence of iron-thiolate clusters. *Biochemistry* **25**, 8353–6.

Goulding, H., Jasani, B., Pereira, H., Reid, A., Galea, M., Bell, J.A., Elston, C.W., Robertson, J. F., Blamey, R. W., Nicholson, R. A.; *et al.* 1995. Metallothionein expression in human breast cancer *British Journal of Cancer* **72**, 968–72.

Griffith, J.K. 1985. Coordinate expression of amplified metallothionein I and II genes in cadmium-resistant Chinese hamster cells. *Molecular and Cellular Biology* **5**, 3525–31.

Haerslev, T., Jacobsen, K., Nedergaard, L. and Zedeler, K. 1994. Immunohistochemical detection of metallothionein in primary breast carcinomas and their axillary lymph node metastases. *Pathology and Research Practice* **190**, 675–81.

Haerslev, T., Jacobsen, G.K. and Zedeler, K. 1995. The prognostic significance of immunohistochemically detectable metallothionein in primary breast carcinomas. *APMIS* **103**, 279–85.

Hamer, D.H. 1986. Metallothionein. *Annual Review of Biochemistry* **55**, 913–51.

Hamer, D. H. 1987. Metallothionein gene regulation in Menkes' syndrome. *Archives of Dermatology* **123**, 1384a–1385a.

Hamer, D.H. and Winge, D.R. (eds) 1989. *Metal Ion Homeostasis*, Alan R. Liss, New York.

Harris, Z.L. and Gitlin, J.D. 1996. Genetic and molecular basis for copper toxicity. *American Journal of Clinical Nutrition* **63**, 836S–841S.

Hashiba, H., Hosoi, J., Karasawa, M., Yamada, S., Nose, K. and Kuroki, T. 1989. Induction of metallothionein mRNA by tumor promoters in mouse skin and its constitutive expression in papillomas. *Molecular Carcinogenesis* **2**, 95–100.

Hayashi, A., Sato, M., Arima, M. and Aoki, T. 1986. Increased cadmium resistance of skin fibroblasts from Menkes disease patients. *Brain Development* **8**, 73–6.

Heguy, A., West, A., Richards, R.I. and Karin, M. 1986. Structure and tissue-specific expression of the human metallothionein IB gene. *Molecular and Cellular Biology* **6**, 2149–57.

Helal, T., Nassiri, M. and Khalifa, A. 1997. Immunohistochemical markers of tumor prognosis in breast cancer in Egypt. *Cancer Detection and Prevention* **21**, 201–6.

Henry, R.B., Liu, J., Choudhuri, S. and Klaassen, C.D. 1994. Species variation in hepatic metallothionein. *Toxicology Letters* **74**, 23–33.

Herzberg, N.H., Wolterman, R.A., van den Berg, G.J., Barth, P.G. and Bolhuis, P.A. 1990. Metallothionein in Menkes disease: Induction in cultured muscle cells. *Journal of Neurological Science* **100**, 50–6.

Hidalgo, J., Castellano, B. and Campbell, I.L. 1997. Regulation of brain metallothioneins. *Current Topics in Neurochemistry* **1**, 1–25.

Hishikawa, Y., Abe, S., Kinugasa, S., Yoshimura, H., Monden, N., Igarashi, M., Tachibana, M. and Nagasue, N. 1997. Overexpression of metallothionein correlates with chemoresistance in esophageal cancer. *Oncology* **54**, 342–7.

Hunziker, P.E. and Kägi, J.H. 1985. Isolation and characterization of six human hepatic isometallothioneins. *Biochemical Journal* **231**, 375–82.

Hunziker, P.E. and Sternlieb, I. 1991. Copper metallothionein in patients with hepatic copper overload. *European Journal of Clinical Investigations* **21**, 466–71.

IARC 1993. *Beryllium, cadmium, mercury and exposures in the glass industry*, International Agency for Research on Cancer Monographs, Volume 58, IARC Press, Lyon, pp. 119–238.

Iszard, R.D., Liu, Y., Dalton, T., Andrews, G.K., Palmiter, R.D. and Klaassen, C.D. 1995. Characterization of metallothionein-I-transgenic mice. *Toxicology and Applied Pharmacology* **133**, 305–12.

Itoh, M., Ebadi, M. and Swanson, S. 1983. The presence of zinc-binding proteins in brain. *Journal of Neurochemistry* **41**, 823–9.

Jin, T. and Frankel, B.J. 1996. Cadmium–metallothionein nephrotoxicity is increased in genetically diabetic as compared with normal Chinese hamsters. *Pharmacology and Toxicology* **79**, 105–8.

Kägi, J.H.R. 1991. Overview of metallothionein, in *Methods in Enzymology*, Vol. 205, Part B, *Metallothionein and Related Molecules*, (eds J.F. Riodan and B.L. Vallee), Academic Press, New York, pp. 613–15.

Kägi, J.H.R. 1993. Evolution, structure and chemical activity of class I metallothioneins: An overview, in *Metallothionein III*, (eds K.T. Suzuki, N. Imura and M. Kimura), Birkhäuser Verlag, Basel, pp. 29–55.

Kägi, J.H.R. and Kojima, Y. 1979. Nomenclature of metallothionein: a proposal. *Experientia* **34**, 141–2.

Kägi, J.H.R. and Kojima, Y. 1987a. Chemistry and biochemistry of metallothionein. *Experientia* **52**, 25–61.

Kägi, J.H.R. and Kojima, Y. (eds) 1987b. *Metallothionein II*, Birkhauser Verlag, Basel.

Kägi, J.H.R. and Schäffer, A. 1988. Biochemistry of metallothionein. *Biochemistry* **27**, 8509–15.

Kägi, J.H.R. and Vallee, B.L. 1961. Metallothionein: a cadmium and zinc-containing protein from equine renal cortex. *Journal of Biological Chemistry* **236**, 2435–42.

Kanno, S., Suzuki, J.S., Aoki, Y. and Suzuki, K.T. 1994a. Selective enhancement of metallothionein mRNA expression by copper in primary cultured liver parenchymal cells of LEC rats. *Research Communications in Chemical Pathology and Pharmacology* **84**, 153–62.

Kanno, S., Aoki, Y., Suzuki, J.S., Takeichi, N., Misawa, S. and Suzuki, K.T. 1994b. Enhanced synthesis of metallothionein as a possible cause of abnormal copper accumulation in LEC rats. *Journal of Inorganic Biochemistry* **56**, 117–25.

Karin, M. and Herschman, H.R. 1979. Dexamethasone stimulation of metallothionein synthesis in HeLa cell cultures. *Science* **204**, 176–7.

Karin, M. and Richards, R.I. 1982. Human metallothionein genes – primary structure of the metallothionein II gene and a related processed gene. *Nature* **299**, 797–802.

Karin, M. and Richards, R.I. 1984. The human metallothionein gene family: structure and expression. *Environmental Health Perspectives* **54**, 111–15.

Karin, M., Slater, E.P and Herschman, H.R. 1981. Regulation of metallothionein synthesis in HeLa cells by heavy metals and glucocorticoids. *Journal of Cellular Physiology* **106**, 63–74.

Karin, M., Eddy, R.L., Henry, W.M., Haley, L.L., Byers, M.G. and Shaws, T.B. 1984. Human metallothionein genes are clustered on chromosome 16. *Proceedings of the National Academy of Sciences, USA*, **81**, 5494–8.

Kelly, E.J. and Palmiter, R.D. 1996. A murine model of Menkes disease reveals a physiological function of metallothionein. *Nature Genetics* **13**, 219–22.

Kelly, E.J., Quaife, C.J., Froelick, G.J. and Palmiter, R. D. 1996. Metallothionein I and II protect against zinc deficiency and zinc toxicity in mice. *Journal of Nutrition* **126**, 1782–90.

Kelly, E.J., Sandgren, E.P., Brinster, R.L. and Palmiter, R.D. 1997. A pair of adjacent glucocorticoid response elements regulate expression of two mouse metallothionein genes. *Proceedings of the National Academy of Sciences, USA* **94**, 10045–50.

Kenaga, C., Cherian, M.G., Cox, C. and Oberdörster, G. 1996. Metallothionein induction and pulmonary responses to inhaled cadmium chloride in rats and mice. *Fundamental and Applied Toxicology* **30**, 204–12.

Kershaw, W.C., Lehman-McKeeman, L.D. and Klaassen, C.D. 1990. Hepatic isometallothioneins in mice: induction in adults and postnatal ontogeny. *Toxicology and Applied Pharmacology* **104**, 267–75.

Klaassen C.D. and Liu, J. 1997. Role of metallothionein in cadmium-induced hepatotoxicity and nephrotoxicity. *Drug Metabolism Reviews* **29**, 79–102.

Klaassen C.D., Liu, J. and Liu, Y-P. 1998. Chronic cadmium chloride-induced nephrotoxicity is not necessarily mediated through the cadmium–metallothionein complex. *Toxicological Sciences (supplement)* **42**, 327.

Klein, D., Greim H. and Summer, K.H. 1986. Stability of metallothionein in gastric-juice. *Toxicology* **41**, 121–9.

Klein, D., Michaelsen, S., Sato, S., Luz, A. Stampfl, A., and Summer, K.H. 1997. Binding of Cu to metallothionein in tissues of the LEC rat with inherited abnormal copper accumulation. *Archives of Toxicology* **71**, 340–3.

Kobune, M., Kato, J., Kohgo, Y. and Niitsu, Y. 1994. Liver cirrhosis in primary hemochromatosis and Wilson disease. *Nippon Rinsho* **52**, 209–14.

Kojima, S., Furukawa, A., Kiyozumi, M. and Shimada, H. 1991. Comparative renal toxicity of metallothioneins with different cadmium/zinc ratios in rats. *Toxicology Letter* **56**, 197–205.

Kondo, Y., Yamagata, K., Satoh, M., Naganuma, A., Imura, N. and Akimoto, M. 1993. Clinical use of a bismuth compound as an adjunct in chemotherapy with *cis*-DDP against human urogenital tumors. In *Metallothionein III*, (eds K.T. Suzuki, N. Imura and M. Kimura), Birkhäuser Verlag, Basel, pp. 269–78.

Kondo, Y., Woo, E.S., Michalaska, A.E., Choo, K.H.A. and Lazo, J.S. 1995. Metallothionein null cells have increased sensitivity to anticancer drugs. *Cancer Research* **55**, 2021–3.

Kontozoglou, T.E., Banerjee, D. and Cherian, M.G. 1989. Immunohistochemical localization of metallothionein in human testicular embryonal carcinoma cells. *Virchows Arch. A: Pathological Anatomy and Histopathology* **415**, 545–9.

Koomagi, R., Mattern, J. and Volm, M. 1995. Up-regulation of resistance-related proteins in human lung tumors with poor vascularization. *Carcinogenesis* **16**, 2129–33.

Krezoski, S. K., Villalobos, J., Shaw III, C.F. and Petering, D. H. 1988. Kinetic liability of zinc bound to Ehrlich cells. *Biochemical Journal* **255**, 483–91.

Kuzuya, T., Hasegawa, T., Shimizu, K. and Nabeshima, T. 1993. Effect of anti-epileptic drugs on serum zinc and copper concentrations in epileptic patients. *International Journal of Clinical Pharmacology, Therapeutics, and Toxicology* **31**, 61–5.

Laird, P.W. and Jaenisch, R. 1994. DNA methylation and cancer. *Human Molecular Genetics* **3**, 1487–95.

Lazo, J.S. and Pitt, B.R. 1995. Metallothioneins and cell death by anticancer drugs. *Annual Reviews of Pharmacology and Toxicology* **35**, 635–53.

Lazo, J.S., Yang, Y.-Y., Woo, E., Kuo, S.-M. and Saijo, N. 1993. Regulation of metallothionein expression and cellular resistance to electrophilic antineoplastic agents, in *Metallothionein III*, (eds K.T. Suzuki, N. Imura and M. Kimura), Birkhäuser Verlag, Basel, pp. 269–78.

Le Beau, M.M., Diaz, M.O., Karin, M. and Rowley, J.D. 1985. Metallothionein gene cluster is split by chromosome 16 rearrangements in myelomonocytic leukaemia. *Nature* **313**, 709–11.

Leber, A.P. and Miya, T.S. 1976. A mechanism for cadmium and zinc induced tolerance to cadmium toxicity: Involvement of metallothionein. *Toxicology and Applied Pharmacology* **37**, 403–14.

Leffler, P.E., Jin, T. and Nordberg, G.F. 1996. Nephrotoxic impact of multiple short-interval cadmium–metallothionein injections in the rat. *Toxicology* **112**, 151–6.

Leone, A. 1986. Metallothionein gene regulation in Menkes disease. *Horizons in Biochemistry and Biophysics* **8**, 207–56.

Leone, A. and Hamer, D. H. 1987. Abnormal copper metabolism and regulation of metallothionein gene expression in Menkes disease. *EXS* **52**, 477–80.

Leone, A., Pavlakis, G.N. and Hamer, D.H. 1985. Menkes disease: abnormal metallothionein gene regulation in response to copper. *Cell* **40**, 301–9.

Levine, A., Cantoni, G.L. and Razin, A.1991. Inhibition of promoter activity by methylation: Possible involvement of protein mediators. *Proceedings of the National Academy of Science USA* **88**, 6515–18.

Li, T.Y., Kraker, A.J., Shaw, C.F. and Petering, D.A. 1980. Ligand substitution reactions of metallothioneins with EDTA and apocarbonic anhydrase. *Proceedings of the National Academy of Science USA* **77**, 6334–9.

Li, Y., Togashi, Y., Sato, S., Emoto, T., Kang, J.H., Takeichi, N., Kobayashi, H., Kojima, Y., Une, Y. and Uchino, J. 1991. Spontaneous hepatic copper accumulation in Long–Evans Cinnamon rats with hereditary hepatitis; A model of Wilson disease. *Journal of Clinical Investigation* **87**, 1858–65.

Liang, L., Fu, K., lee, D.K., Sobieski, R.J., Dalton, T. and Andrews, G.K. 1996. Activation of the complete mouse metallothionein gene locus in the maternal deciduum. *Molecular Reproduction and Development* **43**, 25–37.

Liebermann, M.W., Beach, L.R. and Palmiter, R.D. 1983. Ultraviolet radiation-induced metallothionein-I gene activation is associated with extensive DNA demethylation. *Cell* **35**, 207–14.

Liu, J., Liu, Y., Michalska, A.E., Choo, K.H.A. and Klaassen, C.D. 1996. Distribution and retention of cadmium in metallothionein I and II null mice. *Toxicology and Applied Pharmacology* **136**, 260–8.

Liu, J., Liu, Y., Habeebu, S.S.M. and Klaassen, C.D. 1998. Metallothionein (MT)-null mice are sensitive to cisplatin-induced hepatotoxicity. *Toxicology and Applied Pharmacology* **149**, 24–31.

Liu, Y., Liu, J., Iszard, M.B., Andrews, G.K., Palmiter, R.D. and Klaassen, C.D. 1995. Transgenic mice that overexpress metallothionein-I are protected from cadmium lethality and hepatotoxicity. *Toxicology and Applied Pharmacology* **135**, 222–8.

Luza, S.C. and Speisky, H.C. 1996. Liver copper storage and transport during development implications for cytotoxicity. *American Journal of Clinical Nutrition* **63**, 812S–820S.

MacArthur, C.A. and Lieberman, M.W. 1987. Different types of hypersensitive sites in the mouse metallothionein gene region. *The Journal of Biological Chemistry* **262**, 2161–65.

MacArthur, C.A., Ramabhadran, R., Godwin, A.K., Lebovitz, R.M. and Lieberman, M.W. 1985. Chemical carcinogens induce cadmium resistance and activate metallothionein genes in cadmium sensitive S49 mouse cells. *Carcinogenesis* **6**, 887–92.

MacArthur, C.A., Hawley, T. and Lieberman, M.W. 1986. Coordinate activation and regulation of quiescent metallothionein I and II genes in carcinogen-treated mouse thymic lymphoma cells. *Carcinogenesis* **7**, 1487–95.

McCormick, C.C., Fullmer, C.S. and Garvey, J.S. 1988. Amino acid sequence and comparative antigenicity of chicken metallothionein. *Proceedings of the National Academy of Sciences USA* **85**, 309–13.

McKenna, I.M., Waalkes, M.P., Chen, L.C. and Gordon, T. 1997. Comparison of inflammatory responses in Wistar rats and C57 and DBA mice exposed to cadmium oxide fumes. *Toxicology and Applied Pharmacology* **146**, 196–206.

McKenna, I.M., Gordon, T. Chen, L.C. Anver, M.R. and Waalkes, M.P. 1998. Expression of metallothionein protein in the lungs of Wistar rats and C57 and DBA mice exposed to cadmium oxide fumes. *Toxicology and Applied Pharmacology* **153**, 178–89.

Maier, H., Jones, C., Jasani, B., Ofner, D., Zelger, B., Schmid, K.W. and Budka, H. 1997. Metallothionein overexpression in human brain tumours. *Acta Neuropathol (Berlin)* **94**, 599–604.

Maitani, T., Waalkes, M.P. and Klaassen, C.D. 1984. Distribution of cadmium after oral administration of cadmium–thionein. *Toxicology and Applied Pharmacology* **74**, 237–43.

Margoshes, M. and Vallee, B.L. 1957. A cadmium protein from equine kidney cortex. *Journal of American Chemical Society* **79**, 4813–14.

Markant, A. and Pallauf, J. 1996. Metallothionein and zinc as potential antioxidants in radical-induced lipid peroxidation in cultured hepatocytes. *Journal of Trace Elements in Medicine and Biology* **10**, 88–95.

Masters, B.A., Kelly, E.J., Quaife, C.J., Brinster, R.L. and Palmiter, R.D. 1994. Targeted disruption of metallothionein I and II genes increases sensitivity to cadmium. *Proceedings of the National Academy of Science USA* **91**, 584–8.

Matsumoto, Y., Oka, M., Sakamoto, A., Narasaki, F., Fukuda, M. Takatani, H., Terashi, K., Tsurutani, J., Mayo, K.E. and Palmiter, R.D. 1981. Glucocorticoid regulation of metallothionein-I mRNA synthesis in cultured mouse cells. *The Journal of Biological Chemistry* **256**, 2621–4.

Matsumoto *et al.* 1997. Anticancer Research.

Mayo, K.E. and Palmiter, R.D. 1981. Glucocorticoid regulation of the mouse metallothionein-I gene is selectively lost following amplification of the gene. *The Journal of Biological Chemistry* **257**, 3061–7.

Mayo, K.E., Warren, R. and Palmiter, R.D. 1982. The mouse metallothionein-I gene is transcriptionally regulated by cadmium following transfection into human or mouse cells. *Cell* **29**, 99–108.

Meneghini, R. 1997. Iron homeostasis, oxidative stress, and DNA damage. *Free Radical Biology and Medicine* **23**, 783–92.

Menkes, J.H., Alter, M., Steigleder, G.K., Weakley, D.R. and Sung, J.H. 1962. A sex-linked recessive disorder with retardation of growth, peculiar hair and focal cerebral and cerebellar degeneration. *Pediatrics* **29**, 764–79.

Mercer, J.F. and Grimes, A. 1986. Variation in the amounts of hepatic copper, zinc and metallothionein mRNA during development in the rat. *Biochemical Journal* **238**, 23–7.

Michalska, A.E. and Choo, K.H.A. 1993. Targeting and germ-line transmission of a null mutation at the metallothionein I and II loci in mouse. *Proceedings of the National Academy of Science USA* **90**, 8088–92.

Misra, R.R., Hochadel, J. F., Smith, G.T., Cook, J.C., Waalkes, M.P. and Wink, D.A. 1996. Evidence that nitric oxide enhances cadmium toxicity by displacing the metal from metallothionein. *Chemical Research in Toxicology* **9**, 326–32.

Moffatt, P. and Denizeau, F. 1997. Metallothionein in physiological and physiopathological processes. *Drug Metabolism Reviews* **29**, 261–307.

Monden, N., Abe, S., Sutoh, I., Hishikawa, Y., Kinugasa, S. and Nagasue, N. 1997. Prognostic significance of the expressions of metallothionein, glutathione-*S*-transferase-pi, and P-glycoprotein in curatively resected gastric cancer. *Oncology* **54**, 391–99.

Moussa, M., Kloth, D., Peers, G., Cherian, M.G., Frei, J.V. and Chin, J.L. 1997. Metallothionein expression in prostatic carcinoma: Correlation with Gleason grade, pathologic stage, DNA content and serum level of prostate-specific antigen. *Clinical Investigations and Medicine* **20**, 371–80.

Mulder, T.P., Janssens, A.R., Verspaget, H.W., van Hattum, J. and Lamers, C.B. 1992. Metallothionein concentration in the liver of patients with Wilson disease, primary biliary cirrhosis, and liver metastasis of colorectal cancer. *Journal of Hepatology* **16**, 346–50.

Naganuma, A., Satoh, M. and Imura, N. 1993. Utilization of metallothionein inducer in cancer therapy, in *Metallothionein III*, (eds K.T. Suzuki, N. Imura and M. Kimura), Birkhäuser Verlag, Basel, pp. 255–68.

Naganuma, A., Satoh, M. and Imura, N. 1987. Prevention of lethal and rental toxicity of *cis*-diamminedichloroplatinum(II) by induce of metallothionein synthesis without compromising its antitumor activity in mice. *Cancer Research* **47**, 983–7.

Nakajima, K. and Suzuki, K. 1995. Immunochemical detection of metallothionein in brain. *Neurochemistry International* **27**, 73–97.

Nartey, N.O., Banerjee, D. and Cherian, M.G. 1987a. Immunohistochemical localization of metallothionein in cell nucleus and cytoplasm of fetal human liver and kidney and its changes during development. *Pathology* **19**, 233–8.

Nartey, N.O., Frei, J.V. and Cherian, M.G. 1987b. Hepatic copper and metallothionein distribution in Wilson disease (hepatolenticular degeneration). *Laboratory Investigation* **57**, 397–401.

Nartey, N.O., Cherian, M.G. and Banerjee, D. 1987c. Immunohistochemical localization of metallothionein in human thyroid tumors. *American Journal of Patholology* **129**, 177–82.

Nielson, K.B. and Winge, D.R. 1983. Order of binding in metallothionein. *Journal of Biological Chemistry* **258**, 13063–9.

Nishimura, H., Nishimura, N. and Tohyama, C. 1989a. Immunohistochemical localization of metallothionein in developing rat tissues. *Journal of Histochemistry and Cytochemistry* **37**, 715–22.

Nishimura, H., Nishimura, N. and Tohyama, C. 1989b. Localization of metallothionein in female reproductive organs of rat and guinea pig. *Journal of Histochemistry and Cytochemistry* **37**, 1601–7.

Nordberg, G.F. and Piscator, M. 1975. Influence of long-term cadmium exposure on urinary excretion of protein and cadmium in mice. *Environmental Physiology and Biochemistry* **2**, 37–49.

Nordberg, G.F., Goyer, R.A. and Nordberg, M. 1972. Comparative toxicity of cadmium–metallothionein and cadmium chloride on mouse kidney. *Archives of Pathology* **99**, 192–7.

Nordberg, G.F., Garvey, J.S., and Chang, C.C. 1982. Metallothionein in plasma and urine of cadmium workers. *Environmental Research* **28**, 179–82.

Oberdörster, G., Cherian, M.G. and Baggs, R.B. 1994a. Correlation between cadmium-induced pulmonary carcinogenicity, metallothionein expression, and inflammatory processes: A species comparison. *Environmental Health Perspectives* **102**, 257–63.

Oberdörster, G., Cherian, M.G. and Baggs, R.B. 1994b. Importance of species differences in experimental pulmonary carcinogenicity of inhaled cadmium for extrapolation to humans. *Toxicology Letters* **72**, 339–43.

Ohta, H. and Cherian, M.G. 1991. Gastrointestinal absorption of cadmium and metallothionein. *Toxicology and Applied Pharmacology* **107**, 63–72.

Ohtake, H. and Koga, M. 1979. Purification and characterization of zinc-binding protein from the liver of the partially hepatotectomized rat. *Biochemical Journal* **183**, 683–90.

Olafson, R.W., McCubbin, W.D. and Kay, C.M. 1988. Primary- and secondary-structural analysis of a unique prokaryotic metallothionein from a *Synechococcus* sp. cyanobacterium. *Biochemical Journal* **251**, 691–9.

Onosaka, S. and Cherian, M.G. 1981. The induced synthesis of metallothionein in various tissues of rat in response to metals. I. Effects of repeated injection of cadmium salts. *Toxicology* **22**, 91–101.

Onosaka, S. and Cherian, M.G. 1982. The induced synthesis of metallothionein in various tissues of rats in response to metals. II. Influence of zinc status and specific effect on pancreatic metallothionein. *Toxicology* **23**, 11–20.

Onosaka, S., Min, K.-S., Fukuhara, C., Tanaka, K., Tashiro, S., Shimizu, I., Furuta, M., Yasutomi, T., Kobashi, K. and Yamamoto, K.-I. 1986. Concentrations of metallothionein and metals in malignant and non-malignant tissues in human liver. *Toxicology* **38**, 261–8.

Otvos, J.D., Liu, X., Li, H., Shen, G. and Basti, M. 1993. Dynamic aspects of metallothionein structure, in *Metallothionein III*, (eds K.T. Suzuki, N. Imura and M. Kimura), Birkhäuser Verlag, Basel, pp. 57–74.

Palmiter, R.D. 1987. Molecular biology of metallothionein gene expression. *Experientia* **52**, 63–80.

Palmiter, R.D. 1994. Regulation of metallothionein genes by heavy metals appears to be mediated by a zinc-sensitive inhibitor that interacts with a constitutively active transcription factor, MTF-1. *Proceedings of the National Academy of Science, USA* **91**, 1219–23.

Palmiter, R.D. 1995. Constitutive expression of metallothionein-III (MT-III), but not MT-I, inhibits growth when cells become zinc deficient. *Toxicology and Applied Pharmacology* **135**, 139–46.

Palmiter, R.D., Findley, S.D., Whitmore, T.E. and Durnam, D.M. 1992. MT-III, a brain-specific member of the metallothionein gene family (Alzheimer disease/chromosome assignment/gene expression/gene sequence). *Proceedings of the National Academy of Science, USA* **89**, 6333–7.

Palmiter, R.D., Sandgren, E.P., Koeller, D.A., Findley, S.D. and Brinster, R.L. 1993a. Metallothionein genes and their regulation in transgenetic animals, in *Metallothionein III*, (eds K.T. Suzuki, N. Imura and M. Kimura), Birkhäuser Verlag, Basel, pp. 399–406.

Palmiter, R.D., Sandgren, E.P., Koeller, D.M. and Brinster, R.L. 1993b. Distal regulatory elements from the mouse metallothionein locus stimulate gene expression in transgenic mice. *Molecular and Cellular Biology* **13**, 5266–75.

Pandit, A. and Bhave, S. 1996. Present interpretation of the role of copper in Indian childhood cirrhosis. *American Journal of Clinical Nutrition* **63**(5), 830S–835S.

Payan, J.P., Saillenfait, A.M., Beydon, D., Ban, M. and de Ceaurriz, J. 1990. Pregnancy-associated changes in renal toxicity of cadmium–metallothionein: possible role of intracellular metallothionein. *Toxicology* **65**, 223–32.

Penketh, P.G., Shyam, K. and Sartorelli, A.C. 1996. Mechanisms of resistance to alkylating agents. *Cancer Treatment Research* **87**, 65–81.

Petering, D.H. and Fowler, B.A. 1986. Roles of metallothionein and related proteins in metal metabolism and toxicity, problems and perspectives. *Environmental Health Perspectives* **65**, 217–24.

Petering, D.H., Quesada, A., Dughish, M., Krull, S., Gan, T., Lemkuil, D., Pattanaik, A., Byrnes, R.W., Savas, M., Whelan, H. and Shaw III, C.F. 1993. Metallothionein in tumor and host: intersections of zinc metabolism, the stress response, and tumor therapy, in *Metallothionein III*, (eds K.T. Suzuki, N. Imura and M. Kimura), Birkhäuser Verlag, Basel, pp. 329–46.

Piscator, M. 1964. On cadmium in normal human kidney together with a report on the isolation of metallothionein from livers of cadmium exposed rabbits. *Nord Hyg Tidskr* **45**, 7642–8.

Probst, G.S., Bousquet, W.F. and Miya, T.S. 1977. Correlation of hepatic metallothionein with acute cadmium toxicity in the mouse. *Toxicology and Applied Pharmacology* **39**, 61–9.

Quaife, C.J., Findley, S.D., Erickson, J.C., Froelick, G.J., Kelly, E.J., Zambrowicz, B.P. and Palmiter, R.D. 1994. Induction of a new metallothionein isoform (MT-IV) occurs during differentiation of stratified squamous epithelia. *Biochemistry* **33**, 7250–9.

Rauch, H. 1983. Toxic milk, a new mutation affecting cooper metabolism in the mouse. *Journal of Heredity* **74**, 141–4.

Richards, R.I., Heguy, A., and Karin, M. 1984. Structural and functional analysis of the human metallothionein-IA gene: differential induction by metal ions and glucocorticoids. *Cell* **37**, 263–72.

Riordan, J.R. and Jolicoeur-Paquet, L. 1982. Metallothionein accumulation may account for intracellular copper regulation in Menkes disease. *Journal of Biological Chemistry* **257**, 4639–45.

Robinson, N.J., Tommey, A.M., Kuske, C. and Jackson, P.J. 1993. Plant metallothioneins. *Biochemical Journal* **295**, 1–10.

Saijo, N., Miura, K. and Kasahara, K. 1993. The role of metallothionein for the reduction of adverse effect of cisplatin and the induction of resistance to cisplatin, in *Metallothionein III*, (eds K.T. Suzuki, N. Imura and M. Kimura), Birkhäuser Verlag, Basel, pp. 27992.

Salehi-Ashitiani, K., Widrow, R.J., Markert, C.L. and Goldberg, E. 1993. Testis-specific expression of a metallothionein I-driven transgene correlates with undermethylation of the locus in testicular DNA. *Proceedings of the National Academy of Science, USA* **90**, 8886–90.

Samson, S.L.A. and Gedamu, L. 1998. Molecular analysis of metallothionein gene regulation. *Progress in Nucleic Acid Research* **59**, 257–88.

Sanders, B.M., Goering, P.L. and Jenkins, K. 1996. The role of general and metal-specific cellular responses in protection and repair of metal-induced damage: stress proteins and metallothioneins, in *Toxicology of metals* (eds L.W. Chang, L. Magos and K.T. Suzuki), CRC Lewis Publishing, Boca Raton, pp. 165–87.

Satoh, M., Nishimura, N., Kanayama, Y., Naganuma, A., Suzuki, T. and Tohyama, C. 1997. Enhanced renal toxicity by inorganic mercury in metallothionein-null mice. *Journal of Pharmacology and Experimental Therapeutics* **282**, 1529–33.

Scheinberg, I.H. and Sternlieb, I. 1996. Wilson disease and idiopathic copper toxicosis. *American Journal of Clinical Nutrition* **63**, 842S–845S.

Schmidt, C.J. and Hamer, D.H. 1986. Cell specificity and an effect of ras on human metallothionein gene expression. *Proceedings of the National Academy of Sciences, USA* **83**, 3346–50.

Schmidt, C.J., Hamer, D.H. and McBride, O.W. 1984. Chromosomal location of human metallothionein genes: implications for Menkes disease. *Science* **224**, 1104–6.

Searle, P.F., Davison, B.L., Stuart, G.W., Wilkie, T.M., Norstedt, G. and Palmiter, R.D. 1984. Regulation, linkage, and sequence of mouse metallothionein-I and II genes. *Molecular and Cellular Biology* **4**, 1221–30.

Sendelbach, L.E., Kershaw, W.C., Cuppage, E. and Klaassen, C.D. 1992. Cd-metallothionein nephrotoxicity in inbred strains of mice. *Journal of Toxicology and Environmental Health* **35**, 115–26.

Sewell, A.K., Jensen, L.T., Erickson, J.C., Palmiter, R.D. and Winge, D.R. 1995. Bioactivity of metallothionein-3 correlates with its novel beta domain sequence rather than metal binding properties. *Biochemistry* **34**(14), 4740–7.

Shaikh, Z.A., Ellis, K.J., Subramanian, K.S and Greenberg, A. 1990a. Biological monitoring for occupational cadmium exposure: the urinary metallothionein. *Toxicology* **63**, 53–62.

Shaikh, Z.A., Kido, T., Kito, H., Honda, R. and Nogawa, K. 1990b. Prevalence of metallothioneinuria among the population living in the Kakehashi River basin in Japan – an epidemiological study. *Toxicology* **64**, 59–69.

Shimada, H., Hochadel, J.F. and Waalkes, M.P. 1997. Progesterone pretreatment enhances cellular sensitivity to cadmium despite a marked activation of the metallothionein gene. *Toxicology and Applied Pharmacology* **142**, 178–85.

Shimada, H., Takiguchi, M. and Waalkes, M.P. 1998. Modification of the adverse effects of cadmium exposure by steroid compounds in rodents; A minireview. *Toxic Substance Mechanisms* **17**, 264–7.

Shiraishi, N., Taguchi, T. and Kinebuchi, H.. 1991. Copper-induced toxicity in macular mutant mouse: an animal model for Menkes' kinky-hair disease. *Toxicology and Applied Pharmacology* **110**, 89–96.

Simmers, R.N., Sutherland, G.R., West, A. and Richards, R.I. 1987. Fragile sites at 16q22 are not at the breakpoint of the chromosomal rearrangement in AMMoL. *Science* **236**, 92–4.

Squibb, K.S. and Fowler, B.A. 1984. Intracellular metabolism and effects of circulating cadmium–metallothionein in the kidney. *Environmental Health Perspectives* **54**, 31–35.

Squibb, K.S., Ridlington, J.W., Carmichael, N.G. and Fowler, B.A. 1979. Early cellular effects of circulating cadmium–thionein on kidney proximal tubules. *Environmental Health Perspectives* **28**, 287–96.

Squibb, K.S., Pritchard, J.B. and Fowler, B.A. 1984. Cadmium metallothionein nephropathy: Ultrastructural/biochemical alterations and intracellular cadmium binding. *Journal of Pharamacology and Experimental Therapeutics* **229**, 311–21.

Stennard, F.A., Holloway, A.F., Hamilton, J. and West, A.K. 1994. Characterisation of six additional human metallothioneien genes. *Biochemica et Biophysica Acta* **1218**, 357–65.

Sutherland, G.R., Baker, E., Callen, D.F., Garsion, O.M. and West, A.K. 1990. The human metallothionein gene cluster is not disrupted in myelomonocytic leukemia. *Genomics* **6**, 144–8.

Suzuki, C.A.M. and Cherian, M.G. 1987. Renal toxicity of cadmium–metallothionein and enzymuria in rats. *Journal of Pharmacology and Experimental Therapeutics* **240**, 314–19.

Suzuki, K.T. 1995. Disordered copper metabolism in LEC rats, an animal model of Wilson disease: roles of metallothionein. *Research Communications in Molecular Pathology and Pharmacology* **89**, 221–40.

Suzuki-Kurasaki, M., Okabe, M. and Kurasaki, M. 1997. Copper–metallothionein in the kidney of macular mice: A model for Menkes disease. *Journal of Histochemistry and Cytochemistry* **45**, 1493–501.

Tashiro-Itoh, T., Ichida, T., Matsuda, Y., Satoh, T., Sugiyama, M., Tanaka, Y., Ishikawa, T.,

Itoh, S., Nomoto, M. and Asakura, H. 1997. Metallothionein expression and concentrations of copper and zinc are associated with tumor differentiation in hepatocellular carcinoma. *Liver* **17**, 300–6.

Thomas, G.R., Forbes, J.R., Roberts, E.A., Walshe, J.M. and Cox, D.W. 1995. The Wilson disease gene: spectrum of mutations and their consequence. *Nature Genetics* **9**, 210–17.

Tohyama, C., Suzuki, J. S., Homma, S.T., Nishimura, N. and Nishimura, H. 1993a. Regulation of metallothionein biosynthesis in genital organs of male rats, in *Metallothionein III*, (eds K.T. Suzuki, N. Imura and M. Kimura), Birkhäuser Verlag, Basel, pp. 443–56.

Tohyama, C., Suzuki, J.S., Hemelraad, J., Nishimura, N. and Nishimura, H. 1993b. Induction of metallothionein and its localization in the nucleus of rat hepatocytes after partial hepatectomy. *Hepatology* **18**, 1193–201.

Tsuji, S., Kobayashi, H., Uchida, Y., Ihara, Y. and Miyatake, T. 1992. Molecular cloning of human growth inhibitory factor cDNA and its down-regulation in Alzheimer's disease. *EMBO Journal* **11**, 4843–50.

Tsujikawa, K., Imai, T., Kakutani, M., Kayamori, Y., Mimura, T., Otaki, N., Kimura, M., Fukuyama, R. and Shimizu, N. 1991. Localization of metallothionein in nuclei of growing primary cultured adult rat hepatocytes. *FEBS Letters* **283**, 239–42.

Tsujikawa, K., Sagawa, K., Suzuki, N., Shimaoka, T., Kohama, Y., Otaki, N., Kimura, M. and Mimura, T. 1993. The relationship between nuclear localization of metallothionein and proliferation of rat hepatocytes, in *Metallothionein III*, (eds K.T. Suzuki, N. Imura and M. Kimura), Birkhäuser Verlag, Basel, pp. 427–42.

Tsujikawa, K., Suzuki, N., Sagawa, K., Itoh, M., Sugiyama, T., Kohama, Y., Otaki, N., Kimura, M. and Mimura, T. 1994. Induction and subcellular localization of metallothionein in regenerating rat liver. *European Journal of Cell Biology* **63**, 240–6.

Uchida, Y., Takio, K., Titani, K., Ihari, Y. and Tomonaga, M. 1991. The growth inhibitory factor that is deficient in Alzheimer's disease is a 68 amino acid metallothionein-like protein. *Neuron* **7**, 337–47.

Udom, A.O. and Brady, F.O. 1980. Reactivation of *in vitro* of zinc-requiring apoenzymes by rat liver zinc–thionein. *Biochemical Journal* **187**, 329–35.

Uozaki, H., Horiuchi, H., Ishida, T., Iijima, T., Imamura, T. and Machinami, R. 1997. Overexpression of resistance-related proteins (metallothioneins, glutathione-*S*-transferase, pi, heat shock protein 27, and lung resistance-related protein) in osteosarcoma. Relationship with poor prognosis. *Cancer* **79**, 2336–44.

Vallee, B.L. 1995. The function of metallothionein. *Neurochemistry International* **27**, 23–33.

Vallee, B.L. and Auld, D.S. 1995. Zinc metallochemistry in biochemistry, in *Interface Between Chemistry and Biochemistry*, (eds P. Jollès and P. Jörnvall), Birkhäuser Verlag, Basel, pp. 259–77.

Vallee, B.L. and Maret, W. 1993. The functional potential and potential functions of metallothioneins: A personal perspective, in *Metallothionein III*, (eds K.T. Suzuki, N. Imura and M. Kimura), Birkhäuser Verlag, Basel, pp. 1–27.

Varshney, U. and Gedamu, L. 1984. Human metallothionein MT-I and MT-II processed genes. *Gene* **31**, 135–45

Varshney, U., Hoar, D.I., Starozik, D. and Gedamu, L. 1984. A frequent restriction fragment length polymorphism in the human metallothionein-II processed gene region is evolutionarily conserved. *Molecular Biology and Medicine* **2**, 193–206.

Varshney, U., Jahroudi, N., Foster, R. and Gedamu, L. 1986. Structure, organization, and regulation of human metallothionein IF gene: differential and cell-type-specific expression in response to heavy metals and glucocorticoids. *Molecular and Cellular Biology* **6**, 26–37.

Waalkes, M.P. 1986. Effect of dietary zinc deficiency on the accumulation of cadmium and metallothionein in selected tissues of the rat. *Toxicology and Applied Pharmacology* **18**: 301–13.

Waalkes, M.P. 1993. Biological roles and medical implications, in *Metallothionein III*, (eds K.T. Suzuki, N. Imura and M. Kimura), Birkhaüser Verlag, Basel, Switzerland, pp. 243–54.

Waalkes, M.P. 1995. Metal carcinogenesis, in *Metal Toxicology*, (eds R.A. Goyer, C.D. Klaassen and M.P. Waalkes), Academic Press, San Diego, pp. 47–69.

Waalkes, M.P. 1996. Quantitation of metallothionein, in *Methods in Toxicology – Endocrinology*, (ed. J.A. Thomas), Academic Press, San Diego, pp. 371–91.

Waalkes, M.P. and Bell, J.U. 1980. Depression of metallothionein in fetal rat liver following maternal cadmium exposure. *Toxicology* **18**, 103–10.

Waalkes, M.P. and Goering, P.L. 1990. Metallothionein and other cadmium-binding proteins: Recent developments. *Chemical Research in Toxicology* **3**, 281–8.

Waalkes, M.P. and Klaassen, C.D. 1984. Postnatal ontogeny of metallothionein in various organs of the rat. *Toxicology and Applied Pharmacology* **74**, 314–20.

Waalkes, M.P. and Klaassen, C.D. 1985. Concentration of metallothionein in major organs of rats after administration of various metals. *Fundmental and Applied Toxicology* **5**, 473–7.

Waalkes, M.P. and Rehm, S. 1994. Cadmium and prostate cancer. *Toxicology and Applied Pharmacology* **43**, 251–69.

Waalkes, M.P. and Ward, J.M. 1989. Induction of hepatic metallothionein in male B6C3F1 mice exposed to hepatic tumor promoters: Effects of phenobarbital, acetaminophen, sodium barbital, and di(2-ethylhexyl)phthalate. *Toxicology and Applied Pharmacology* **100**, 217–26.

Waalkes, M.P., Ross, S.M., Craig, C.R. and Thomas, J.A. 1982. Induction of metallothionein in the rat brain by copper-implantation but not by cobalt-implantation. *Toxicology Letters* **12**, 137–42.

Waalkes, M.P., Hjelle, J.J. and Klaassen, C.D. 1984. Transient induction hepatic metallothionein following oral ethanol administration. *Toxicology and Applied Pharmacology* **74**, 230–6.

Waalkes, M.P., Wilson, M.J. and Poirier, L.A. 1985. Reduced cadmium-induced cytotoxicity in cultured liver cells following 5-azacytidine pretreatment. *Toxicology and Applied Pharmacology* **81**, 250–7.

Waalkes, M.P., Miller, M.S., Wilson, M.J., Bare, R.M. and McDowell, A.E. 1988. Increased metallothionein gene expression in 5-aza-2′-deoxycytidine-induced resistance to cadmium cytotoxicity. *Chemical-Biological Interactions* **66**, 189–204.

Waalkes, M.P., Diwan, B.A., Weghorst, C.M., Bare, R.M., Ward, J.M. and Rice, J.M. 1991. Anti-carcinogenic effects of cadmium in B6C3F1 mouse liver and lung. *Toxicology and Applied Pharmacology* **110**, 327–35.

Waalkes, M.P., Diwan, B.A., Weghorst, C.M., Ward, J.M., Rice, J.M., Cherian, M.G. and Goyer, R.A. 1993. Further evidence of the tumor-suppressive effects of cadmium in the B6C3F1 mouse liver and lung: Late stage vulnerability of tumors to cadmium and the role of metallothionein. *Journal of Pharmacology and Experimental Therapeutics* **266**, 1656–63.

Waalkes, M.P., Diwan, B.A., Rehm, S., Moussa, M., Cherian, M.G. and Goyer, R.M. 1996. Down regulation of metallothionein expression in human and murine hepatocellular tumors: Association with the tumor necrotizing and antineoplastic effects of cadmium in mice. *Journal of Pharmacology and Experimental Therapeutics* **277**, 1026–33.

Waalkes, M.P., Rehm, S., Coogan, T.P. and Ward, J.M. 1997. Role of cadmium in the etiology of prostate cancer, in *Target Organ Toxicology Series: Endocrine Toxicology*, (eds J.A. Thomas and H.D. Colby), Raven Press, New York, pp. 227–44.

Waalkes, M.P., Nagamine, T., Shimada, H., Diwan, B.A., Coogan, T.P., Shiraishi, N., Cherian, M. G. and Goyer, R. A. 1998. Further studies on the role of metallothionein in the antitumor effects of cadmium, in *Metallothionein IV*, (ed. C.D. Klaassen), Birkhäuser Verlag, Basel, pp. 569–72.

Waldrop, G.L. and Ettinger, M.J. 1990. The relationship of excess copper accumulation by fibroblasts from the brindled mouse model of Menkes disease to the primary defect. *Biochemistry Journal* **15**, 417–22.

Webb, M. 1972. Protection by zinc against cadmium toxicity. *Biochemical Pharmacology* **21**, 2767–71.

Webb, M. and Verschoyle, R.D. 1976. An investigation of the role of metallothioneins in protection against the acute toxicity of the cadmium ion. *Biochemical Pharmacology* **25**, 673–9.

West, A.K., Stallings, R., Hildebrand, C.E., Chiu, R., Karin, M. and Richards, R.I. 1990. Human metallothionein genes: structure of the functional locus at 16q13. *Genomics* **8**, 513–18.

Wilson, S.A.K. 1912. Progressive lenticular degeneration: a familial nervous disease associated with cirrhosis of the liver. *Brain* **34**, 295–509.

Wong, K.-L. and Klaassen, C.D. 1979. Isolation and characterization of metallothionein which is highly concentrated in newborn rat liver. *Journal of Biological Chemistry* **254**, 12399–403.

Woo, E.S., Kondo, Y., Watkins, S.C., Hoyt, D.G. and Lazo, J.S. 1996. Nucleophilic distribution of metallothionein in human tumor cells. *Experimental Cell Research* **224**, 365–71.

Woo, E.S., Monks, A., Watkins, S.C., Wang, A.S. and Lazo, J.S. 1997. Diversity of metallothionein content and subcellular localization in the National Cancer Institute tumor panel. *Cancer Chemotherapy and Pharmacology* **41**, 61–8.

Wu, J., Forbes, J.R., Chen, H.S. and Cox, D.W. 1994. The LEC rat has a deletion in the copper transporting ATPase gene homologous to the Wilson disease gene. *Nature Genetics* **7**, 541–5.

Yang, X.L., Miura, N., Kawarada, Y., Terada, K., Petrukhin, K., Gilliam, T. and Sugiyama, T. 1997. Two forms of Wilson disease protein produced by alternative splicing are localized in distinct cellular compartments. *Biochemical Journal* **326**, 897–902.

Yoshida, Y., Tokusashi, Y., Lee, G.H. and Ogawa, H. 1996. Intrahepatic transplantation of normal hepatocytes prevents Wilson disease in Long–Evans cinnamon rats. *Gastroenterology* **111**, 1654–60.

Yoshikawa, H. 1970. Preventive effect of pretreatment with low dose of metals on the acute toxicity of metals in mice. *Industrial Health* **8**, 184–91.

Yoshikawa, H. 1973. Preventive effects of pretreatment with cadmium on acute cadmium poisoning in rats. *Industrial Health* **11**, 113–19.

Zalups, R.K. and Cherian, M.G. 1992. Renal metallothionein metabolism after a reduction of renal mass. II. Effect of zinc pretreatment on the renal toxicity and intrarenal accumulation of inorganic mercury. *Toxicology* **71**, 103–17.

Zalups, R.K., Gelein, R.M. and Cherian, M.G. 1992. Shifts in the dose–effect relationship for the nephropathy induced by cadmium–metallothionein in rats after a reduction in renal mass. *Journal of Pharmacology and Experimental Therapeutics* **262**, 1256–66.

Zalups, R.K., Fraser, J. and Koropatnick, J. 1995. Enhanced transcription of metallothionein genes in rat kidney: effect of uninephrectomy and compensatory renal growth. *American Journal of Physiology* **268**, F643–F650.

Zeng, J., Vallee, B.L. and Kagi, J.H.R. 1991. Zinc transfer from transcription factor IIIA fingers to thionein clusters. *Proceedings of the National Academy of Sciences, USA* **88**, 9984–8.

Zhang, X.H. and Takenaka, I. 1998. Incidence of apoptosis and metallothionein expression in renal cell carcinoma. *British Journal of Urology* **81**, 9–13.

Zhao, C.Q., Young, M.R., Diwan, B.A., Coogan, T.P. and Waalkes, M.P. 1997. Association of arsenic-induced malignant transformation with DNA hypomethylation and aberrant gene expression. *Proceedings of the National Academy of Sciences, USA* **94**, 10907–12.

Zheng, H., Berman, N.E. and Klaassen, C.D. 1995. Chemical modulation of metallothionein I and III mRNA in mouse brain. *Neurochemistry International* **27**, 43–58.

Zheng, H., Liu, J., Choo, K.H.A., Michalska, A.E. and Klaassen, C.D. 1996. Metallothionein-I and -II knock-out mice are sensitive to cadmium-induced liver mRNA expression of c-jun and p53. *Toxicology and Applied Pharmacology* **136**, 229–235.

Zhu, Z. and Thiele, D.J. 1996. Toxic metal-responsive gene transcription, in *Stress-inducible cellular responses*, (eds U. Feige, R.I. Morimoto, I. Yahara and B. Polla), Birkhäuser Verlag, Basel, pp. 307–20.

15

METAL IONS AND THE CYTOSKELETON

Douglas M. Templeton

15.1 Components of the cytoskeleton

At an ultrastructural level, the cytoskeleton consists of microfilaments of 6–8 nm diameter, microtubules of 25 nm diameter and filaments of an intermediate size known collectively as intermediate filaments. The microfilaments represent actin polymers and the microtubules consist of polymerized tubulin. Both have associated motor proteins and are involved in motility and shape changes as well as maintaining cell structure, and both are found in all eucaryotic cells. The intermediate filaments serve in a true skeletal role, are expressed in a cell-specific pattern, and include members of the keratin, desmin/vimentin, neurofilament and lamin families of proteins. On the other hand, polymerization/depolymerization of actin and tubulin are highly dynamic processes that are regulated in part by divalent metal-ion-binding sites, generally for Ca^{2+} and Mg^{2+} *in vivo*, and this means that other divalent metals can act both as probes for structural investigations and as potential toxic interferents with microfilaments and microtubules. Less is known about the assembly of intermediate filaments, but in general they assemble spontaneously into stable, long-lived, protease-resistant filaments. An extensive compilation of the literature on intermediate filaments has been presented (Quinlan *et al.*, 1994). Because metal ions are not prominent in intermediate filament binding or assembly, these filaments will not be considered further here.

15.2 Actin

Actin is a highly conserved protein of 375 amino acids that occurs in at least six different isoforms representing distinct gene products in mammals (Vandekerchove and Weber, 1978; Sheterline and Sparrow, 1994). There are two smooth-muscle actins, two striated-muscle actins and two nonmuscle actins. Alternatively, actins may be grouped as α, β and γ, based upon differences in electrophoretic mobility. Thus, the six mammalian actins are designated α-SM (smooth muscle), γ-SM, α_c (cardiac), α_{sk} (skeletal), β_{cyto} (cytoplasmic), and γ_{cyto}. β_{cyto} and γ_{cyto} are coexpressed in nonmuscle cells. α-SM and γ-SM are expressed in smooth muscle, with α-SM predominating in vascular smooth muscle and γ-SM in visceral smooth muscle. The remaining two α actins are expressed in both cardiac and skeletal muscle, α_c predominating in heart and α_{sk} in skeletal muscle. Gene sequences have been identified that give rise to the tissue-specific patterns of expression (Petropoulos *et al.*, 1989). Structural diversity is further achieved with different patterns of post-translational modification (Vandekerchove, 1993) but the consequences for polymerization and protein interactions are not generally understood (Ayscough and Drubin, 1996).

The actin monomer has a flattened, globular structure (G-actin) with four domains and a cleft between domains I/II and III/IV (Kabsch *et al.*, 1990; Sheterline and Sparrow, 1994). The cleft accommodates a divalent metal ion and ATP. G-actin with bound ATP associates noncovalently to form a growing double-helical actin filament (F-actin). When bound to the filament, ATP is hydrolyzed to produce ADP–actin and the cleft narrows (Lorenz *et al.*, 1993; Tirion *et al.*, 1995). Domains III and IV lie along the axis of the filament, and residues of domains I (containing the N and C termini) and II are exposed and probably involved in interactions of filament stabilization (Ayscough and Drubin, 1996). At the critical concentration, polymerization of actin is spontaneous, and below this concentration depolymerization occurs. A polymerization cycle consists of binding of ATP–actin to the filament with an association constant of about 10^{10}/M in the presence of Ca^{2+}, hydrolysis and release of P_i, dissociation of ADP–actin, and exchange of ATP for ADP. Exchange occurs only on the monomeric G-actin. The actin filament is polarized, the two ends being referred to as barbed and pointed, based on their appearance following decoration with heavy meromyosin. The kinetics of association and dissociation of ADP–actin and ATP–actin differ at the two ends; dissociation of ADP–actin from the pointed end and association of ATP–actin with the barbed end are the most rapid processes, and the critical concentrations for addition of Ca^{2+}–ATP–G-actin *in vitro* are 0.1 μM and 0.5 μM at the barbed and pointed ends, respectively. This gives rise to more rapid growth at the barbed end, or to growth at the barbed end with dissolution from the pointed end, a process referred to as treadmilling (Fechheimer and Zigmond, 1993; Sheterline and Sparrow, 1994). By this mechanism, rapid changes in cell shape and extension of cell processes can be achieved. The critical concentration for ADP–actin polymerization is 25-fold higher than for ATP–actin, and the rate constant for association of ATP–actin is 2.5-fold higher than for ADP–actin (Carlier *et al.*, 1994).

The state of polymerization of F-actin is further regulated by a large number of actin-binding proteins that can sequester G-actin, cap the growing filament, sever F-actin, or organize and stabilize the filament structure. At least 48 classes of such proteins have been considered (Vandekerchove, 1993). That actin binding is the primary function of all these proteins seems unlikely, and abundant actin filaments may serve as secondary binding sites for some of them. Nevertheless, there is a large degree of redundancy in the effects of these proteins on actin. Among the actin-binding proteins are the myosins that, in muscle cells, bind weakly in rapid equilibrium in the absence of ATP. In the presence of ATP, the actin–myosin complex hydrolyzes the nucleotide and, in the presence of the resulting ADP, a transition to a strongly bound form occurs that is coupled to movement of the myosin head along the filament, the production of force and muscle contraction. Myosin is also involved in the translocation of endocytic vesicles along actin filaments (Cheney and Mooseker, 1992).

15.2.1 Divalent metals and actin polymerization

Actin has a single, high-affinity divalent metal-binding site that is generally thought to be occupied by Mg^{2+} *in vivo* and readily replaced by Ca^{2+} *in vitro* (Carlier, 1991;

Gershman *et al.*, 1994). A dissociation constant for Ca^{2+} of 1–2 × 10^{-9} M has been estimated (Nowak *et al.*, 1988; Carlier, 1991; Gershman *et al.*, 1994) For Mg^{2+} the value is 5 × 10^{-9} M. These are higher affinities than for all other known Ca^{2+}- and Mg^{2+}-binding proteins (Gershman *et al.*, 1994). In addition, there are several weak cation-binding sites for mono- and divalent ions. These low-affinity sites appear to be involved in activating the monomer for the initiation of nucleation (Carlier, 1991). Four such Ca^{2+} sites were found in rabbit skeletal muscle and bovine aorta actins, and three in chicken gizzard actin (Strzelecka-Golaszewska *et al.*, 1989). This was thought to reflect involvement of an acidic residue in the N-terminus of α_{sk} and α-SM actins that is missing in γ-SM actin. The X-ray structure reveals only one bound Ca^{2+} in skeletal muscle actin, which presumably represents the high-affinity site (Kabsch *et al.*, 1990). This is located in a deep, hydrophilic pocket in the interdomain nucleotide-binding cleft. Asp11, Gln137 and Asp154 contribute to this pocket, but do not appear to be in direct coordination distance to the ion. Rather, Ca^{2+} is bound to the β- and γ-phosphate oxygens of ATP in the actin–ATP complex and to the β-phosphate oxygen in the ADP complex.

Although ATP will bind to G-actin in the absence of metal, the presence of the divalent ion increases the affinity by several orders of magnitude (Valentin-Ranc and Carlier, 1991). Furthermore, the nature of the bound ion influences both the nucleotide exchange kinetics and ATPase activity of actin. On binding of the ATP–G-actins to the filament, the rates of hydrolysis of ATP and Ca^{2+}–ATP are similar ($k_{hydrolysis}$ = 0.01/s) whereas Mg^{2+}–ATP is hydrolyzed about three orders of magnitude faster ($k_{hydrolysis}$ = 13/s), implying a catalytic role for Mg^{2+} (Carlier *et al.*, 1994). A comparison of the rates of dissociation of metal, nucleotide and metal–nucleotide complexes from G-actin (Carlier, 1991; Sheterline and Sparrow, 1994) indicates that rapid dissociation of Ca^{2+} is followed by release of ATP, and the rate of ATP dissociation is therefore dependent upon [Ca^{2+}]. On the other hand, Mg^{2+}–ATP and Mg^{2+} dissociate much more slowly than the rate of nucleotide exchange in Mg^{2+}–ATP–actin, implying that ATP dissociates with Mg^{2+} remaining bound to actin. One might speculate that the hydrophilic residues in the binding cleft are better able to coordinate Mg^{2+} (ionic radius, r = 0.65 Å) than Ca^{2+} (r = 0.99 Å) as the smaller ion may penetrate deeper into the cleft, and this might account, in part, for the different properties of Ca^{2+}–actin and Mg^{2+}–actin. Validation of this model probably awaits determination of the crystal structure of Mg^{2+}–ATP–actin. It might also be questioned whether ionic or hydrated radii are more relevant to this discussion. Gershman *et al.* (1994) have pointed out that whereas the affinity constant of Mg^{2+} at the high-affinity actin-binding site is only about one-fifth that for Ca^{2+} (i.e. 5 nM versus about 1 nM), the dissociation rate constant for Mg^{2+} is an order of magnitude smaller and the association rate constant is, as required by these numbers, about 50 times smaller than for Ca^{2+}. The explanation offered is that while association of Ca^{2+} ($k \simeq 2 \times 10^7$/M/s) is near the diffusion limit, the rate for Mg^{2+} ($k \simeq 2.5 \times 10^5$/M/s) reflects the rate of removal of a water molecule from the inner hydration sphere of Mg^{2+} ($k_{-H2O} = 1.5 \times 10^5$/s). For Ca^{2+}, $k_{-H2O} = 5 \times 10^8$/s (Williams, 1985). The significance of the relative water exchange rates of metal ions in determining their biological roles has been discussed by Williams (1985). The present example suggests the importance of dehydration of Ca^{2+} and Mg^{2+} aquo ions in binding to actin, and is in keeping with the strong correlation between the ionic radii of a number of elements and their ability to activate calmodulin by substituting at its Ca^{2+}-binding sites (Chao *et al.*, 1984).

The nature of the divalent cation influences the conformation of actin in a number of ways, the significance of which is beginning to be appreciated. Both metal exchange and polymerization have been studied frequently by observing changes in the fluorescence of probes attached to the C-terminal region of G-actin. Noting that this implies conformational changes occurring in this region in both processes, Selden *et al.* (1994) monitored the intrinsic fluorescence of tryptophan residues near the C terminus to demonstrate different effects of Ca^{2+} and Mg^{2+} on the conformation of the native monomer, and on conformational changes following polymerization. Replacement of Ca^{2+} with Mg^{2+} in native ATP–G-actin changed susceptibility of additional regions of the protein to proteolytic cleavage (Strzelecka-Golaszewska *et al.*, 1993), indicating more global conformational changes than those affecting only the C terminus. Removal of metal ion altogether caused the protein to become unfolded and exposed all Cys residues for titration (Valentin-Ranc and Carlier, 1991).

Structural changes in actin depending on the nature of the bound metal cation have biomechanical consequences for actin filaments. Both the flexural and torsional rigidity of actin filaments *in vitro* can be assessed by optical microscopic observation of their Brownian motion. Whereas Ca^{2+}–ADP filaments and Mg^{2+}–ADP filaments have similar flexural rigidity (Yasuda *et al.*, 1996), in fact similar to metal-free ADP–actin in this regard (Isambert *et al.*, 1995), the torsional rigidity of the Ca^{2+}-containing filament is about three times that of Mg^{2+}–ADP–F-actin (Yasuda *et al.*, 1996). Heavy meromyosin binds to F-actin with two heads on the same helical strand. Heavy meromyosin binding to Ca^{2+}–F-actin induces structural changes along the strand that lead to highly co-operative binding of additional heavy meromyosin molecules to the same strand (Orlova and Egelman, 1997). With Mg^{2+} at the high-affinity metal-binding site, this cooperativity is lost (Orlova and Egelman, 1997). The significance of these structural differences between Ca^{2+}–F-actin and Mg^{2+}–F-actin is not yet understood.

15.2.2 Metal ions as probes of actin structure and dynamics

Other metal ions can bind with low affinity to actin, and in particular some can occupy the high-affinity binding cleft. Thus, they have occasionally been exploited as probes of actin structure and function. Calcium, Sr^{2+}, Mn^{2+}, and Ni^{2+} appear to compete for the same low-affinity binding sites, whereas Zn^{2+} binds to these and additional sites (Strzelecka-Golaszewska *et al.*, 1978b). Polymerization of Ca^{2+}–ATP–G-actin was metal-dependent when other divalent cations were added to the *in vitro* polymerization buffer. Polymerization was complete with four moles of either Ca^{2+}, Sr^{2+}, Ni^{2+}, or Zn^{2+} bound per monomer (Strzelecka-Golaszewska *et al.*, 1978b). This is consistent with the notion that the low-affinity sites activate the monomer for nucleation (Carlier, 1991), and suggests that masking negative charges on G-actin may be more important for this event than specific site chemistry. Physiologic concentrations of K^+ are usually included to enhance polymerization in *in vivo* experiments. $LiCl_2$ alone is able to induce actin polymerization *in vitro*, and replacement of K^+ in the Mg^{2+}-containing polymerization buffer with Li^+ decreases the critical concentration for polymerization, enhances nucleation and stabilizes F-actin (Colombo *et al.*, 1991). The polyelectrolyte nature of the resulting actin filaments can be demonstrated by using metal ions to cause filaments to bundle together, as occurs with DNA. In the Manning polyelectrolyte model (Manning, 1979), condensation of delocalized counter-ions into the domain of the polymer achieves partial charge neutralization and overcomes interpolymer electrostatic repulsion. In Manning's theory, the extent of ion binding, and hence charge neutralization, is determined only by the charge of the ion and not its specific identity. This model works well for DNA, but only qualitatively for actin filaments. $Co(NH_3)_6^{3+}$, Co^{2+}, Mn^{2+}, Ca^{2+} and Mg^{2+} all caused bundling of filaments, but did so in an ion-specific manner, at concentrations of 2, 5.5, 7, 20 and 27 mM, respectively (Tang and Janmey, 1996). Cu^{2+} has additional effects on bundling, causing the phenomenon at only 0.4 mM, probably due to additional charge reduction from specific binding to the C terminus of actin (Tang and Janmey, 1996).

Cr(III)–ATP was used to deduce the involvement of ATP/ADP phosphate groups and their stereochemistry in the high-affinity binding site before the crystal structure had been solved (Valentin-Ranc and Carlier, 1989). Cr(III) forms a stable chelate with ATP; the metal exchange rate is on the order of days (Cleland, 1982). The chelate is bidentate with coordination through two oxygens of the β- and γ-phosphates. This renders the β-phosphate chiral, with diastereomers designated in the nomenclature of Cleland as Λ and Δ (Cleland, 1982), and so the molecules can serve as stable probes of the geometry of Mg^{2+}–ATP-binding sites. ^{45}Ca–ADP–actin was prepared with ^{45}Ca occupying the high-affinity site (Valentin-Ranc and Carlier, 1989). ATP then exchanged for ADP with ^{45}Ca remaining bound. However, Cr(III)–ATP displaced the label, suggesting that the stable nucleotide substituted both the metal and nucleotide site, which were also therefore adjacent. Displacement occurred only with the Λ isomer (Valentin-Ranc and Carlier, 1989). This configuration of metal and nucleotide was confirmed in the crystal structure of Ca^{2+}–ATP–G-actin (Kabsch *et al.*, 1990).

15.2.3 Actin filaments and metal ion toxicity

Toxic metals are unlikely to achieve sufficient concentrations to compete with Ca^{2+} and Mg^{2+} at the low-affinity binding sites, and some nonspecificity at these sites is predicted by polyelectrolyte theory. There is the possibility of competition/displacement at the high-affinity site, but no clear evidence of this occurring *in vivo* has been presented. Nevertheless, a number of metal ions disrupt F-actin in intact cells. In MDCK renal epithelial cells, exposure to 10 μM Zn^{2+} for 4 h led to a loss of stress fibers with reorganization of actin into plaque-like structures and an overall increase in total F-actin as determined by fluorescence of rhodamine-conjugated phalloidin (Mills *et al.*, 1992). This increase in filaments is reminiscent of the stabilization of F-actin by four atoms of bound Zn^{2+} (Strzelecka-Golaszewska *et al.*, 1978b), but it is highly unlikely that sufficient local concentrations of Zn^{2+} could be achieved to account for the formation of plaques by the bundling phenomenon noted above. On the other hand, toxic effects of metals on the actin cytoskeleton often appear to be secondary to effects on other regulatory systems. As an example of the complexities that must be considered in interpreting observations of effects of metals, it was shown that the metal chelator *N*, *N*, *N′*, *N′*-tetrakis (2-pyridylmethyl)-ethylene diamine (TPEN) could protect C6 glioma cells from phorbol ester-induced microfilament disruption, and Zn^{2+} could overcome this protection (Hedberg *et al.*, 1994). TPEN showed some specificity for Zn^{2+} and a Zn^{2+}-dependent event was implied. Phorbol esters translocate and activate protein kinase C (PKC), and activation of PKC has a profound influence on microfilament organization. For example, the myristoylated alanine-rich C kinase substrate

(MARCKS) appears to mediate the reversible association of actin with the plasma membrane, dependent upon its state of phosphorylation by PKC (Thelen *et al.*, 1991). In the study cited, TPEN and Zn^{2+} were found to affect phosphorylation of downstream targets of PKC, indicating a Zn^{2+}-dependent kinase activity ultimately leading to microfilament disassembly (Hedberg *et al.*, 1994).

Exposure of Swiss 3T3 cells to only 2.5 μM As^{3+} resulted in a loss of thick actin fibers with retention of F-actin microfilaments (Li and Chou, 1992). Again the mechanism is unknown, but As^{3+} increased levels of cellular glutathione by stimulating its synthesis, and glutathione level is an important determinant of cytoskeletal organization. Following increased glutathione levels in response to As^{3+}, the content of exposed thiols in the total cytoskeletal fraction was increased (Li and Chou, 1992). If this was in part due to actin, it could represent a conformational change, perhaps dependent on cellular redox status, similar to the exposure of thiols observed with binding of metal-free ATP (Valentin-Ranc and Carlier, 1991). However, exposed sulfhydryls in actin were not specifically measured, and any conclusions remain speculative.

Neutrophil chemotaxis requires polymerization of actin for motility. The ability of a series of metal ions to inhibit chemotaxis was found to correlate strongly with their ability to prevent stimulated actin polymerization (Simchowitz and Cragoe, 1990). K_i values for F-actin formation ranged from less than 10 μM for Ni^{2+} and Co^{2+} to 2.9 mM for Mg^{2+}, with intermediate values for Sr^{2+}, Zn^{2+} and La^{3+} (10–100 μM), and Cd^{2+}, Mn^{2+}, and Ba^{2+} (100–350 μM). Noting that Ni^{2+}, Zn^{2+}, or Sr^{2+} alone sustain actin polymerization *in vitro* (Strzelecka-Golaszewska *et al.*, 1978a), the authors again conclude that the effects of the metals must be on an unidentified regulatory component of activation.

In some instances, it appears that the metal ion may influence F-actin through activation of actin-associated proteins. $HgCl_2$ at 10 μM caused a loss of actin stress fibers in cultured proximal tubule epithelia that became apparent as early as 2 min after exposure (Elliget *et al.*, 1991). Interestingly, this effect was prevented by the protease inhibitor, leupeptin, indicating that active enzymatic disassembly may be occurring. The authors suggest that a Hg^{2+}-dependent rise in intracellular Ca^{2+} may have activated a Ca^{2+}-dependent protease such as calpain. A number of other candidate actin-associated molecules are influenced by Ca^{2+}, including capping and severing proteins, such as gelsolin and severin, that are activated by Ca^{2+} at a concentration greater than 10 μM, and a 30 kDa bundling protein that dissociates from F-actin in 0.1 μM Ca^{2+}. In addition, Ca^{2+}/calmodulin causes dissociation of some proteins such as MARCKS from F-actin and inhibits an actin cross-linking protein, spectrin (Janmey, 1994).

Perhaps the most thoroughly studied example of actin cytoskeletal disruption by a metal ion is Cd^{2+}. A marked disorganization and loss of F-actin was caused by 10 μM Cd^{2+} over 6 hours in MDCK cells (Mills and Ferm, 1989). Upon removal of Cd^{2+}, stress fibers were re-established over 18 hours. The same concentration for 16 hours caused disorganization of actin filaments and an increase in total cytoskeletal exposed sulfhydryl groups in Swiss 3T3 cells, while 2 mM Ni^{2+} caused peripheral bundling of actin and loss of cytoskeletal sulfhydryls (Li *et al.*, 1993).[1] A decrease in stress fibers

1 As Cd^{2+} exposure was in medium containing 10% serum, the effective concentration of available Cd^{2+} is substantially lower, due to binding to albumin (Templeton, 1990).

and loss of membrane-associated actin in LLC-PK_1 renal epithelial cells was caused by 60 μM $CdCl_2$, and was associated with cell–cell separation and loss of intercellular tight junctions (Prozialeck and Niewenhuis, 1991) (i.e. a loss of barrier function). In MDCK cells equimolar amounts of Cd^{2+} and Zn^{2+} produced the appearance of F-actin plaques that were seen with Zn^{2+} alone, while the same concentration of Cd^{2+} alone caused loss of stress fibers and peripheral actin without the appearance of plaques (Mills *et al.*, 1992). Swiss 3T3 cells showed a loss of actin filaments in 5 μM Cd^{2+} that were notably absent in cells selected for cadmium resistance by growth in incremental Cd^{2+} concentrations, a maneuver that results in increased cellular metallothionein, glutathione and protein sulfhydryls (Li *et al.*, 1994). Actin filament density was increased and somewhat disorganized in the resistant cells, but unaffected by 50 μM Cd^{2+} over 16 h. Again, in these examples a number of mechanisms are possible. A significant portion of cellular cadmium is associated with the cytoskeletal fraction (Kunimoto *et al.*, 1986). Also, the ionic radius of Cd^{2+} (0.97 Å) is extremely close to that of Ca^{2+}, and Cd^{2+} can substitute for Ca^{2+} in a number of Ca^{2+}-binding proteins, including calmodulin (Chao *et al.*, 1984), Ca^{2+}-dependent PKC isoforms (Mazzei *et al.*, 1984) and, potentially, actin-binding proteins. Therefore, we undertook to determine whether the effects of Cd^{2+} on F-actin involved direct Cd^{2+}–actin interaction and whether they were dependent upon, or competitive with, Ca^{2+}.

We have studied the effect of Cd^{2+} on actin in both cultured vascular smooth-muscle cells and smooth-muscle-like mesangial cells from the kidney (Wang *et al.*, 1996; Wang and Templeton, 1996). In these cells, exposure to 10 μM Cd^{2+} for 8 h causes extensive disruption of the cytoskeleton and loss of F-actin over 4 h (Wang *et al.*, 1996), a treatment that is without effect on cell viability and leads to activation of mitogen-activated protein kinases and c-*fos* proto-oncogene induction (Wang and Templeton, 1998). The phenomenon is highly specific for Cd^{2+} in these cells. For example, in contrast to neutrophils where, as noted above, Ni^{2+}, Co^{2+}, Sr^{2+}, Zn^{2+} and La^{3+} were much more potent inhibitors of F-actin formation than Cd^{2+} (Simchowitz and Cragoe, 1990), no other metals tested (Mn^{2+}, Co^{2+}, Ni^{2+}, Cu^{2+}, Zn^{2+}, Hg^{2+} or Mg^{2+}) in our smooth-muscle cells had a discernible effect on F-actin under the same conditions, even at 30 μM. The lack of effect of Hg^{2+}, in particular, seems to rule out generalized toxicity or altered cellular thiol status as participating in the effect of Cd^{2+}, as Hg^{2+} is quite toxic to the cells under these circumstances and is a very soft metal ion with similar effects on cellular thiols to Cd^{2+}. When fluorescent-labeled G-actin was allowed to polymerize *in vitro* in the presence of Mg^{2+} and ATP, the rate of F-actin formation was actually increased by Cd^{2+}, although significant effects were only observed at concentrations of 100 μM or greater. When F-actin was diluted below its critical concentration and allowed to depolymerize, concentrations of Cd^{2+} up to 10 μM were actually found to stabilize the polymer; concentrations of 100 μM and higher were added before enhanced depolymerization was seen (Wang and Templeton, 1996). Cadmium had no effect on the critical concentration at which polymerization was initiated. Quite recently, DalleDonne *et al.* (1997) have used very high Cd^{2+} concentrations *in vitro* to examine interactions of the cation with actin. In the range 0.25–0.5 mM, Cd^{2+} increased the rate of polymerization compared to Ca^{2+} and Mg^{2+}, but even higher concentrations apparently denatured the protein. At 0.5 mM, Cd^{2+} displaced Ca^{2+} from the high-affinity binding site and enhanced nucleation/activation (DalleDonne *et al.*, 1997). Consistent with measurements in epithelial cells (Quamme, 1992), Cd^{2+} reached concentrations only in the picomolar range in our smooth-muscle cells[2] (Wang *et al.*, 1996). Therefore, a direct effect of Cd^{2+} on actin can be eliminated as an explanation for cytoskeletal disruption *in vivo*.

It has been reported that Cd^{2+} binding to an orphan receptor on fibroblasts can elicit an intracellular Ca^{2+} signal (Smith *et al.*, 1989). No such signal was observed in response to 10 μM Cd^{2+} in smooth-muscle cells, but exposures of several hours caused an increase in $[Ca^{2+}]_i$ to 100–150 nM above basal values (Wang *et al.*, 1996). To investigate the role of Ca^{2+} further, we used the Ca^{2+}–ATPase inhibitor, thapsigargin. This agent inhibits the resequestration of cytosolic Ca^{2+} into the endoplasmic reticulum, with the consequence that cellular Ca^{2+} stores become depleted. Thapsigargin-treated cells had lower basal levels of $[Ca^{2+}]_i$ than untreated cells and, in contrast to the untreated cells, did not show any increase in $[Ca^{2+}]_i$ upon exposure to Cd^{2+}. However, thapsigargin treatment had no effect on F-actin levels and did not diminish the decrease in F-actin observed in response to Cd^{2+} (Wang *et al.*, 1996). Therefore, it appears that increased cytosolic Ca^{2+} levels do not mediate the effect of Cd^{2+} on F-actin. In fact, when the above experiments were repeated in medium containing 10 μM Ca^{2+} instead of Ca^{2+}-replete medium (1 mM), the combined effect of Cd^{2+} and thapsigargin on F-actin was enhanced (Wang *et al.*, 1996).

Because Cd^{2+} does not affect actin polymerization *in vitro* in the same way as Cd^{2+} exposure does *in vivo*, other cellular factors must be involved in the *in vivo* response. Indeed, the *in vivo* response could be replicated *in vitro* by adding cytosolic preparations from Cd^{2+}-treated cells (Wang and Templeton, 1996). That is, when the *in vitro* polymerization and depolymerization reactions were supplemented with cytosol from either Cd^{2+}-treated or control cells, cytosol from the Cd^{2+}-treated cells decreased the rate of polymerization and increased the rate of depolymerization. Several observations favored the involvement of a cellular regulatory factor(s) affected by Cd^{2+}:

1. Differences in Cd^{2+} or Ca^{2+} concentration between cytosols of Cd^{2+}-treated and control cells were insufficient to influence the *in vitro* reactions.
2. EGTA partially restored the rates of polymerization and depolymerization in the presence of Cd^{2+}-treated cytosol to control levels, consistent with regulation by a Ca^{2+}-sensitive factor.
3. The effect on polymerization of cytosol from Cd^{2+}-exposed cells was selective for elongation from the barbed end of the filament, suggesting the involvement of capping proteins.

Therefore, we examined the effect of Cd^{2+} on actin-binding proteins by overlaying protein electrophoretic gels with ^{125}I-labeled actin (Wang and Templeton, 1996). We found that Cd^{2+} caused the rapid disappearance of an actin-binding protein of 93 kDa from both cytosolic and cytoskeletal fractions, while a 42 kDa actin-binding protein present in cytosol became associated with the cytoskeleton after Cd^{2+} treatment. While the identity of these proteins and the basis for changes in the level of actin-binding activity remain to be determined, it seems reasonable to conclude that the effects of

2 Although uptake of Cd^{2+} continues at a constant rate for many more hours (Templeton, 1990), subsequent induction of and binding to metallothionein presumably buffers $[Cd^{2+}]_i$ at about 1 pM.

Cd^{2+} on F-actin in smooth-muscle cells are secondary to changes in the concentration or activity of Ca^{2+}-dependent actin-binding proteins rather than to direct effects of Cd^{2+} on actin or secondary effects of cell Ca^{2+} concentration.

15.3 Tubulin

Microtubules are hollow tubes that, in cross-section have 13 or 14 subunits staggered by about 0.9 nm that can be viewed longitudinally as being arranged in helical protofilaments (Desai and Mitchison, 1997). The subunits consist of α-tubulin and β-tubulin, each with about half a dozen isoforms coded by distinct genes in vertebrates (Cleveland, 1993), and all having about 450 amino acids. The microtubule assembles from α, β-heterodimers that are preformed and folded through a complex pathway involving chaperonin, cycles of ATP-dependent modifications and a number of specific protein cofactors (Lewis *et al.*, 1997). Microtubules form the major structural component of the mitotic spindle and motion between microtubules is the basis of flagellar beating. They form structures along which intracellular organelles are transported. These functions depend on association with two families of motor proteins, the kinesins (Vale and Fletterick, 1997) and dyneins (Porter, 1996). Microtubule assembly can occur spontaneously *in vitro*, but *in vivo* arises from microtubule organizing centers (MTOCs), (e.g. around centrosomes or flagellar basal bodies). A third tubulin, γ-tubulin, occurs in lower amounts and may be involved in nucleation at the MTOC (Cleveland, 1993).

Microtubules are polarized with the ends referred to as plus and minus, β-tubulin being exposed at the plus end and α-tubulin at the minus end (Desai and Mitchison, 1997). Growth is fastest at the plus end, which is usually directed away from the MTOC. The heterodimer binds two moles of GTP, one nonexchanging (N-site) in the α-subunit and one bound exchangeably (E-site) as Mg^{2+}-GTP in the β-subunit. Upon polymerization, GTP is hydrolyzed to GDP at the E-site and remains bound. Only the free heterodimer exchanges GDP for GTP. The polymer undergoes disassembly by disassociating to oligomers and then heterodimers, and re-assembly by stepwise addition of heterodimers (Mandelkow and Mandelkow, 1993). Hydrolysis of GTP favors dissociation, delayed hydrolysis producing a GTP-rich 'cap' at the plus end that may stabilize the polymer. Upon GTP hydrolysis, the protofilaments appear to undergo a conformational change, resulting in bending that accompanies filament loss (Tran *et al.*, 1997). At steady state, microtubules are in a state referred to as dynamic instability, with prolonged periods of polymerization followed by rapid and extensive depolymerization. Because dynamic instability is a nonequilibrium situation it requires an energy source, and this is provided by GTP hydrolysis (Desai and Mitchison, 1997). In cultured cells in interphase, microtubule half-life is on the order of minutes (Gelfand and Bershadsky, 1991). In the mitotic spindle it is much shorter. Assembly is favoured by Mg^{2+}, GTP and association with a variety of microtubule-associated proteins (MAPs); disassembly by Ca^{2+} and GDP.

15.3.1 Metal ions and tubulin polymerization

Microtubule polymerization *in vitro* is generally performed in the presence of Mg^{2+} with EGTA included in the buffer to sequester Ca^{2+}; Mg^{2+} is required for assembly and Ca^{2+} inhibits polymerization. However, as Mg^{2+} is increased above physiologic concentrations to 20 mM, it has a comparable effect to physiologic Ca^{2+} concentrations with respect to the rate of shortening of filaments *in vitro* (Tran *et al.*, 1997). By analogy with actin and Mg^{2+}–ATP, Mg^{2+} has a high-affinity binding site at the exchangeable GTP nucleotide-binding site on β-tubulin. Mn^{2+} can displace the tightly bound Mg^{2+} and also supports polymerization (Gaskin, 1981), and was used as an electron paramagnetic resonance probe to demonstrate the existence of a bound metal–GTP complex (Jemiolo and Grisham, 1982). The effect of Mn^{2+} on the ^{19}F and ^{31}P nuclear magnetic resonance spectra of guanosine 5′-(γ-fluorotriphosphate) bound to tubulin confirmed the participation of the nucleotide in the metal-binding site (Monasterio, 1987). Therefore, Mg^{2+} is positioned to catalyze GTP hydrolysis as it does ATP hydrolysis in actin. Loss of the γ-phosphate upon hydrolysis changes the coordination of Mg^{2+}, and it is proposed that this is the switch that converts tubulin between the straight tubule and the bent filaments that precede depolymerization (Shearwin and Timasheff, 1992). Similar changes in Mg^{2+} coordination with nucleotide triphosphate hydrolysis are thought to underlie inactivation of other GTP-binding proteins, such as $p21^{ras}$.

Cr(III)-GTP substitutes for Mg^{2+} + GTP in supporting polymerization (Gaskin, 1981), and this further suggests that the metal–nucleotide complex is present at the high-affinity binding site. It was subsequently shown that the Δ isomer of Cr(III)–GTP binds preferentially to tubulin, is efficiently hydrolyzed, and then Cr^{3+} and P_i are released (Carlier *et al.*, 1991). The release of Cr^{3+} from the stable nucleotide diphosphate–chromium complex is in contrast to Cr(III)–ADP binding to actin, discussed above, or binding of Cr(III)–NDP to a number of other nucleotide triphosphatases (e.g. $p21^{ras}$). This has been taken as evidence for strongly nucleophilic residues near the tubulin–nucleotide bond (Carlier *et al.*, 1991), a suggestion that awaits careful analysis of the recently determined crystal structure (Nogales *et al.*, 1998) for confirmation. However, it has been noted that Asp205 is in a sequence identified by homology to Mg^{2+} coordination sites in other nucleotide-binding proteins (Sternlicht *et al.*, 1987).

Whereas Mn^{2+} and Mg^{2+} bind with comparable affinity to this site, Ca^{2+} binds only weakly. Zn^{2+} and Co^{2+} also bind with high affinity[3] and support polymerization (Jemiolo and Grisham, 1982), but the polymer structure in the case of these ions is altered to form sheets rather than tubules (Gaskin, 1981). Microtubules induced with Mg^{2+}–GTP revert to sheets on treatment with Zn^{2+}. Beryllium as Be^{2+} can substitute for Mg^{2+} in supporting polymerization *in vitro*, but microtubules formed in the presence of Be^{2+} contain a higher proportion of GTP and are more stable than those formed with Mg^{2+}–GTP (Hamel *et al.*, 1992). Apparently Be^{2+} is a poor catalyst for GTP hydrolysis and the increased GTP content at the E-site may account for the increased stability. However, Al^{3+}, Co^{2+}, Mn^{2+}, Zn^{2+}, Cd^{2+} and Sn^{2+} also supported polymerization and enhanced polymer stability did not correlate with GTP content (Hamel *et al.*, 1992). It was suggested that decreased GTP hydrolysis with Be^{2+}, Al^{3+} and Sn^{2+} is a result of tight binding of the metal to the nucleotide in the E-site, while stability is a consequence of direct binding of the metals to tubulin or polymer. Zinc, for instance, binds to amino-

3 Colchicine was found to decrease incorporation of Co^{2+} into the E-site, while energy transfer experiments showed the interaction between Co^{2+} and colchicine to be long range (Ward and Timasheff, 1988). This demonstrates long-range conformational effects on the E-site by colchicine, an effect that could not be observed easily with the native Mg^{2+} bound molecule.

terminal regions of both α- and β-tubulin (Serrano *et al.*, 1988). Al^{3+} binds to GTP–tubulin in a ternary complex with an association constant 10^7 times higher than Mg^{2+} and supports polymerization at a concentration of only 4×10^{-11} M, even in the presence of Ca^{2+} (Macdonald *et al.*, 1987). In addition to decreased GTP hydrolysis, therefore, decreased sensitivity to Ca^{2+}-induced depolymerization may contribute to the increased stability of the microtubule in the presence of this ion. In contrast to these observations, Miura *et al.* (1984) reported that Hg^{2+} and CH_3Hg^+, Cr^{3+}, Cu^{2+} and Cd^{2+} all inhibited *in vitro* polymerization at concentrations in the range of 10^{-4}–10^{-5} M (Miura *et al.*, 1984). Micromolar concentrations of Hg^{2+} chelates inhibit the binding of GTP to the E-site of tubulin (Duhr *et al.*, 1993).

Mg^{2+} also binds to the nonexchangeable N-site on α-tubulin. After hydrolysis of GTP at the E-site the affinity of Mg^{2+} there is decreased, and P_i and Mg^{2+} dissociate. However, Mg^{2+}–GTP remains bound to the N-site. The association constants for Mg^{2+} with GTP–tubulin are about 1×10^6 M at both sites (Correia *et al.*, 1988). Mn^{2+} displaces Mg^{2+} at the N-site as well, and association constants with Mn^{2+} are about 5×10^7 M at both sites. The association constants with both metals drop by about three orders of magnitude at the E-site when GTP is hydrolyzed. Electron paramagnetic resonance of the Mn^{2+}-substituted protein was used to demonstrate the very slow exchange (k_{app} = 0.020/min) of metal at the N-site (Correia *et al.*, 1988).

15.3.2 Microtubules and metal ion toxicity

The above *in vitro* studies indicate that a variety of metal ions can bind to tubulin, in general favoring polymerization while affecting stability and conformation of the resulting polymer. However, as noted above for microfilaments, whether concentrations of other metal ions can be achieved *in vivo* to compete with Mg^{2+} and Ca^{2+} in regulating microtubule dynamics remains uncertain. Aluminum may be a notable exception (Macdonald *et al.*, 1987), although effects of Al^{3+} on neuronal cytoskeleton *in vivo* or on whole cytoskeletal preparations appear to be mediated through aggregation of phosphorylated cytoskeletal-associated proteins rather than involvement of tubulin (Diaz-Nido and Avila, 1990; Hewitt *et al.*, 1991). Nevertheless, microtubules appear to be important targets for metal toxicity, and a number of studies have demonstrated disruption of microtubules in response to various metal ions. As with microfilaments, these effects would generally seem to be secondary to other effects on regulatory processes, but mechanistic data are, on the whole, lacking.

Choi and co-workers showed a number of years ago that 20 μM CH_3Hg^+ caused complete disappearance of microtubules in cultured fetal neuronal cells and astrocytes over several hours (Choi *et al.*, 1981) and concluded that this might contribute to the neurotoxicity of CH_3Hg^+ observed in accidental human exposures. Subsequently numerous authors observed a similar phenomenon. Miura *et al.* (1984) found that CH_3Hg^+ disrupted the microtubule network in cultured glioma cells, while several other metal ions at concentrations that inhibited cell growth had no such effect. Clarkson (1991) concluded that developmental toxicity of CH_3Hg^+ is due to effects on both the cell cycle and cell migration, with microtubule depolymerization as a common denominator. Primary rat neurons were arrested in G_2/M by 2 μM CH_3Hg^+ (Ponce *et al.*, 1994), consistent with disruption of the mitotic spindle. However, CH_3Hg^+ may alternatively have effects on cell-cycle-regulatory gene expression (Ou *et al.*, 1997), and the effects observed on microtubules might also be secondary to a general effect on cell thiols; thiol-blocking agents are known to inhibit microtubule polymerization (Kuriyama, 1976). Although CH_3Hg^+ was found to bind to 15 free thiols per tubulin dimer *in vitro*, it only caused depolymerization, accompanied by protein aggregation, above 10 μM (Vogel *et al.*, 1985).

Microtubule assembly appeared to be impaired in the brains of zinc-deficient animals (Hesketh, 1981), perhaps consistent with the changes in structure of the polymer caused by Zn^{2+} *in vitro* (Gaskin, 1981). However, microtubules were unaffected by Zn^{2+} in MDCK cells at concentrations that disrupted the actin cytoskeleton, but depolymerized at Zn^{2+} concentrations at or above 50 μM (Mills *et al.*, 1992). Likewise, higher concentrations of As^{3+} were required to disrupt microtubules than actin, but marked effects on microtubules were observed at or above 20 μM. Ni^{2+} caused reorganization and bundling of microtubules in Swiss 3T3 cells at a concentration of 2 mM, and this effect was antagonized by supplementation of the culture medium with Mg^{2+} (Li *et al.*, 1996). This effect, too, may be mediated through cellular thiol status rather than competitive binding of metal to tubulin; Ni^{2+} caused a decrease in cellular glutathione levels that were restored upon Mg^{2+} supplementation. However, the relationship is far from straightforward. A concentration of Cd^{2+} (10 μM) that caused an increase in cellular glutathione over 16 hours also induced microtubule depolymerization (Li *et al.*, 1993) and Cd^{2+}-resistant cells with increased cellular thiol status showed disturbances in their microtubules (Li *et al.*, 1994). Cadmium depolymerized microtubules in MDCK cells, again after changes in actin had begun (Mills and Ferm, 1989). The prevention of microtubule disruption by 50 μM Cd^{2+} in the presence of trifluoperazine, a calmodulin inhibitor, may implicate activation of calmodulin by Cd^{2+} acting as a Ca^{2+} analog (Perrino and Chou, 1986). However, activation of calmodulin by Cd^{2+} has not been demonstrated *in vivo* (cf. Wang and Templeton, 1998) and other effects of trifluoperazine cannot be ruled out.

15.4 Conclusions and future prospects

The major components of the dynamic cytoskeleton, microtubules and microfilaments, are both composed of subunits that bind GTP or ATP, respectively, in a Mg^{2+}-dependent manner *in vivo*. Polymerization is linked to Mg^{2+}-catalyzed hydrolysis of the nucleotide triphosphate at a high-affinity Mg^{2+}-binding site. Whereas Ca^{2+} can substitute for Mg^{2+} in actin polymerization, it inhibits tubulin polymerization. A number of other metal ions can displace Mg^{2+} from its high-affinity binding site and support polymerization *in vitro*, the latter being especially true for tubulin. However, these *in vitro* effects generally occur at comparatively high concentrations of metal ion, and are frequently in the opposite direction to the effect of the metal on microfilaments and microtubules in intact cells. Nevertheless, a variety of metal ions have been shown to cause F-actin and (or) microtubule disorganization or depolymerization in cells. In most cases, then, these effects appear to be secondary effects on regulation of cytoskeletal assembly rather than direct effects of metal-ion interactions on polymer dynamics.

The molecular mechanisms modulating polymerization of actin and tubulin are exceedingly complex, involving interactions with many dozens of other proteins, and isolated aspects of the overall picture are just beginning to be understood. Many of these interactions are regulated by phosphorylation/dephosphorylation events, suggesting that the effects of metal ions on catalytic kinase cascades may be one important means by which they influence the cytoskeleton. For instance, cutting of F-actin by cofilin and destrin is inhibited by phosphorylation, and the vasodilator-stimulated phosphoprotein (VASP) binds profilin and recruits actin–profilin complexes (reviewed by Zigmond, 1996). The Rho GTPases, Rho, Rac and cdc42, are emerging as key regulators of the cytoskeleton (Zigmond, 1996; Tapon and Hall, 1997). The Rho-associated kinase, Rho-K, inhibits myosin light-chain phosphatase, increasing the level of phosphorylation of myosin light-chain and favoring bundling of actin filaments. Rac increases phosphotidyl inositol bisphosphate (PIP_2) levels and PIP_2 inhibits the barbed-end actin-capping protein, gelsolin. The ezrin/radixin/moesin family of proteins are key linkers of actin with the plasma membrane and their linking activity now appears to be, in part, under control of Rho and modified by PIP_2 (Tsukita *et al.*, 1997). Future studies with metal ions will probably focus more on understanding their effects on these regulatory pathways than on metal-binding properties of purified cytoskeletal components.

References

Ayscough, K.R. and Drubin, D.G. 1996. Actin: General principles from studies in yeast. *Annual Review of Cell Developmental Biology* **12**, 129–60.

Carlier, M.-F. 1991. Actin: Protein structure and filament dynamics. *Journal of Biological Chemistry* **266**, 1–4.

Carlier, M.-F., Didry, D. and Valentin-Ranc, C. 1991. Interaction between chromium GTP and tubulin. Stereochemistry of GTP binding, GTP hydrolysis, and microtubule stabilization, *Journal of Biological Chemistry* **266**, 12361–8.

Carlier, M.-F., Valentin-Ranc, C., Combeau, C., Fievez, S. and Pantoloni, D. 1994. Actin polymerization: Regulation by divalent metal ion and nucleotide binding, ATP hydrolysis and binding of myosin. *Advances in Experimental Biology and Medicine* **358**, 71–81.

Chao, S.-H., Suzuki, Y., Zysk, J.R. and Cheung, W.Y. 1984. Activation of calmodulin by various metal cations as a function of ionic radius. *Molecular Pharmacology* **26**, 75–82.

Cheney, R.E. Mooseker, M. 1992. Unconventional myosins. *Current Opinion in Cell Biology* **4**, 27–35.

Choi, B.H., Cho, K.H. and Lapham, L.W. 1981. Effects of methylmercury on human fetal neurons and astrocytes *in vitro*: a time-lapse cinematographic, phase and electron microscopic study. *Environmental Research* **24**, 61–74.

Clarkson, T. 1991. Methylmercury. *Fundamental and Applied Toxicology* **16**, 20–1.

Cleland, W.W. 1982. Preparation of chromium(III) and cobalt(III) nucleotides as chirality probes and inhibitors. *Methods in Enzymology* **87**, 159–79.

Cleveland, D.W. 1993. Tubulin and associated proteins, in *Guidebook to the Cytoskeletal and Motor Proteins*, (eds T. Kreis and R. Vale), Oxford University Press, Oxford, pp. 101–5.

Colombo, R., Milzani, A., Contini, P. and Dalle Donne, I. 1991. Effects of lithium ions on actin polymerization in the presence of magnesium ions. *Biochemical Journal* **274**, 421–5.

Correia, J.J., Beth, A.H., Williams, R.C. Jr 1988. Tubulin exchanges divalent cations at both nucleotide-binding sites. *Journal of Biological Chemistry* **263**, 10681–6.

DalleDonne, I., Milzani, A. and Colombo, R. 1997. Actin assembly by cadmium ions. *Biochimica et Biophysica Acta* **1357**, 5–17.

Desai, A. and Mitchison, T.J. 1997. Microtubule polymerization dynamics. *Annual Review of Cell Biology* **13**, 83–117.

Diaz-Nido, J. and Avila, J. 1990. Aluminum induces the *in vitro* aggregation of bovine brain cytoskeletal proteins. *Neuroscience Letters* **110**, 221–6.

Duhr, E.F., Pendergrass, J.C., Slevin, J.T. and Haley, B.E. 1993. HgEDTA complex inhibits GTP interactions with the E-site of brain β-tubulin. *Toxicology and Applied Pharmacology* **122**, 273–80.

Elliget, K.A., Phelps, P.C. and Trump, B.F. 1991. $HgCl_2$-induced alteration of actin filaments in cultured primary rat proximal tubule epithelial cells labelled with fluorescent phalloidin. *Cell Biology and Toxicology* 7, 263–80.

Fechheimer, M. and Zigmond, S.H. 1993. Focusing on unpolymerized actin. *Journal of Cell Biology* **123**, 1–5.

Gaskin, F. 1981. *In vitro* microtubule assembly regulation by divalent cations and nucleotides. *Biochemistry* **20**, 1318–22.

Gelfand, V.I. and Bershadsky, A.D. 1991. Microtubule dynamics: Mechanism, regulation, and function. *Annual Review of Cell Biology* 7, 93–116.

Gershman, L.G., Selden, L.A., Kinosian, H.J. and Estes, J.E. 1994. Actin-bound nucleotide/divalent cation interactions. *Advances in Experimental Biology and Medicine* **358**, 35–49.

Hamel, E., Lin, C.M., Kenney, S., Skehan, P. and Vaughns, J. 1992. Modulation of tubulin-nucleotide interactions by metal ions: Comparison of beryllium with magnesium and initial studies with other cations. *Archives of Biochemistry and Biophysics* **295**, 327–39.

Hedberg, K.K., Birrell, B., Mobley, P.L. and Griffith, O.H. 1994. Transition metal chelator TPEN counteracts phorbol ester-induced actin cytoskeletal disruption in C6 rat glioma cells without inhibiting activation or translocation of protein kinase C. *Journal of Cell Physiology* **158**, 337–46.

Hesketh, J.E. 1981. Impaired microtubule assembly in brain from zinc-deficient pigs and rats. *International Journal of Biochemistry* **13**, 921–6.

Hewitt, C.D., Herman, M.M., Lopes, M.B., Savory, J. and Wills, M.R. 1991. Aluminium maltol-induced neurocytoskeletal changes in fetal rabbit midbrain in matrix culture. *Neuropathology and Applied Neurobiology* **17**, 47–60.

Isambert, H., Vrnier, P., Maggs, A.C., Fattoum, A., Kassab, R., Pantaloni, D. and Carlier, M.-F. 1995. Flexibility of actin filaments derived from thermal fluctuations: Effect of bound nucleotide, phalloidin, and muscle regulatory proteins. *Journal of Biological Chemistry* **270**, 11437–44.

Janmey, P.A. 1994. Phosphoinositides and calcium as regulators of cellular actin assembly and disassembly. *Annual Review of Physiology* **56**, 169–91.

Jemiolo, D.K. and Grisham, C.M. 1982. Divalent cation-nucleotide complex at the exchangeable nucleotide binding site of tubulin. *Journal of Biological Chemistry* **257**, 8148–52.

Kabsch, W., Mannherz, H., Suck, D., Pai, E. and Holmes, K. 1990. Atomic structure of the actin : DNase1 complex. *Nature* **347**, 37–44.

Kunimoto, M., Miyasaka, K. and Miura, T. 1986. Changes in membrane properties of rat red blood cells induced by cadmium accumulation in the membrane fraction. *Journal of Biochemistry* **99**, 397–406.

Kuriyama, R. 1976. *In vitro* polymerization of flagellar and ciliary outer fiber tubulin into microtubules. *Journal of Biochemistry* **80**, 153–65.

Lewis, S.A., Tian, G. and Cowan, N.J. 1997. The α- and β-tubulin folding pathways. *Trends in Cell Biology* 7, 479–84.

Li, W. and Chou, I.-N. 1992. Effects of sodium arsenite on the cytoskeleton and cellular glutathione levels in cultured cells. *Toxicology and Applied Pharmacology* **114**, 132–9.

Li, W., Zhao, Y. and Chou, I.-N. 1993. Alterations in cytoskeletal protein sulfhydryls and cellular glutathione in cultured cells exposed to cadmium and nickel ions. *Toxicology* 77, 65–79.

Li, W., Kagan, H.M. and Chou, I.-N. 1994. Alterations in cytoskeletal organization and homeostasis of cellular thiols in cadmium-resistant cells. *Toxicology and Applied Pharmacology* **126**, 114–23.

Li, W., Zhao, Y. and Chou, I.-N. 1996. Mg^{2+} antagonism on Ni^{2+}-induced changes in microtubule assembly and cellular thiol homeostasis. *Toxicology and Applied Pharmacology* **136**, 101–11.

Lorenz, M., Popp, D. and Holmes, K. 1993. Refinement of the F-actin model against X-ray fiber diffraction data by use of a directed mutation algorithm. *Journal of Molecular Biology* **234**, 826–36.

Macdonald, T.L., Humphreys, W.G. and Martin, R.B. 1987. Promotion of tubulin assembly by aluminum ion *in vitro*. *Science* **236**, 183–6.

Mandelkow, E.-M. and Mandelkow, E. 1993. α/β-Tubulin, in *Guidebook to the Cytoskeletal and Motor Proteins*, (eds T. Kreis and R. Vale), Oxford University Press, Oxford, pp. 127–130.

Manning, G.S. 1979. Counterion binding in polyelectrolyte theory. *Accounts of Chemical Research* **12**, 443–9.

Mazzei, G.J., Girard, P.R. and Kuo, J.F. 1984. Environmental pollutant Cd^{2+} biphasically and differentially regulates myosin light chain kinase and phospholipid/Ca^{2+}-dependent protein kinase. *FEBS Letters* **173**, 124–8.

Mills, J.W. and Ferm, V.H. 1989. Effect of cadmium on F-actin and microtubules of Madin–Darby canine kidney cells. *Toxicology and Applied Pharmacology* **101**, 245–54.

Mills, J.W., Zhou, J.-H., Cardoza, L. and Ferm, V.H. 1992. Zinc alters actin filaments in Madin–Darby canine kidney cells. *Toxicology and Applied Pharmacology* **116**, 92–100.

Miura, K., Inokawa, M. and Imura, N. 1984. Effects of methylmercury and some metal ions on microtubule networks in mouse glioma cells and in vitro tubulin polymerization. *Toxicology and Applied Pharmacology* **73**, 218–31.

Monasterio, O. 1987. ^{19}F Nuclear magnetic resonance measurement of the distance between the E-site GTP and the high-affinity Mg^{2+} in tubulin. *Biochemistry* **26**, 6099–106.

Nogales, E., Wolf, S.G and Downing, K.H. 1998. Structure of the $\alpha\beta$tubulin dimer by electron crystallography. *Nature* **391**, 199–203.

Nowak, E., Strzelecka-Golaszewska, H. and Goody, R.S. 1988. Kinetics of nucleotide and metal ion interaction with G-actin. *Biochemistry* **27**, 1785–92.

Orlova, A. and Egelman, E.H. 1997. Cooperative rigor binding of myosin to actin is a function of Factin structure. *Journal of Molecular Biology* **265**, 469–74.

Ou, Y.C., Thompson, S.A., Kirchner, S.C., Kavanagh, T.J. and Faustman, E.M. 1997. Induction of growth arrest and DNA damage-inducible genes Gadd45 and Gadd153 in primary rodent embryonic cells following exposure to methylmercury. *Toxicology and Applied Pharmacology* **147**, 31–8.

Perrino, B.A. and Chou, I.-N. 1986. Role of calmodulin in cadmium-induced microtubule disassembly. *Cell Biology International Reports* **10**, 565–73.

Petropoulos, C.J., Rosenberg, M.P., Jenkins, N.A., Copeland, N.G. and Hugnes, S.H. 1989. The chicken skeletal muscle α-actin promoter is tissue specific in transgenic mice. *Molecular and Cellular Biology* **9**, 3785–92.

Ponce, R.A., Kavanagh, T.J., Mottet, N.K., Whittaker, S.G. and Faustman, E.M. 1994. Effects of methyl mercury on the cell cycle of primary rat CNS cells in vitro. *Toxicology and Applied Pharmacology* **127**, 83–90.

Porter, M. 1996. Axonemal dyneins: assembly, organization, and regulation. *Current Opinion in Cell Biology* **8**, 10–17.

Prozialeck, W.C. and Niewenhuis, R.J. 1991. Cadmium (Cd^{2+}) disrupts intercellular junctions and actin filaments in LLC-PK1 cells. *Toxicology and Applied Pharmacology* **107**, 81–97.

Quamme, G.A. 1992. Free cadmium activity in renal epithelial cells is enhanced by Mg^{2+} depletion. *Kidney International* **41**, 1237–44.

Quinlan, R., Hutchison, C. and Lane, B. 1994. Intermediate filament proteins. *Protein Profile* **1**, 779–911.

Selden, L.A., Kinosian, H.J., Estes, J.E. and Gershman, L.C. 1994. Influence of the high affinity divalent cation on actin tryptophan fluorescence. *Advances in Experimental Biology and Medicine* **358**, 51–7.

Serrano, L., Dominguez, J.E. and Avila, J. 1988. Identification of zinc-binding sites of proteins: Zinc binds to the amino-terminal region of tubulin. *Analytical Biochemistry* **172**, 210–18.

Shearwin, K.E. and Timasheff, S.N. 1992. Linkage between ligand binding and control of tubulin conformation. *Biochemistry* **31**, 8080–9.

Sheterline, P. and Sparrow, J.C. 1994. Actin. *Protein Profile* **1**, 1–121.

Simchowitz, L. and Cragoe E.J. Jr 1990. Polyvalent cations inhibit human neutrophil chemotaxis by interfering with the polymerization of actin. *Journal of Biological Chemistry* **265**, 13457–63.

Smith, J.B., Dwyer, S.D. and Smith, L. 1989. Cadmium evokes inositol polyphosphate formation and calcium mobilization. *Journal of Biological Chemistry* **264**, 7115–18.

Sternlicht, H., Yaffe, M.B. and Farr, G.W. 1987. A model of the nucleotide-binding site in tubulin. *FEBS Letters* **214**, 226–35.

Strzelecka-Golaszewska, H., Prochniewicz, E. and Drabikowski, W. 1978a. Interaction of actin with divalent cations. 1. The effect of various cations on the physical state of actin. *European Journal of Biochemistry* **88**, 219–27.

Strzelecka-Golaszewska, H., Prochniewicz, E. and Drabikowski, W. 1978b. Interaction of actin with divalent cations. 2. Characterization of protein-metal complexes. *European Journal of Biochemistry* **88**, 229–237.

Strzelecka-Golaszewska, H., Boguta, G., Zmorzynski, S. and Moraczewska, J. 1989. Biochemical and theoretical approach to localization of metal-ion-binding sites in the actin primary structure. *European Journal of Biochemistry* **182**, 299–305.

Strzelecka-Golaszewska, H., Moraczewska, J., Khaitlina, S.Y. and Mossakowska, M. 1993. Localization of the tightly bound divalent-cation-dependent and nucleotide-dependent conformation changes in G-actin using limited proteolytic digestion. *European Journal of Biochemistry* **211**, 731–42.

Tang, J.X. and Janmey, P.A. 1996. The polyelectrolyte nature of F-actin and the mechanism of actin bundle formation. *Journal of Biological Chemistry* **271**, 8556–63.

Tapon, N. and Hall, A. 1997. Rho, Rac, and Cdc42 GTPases regulate the organization of the actin cytoskeleton. *Current Opinion in Cell Biology* **9**, 86–92.

Templeton, D.M. 1990. Cadmium uptake by cells of renal origin. *Journal of Biological Chemistry* **265**, 21764–70.

Thelen, M., Rosen, A., Nairn, A.C. and Aderem, A. 1991. Regulation by phosphorylation of reversible association of a myristoylated protein kinase C substrate with the plasma membrane. *Nature* **351**, 320–2.

Tirion, M., ben-Avraham, D., Lorenz, M. and Holmes, K. 1995. Normal modes as refinement parameters for the F-actin model. *Biophysical Journal* **68**, 5–12.

Tran, P.T., Joshi, P. and Salmon, E.D. 1997. How tubulin subunits are lost from the shortening ends of microtubules. *Journal of Structural Biology* **118**, 107–18.

Tsukita, S., Yonemura, S. and Tsukita, S. 1997. ERM (ezrin/radixin/moesin) family: from cytoskeleton to signal transduction. *Current Opinion in Cell Biology* **9**, 70–5.

Vale, R.D. and Fletterick, R.J. 1997. The design plan of kinesin motors. *Annual Review of Cell Biology* **13**, 745–77.

Valentin-Ranc, C. and Carlier, M.-F. 1989. Evidence for the direct interaction beteen tightly bound divalent metal ion and ATP on actin. Binding of the Λ isomers of $\beta\alpha$-bidentate CrATP to actin. *Journal of Biological Chemistry* **264**, 20871–80.

Valentin-Ranc, C. and Carlier, M.-F. 1991. Role of ATP-bound divalent metal ion in the conformation and function of actin. Comparison of Mg-ATP, Ca-ATP, and metal ion-free ATP-actin. *Journal of Biological Chemistry* **266**, 7668–75.

Vandekerchove, J. 1993. Actins, in *Guidebook to the Cytoskeletal and Motor Proteins*, (eds T. Kreis and R. Vale), Oxford University Press, Oxford, pp. 13–15.

Vandekerchove, J. and Weber, K. 1978. At least six different actins are expressed in a higher mammal: an analysis based on the amino acid sequence of the amino-terminal tryptic peptide. *Journal of Molecular Biology* **126**, 783–802.

Vogel, D.G., Margolis, R.L. and Mottet, N.K. 1985. The effects of methyl mercury binding to microtubules. *Toxicology and Applied Pharmacology* **80**, 473–86.

Wang, Z. and Templeton, D.M. 1996. Cellular factors mediate cadmium-dependent actin depolymerization. *Toxicology and Applied Pharmacology* **139**, 115–21.

Wang, Z. and Templeton, D.M. 1998. Induction of c-*fos* proto-oncogene in mesangial cells by cadmium. *Journal of Biological Chemistry* **273**, 73–9.

Wang, Z., Chin, T.A. and Templeton, D.M. 1996. Calcium-independent effects of cadmium on actin assembly in mesangial and vascular smooth muscle cells. *Cell Motility and the Cytoskeleton* **32**, 208–22.

Ward, L.D. and Timasheff, S.N. 1988. Energy-transfer studies of the distance between the high-affinity metal binding site and the colchicine and 8-anilino-naphthalenesulfonic acid binding sites on calf brain tubulin. *Biochemistry* **27**, 1508–14.

Williams, R.J.P. 1985. The symbiosis of metal and protein functions. *European Journal of Biochemistry* **150**, 231–48.

Yasuda, R., Miyata, H. and Kinosita, K. Jr 1996. Direct measurement of the torsional rigidity of single actin filaments. *Journal of Molecular Biology* **263**, 227–36.

Zigmond, S.H. 1996. Signal transduction and actin filament organization. *Current Opinion in Cell Biology* **8**, 66–73.

16

THE ROLE OF METALS IN OXIDATIVE DAMAGE AND REDOX CELL-SIGNALING DERANGEMENT

Kazimierz S. Kasprzak and Gregory S. Buzard

16.1 Introduction

Exposure of humans and animals to certain metal derivatives by various routes results in a plethora of more or less specific toxic effects, ranging from acute to chronic, including immediate lethality or the slow process of carcinogenesis. Owing to their rich chemistry, several transition metals (e.g. zinc, iron, manganese, copper and cobalt in mammals) are 'essential' for life. Cells employ metals to control many metabolic and signaling functions. However, the very richness of their chemistry makes metals capable of evading the established controls of their cellular uptake, transport, compartmentalization, and binding to tissue or cell constituents. Disruption of the cell's well-being occurs when exogenous metals bind to proteins other than those designed to detoxify them, or they displace the physiologic metals from their natural carriers. From the molecular perspective, binding of a metal cation to an adventitious ligand may result in steric rearrangement of the latter (or 'cross-linking' of several ligands in binary or ternary complexes) that may impair or abrogate its physiologic function(s) and thus damage the ligand. Such damage may be reversed, for example, by sequestration of the toxic metal with stronger ligands, as is done in chelation therapy, or by supplementation with a physiologic metal.

Besides imposing conformational changes, a bound transition metal may also act as a catalytic center for redox reactions with molecular oxygen or endogenous peroxides. The high pathogenic importance of such reactions results from the spatial extent and chemical irreversibility of the damage they may cause in the cell. The damage may include oxidative modification of both the metal-complexing ligand itself and the neighboring molecules that are attacked by reactive intermediates of the original reaction. The reactive intermediates consist of a wide spectrum of oxygen-, carbon- or

This project has been funded in whole with Federal funds from the National Cancer Institute, National Institutes of Health, in part under Contract No. NO1-CO-56000. The content of this publication does not necessarily reflect the views or policies of the Department of Health and Human Services, nor does mention of trade names, commercial products, or organizations imply endorsement by the US Government. The publisher or recipient acknowledges the right of the US Government to retain a nonexclusive, royalty-free licence in and to any copyright covering the article.

sulfur-centered radicals, originating from dioxygen, hydrogen peroxide, lipid peroxides, as well as the amino acid, peptide, protein and other ligands carrying the toxic metal. The spectrum of the damage the radicals may cause in a cell and the resulting pathology at an organismal level may be very broad. Over 40 clinical conditions in which oxygen radicals may play pathogenic roles have been listed by Cross *et al.* (1987).

Reviewing the existing literature on all the facets of metal-mediated oxidative damage and its pathogenicity is too formidable a task. Therefore, this chapter is not such a comprehensive nor historical review, but rather a summary of what we feel are important current facts and mechanistic concepts on the diverse roles of selected metals, mainly Ni, Cr, Cu, Fe, Cd, As, Pb, and briefly several others. These metals are known to be toxic or carcinogenic contaminants of the occupational or general environments, and to play a role in the activation of various oxygen species and the mediation of oxidative damage to major cell constituents, such as lipids, proteins and nucleic acids. Our discussion of the biologic importance of this type of damage is likewise limited to selected issues of genetic and epigenetic toxicity and the possible derangement of redox regulation of cell signaling. All these effects are integrated into our understanding of the multi-step process of metal-mediated carcinogenesis.

16.2 Mechanistic background: direct versus indirect action

16.2.1 Basic redox biochemistry of toxic metals

Although most metals have a rich redox chemistry, not all of them, especially when complexed with natural tissue and cellular ligands, can react with certain oxygen and other substrates and activate them to highly oxidizing species, such as the hydroxyl radical, $^{\cdot}OH$, under physiologic conditions (Spiro, 1980; Sawyer, 1987). The most important biologically relevant mechanisms of oxygen activation by metal ions involve Fenton/Haber–Weiss chemistry and autoxidation. The substrates are molecular oxygen and two of its metabolic products, H_2O_2 and superoxide, $O_2^{-\cdot}$ (Halliwell and Gutteridge, 1986; Peskin and Shlyahova, 1986; Imlay and Linn, 1988; Szatrowski and Nathan, 1991). Equation 16.1 illustrates the conversion of H_2O_2 into hydroxyl radical ($^{\cdot}OH$) via the oxidation of a metal cation:

$$M^{n+} + H_2O_2 \rightarrow M^{(n+1)+} + OH^- + {}^{\cdot}OH \quad \text{(Fenton reaction)} \qquad (16.1)$$

The metal cation, just oxidized, may be turned back to reaction 16.1 by reduction with $O_2^{-\cdot}$:

$$M^{(n+1)+} + O_2^{-\cdot} \rightarrow M^{n+} + O_2 \qquad (16.2)$$

The balance of these two reactions is:

$$H_2O_2 + O_2^{-\cdot} \rightarrow O_2 + OH^- + {}^{\cdot}OH \quad \text{(Haber–Weiss reaction)} \qquad (16.3)$$

Thus, the two oxidation states of a metal cation (M^{n+} and $M^{(n+1)+}$) can form a catalytic electron transfer (redox) couple. In the absence of chelators, the Fenton reaction is driven by Cu(I), Fe(II), Co(II), Ti(III) and Cr(V) ions (Walling, 1975; Gutteridge, 1985;

Shi and Dalal, 1992). Some other metal ions become reactive after proper chelation (see below). H_2O_2 and $O_2^{-\cdot}$ needed for those reactions may originate not only from cell metabolism but also from oxidation of some metal ions with molecular oxygen (a process known as autoxidation) (Spiro, 1980; Sawyer, 1987). The latter is illustrated below:

$$M^{n+} + O_2 \rightarrow M^{(n+1)+} + O_2^{-\cdot} \tag{16.4}$$

$$M^{n+} + O_2^{-\cdot} + 2H^+ \rightarrow M^{(n+1)+} + H_2O_2 \tag{16.5}$$

The yield of $O_2^{-\cdot}$ and H_2O_2 in the above reactions strongly depends on pH and the presence or absence of chelators (Spiro, 1980; Sawyer, 1987). Equations 16.2 and 16.5 indicate that, depending on the reaction conditions, $O_2^{-\cdot}$ may serve as either the reductant or oxidant of a metal ion. Owing to this, $O_2^{-\cdot}$ may oxidize and release iron from the enzymatic [4Fe–4S] clusters; this labilized Fe(III) may then bind to other cellular ligands (e.g. DNA) where $O_2^{-\cdot}$ can reduce it to Fe(II) and thus make it Fenton-reactive (Keyer and Imlay, 1996).

Metal compounds interacting with H_2O_2 produce not only free $^{\cdot}OH$ radicals but also other strong oxidants, such as singlet oxygen (Kawanishi *et al.*, 1989a), and metallo-oxo and -peroxo species, all capable of damaging DNA and proteins (Kawanishi *et al.*, 1986; Yamamoto and Kawanishi, 1989, 1991, 1992; Yamamoto *et al.*, 1989, 1993; Rush *et al.*, 1990; Kawanishi and Yamamoto, 1991; Stadtman and Berlett, 1991; Ito *et al.*, 1992; Wink *et al.*, 1994). The formation of metallo-oxo (equation 16.6) and -peroxo (equation 16.7) species is suggested as an alternative to the classical Fenton reaction (reviewed by Wink *et al.*, 1994):

$$M^{n+} + H_2O_2 \rightarrow [MO]^{n+} + H_2O \tag{16.6}$$

$$M^{n+} + H_2O_2 \rightarrow [MO_2]^{(n-2)+} + 2H^+ \tag{16.7}$$

The formation and action of metal-associated oxidants at sites of metal binding to the ligand (protein, DNA) allow for an explanation of the often observed site specificity of metal-mediated oxidative DNA damage (Kawanishi *et al.*, 1986; Yamamoto and Kawanishi, 1989, 1991, 1992; Yamamoto *et al.*, 1989, 1993; Rush *et al.*, 1990; Kawanishi and Yamamoto, 1991; Mouret *et al.*, 1991; Stadtman and Berlett, 1991; Ito *et al.*, 1992; Wink *et al.*, 1994; Bal *et al.*, 1997a,b).

Although it is the metal cation itself that drives electron transfer in the above reactions, the ligand with which the cation is complexed has a profound effect on that transfer. It may either enhance or inhibit the redox reactions. For example, autoxidation of Fe(II) to Fe(III) is enhanced by ethylenediaminetetraacetic acid (EDTA) and nitrilotriacetic acid (NTA) (Hamazaki *et al.*, 1989), but inhibited by *o*-phenanthroline (Nassi-Calo *et al.*, 1989) and deferoxamine (Loeb *et al.*, 1988). NTA enables Fe(III) to react with H_2O_2 and produce $^{\cdot}OH$ (Inoue and Kawanishi, 1987). Deferoxamine inhibits free-radical generation and 2′-deoxyguanosine (dG) hydroxylation mediated by Cr(V) (Shi *et al.*, 1992b). Ni(II) aquo-cation does not react with O_2 and H_2O_2. However, its reactivity towards these oxygen species at physiologic pH range may be facilitated by several bioligands, such as oligopeptides or proteins (reviewed by Kasprzak, 1991, 1995, 1996, 1997a). Likewise, inorganic Co(II) cation, which is stable under air, becomes sensitive to oxidation with atmospheric O_2 in the presence of organic ligands

(Moorhouse *et al.*, 1985). Quite often, the chelator participates in the redox reaction and becomes itself a source of radical intermediates capable of damaging other molecules, including proteins and DNA (Kasprzak and Bare, 1989; Standeven and Wetterhahn, 1991).

Some metal ions can catalyze radical decomposition of hypochlorous acid (HOCl) and organic peroxides produced by phagocytes (Halliwell and Gutteridge, 1984; Sunderman, 1986; Tofigh and Frenkel, 1989; Akman *et al.*, 1992; Prutz, 1996). HOCl yields $^{\cdot}OH$ and Cl^-, while the peroxides (e.g. lipid peroxides and nucleoside hydroperoxides produced by ionizing radiation) are degraded through chain radical reactions with the formation of transient radical species, including $^{\cdot}OH$. Thus, decomposition of 5-hydroperoxy-methyl-2′-deoxyuridine (a relatively stable product of deoxythymidine (dT) oxidation) to 5-hydroxymethyl-2′-deoxyuridine (5-OHMe-dU) and 5-formyl-2′-deoxyuridine proceeds rapidly with Cu(I), Cu(II), Fe(II) and Sn(II); slowly with Co(II) and Ni(II), but not at all in the presence of Fe(III), Mn(II), Mn(III), Al(III), Sn(IV) or Ca(II) (Tofigh and Frenkel, 1989). Owing to those reactions, 5-hydroperoxymethyl-2′-deoxyuridine in DNA appears to be capable of oxidizing neighboring bases (Patel *et al.*, 1992). Benzoyl peroxide, a commercial oxidant widely used for industrial, household and medicinal purposes, was found to decompose and damage DNA in the presence of Cu(I) (Akman *et al.*, 1992). Chelating agents, such as EDTA, diethylenetriaminepentaacetic acid, deferoxamine and proteins may either inhibit or enhance that degradation (Frenkel and Tofigh, 1989; Tofigh and Frenkel, 1989).

Certain metal sulfides display higher carcinogenic activity than other derivatives of the same metal (Lin *et al.*, 1991; Kargacin *et al.*, 1993; Lee *et al.*, 1993). A possible reason for this is the potential of both the metal cation and the sulfide anion (S^{2-}) to undergo oxidation and thus activate O_2 or other oxygen species (Kasprzak and Sunderman, 1977; Lin *et al.*, 1991; Shi *et al.*, 1994). One product of S^{2-} autoxidation, the sulfite anion (SO_3^{2-}), has been demonstrated to damage chromatin by deamination of DNA bases (Singer and Grunberger, 1983) and degradation of proteins (Ito and Kawanishi, 1991). In addition, SO_3^{2-} is able to activate O_2 to free-radical species during autoxidation to the sulfate (SO_4^{2-}) and thus promote oxidation of other molecules. For example, nickel subsulfide (Ni_3S_2) in aerobic solution caused oxidation of 2′-deoxyguanosine (dG) to 8-oxo-2′-deoxyguanosine (8-oxo-dG) (Kasprzak and Hernandez, 1989) and deamination of 5-methyl-2′-deoxycytidine to thymidine (Kasprzak *et al.*, 1991b). Ni_3S_2-rich nickel mattes, to which nickel refinery workers are exposed, were found to catalyze oxidation by O_2 of formate to CO_2^- radical (Costa *et al.*, 1989a). Oxygen-activating capacity was observed for particles of various sulfides (Costa *et al.*, 1989a,b,c; Zalma *et al.*, 1989) and other classes of insoluble metal compounds, including asbestos (Costa *et al.*, 1989a,b,c; Zalma *et al.*, 1989; Vallyathan *et al.*, 1992; Berger *et al.*, 1993; Fung *et al.*, 1998).

Because of the high carcinogenic potency of nickel and chromium, the redox biochemistry of these two metals has been discussed in detail in several publications (Lancaster, 1988; Wetterhahn *et al.*, 1989; Kasprzak, 1991, 1995, 1996, 1997a; Klein *et al.*, 1991; Magos, 1991; Standeven and Wetterhahn, 1991; Snow, 1992; Kortenkamp *et al.*, 1996, 1997). Briefly, biological oxidative effects of nickel depend on its ability to form an electron transfer couple, Ni(III)/Ni(II) (compare equations 16.1 and 16.2), but only when complexed by certain oligopeptides (Margerum and Anliker, 1988; Cotelle *et al.*, 1992). An important result of the reactions of certain Ni(II) complexes with O_2 or H_2O_2 is the production of not only $^{\cdot}OH$ (or an oxo-cation NiO^{2+}), but also other oxygen-, carbon- and, perhaps, sulfur-centered radicals originating from the ligands. All these species may damage neighboring molecules (Bossu *et al.*, 1978; Kasprzak and Bare, 1989; Torreilles and Guerin, 1990). The redox biochemistry of chromium is richer and involves several oxidation states: Cr(VI), Cr(V), Cr(IV), Cr(III) and, possibly, Cr(II). Chromium is more toxic and carcinogenic when administered as chromate or dichromate anion [Cr(VI)] than as the Cr(III) cation, since Cr(VI) is much more easily taken up by cells. The intracellular Cr(VI) is reduced by glutathione (GSH) reductase, carbohydrates, ascorbic acid, GSH and other molecules to more reactive products, including Cr(V), Cr(IV) and Cr(III). The by-products of the Cr(VI) reduction processes, singlet oxygen and oxygen- and sulfur-centered radicals (e.g. lipid hydroperoxide-derived, or GSH-thiyl radicals, and $^{\cdot}OH$), can damage DNA and other molecules. Hence, the ability of Cr(VI) to damage cellular components depends on its gradual reduction and oxygen activation (Goodgame and Joy, 1986; Wetterhahn *et al.*, 1989; Shi and Dalal, 1989, 1990a,b; Sugiyama *et al.*, 1991a,b; Shi *et al.*, 1993c; Kortenkamp *et al.*, 1996, 1997) and the 'ultimate' pathogenic species in acute toxicity and carcinogenesis by chromium may be $^{\cdot}OH$ and other radicals (Shi and Dalal, 1989, 1990a; Shi *et al.*, 1993c).

Growing evidence suggests that oxidative DNA damage may be involved also in cobalt-induced toxicity and carcinogenesis (Nackerdien *et al.*, 1991; Beyersmann and Hartwig, 1992; Shi *et al.*, 1993a; Kasprzak *et al.*, 1994b). That evidence is strongly supported by the redox biochemistry of cobalt (Kadiiska *et al.*, 1989; Hanna *et al.*, 1992; Shi *et al.*, 1993a). This metal, in the form of Co(II) complexes with some ligands (e.g. GSH and L-cysteine (Cys)), has been found to generate $^{\cdot}OH$ and other oxygen- and carbon-centered radicals from model lipid peroxides at physiologic pH (Shi *et al.*, 1993a). Likewise, NADH, GSH and anserine render Co(II) reactive with H_2O_2 to produce $\cdot OH$ (Kadiiska *et al.*, 1989; Shi *et al.*, 1993a).

Experimental data also indicate that nonredox active carcinogenic metals, cadmium and lead, promote oxidative damage (Frenkel *et al.*, 1994; Hartwig, 1994, 1995; Hassoun and Stohs, 1996; Hiraku and Kawanishi, 1996; Yang *et al.*, 1996, 1997; Mikhailova *et al.*, 1997; Saplakoglu *et al.*, 1997; Sieg and Billings, 1997). Such metals are likely to inflict the damage indirectly, for example, by triggering inflammation and lipid peroxidation, interfering with calcium, copper and iron metabolism, or inhibiting antioxidant defenses, as discussed below.

16.2.2 *Effect on cellular active oxygen-handling systems*

In addition to the direct generation of oxygen radicals, toxic metals may also sustain oxidative damage indirectly through enhancement of physiologic oxygen activation systems at various molecular and cellular levels, including stimulation of inflammatory cells and inactivation of cellular defenses against endogenous (metabolic) oxidants (reviewed by Kasprzak, 1991, 1995, 1996, 1997a).

16.2.2.1 *Inflammation*

Various metal derivatives administered parenterally trigger an inflammatory response which results in infiltration of the affected tissue by neutrophils and phagocytes with all their excretory oxidative weapons, including hydrogen peroxide and hypochlorous acid (Babior, 1978; Fleet *et al.*, 1990; Landino *et al.*, 1996). Chronic inflammation has been long associated with carcinogenesis (reviewed by Weitzman and Gordon, 1990). Activated neutrophils are known to damage the base residues in isolated DNA (Frenkel, 1989; Jackson *et al.*, 1989; Reid and Loeb, 1992). Although it seems unlikely that phagocyte-derived active oxygen products can diffuse far enough to attack DNA in neighboring cells, experimental evidence suggests that at least one such product does, in fact, achieve that goal. Frenkel and Chrzan (1987) incubated HeLa cells with stimulated polymorphonuclear leukocytes and found increased amounts of 5-OHMe-dU (a dT oxidation product) in DNA of the cells. Formation of this derivative could be prevented with catalase, indicating that H_2O_2 was the diffusible agent (or at least its precursor) responsible for dT oxidation catalyzed, perhaps, by DNA-bound iron. Phagocytes are also known to cause bacterial cell mutations. Thus, His-requiring mutants of *Salmonella typhimurium* TA100 were reverted into His-independence following incubation with human blood leukocytes. Heat-killed leukocytes or leukocytes from a patient with chronic granulomatous disease ($O_2^{-\cdot}$ production defect) were mutationally inactive (Weitzman and Stossel, 1981).

16.2.2.2 Inactivation of antioxidants

Metal ions are capable of reacting with a variety of cellular components serving as active oxygen scavengers. Thus, Ni(II) was found to transiently inhibit catalase and GSH peroxidase activity in rats *in vitro* (Misra *et al.*, 1990; Rodriguez *et al.*, 1990; Rodriguez and Kasprzak, 1992). Superoxide dismutase appeared to be less sensitive to Ni(II) inhibition (Misra *et al.*, 1990). However, the limited extent and great variability of the effects of Ni(II) on these and other antioxidant enzymes in different species, strains and tissues do not allow us to fully assess possible significance of the inhibition to oxidative damage caused by this metal. Unlike Ni(II), Cd(II) appears to be a potent inhibitor of superoxide dismutase, thus promoting metabolic $O_2^{-\cdot}$ build-up in affected tissues (Amoruso *et al.*, 1982; Ochi *et al.*, 1983; Shukla *et al.*, 1987). Moreover, published evidence indicates that Cd(II) may enhance production of $O_2^{-\cdot}$ and nitric oxide, which may combine into the powerful oxidant peroxynitrite (Amoruso *et al.*, 1982; Ochi *et al.*, 1983; Hassoun and Stohs, 1996). Cantoni *et al.* (1984) found that the addition of superoxide dismutase or catalase to cells treated with Hg(II) chloride markedly reduced the extent of DNA single-strand breaks without decreasing Hg(II) uptake. This would signify increased accumulation of intracellular $O_2^{-\cdot}$ and H_2O_2, possibly through inhibition of these two enzymes by Hg(II).

The metals may also enhance cellular levels of metabolic active oxygen species through stimulation or mimicking of certain oxygen-activating enzymes, or generation of specific redox-active metabolites. For example, Ni(II) was found to augment the activity of myeloperoxidase (Rodriguez and Kasprzak, 1989), whereas Cu(II) was found to mimic that enzyme (Frenkel *et al.*, 1986). Under certain conditions, the copper center in Cu/Zn superoxide dismutase may catalyze the formation of $^{\cdot}OH$ from H_2O_2 via a Fenton-like reaction and the nitration of aromatic amino acids by peroxynitrite (Ischiropoulos *et al.*, 1992). Oxidative DNA-damaging activity was also found for Mn-superoxide dismutase (Dowjat *et al.*, 1996). Arsenite [As(III)] has the potential to modulate oxidative damage by a mechanism involving xanthine oxidase (Stewart *et al.*,

1985). Pb(II), in turn, provides an example of a metal that may induce oxidative cell damage through a product of its toxicity. This metal causes accumulation of δ-aminolevulinic acid (a heme precursor) which undergoes autoxidation accompanied by formation of active species, such as H_2O_2, $O_2^{-\cdot}$, and eventually •OH (with participation of cellular iron or copper) (Hermes-Lima *et al.*, 1991; Hiraku and Kawanishi, 1996).

Some metal cations reacting with cellular antioxidants, such as ascorbate, Cys, His, GSH and others, can exhaust their pools or modify their reactions with oxygen species to produce free-radicals (Kasprzak and Hernandez, 1989; Datta *et al.*, 1992; Kasprzak *et al.*, 1992b; Shi *et al.*, 1992a,1993b). Thus arsenate [As(V)] can oxidize GSH to glutathione disulfide (GSSG) and deplete cells from the antioxidant form of glutathione (Carter, 1995). Chromate [Cr(VI)] does the same and generates also the damaging thiyl radical from GSH (Carter, 1995). Ni(II)–His complexes enhance *in vitro* oxidation of the guanine (G) moiety in free dG and in DNA (Datta *et al.*, 1992) and *in vivo* oxidation of DNA bases in the rat kidney (Misra *et al.*, 1993). However, the presumed histidyl oligopeptide antioxidants, carnosine, homocarnosine and anserine, increase Ni(II)-mediated *in vitro* oxidation by H_2O_2 only of free dG, but not of DNA- or nucleohistone-bound dG (Datta *et al.*, 1993). Similar enhancement by His of dG oxidation with H_2O_2 has been observed for Fe(III), Cr(III) and Cu(II), but not Co(II). In the case of DNA-bound dG, however, the enhancement was noticed only for Cu(II), while the effect of other metals remained unchanged [Cr(III), Co(II)], or was inhibited [Fe(III)] by His (Kasprzak, 1997a and unpublished results).

16.2.3 *Inhibition of DNA repair*

While oxidatively altered proteins and lipids are usually marked for prompt enzymatic degradation, oxidative DNA damage can be repaired (or prevented) in living cells by various enzymatic mechanisms (Kasai *et al.*, 1986; Chung *et al.*, 1991a,b; Demple and Amabile-Cuevas, 1991; Vos and Wauthier, 1991; Beyersmann and Hartwig, 1992; Klein *et al.*, 1992; Yamamoto *et al.*, 1992; Bessho *et al.*, 1993; Grollman and Moriya, 1993; Hartwig, 1995), the efficiency of which may differ among cells and tissues (Kasprzak *et al.*, 1997). The cellular protection against the 8-oxo-dG lesion includes enzymes capable of removing 8-oxo-2′-deoxyguanosine triphosphate (8-oxo-dGTP) from the deoxynucleotide pool and 8-oxo-dG from the DNA molecule. Data on these enzymes have been reviewed previously (Grollman and Moriya, 1993; Demple and Harrison, 1994; Sancar, 1995). The first enzyme, 8-oxo-dGTP pyrophosphatase (8-oxo-dGTPase), called MutT in *E. coli* or MTH1 in humans, can be inhibited *in vitro* by Cu(II), Cd(II) and to a lesser degree by Ni(II) and Co(II) (Porter *et al.*, 1996, 1997). The inhibitory effect of Cd(II) in this enzyme's activity has been confirmed recently in cultured cells (Bialkowski and Kasprzak, 1998). The second enzyme, the Fpg glycosylase, is a zinc-finger protein in which Zn(II) may be replaced by Cu(II), Cd(II) or Hg(II), but not Mn(II), Fe(II), Co(II) or Ni(II) (O'Connor *et al.*, 1993). The replacement with Cu(II) may increase sensitivity of Fpg protein to oxidative damage, but this, as well as the effects of Zn(II) replacement by the other metals on Fpg enzymatic function, awaits further studies. Wink and Laval (1994) have shown that nitric oxide inactivates Fpg protein through oxidation of its zinc-finger cysteines. We may expect, therefore, that metals may enhance this inhibitory effect by inciting inflammatory cell responses.

Several transition metal cations are known to inhibit certain steps in the DNA excision repair process. It is not surprising, therefore, that the repair of damaged DNA in some metal-exposed cells is found to be impaired. Thus, Ni(II) and Co(II) inhibited repair of DNA strand breaks in cultured mammalian cells after UV or X-irradiation, apparently by affecting the polymerization or ligation steps (Hartwig and Beyersmann, 1989; Hartwig *et al.*, 1991, 1992, 1994; Lee-Chen *et al.*, 1993). In the case of Ni(II), the DNA repair inhibition could be prevented by Mg(II) (Hartwig *et al.*, 1994), a known antagonist of nickel toxicity and carcinogenesis (reviewed by Kasprzak, 1997b). Co(II) was found to suppress repair of the UV-generated pyrimidine dimer lesion by blocking its excision step (Hartwig *et al.*, 1991). There are indications that the repair of oxidatively damaged DNA may also be obstructed by Pb(II), Cr(VI), Cd(II) and As(III) (Hartwig and Beyersmann, 1989; Lee-Chen *et al.*, 1992; Hartwig, 1994, 1995).

16.3 Lipid peroxidation

Metal ions promote lipid peroxidation in living cells (Aust and Svingen, 1982; Sunderman, 1986; Athar *et al.*, 1987; Andersen and Andersen, 1989; Knight and Voorhees, 1990; Sole *et al.*, 1990; Buettner, 1993; Dix and Aikens, 1993). The lipid peroxidation process constitutes a chain reaction in which polyunsaturated fatty acids of cell membranes are oxidized through C- and O-centered radicals and hydroperoxy intermediates to yield various products, including epoxy-fatty acids, alkanes, alkenes and aldehydes (e.g. malondialdehyde) (Sunderman, 1986). Metals may sustain lipid peroxidation through generation of other radical species ($O_2^{-\cdot}$, $^{\cdot}OH$) that initiate the chain reaction, and then further propagate the reaction through reduction of the relatively stable lipid hydroperoxide intermediates (Buettner, 1993; Halliwell, 1995). Enhancement of lipid peroxidation *in vitro* and *in vivo* was observed for Cd(II), Co(II), Cu(II), Hg(II), Ni(II), Pb(II), Sn(II) and V(V) (Sunderman, 1986; Byczkowski *et al.*, 1988; Sole *et al.*, 1990; Sarkar *et al.*, 1995; Xie *et al.*, 1995), as well as for Fe(III) and Fe(II) compounds (Aust and Svingen, 1982; Winkler *et al.*, 1984; Okada *et al.*, 1987; El Din *et al.*, 1988; Tillian *et al.*, 1989; Umemura *et al.*, 1990a; Hartwig *et al.*, 1993; Minotti, 1993; Glass and Stark, 1997). Lipid peroxidation products were found to cause the formation of 8-oxo-dG (Park and Floyd, 1992), strand breaks (Hartwig *et al.*, 1993) and possibly larger adducts in DNA (Brambilla *et al.*, 1989; Vaca *et al.*, 1988).

Another indication of a possible causal relationship between peroxidation and DNA damage is that metal-induced lipid peroxidation and the concurrent chromosomal aberrations *in vivo* and *in vitro* could be inhibited by natural antioxidants (e.g. GSH) (Sunderman, 1986; Srivastava *et al.*, 1990; Umemura *et al.*, 1991). However, vitamin E, a well-established antioxidant, could actually enhance lipid peroxidation caused by nickel (Husain *et al.*, 1995), but still inhibit chromosomal aberrations caused by this metal (Lin *et al.*, 1991). Lipid peroxidation by Ni(II) in mice was found to be high in strains that are low in GSH and GSH peroxidase (e.g., BALB/c) compared to other strains (e.g. B6C3F1 or C3H) (Misra *et al.*, 1991; Rodriguez *et al.*, 1991). Metallothionein, the highly inducible metal-binding protein and free-radical scavenger, may or may not protect cells against lipid peroxidation driven by copper and iron, depending on the concentrations of other natural ligands and antioxidants, such as ceruloplasmin, GSH, ascorbate and others (Sato and Bremner, 1993; Stephenson *et al.*, 1994; Halliwell, 1995; Oikawa *et al.*, 1995; Glass and Stark, 1997). The common antioxidant ascorbic acid can actually assist in metal-driven lipid peroxidation by reducing the metal cations to lower, more reactive valences (Buettner, 1993; Halliwell, 1995; Glass and Stark, 1997; Nappi and Vass, 1997). Cysteine, homocysteine and glutathione enhanced Fe(III)-mediated, but inhibited Cu(II)-mediated lipid peroxidation (Lynch and Frei, 1997). Heparin, in turn, was found to prevent Fe(II)- and Cu(II)- dependent lipid peroxidation by H_2O_2 (Albertini *et al.*, 1996).

16.4 Protein oxidation

The oxidation of amino acids and proteins by oxygen radicals and the roles of toxic metals in this process have been reviewed by several authors (Pacifici and Davies, 1990; Stadtman,1993; Vilar-Rojas *et al.*, 1996; Stadtman and Berlett, 1997). Along with the other types of oxidative damage, protein oxidation is believed to be mechanistically involved in aging, the development of ischemia/reperfusion injury, Alzheimer's disease, Parkinson's disease, atherosclerosis, liver cirrhosis, cancer and other pathologic effects (Stadtman, 1993; Crawford, 1995; Kienzl *et al.*, 1995; Schubert and Chevion, 1995; Vile and Tyrrell, 1995; Arora and Gores, 1996; Darley-Usmar and Halliwell, 1996; Sokol, 1996; Butterfield, 1997; Leeuwenburgh *et al.*, 1997; Smith *et al.*, 1997). Metal-catalyzed oxidation may also adversely affect the quality of some commercially produced proteins (Zhao *et al.*, 1997).

Free amino acids are oxidized in aqueous solutions by H_2O_2 in the presence of Fe(II), Cu(II) and other Fenton-reactive metal cations, predominantly to α-ketoacids, oximes and carboxylic acids having one less carbon atom, with the release of ammonia and carbon dioxide (Stadtman, 1993). Thus, deamination and decarboxylation are the major effects. In addition, the sulfhydryl, aromatic and heterocyclic groups of amino acids are oxidized to more specific derivatives which are mentioned below. Redox-active metals may also promote oxidative modification of the amino-acid residues in proteins and peptides. In this case, the major targets are Cys, His, Arg, Lys and Pro residues. These differ from the targets attacked by ionizing radiation-generated oxygen radicals. The sulfhydryls are commonly oxidized to disulfides, but they may be turned into sulfino-, sulfeno- and solfono-derivatives as well; the latter is also true for Met (Li *et al.*, 1995a; Bal *et al.*, 1996). Histidine residues may be converted into aspartic acid, asparagine, or 2′-OH-His (Stadtman, 1993; Lewisch and Levine, 1995; Bal *et al.*, 1997b). Arg is converted to glutamic-semialdehyde; Lys is oxidized to 2-amino-adipic-semialdehyde, and Pro is turned into glutamic acid, pyroglutamic acid, aminobutyric acid, and glutamic-semialdehyde (Stadtman, 1993). The formation and rearrangements of radical intermediates arising in the oxidation process of protein molecules also result in protein fragmentation and intra- and interprotein cross-linking (aggregation, denaturation) (Pacifici and Davies, 1990; Kito *et al.*, 1995; Li *et al.*, 1995b; Gill *et al.*, 1997). For the same reason, proteins may become cross-linked with DNA (Zhuang *et al.*, 1994; Olin *et al.*, 1996). It seems obvious that the variety of oxidative damage may result in a wide spectrum of more or less pathogenic events, ranging from functional impairment to quick degradation of the affected protein by cellular proteases (Climent *et al.*, 1989; Pacifici and Davies, 1990; Stadtman, 1993).

The site-specificity of metal-mediated oxidative damage depends on the binding of metal ions by proteins and peptides predominantly through the sulfhydryl, imidazole and the deprotonated peptide-bond and side-chain nitrogens, and generation of metal-associated oxidizing species at these particular sites. Therefore, Trp, Tyr, Phe and Met, which are also sensitive to free oxygen radicals (e.g. those generated by ionizing radiation), but do not coordinate metal cations under physiological conditions, are less likely to be targeted by metal-catalyzed oxidation (Stadtman, 1993; Vilar-Rojas *et al.*, 1996). They may, none the less, be damaged if they are located close to the metal-binding site, either in the amino-acid backbone, or due to conformation of the protein molecule. For example, in a 15-mer peptide modeling the N-terminal sequence of human protamine P2, Ni(II) and Cu(II) catalyzed oxidation by H_2O_2 of not only Arg1 and His3, which coordinate the metals, but also (and very strongly) of Tyr8, located away from the metal-binding site, but very likely brought close to it due to conformational effect of the bound metal on peptide structure (Bal *et al.*, 1997a). Owing to different affinities of various metal cations for different side-chain donors in peptides and proteins, the extent and character of the damage may strongly depend on the metal. This, for example, was observed for Fe(II) and Cu(II) catalysis of collagen oxidation by H_2O_2 (Hawkins and Davies, 1997). Because of the site-specific generation of oxidizing species, the metal-catalyzed oxidative damage of proteins, unlike the damage caused by diffusible oxygen radicals, is resistant to radical scavengers, but may be quenched by certain strong metal chelators. On the other hand, quite importantly, oxidative modification of free amino acids and amino-acid residues in proteins in the presence of Fe(II) and Mn(II) was found to be stimulated at alkaline pH by the ubiquitous bicarbonate anion (Stadtman, 1993).

In addition to mediation of oxidative damage to protein ligands, protein-bound metals may promote oxidation of neighboring macromolecules. Relevant examples would include catalysis of lipoprotein oxidation by ceruloplasmin-bound copper (Mukhopadhyay *et al.*, 1997), DNA damage by copper/zinc superoxide dismutase (Dowjat *et al.*, 1996) and by nickel bound to protamine- and histone-modeling peptides, isolated chromatin or nucleohistone (Kasprzak and Bare, 1989; Nackerdien *et al.*, 1992; Bal *et al.*, 1996, 1997b; Liang *et al.*, 1998). Redox-active metals may also assist in propagation of oxidative damage caused in proteins by ionizing radiation or endogenous oxidants, for example by degradation of protein hydroperoxides to various reactive radical species capable of damaging other biomolecules (Fu *et al.*, 1995).

16.5 Oxidative DNA damage

The DNA molecule, having an abundance of phosphate anions and nitrogen and oxygen donor groups, is an ideal binding partner for metal cations (Kazakov, 1996). The nuclear proteins can bind metals as well (Bal *et al.*, 1995, 1996, 1997a, 1998). This may explain why, following *in vivo* exposure, metals have been found in cell nuclei (Bryan, 1981; Kasprzak and Poirier, 1985). It is also possible that endogenous Fenton/Haber–Weiss-reactive metals, iron and copper, released from their natural carriers under certain pathologic conditions (e.g. by $O_2^{-\cdot}$ in inflammation) may become associated with chromatin (Keyer and Imlay, 1996). Iron is considered to be a natural component of cell nuclei (Meneghini, 1997). The presence of $O_2^{-\cdot}$ and H_2O_2, the major substrates for oxygen activation to $^{\cdot}OH$ and metal-associated oxidants, was also detected in the cell nucleus (Peskin and Shlyahova, 1986; Szatrowski and Nathan, 1991). The pool of substrates is further enriched by lipid peroxides arising in cells in response to metal-induced stress. Hence, the reactions leading to oxygen activation can proceed *in vivo* around and inside the cell nucleus (Kadiiska *et al.*, 1992; Huang *et al.*, 1993).

The major oxidative DNA lesions associated with administration of toxic metals, such as nickel, chromium, and cobalt, to experimental animals and cultured cells, include strand scission, depurination, cross-linking and base modifications (Hamilton-Koch *et al.*, 1986; Oleinick *et al.*, 1987; Wetterhahn *et al.*, 1989; Kasprzak *et al.*, 1992a, 1994b; Misra *et al.*, 1993). Besides nuclear DNA, an important target for metals is also mitochondrial DNA (Rossi *et al.*, 1988; Rossi and Wetterhahn, 1989). The techniques of detection and quantification of DNA damage have been described by several authors (Berger *et al.*, 1990; Dizdaroglu, 1991; Frenkel *et al.*, 1991; Halliwell and Dizdaroglu, 1992; Cadet and Weinfeld, 1993; Randerath *et al.*, 1996).

16.5.1 DNA base damage

The spectrum of chemical changes produced by $^{\cdot}OH$, either diffusible or associated with a metal ion, in the base moiety of DNA, is well established (von Sonntag, 1987; Dizdaroglu, 1991, 1992; Halliwell and Aruoma, 1991; Henle *et al.*, 1996; Luo *et al.*, 1996). The most commonly assayed base products are presented in Fig. 16.1 (Dizdaroglu, 1991). Many bases from that spectrum have been found in isolated chromatin or DNA exposed to H_2O_2 plus Ni(II), Co(II) (Nackerdien *et al.*, 1991), Cu(II) or Fe(III) (Dizdaroglu *et al.*, 1991). Ni(II) generated such bases even when H_2O_2 was not added, an indication that Ni(II) bound to chromatin components (but not to pure DNA) was able to activate ambient O_2 (Nackerdien *et al.*, 1991). Recently, Ni(II)- and Cu(II)-binding amino acid motifs were found in core histones H3 and H2A (Bal *et al.*, 1995, 1998), and also in protamine P2 (Bal *et al.*, 1997a), some of them having the potential to promote DNA base oxidation and other forms of oxidative damage (Bal *et al.*, 1996, 1997b; Liang *et al.*, 1998). Base damage in isolated DNA by active oxygen arising from the reactions of Cr(VI) with H_2O_2 (Wetterhahn *et al.*, 1989), GSH reductase (Shi and Dalal, 1989) and GSH (Kortenkamp *et al.*, 1996) was also observed. In the latter case, the presence of molecular oxygen was essential. Elevated levels of base damage have been found also in DNA exposed to Fe(II) and Fe(III) in the presence of H_2O_2, $O_2^{-\cdot}$, or stimulated neutrophils; EDTA and NTA enhanced the effects (Kasai and Nishimura, 1984b; Inoue and Kawanishi, 1987; Aruoma *et al.*, 1989; Jackson *et al.*, 1989). Several brands of iron-rich asbestos mediated oxidative DNA base damage proportionally to their iron contents (Kasai and Nishimura, 1984a). Interestingly, various copper, iron and chromium ions, expected to have different affinities for coordination by particular DNA bases (Kazakov, 1996) and thus catalyze the damage in metal-characteristic sites, nevertheless produced very similar patterns of DNA base oxidation with H_2O_2 plus ascorbate. Moreover, those patterns were remarkably similar to patterns of DNA base damage observed in cells exposed to H_2O_2 alone (with an unidentified endogenous transition metal, or metal-like ligand, catalyzing the reaction), indicating that metal-dependent H_2O_2-generated DNA damage is more sequence-specific than metal-specific (Rodriguez *et al.*, 1997).

Increased amounts of at least one of the many detectable damaged DNA bases were found in organs of Fischer rats (Kasprzak *et al.*, 1990, 1992a, 1997; Misra *et al.*, 1993) and BALB/c mice (Kasprzak *et al.*, 1991a) following *in vivo* exposure to Ni(II). Moreover, the distribution and persistence of damaged bases in the rat kidney, which is a target organ for nickel carcinogenesis, differed significantly from that in liver, a nontarget organ (Kasprzak *et al.*, 1990, 1997). In Ni(II)-treated mice, renal 8-oxo-dG

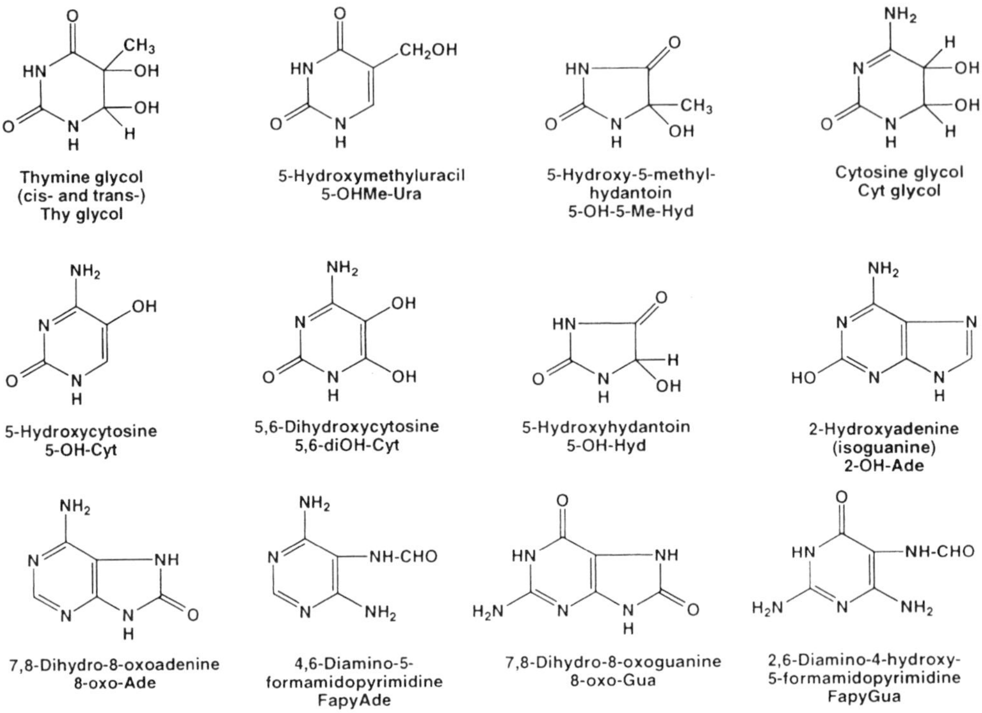

Figure 16.1 Structures of free-radical-induced products of pyrimidines and purines in DNA, which were identified by the use of gas chromatography/mass spectrometry (according to Dizdaroglu, 1991).

levels were increased only in the BALB/c strain, which had low GSH and GSH peroxidase levels compared to two other strains investigated, the B6C3F1 and C3H (Kasprzak *et al.*, 1991a; Misra *et al.*, 1991; Rodriguez *et al.*, 1991). Oxidative DNA base damage was also observed in organs of rats injected with Co(II) (Kasprzak *et al.*, 1994b) and Fe(III) complexed with NTA (Umemura *et al.*, 1990a,b, 1991; Zhang *et al.*, 1997), and in cultured cells exposed to Cd(II) (Mikhailova *et al.*, 1997) and Fe(III)-NTA (but not Fe(III) citrate) (Hartwig and Schlepegrell, 1995).

16.5.2 Cross-linking

The most common nonlethal effect of toxic metals on nuclear chromatin observed *in vitro* and *in vivo* is DNA–protein cross-linking (Wedrychowski *et al.*, 1986a; Borges and Wetterhahn, 1989; Wetterhahn *et al.*, 1989; Chang *et al.*, 1993). Generally, metals can generate DNA–protein cross-links in two ways: polyvalent metal cations may simply bring DNA and a protein together in a mixed-ligand complex, or they can induce the formation of strong covalent bonds between DNA and protein. Cross-links of the first type are relatively weak and dissociate when the ionic strength of the solution is increased (Wedrychowski *et al.*, 1986b; Patierno *et al.*, 1987) or metal chelators are added (Wedrychowski *et al.*, 1986a; Borges and Wetterhahn, 1989). Cross-links of the second type do not break under such conditions. They are typical for free-radical attack on chromatin (Lesko *et al.*, 1982; Oleinick *et al.*, 1987; Dizdaroglu, 1991; Misra *et al.*, 1993; Mattagajasingh and Misra, 1996). A good example of this type of very strong cross-linking is that produced in isolated nuclear chromatin by Fe(II)EDTA, but only in the presence of H_2O_2; the latter alone does not cause cross-link formation (Lesko *et al.*, 1982). Another example comes from an *in vivo* experiment in which DNA–protein cross-linking induced by Ni(II) was greatly enhanced by His, known to promote oxidation-mediating capability of Ni(II) and some other toxic metals (Datta *et al.*, 1992; Kasprzak, 1997a). Likewise, tetraglycine that renders Ni(II) redox-active, facilitated DNA–protein cross-linking by Ni(II) in isolated calf thymus nucleohistone and protein–protein cross-linking among histones (Kasprzak and Bare, 1989).

As mentioned earlier, carcinogenic Cr(VI) reacts with DNA only after (or in the course of) metabolic reduction of the metal to lower oxidation states. Oxygen- and sulfur-centered radicals derived from the reductants may attack DNA and become cross-linked to it. The resulting adducts with GSH (Goodgame and Joy, 1986; Wetterhahn *et al.*, 1989), Cys (Borges and Wetterhahn, 1989) and proteins of different molecular weights (Borges and Wetterhahn, 1989; Miller and Costa, 1989) were identified. Increased DNA–protein cross-linking in lymphocytes has been proposed as a potential biomarker of genotoxic effects of chemicals, including several metals. Potassium permanganate, mercury chloride, lead nitrate, cadmium chloride, aluminum chloride and magnesium chloride did not produce DNA–protein cross-links, while copper sulfate, potassium chromate, *cis*-platinum and arsenic trioxide induced the cross-links (Costa *et al.*, 1993, 1996; Kuykendall *et al.*, 1996). Unfortunately, little is known about the influence of specific metals on the type and extent of the covalent cross-links and the exact chemistry of their formation (Dizdaroglu, 1992; Salnikow *et al.*, 1992; Zhitkovich and Costa, 1992; Costa *et al.*, 1993; Lloyd *et al.*, 1997). Cross-links between thymine and tyrosine were identified in isolated chromatin treated with H_2O_2 in the presence of Fe(III) or Cu(II) (Dizdaroglu, 1992). DNA–protein cross-links produced in calf thymus nucleohistone by $^{\cdot}OH$ generated by ionizing radiation involved formation of covalent bonds mainly between the pyrimidine bases and glycine, alanine, valine, leucine, isoleucine, threonine, lysine and tyrosine (Dizdaroglu, 1991).

Intra- and interstrand cross-linking between neighboring DNA nucleotides is also possible, but the effect of metals on its formation has not been fully explored. Randerath *et al.* (1996) observed formation of intrastrand cross-link between two adjacent nucleotides in a series of single-stranded oligodeoxyribonucleotides incubated with H_2O_2 and Fe(II) or Ni(II). Formation of this type of cross-link between two cytosines or two thymines in the same strand is characteristic of DNA damage produced by free-radicals generated by γ- and UV-radiation (Dizdaroglu and Simic, 1984a,b). Such cross-links are believed to result in tandem double CC $\rightarrow$ TT mutations that result from DNA exposure to radiation or to iron, nickel and copper plus oxidant (Dizdaroglu and Simic, 1984a,b; Tkeshelashvili *et al.*, 1993). The formation of interstrand cross-links between guanines in DNA of cultured fibroblasts treated with Cr(VI) was found to be the most likely cause of a block in DNA replication (Xu *et al.*, 1996).

The formation of cross-links in chromatin may manifest itself in morphologic aberrations of chromosomes. Such aberrations were observed, for example, in lymphocytes of workers exposed to nickel and chromium compounds (Deng *et al.*, 1988; Sunderman, 1989; Costa *et al.*, 1993). In cultured Chinese hamster ovary cells, chromosomal alterations caused by Ni(II) were predominantly localized in the protein-rich, heterochromatic region of the X chromosome (Sen and Costa, 1985). This is consistent with more recent findings that heterochromatic proteins specifically enhance Ni(II)-induced oxidative damage (Huang *et al.*, 1995).

16.5.3 *Strand scission*

DNA single-strand scission was found in kidneys and lungs, but not in livers of rats after parenteral administration of Cd(II) and Ni(II) chlorides (Saplakoglu *et al.*, 1997). Likewise, cells cultured with Cd(II) (Hassoun and Stohs, 1996; Mikhailova *et al.*, 1997) or Fe(III)-NTA, but not Fe(III) citrate (Hartwig and Schlepegrell, 1995), had their nuclear DNA fragmented in the metal dose-dependent manner. *In vitro*, single-strand breaks were produced in DNA by Cr(VI) in the presence of H_2O_2 or GSH plus ambient oxygen (Aiyar *et al.*, 1991; Kortenkamp *et al.*, 1996), and by Cu(II) and Ni(II) plus H_2O_2. The latter effect was enhanced for Ni(II), but abolished for Cu(II), by an oligopeptide modeling the N-terminal metal-binding amino-acid sequence of protamine P2 (Bal *et al.*, 1997b; Liang *et al.*, 1998). Kawanishi *et al.* (1989b) have found that Ni(II) facilitates *in vitro* DNA cleavage by H_2O_2 in a site-specific way characteristic for action of an active nickel–oxygen complex rather than ˙OH or singlet oxygen. The most sensitive sites appear to be at the C, T and G residues, but rarely at A residues. This pattern is surprising, especially in view of the 'hole migration' phenomenon (Candeias and Steenken, 1993; Melvin *et al.*, 1995) and the stronger interaction of Ni(II) with the purines (mainly guanine) than with the pyrimidines (Datta *et al.*, 1991). The results of Kawanishi *et al.* (1989b) might indicate that interaction of Ni(II) with the DNA molecule as a whole does not follow affinity rules established for separate nucleotides and nucleosides, a phenomenon observed also for iron (Luo *et al.*, 1996). Perhaps the difference results from the participation of other ligands, such as proteins, which may form ternary (mixed ligand) complexes with Ni(II) and DNA. This would explain the site specificity of DNA–protein cross-linking observed by Dizdaroglu (1992) and Salnikow *et al.* (1992). DNA cleavage mediation by Ni(II) complexes and ligand effects on the selectivity of DNA oxidation with various oxidants, including O_2 and H_2O_2, were studied in detail by Mack and Dervan (1992), and Burrows, Rokita and co-workers (Li and Rokita, 1991; Muller *et al.*, 1992; Cheng *et al.*, 1993; Wey *et al.*, 1993). Similar effects of Fe(III)-substituted DNA-binding zinc-finger proteins and Cu(II)- and Ni(II)-binding peptides have been investigated by Sarkar *et al.* (reviewed by Sarkar, 1997).

16.5.4 *Depurination*

A more general DNA-damaging effect that results from the exposure to toxic and carcinogenic metals is depurination. For example, Cr(VI) and Cu(II) were found to release guanine (G) and adenine (A), respectively, and Ni(II) appeared to release A (the product was not definitely identified) from the DNA molecule. The authors, Schaaper *et al.* (1987), speculated that the mechanism of depurination involves oxygen free-radicals. Depurination occurs concurrently with DNA strand scission and both effects can result from ˙OH attack on the DNA sugar moiety (Dizdaroglu, 1991). Modified sugars that remain in the DNA backbone constitute alkali-labile sites that are frequently observed in DNA from metal-treated cells (Hamilton-Koch *et al.*, 1986; Dizdaroglu,

1991; Sugiyama *et al.*, 1991a,b). Thus, the presence of such sites may also indicate possible depurination. A great enhancement by Fe(II) and Ni(II) of *in vitro* depurination was observed in dG treated with an oxidizing mixture of H_2O_2 plus ascorbate (Kasprzak and Hernandez, 1989; Littlefield *et al.*, 1991).

16.6 Metal effects on redox cell signaling

Reactive oxygen species may serve as physiologic signal transduction messengers. The literature on this subject has been reviewed by Colburn (1992) and Sen and Packer (1996). The original messenger molecules were the superoxide anion and hydrogen peroxide, with NADPH oxidase and superoxide dismutase serving as their physiologic generators. More recently, the free-radical nitric oxide, NO, generated by nitric oxide synthases (NOS and iNOS), has been added to the list (Felley-Bosco *et al.*, 1998). The evidence supporting the signaling functions of radical species includes redox activation/inactivation of transcription factors such as nuclear factor κB (NF-κB) and activating protein-1 (AP-1) (exemplified in Table 16.1) which, among others, control genes involved in the immune, inflammatory and acute-phase responses, as well as oncogene and tumor suppressor gene products, such as $p21^{ras}$ or p53, and many other proteins. Generally speaking, the cysteine-rich motifs of metal-binding proteins and the metal clusters of metalloregulatory proteins, such as p53, Ref-1, HIF-1, zinc-finger transcription factors, *E. coli* SoxR/S proteins, and iron regulatory proteins, are all natural sensors of oxygen radicals and NO (Drapier and Bouton, 1996). Some of these responses (e.g. cellular immunity) have already been shown to be deregulated by metal treatment (Kasprzak *et al.*, 1987; Goebeler *et al.*, 1995; Ye *et al.*, 1995; Zeromski *et al.*, 1995; Beyersmann and Hechtenberg, 1997). It seems obvious that the Fenton/Haber–Weiss chemistry driven by adventitious cellular metals will disrupt the orderly

Table 16.1 Two-level redox control of NF-κB and AP-1 and the mediators involved (according to Galter *et al.*, 1994; Xanthoudakis *et al.*, 1996; Hirota *et al.*, 1997; Jin *et al.*, 1997)

Transcription factor	*Level 1: Redox modulation of induction/activation*	*Level 2: Redox modulation of DNA binding*
NF-κB	**Positive mediators:** *oxidants*; e.g. H_2O_2, GSSG (induction of NF-κB activation and nuclear translocation)	**Negative mediators:** *oxidants*; e.g. GSSG (inhibition by mixed disulfide formation at the Cys residue in the DNA-binding region?)
	Negative mediators: *reductants*; reduced thioredoxin	**Positive mediators:** *reductants*; e.g. Ref-1, reduced thioredoxin
AP-1	**Positive mediators:** *oxidants*; reactive oxygen species, GSSG (oxidative activation via Jun phosphorylation); *reductants*; thioredoxin (activation of c-*fos* gene transcription)	**Negative mediators:** *oxidants*; e.g. oxidized thioredoxin (oxidative inhibition of the Cys residue, as above) **Positive mediators:** *reductants*; e.g. reduced thioredoxin, Ref-1

physiologic production and functions of the main redox signaling molecules $O_2^{-\cdot}$, H_2O_2 and NO. Therefore, generation of radical species by metals, and especially metal-containing particles, may be considered as a unifying factor in determining their pathogenicity (Donaldson *et al.*, 1996). However, the enormous complexity of the redox signaling network precludes a full discussion of the observed or possible interference by metals in its pathways. The discussion below is thus limited to a few examples of basic interactions of metals and metal-generated oxidants with selected signaling molecules (or molecule systems), including NO, NF-κB, AP-1, p53, $p21^{ras}$, and redox factor-1 (Ref-1), and a few examples of more complex interplay of metals and signaling in oxidative stress.

16.6.1 *Interactions with major signal molecules*

16.6.1.1 *Nitric oxide*

The ubiquitous signaling radical molecule nitric oxide, NO, regulates the expression and function of many diverse genes and their products. Proteins that contain thiol or metal groups located in active or allosteric regulatory sites are targets for, and transducers of, the NO signal molecule. These are carrier proteins, enzymes, receptors, ion channels and structural elements such as albumin, P_{450}s, soluble guanylate cyclase, cytochrome *c* oxidase, glyceraldehyde-3-phosphate dehydrogenase, glutathione peroxidase, metallothioneins, cardiac calcium-release channels (Xu *et al.*,1998) and heme proteins. Metals often help regulate gene activity as transcription cofactors that form metal–thiolate polymetallic clusters. Disruption of metallo-thiolate clusters by NO and NO-derived species causes allosteric changes in the protein structures that alter their enzymatic and regulatory functions. NO reacts readily with metal centers in proteins, including heme-iron, iron–sulfur clusters, zinc–sulfur clusters and copper (Radi, 1996). Either stable nitrosyl–metal complexes or unstable redox-active protein derivatives are formed. It is often not only NO, but rather NO-derived species, such as the NO and $O_2^{-\cdot}$ derivative peroxynitrite, that are the ultimate species attacking the metalloprotein, so there are variations in their reactive rates, allosteric effects and product stabilities.

NO is capable of reacting with transition metals and their protein carrier ligands very rapidly (Felley-Bosco *et al.*, 1998) and thus disrupting their physiologic functions. And vice versa: owing to this high reactivity, some metals may modulate NO's biological activities, switching them from regulatory to damaging to a cell. NO-mediated damage may occur by disruption of mitochondrial respiration, enzyme inhibition, lipid peroxidation, alteration of gene expression and genetic mutations via intermediates such as N_2O_3 and peroxynitrites from $O_2^{-\cdot}$ or other oxygen radicals (Gordge, 1998).

NO's reactivity with the metal centers of metalloproteins is governed by its unique redox properties. Because NO is neither a strong oxidant nor a strong reductant, the one-electron switch from $NO^{\cdot}$ to NO^+ or NO^- occurs with ease. NO reacts with iron and copper metalloproteins with redox chemistries that depend on the charge properties of the bound metal group, and frequently electron transfer occurs as needed (Radi, 1996). Albumin acts as a carrier for Cu(II) in the blood. *S*-Nitrosation by NO releases the Cu(II), suggesting that NO plays a vital role in transport of organic ions and heavy metals in the circulation (Kashiba-Iwatsuki *et al.*, 1997). Furthermore, the temporary association of NO with one metalloprotein can change its properties so that the

NO–protein complex can then undergo a different type of NO chemistry with another metalloprotein or target molecule in a sort of a 'bait-and-switch' exchange. For example, NO's interaction with hemoglobin in the oxygen-rich environment of the lung alters its oxygen-binding capacity, and is reversed in peripheral tissues (Gow and Stamler, 1998). In addition, the site left behind when NO is released from a NO carrier creates a reactive site that can have important biological consequences. For example, metallothionein, which has 20 cysteines that carry Zn(II), forms transient *S*-nitrosyl groups, but then releases the NO with formation of disulfide bonds and loss of the metal, leading to disruption of zinc homeostasis. These disulfide bonds must be enzymatically reduced back to the -SH groups before the metallothionein protein can regain its zinc (or certain other metal cations) chelating function.

Thus, the essential metals iron, copper and zinc on one side, and the endogenous free-radicals, superoxide and NO, on the other side, are entangled in a complex network of mutual regulation, disruption of which, for example, through supplementation of excessive amounts of one of the reactants, may lead to oxidative damage.

16.6.1.2 Nitric oxide and iron homeostasis

Usually, only a very small pool of free redox-active iron moves between the transferin/ferritin carriers and the cell's mitochondria. Excesses in this free iron pool, in the absence of adequate antioxidants, catalyzes deleterious Fenton-type free-radical reactions. When increased levels of iron are needed, protein synthesis of transferin receptor is upregulated and ferritin synthesis is inhibited by iron regulatory proteins (IRPs). IRP-1 and IRP-2 are redox-sensitive iron–sulfur proteins whose regulatory activity is changed by iron depletion. Binding of the IRPs to the RNA stem–loop structures of the iron-responsive elements (IRE) of the ferritin mRNA suppresses its translation to protein, while IRPs binding to the IRE of the normally short-lived transferin receptor mRNA stabilizes the message and results in a rapid accumulation of the transferin receptor protein. The subsequent increases in intracellular iron levels cause the IRPs to switch back to an iron–sulfur structure that no longer binds mRNAs, but instead now acts as the Krebs cycle aconitase enzyme. Oxidative stress from NO, or its product, peroxynitrite, interacts with a Cys residue to predispose IRP-1 to bind to IREs and reduces IRP-2 binding to IREs, showing that both IRP proteins are regulated by post-translational redox events (Bouton *et al.*, 1997). Thus, exposure of cells to exogenous NO mimics iron deficiency in regulating the processes of iron homeostasis (Drapier *et al.*, 1993; Klausner *et al.*, 1993; Weiss *et al.*, 1993). NO cleaves the iron cation from aconitase, converting the protein to the IRP-1 function that increases transferin receptor synthesis (Phillips *et al.*, 1996) and downregulates ferritin expression (Weiss *et al.*, 1994). The overall reaction may require the presence of superoxide for NO to form peroxynitrite (Gardner and Fridovich, 1991; Hausladen and Fridovich, 1994). If an NO-induced excess of free iron relative to ferritin develops, DNA-damaging Fenton oxidants can be formed (Keyer and Imlay, 1996) (Fig. 16.2).

Upregulation of the inducible nitric oxide synthase (iNOS) protein within a cell slightly reduces transferin receptor synthesis, in contrast to the increased receptor expression seen when NO is generated exogenously for the cells by chemical NO donors (Phillips *et al.*, 1996). Complexes are formed by reaction of NO with free iron(II) species (Mulsch *et al.*, 1993). Interestingly, addition of iron salts to murine macrophage cultures

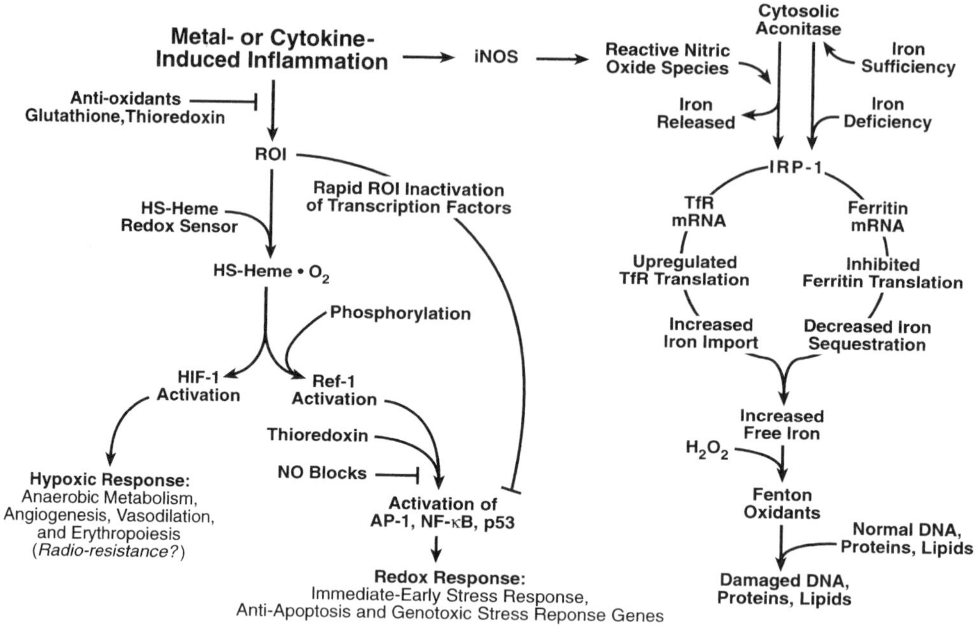

Figure 16.2 Examples of effects of metals on iron metabolism and intracellular signaling. The metal-induced inflammatory response produces cytokines, reactive nitric oxide species (RNOS) and other reactive oxygen intermediates (ROI). RNOS nitrosate aconitase, converting it to its iron regulatory protein-1 (IRP-1) function, which in turn leads to an eventual increase in the levels of free intracellular iron. This iron can undergo Fenton chemistry to cause oxidative damage to cellular constituents. ROI which are not neutralized by antioxidants or reducing peptides and proteins can oxidize and inactivate a number of transcription factors involved in various types of stress response, by blocking their DNA-binding ability. Ref-1 and its reducer thioredoxin are required to reactivate the transcription factors.

represses cytokine induction of the inducible nitric oxide synthase gene (*iNOS*), while the iron chelator desferrioxamine stimulates iNOS expression. This feedback mechanism for counteracting nitric oxide's contribution to intracellular iron overload may protect the cell's DNA from NO-induced Fenton oxidation (Weiss *et al.*, 1994). To add complexity to this story, NO can also directly inactivate reactive Fenton intermediates (Kanner *et al.*, 1991).

16.6.1.3 S-Nitrosothiols and metals

S-Nitrosothiols, the product of NO's reaction with protein thiol groups, serve as physiologic and pharmaceutic storage and transport molecules for NO delivery. *S*-Nitrosothiols are quite stable, but only in the absence of certain metal ions. Metals rapidly decompose *S*-nitrosothiols to NO and disulfide (Singh *et al.*, 1996). Likewise, superoxide causes decomposition of *S*-nitrosothiols, with the release of NO and the formation of peroxynitrite (Trujillo *et al.*, 1998). The association–dissociation kinetics of NO in metalloproteins, and in *S*-nitrosylated proteins, have drastic implications for the role of NO in biologic signaling systems. It is plausible that enzymes or scavenger proteins exist to accelerate the transfer or release of more tightly bound NO-like groups from signaling proteins. Rapid *trans*-nitrosation might be required to recycle proteins that are meant to be only transiently *S*-nitrosated.

The primary metal candidate to form part of a *S*-nitrosothiol NO-releaser is copper. Numerous proteins have various types of copper centers. Cytochrome *c* oxidase (which releases cytochrome *c* subunits from the mitochondria, a reaction that figures prominently in inducing apoptosis) contains an NO-reactive copper center. On the other hand, Cu/Zn-SOD has a copper center that does not react with NO, allowing it to continue its critical role of removing superoxide even in the presence of high amounts of NO, thus preventing peroxynitrite formation and cytotoxicity (Radi, 1996). Peroxynitrite reacts with all of the Cu/Zn-, Mn- and Fe-containing types of SOD, but only the Mn- and Fe-types are inactivated (reviewed by Radi, 1996). In other metalloproteins, Cu(I) and Cu(III) seem to be more reactive with NO than Cu(II) centers. More importantly, Cu(I) is able to scavenge NO from *S*-nitrosothiols.

16.6.1.4 Redox active peptides and proteins

Functional activation and deactivation of peptides and proteins may occur through various mechanisms, including phosphorylation/dephosphorylation and oxidation/reduction reactions. Phosphorylation of proteins occurs at serines, threonines, tyrosines or histidines. There is as yet no evidence that toxic metals and radical species may affect any of the phosphorylation events directly. Redox modifications of proteins occur mainly at cysteines, for example through *S*-nitrosation, oxidation of thiols and reduction of the disulfide bonds. It is here that redox active metals, NO, and other radicals have much opportunity for interference. The redox status of a protein can affect its molecular conformation and thus alter its functions. The protein pairs that safeguard the cellular redox 'mean streets' are glutathione/glutathione reductase and thioredoxin/thioredoxin reductase. Two additional proteins, glutaredoxin and Ref-1, seem to partner more freely with other members of the team (Powis *et al.*, 1995). Ref-1, in particular, seems to be an integral regulator of the transcription factors and genes responsible for responding to oxidative stress (discussed below). The most interesting transcription factors whose functions are regulated by redox interactions are the AP-1, NF-κB, p53 and $p21^{ras}$ proteins.

16.6.1.5 AP-1 and NF-κB

The DNA-binding sites for AP transcription factors are located in the promoter regions of a variety of genes that are engaged directly in initiating the inflammatory cascade and in the ensuing cellular responses to oxidative stress (Sen and Packer, 1996). Oxidation of Cys residues in the evolutionarily conserved –Lys–Cys–Arg– sequence of each subcomponent of AP-1, Fos, and Jun, has an inhibitory effect on AP-1 binding to DNA sequences (Xanthoudakis *et al.*, 1996). The oxidative suppression of AP-1-triggered genes is a model mechanism for negative feedback control of the inflammatory response.

The Ref-1 protein, in conjunction with thioredoxin, operates as an essential physiologic reducer of specific transcription factors to stimulate their DNA binding. The list of known transcription factors activated by Ref-1 includes AP-1, NF-κB, and p53 (Xanthoudakis *et al.*, 1996; Parks *et al.*, 1997; Verhaegh *et al.*, 1997). In addition, Ref-1 has a second function: it is a class II A/P endonuclease that repairs oxidative lesions (mainly abasic sites) in DNA (Xanthoudakis *et al.*, 1996; Parks *et al.*, 1997). Ref-1 emerges, therefore, as a critical component of cellular response to oxidative stress in general, and DNA damage in particular.

The functional dependence of transcription factors AP-1, NF-κB and other nuclear proteins (Galter *et al.*, 1994; Lickteig *et al.*, 1996; Xanthoudakis *et al.*, 1996; Parks *et al.*, 1997), as well as their mediators GSSG and Ref-1, on the oxidation status of Cys residues, on one hand, and the known ability of some metals to catalyze Cys oxidation, on the other hand, give the metals a mechanism of interference with gene expression. For example gold(I) directly inhibited AP-1 DNA binding by interacting with the thiol groups on Jun and Fos proteins; a similar action was observed for the metalloid selenium (as the redox-active selenite and selenodiglutathione) (Handel *et al.*, 1995; Bjornstedt *et al.*, 1997). Suppression of AP-1 and other redox-regulated transcription factors' functions may thus contribute to the antiarthritic (anti-inflammatory) and anticancer effects of gold and selenium, respectively. Zinc is similarly able to attenuate oxidative stress damage to endothelial cells by interfering with certain aspects of the signaling and inflammatory responses, specifically by lowering NF-κB and AP-1 DNA-binding levels (Connell *et al.*, 1997). When 3T3 cells were selected for resistance to toxic amounts of Ni(II), they had diminished DNA binding of NF-κB and AP-1 (Salnikow *et al.*, 1994). In contrast, Ni(II) and Co(II) enhanced NF-κB and AP-1 DNA binding in vascular endothelial cells, whereas Co(II) and Cr(III) did not (Wagner *et al.*, 1997). Very interestingly, two chemically very different metals, lithium and lead, cause similar increased AP-1 DNA binding in brain cells (Ozaki and Chuang, 1997; Pennypacker *et al.*, 1997). Thus, the enhancement or suppression of AP-1 and NF-κB may depend on the metal, its cellular concentration and the target cells.

The NF-κB transcription complex is critically important to regulation of iNOS induction and to the cell's responses to many types of reactive oxygen species and genotoxic stresses. Activating stimuli for NF-κB include tumor necrosis factor-α (TNF-α), lipopolysaccharide, phorbol ester, viral infection and oxidative stress of various origins (reviewed by Schreck *et al.*, 1992; Baeuerle and Baltimore, 1996; Cheshire and Baldwin, 1997). It is therefore not surprising that activating stimuli also include metal ions, such as Ni(II), Co(II) (Goebeler *et al.*, 1995) and Cr(VI) (Ye *et al.*, 1995).

The NF-κB complex is composed of a heterodimer of two proteins of the Rel family, often the p50 and p65 subunits. The NF-κB transcription activation complex is held in the cytoplasm in a latent form bound to at least one member of the IκB family of inhibitors. In response to specific activating stimuli, site-specific serine phosphorylation by an unidentified kinase targets the IκB subunit for the ubiquitin–proteosome degradation, unmasking the NF-κB nuclear translocation signal. The NF-κB subunit complex is then actively transported to the nucleus. This IκB-kinase is activated by MEKK1, a kinase in the Jun activation pathway (reviewed in Li and Nabel, 1997). Tyrosine phosphorylation inactivates IκB without causing its degradation, allowing a more transient NF-κB signal to be sent (reviewed by Jin *et al.*, 1997).

Just as kinases and phosphatases regulate signal pathways by changing the phosphorylation of the pathways' proteins, the oxidants and antioxidants play a leading role in the pleiotropic signaling pathways affected by redox status. Redox status directly controls NF-κB function in the nucleus (Table 16.1). Following enzymatic reduction of oxidized NF-κB by Ref-1, the now activated NF-κB complex transactivates a number of genes, including the *iNOS* gene in appropriate cell types. Once the critical cysteine residues of NF-κB are activated by Ref-1 reduction, the reduced state is maintained by thioredoxin and thioredoxin reductase levels in the nucleus. The metalloid selenium at low (nM) levels is essential for redox-active seleno-enzymes such as thioredoxin reductase and glutathione peroxidases. However, at higher (μM) concentrations, selenium forms RS–Se–SR adducts with thiol groups, inactivating proteins such as NF-κB, and thus blocking stimulation of genes like *iNOS* and *TNF-α* (Kim and Stadtman, 1997).

16.6.1.6 IκB

Redox reactions also regulate NF-κB's access to the nucleus by modulating the phosphorylation of IκB (Jin *et al.*, 1997). Of the three major reducing antioxidants, glutathione, glutaredoxin and thioredoxin, it is the thioredoxin pathway that mostly controls IκB phosphorylation. The thioredoxin pathway has three enzymes, thioredoxin reductase, thioredoxin, and thioredoxin peroxidase; the latter directly reduces the target molecule. A newly discovered cytoplasmic human thioredoxin peroxidase, AOE372, is also responsible for modulating IκB phosphorylation, perhaps by affecting the IκB kinase (Jin *et al.*, 1997).

16.6.1.7 p53

The tumor suppressor protein p53 is a transcription factor regulating cell cycle and DNA repair. Responding to oxidative cell damage, p53 binds to DNA and activates several genes whose expression products arrest cell cycle in G_1 or at the G_2/M border (for DNA repair), or may eventually trigger apoptosis. These events halt proliferation of cells with damaged DNA and thus prevent mutations. The DNA-binding activity of p53 depends on an essential metal and redox status of its Cys residues. The essential metal is zinc(II) coordinated by the Cys176, His179, Cys238 and Cys242 residues of p53. The removal of zinc(II) by chelators promotes oxidation of the Cys residues and abrogates p53's DNA-binding ability (Verhaegh *et al.*, 1997). The p53 protein can also bind copper, the redox status of which [Cu(I) or Cu(II)] may exert either a stimulatory or inhibitory effect on p53 activity (Hainaut *et al.*, 1995).

16.6.1.8 $p21^{ras}$

The role of $p21^{ras}$ in the mediation of free-radical signaling was studied by Lander *et al.* (1995, 1997). They found that signaling by agents causing oxidative stress, including Hg(II), NO and H_2O_2, was prevented in cells with blocked $p21^{ras}$ activity. The site on $p21^{ras}$ responsible for molecular interaction with NO was identified as Cys118; it was activated by *S*-nitrosylation. The authors hypothesized that modification of the same Cys by redox agents other than NO (e.g. transition metals or hydroxyl radical), although chemically different, may result in similar structural alterations of the $p21^{ras}$ molecule and thus produce similar effects (Lander *et al.*, 1997).

16.6.2 Hypoxia and redox regulation of signaling

Oxygen is such an essential substrate in aerobic metabolism that organisms and cells have developed both immediate and long-term compensatory mechanisms for dealing with oxygen deprivation, either anoxia or hypoxia. Oxygen transport and metabolism are tightly dependent on iron; hence, lowering of iron levels triggers cellular responses resembling those induced by hypoxia. Hypoxia is known to play a key role in the pathophysiology of cancer, but also in diseases with a link to dysfunction of the redox signaling pathways. Adaptation to hypoxia includes gene expression changes which lead to increased O_2 delivery to the tissues. Adaptation at the cellular level involves expression of hypoxic stress-related proteins. Hypoxia activates the Ref-1 reduction activity for activation of p53, Fos, Jun, NF-κB and other transcription activator proteins. The induction of the potent vasoconstrictor endothelin-1 by hypoxia, and previously described hypoxic responses, are mimicked by treatment of cells with divalent transition metals such as colbaltous ion or by iron chelators such as deferoxamine (Bodi *et al.*, 1995).

Regulation of many of the proteins participating in adaptation to hypoxia occurs at the level of gene expression. The molecular mechanism of hypoxia-dependent regulation is transcriptional induction (and repression) via the binding of a growing family of transcription factors related to hypoxia-inducible factor-1 (HIF-1), including EPAS-1 and HIF-1-related factor (HLF). The HIF-1 family are the only known mammalian transcription factors expressed uniquely in response to physiologically relevant levels of hypoxia. HIF-1, as a critical component of the oxygen-signaling pathway, is a prime candidate regulator molecule for the role of coordinating vascular oxygen supply with cellular growth and energy metabolism.

HIF-1 responds to oxygen concentration via a pathway involving an unknown specific oxygen sensor, which appears to be a heme protein. In one model, the putative heme protein adopts different conformational states, depending on whether O_2 is bound. Another model proposes that the heme protein converts O_2 to H_2O_2. Ref-1 is a strong candidate for being this oxy-sensor. The phosphorylation and redox state of HIF-1 regulates its transcription activation functions. An intact redox-dependent signaling pathway involving Ref-1 is required for destabilizing the HIF-1-α mRNA and protein and allowing the DNA-binding activity of HIF-1 (Tian *et al.*, 1997). The HIF-1 binding is reversibly blocked by oxidation of HIF-1. Thioredoxin and Ref-1 can rescue/reactivate oxidized HIF-1 (Huang *et al.*, 1996).

HIF-1 is a heterodimer composed of HIF-1-α and usually HIF-1-β subunits. A major control of HIF-1 activity by oxygen tension is achieved by changes in the protein level of the HIF-1-α subunit, which complexes with the constitutively expressed HIF-1-β subunit. Both the HIF-1-α protein and mRNA have a very short half-life in normal cells. Changes in HIF-1-α protein abundance occur largely via regulated stability involving proteolysis, but also at the level of transcription. Long-term induction of many of the other hypoxic-response proteins by hypoxia also requires an increase in mRNA stability, which is mediated by the binding of regulatory proteins to specific sequences within the mRNAs. Hypoxia, low iron levels, or a number of other cell stresses result in inactivation of the specific protease and the ribonuclease that are responsible for suppressing HIF-1-α protein expression. Like another more familiar oxidative stress-response protein, p53, this results in rapid HIF-1 protein accumulation, transport to the nucleus and altered regulation of a battery of genes which will, first, immediately protect the cell from the hypoxic stress response, including temporary suppression of the apoptotic response, and secondly, initiate long-term physiologic responses that will alleviate the hypoxia.

16.6.3 Apoptosis

Cells can accumulate when either proliferation rates are higher or apoptosis rates are lower than normal. Diminished apoptosis is characteristic for some cancers and autoimmune disorders. The inhibition of apoptosis in human cancer has been associated with overexpression of thioredoxin (Baker *et al.*, 1997). Undesired apoptosis is the cause of some neurodegenerative disorders and ischemic injuries. Apoptosis is achieved by two mechanisms. Positive induction often occurs by ligand binding to a plasma membrane receptor, such as Fas/FasL or TNF/TNFR interactions. Alternatively, or coincidentally, it occurs by negative induction via loss of apoptosis suppressor functions, as happens when Bcl-2 family members are underexpressed or inactivated by heterodimerization with Bad/Bax family members. Each has several sequelae in common; chief among them is the activation of a cascade of cysteine proteases of the ICE/Caspase family. This in turn has the dual role of cleavage activation of several enzymes and transcription factors that lead to proteolysis of proteins essential for cell survival and the irrevocable initiation of the transcriptional programmed cell-death pathway.

Cellular cytokines, such as TNF-α and TGF-β, play critical roles in the delicate balance between cell-death and cell-proliferation signals. Both ligands bind to receptors that result in a cascade of signals that eventually activate/induce or inactivate/suppress many genes and gene products. Common to both these ligands is the activation of the transcription factor NF-κB, which, as discussed previously, acts in many noxious stimuli scenarios, such as heavy metal exposures, to prevent apoptosis from occurring. It is the balance between the simultaneous activation of the 'death domain' pathways of cysteine-aspartic acid (CASP) proteases, Bcl-family inactivation and sphigomylinase/ceramide signaling through the JNK/SAPK and Ras/Raf pathways that determines whether a given cell will proliferate, get into a cell-cycle arrest, or die. In macrophages and some other cells there is a paracrine and autocrine negative feedback regulatory loop between the production of NO in response to combinations of cytokines and the expression of inducing and suppressing cytokines. This, in turn, regulates the inflammatory and apoptotic responses, important components of the metal-mediated carcinogenesis process. Lead, mercury and cadmium, and less so gold and copper, were found to modulate the cellular inflammatory response through inhibition of iNOS-dependent production of NO in murine macrophages in response to TNF-α. Nickel(II) has been known to cause contact dermatitis, again indicating an effect on inflammation, while chromium(III) and zinc(II) have no effect (Tian and Lawrence, 1996).

Several toxic metal compounds may cause apoptotic cell death, though the underlying mechanistic pathways are not well defined. They include organomercurial derivatives (Shenker *et al.*, 1997), chromium(VI) (Blankenship *et al.*, 1994), iron–NTA (Zhang *et al.*, 1997) and nickel (Shiao *et al.*, 1998). The effect of iron–NTA was inhibited by the antioxidant vitamin E. Interestingly, chromium(VI)-induced apoptosis in CHO cells could be inhibited by Cd(II), possibly via a Bcl-2-dependent pathway; Cd(II) alone did not cause apoptosis (Shimada *et al.*, 1998; Shiao, personal communication).

16.6.3.1 The roles of mitochondria and calcium

The mitochondrion is where the bulk of the cell oxygen is reduced. Mitochondrial function is potently affected by NO and its congeners. These, in turn, may be modulated by metals, as discussed before. NO and peroxynitrite may target and damage mitochondrial membranes, enzymes and DNA, and stimulate calcium release from the mitochondrial stores (Richter, 1997). It is this loss of mitochondrial function and the resulting change in cellular redox potential that may connect various aspects of metal-induced oxygen signaling, NO signaling and inflammation to cancer cell promotion, proliferation, escape from immune surveillance, induction of angiogenesis and response to radiation therapy (Guillemin and Krasnow, 1997). The central players in this drama may be HIF-1 and Ref-1.

Mitochondrial integrity in response to hypoxia plays a large role in maintaining proper sequestration of the calcium ions needed for activation of many of the CASPs. Any reduction of membrane potential and release of calcium ions from damaged mitochondria thus play critical roles in inducing apoptosis. Calcium efflux from mitochondrial and endoplasmic reticulum stores is sensitive to cellular redox states. Toxic metals may interfere with calcium homeostasis through various mechanisms involving calcium channels, pumps and the function of calcium-dependent enzymes such as kinases and caspases, and disrupt calcium signaling systems that may eventually result in aberrant cell growth control and transformation. Thus, cadmium ions are known to pass through membranes via calcium channels and stimulate the immediate early genes for Fos, Jun and c-Myc, and the oxidative stress response genes for p53, metallothionein, GSH and HSPs (Beyersmann and Hechtenberg, 1997). Nickel can interfere with calcium functions by specifically stimulating a calcium-dependent expression of the *Cap43* gene (Salnikow *et al.*, 1998). The increase in intracellular calcium concentration by Ni(II) is also consistent with the observed nickel induction of the activating transcription factor ATF-1 (Sun *et al.*, 1996; Salnikow *et al.*, 1997). The many regulatory functions of ATF-1 include participation in cell signaling in response to hypoxia and enhancing the erythropoietin gene expression (Kvietikova *et al.*, 1997). The latter, in turn, has been known to be upregulated in rat kidneys in the early stages of nickel-induced renal carcinogenesis (Sunderman *et al.*, 1984; Kasprzak *et al.*, 1994a). Erythropoietin was also induced by cobalt in mice and this induction greatly enhanced the micronuclei response to mutagens (Suzuki *et al.*, 1993).

16.7 Carcinogenesis as an example of pathogenicity of metal-mediated oxidative damage

Several metals, including nickel, chromium, cadmium, beryllium and arsenic, have been found to be carcinogenic in humans, and many more appear to be carcinogenic in experimental animals (reviewed by Kasprzak, 1995, 1996; Hayes, 1997). However, the underlying mechanisms remain unclear (Nieboer *et al.*, 1986; Snow, 1992; Lee *et al.*, 1995; Stohs and Bagchi, 1995; Klein and Costa, 1997; Kasprzak *et al.*, 1999). Carcinogenesis is a lengthy multi-step process that may be arbitrarily divided into initiation, promotion and progression of tumors. The genomic instability resulting from genotoxic and epigenetic insults of a carcinogen in somatic cells, consisting of multiple (sequential) DNA mutations and alterations in cell signaling, is presumed to play a major role in the entire process. Thus, the potential of metals to mediate oxidative damage that may lead to both genotoxic and epigenetic consequences, along with other nonredox effects of metals, makes them capable of advancing carcinogenesis in most, if not all, of its stages.

16.7.1 Genotoxicity

There is strong evidence that at least some types of oxidative DNA damage are potentially mutagenic. Thus far, the damage/mutation relationship has been best established for mutations produced by damaged DNA bases; among them, for 8-oxo-G, frequently assayed as 8-oxo-dG. As found by many groups, in addition to pairing with C, 8-oxo-G (7, 8-dihydro-8-oxoguanine) may also mispair with A. If the lesion is present in a DNA template, the mispairing results in a G → T transversion mutation (Kuchino *et al.*, 1987; Wood *et al.*, 1990, 1992; Moriya *et al.*, 1991; Shibutani *et al.*, 1991; Cheng *et al.*, 1992; Kamiya *et al.*, 1992; Le Page *et al.*, 1995, 1998). The same damaged base in the form of substrate for DNA synthesis, 8-oxo-dGTP, causes A → C substitutions (Cheng *et al.*, 1992; Purmal *et al.*, 1994a). The 8-oxo-G : A mismatch is repaired in cells less efficiently than the 8-oxo-G : C mismatch that increases the mutagenic frequency of the former (Le Page *et al.*, 1998).

Besides 8-oxo-dG, several more products of DNA base oxidation have been thus far identified as potential mutagenic DNA lesions. Thus, 8-oxo-dA (7,8-dihydro-8-oxo-2′-deoxyadenosine) caused A → G transitions and A → C transversions (Kamiya *et al.*, 1995a); 2-OH-dA induced A → T and A → C transversions, and possibly also A → G transitions (Kamiya *et al.*, 1995b; Kamiya and Kasai, 1996); 5-OH-dC and 5-OH-dU had the potential to induce C → T transitions and C → G transversions (Purmal *et al.*, 1994b); and *β*-ureidoisobutyric acid (a thymine ring fragmentation product) causes T → A transversions (Maccabee *et al.*, 1994). 5-OH-dCTP was found to have a relatively high potential of inducing T → C transitions through mispairing with template dA (Purmal *et al.*, 1994a). In contrast, 5-OH-dUTP, dU-glycol tri-phosphate and 8-oxo-dATP are thought to be rather unimportant oxidative lesions when present in the metabolic pools (Purmal *et al.*, 1994a, 1998). Two more products of DNA base oxidation, 5-OHMe-dU and thymidine glycol, are also potentially mutagenic (Billimoria and Gupta, 1986; Shirname-More, 1987; Hayes *et al.*, 1988; Basu *et al.*, 1989; Akman *et al.*, 1991). Mutagenicity of other oxidatively damaged nucleobases is still debated or unexplored (Ames, 1989; Breimer, 1990; Akman *et al.*, 1991, 1993; Wood *et al.*, 1992).

The mutation spectra produced by metal-mediated oxidation of DNA (Table 16.2) have been established *in vitro* for Fe(II), Cu(I), Cu(II), and Ni(II) (Moraes *et al.*, 1990; Loeb *et al.*, 1991; McBride *et al.*, 1991; Tkeshelashvili *et al.*, 1991). However, the underlying types of the initial DNA lesions were not always identified. The mutations induced by these metals were predominantly single-base substitutions which clustered characteristically for each agent. The most frequent mutations observed for single-stranded DNA after aerobic incubation with Fe(II) were G → C transversions followed by C → T transitions and G → T transversions (McBride *et al.*, 1991). With Cu(I) and Cu(II), the mutations were predominantly C → T transitions, followed by G → T transversions (Tkeshelashvili *et al.*, 1991). Reid and Loeb (1993) and Tkeshelashvili *et al.* (1993) have established that the tandem double CC → TT mutations, known to occur via UV damage to DNA, can also be produced by other treatments generating

Table 16.2 Mutations resulting from metal exposure

Mutation found	*Possible underlying lesion*	*Suspected cause of lesion*
G → T[1, 2, 3]	8-oxo-dG[9, 10] abasic site[2, 11]	Oxidation and/or depurination of dG
C → T[1, 2, 4]	5-OH-dC[12] dU or dT mismatch[1]	Oxidation of dC; deamination of dC and 5-Me-dC
G → C[2], G → A[4]	dX[13]	Deamination of dG
CC → TT[4, 5]	dC-dC cross-link[4, 5, 14]	Cross-linking of neighboring C residues in the same DNA strand
T : A → A : T[6] T → A[7]	β-ureidobutyric acid[8]	Ring fragmentation of dT
G : C → C : G[6]	5-OH-dU[6]	Oxidation/deamination of dC
T : A → G : C[6]	8-oxo-dGTP[9, 10]	Oxidation of dGTP and misincorporation of 8-oxo-dGTP
G : C → A : T[15] G : C → T : A[15]	DNA-Cr(III)-GSH, -His, or -Cys cross-link[15]	Formation of ternary Cr(III) complexes with DNA, GSH or amino acids

[1] Tkeshelashvili *et al.* (1991); [2] McBride *et al.* (1991); [3] Higinbotham *et al.* (1992); [4] Tkeshelashvili *et al.* (1993); [5] Reid and Loeb (1993); [6] Yang *et al.* (1996); [7] Loeb *et al.* (1988); [8] Maccabee *et al.* (1994); [9] Cheng *et al.* (1992); [10] Le Page *et al.* (1995); [11] Schaaper *et al.* (1987); [12] Feig *et al.* (1994); [13] Eritja *et al.* (1986); [14] Dizdaroglu and Simic (1984a); [15] Voitkun *et al.* (1998).

active oxygen species, for example by Fe(II), Cu(I), Cu(II) or Ni(II) compounds plus O_2 or H_2O_2. The authors speculated that the mutations occurred because of base mispairing at sites of cytosine dimers produced by ·OH (Dizdaroglu and Simic, 1984a).

Metal exposure-related mutations were also determined *in vivo*. The G → T transversion mutation characteristic for the 8-oxo-dG or G apurinic lesions (McBride *et al.*, 1991) was found by our laboratory to occur exclusively in the K-*ras* oncogene isolated from nickel subsulfide (Ni_3S_2)-induced rat renal tumors (Higinbotham *et al.*, 1992). The latency of tumors was shortened and the incidence of codon 12 mutations in K-*ras* was greatly increased by the addition of metallic iron powder to Ni_3S_2. Since the iron powder did not produce tumors by itself and did not affect the final tumor incidence by Ni_3S_2, the enhancement was most likely due to assistance in tumor progression. The latter is consistent with *ras* involvement in tumor latency (Higinbotham *et al.*, 1992). Yang *et al.* (1996) established the mutation frequencies in the *hprt* gene of CHO K-1 cells cultured with Cd(II), known to induce oxidative damage. Base transversions occurred more frequently than transitions. The most frequent mutations found were G : C → C : G (30%), T : A → A : T (26%), and T : A → G : C (19%) transversions, followed by G : C → A : T (11%) and A : T → G : C (7%) transitions. Hence, the most likely base lesions responsible for Cd(II) mutagenesis in

CHO K-1 cells could be 5-OH-dU and β-ureidoisobutyric acid. The frequency of G : C → T : A transversions typical for 8-oxo-dG or apurinic site-caused mutagenesis was low, indicating that direct generation of these lesions in DNA could not account for Cd(II) mutagenicity. However, the A : T → C : G transversion mutation, associated with misincorporation of 8-oxo-dGTP into DNA, seemed to be high enough to make this event likely to occur in Cd(II)-exposed cells in concordance with the observed inhibition of 8-oxo-dGTPases by this metal (Porter *et al.*, 1996, 1997).

The Fe(III)–NTA complex is strongly carcinogenic to the rat and mouse kidney (Okada *et al.*, 1983; Li *et al.*, 1987). Since a significant increase of 8-oxo-dG production was found in renal DNA of male Wistar rats given a single intraperitoneal injection of Fe(III)–NTA, we might suspect that the mutagenicity of this damaged base contributed to tumor development; but the related mutations were not determined. Likewise, elevated levels of 8-oxo-dG, but not mutations, were also observed in V79 cells cultured with Fe(III)–NTA (Hartwig and Schlepegrell, 1995).

Possible promutagenic effects of metal-induced DNA–protein (or amino acid) cross-linking, DNA strand scission and depurination are not fully recognized. Iron ions, but not copper ions, may cause thymine–tyrosine cross-links between DNA and proteins (Altman *et al.*, 1995), but mutagenicity, if any, of this event is unknown. Strand scissions have been found to be mutagenic by causing deletions (Runger and Kraemer, 1989) or insertions (Chiocca *et al.*, 1991). For example, treatment of cultured cells with Ni(II) chloride resulted in a mutation consisting of duplication of a 70-base DNA strand, an indication of strand breakage followed by faulty repair, or erratic DNA replication (Chiocca *et al.*, 1991). Depurination of the G site may result in G → T transversions (McBride *et al.*, 1991). Most recently, Voitkun *et al.* (1998) determined that the Cr(III)–GSH–, Cr(III)–His– and Cr(III)–Cys–DNA adducts were mutagenic in human cells, with G : C → A : T transitions and G : C → T : A transversions being almost equally frequent for the –His– and –Cys– adducts, and G : C → T : A transversions being prevalent for the Cr(III)–GSH–DNA adduct.

16.7.2 Epigenetic effects

Mechanistically, various epigenetic effects of metals contributing to tumor promotion and progression may be expressed by changing the oxidation status and reactivity of cellular signaling molecules, such as NF-κB, AP-1, p53, K-ras, Bcl-2, NO and others, described in more detail in Section 16.6 (the multifaceted roles of NO in cancer have been most recently reviewed by Wink *et al.*, 1998). However, as can be seen in Table 16.1, the metal-mediated oxidation may have opposite effects on a given signaling molecule (e.g. oxidation may activate NF-κB and AP-1, but at the same time prevent their binding to DNA). Therefore, by considering only a single target molecule, it is impossible to predict the final result, which may depend on many factors, including spatial and temporal distribution of the metal among subcellular and cellular tissue structures, its redox activity, as well as the wide variety of interdependent signaling pathways. Very likely for these reasons, iron, known to greatly enhance oxidative damage, administered together with the nickel carcinogen Ni_3S_2, promoted carcinogenesis in the rat kidney (see above), but prevented it in skeletal muscle (Kasprzak, 1997b). Similarly, cadmium has been shown to both induce and prevent carcinogenesis, very likely through interplay with metallothionein and nitric oxide (Shiraishi *et al.*,

1995; Misra *et al.*, 1996; Waalkes *et al.*, 1996). Clearly, despite our knowledge of some general phenomena associated with epigenetic toxicology of metals, such as inhibition of cell–cell communication and DNA methylation derangement, the mechanisms underlying the epigenetic effects of metals are poorly understood.

16.7.2.1 Intercellular communication

The gap-junctional intercellular communication is essential for maintaining the homeostasis of multicellular organisms. Its disruption is thought to be mechanistically involved in various pathogenic processes, including tumor promotion in which oxidative stress has also been implicated (Upham *et al.*, 1997; Klaunig *et al.*, 1998; Trosko, 1998; Trosko and Ruch, 1998). Since certain antioxidants have been found to prevent the inhibition of intercellular communication by tumor promoters (Ruch *et al.*, 1989; Morrow and McGlynn, 1997), one could presume that the ultimate inhibitors are reactive oxygen radicals. However, the mechanisms are not that simple. For example, H_2O_2, a tumor promoter and potent oxidant, inhibited gap junctional communication only in glutathione-sufficient cells. Glutathione-deficient cells were not affected. Moreover, the inhibition was insensitive to antioxidants, indicating lack of involvement of free radicals (Upham *et al.*, 1997). This seems to give a signaling role to molecular H_2O_2 in regulation of cell–cell communication in concert with glutathione. Thus, toxic transition metals, for which these two molecules are easy targets, may interfere with such regulation. Toxic metals may also derange intercellular communication by interactions with the essential metals, calcium and magnesium, which are critical to the process of establishing communication among cells through new permeable junctions (Davidson *et al.*, 1984). Disruption of intercellular communication was, indeed, observed in NIH 3T3 cells exposed to Ni(II) (Miki *et al.*, 1987) and in cultured renal epithelial cells exposed to Fe(II), bromate or dimethylnitrosamine (Noguchi *et al.*, 1998). However, Ni(II) did increase cell–cell communication in a cultured uterine cell line (Marty and Loch-Caruso, 1993), indicating a possible tissue and metal concentration dependence of the effect. It is worthy of mention that in both of the above Ni(II) experiments, Ni(II) action was prevented by Mg(II), an illustration of metal involvement and antagonisms in regulating the gap junctions.

16.7.2.2 DNA methylation derangement

The preconceptional carcinogenesis phenomenon, studied by several groups, including our own, has been linked to the derangement of imprinting (reviewed by Sipowicz *et al.*, 1997; Yu *et al.*, 1999). Genetic imprinting is associated, in part, with methylation of DNA cytosines that occurs within the pre-implantation blastocyst based upon unknown imprinting markers (Barletta *et al.*, 1997). This methylation, in turn, was also found to be deregulated by the presence of 8-oxo-dG in template DNA (Weitzman *et al.*, 1994; Turk *et al.*, 1995; Cerda and Weitzman, 1997). After fertilization, the DNA in the early zygote is subject to sequential processes of extensive demethylation then remethylation (Buckley, 1994; Yeivin *et al.*, 1996); therefore, any damaged and unrepaired DNA base, in addition to bearing a mutagenic potential, may alter the methylation-dependent imprinting and thus lead to improper gene expression, and eventually cancer, in the affected offspring.

Nickel has been shown previously to produce hypermethylation of DNA in cultured cells (Klein and Costa, 1997). It remains, however, an open question as to whether this methylation effect is specific to this metal (by unknown mechanism), or causally associated with nickel-induced oxidative DNA damage. Clearly, both transient and chronic metal-mediated alterations in somatic-cell methylation patterns can lead to inappropriate overexpression of oncogenes or reduction in tumor suppressor and mutation suppressor gene expressions, leading to neoplastic cell transformation.

16.8 Conclusion

The overall picture that emerges from the above-discussed findings indicates that the toxic metals are capable of deleteriously interacting with almost all cell constituents, eventually causing a wide variety of adverse effects in humans and animals, including aging, allergies, Alzheimer's and Parkinson's diseases, atherosclerosis, liver cirrhosis, cancer and many other pathologies (Kon, 1978; Cross *et al.*, 1987). Mechanistically, the action of metals may be due to structural and functional alterations of biomolecules caused by the bound metal itself (conformational effects) and by oxygen radicals and other reactive intermediates arising from metal-catalyzed redox reactions with endogenous oxidants. The catalytic effects of metals stem from their ability of redox cycling under physiologic conditions or inhibition of antioxidant cellular defenses. Support for the importance of oxidative damage in metal toxicity in general is very strong for two essential (but toxic when overdosed) metals, iron and copper, and three well-established human metal carcinogens, nickel, chromium and cadmium. However, along with more specific effects (Magos, 1991; Snow, 1992), oxidative damage seems to be slowly taking the leading role in explaining mechanisms of pathogenic effects of other metals as well.

Acknowledgments

The authors grateful to Dr Paul J. Donovan for valuable critical comments on this chapter and to Ms Maritta Perry Grau and Ms Kathy Breeze for editorial help.

References

Aiyar, J., Berkovits, H.J., Floyd, R.A. and Wetterhahn, K.E. 1991. Reaction of chromium(VI) with glutathione or with hydrogen peroxide: Identification of reactive intermediates and their role in chromium(VI)-induced DNA damage. *Environmental Health Perspectives* **92**, 53–62.

Akman, S.A., Forrest, G.P., Doroshow, J.H. and Dizdaroglu, M. 1991. Mutation of potassium permanganate- and hydrogen peroxide-treated plasmid pZ189 replicating in CV-1 monkey kidney cells. *Mutation Research* **261**, 123–30.

Akman, S.A., Doroshow, J.H. and Kensler, T.W. 1992. Copper-dependent site-specific mutagenesis by benzoyl peroxide in the *supF* gene of the mutation reporter plasmid pS189. *Carcinogenesis* **13**, 1783–7.

Akman, S.A., Kensler, T.W., Doroshow, J.H. and Dizdaroglu, M. 1993. Copper ion-mediated modification of bases in DNA *in vitro* by benzoyl peroxide. *Carcinogenesis* **14**, 1971–4.

Albertini, R., Rindi, S., Passi, A., Pallavicini, G. and De Luca, G. 1996. Heparin protection against Fe^{2+}- and Cu^{2+}-mediated oxidation of liposomes. *FEBS Letters* **383**, 155–8.

Altman, S.A., Zastawny, T.H., Randers-Eichhorn, L., Cacciuttolo, M.A., Akman, S.A.,

Dizdaroglu, M. and Rao G. 1995. Formation of DNA-protein cross-links in cultured mammalian cells upon treatment with iron ions. *Free Radicals in Biology and Medicine* **19**, 897–902.

Ames, B.N. 1989. Mutagenesis and carcinogenesis. *Environmental and Molecular Mutagenesis* **14** (Suppl. 16), 66–77.

Amoruso, M.A., Witz, G. and Goldstein, B.D. 1982. Enhancement of rat and human phagocyte superoxide anion radical production by cadmium *in vitro*. *Toxicology Letters* **10**, 133–8.

Andersen, H.R and Andersen, O. 1989. Effect of nickel chloride on hepatic lipid peroxidation and glutathione concentration in mice. *Biological Trace Element Research* **21**, 255–61.

Arora, A.S. and Gores, G.J. 1996. The role of metals in ischemia/reperfusion injury of the liver. *Seminars in Liver Disease* **16**, 31–8.

Aruoma, O.I., Halliwell, B. and Dizdaroglu, M. 1989. Iron ion-dependent modification of bases in DNA by the superoxide radical-generating system hypoxanthine/xanthine oxidase. *Journal of Biological Chemistry* **264**, 13024–8.

Athar, M., Hasan, S.K. and Srivastava, R.C. 1987. Evidence for the involvement of hydroxyl radicals in nickel-mediated enhancement of lipid peroxidation: Implications for nickel carcinogenesis. *Biochemical Biophysical Research Communications* **147**, 1276–81.

Aust, S.D. and Svingen, B.A. 1982. The role of iron in enzymatic lipid peroxidation, in *Free Radicals in Biology*, (ed. W.A. Pryor), Academic Press, New York, Vol. 5, pp. 1–28.

Babior, B.M. 1978. Oxygen-dependent microbial killing by phagocytes. *New England Journal of Medicine* **298**, 659–68.

Baker, A., Payne, C.M., Briehl, M.M. and Powis, G. 1997. Thioredoxin, a gene found overexpressed in human cancer, inhibits apoptosis *in vitro* and *in vivo*. *Cancer Research* **57**, 5162–7.

Bal, W., Lukszo, J., Jezowska-Bojczuk, M. and Kasprzak, K.S. 1995. Interactions of nickel(II) with histones: Stability and solution structure of complexes with CH_3CO-Cys-Ala-Ile-His-NH_2, a putative metal binding sequence of histone H3. *Chemical Research in Toxicology* **8**, 683–92.

Bal, W., Lukszo, J. and Kasprzak, K.S. 1996. Interactions of nickel(II) with histones: Enhancement of 2′-deoxyguanosine oxidation by Ni(II) complexes with CH_3CO-Cys-Ala-Ile-His-NH_2, a putative metal binding sequence of histone H3. *Chemical Research in Toxicology* **9**, 535–40.

Bal, W., Lukszo, J., Jezowska-Bojczuk, M. and Kasprzak, K.S. 1997a. Binding of nickel(II) and copper(II) to the N-terminal sequence of human protamine HP2. *Chemical Research in Toxicology* **10**, 906–14.

Bal, W., Lukszo, J. and Kasprzak, K.S. 1997b. Mediation of oxidative DNA damage by nickel(II) and copper(II) complexes with the N-terminal sequence of human protamine HP2. *Chemical Research in Toxicology* **10**, 915–21.

Bal, W., Lukszo, J., Bialkowski, K. and Kasprzak, K.S. 1998. Interactions of nickel(II) with histones: Interactions of Ni(II) with CH_3CO-Thr-Glu-Ser-His-His-Lys-NH_2, a peptide modeling the potential metal binding site in the 'C-tail' region of histone H2A. *Chemical Research inToxicology* **11**, 1014–23.

Barletta, J.M., Rainier, S. and Feinberg, A.P. 1997. Reversal of loss of imprinting in tumor cells by 5-aza-2′-deoxycytidine. *Cancer Research* **57**, 48–50.

Basu, A.K., Loechler, E.L., Leadon, S.A. and Essigman, J.M. 1989. Genetic effects of thymine glycol: Site-specific mutagenesis and molecular modeling studies. *Proceedings of the National Academy of Sciences (U.S.A.)* **86**, 7677–81.

Baeuerle, P.A. and Baltimore, D. 1996. NF-κB: Ten years after. *Cell* **87**, 13–20.

Berger, M., Anselmino, C., Mouret, J.F. and Cadet, J. 1990. High performance liquid chromatography – electrochemical assay for monitoring the formation of 8-oxo-7,8-dihydroadenine and its related 2′-deoxyribonucleoside. *Journal of Liquid Chromatography* **13**, 929–40.

Berger, M., de Hazen, M., Nejjari, A., Fournier, J., Guignard, J., Pezerat, H. and Cadet, J. 1993. Radical oxidation reactions of the purine moiety of 2′-deoxyribonucleosides and DNA by iron-containing minerals. *Carcinogenesis* **14**, 41–6.

Bessho, T., Roy, R., Yamamoto, K., Kasai, H., Nishimura, S., Tano, K. and Mitra, S. 1993. Repair of 8-hydroxyguanine in DNA by mammalian *N*-methylpurine-DNA glycosylase. *Proceedings of the National Academy of Sciences (U.S.A.)* **90**, 8901–4.

Beyersmann, D. and Hartwig, A. 1992. The genetic toxicology of cobalt. *Toxicology and Applied Pharmacology* **115**, 137–45.

Beyersmann, D. and Hechtenberg, S. 1997. Cadmium, gene regulation, and cellular signalling in mammalian cells. *Toxicology and Applied Pharmacology* **144**, 247–61.

Bialkowski, K. and Kasprzak, K.S. 1998. A novel assay of 8-oxo-2′-deoxyguanosine 5′-triphosphate pyrophosphohydrolase (8-oxo-dGTPase) activity in cultured cells and its use for evaluation of cadmium(II) inhibition of this activity. *Nucleic Acids Research* **26**, 3194–201.

Bilimoria, M.H. and Gupta, S.V. 1986. Comparison of the mutagenic activity of 5-hydroxy-methyldeoxyuridine with 5-substituted 2′-deoxyuridine analogs in the Ames *Salmonella/* microsome test. *Mutation Research* **169**, 123–7.

Bjornstedt, M., Kumar, S., Bjorkhem, L., Spyrou, G. and Holmgren, A. 1997. Selenium and the thioredoxin and glutaredoxin systems. *Biomedical and Environmental Sciences* **10**, 271–9.

Blankenship, L.J., Manning, F.C.R., Orenstein, J.M. and Patierno, S.R. 1994. Apoptosis is the mode of cell death caused by carcinogenic chromium. *Toxicology and Applied Pharmacology* **126**, 75–83.

Bodi, I., Bishopric, N.H., Discher, D.J., Wu, X. and Webster, K.A. 1995. Cell-specificity and signaling pathway of endothelin-1 gene regulation by hypoxia. *Cardiovascular Research* **30**, 975–84.

Borges, K.M. and Wetterhahn, K.E. 1989. Chromium cross-links glutathione and cysteine to DNA. *Carcinogenesis* **10**, 2165–8.

Bossu, F.P., Paniago, E.B., Margerum, D.W., Kirksey, S.T. and Kurtz, J.L. 1978. Trivalent nickel catalysis of the autoxidation of nickel(II) tetraglycine. *Inorganic Chemistry* **17**, 1034–42.

Bouton, C., Hirling H. and Drapier, J.C. 1997. Redox modulation of iron regulatory proteins by peroxynitrite. *Journal of Biological Chemistry* **272**, 19969–75.

Brambilla, G., Martelli, A. and Marinari, U.M. 1989. Is lipid peroxidation associated with DNA damage? *Mutation Research* **214**, 123–7.

Breimer, L.H. 1990. Molecular mechanisms of oxygen radical carcinogenesis. *Molecular Carcinogenesis* **3**, 188–97.

Bryan, S.E. 1981. Heavy metals in the cell's nucleus, in *Metal Ions in Genetic Information Transfer*, (eds G.L. Eichhorn and L.G. Marzilli), Elsevier, New York, pp. 87–101.

Buckley, J. 1994. Male-mediated developmental toxicity: Paternal exposures and childhood cancer, in *Male-Mediated Developmental Toxicity*, (eds A.F. Olshan and D.R. Mattison), Plenum Press, New York, pp. 169–75.

Buettner, G.R. 1993. The pecking order of free radicals and antioxidants: Lipid peroxidation, α-tocopherol, and ascorbate. *Archives of Biochemistry and Biophysics* **300**, 535–43.

Butterfield, D.A. 1997. *β*-Amyloid-associated free radical oxidative stress and neurotoxicity: Implications for Alzheimer's disease. *Chemical Research in Toxicology* **10**, 495–506.

Byczkowski, J.Z., Wan, B. and Kulkarni, A. 1988. Vanadium-mediated lipid peroxidation in microsomes from human term placenta. *Bulletin of Environmental Contamination and Toxicology* **41**, 696–703.

Cadet, J. and Weinfeld, M. (1993). Detecting DNA damage. *Analytical Chemistry* **65**, 675A–682A.

Candeias, L.P. and Steenken, S. 1993. Electron transfer in di(deoxy)nucleoside phosphates in aqueous solution: Rapid migration of oxidative damage (via adenine) to guanine. *Journal of the American Chemical Society* **115**, 2437–40.

Cantoni, O., Christie, N.T., Swann, A., Drath, D.B. and Costa, M. 1984. Mechanism of $HgCl_2$ cytotoxicity in cultured mammalian cells. *Molecular Pharmacology* **26**, 360–8.
Carter, D.E. 1995. Oxidation–reduction reactions of metal ions. *Environmental Health Perspectives* **103** (Suppl. 1), 17–19.
Cerda, S. and Weitzman, S.A. 1997. Influence of oxygen radical injury on DNA methylation. *Mutation Research* **386**, 141–52.
Chang, J., Watson, W., Randerath, E. and Randerath, K. 1993. Bulky DNA adduct formation induced by Ni(II) *in vitro* and *in vivo* as assayed by ^{32}P-postlabeling. *Mutation Research* **291**, 147–59.
Cheng, C.C., Rokita, S.E. and Burrows, C.J. 1993. Nickel(III)-promoted DNA cleavage with ambient dioxygen. *Angewandte Chemie International* **32**, 277–8 [ed. in English].
Cheng, K.C., Cahill, D.S., Kasai, H., Nishimura, S. and Loeb, L.A. 1992. 8-Hydroxy-guanine, an abundant form of oxidative DNA damage, causes G → T and A → C substitutions. *Journal of Biological Chemistry* **267**, 166–72.
Cheshire, J.L. and Baldwin, A.S. 1997. Synergistic activation of NF-κB by tumor necrosis factor alpha and gamma interferon via enhanced IκBα degradation and *de novo* IκBβ degradation. *Molecular and Cellular Biology* **17**, 6746–54.
Chiocca, S.M., Sterner, D.A., Biggart, N.W. and Murphy, E.C. Jr 1991. Nickel mutagenesis: Alteration of the MuSVts110 thermosensitive splicing phenotype by a nickel-induced duplication of the 3′ splice site. *Molecular Carcinogenesis* **4**, 61–71.
Chung, M.H., Kasai, H., Jones, D.S., Inoue, H., Ishikawa, H., Ohtsuka, E. and Nishimura, S. 1991a. An endonuclease activity of *Escherichia coli* that specifically removes 8-hydroxyguanine residues from DNA. *Mutation Research* **254**, 1–12.
Chung, M.H., Kim, H.K., Ohtsuka, E., Kasai, H., Yamamoto, F. and Nishimura, S. 1991b. An endonuclease activity in human polymorphonuclear neutrophils that removes 8-hydroxyguanine residues from DNA. *Biochemical and Biophysical Research Communications* **178**, 1472–8.
Climent, I., Tsai, L. and Levine, R.L. 1989. Derivatization of gamma-glutamyl semialdehyde residues in oxidized proteins by fluoresceinamine. *Analytical Biochemistry* **182**, 226–32.
Colburn, N. 1992. Gene regulation by active oxygen and other stress inducers: Role in tumor promotion and progression, in *Biological Consequences of Oxidative Stress. Implications for Cardiovascular Disease and Carcinogenesis*, (eds L. Spatz and A.D. Bloom), Oxford University Press, Oxford, pp. 122–37.
Connell, P., Yang, V.M., Toborek, M., Cohen, D.A., Barve, S., McClain, C.J. and Henning, B. 1997. Zinc attenuates tumor necrosis factor-mediated activation of transcription factors in endothelial cells. *Journal of the American College of Nutrition* **16**, 411–17.
Costa, D., Guignard, J. and Pezerat, H. 1989a. Production of free radicals arising from the surface activity of minerals and oxygen. Part II. Arsenides, sulfides, and sulfoarsenides of iron, nickel, and copper. *Toxicology and Industrial Health* **5**, 1079–97.
Costa, D., Guignard, J. and Pezerat, H. 1989b. Production of radicals by non-fibrous materials in a cell-free buffer medium. *NATO ASI Series* **H30**, 189–96.
Costa, D., Guignard, J., Zalma, R. and Pezerat, H. 1989c. Production of free radicals arising from the surface activity of minerals and oxygen. Part I. Iron mine ores. *Toxicology and Industrial Health* **5**, 1061–78.
Costa, M., Zhitkovich, A. and Toniolo, P. 1993. DNA–protein cross-links in welders: Molecular implications. *Cancer Research* **53**, 460–3.
Costa, M., Zhitkovich, A., Gargas, M., Paustenbach, D., Finley, B., Kuykendall, J., Billings, R., Carlson, T.J., Wetterhahn, K., Xu, J., Patierno, S. and Bogdanffy M. 1996. Interlaboratory validation of a new assay for DNA–protein crosslinks. *Mutation Research* **369**, 13–21.
Cotelle, N., Tremolieres, E., Bernier, J.L., Catteau, J.P. and Henichart, J.P. 1992. Redox chemistry

of complexes of nickel(II) with some biologically important peptides in the presence of reduced oxygen species. An ESR study. *Journal of Inorganic Chemistry* **46**, 7–15.

Crawford, R.D. 1995. Proposed role for a combination of citric acid and ascorbic acid in the production of dietary iron overload: A fundamental cause of disease. *Biochemistry and Molecular Medicine* **54**, 1–11.

Cross, C.E., Halliwell, B., Borish, E.T., Pryor, W.A., Ames, B.N., Saul, R.L., McCord, J.M. and Harman, D. 1987. Oxygen radicals and human disease. *Annals of Internal Medicine* **107**, 526–45.

Darley-Usmar, V. and Halliwell, B. 1996. Reactive nitrogen species, reactive oxygen species, transition metal ions, and the vascular system. *Pharmaceutical Research* **13**, 649–62.

Datta, A.K., Riggs, C.W., Fivash, M.J. and Kasprzak, K.S. 1991. Mechanisms of nickel carcinogenesis: Interaction of Ni(II) with 2′-deoxynucleosides and 2′-deoxynucleotides. *Chemico-Biological Interactions* **79**, 323–34.

Datta, A.K., Misra, M., North, S.L. and Kasprzak, K.S. 1992. Enhancement by nickel(II) and L-histidine of 2′-deoxyguanosine oxidation with hydrogen peroxide. *Carcinogenesis* **13**, 283–7.

Datta, A.K., Shi, X. and Kasprzak, K.S. 1993. Effect of carnosine, homocarnosine, and anserine on hydroxylation of the guanine moiety in 2′-deoxyguanosine, DNA, and nucleohistone with hydrogen peroxide in the presence of nickel(II). *Carcinogenesis* **14**, 417–22.

Davidson, J.S., Baumgarten, I.M. and Harley, E.H. 1984. Effects of extracellular calcium and magnesium on junctional intercellular communication in human fibroblasts. *Experimental Cell Research* **155**, 406–12.

Demple, B. and Amabile-Cuevas, C.F. 1991. Redox-redux: The control of oxidative stress responses. *Cell* **67**, 837–9.

Demple, B. and Harrison, L. 1994. Repair of oxidative damage to DNA: Enzymology and biology. *Annual Review of Biochemistry* **63**, 915–48.

Deng, C, Lee, H.H., Xian, H., Yao, M., Huang, J. and Ou, B. 1988. Chromosomal aberrations and sister chromatid exchanges of peripheral blood lymphocytes in Chinese electroplating workers: Effect of nickel and chromium. *Journal of Trace Elements in Experimental Medicine* **1**, 57–62.

Dix, T.A. and Aikens, J. 1993. Mechanisms and biological relevance of lipid peroxidation initiation. *Chemical Research in Toxicology* **6**, 2–18.

Dizdaroglu, M. 1991. Chemical determination of free radical-induced damage to DNA. *Free Radicals in Biology and Medicine* **10**, 225–42.

Dizdaroglu, M. 1992. Oxidative damage to DNA in mammalian chromatin. *Mutation Research* **275**, 331–42.

Dizdaroglu, M. and Simic, M.G. 1984a. Radiation-induced crosslinking of cytosine. *Radiation Research* **100**, 41–6.

Dizdaroglu, M. and Simic, M.G. 1984b. Radiation-induced formation of thymine–thymine crosslinks. *International Journal of Radiation Biology* **46**, 241–6.

Dizdaroglu, M., Rao, G., Halliwell, B. and Gajewski, E. 1991. Damage to the DNA bases in mammalian chromatin by hydrogen peroxide in the presence of ferric and cupric ions. *Archives of Biochemistry and Biophysics* **285**, 317–24.

Donaldson, K., Beswick, P.H. and Gilmour, P.S. 1996. Free radical activity associated with the surface of particles: A unifying factor in determining biological activity? *Toxicology Letters* **88**, 293–8.

Dowjat, K.W., Kharatishvili, M. and Costa, M. 1996. DNA and RNA strand scission by copper, zinc and manganese superoxide dismutases. *BioMetals* **9**, 327–35.

Drapier, J.-C. and Bouton, C. 1996. Modulation by nitric oxide of metallo-protein regulatory activities. *BioEssays* **18**, 549–56.

Drapier, J.-C., Hirling, H., Wietzerbin, J., Kaldy, P. and Kühn, L.C. 1993. Biosynthesis of nitric oxide activates ion regulatory factor in macrophages. *EMBO Journal* **12**, 3643–9.

El Din, M.S., Schaur, R.J. and Schauenstein, E. 1988. Uptake of ferrous iron histidinate, a promoter of lipid peroxidation, by Ehrlich ascites tumor cells. *Biochimica et Biophysica Acta* **962**, 37–41.

Eritja, R., Horowitz, D.M., Walker, P.A., Ziehler-Martin, J.P., Boosalis, M.S., Goodman, M.F., Itakura, K. and Kaplan, B.E. 1986. Synthesis and properties of oligonucleotides containing 2′-deoxynebularine and 2′-deoxyxanthosine. *Nucleic Acids Research* **14**, 8135–53.

Feig, D.I., Sowers, L.C. and Loeb, L.A. 1994. Reverse chemical mutagenesis: Identification of the mutagenic lesions resulting from reactive oxygen species-mediated damage to DNA. *Proceedings of the National Academy of Sciences (U.S.A.)* **91**, 6609–13.

Felley-Bosco, E., Buzard, G.S. and Keefer, L.K. 1998. Nitric oxide, altered DNA, and mammalian disease, in *Molecular Biology of Free Radicals in Human Diseases*, (eds O.I. Aruoma and B. Halliwell), OICA International, Castries, St. Lucia, pp. 287–325.

Fleet, J.C., Golemboski, K.A., Dietert, R.R., Andrews, G.K. and McCormick, C.C. 1990. Induction of hepatic metallothionein by intraperitoneal metal injection: An associated inflammatory response. *American Journal of Physiology* **258**, G926–G933.

Frenkel, K. 1989. Oxidation of DNA bases by tumor promoter-activated processes. *Environmental Health Perspectives* **81**, 45–54.

Frenkel, K. and Chrzan, K. 1987. Radiation-like modification of DNA and H_2O_2 formation by activated human polymorphonuclear leukocytes (PMNs), in *Anticarcinogenesis and Radiation Protection*, (eds P.A. Cerutti, O.F. Nygaard and M.G. Simic), Plenum Press, New York, pp. 97–102.

Frenkel, K. and Tofigh, S. 1989. Decomposition of nucleoside hydroperoxide by metals and metalloproteins. *Biological Trace Element Research* **21**, 351–7.

Frenkel, K., Blum, F. and Troll, W. 1986. Copper ions and hydrogen peroxide form hypochlorite from NaCl thereby mimicking myeloperoxidase. *Journal of Cellular Biochemistry* **30**, 181–93.

Frenkel, K., Zhong, Z., Wei, H., Karkoszka, J., Patel, U., Rashid, K., Georgescu, M. and Solomon, J.J. 1991. Quantitative high performance liquid chromatography analysis of DNA oxidized *in vitro* and *in vivo*. *Analytical Biochemistry* **196**, 126–36.

Frenkel, K., Karkoszka, J., Cohen, B., Baranski, B., Jakubowski, M., Cosma, G., Taioli, M. and Toniolo, P. 1994. Occupational exposures to Cd, Ni and Cr modulate titers of anti-oxidized DNA base antibodies. *Environmental Health Perspectives* **102** (Suppl. 3), 221–5.

Fu, S., Gebicki, S., Jessup, W., Gebicki, J.M. and Dean, R.T. 1995. Biological fate of amino acid, peptide and protein hydroperoxides. *Biochemical Journal* **311**, 821–7.

Fung, H., Kow., Y.W., Van Houten, B., Taatjes, D.J., Hatahet, Z., Janssen, Y.M.W., Vacek, P., Faux, S.P. and Mossman, B.T. 1998. Asbestos increases mammalian AP-endonuclease gene expression, protein levels, and enzyme activity in mesothelial cells. *Cancer Research* **58**, 189–94.

Galter, D., Mihm, S. and Droge, W. 1994. Distinct effects of glutathione disulphide on the nuclear transcription factors κB and the activator protein-1. *European Journal of Biochemistry* **221**, 639–48.

Gardner, P.R. and Fridovich, I. 1991. Superoxide sensitivity of the *Escherichia coli* aconitase. *Journal of Biological Chemistry* **266**, 19328–33.

Gill, G., Richter-Rusli, A.A., Ghosh, M., Burrows, C.J. and Rokita, S. 1997. Nickel-dependent oxidative cross-linking of a protein. *Chemical Research in Toxicology* **10**, 302–9.

Glass, G.A. and Stark, A.-A. 1997. Promotion of glutathione–γ–glutamyl transpeptidase-dependent lipid peroxidation by copper and ceruloplasmin: The requirement for iron and the effects of antioxidants and antioxidant enzymes. *Environmental and Molecular Mutagenesis* **29**, 73–80.

Goebeler, M., Roth, J., Broker, E.B., Sorg, C. and Schultze-Osthoff, K. 1995. Activation of nuclear factor-κB and gene expression in human endothelial cells by the common haptens nickel and cobalt. *Journal of Immunology* **155**, 2459–67.

Goodgame, D.M.L and Joy, A.M. 1986. Relatively long-lived chromium(V) species are produced by the action of glutathione on carcinogenic chromium(VI). *Journal of Inorganic Biochemistry* **26**, 219–24.

Gordge, M.P. 1998. How cytotoxic is nitric oxide? *Experimental Nephrology* **6**, 12–16.

Gow, A.J. and Stamler, J.S. 1998. Reactions between nitric oxide and hemoglobin under physiological conditions. *Nature* **391**, 169–73.

Grollman, A.P. and Moriya, M. 1993. Mutagenesis by 8-oxoguanine: An enemy within. *Trends in Genetics* **9**, 246–9.

Guillemin, K. and Krasnow, M.A. 1997. The hypoxic response: Huffing and HIFing. *Cell* **89**, 9–12.

Gutteridge, J.M. 1985. Superoxide dismutase inhibits the superoxide-driven Fenton reaction at two different levels: Implications for a wider protective role. *FEBS Letters* **185**, 19–23.

Hainaut, P., Rolley, N., Davies, M. and Milner, J. 1995. Modulation by copper of p53 conformation and sequence-specific DNA binding: Role for Cu(II)/Cu(I) redox mechanism. *Oncogene* **10**, 27–32.

Halliwell, B. 1995. Oxidation of low-density lipoproteins: Questions of initiation, propagation, and the effect of antioxidants. *American Journal of Clinical Nutrition* **61** (Suppl.), 670S–677S.

Halliwell, B. and Aruoma, O.I. 1991. DNA damage by oxygen-derived species: Its mechanism and measurement in mammalian system. *FEBS Letters* **281**, 9–19.

Halliwell, B. and Dizdaroglu, M. 1992. The measurement of oxidative damage to DNA by HPLC and GC/MS techniques. *Free Radical Research Communications* **16**, 75–87.

Halliwell, B. and Gutteridge, J.M.C. 1984. Oxygen toxicity, oxygen radicals, transition metals, and disease. *Biochemical Journal* **219**, 1–14.

Halliwell, B. and Gutteridge, J.M.C. 1986. Oxygen free radicals and iron in relation to biology and medicine: Some problems and concepts. *Archives of Biochemistry and Biophysics* **246**, 501–14.

Hamazaki, S., Okada, S., Li, J.L., Toyokuni, S. and Midorikawa, O. 1989. Oxygen reduction and lipid peroxidation by iron chelates with special reference to ferric-nitrilotriacetate. *Archives of Biochemistry and Biophysics* **272**, 10–17.

Hamilton-Koch, W., Snyder, R.D. and Lavelle, J.M. 1986. Metal-induced DNA damage and repair in human diploid fibroblasts and Chinese hamster ovary cells. *Chemico-Biological Interactions* **59**, 17–28.

Handel, M.L., Watts, C.K., deFazio, A., Day, R. O. and Sutherland, R.L. 1995. Inhibition of AP-1 binding and transcription by gold and selenium involving conserved cysteine residues in Jun and Fos. *Proceedings of the National Academy of Sciences (U.S.A.)* **92**, 4497–501.

Hanna, P.M., Kadiiska, M.B. and Mason, R.P. 1992. Oxygen-derived free radical and active oxygen complex formation from cobalt(II) chelates *in vitro*. *Chemical Research in Toxicology* **5**, 109–15.

Hartwig, A. 1994. Role of DNA repair inhibition in lead- and cadmium-induced genotoxicity: A review. *Environmental Health Perspectives* **102** (Suppl. 3), 45–50.

Hartwig, A. 1995. Current aspects in metal genotoxicity. *BioMetals* **8**, 3–11.

Hartwig, A. and Beyersmann, D. 1989. Comutagenicity and inhibition of DNA repair by metal ions in mammalian cells. *Biological Trace Element Research* **21**, 359–65.

Hartwig, A. and Schlepegrell, R. 1995. Induction of oxidative DNA damage by ferric iron in mammalian cells. *Carcinogenesis* **16**, 3009–13.

Hartwig, A., Snyder, R.D., Schlepegrell, R. and Beyersmann, D. 1991. Modulation by Co(II) of UV-induced DNA repair, mutagenesis, and sister chromatid exchanges in mammalian cells. *Mutation Research* **248**, 177–85.

Hartwig, A., Schlepegrell, R. and Beyersmann, D. 1992. Interactions in nickel mutagenicity: Indirect mechanisms in genotoxicity and interference with DNA repair, in *Metal Compounds in Environment and Life*, (eds E. Merian and W. Haerdi), Science Reviewers, Inc, Wilmington, Vol. 4, pp. 475–80.

Hartwig, A., Klyszcz-Nasko, H., Schlepegrell, R. and Beyersmann, D. 1993. Cellular damage by ferric nitrilotriacetate and ferric citrate in V79 cells: Interrelationship between lipid peroxidation, DNA strand breaks and sister chromatid exchanges. *Carcinogenesis* **14**, 107–12.

Hartwig, A., Mullenders, L.H.F., Schlepegrell, R., Kasten, U. and Beyersmann, D. 1994. Nickel(II) interferes with the incision step in nucleotide excision repair in mammalian cells. *Cancer Research* **54**, 4045–51.

Hassoun, E.A. and Stohs, S.J. 1996. Cadmium-induced production of superoxide anion and nitric oxide, DNA single strand breaks and lactate dehydrogenase leakage in J774A.1 cell cultures. *Toxicology* **112**, 219–26.

Hausladen, A. and Fridovich, I. 1994. Superoxide and peroxynitrite inactivate aconitases, but nitric oxide does not. *Journal of Biological Chemistry* **269**, 29405–8.

Hawkins, C.L. and Davies, M.J. 1997. Oxidative damage to collagen and related substrates by metal ion/hydrogen peroxide systems: Random attack or site-specific damage? *Biochimica et Biophysica Acta* **1360**, 84–96.

Hayes, R.B. 1997. The carcinogenicity of metals in humans. *Cancer Causes and Control* **8**, 371–85.

Hayes, R.C., Petrullo, L.A., Huang, H., Wallace, S.S. and LeClerc, J.E. 1988. Oxidative damage in DNA: Lack of mutagenicity of thymine glycol lesions. *Journal of Molecular Biology* **201**, 239–46.

Henle, E.S., Luo, Y., Gassmann, W. and Linn, S. 1996. Oxidative damage to DNA constituents by iron-mediated Fenton reactions. The deoxyguanosine family. *Journal of Biological Chemistry* **271**, 21177–86.

Hermes-Lima, M., Valle, V.G.R, Vercesi, A.E. and Bechara, E.J.H. 1991. Damage to rat liver mitochondria promoted by δ-aminolevulinic acid-generated reactive oxygen species: Connections with acute intermittent porphyria and lead poisoning. *Biochimica et Biophysica Acta* **1056**, 57–63.

Higinbotham, K.G., Rice, J.M., Diwan, B.A., Kasprzak, K.S., Reed, C.D. and Perantoni, A. 1992. GGT to GTT transversions in codon 12 of the K-*ras* oncogene in rat renal sarcomas induced with nickel subsulfide or nickel subsulfide/iron are consistent with oxidative damage to DNA. *Cancer Research* **52**, 4747–51.

Hiraku, Y. and Kawanishi, S. 1996. Mechanism of oxidative DNA damage induced by δ-aminolevulinic acid in the presence of copper ion. *Cancer Research* **56**, 1786–93.

Hirota, K., Matsui, M., Iwata, S., Nishiyama, A., Mori, K. and Yodoi, J. 1997. AP-1 transcriptional activity is regulated by a direct association between thioredoxin and Ref-1. *Proceedings of the National Academy of Science (U.S.A.)* **94**, 3633–8.

Huang, X., Frenkel, K., Klein, C.B. and Costa, M. 1993. Nickel induces increased oxidants in intact cultured mammalian cells as detected by dichlorofluorescein fluorescence. *Toxicology and Applied Pharmacology* **120**, 29–36.

Huang, X., Kitahara, J., Zhitkovich, A., Dowjat, K. and Costa, M. 1995. Heterochromatic proteins specifically enhance nickel-induced 8-oxo-dG formation. *Carcinogenesis* **16**, 1753–9.

Huang, L.E., Arany, Z., Livingston, D.M. and Bunn, H.F. 1996. Activation of hypoxia-inducible transcription factor depends primarily upon redox-sensitive stabilization of its alpha subunit. *Journal of Biological Chemistry* **271**, 32253–9.

Husain, M.M., Ahmad, N., Gupta, S., Behari, J.R., Hasan, S.K., Srivastava, S.K. and Srivastava, R.C. 1995. Exacerbation of nickel-induced oxidative response by vitamin E. *Industrial Health* **33**, 143–52.

Imlay, J.A. and Linn, S. 1988. DNA damage and oxygen radical toxicity. *Science* **240**, 1302–9.

Inoue, S. and Kawanishi, S. 1987. Hydroxyl radical production and human DNA damage induced by ferric nitrilotriacetate and hydrogen peroxide. *Cancer Research* **47**, 6522–7.

Ischiropoulos, H., Zhou, L., Chen, J., Tsai, M., Martin, V.C., Smith, C.D. and Beckman, J.S. 1992. Peroxynitrite-mediated tyrosine nitration catalyzed by superoxide dismutase. *Archives of Biochemistry and Biophysics* **298**, 431–7.

Ito, K. and Kawanishi, S. 1991. Site-specific fragmentation and modification of albumin by sulfite in the presence of metal ions or peroxidase/H_2O_2: Role of sulfate radical. *Biochemical and Biophysical Research Communications* **176**, 1306–12.

Ito, K., Yamamoto, K. and Kawanishi, S. 1992. Manganese-mediated oxidative damage of cellular and isolated DNA by isoniazid and related hydrazines: Non-Fenton type hydroxyl radical formation. *Biochemistry* **31**, 11606–13.

Jackson, J.H., Gajewski, E., Schraufstatter, I., Hyslop. P.A., Fuciarelli, A.F., Cochrane, C.G. and Dizdaroglu, M. 1989. Damage to the bases in DNA induced by stimulated human neutrophils. *Journal of Clinical Investigation* **84**, 1644–9.

Jin, D.Y., Chae, H.Z., Rhee, S.G. and Jeang, K.T. 1997. Regulatory role of a novel human thioredoxin peroxidase in NF-κB activation. *Journal of Biological Chemistry* **272**, 30952–61.

Kadiiska, M.B., Maples, K.R. and Mason, R.P. 1989. A comparison of cobalt(II) and iron(II) hydroxyl and superoxide free radical formation. *Archives of Biochemistry and Biophysics* **275**, 98–111.

Kadiiska, M.B., Hanna, P.M., Hernandez, L. and Mason, R.P. 1992. *In vivo* evidence of hydroxyl radical formation after acute copper and ascorbic acid intake: Electron spin resonance spin-trapping investigation. *Molecular Pharmacology* **42**, 723–9.

Kamiya, H. and Kasai, H. 1996. Effects of sequence contexts on misincorporation of nucleotides opposite 2-hydroxyadenine. *FEBS Letters* **391**, 113–16.

Kamiya, H., Miura, H., Ishikawa, H., Inoue, H., Nishimura, S. and Ohtsuka, E. 1992. c-Ha-*ras* containing 8-hydroxyguanine at codon 12 induces point mutations at the modified and adjacent positions. *Cancer Research* **52**, 3483–5.

Kamiya, H., Miura, H., Murata-Kamiya, N., Ishikawa, H., Sakaguchi, T., Inoue, H., Sasaki, T., Masutani, C., Hanaoka, F., Nishimura, S. and Ohtsuka, E. 1995a. 8-Hydroxyadenine (7,8-dihydro-8-oxoadenine) induces misincorporation in *in vitro* DNA synthesis and mutations in NIH 3T3 cells. *Nucleic Acids Research* **23**, 2893–9.

Kamiya, H., Ueda, T., Ohgi, T., Matsukage, A. and Kasai, H. 1995b. Misincorporation of dAMP opposite 2-hydroxyadenine, an oxidative form of adenine. *Nucleic Acids Research* **23**, 761–6.

Kanner, J., Harel, S. and Granit, R. 1991. Nitric oxide as an antioxidant. *Archives of Biochemistry and Biophysics* **289**, 130–6.

Kargacin, B., Klein, C.B. and Costa, M. 1993. Mutagenic responses of nickel oxides and sulfides in Chinese hamster V79 cell lines at the xanthine–guanine phosphoribosyl transferase locus. *Mutation Research* **300**, 63–72.

Kasai, H. and Nishimura, S. 1984a. DNA damage induced by asbestos in the presence of hydrogen peroxide. *Gann* **75**, 841–4.

Kasai, H. and Nishimura, S. 1984b. Hydroxylation of deoxyguanosine at the C-8 position by polyphenols and aminophenols in the presence of hydrogen peroxide and ferric ion. *Gann* **75**, 565–6.

Kasai, H., Crain, P.F., Kuchino, Y., Nishimura, S., Ootsuyama, A. and Tanooka, H. 1986. Formation of 8-hydroxyguanine moiety in cellular DNA by agents producing oxygen radicals and evidence for its repair. *Carcinogenesis* **7**, 1849–51.

Kashiba-Iwatsuki, M., Miyamoto, M. and Inoue, M. 1997. Effect of nitric oxide on the ligand-binding activity of albumin. *Archives of Biochemistry and Biophysics* **345**, 237–42.

Kasprzak, K.S. 1991. The role of oxidative damage in metal carcinogenicity. *Chemical Research in Toxicology* **4**, 604–15.

Kasprzak, K.S. 1995. Possible role of oxidative damage in metal-induced carcinogenesis. *Cancer Investigation* **13**, 411–30.

Kasprzak, K.S. 1996. Oxidative DNA damage in metal-induced carcinogenesis, in *Toxicology of Metals*, (eds L.W. Chang, L. Magos and T. Suzuki), Lewis Publishers, Boca Raton, pp. 299–320.

Kasprzak, K.S. 1997a. The oxidative damage hypothesis of metal-induced genotoxicity and carcinogenesis, in *Cytotoxic, Mutagenic and Carcinogenic Potential of Heavy Metals Related to Human Environment*, (ed. N.D. Hadjiliadis), Kluwer Academic Publishers, Dordrecht, NATO-ASI Series, Vol. 26, pp. 73–92.

Kasprzak, K.S. 1997b. Effects of calcium, magnesium, zinc, and iron on nickel carcinogenesis: Inhibition versus enhancement, in *Cytotoxic, Mutagenic and Carcinogenic Potential of Heavy Metals Related to Human Environment*, (ed. N.D. Hadjiliadis, Kluwer Academic Publishers, Dordrecht, NATO-ASI Series, Vol. 26, pp. 93–106.

Kasprzak, K.S. and Bare, R.M. 1989. *In vitro* polymerization of histones by carcinogenic nickel compounds. *Carcinogenesis* **10**, 621–4.

Kasprzak, K.S. and Hernandez, L. 1989. Enhancement of hydroxylation and deglycosylation of 2′-deoxyguanosine by carcinogenic nickel compounds. *Cancer Research* **49**, 5964–8.

Kasprzak, K.S. and Poirier, L.A. 1985. Effects of calcium(II) and magnesium(II) on nickel(II) uptake and stimulation of thymidine incorporation into DNA in the lungs of strain A mice. *Carcinogenesis* **6**, 1819–21.

Kasprzak, K.S. and Sunderman, F.W. Jr 1977. Mechanisms of dissolution of nickel subsulfide in rat serum. *Research Communications in Chemical Pathology and Pharmacology* **16**, 95–108.

Kasprzak, K.S., Ward, J.M., Poirier, L.A., Reichardt, D.A., Denn, A.C. III and Reynolds, C.W. 1987. Nickel–magnesium interactions in carcinogenesis: dose-effects and involvement of natural killer cells. *Carcinogenesis* **8**, 1005–11.

Kasprzak, K.S., Diwan, B.A., Konishi, N, Misra, M. and Rice, J.M. 1990. Initiation by nickel acetate and promotion by sodium barbital of renal cortical epithelial tumors in male F344 rats. *Carcinogenesis* **11**, 647–52.

Kasprzak, K.S., Misra, M., Rodriguez, R.E. and North, S.L. 1991a. Nickel-induced oxidation of renal DNA guanine residues *in vivo* and *in vitro*. *Toxicologist* **11**, 233.

Kasprzak, K.S., North, S.L. and Keefer, L.K. 1991b. Deamination of 5-methyl-2′-deoxycytidine with carcinogenic nickel subsulfide and its metabolite nickel(II) sulfite. *Proceedings of the American Association for Cancer Research* **32**, 108.

Kasprzak, K.S., Diwan, B.A., Rice, J.M., Misra, M., Riggs, C.W., Olinski, R. and Dizdaroglu, M. 1992a. Nickel(II)-mediated oxidative DNA base damage in renal and hepatic chromatin of pregnant rats and their fetuses: Possible relevance to carcinogenesis. *Chemical Research in Toxicology* **5**, 809–15.

Kasprzak, K.S., North, S.L. and Hernandez, L. 1992b. Reversal by nickel(II) of inhibitory effects of some scavengers of active oxygen species upon hydroxylation of 2′-deoxyguanosine *in vitro*. *Chemico-Biological Interactions* **84**, 11–19.

Kasprzak, K.S., Diwan, B.A. and Rice, J.M. 1994a. Iron accelerates while magnesium inhibits nickel-induced carcinogenesis in the rat kidney. *Toxicology* **90**, 129–40.

Kasprzak, K.S., Zastawny, T.H., North, S.L., Riggs, C.W., Diwan, B.A., Rice, J.M. and Dizdaroglu, M. 1994b. Oxidative DNA base damage in renal, hepatic, and pulmonary chromatin of rats after intraperitoneal injection of cobalt(II) acetate. *Chemical Research in Toxicology* **7**, 329–35.

Kasprzak, K., Jaruga, P., Zastawny, T.H., North, S.L., Riggs, C.W., Olinski, R. and Dizdaroglu, M. 1997. Oxidative DNA damage and its repair in kidneys and livers of nickel(II)-treated male F344 rats. *Carcinogenesis* **18**, 271–7.

Kasprzak, K.S., Bal, W., Porter, D.W. and Bialkowski, K. 1999. Studies on oxidative mechanisms of metal-induced carcinogenesis: Recent developments, in *Advances in DNA Damage and Repair*, (ed. M. Dizdaroglu and A. Karakaya), Kluwer Academic/Plenum Publishers, New York, pp. 193–208.

Kawanishi, S. and Yamamoto, K. 1991. Mechanism of site-specific DNA damage induced by methyl hydrazines in the presence of copper(II) and manganese(II). *Biochemistry* **30**, 3069–75.

Kawanishi, S., Inoue, S. and Sano, S. 1986. Mechanism of DNA cleavage induced by sodium chromate(VI) in the presence of hydrogen peroxide. *Journal of Biological Chemistry* **261**, 5952–8.

Kawanishi, S., Inoue, S. and Yamamoto, K. 1989a. Hydroxyl radical and singlet oxygen production and DNA damage induced by carcinogenic metal compounds and hydrogen peroxide. *Biological Trace Element Research* **21**, 367–72.

Kawanishi, S., Inoue, S. and Yamamoto, K. 1989b. Site-specific DNA damage by nickel(II) ion in the presence of hydrogen peroxide. *Carcinogenesis* **12**, 2231–5.

Kazakov, S.A. 1996. Nucleic acid binding and catalysis by metal ions, in *Bioorganic Chemistry: Nucleic Acids*, (ed. S.M. Hecht), Oxford University Press, New York, pp. 244–87.

Keyer, K. and Imlay, J.A. 1996. Superoxide accelerates DNA damage by elevating free-iron levels. *Proceedings of the National Academy of Sciences (U.S.A.)* **93**, 13635–40.

Kienzl, E., Puchinger, L., Jellinger, K., Linert, W., Stachelberger, H. and Jameson, R.F. 1995. The role of transition metals in the pathogenesis of Parkinson's disease. *Journal of Neurological Sciences* **134** (Suppl.), 69–78.

Kim, I.Y. and Stadtman, T.C. 1997. Inhibition of NF-κB DNA binding and nitric oxide induction in human T cells and lung adenocarcinoma cells by selenite treatment. *Proceedings of the National Academy of Sciences (U.S.A.)* **94**, 12904–7.

Kito, M., Takenaka, Y. and Urade, R. 1995. The 1,10-phenantroline micelles-copper(I) complex catalyzes protein degradation. *FEBS Letters* **362**, 39–42.

Klaunig, J.E., Xu, Y., Isenberg, J.S., Bachowski, S., Kolaja K.L., Jiang, J., Stevenson, D.E. and Walborg, E.F. Jr 1998. The role of oxidative stress in chemical carcinogenesis. *Environmental Health Perspectives* **106** (Suppl. 1), 289–95.

Klausner, R.D., Rouault, T.A. and Harford, J.B. 1993. Regulating the fate of mRNA: The control of cellular iron metabolism. *Cell* **72**, 19–28.

Klein, C.B. and Costa, M. 1997. DNA methylation, heterochromatin and epigenetic carcinogens. *Mutation Research* **386**, 163–80.

Klein, C.B., Frenkel, K. and Costa, M. 1991. The role of oxidative processes in metal carcinogenesis. *Chemical Research in Toxicology* **4**, 592–604.

Klein, J.C., Bleeker, M.J., Saris, C.P., Roelen, H.C.P.F., Brugghe, H.F., van den Elst, H., van der Marel, G.A., van Boom, J.H., Westra, J.G., Kriek, E. and Berns, A.J.M. 1992. Repair and replication of plasmids with site-specific 8-oxodG and 8-AAFdG residues in normal and repair-deficient human cells. *Nucleic Acids Research* **20**, 4437–43.

Knight, J.A. and Voorhees, R.P. 1990. Peroxidation of linolenic acid-catalysis by transition metal ions. *Annals of Clinical and Laboratory Science* **20**, 347–52.

Kon, S.H. 1978. Biological autoxidation: 1. Decontrolled iron: An ultimate carcinogen and toxicant: A hypothesis. *Medical Hypotheses* **4**, 445–63.

Kortenkamp, A., Casadevall, M., Faux, S.P., Jenner, A., Shayer, R.O.J., Woodbridge, N. and O'Brien, P. 1996. A role for molecular oxygen in the formation of DNA damage during the reduction of the carcinogen chromium(VI) by glutathione. *Archives of Biochemistry and Biophysics* **329**, 199–207.

Kortenkamp, A., Casadevall, M., Da Cruz Fresco, P. and Shayer, R.O.J. 1997. In *Cytotoxic, Mutagenic and Carcinogenic Potential of Heavy Metals Related to Human Environment*, (ed. N.D. Hadjiliadis), Kluwer Academic Publishers, Dordrecht, NATO-ASI Series, Vol. 26, pp. 15–34.

Kuchino, Y., Mori, F., Kasai, H., Inoue, H., Iwai, S., Miura, K., Ohtsuka, E. and Nishimura, S. 1987. Misreading of DNA templates containing 8-hydroxydeoxyguanosine at the modified base and at adjacent residues. *Nature* **327**, 77–9.

Kuykendall, J.R., Kerger, B.D., Jarvi, E.J., Corbett, G.E. and Paustenbach, D.J. 1996. Measurement of DNA-protein cross-links in human leukocytes following acute ingestion of chromium in drinking water. *Carcinogenesis* **17**, 1971–7.

Kvietikova, I., Wenger, R.H., Marti, H.H. and Gassmann, M. 1997. The hypoxia-inducible factor-1 DNA recognition site is cAMP-responsive. *Kidney International* **51**, 564–6.

Lancaster, J.R. Jr (ed.) 1988. *The Bioinorganic Chemistry of Nickel*, VCH, New York.

Lander, H.M., Ogiste, J.S., Teng, K.K. and Novogrodsky, A. 1995. p21^{ras} as a common signaling target of reactive free radicals and cellular redox stress. *Journal of Biological Chemistry* **270**, 21195–8.

Lander, H.M., Hajjar, D.P., Hempstead, B.L., Mirza, U.A., Chait, B.T., Campbell, S. and Quilliam, L.A. 1997. A molecular redox switch on p21^{ras}. *Journal of Biological Chemistry* **272**, 4323–6.

Landino, L.M., Crews, B.C., Timmons, M.D., Morrow, J.D. and Marnett, L.J. 1996. Peroxynitrite, the coupling product of nitric oxide and superoxide, activates prostaglandin biosynthesis. *Proceedings of the National Academy of Sciences (U.S.A.)* **93**, 15069–74.

Lee, Y.W., Pons, C., Tummolo, D.M., Klein, C.B., Rossman, T.G. and Christie, N.T. 1993. Mutagenicity of soluble and insoluble nickel compounds at the *gpt* locus in G12 Chinese hamster cells. *Environmental and Molecular Mutagenesis* **21**, 365–71.

Lee, Y.W., Klein, C.B., Kargacin, B., Salnikow, K., Kitahara, J., Dowjat, K., Zhitkovich, A., Christie, N.T. and Costa, M. 1995. Carcinogenic nickel silences gene expression by chromatin condensation and DNA methylation: A new model for epigenetic carcinogenesis. *Molecular and Cellular Biology* **15**, 2547–57.

Lee-Chen, S.F., Yu, C.T. and Jan, K.Y. 1992. Effect of arsenite on the DNA repair of UV-irradiated Chinese hamster ovary cells. *Mutagenesis* **7**, 51–5.

Lee-Chen, S.F., Wang, M.C., Yu, C.T., Wu, D.R. and Jan, K.Y. 1993. Nickel chloride inhibits the DNA repair of UV-treated but not methyl methanesulfonate-treated Chinese hamster ovary cells. *Biological Trace Element Research* **37**, 39–50.

Leeuwenburgh, C., Rasmussen, J.E., Hsu, F.F., Mueller, D.M., Pennathur, S. and Heinecke, J.W. 1997. Mass spectrometric quantification of markers for protein oxidation by tyrosyl radical, copper, and hydroxyl radical in low density lipoprotein isolated from human atherosclerotic plaques. *Journal of Biological Chemistry* **272**, 3520–6.

Le Page, F., Margot, A., Grollman, A.P., Sarasin, A. and Gentil, A. 1995. Mutagenicity of a unique 8-oxoguanine in a human Ha-*ras* sequence in mammalian cells. *Carcinogenesis* **16**, 2779–84.

Le Page, F., Guy, A., Cadet, J., Sarasin, A. and Gentil, A. 1998. Repair and mutagenic potency of 8-oxo-G:A and 8-oxo-G:C base pairs in mammalian cells. *Nucleic Acids Research* **26**, 1276–81.

Lesko, S.A., Drocourt, J.-L. and Yang, S.-U. 1982. Deoxyribonucleic acid–protein and deoxyribonucleic acid interstrand cross-links induced in isolated chromatin by hydrogen peroxide and ferrous ethylenediamine-tetraacetate chelates. *Biochemistry* **21**, 5010–15.

Lewisch, S.A. and Levine, R.L. 1995. Determination of 2-oxohistidine by amino acid analysis. *Analytical Biochemistry* **231**, 440–6.

Li, Z. and Nabel, G.J. 1997. A new member of the IκB protein family, IκBε inhibits RelA (p65)-mediated NF-κB transcription. *Molecular and Cellular Biology* **17**, 6184–90.

Li, T. and Rokita, S.E. 1991. Selective modification of DNA controlled by an ionic signal. *Journal of the American Chemical Society* **113**, 7771–3.

Li, J.-L., Okada, S., Hamazaki, S., Ebina, Y. and Midorikawa, S. 1987. Subacute nephrotoxicity and induction of renal cell carcinoma in mice treated with ferric nitrilotriacetate. *Cancer Research* **47**, 1867–9.

Li, S., Schoneich, C. and Borchardt, R.T. 1995a. Chemical pathways of peptide degradation. VIII. Oxidation of methionine in small model peptides by prooxidant/transition metal ion systems: Influence of selective scavengers for reactive oxygen intermediates. *Pharmaceutical Research* **12**, 348–55.

Li, S., Nguyen, T.H., Schoneich, C. and Borchardt, R.T. 1995b. Aggregation and precipitation of human relaxin induced by metal-catalyzed oxidation. *Biochemistry* **34**, 5762–72.

Liang, R., Jaruga, P., Dizdaroglu, M. and Kasprzak, K.S. 1998. Effect of a pentadecapeptide modeling the N-terminal metal-binding motif of human protamine P2 on Ni(II)- and Cu(II)-mediated oxidative strand scission and base damage in pUC19 DNA. *Toxicologist* **42**, 127.

Lickteig, K., Lamb, K. and Rizzino, A. 1996. Effect of oxidation and reduction on the binding of transcription factors to *cis*-regulatory elements located in the FGF-4 gene. *Molecular Reproduction and Development* **44**, 146–52.

Lin, X., Sugiyama, M. and Costa, M. 1991. Differences in the effect of vitamin E on nickel sulfide or nickel chloride-induced chromosomal aberrations in mammalian cells. *Mutation Research* **260**, 159–64.

Littlefield, N.A., Fullerton, F.R. and Poirier, L.A. 1991. Hydroxylation and deglycosylation of 2′-deoxyguanosine in the presence of magnesium and nickel. *Chemico-Biological Interactions* **79**, 217–28.

Lloyd, D.R., Phillips, D.H. and Carmichael, P.L. 1997. Generation of putative intrastrand cross-links and strand breaks in DNA by transition metal ion-mediated oxygen radical attack. *Chemical Research in Toxicology* **10**, 393–400.

Loeb, L.A., James, E.A., Waltersdorph, A.M. and Klebanoff, S.J. 1988. Mutagenesis by the autoxidation of iron with isolated DNA. *Proceedings of the National Academy of Sciences (U.S.A.)* **85**, 3918–22.

Loeb, L.A., McBride, T.J., Reid, T.M. and Cheng, K.C. 1991. Mutagenic spectra of oxygen free radicals, in *New Horizons in Molecular Toxicology*, (eds G.S. Probst, M.J. Vodicnik and M.A. Dorato), FASEB Publ, Bethesda, pp. 35–9.

Luo, Y., Henle, E.S. and Linn, S. 1996. Oxidative damage to DNA constituents by iron-mediated Fenton reactions. The deoxycytidine family. *Journal of Biological Chemistry* **271**, 21167–6.

Lynch, S.M. and Frei, B. 1997. Physiological thiol compounds exert pro- and anti-oxidant effects, respectively, on iron- and copper-dependent oxidation of human low-density lipoprotein. *Biochimica et Biophysica Acta* **1345**, 215–21.

McBride, T.J., Preston, B.D. and Loeb, L.A. 1991. Mutagenic spectrum resulting from DNA damage by oxygen free radicals. *Biochemistry* **30**, 207–13.

Maccabee, M., Evans, J.S., Glackin, M.P., Hatahet, Z. and Wallace, S.S. 1994. Pyrimidine ring fragmentation products: Effects of lesion structure and sequence context on mutagenesis. *Journal of Molecular Biology* **236**, 514–30.

Mack, D.P. and Dervan, P.B. 1992. Sequence-specific oxidative cleavage of DNA by a designated metalloprotein, nickel(II)GGH(Hin139–190). *Biochemistry* **31**, 9399–405.

Magos, L. 1991. Epidemiological and experimental aspects of metal carcinogenesis: Physicochemical properties, kinetics, and the active species. *Environmental Health Perspectives* **95**, 157–89.

Margerum, D.W. and Anliker, S.L. 1988. Nickel(III) chemistry and properties of the peptide complexes of Ni(II) and Ni(III), in *The Bioinorganic Chemistry of Nickel*, (ed. J.R. Lancaster Jr), VCH, New York, pp. 29–51.

Marty, M.S. and Loch-Caruso, R. 1993. Nickel-induced increases in gap junctional communication in the uterine cell line SK-UT-1. *In Vitro Cellular and Developmental Biology* **29A**, 215–20.

Mattagajasingh, S. and Misra, H.P. 1996. Mechanisms of the carcinogenic chromium(VI)-induced DNA–protein cross-linking and their characterization in cultured intact human cells. *Journal of Biological Chemistry* **271**, 33550–60.

Melvin, T., Botchway, S., Parker, A.W. and O'Neill, P. 1995. Migration of photoinduced oxidative damage in models for DNA. *Journal of the Chemical Society, Chemical Communications* 653–4.

Meneghini, R. 1997. Iron homeostasis, oxidative stress, and DNA damage. *Free Radicals in Biology and Medicine* **23**, 783–92.

Mikhailova, M.V., Littlefield, N.A., Hass, B.S., Poirier, L.A. and Chou, M.W. 1997. Cadmium-induced 8-hydroxydeoxyguanosine formation, DNA strand breaks and antioxidant enzyme activities in lymphoblastoid cells. *Cancer Letters* **115**, 141–8.

Miki, H., Kasprzak, K.S., Kenney, S. and Heine, U.I. 1987. Inhibition of intercellular communication by nickel(II): Antagonistic effect of magnesium. *Carcinogenesis* **8**, 1757–60.

Miller, C.A. III and Costa, M. 1989. Immunological detection of DNA–protein complexes induced by chromate. *Carcinogenesis* **10**, 667–72.

Minotti, G. 1993. Sources and role of iron in lipid peroxidation. *Chemical Research in Toxicology* **6**, 134–46.

Misra, M., Rodriguez, R.E. and Kasprzak, K.S. 1990. Nickel-induced lipid peroxidation in the rat: Correlation with nickel effect on antioxidant defense systems. *Toxicology* **64**, 1–17.

Misra, M., Rodriguez, R.E., North, S.L. and Kasprzak, K.S. 1991. Nickel-induced renal lipid peroxidation in different strains of mice: Concurrence with nickel effect on antioxidant defense systems. *Toxicology Letters* **59**, 121–33.

Misra, M., Olinski, R., Dizdaroglu, M. and Kasprzak, K.S. 1993. Enhancement by L-histidine of nickel(II)-induced DNA–protein cross-linking and oxidative DNA base damage in the rat kidney. *Chemical Research in Toxicology* **6**, 33–7.

Misra, R.R., Hochadel, J.F., Smith, G.T., Cook, J.C., Waalkes, M.P. and Wink, D.A. 1996. Evidence that nitric oxide enhances cadmium toxicity by displacing the metal from metallothionein. *Chemical Research in Toxicology* **9**, 326–32.

Moorhouse, C.P., Halliwell, B., Grootveld, M. and Gutteridge, J.M.C. 1985. Cobalt(II) ion as a promoter of hydroxyl radical and possible 'crypto-hydroxyl' radical formation under physiological conditions: Differential effects of hydroxyl radical scavengers. *Biochimica et Biophysica Acta* **843**, 261–8.

Moraes, E.C., Keyse, S.M. and Tyrrell, R.M. 1990. Mutagenesis by hydrogen peroxide treatment of mammalian cells: A molecular analysis. *Carcinogenesis* **11**, 283–93.

Moriya, M., Ou, C., Bodepudi, C., Johnson, F., Takeshita, M. and Grollman, A.P. 1991. Site specific mutagenesis using a gapped duplex vector: A study of translesion synthesis past 8-oxo-deoxyguanosine in *E. coli*. *Mutation Research* **254**, 281–8.

Morrow, D.M. and McGlynn, H. 1997. Antioxidant prevention of tumor promoter induced inhibition of intercellular communication in renal cell lines. *Biochemical Society Transactions* **25**, S561.

Mouret, J.F., Berger, M., Anselmino, C., Polverelli, M. and Cadet, J. 1991. Étude comparative de l'oxydation radicalaire de l'ADN et de ses nucléosides par les radicaux hydroxyles et les ions ferryles issus de la réaction de Fenton. *Journal of Chemical Physics* **88**, 1053–61.

Mukhopadhyay, C.K, Mazumder, B., Lindley, P.F. and Fox, P.L. 1997. Identification of the prooxidant site of human ceruloplasmin: A model for oxidative damage by copper bound to protein surfaces. *Proceedings of the National Academy of Sciences (U.S.A.)* **94**, 11546–51.

Muller, J.G., Chen, X., Dadiz, A.C., Rokita, S.E. and Burrows, C.J. 1992. Ligand effects associated with the intrinsic selectivity of DNA oxidation promoted by nickel(II) macrocyclic complexes. *Journal of the American Chemical Society* **114**, 6407–11.

Mulsch, A., Mordvintcev, P.I., Vanin, A.F. and Busse, R. 1993. Formation and release of dinitrosyl iron complexes by endothelial cells. *Biochemical and Biophysical Research Communications* **196**, 1303–8.

Nackerdien, Z., Kasprzak, K.S., Rao, G., Halliwell, B. and Dizdaroglu, M. 1991. Nickel(II)-and cobalt(II)-dependent damage by hydrogen peroxide to the DNA bases in isolated human chromatin. *Cancer Research* **51**, 5837–42.

Nappi, A.J. and Vass, E. 1997. Comparative studies of enhanced iron-mediated production of hydroxyl radical by glutathione, cysteine, ascorbic acid, and selected catechols. *Biochimica et Biophysica Acta* **1336**, 295–301.

Nassi-Calo, L., Mello-Filho, A.C. and Meneghini, R. 1989. *o*-Phenanthroline protects mammalian cells from hydrogen peroxide-induced gene mutation and morphological transformation. *Carcinogenesis* **10**, 1055–7.

Nieboer, E., Maxwell, R.I., Rosetto, F.E., Stafford, A.R. and Stetsko, P.I. 1986. Concepts in nickel carcinogenesis, in *Frontiers in Bioinorganic Chemistry*, (ed. A.V. Xavier), VCH, Heidelberg, pp. 142–51.

Noguchi, M., Nomata, K., Watanabe, J., Kanetake, H. and Saito, Y. 1998. Changes in the gap junctional intercellular communication in renal tubular epithelial cells *in vitro* treated with renal carcinogens. *Cancer Letters* **122**, 77–84.

Ochi, T., Ishiguro, T. and Ohsawa, M. 1983. Participation of active oxygen species in the induction of DNA single-strand scissions by cadmium chloride in cultured Chinese hamster cells. *Mutation Research* **122**, 169–75.

O'Connor, T.R., Graves, R.J., de Murcia, G., Castaing, B. and Laval, J. 1993. Fpg protein of *Escherichia coli* is a zinc finger protein whose cysteine residues have a structural and/or functional role. *Journal of Biological Chemistry* **268**, 9063–70.

Oikawa, S., Kurosaki, M., Kojima, Y. and Kawanishi, S. 1995. Oxidative and nonoxidative mechanisms of site-specific DNA cleavage induced by copper-containing metallothioneins. *Biochemistry* **34**, 8763–70.

Okada, S., Hamazaki, S., Ebina, Y., Fujioka, M. and Midorikawa, O. 1983. Nephrotoxicity and induction of renal adenocarcinoma by ferric nitrilotriacetate (Fe-NTA) in rats, in *Structure and Function of Iron Storage and Transport Proteins*, (eds I. Urushizaki, P. Aisen, I. Litowsky and J.W. Drysdale), Elsevier, New York, pp. 473–8.

Okada, S., Hamazaki, S., Ebina, Y., Li, J.L. and Midorikawa, O. 1987. Nephrotoxicity and its prevention by vitamin E in ferric nitrilotriacetate prompted lipid peroxidation. *Biochimica et Biophysica Acta* **922**, 28–33.

Oleinick, N.L, Chiu, S., Ramakrishnan, N. and Xue, L. 1987. The formation, identification, and significance of DNA–protein cross-links in mammalian cells. *British Journal of Cancer* **55** (Suppl. VIII), 135–40.

Olin, K.L., Cherr, G.N., Rifkin, E. and Keen, C.L. 1996. The effects of some redox-active metals and reactive aldehydes on DNA–protein cross-links *in vitro*. *Toxicology* **110**, 1–8.

Ozaki, N. and Chuang, D.M. 1997. Lithium increases transcription factor binding to AP-1 and cyclic AMP-responsive element in cultured neurons and rat brain. *Journal of Neurochemistry* **69**, 2336–44.

Pacifici, R.E. and Davies, K.J.A. 1990. Protein degradation as an index of oxidative stress. *Methods in Enzymology* **186**, 485–502.

Park, J.W. and Floyd, R.A. 1992. Lipid peroxidation products mediate the formation of 8-hydroxydeoxyguanosine in DNA. *Free Radicals in Biology and Medicine* **12**, 245–50.

Parks, D., Bolinger, R. and Mann, K. 1997. Redox state regulates binding of p53 to sequence-specific DNA, but not to non-specific or mismatched DNA. *Nucleic Acids Research* **26**, 1289–95.

Patel, U., Bhimani, R. and Frenkel, K. 1992. Mechanism of mutagenicity by 5-hydroperoxymethyl-2′-deoxyuridine, an intermediate product of ionizing radiation, in bacteria. *Mutation Research* **283**, 145–56.

Patierno, S.R., Sugiyama, M. and Costa, M. 1987. Effect of nickel(II) on DNA–protein binding, thymidine incorporation, and sedimentation pattern of chromatin fractions from intact mammalian cells. *Journal of Biochemical Toxicology* **2**, 13–23.

Pennypacker, K.R., Xiao, Y., Xu, R.H. and Harry, G.J. 1997. Lead-induced developmental changes in AP-1 DNA binding in rat brain. *International Journal of Developmental Neuroscience* **15**, 321–8.

Peskin, A.V. and Shlyahova, L. 1986. Cell nuclei generate DNA-nicking superoxide radicals. *FEBS Letters* **194**, 317–21.

Phillips, J.D., Kinikini, D.V., Yu, Y., Guo, B. and Leibold, E.A. 1996. Differential regulation of IRP1 and IRP2 by nitric oxide in rat hepatoma cells. *Blood* **87**, 2983–92.

Porter, D.W., Nelson, V.C., Fivash, M.J. Jr and Kasprzak, K.S. 1996. Mechanistic studies of the inhibition of MutT dGTPase by the carcinogenic metal Ni(II). *Chemical Research in Toxicology* **9**, 1375–81.

Porter, D.W., Yakushiji, H., Nakabeppu, Y., Sekiguchi, M., Fivash, M.J. Jr and Kasprzak, K.S.

1997. Sensitivity of *Escherichia coli* (MutT) and human (MTH1) 8-oxo-dGTPases to *in vitro* inhibition by the carcinogenic metals, nickel(II), copper(II), cobalt(II) and cadmium(II). *Carcinogenesis* **18**, 1785–91.

Powis, G., Briehl, M. and Oblong, J. 1995. Redox signalling and the control of cell growth and death. *Pharmacology and Therapeutics* **68**, 149–73.

Prutz, W.A. 1996. Measurement of copper-dependent oxidative DNA damage by HOCl and H_2O_2 with the ethidium-binding assay. *Journal of Biochemical and Biophysical Methods* **32**, 125–35.

Purmal, A.A., Kow, Y.W. and Wallace, S.S. 1994a. 5-Hydroxypyrimidine deoxynucleoside triphosphates are more efficiently incorporated into DNA by exonuclease-free Klenow fragment than 8-oxopurine deoxynucleoside triphosphates. *Nucleic Acids Research* **22**, 3930–5.

Purmal, A.A., Kow, Y.W. and Wallace, S.S. 1994b. Major oxidative products of cytosine, 5-hydroxycytosine and 5-hydroxyuracil, exhibit sequence context-dependent mispairing *in vitro*. *Nucleic Acids Research* **22**, 72–8.

Purmal, A.A., Bond, J.P., Lyons, B.A., Kow, Y.W. and Wallace, S.S. 1998. Uracil glycol deoxynucleoside triphosphate is a better substrate for DNA polymerase Klenow fragment than thymine glycol deoxynucleoside triphosphate. *Biochemistry* **37**, 330–8.

Radi, R. 1996. Reactions of nitric oxide with metallo-proteins. *Chemical Research in Toxicology* **9**, 828–35.

Randerath, K., Randerath, E., Smith, C.V. and Chang, J. 1996. Structural origins of bulky oxidative DNA adducts (Type III-compounds) as deduced by oxidation of oligonucleotides of known sequence. *Chemical Research in Toxicology* **9**, 247–54.

Reid, T.M. and Loeb, L.A. 1992. Mutagenic specificity of oxygen radicals produced by human leukemia cells. *Cancer Research* **52**, 1082–6.

Reid, T.M. and Loeb, L.A. 1993. Tandem double CC → TT mutations are produced by reactive oxygen species. *Proceedings of the National Academy of Sciences (U.S.A.)* **90**, 3904–7.

Richter, C. 1997. Reactive oxygen and nitrogen species regulate mitochondrial Ca^{2+} homeostasis and respiration. *Biosciences Reports* **17**, 53–66.

Rodriguez, R.E. and Kasprzak, K.S. 1989. Glutathione peroxidase and myeloperoxidase activity *in vitro* in the presence of Ni(II). *Proceedings of the American Association for Cancer Research* **30**, 204.

Rodriguez, R.E. and Kasprzak, K.S. 1992. Effects of nickel on catalase and the glutathione peroxidase-reductase system *in vitro*, in *Nickel and Human Health: Current Perspectives*, (eds E. Nieboer and J.O. Nriagu), J. Wiley & Sons, New York, pp. 375–85.

Rodriguez, R.E., Misra, M. and Kasprzak, K.S. 1990. Effects of nickel on catalase activity *in vitro* and *in vivo*, *Toxicology* **63**, 45–52.

Rodriguez, R.E., Misra, M., North, S.L. and Kasprzak, K.S. 1991. Nickel-induced lipid peroxidation in the liver of different strains of mice and its relation to nickel effects on antioxidant systems. *Toxicology Letters* **57**, 269–81.

Rodriguez, H., Holmquist, G.P., D'Agostino, R. Jr, Keller, J. and Akman, S.A. 1997. Metal ion-dependent hydrogen peroxide-induced DNA damage is more sequence specific than metal specific. *Cancer Research* **57**, 2394–403.

Rossi, S.C. and Wetterhahn, K.E. 1989. Chromium(V) is produced upon reduction of chromate by mitochondrial electron transport chain complexes. *Carcinogenesis* **10**, 913–20.

Rossi, S.C., Gorman, N. and Wetterhahn, K.E. 1988. Mitochondrial reduction of the carcinogen chromate: Formation of chromium(V). *Chemical Research in Toxicology* **1**, 101–7.

Ruch, R.J., Cheng, S.J. and Klaunig, J.S. 1989. Prevention of cytotoxicity and inhibition of intercellular communication by antioxidant catechins isolated from Chinese green tea. *Carcinogenesis* **10**, 1003–8.

Runger, T.M. and Kraemer, K.H. 1989. Joining of linear plasmid DNA is reduced and error-prone in Bloom's syndrome cells. *EMBO Journal* **8**, 1419–25.

Rush, J.D., Maskos, Z. and Koppenol, W.H. 1990. Reactions of iron(II) nucleotide complexes with hydrogen peroxide. *FEBS Letters* **261**, 121–3.

Salnikow, K., Zhitkovich, A. and Costa, M. 1992. Analysis of the binding sites of chromium to DNA and protein *in vitro* and in intact cells. *Carcinogenesis* **13**, 2341–6.

Salnikow, K., Gao, M., Voitkun, V., Huang, X. and Costa, M. 1994. Altered oxidative stress responses in nickel-resistant mammalian cells. *Cancer Research* **54**, 6407–12.

Salnikow, K., Wang, S. and Costa, M. 1997. Induction of activating transcription factor 1 by nickel and its role as a negative regulator of thrombospondin I gene expression. *Cancer Research* **57**, 5060–6.

Salnikow, K., Zhou, D., Whang, C. and Costa, M. 1998. Carcinogenic nickel targeting calcium-dependent signaling pathway. *Proceedings of the American Association for Cancer Research* **39**, 21.

Sancar, A. 1995. DNA repair in humans. *Annual Review of Genetics* **29**, 69–105.

Saplakoglu, U., Iscan, M. and Iscan, M. 1997. DNA single-strand breakage in rat lung, liver and kidney after single and combined treatments of nickel and cadmium. *Mutation Research* **394**, 133–40.

Sarkar, B. 1997. Zinc finger–DNA interaction: Effect of metal replacement, free radical generation and DNA damage and its relevance to carcinogenesis, in *Cytotoxic, Mutagenic and Carcinogenic Potential of Heavy Metals Related to Human Environment*, (ed. N.D. Hadjiliadis), Kluwer Academic Publishers, Dordrecht, NATO-ASI Series, Vol. 26, pp. 1–14.

Sarkar, S., Yadav, P., Trivedi, R., Bansal, A.K. and Bhatnagar, D. 1995. Cadmium-induced lipid peroxidation and the status of the antioxidant system in rat tissues. *Journal of Trace Elements in Medicine and Biology* **9**, 144–9.

Sato, M. and Bremner, I. 1993. Oxygen free radicals and metallothionein. *Free Radicals in Biology and Medicine* **14**, 325–37.

Sawyer, D.T. 1987. The chemistry and activation of dioxygen species (O_2, $O_2^{-\cdot}$, and HOOH) in biology, in *Oxygen Complexes and Oxygen Activation by Transition Metals*, (eds A.E. Martell and D.T. Sawyer), Plenum Press, New York, pp. 131–48.

Schaaper, R.M., Koplitz, R.M., Tkeshelashvili, L.K. and Loeb, L.A. 1987. Metal-induced lethality and mutagenesis: Possible role of apurinic intermediates. *Mutation Research* **177**, 179–88.

Schreck, R., Alberman, N.K. and Baeuerle, P.A. 1992. Nuclear factor κB: An oxidative stress-responsive transcription factor of eukaryotic cells (a review). *Free Radical Research Communications* **17**, 221–37.

Schubert, D. and Chevion, M. 1995. The role of iron in beta amyloid toxicity. *Biochemical and Biophysical Research Communications* **216**, 702–7.

Sen, C.K. and Packer, L. 1996. Antioxidant and redox regulation of gene transcription. *FASEB Journal* **10**, 709–20.

Sen, P. and Costa, M. 1985. Induction of chromosomal damage in Chinese hamster ovary cells by soluble and particulate nickel compounds: Preferential fragmentation of the heterochromatic long arm of the X-chromosome by carcinogenic crystalline NiS particles. *Cancer Research* **45**, 2320–5.

Shenker, B.J., Datar, S., Mansfield, K. and Shapiro, I.M. 1997. Induction of apoptosis in human T-cells by organomercuric compounds: A flow cytometric analysis. *Toxicology and Applied Pharmacology* **143**, 397–406.

Shi, X. and Dalal, N.S. 1989. Chromium(V) and hydroxyl radical formation during the glutathione reductase-catalyzed reduction of chromium(VI). *Biochemical and Biophysical Research Communications* **163**, 627–34.

Shi, X. and Dalal, N.S. 1990a. On the hydroxyl radical formation in the reaction between hydrogen peroxide and biologically-generated chromium(V) species. *Archives of Biochemistry and Biophysics* **277**, 342–50.

Shi, X. and Dalal, N.S. 1990b. One-electron reduction of chromate by NADPH-dependent glutathione reductase. *Journal of Inorganic Biochemistry* **40**, 1–12.

Shi, X. and Dalal, N.S. 1992. The role of superoxide radical in chromium(VI)-generated hydroxyl radical: The Cr(VI) Haber–Weiss cycle. *Archives of Biochemistry and Biophysics* **292**, 323–7.

Shi, X., Dalal, N.S. and Kasprzak, K.S. 1992a. Generation of free radicals from lipid hydroperoxides by Ni^{2+} in the presence of oligopeptides. *Archives of Biochemistry and Biophysics* **299**, 154–62.

Shi, X., Sun, X., Gannett, P.M. and Dalal, N.S. 1992b. Deferoxamine inhibition of Cr(V)-mediated radical generation and deoxyguanine hydroxylation: ESR and HPLC evidence. *Archives of Biochemistry and Biophysics* **293**, 281–6.

Shi, X., Dalal, N.S. and Kasprzak, K.S. 1993a. Generation of free radicals from model lipid hydroperoxides and H_2O_2 by Co(II) in the presence of cysteinyl and histidyl chelators. *Chemical Research in Toxicology* **6**, 277–83.

Shi, X., Dalal, N.S. and Kasprzak, K.S. 1993b. Generation of free radicals in reactions of Ni(II)–thiol complexes with molecular oxygen and model lipid hydroperoxides. *Journal of Inorganic Biochemistry* **50**, 211–25.

Shi, X., Dalal, N.S. and Kasprzak, K.S. 1993c. Generation of free radicals from hydrogen peroxide and lipid hydroperoxides in the presence of Cr(III). *Archives of Biochemistry and Biophysics* **302**, 294–9.

Shi, X., Dalal, N.S. and Kasprzak, K.S. 1994. Enhanced generation of hydroxyl and sulfur trioxide anion radicals from oxidation of sodium sulfite, nickel(II) sulfite, and nickel subsulfide in the presence of nickel(II) complexes. *Environmental Health Perspectives* **102** (Suppl. 3), 91–6.

Shiao, Y.H., Lee, S.H. and Kasprzak, K.S. 1998. Cell cycle arrest, apoptosis, and p53 expression in nickel(II) acetate-treated Chinese hamster ovary cells. *Carcinogenesis* **19**, 1203–7.

Shibutani, S., Takeshita, M. and Grollman, A.P. 1991. Insertion of specific bases during DNA synthesis past the oxidation-damaged base 8-oxodG. *Nature* **349**, 431–4.

Shimada, H., Shibata, M.A., Shiao, Y.H. and Waalkes, M.P. 1998. Cadmium suppresses apoptosis induced by chromium. *Journal of Toxicology and Environmental Health* **54**, 159–68.

Shiraishi, N., Hochadel, J.F., Coogan, T.P., Koropatnick, J. and Waalkes, M.P. 1995. Sensitivity to cadmium-induced genotoxicity in rat testicular cells is associated with minimal expression of the metallothionein gene. *Toxicology and Applied Pharmacology* **130**, 229–36.

Shirname-More, L., Rossman, T.G., Troll, W., Teebor, G.W. and Frenkel, K. 1987. Genetic effects of 5-hydroxymethyl-2′-deoxyuridine, a product of ionizing radiation. *Mutation Research* **178**, 177–86.

Shukla, G.S., Hussain, T. and Chandra, S.V. 1987. Possible role of regional superoxide dismutase activity and lipid peroxide levels in cadmium neurotoxicity: *In vivo* and *in vitro* studies in growing rats. *Life Sciences* **41**, 2215–21.

Sieg, D.J. and Billings, R.E. 1997. Lead/cytokine-mediated oxidative DNA damage in cultured mouse hepatocytes. *Toxicology and Applied Pharmacology* **142**, 106–15.

Singer, B. and Grunberger, D. 1983. *Molecular Biology of Mutagens and Carcinogens*, Plenum Press, New York.

Singh, R.J., Hogg, N., Joseph, J. and Kalyanaraman, B. 1996. Mechanism of nitric oxide release from S-nitrosothiols. *Journal of Biological Chemistry* **271**, 18596–603.

Sipowicz, M., Diwan, B.A., Ramljak, D., Buzard, G.S., Yu, W., Waalkes, M.P., Rice, J.M., Kasprzak, K.S. and Anderson, L.M. 1997. Perinatal effects of metals and cancer in offspring, in *Cytotoxic, Mutagenic and Carcinogenic Potential of Heavy Metals Related to Human Environment*, (ed N.D. Hadjiliadis), Kluwer Academic Publishers, Dordrecht, NATO-ASI Series, Vol. 26, pp. 123–38.

Smith, M.A., Harris, P.R.L., Sayre, L.M. and Perry, G. 1997. Iron accumulation in Alzheimer's disease is a source of redox-generated free radicals. *Proceedings of the National Academy of Sciences (U.S.A.)* **94**, 9866–8.

Snow, E.T. 1992. Metal carcinogenesis: Mechanistic implications. *Pharmacology and Therapeutics* **53**, 31–65.

Sokol, R.J. 1996. Antioxidant defenses in metal-induced liver damage. *Seminars on Liver Disease* **16**, 39–46.

Sole, J., Huguet, J., Arola, L. and Romeu, A. 1990. *In vivo* effects of nickel and cadmium in rats on lipid peroxidation and ceruloplasmin activity. *Bulletin of Environmental Contamination and Toxicology* **44**, 686–91.

Spiro, T.G. (ed.) 1980. *Metal Ion Activation of Dioxygen*, J. Wiley & Sons, New York.

Srivastava, R.C., Kumar, A., Srivastava, S.K., Gupta, S., Hasan, S.K. and Athar, H.M. 1990. Nickel-mediated inhibition in the glutathione-dependent protection against lipid peroxidation. *Biochemistry International* **20**, 495–501.

Stadtman, E.R. 1993. Oxidation of free amino acids and amino acid residues in proteins by radiolysis and by metal-catalyzed reactions. *Annual Review of Biochemistry* **62**, 797–821.

Stadtman, E.R. and Berlett, B.S. 1991. Fenton chemistry: Amino acid oxidation. *Journal of Biological Chemistry* **266**, 17201–11.

Stadtman, E.R. and Berlett, B.S. 1997. Reactive oxygen-mediated protein oxidation in aging and disease. *Chemical Research in Toxicology* **10**, 485–94.

Standeven, A.M. and Wetterhahn, K.E. 1991. Is there a role for reactive oxygen species in the mechanism of chromium(VI) carcinogenesis? *Chemical Research in Toxicology* **4**, 616–25.

Stephenson, G.F., Chan, H.M. and Cherian, M.G. 1994. Copper-metallothionein from the toxic milk mutant mouse enhances lipid peroxidation initiated by an organic hydroperoxide. *Toxicology and Applied Pharmacology* **125**, 90–6.

Stewart, R.C., Hille, R. and Massey, V. 1985. The reaction of arsenite-complexed xanthine oxidase with oxygen. *Journal of Biological Chemistry* **260**, 8892–904.

Stohs, S.J. and Bagchi, D., 1995. Oxidative mechanisms in the toxicity of metal ions. *Free Radicals in Biology and Medicine* **18**, 321–36.

Sugiyama, M., Tsuzuki, K. and Ogura, R. 1991a. Effect of ascorbic acid on DNA damage, cytotoxicity, glutathione reductase, and formation of paramagnetic chromium in Chinese hamster V-79 cells treated with sodium chromate(VI). *Journal of Biological Chemistry* **266**, 3383–6.

Sugiyama, M., Lin, X., and Costa, M. 1991b. Protective effect of vitamin E against chromosomal aberrations and mutation induced by sodium chromate in Chinese hamster V79 cells. *Mutation Research* **260**, 19–23.

Sun, P., Lou, L. and Maurer, R.A. 1996. Regulation of activating transcription factor-1 and the cAMP response element-binding protein by Ca^{2+}/calmodulin-dependent protein kinase type I, II, and IV. *Journal of Biological Chemistry* **271**, 3066–73.

Sunderman, F.W. Jr 1986. Metals and lipid peroxidation. *Acta Pharmacologica et Toxicologica* **59** (Suppl. VII), 248–55.

Sunderman, F.W. Jr 1989. Mechanisms of nickel carcinogenesis. *Scandinavian Journal of Work and Environmental Health* **15**, 1–12.

Sunderman, F.W. Jr, McCully, K.S. and Hopfer, S.M. 1984. Association between erythrocytosis and renal cancers in rats following intrarenal injection of nickel compounds. *Carcinogenesis* **5**, 1511–16.

Suzuki, Y., Shimizu, H., Nagae, Y., Fukumoto, M., Okonogi, H. and Kadokura, M. 1993. Micronucleus test and erythropoiesis: Effect of cobalt on the induction of micronuclei by mutagens. *Environmental and Molecular Mutagenesis* **22**, 101–6.

Szatrowski, T.P. and Nathan, C.F. 1991. Production of large amounts of hydrogen peroxide by human tumor cells. *Cancer Research* **51**, 794–8.

Tian, L. and Lawrence, D.A. 1996. Metal-induced modulation of nitric oxide production *in vitro* by murine macrophages: Lead, nickel, and cobalt utilize different mechanisms. *Toxicology and Applied Pharmacology* **141**, 540–7.

Tian, H., McKnight, S.L. and Russel, D.W. 1997. Endothelial PAS domain 1 (EPAS1), a transcription factor selectively expressed in endothelial cells. *Genes and Development* **11**, 72–82.

Tillian, H.M., Hammer, A., Kink, E., Schaur, R.J. and Schauenstein, E. 1989. Iron-induced lipid peroxidation and inhibition of proliferation in Ehrlich ascites tumor cells. *Journal of Cancer Research and Clinical Oncology* **115**, 79–83.

Tkeshelashvili, L.K., McBride, T.J., Spence, K. and Loeb, L.A. 1991. Mutation spectrum of copper-induced DNA damage. *Journal of Biological Chemistry* **266**, 6401–6.

Tkeshelashvili, L.K., Reid, T.M., McBride, T.J. and Loeb, L.A. 1993. Nickel induces a signature mutation for oxygen free radical damage. *Cancer Research* **53**, 4172–4.

Tofigh, S. and Frenkel, K. 1989. Effect of metals on nucleoside hydroperoxide, a product of ionizing radiation in DNA. *Free Radicals in Biology and Medicine* **7**, 131–43.

Torreilles, J. and Guerin, M.-C. 1990. Nickel(II) as a temporary catalyst for hydroxyl radical generation. *FEBS Letters* **272**, 58–60.

Trosko, J.E. 1998. Hierarchical and cybernetic nature of biologic systems and their relevance to homeostatic adaptation to low-level exposures to oxidative stress-inducing agents. *Environmental Health Perspectives* **106** (Suppl. 1), 331–9.

Trosko, J.E. and Ruch, R.J. 1998. Cell–cell communication in carcinogenesis. *Frontiers in Bioscience* **3**, D208–D236.

Trujillo, M., Alvarez, M.N., Peluffo, G., Freeman, B.A. and Radi, R. 1998. Xanthine oxidase-mediated decomposition of *S*-nitrosothiols. *Journal of Biological Chemistry* **273**, 7828–34.

Turk, P.W., Laayoun, A., Smith, S.S. and Weitzman, S.A. 1995. DNA adduct 8-hydroxyl-2′-deoxyguanosine (8-hydroxyguanosine) affects function of human DNA methyltransferase. *Carcinogenesis* **16**, 1253–5.

Umemura, T., Sai, K., Takagi, A., Hasegawa, R. and Kurokawa, Y. 1990a. Oxidative DNA damage, lipid peroxidation, and nephrotoxicity induced in the rat kidney after ferric nitrilotriacetate administration. *Cancer Letters* **54**, 95–100.

Umemura, T., Sai, K., Takagi, A., Hasegawa, R. and Kurokawa, Y. 1990b. Formation of 8-hydroxydeoxyguanosine (8-OH-dG) in rat. *Carcinogenesis* **11**, 345–7.

Umemura, T., Sai, K., Takagi, A., Hasegawa, R. and Kurokawa, Y. 1991. The effects of exogenous glutathione and cysteine on oxidative stress induced by ferric nitrilotriacetate. *Cancer Letters* **58**, 49–56.

Upham, B.L., Kang, K.S., Cho, H.Y. and Trosko, J.E. 1997. Hydrogen peroxide inhibits gap junctional intercellular communication in glutathione sufficient, but not glutathione deficient cells. *Carcinogenesis* **18**, 37–42.

Vaca, C.E., Wilhelm, J. and Harms-Ringdahl, M. 1988. Interaction of lipid peroxidation products with DNA. A review. *Mutation Research* **195**, 137–49.

Vallyathan, V., Mega, J.F., Shi, X. and Dalal, N.S. 1992. Enhanced generation of free radicals from phagocytes induced by mineral dusts. *American Journal of Respiratory and Molecular Biology* **6**, 404–13.

Verhaegh, G.W., Richard, M.J. and Hainaut, P. 1997. Regulation of p53 by metal ions: Dithiocarbamate down-regulates p53 DNA-binding activity by increasing the intracellular level of copper. *Molecular and Cellular Biology* **17**, 5699–706.

Vilar-Rojas, C., Guzman-Grenfell, A.M. and Hicks, J.J. 1996. Participation of oxygen-free radicals in the oxido-reduction of proteins. *Archives of Medical Research* **27**, 1–6.

Vile, G.F. and Tyrrell, R.M. 1995. UVA radiation-induced oxidative damage to lipids and proteins *in vitro* and in human skin fibroblasts is dependent on iron and singlet oxygen. *Free Radicals in Biology and Medicine* **18**, 721–30.

Voitkun, V., Zhitkovich, A. and Costa, M. 1998. Cr(III)-mediated crosslinks of glutathione or amino acids to the DNA–phosphate backbone are mutagenic in human cells. *Nucleic Acids Research* **26**, 2024–30.

von Sonntag, C. 1987. *The Chemical Basis of Radiation Biology*, Taylor and Francis, London.

Vos, J.M. and Wauthier, E.L. 1991. Differential introduction of DNA damage and repair in mammalian genes transcribed by RNA polymerases I and II. *Molecular and Cellular Biology* **11**, 2245–52.

Waalkes, M.P., Diwan, B.A., Rehm, S., Ward, J.M., Moussa, M., Cherian, M.G. and Goyer, R.A. 1996. Down-regulation of metallothionein expression in human and murine hepatocellular tumors: Association with the tumor necrotizing and antineoplastic effects of cadmium in mice. *Journal of Pharmacology and Experimental Therapeutics* **277**, 1026–33.

Wagner, M., Klein, C.L., Kleinert, H., Euchenhofer, C., Forstermann, U. and Kirkpatrick, C.J. 1997. Heavy metal ion induction of adhesion molecules and cytokines in human endothelial cells: the role of NF-kappaB, I kappaB-alpha and AP-1. *Pathobiology* **65**, 241–52.

Walling, C. 1975. Fenton's reagent revisited. *Acc. Chemical Research* **8**, 125–31.

Wedrychowski, A., Schmidt, W.N. and Hnilica, L.S. 1986a. The *in vivo* cross-linking of proteins and DNA by heavy metals. *Journal of Biological Chemistry* **261**, 3370–6.

Wedrychowski, A., Schmidt, W.N. and Hnilica, L.S. 1986b. DNA–protein crosslinking by heavy metals in Novikoff hepatoma. *Archives of Biochemistry and Biophysics* **251**, 397–402.

Weiss, G., Goossen, B., Doppler, W., Fuchs, D., Pantopoulos, K., Werner-Felmayer, G., Wachter, H. and Hentze, M.W. 1993. Translational regulation via iron-responsive elements by the nitric oxide/NO-synthase pathway. *EMBO Journal* **12**, 3651–7.

Weiss, G., Werner-Felmayer, G., Werner, E.R., Grünewald, K., Wachter, H. and Hentze, M.W. 1994. Iron regulates nitric oxide synthase activity by controlling nuclear transcription. *Journal of Experimental Medicine* **180**, 969–76.

Weitzman, S.A. and Gordon, L.I. 1990. Inflammation and cancer: Role of phagocyte-generated oxidants in carcinogenesis. *Blood* **76**, 655–63.

Weitzman, S.A. and Stossel, T.P. 1981. Mutation caused by human phagocytes. *Science* **212**, 546–7.

Weitzman, S.A., Turk, P.W., Milkowski, D.H. and Kozlowski, K. 1994. Free radical adducts induce alterations in DNA cytosine methylation. *Proceedings of the National Academy of Sciences (U.S.A.)* **91**, 1261–4.

Wetterhahn, K.E., Hamilton, J.W., Aiyar, J., Borges, K.M. and Floyd, R. 1989. Mechanisms of chromium(VI) carcinogenesis. Reactive intermediates and effect on gene expression. *Biological Trace Element Research* **21**, 405–11.

Wey, S.J., O'Connor, K.J. and Burrows, C.J. 1993. Preparation of primary vicinal diamines from amino acid esters and crystal structure of a chiral nickel salen complex. *Tetrahedron Letters* **34**, 1905–8.

Wink, D.A. and Laval, J. 1994. The Fpg protein, a DNA repair enzyme, is inhibited by the biomediator nitric oxide *in vitro* and *in vivo*. *Carcinogenesis* **15**, 2125–9.

Wink, D.A., Wink, C.B., Nims, R.W. and Ford, P.C. 1994. Oxidizing intermediates generated in the Fenton reagent: Kinetic arguments against intermediacy of the hydroxyl radical. *Environmental Health Perspectives* **102** (Suppl. 3), 11–15.

Wink, D.A., Vodovoz, Y., Laval, J., Laval, F., Dewhirst, M.W. and Mitchell, J.B. 1998. The multifaceted roles of nitric oxide in cancer. *Carcinogenesis* **19**, 711–21.

Winkler, P., Schaur, R.J. and Schauenstein, E. 1984. Selective promotion of ferrous ion-dependent lipid peroxidation in Ehrlich ascites tumor cells by histidine as compared with other amino acids. *Biochimica et Biophysica Acta* **796**, 226–31.

Wood, M.L., Dizdaroglu, M., Gajewski, E. and Essigmann, J.M. 1990. Mechanistic studies of ionizing radiation and oxidative mutagenesis: Genetic effects of a single 8-hydroxyguanine (7-hydro-8-oxoguanine) residue inserted at a unique site in a viral genome. *Biochemistry* **29**, 7024–32.

Wood, M.L., Esteve, A., Morningstar, M.L., Kuziemko, G.M. and Essigmann, J.M. 1992. Genetic effects of oxidative DNA damage: Comparative mutagenesis of 7,8-dihydro-8-oxoguanine and 7,8-dihydro-8-oxoadenine in *Escherichia coli*. *Nucleic Acids Research* **20**, 6023–32.

Xanthoudakis, S., Smeyne, R.J., Wallace, J.D. and Curran, T. 1996. The redox/DNA repair protein, Ref-1, is essential for early embryonic development in mice. *Proceedings of the National Academy of Sciences (U.S.A.)* **93**, 8919–23.

Xie, J., Funakoshi, T., Shimada, H. and Kojima, H. 1995. Effects of chelating agents on testicular toxicity in mice caused by acute exposure to nickel. *Toxicology* **103**, 147–55.

Xu, J., Bubley, G.J., Detrick, B., Blankenship, L.J. and Patierno, S.R. 1996. Chromium(VI) treatment of normal human lung cells results in guanine-specific DNA polymerase arrest, DNA–DNA cross-links and S-phase blockade of cell cycle. *Carcinogenesis* **17**, 1511–17.

Xu, L., Eu, J.P., Meissner, G. and Stamler, J.S. 1998. Activation of the cardiac calcium release channel (ryanodine receptor) by poly-*S*-nitrosylation. *Science* **279**, 234–7.

Yamamoto, K. and Kawanishi, S. 1989. Hydroxyl free radical is not the main active species in site-specific DNA damage induced by copper(II) ion and hydrogen peroxide. *Journal of Biological Chemistry* **264**, 15435–40.

Yamamoto, K. and Kawanishi, S. 1991. Site-specific DNA damage induced by hydrazine in the presence of manganese and copper ions. *Journal of Biological Chemistry* **266**, 1509–15.

Yamamoto, K. and Kawanishi, S. 1992. Site-specific DNA damage by phenylhydrazine and phenelzine in the presence of Cu(II) ion or Fe(III) complexes: Roles of active oxygen species and carbon radicals. *Chemical Research in Toxicology* **5**, 440–6.

Yamamoto, K., Inoue, S., Yamazaki, A., Yoshinaga, T. and Kawanishi, S. 1989. Site-specific DNA damage induced by cobalt(II) ion and hydrogen peroxide: Role of singlet oxygen. *Chemical Research in Toxicology* **2**, 234–9.

Yamamoto, F., Kasai, H., Bessho, T., Chung, M.H., Inoue, H., Ohtsuka, E., Hori, T. and Nishimura, S. 1992. Ubiquitous presence in mammalian cells of enzymatic activity specifically cleaving 8-hydroxyguanine-containing DNA. *Japanese Journal of Cancer Research* **83**, 351–7.

Yamamoto, K., Inoue, S. and Kawanishi, S. 1993. Site-specific DNA damage and 8-hydroxydeoxyguanosine formation by hydroxylamine and 4-hydroxyaminoquinoline-1-oxide in the presence of Cu(II): Role of active oxygen species. *Carcinogenesis* **14**, 1397–401.

Yang, C.F., Shen, H.-M., Shen, Y., Zhuang, Z.-X. and Ong, C.-N. 1997. Cadmium-induced oxidative cellular damage in human fetal lung fibroblasts (MRC-5 cells). *Environmental Health Perspectives* **105**, 712–16.

Yang, J.-L., Chao, J.-I. and Lin, J.-G. 1996. Reactive oxygen species may participate in the mutagenicity and mutational spectrum of cadmium in Chinese hamster ovary-K1 cells. *Chemical Research in Toxicology* **9**, 1360–7.

Ye, J., Zhang, X., Young, H.A., Mao, Y. and Shi, X. 1995. Chromium(VI)-induced nuclear factor-κB activation in intact cells via free radical reactions. *Carcinogenesis* **16**, 2401–5.

Yeivin, A., Levine, A. and Razin, A. 1996. DNA methylation patterns in tumors derived from F9 cells resemble methylation at the blastula stage. *FEBS Letters* **395**, 11–16.

Yu, W., Sipowicz, M.A., Diwan, B.A., Kasprzak, K.S., Haines, D.C. and Anderson, L.M. 1999. Preconception exposure of male mice to urethane or chromium: Increased tumors in multiple organs of offspring. *Toxicology and Applied Pharmacology* **158**, 161–76.

Zalma, R., Guignard, J. and Pezerat, H. 1989. Production of radicals arising from surface activity of fibrous materials. *NATO ASI Series* **H30**, 257–64.

Zeromski, J., Jezewska, E., Sikora, J. and Kasprzak, K.S. 1995. The effect of nickel compounds on immunophenotype and natural killer cell function of normal human lymphocytes. *Toxicology* **97**, 39–48.

Zhang, D., Okada, S., Yu, Y., Zheng, P., Yamaguchi, R. and Kasai, H. 1997. Vitamin E inhibits apoptosis, DNA modification, and cancer incidence induced by iron-mediated peroxidation in Wistar rat kidney. *Cancer Research* **57**, 2410–14.

Zhao, F., Ghezzo-Schoneich, E., Aced, G.I., Hong, J., Milby, T. and Schoneich, C. 1997. Metal-catalyzed oxidation of histidine in human growth hormone. *Journal of Biological Chemistry* **272**, 9019–29.

Zhitkovich, A. and Costa, M. 1992. A simple, sensitive assay to detect DNA–protein crosslinks in intact cells and *in vivo*. *Carcinogenesis* **13**, 1485–9.
Zhuang, Z., Huang, X. and Costa, M. 1994. Protein oxidation and amino acid–DNA cross-linking by nickel compounds in intact cultured cells. *Toxicology and Applied Pharmacology* **126**, 319–25.

17

COPPER ION REGULATION OF GENE EXPRESSION IN YEAST

Laran T. Jensen and Dennis R. Winge

17.1 Introduction

Cells in the natural world encounter a changing environment with fluctuations in nutrient concentrations. Alkali and alkaline earth metal ions (sodium, potassium, magnesium and calcium) are the most abundant metal ions in biological systems. Sodium, potassium, magnesium and calcium are all present in high concentrations. Environmental changes in the concentrations of these metal ions do not typically present a nutritional problem for cells. In contrast, other physiologically important metals are present at low levels and are designated as trace or ultratrace metals. This group includes the 3d transition metal ions from vanadium to copper, zinc, and the second row transition metal molybdenum. In addition to being present at low concentrations, the bioavailability of these metal ions depends on the complexes in which they exist in the environment. This combination can result in nutritional deprivation of these trace metal ions for cells. Deficiency of an essential metal ion can impair physiology through loss of critical functions of metalloenzymes.

Many of these trace metal ions can also be toxic. Excessive concentrations of metal ions may affect cellular metabolism by a myriad of ways, including disruption of cell membrane integrity, alteration of the cellular redox state, as well as inhibition of respiration, DNA replication and protein synthesis (Agarwal *et al.*, 1989; Gadd and White, 1989). These toxic effects can arise from direct inhibition of protein function or indirect mechanisms such as metal-induced free-radical reactions. The multiple valence states in which many trace metal ions can exist yields catalytic properties that are utilized by enzymes. However, the same properties of these trace metal ions can also catalyze formation of highly reactive hydroxyl radicals through Fenton chemistry (Fridovich, 1978; Henle and Linn, 1997). The high reactivity of biomolecules to radicals results in damage to DNA, RNA, proteins and lipids (Imlay *et al.*, 1988; Farr and Kogoma, 1991).

Thus, cells may be challenged with a range of trace metal-ion concentrations. Physiologic responses to fluctuations in the concentration of these metal ions are necessary for cell survival. Cells confront at least three problems with regard to metal-ion levels to ensure that a positive balance is maintained for each physiologically important metal ion. The discrimination between metal ions in biology is complex, since simple chemical chelators often fail to exhibit the exquisite metal-ion selectivity that is seen in biological systems. Secondly, the intracellular concentration of essential metal ions must be controlled, ensuring adequate but not toxic intracellular quantities. Several homeostatic mechanisms have been identified that regulate metal-ion concentration, including regulated uptake and efflux of metal ions as well as sequestration of metal ion by proteins and intracellular compartments. Thirdly, some, and perhaps all, metal ions need to be routed to sites of utilization. Metal-ion chaperones can prevent the formation of hydroxyl radicals by transporting redox-active metal ions to intracellular compartments or metal-binding proteins. These chaperones may also assist in selection of the metal ion, increasing specificity and preventing mis-metallation of proteins.

The budding yeast *Saccharomyces cerevisiae* has been used extensively as a model organism to study metal-ion homeostasis. Insights have emerged into the mechanisms of uptake, intracellular distribution and detoxification of metal ions. This chapter highlights mechanisms in *S. cerevisiae* that have evolved to cope with fluctuations in the environmental concentration of copper ions.

Copper ions are essential for normal cellular physiology in *S. cerevisiae*. At least three critical copper-containing enzymes have been identified in yeast that are required for normal growth. The mitochondrial cytochrome oxidase complex requires three copper ions as cofactors. The ability of yeast cells to grow on nonfermentable carbon sources, such as glycerol and ethanol, is dependent on cytochrome oxidase activity. In addition, oxidative growth requires defense against reactive oxygen intermediates. Cytoplasmic superoxide dismutase, which contains both copper and zinc ions, functions to convert reactive superoxide anions to hydrogen peroxide which is subsequently converted to water and oxygen by catalase. A third key, copper-containing enzyme is Fet3, which contains a trinuclear copper center. Fet3 is ferro-oxidase and has been shown to be critical for iron uptake (Askwith *et al.*, 1994; De Silva *et al.*, 1995; Yuan *et al.*, 1995). Preliminary data suggest that hydroxylation of inositol phosphorylceramide-c requires copper, suggesting a fourth important copper metalloenzyme in yeast (Beeler *et al.*, 1997). Other organisms produce a myriad of oxidases and oxygenases that require copper as a cofactor, but the presence of these enzymes in the yeast has not been demonstrated.

Two copper-regulated transcription factors exist in *S. cerevisiae*, Mac1 (metal binding activator) and Ace1 (activator of *CUP1* expression). Transcription factors serve as integration centers for the different signal transduction pathways that can affect a gene. The regulation of these regulators themselves is of crucial importance for differential gene expression.

In addition to the ease of culturing and genetic manipulations in yeast, the completion of the *S. cerevisiae* genome sequence has increase its value as a model organism for interpreting animal DNA sequences. Nearly 31% of potential yeast DNA open reading frames (ORFs) have statistically robust homologs among mammalian protein sequences (Botstein *et al.*, 1997). Human homologs of several key copper and iron homeostasis genes in yeast have already identified (Yuan *et al.*, 1995; Amaravadi *et al.*, 1997; Babcock *et al.*, 1997; Klomp *et al.*, 1997; Zhou and Gitschier, 1997). It remains to be determined how far the similarities between yeast and animal metal homeostatic systems may extend.

17.2 Nutritional response to copper deprivation

The essential nature of copper for normal yeast physiology is consistent with the fact that *S. cerevisiae* possesses mechanisms to ensure that adequate cellular copper levels

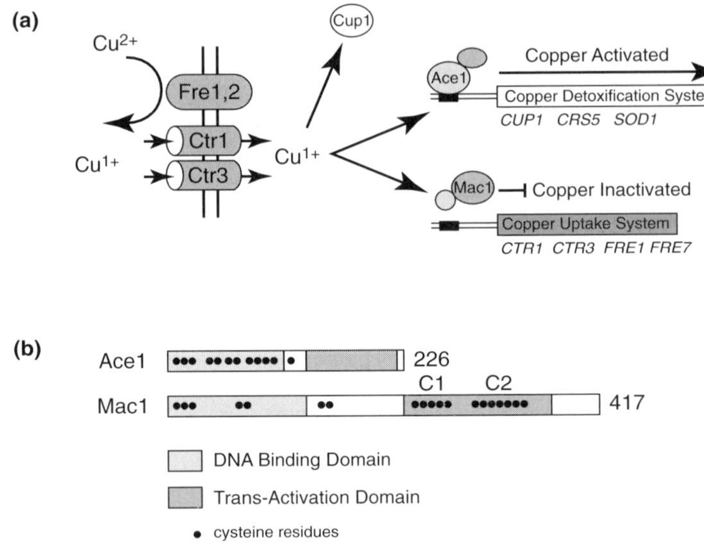

Figure 17.1 (a) Scheme of copper regulation of gene expression in yeast. Two copper-regulated transcription factors Ace1 and Mac1 mediate expression of genes involved in copper ion buffering and copper ion uptake, respectively. (b) The two functional domains of each protein are illustrated with the positions of cysteines specified by black dots.

are attained. Copper deficiency results in the expression of four genes, three of which have been demonstrated to be involved in cellular uptake of copper ions (Dancis *et al.*, 1994; Hassett and Kosman, 1995; Georgatsou *et al.*, 1997; Labbe *et al.*, 1997; Yamaguchi-Iwai *et al.*, 1997). Expression of genes in the high-affinity copper-uptake system occurs in *S. cerevisiae* cultured in medium containing less than 1 nM Cu(II) (Fig. 17.1a). The copper ion uptake system consists of specific permeases and a cell-surface metalloreductase. Two of the genes encode the high-affinity plasma-membrane copper ion permeases, Ctr1 and Ctr3 (Dancis *et al.*, 1994; Labbe *et al.*, 1997). Ctr1 and Ctr3 are functionally redundant, in that a copper deficient state arises only when both permeases are nonfunctional (Knight *et al.*, 1996). In addition to enhanced susceptibility to copper deficiency, cells lacking a functional permease are more resistant to copper toxicity (Dancis *et al.*, 1994).

Ctr1 is oligomeric, with three candidate transmembrane segments per subunit (Dancis *et al.*, 1994). The candidate ectodomain is characterized by an unusual abundance of Met and Ser residues. A Met–X–X–Met sequence motif is repeated eleven times in the N-terminal region. Multiple Met–X–X–Met sequences may function as low-affinity Cu(II) scavenging motifs, gathering Cu(II) ions for transport across the plasma membrane. Transport across the lipid bilayer may be facilitated in part by binding of copper ions to two Cys–X–Cys sequence motifs within the candidate cytoplasmic domain. Copper uptake in yeast is energy dependent, although the mechanism is unresolved (Lin and Kosman, 1990).

The *CTR3* permease gene is not expressed in most laboratory yeast strains due to the insertion of a transposable element (Knight *et al.*, 1996). Ctr3 (241 residues) is a smaller molecule than Ctr1 (406 residues) and exhibits limited similarity to Ctr1. The presence of 11 cysteinyl residues in CC or CXXC motifs are thought to be important in copper transport (Knight *et al.*, 1996).

A third gene encodes Fre1 metalloreductase, which generates a diffusible reductant into the growth medium to mobilize Cu(II) ions (Lesuisse *et al.*, 1996) (Fig. 17.1a). Fre1 appears to be a flavocytochrome-containing NADPH oxidase (Finegold *et al.*, 1996; Lesuisse *et al.*, 1996; Shatwell *et al.*, 1996). Four essential histidines are predicted to coordindate two heme moieties that are expected to function in transmembrane electron transfer (Finegold *et al.*, 1996). Fre1 and a second metalloreductase, Fre2, function in both copper and iron ion uptake. *FRE1* is derepressed by either copper- or iron-deficient growth conditions, whereas *FRE2* is uniquely expressed under iron-limiting conditions (Dancis *et al.*, 1992; Georgatsou and Alexandraki, 1994; Shatwell *et al.*, 1996). The significance of cell-surface metalloreductases for iron and copper homeostasis is that iron and copper ions in the environment are largely present in insoluble, oxidized valent states. Bioavailability of these ions is increased via reduction.

The fourth gene derepressed is homologous to Fre1. The function of this gene product, designated Fre7, remains unresolved. Fre7 exhibits sequence similarity to Fre1 in regions expected to bind FAD, NADPH and the two hemes (Finegold *et al.*, 1996). Thus, Fre7 is expected to be a NADPH oxidase, although its cellular localization has not been identified.

The expression of *CTR1*, *CTR3*, *FRE1* and *FRE7* is a cellular response to inadequate intracellular copper levels. Copper-deficient cells upregulate these genes efficiently. The kinetics of *CTR3* derepression has been investigated (Labbe *et al.*, 1997). Cells shifted to copper-limiting conditions exhibit a rapid derepression of *CTR3* expression, with maximal response occurring within 10 min (Labbe *et al.*, 1997). In this experiment, cells were shifted from cultures containing 16 nM Cu(II) to a medium containing <1 nM Cu(II). In contrast, expression of these genes is attenuated in cells cultured in medium containing greater than 1 nM Cu(II). The addition of 1 μM Cu(II) to copper-depleted cells resulted in a time-dependent attenuation of *CTR3* expression, with maximal inhibition requiring 60 min (Labbe *et al.*, 1997). Copper-replete cells have no need for the high-affinity copper-uptake system. Copper uptake can occur through low-affinity permeases. Therefore, one cellular response to sufficient intracellular copper is repression of high-affinity copper uptake.

A second response is the copper-dependent removal of Ctr1 and Ctr3 transporters from the plasma membrane (Knight *et al.*, 1986; Ooi *et al.*, 1996). Immunofluorescence analyses of epitope-tagged Ctr1 revealed copper-induced internalization of Ctr1 from the cell surface (Ooi *et al.*, 1996). In addition, Western analysis revealed a copper-induced degradation of Ctr1 (Ooi *et al.*, 1996). Copper-induced Ctr1 degradation occurs independent of internalization and requires a higher exogenous copper level [10 μM Cu(II)] than that required for inhibition of *CTR1* expression [1 nM Cu(II)] (Ooi *et al.*, 1996).

In addition to the high-affinity permease system, yeast contains at least two low-affinity copper ion transporters, Ctr2 and Fet4 (Kampfenkel *et al.*, 1995; Eide and Kosman, unpublished observation). Disruption of *CTR2* conferred limited copper resistance and overexpression of *CTR2* resulted in an increased sensitivity to copper toxicity (Kampfenkel *et al.*, 1995). Neither *FET4* nor *CTR2* are copper specific or copper regulated.

17.3 Transcriptional regulator of copper nutritional response

Genes transcriptionally expressed under copper-deficient conditions are regulated by the transcription factor, Mac1 (metal-binding activator) (Jungmann *et al.*, 1993; Hassett and Kosman, 1995; Georgatsou *et al.*, 1997; Labbe *et al.*, 1997; Yamaguchi-Iwai *et al.*, 1997) (Fig. 17.1). *MAC1* was originally identified as a partially dominant *MAC1* mutation, *MAC1*^{up1} (Jungmann *et al.*, 1993). Cells harboring the *MAC1*^{up1} allele are incapable of repression of copper-uptake genes (Hassett and Kosman, 1995; Yamaguchi-Iwai *et al.*, 1997). As a result, *MAC1*^{up1} cells are hypersensitive to copper salts in the growth medium (Jungmann *et al.*, 1993). The copper sensitivity of *MAC1*^{up1} cells demonstrates the importance of downregulating the high-affinity uptake system in copper-replete cells. In contrast, a frame-shift mutation in *MAC1* designated *mac1-1* results in substantial loss of both Cu(II) and Fe(III) reduction as well as uptake (Jungmann *et al.*, 1993; Hassett and Kosman, 1995). These cells have undetectable *CTR1*, *CTR3* or *FRE1* transcripts (Labbe *et al.*, 1997; Yamaguchi-Iwai *et al.*, 1997). The phenotypes of *mac1-1* cells include respiratory deficiency and sensitivity to a myriad of stresses. These phenotypes are reversed upon addition of exogenous copper salts, suggesting that *mac1-1* cells are copper deficient (Jungmann *et al.*, 1993).

The Mac1 binding site is a conserved DNA element [TTTGC(T,G)CA] repeated in the 5′ regions of *CTR1*, *CTR3*, *FRE1* and *FRE7* (Labbe *et al.*, 1997; Yamaguchi-Iwai *et al.*, 1997; Martins *et al.*, 1998) (Fig. 17.2a). Mac1-mediated transcription requires two functional elements (Labbe *et al.*, 1997; Yamaguchi-Iwai *et al.*, 1997; Martins *et al.*, 1998). The elements, designated copper-regulatory elements or CuRE (Labbe *et al.*, 1997), can exist as inverted or direct repeats with variable spacing between the elements ranging from 7 to 40 basepairs (Labbe *et al.*, 1997; Yamaguchi-Iwai *et al.*, 1997; Martins *et al.*, 1998) (Fig. 17.2a). In addition, expansion of the spacing between the two elements upstream of *CTR1* revealed attenuated expression (Martins *et al.*, 1998).

Transcriptional activators of genes transcribed by RNA polymerase II contain at least two domains, one for specific DNA binding and another for transactivation (Ptashne, 1988; Mitchell and Tjian, 1989; Zawel and Reinberg, 1995). These two activities are typically separable and functionally independent. Interaction of the activation domain (AD) with proteins that bind the TATA element allow for RNA polymerase binding, thus creating the pre-initiation complex (Mitchell and Tjian, 1989; Hahn, 1993; Zawel and Reinberg, 1995). Yeast activation domains are typified by an unusual abundance of acidic residues, prolyl residues or glutamines (Mitchell and Tjian, 1989; Hahn, 1993). The acidic nature of the AD is more a descriptive feature of this class than a functional feature, since no strict correlation exists between net charge and activation potential (Hahn, 1993; Leuther *et al.*, 1993; Regier *et al.*, 1993; Dahlman-Wright *et al.*, 1995).

DNA-binding and activation domains have been characterized in Mac1 (Fig. 17.1b). The DNA-binding domain of Mac1 was mapped by fusing the herpes simplex VP16 activation domain to C-terminal truncations of Mac1 (Jensen *et al.*, 1998). The minimal AD from the herpes simplex virus-1 VP16 has been used extensively as a heterologous activator (Sadowski *et al.*, 1988; Lefstin *et al.*, 1994; Silverman *et al.*, 1994; Wang *et al.*, 1997). Fusion of the N-terminal 194 residues of Mac1 to residues 417–491 of VP16 generated a functional transcription factor. Cells harboring the Mac1/VP16 fusion were methionine prototrophs, unlike *mac1-1* cells containing the control plasmid which are

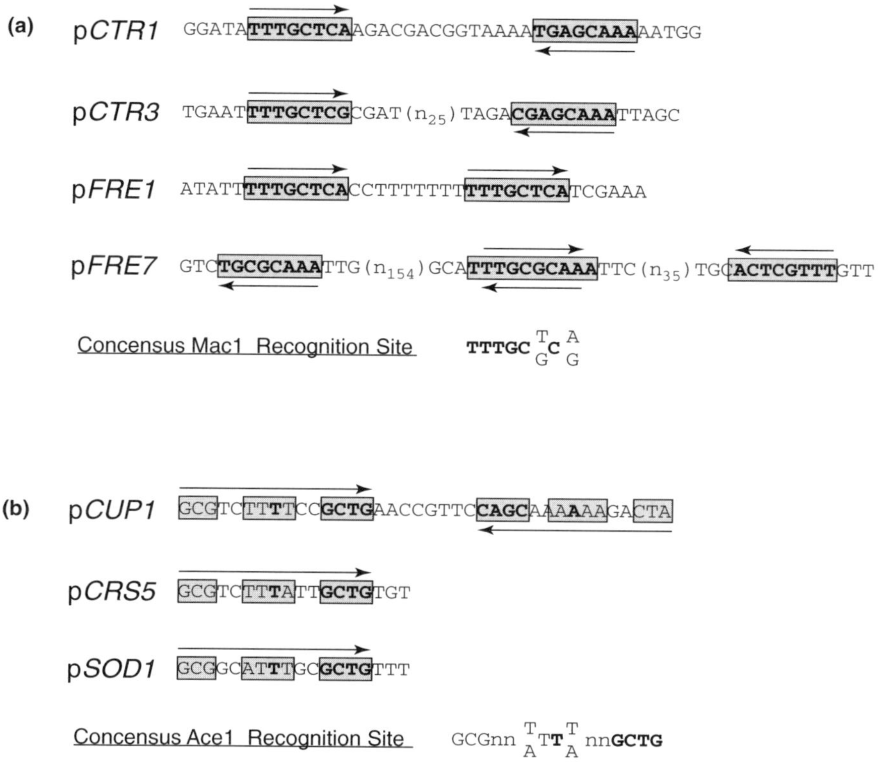

Figure 17.2 Copper-responsive promoter elements in genes regulated by (a) Mac1 and (b) Ace1.

Met auxotrophs due to depressed NADPH levels arising from copper deprivation (Slekar *et al.*, 1996). In addition, the Mac1/VP16 fusion was active in promoting expression of a *pCTR1/lacZ* reporter construct (Jensen *et al.*, 1998). A shorter fusion containing only the N-terminal 159 residues was equally active as the 194mer, although further truncations to the N-terminal 125 residues or fewer resulted in substantially less activity compared to Mac1(1–159). Mac1(1–125) was also less active in *in vitro* DNA binding than Mac1(1–194) or Mac1(1–159). Thus, the DNA-binding domain of Mac1 appears to be contained within the N-terminal 159 residues (Fig. 17.1b). The activities of the Mac1/VP16 fusions were not modulated by the addition of copper salts.

The N-terminal 40 residues of Mac1 are highly homologous with the Ace1 and Amt1 copper-regulated transcription factors from *S. cerevisiae* and *Candida glabrata*, respectively (Jungmann *et al.*, 1993). The first 40 residues of Amt1 enfold a nonclassical, zinc finger with 3 Cys 1 His Zn ligation (Turner *et al.*, 1998). These residues, Cys11, Cys14, Cys23 and His25 are conserved in Mac1. Thus, the N-terminal 40 residues of Mac1 probably consist of an Amt1-like Zn motif. The Zn modules in Ace1 and Amt1 are a subdomain of the DNA-binding region responsible for base-specific minor groove contacts (Buchman *et al.*, 1989; Dobi *et al.*, 1995; Dixon *et al.*, 1996; Farrell *et al.*, 1996; Koch and Thiele, 1996).

The N-terminal 159 residues of Mac1 bind two Zn(II) ions (Jensen *et al.*, 1998). One Zn(II) ion is presumably bound within the conserved Zn module. Since the Zn module in Ace1 contacts an A/T-minor groove sequence (Buchman *et al.*, 1990; Dobi *et al.*, 1995; Koch and Thiele, 1996), the corresponding Zn module in Mac1 would be expected to contact the TTT segment of the TTTGCTCA site (Yamaguchi-Iwai *et al.*, 1997). The segment of Mac1 ranging from residues 40 to 159 is predicted to contact the GCTCA subsite of the Mac1-responsive element. The ligands for the second bound Zn(II) ion in Mac1 are unresolved. If the CysCys pairs at positions 69, 70 are involved, the Zn(II) would not be critical for DNA binding, since Mac1(1–194) containing Cys → Ser substitutions was competent to bind DNA. Mac1 synthesized by an *in vitro* transcription/translation system was shown to bind specifically to the TTTGCTCA sequence (Yamaguchi-Iwai *et al.*, 1997). Purified Mac1(1–159) binds to a DNA probe containing a single repeat of the TTTGCTCA sequence with an affinity of 4 nM, but not two mutant duplexes (Jensen *et al.*, 1998). This confirms that Mac1(1–159) binds specifically to the TTTGCTCA element. While Mac1 can bind to a single TTTGCTCA element *in vitro*, two repeats of this element are necessary for *in vivo* activity (Labbe *et al.*, 1997; Yamaguchi-Iwai *et al.*, 1997; Martins *et al.*, 1998). Addition of Mac1(1–159) to a DNA probe containing an inverted repeat of the TTTGCTCA element yielded two gel-retarded complexes consistent with binding of two protein molecules. The binding affinity of the 1 : 1 complex with DNA containing the inverted repeat was 4 nM – a value similar to that observed for the isolated half-site. The affinity of the 2 : 1 Mac1 : DNA complex was approximately 13 nM. Each site appears to bind independently.

Repeated TTTGCTCA elements are not necessary for high-affinity DNA binding by Mac1 *in vitro* (Jensen *et al.*, 1998). The requirement for two Mac1 binding sites *in vivo* suggests synergism between the sites. The Mac1(1–194)/VP16 fusion stimulates transcription from a promoter containing a single TTTGCTCA element, but expression of the reporter gene was nearly fivefold greater with two TTTGCTCA promoter elements. The difference in activity between one and two elements with Mac1(194)/VP16 is suggestive of synergism between the sites.

Mac1 contains a copper-regulated transactivation domain in the C-terminal region of the protein (Georgatsou *et al.*, 1997; Graden and Winge, 1997). The C-terminal segment of Mac1 contains two Cys-rich motifs within an acidic region that typifies fungal activation domains. The known *MAC1*^{up} mutations which abolish copper regulation of Mac1 function lie within the first Cys-rich region (Jungmann *et al.*, 1993; Yamaguchi-Iwai *et al.*, 1997). The demonstration of transactivation activity in the Cys-rich region came from Gal4/Mac1 and LexA/Mac1 fusion proteins in which heterologous DNA-binding domains were fused in frame to segments of Mac1 (Georgatsou *et al.*, 1997; Graden and Winge, 1997). Expression of the Gal4/Mac1 (Mac1 residues 42–417) fusion protein in a cell containing *pGAL1/lacZ* resulted in expression of β-galactosidase, implying that residues 42–417 of Mac1 contain the AD (Graden and Winge, 1997). In addition, expression of β-galactosidase was copper regulated (Graden and Winge, 1997). Under conditions of limited copper bioavailability, β-galactosidase activity was markedly elevated; whereas cells from copper-supplemented medium exhibited low β-galactosidase levels.

Transactivation activity was mapped to the two C-terminal Cys-rich motifs (labeled C1 and C2 in Fig. 17.1b). The first Cys-rich motif (C1; residues 240–311) contains modest activation activity, whereas the second motif (C2; residues 311–350) possesses potent activation activity (Zu *et al.*, 1998; Keller and Winge, unpublished observation). Copper regulation of transactivation activity in Mac1 requires sequences upstream of the C1,C2 motifs (Graden and Winge, 1997). The importance of the C1 motif in copper regulation of Mac1 function is highlighted by the loss of copper regulation in mutant Mac1 molecules with Cys → Ser substitutions at each of the five Cys residues in this motif (Graden and Winge, 1997, Yamaguchi-Iwai *et al.*, 1997). Cys → Ser substitutions in C2 had no effect on copper attenuation of transactivation (Graden and Winge, 1997).

The Cys-rich motifs, consisting of CXCXXXXCXCXXCXXH sequences, resemble Cu(I)-binding cysteinyl sequence motifs found in Ace1 and metallothioneins (Jungmann *et al.*, 1993). Mac1 is likely to be the copper sensor itself. Mac1 binds both Zn(II) and Cu(I) ions. Metal repression of Mac1 function is specific for Cu(I) ions among the physiologically important metal ions. The fact that Ag(I) attenuates Mac1 function is consistent with Mac1 regulation by cuprous ions because Cu(I) and Ag(I) ions are known to form similar metal–thiolate complexes (Labbe *et al.*, 1997; Winge *et al.*, 1997). Cu(I) ion specificity may be achieved through formation of a polycopper–thiolate cluster in the repressed Mac1 conformer analogous to the Cu(I)–thiolate clusters that exist in metallothioneins (Winge *et al.*, 1997).

In addition to the observed copper repression of Mac1 transactivation, the Mac1 DNA-binding function is impaired in copper-replete cells (Labbe *et al.*, 1997). Copper-replete cells show a loss of an *in vivo* Mac1 footprint on the *CTR3* copper regulatory element, consistent with copper-mediated loss of DNA binding (Labbe *et al.*, 1997). The loss of an *in vivo* footprint in copper-treated cells is in contrast to *in vitro* studies (Jensen *et al.*, 1998).

A variety of mechanisms may account for the observed loss of an *in vivo* Mac1 footprint. These include copper-dependent nuclear export of Mac1, copper-dependent Mac1 degradation, or masking of DNA-binding activity through an intramolecular interaction between the N-terminal DNA-binding domain and the C-terminal copper-binding activation domain. Neither copper-dependent nuclear export nor copper-dependent degradation of Mac1 is responsible for copper repression of Mac1 function. Mac1 is localized within the nucleus independent of the copper status of the cell. Copper-dependent degradation of Mac1 occurs only in cells exposed to elevated copper levels above physiologically relevant levels (Zu *et al.*, 1998).

An intramolecular interaction exists in Mac1 in copper-replete cells between the C1, C2 AD and the N-terminal DNA-binding domain (Jensen and Winge, 1998). The intramolecular interaction was demonstrated in a two-hybrid experiment in which the Mac1 DNA-binding domain (residues 1–200) was fused to the Gal4 DNA-binding domain in the first fusion gene. The second fusion gene contained the Mac1 AD (residues 240–417) fused to the Gal4 AD. Cells transformed with vectors encoding both fusions were found to express a *GAL1/lacZ* fusion gene only in copper-treated cells, implying a copper-mediated oligomerization of the two hybrid proteins. This interaction was precluded when the Mac1 AD sequence contained the *MAC1*^{up1} mutation.

In summary, Mac1 is a transcriptional activator in copper-deficient cells, but not in copper-replete cells. Cells starved for copper contain a derepressed Mac1 and show enhanced expression of genes specific for copper ion uptake. Activation of Mac1 is a copper starvation response. A model for copper repression of Mac1 function emerges. Cu(I) binding to the Mac1 AD stabilizes a conformation, exposing key residues in the C1 region that form an intramolecular interface with residues in the DNA-binding domain. This interaction masks the Mac1 AD and attenuates DNA binding. No information is available on why copper repression of Mac1 occurs at an extracellular copper concentration three orders of magnitude lower than the concentration required for Ace1 activation.

An understanding of the copper regulation of Mac1 function may shed insight on two other known metal-responsive yeast transcription factors, Zap1 and Aft1. No detailed mechanistic information is available on how Fe and Zn ions repress Aft1 and Zap1 function, respectively, in yeast (Yamaguchi-Iwai *et al.*, 1995, Yamaguchi-Iwai *et al.*, 1996; Zhao and Eide, 1997).

A Mac1-like factor exists in *Posospora anserina* (Osiewacz and Nuber, 1996). The factor, designated Grisea, is important in cell differentiation and senescence. Cells lacking a functional *GRISEA* have an increased life span and produce no female gametangia (Osiewacz and Nuber, 1996). These effects can be reversed by supplementation of copper to the cultures, suggesting that the phenotype arises from copper deficiency (Marbach *et al.*, 1994). Grisea resembles Mac1 with the conserved N-terminal Zn module and two homologous C-terminal Cys-rich motifs.

17.4 Copper-induced stress responsiveness

Cells exposed to an environment with elevated Cu(II) levels exhibit protective responses to counteract the cytotoxicity of copper ions. Excess metal ions in the environment result in the downregulation of metal ion uptake genes as well as induced expression of a subset of genes that have protective roles (Furst *et al.*, 1988; Gralla *et al.*, 1991; Culotta *et al.*, 1994). Growth conditions in which the extracellular Cu(II) concentration exceeds 1 μM lead to the expression of at least three genes, *CUP1*, *CRS5* and *SOD1* (Furst *et al.*, 1988; Gralla *et al.*, 1991; Culotta *et al.*, 1994) (Fig. 17.1a). *CUP1* and *CRS5* encode cysteinyl-rich polypeptides in the metallothionein (MT) family (Culotta *et al.*, 1994; Jensen *et al.*, 1996). *CUP1* is the dominant locus that confers copper resistance (Fogel and Welch, 1982; Hamer *et al.*, 1985; Jensen *et al.*, 1996).

The importance of *CUP1* in copper tolerance was demonstrated in genetic crosses of copper-resistant strains (Cu^R) and copper-sensitive strains (Cu^S) which showed that Cu^R was mediated by *CUP1* (Brenes-Pomales *et al.*, 1955). Targeted disruption of the *CUP1* gene reduced the copper resistance of cells from 100 μM to less than 10 μM (Hamer *et al.*, 1985). In addition, gene amplification of *CUP1* results in Cu^R (Fogel and Welch, 1982). The repeating DNA element within an amplified *CUP1* locus is a 2 kb sequence and contains two transcription units, one of which encodes MT (Fogel and Welch, 1982; Karin *et al.*, 1984). The second open reading frame is nonessential for the copper-resistance phenotype. Cu^S cells contain a single 2 kb unit, whereas Cu^R strains contain multiple tandemly iterated units, from 2 to 14 copies of *CUP1* on chromosome 8 (Fogel and Welch, 1982; Fogel *et al.*, 1988). Thus, copper resistance in *S. cerevisiae* is related to the gene copy number of *CUP1*.

MT genes are the dominant determinant of copper resistance in other species. The imperfect yeast, *C. glabrata*, contains three distinct chromosomal MT loci, one of which is present as tandemly duplicated genes (Mehra *et al.*, 1990). Targeted disruption of the amplified MT locus and a second MT gene confers a Cu^S phenotype (Mehra *et al.*, 1992). Targeted MT disruptions have also been reported in cyanobacteria and the mouse. In each case the only observed phenotype is metal sensitivity (Michalska and Choo, 1993; Turner *et al.*, 1993; Masters *et al.*, 1994).

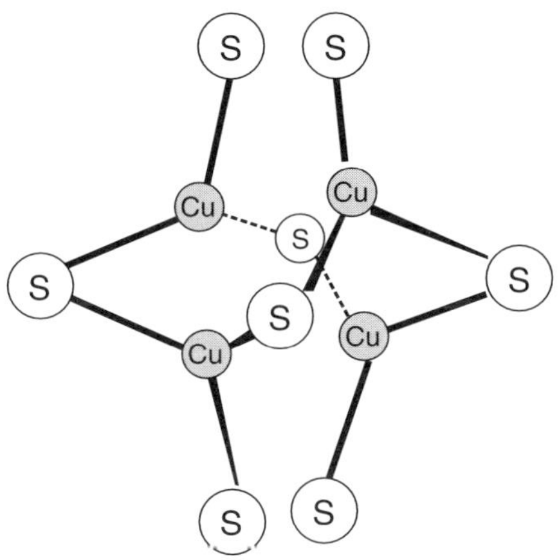

Figure 17.3 Model for the tetracopper–thiolate cluster in the copper-regulatory domain of Ace1 involving residues 40–100. The tetracopper cluster is believed to be stabilized by four bridging thiolate sulfur atoms.

The Cup1 MT buffers the intracellular copper ion concentration by chelating Cu(I) ions within a heptacopper–thiolate cluster (Narula *et al.*, 1991; Peterson *et al.*, 1996) (Fig. 17.3). The effectiveness of Cup1 MT in copper-ion buffering is dramatically enhanced by the coupling of Cup1 biosynthesis to intracellular Cu(I) levels (Furst *et al.*, 1988). Copper regulation of Cup1 biosynthesis occurs at the level of transcriptional activation (Butt *et al.*, 1984; Karin *et al.*, 1984).

Saccharomyces cerevisiae contains a second MT gene, *CRS5*, whose expression is also copper regulated (Culotta *et al.*, 1994). *CRS5* is present as a single-copy gene, unlike the tandem array of *CUP1* genes (Culotta *et al.*, 1994). Targeted disruption of *CUP1* in cells with a wild-type *CRS5* locus confers hypersensitivity to copper salts (Culotta *et al.*, 1994). An additional targeted disruption of *CRS5* exacerbates the hypersensitivity (Culotta *et al.*, 1994). The dominance of Cup1 MT over Crs5 in copper buffering arises from the amplification of the *CUP1* locus, the greater copper responsiveness of the *CUP1* promoter and the greater stability of the CuCup1 complex compared to the CuCrs5 complex (Jensen *et al.*, 1996).

Yeast MT confers resistance to metals other than copper when chimeric yeast MT are constructed in which *CUP1* is placed under a constitutive promoter (Ecker *et al.*, 1986; Yu *et al.*, 1994). *CUP1* MT is capable of binding Cd(II) ions, yet *CUP1* does not impart cadmium resistance unless the gene is induced (Yu *et al.*, 1994). MT confers resistance to only those metal ions capable of inducing the transcriptional activation of MT genes. In *S. cerevisiae* and *C. glabrata*, Ag(I) is the only other metal ion known to induce MT expression (Furst *et al.*, 1988).

Additional functions may exist for Cup1 and Crs5 since both genes are highly basally expressed. SAGE (serial analysis of gene expression) revealed that *S. cerevisiae* cells contain at least 4665 genes, with expression levels ranging from 0.3 to 200 mRNA transcripts per cell (Velculescu *et al.*, 1997). The *CUP1* transcript level was near 75, which is similar in abundance to many ribosomal protein genes (Velculescu *et al.*, 1997). DNA microchip analysis of yeast reveals that *CRS5* is also highly basally expressed. The high basal expression of *CUP1* and *CRS5* suggests that they may have additional housekeeping roles. One additional role for Cup1 is its limited superoxide dismutase activity (Tamai *et al.*, 1993). *CUP1* can functionally suppress the oxygen sensitivity of a *sod1δ* strain (Tamai *et al.*, 1993). The mechanism of superoxide dismutation by CuCup1 involves a thiyl radical species (Sievers *et al.*, 1996).

The third known gene activated by copper ions is *SOD1*, encoding Cu, Zn-superoxide dismutase (Tamai *et al.*, 1993). Copper-induced expression of Sod1 may impart greater defense against copper-catalyzed free-radical generation, yet the antioxidant function of Sod1 is not dependent on copper regulation (Culotta *et al.*, 1995). Sod1 also contributes to copper buffering in *S. cerevisiae*, similar to Cup1 (Culotta *et al.*, 1995). Thus, Cup1 and Sod1 each contribute to copper and oxygen-radical homeostasis.

17.5 Transcriptional regulator of copper-stress response

The *trans*-acting factor that mediates Cu(I) activation in *S. cerevisiae* was identified as Ace1 (also designated Cup2) (Thiele, 1988; Buchman *et al.*, 1989) (Fig. 17.1a). Cu(I) activation of Ace1 occurs through copper-dependent Ace1 binding to 5′ promoter elements (designated UAS_{Cu}) within the *CUP1* gene (Furst *et al.*, 1988; Buchman *et al.*, 1990) (Fig. 17.2b). Ace1 forms a specific complex with UAS_{Cu} DNA only in the presence of Cu(I) or Ag(I) ions (Furst *et al.*, 1988). The DNA-binding domain of Ace1 maps to the N-terminal 122 residues (Furst *et al.*, 1988) (Fig. 17.1b). Twelve cysteinyl residues are present in this domain, 11 of which are critical for copper-induced expression of *CUP1* (Hu *et al.*, 1990). Ten of the 11 critical cysteinyl residues are present in Cys–$X_{1,2}$–Cys motifs that are commonly found in copper-binding metallothioneins. These results led Furst *et al.* to postulate in 1988 that Cu(I) binding to Ace1 triggered a conformational change to a fold that was poised for DNA binding (Furst *et al.*, 1988).

A related transcription factor, Amt1, is responsible for copper metalloregulation in *Candida glabrata* (Zhou and Thiele, 1991). Ace1 and Amt1 are structurally homologous. The N-terminal half of Amt1 is 50% identical to the corresponding region of Ace1, with complete conservation of the 11 critical cysteinyl residues (with one added gap). Homology between Ace1 and Amt1 ends after residue 100 in Ace1. The C-terminal region (residues 120–225) of Ace1 contains the transactivation domain (Munder and Furst, 1992). Thus, Ace1 and Amt1, like many yeast transcription factors, are modular in nature, with a specific DNA-binding domain and a separate and independently acting transactivation domain (Hahn, 1993).

Footprinting analyses of Ace1 binding to UAS_{Cu} revealed major groove base contacts at the two ends of UAS_{Cu} and minor groove contacts in the middle A/T-rich region (Buchman *et al.*, 1990; Dobi *et al.*, 1995; Dixon *et al.*, 1996). The prediction was made that Ace1 lies atop the minor groove, contacting the major groove on both sides (Buchman *et al.*, 1990). The DNA-binding site for Amt1 is more restricted, involving only the core GCTG sequences and the A/T-rich region (Zhou and Thiele, 1993; Koch and Thiele, 1996). Amt1 shows no major groove contacts in the region corresponding to the Ace1 distal site.

The DNA-binding activity of Ace1 and Amt1 is dependent on the presence of copper ions (Furst *et al.*, 1988; Dameron *et al.*, 1991; Thorvaldsen *et al.*, 1994). The addition of Cu(I) or Ag(I) to Ace1 or Amt1 polypeptides yields a protein capable of specific DNA binding (Furst *et al.*, 1988; Dameron *et al.*, 1991). Expression of recombinant Ace1 or Amt1 in bacteria enabled the isolation of Cu_4Zn_1 protein complexes (Dameron *et al.*, 1991; Thorvaldsen *et al.*, 1994; Farrell *et al.*, 1996).

Biophysical analyses of the Cu_4Zn_1Ace1 and Cu_4Zn_1Amt1 complexes revealed ultraviolet transitions consistent with Cu–thiolate charge-transfer bands and luminescence in aqueous solutions, indicative of trigonal Cu(I) binding in an environment shielded from solvent interactions (Dameron *et al.*, 1991; Thorvaldsen *et al.*, 1994; Farrell *et al.*, 1996). Copper reconstitution studies of apo–Amt1 revealed an all-or-nothing formation of a tetracopper species (Thorvaldsen *et al.*, 1994). X-ray absorption spectroscopy of Ace1 and Amt1 confirmed the trigonal coordination of Cu(I) ions and revealed the existence of a Cu(I) polynuclear cluster (Pickering *et al.*, 1993; Thorvaldsen *et al.*, 1994).

The CuS cluster in Ace1 and Amt1 resembles synthetic tetracopper clusters formed with thiolphenolate ligands and polycopper clusters in MTs (Dance, 1986; Pickering *et al.*, 1993) (Fig. 17.3). The Cup1 MT enfolds a single heptacopper–thiolate cluster (Narula *et al.*, 1991; Peterson *et al.*, 1996) (Fig. 17.4a). The polycopper cluster in Cup1 is buried within the core of Cup1. Figure 17.4b shows a space-filling model of CuCup1 in which the seven black-colored Cu(I) ions are largely shielded by the polypeptide sheath. It is expected that the tetracopper cluster in Ace1 is likewise buried within the core of the Ace1 protein, stabilizing the tertiary fold. The regulatory factor Ace1 resembles the gene product it modulates (Cup1) in containing a polycopper cluster.

Mapping studies on Amt1 and Ace1 revealed the cysteinyl thiolates that serve as ligands for Zn(II) and the tetracopper cluster (Farrell *et al.*, 1996). The single Zn(II) ion is coordinated within the N-terminal 40 residues through the thiolates of Cys11, Cys14, Cys23 and the imidazole of His25 (Farrell *et al.*, 1996; Posewitz *et al.*, 1996). The spacing of Zn(II) ligands is $C–X_2–C–X_8–C–X–H$. Synthetic peptides encompassing residues 1–40 of Amt1 and Ace1 form stably folded complexes in the presence of 1 mol eq. Zn(II). The solution structure of the Amt1 Zn module of Amt1 was shown to be distinct from previously characterized zinc-finger motifs. The Zn module consists of a three-stranded antiparallel β-sheet with two short helical segments that project from one end of the β-sheet (Turner *et al.*, 1998). Conserved Arg and Lys residues form a basic patch predicted to be important for DNA binding. The Zn(II) module in Ace1 and Amt1 is conserved in Mac1, Grisea, an additional *S. cerevisiae* protein, Lpz8, and two additional fungal ORFs (Farrell *et al.*, 1996).

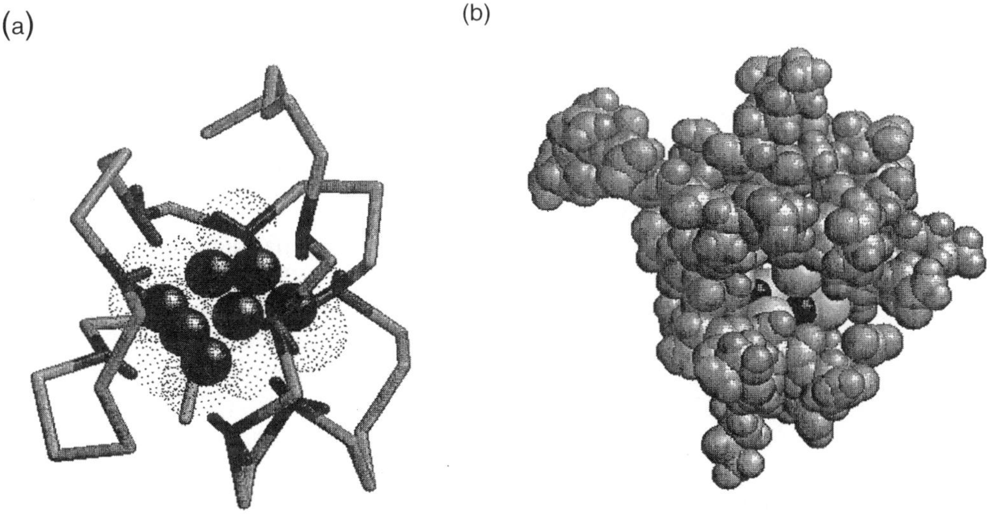

Figure 17.4 Structure of the heptacopper–thiolate cluster in Cup1. (a) The polypeptide backbone chain with the positions of cysteinyl residues darkened. The seven Cu(I) ions and cysteinyl sulfurs are shown in space-filling display with Cu(I) ions shown as dark spheres and sulfurs as dotted spheres. (b) The Cup1 structure in space-filling display with the seven Cu(I) ions again shown by dark spheres. Notice that the Cu ions are shielded by the polypeptide fold.

The tetracopper domain of Amt1 is enfolded by residues 41–110 (Graden *et al.*, 1996). Expression of a truncated Amt1 (residues 37–110) in bacteria resulted in the isolation of a Cu(I)-containing complex with 4 mol eq. Cu(I) bound (Graden *et al.*, 1996). Spectral analyses of the truncated Cu_4Amt1 complex suggested that Cu(I) binding was analogous to the intact CuZnAmt1 complex. X-ray absorption spectroscopy of the truncated CuAmt1 complex yielded a similar mean Cu–S bond distance and Cu–Cu separation as the intact CuZnAmt1 complex (Graden *et al.*, 1996). The truncated CuAmt1 complex retained high affinity DNA-binding activity, with the actual affinity reduced only a factor of 10 relative to the intact Amt1 (Graden *et al.*, 1996).

Thus, Amt1 and Ace1 consist of three separate domains: residues 1–40, the Zn(II) domain; residues 41–110 (100 in Ace1), the tetracopper domain; and residues 110–C-terminal end, the transactivation domain. A model for the CuS cluster in Ace1 and Amt1 is shown in Fig. 17.3. The DNA-binding function of each protein consists of both the N-terminal Zn module and Cu-regulatory domain.

The Zn(II) module is essential for *in vivo* function of both Ace1 and Amt1 (Thiele, 1988; Buchman *et al.*, 1990; Koch and Thiele, 1996). One function involves DNA binding in the A/T-rich region of UAS_{Cu}. A short sequence [(R/K)GRP] exists in the Zn module that is homologous to a minor-groove DNA-binding motif found in various nuclear proteins from animals, plants, insects, yeast and bacteria (Reeves and Nissen, 1990; Geierstanger *et al.*, 1994; Bustin and Reeves, 1996). The RGRP motif was originally described in the high mobility group HMG-I/Y protein family (Reeves and Nissen, 1990; Geierstanger *et al.*, 1994; Bustin and Reeves, 1996). In the HMG-I/Y : DNA complex, the arginine side-chains are buried deep within the minor groove making specific A/T contacts (Geierstanger *et al.*, 1994).

The RGRP sequence (residues 36–39) in Amt1 is responsible for minor groove DNA binding within the A/T-rich region (Koch and Thiele, 1996). An Arg_{38}Lys codon mutation in Amt1 greatly attenuated its *in vivo* function (Koch and Thiele, 1996). Likewise, a Gly_{37}Glu codon mutation in Ace1 compromised the ability of Ace1 to mediate copper-induced expression of *CUP1* (Buchman *et al.*, 1989; Buchman *et al.*, 1990). The substitutions in each protein diminished *in vitro* DNA-binding affinity four- to fivefold (Buchman *et al.*, 1990; Koch and Thiele, 1996).

Since the RGRP motif in Ace1 and Amt1 is part of the conserved Zn module, one function of the Zn module may be to stabilize a conformation that presents the RGRP sequence in proper juxtaposition to contact DNA. An additional function of the Zn(II) module in Ace1 is its role in major groove contacts with DNA bases (Buchman *et al.*, 1990; Dobi *et al.*, 1995). As mentioned, CuZnAce1 contacts bases separated by one and one-half turns of the DNA helix (Buchman *et al.*, 1990; Dobi *et al.*, 1995). The Zn(II) module of Ace1 appears to be responsible for the major groove contacts in the distal GCG region of UAS_{Cu} (Buchman *et al.*, 1990; Dobi *et al.*, 1995) (Fig. 17.2b). The evidence for such major groove contacts is twofold. First, footprinting studies with the Ace1-1 mutant molecule containing a Cys_{11}Tyr substitution revealed a loss of contacts in the distal region (Buchman *et al.*, 1990). Secondly, these same contacts were also lost in footprinting studies using a modified Ace1 in which the three cysteinyl residues in the N-terminal 40 residues were alkylated (Dobi *et al.*, 1995).

As mentioned, the DNA-binding site for Amt1 involves only the core GCTG sequences and the A/T-rich region (Zhou and Thiele, 1993; Koch and Thiele, 1996). Mutations within codons of the Zn module of Amt1 (Cys_{11}Tyr, Cys_{14}Ser, Cys_{23}Ser) reduce *in vivo* function of Amt1 (Posewitz *et al.*, 1996). The Zn module of Amt1 is therefore expected to contact only the A/T-sequence.

The fact that the copper regulatory domain (CuRD) binds DNA with high affinity is consistent with this domain dominating the protein : DNA interface. The truncated CuAmt1 complex may be expected to contact only the core GCTG sequence and not the A/T-rich region. The reduced DNA-binding affinity of the truncated CuAmt1 is likely related to the loss of minor groove contacts. The DNA-binding affinity of a mutant Amt1 complex with an R → K substitution in the Zn module shows a similar impact (Koch and Thiele, 1996).

The activation of Amt1 and Ace1 to become DNA-binding proteins occurs by formation of the tetracopper cluster. The tetracopper domain is functionally a CuRD. The CuRD contains eight cysteinyl residues. This domain is also found in the *S. cerevisiae* Lpz8p and *Yarrowia lipolyitica* Crf1 sequences. A consensus sequence can be derived from the four CuRD sequences: $C–X_2–C–(X)_{12,14}–C–X–C–(X)_{10–27}–C–X–C–X_5–C–X–C$. The only site of significant variability is the segment separating the two halves, each containing four cysteines. The variable spacing between the segments suggests that the CuRD exists as a two-lobe structure with the tetracopper center located between them. This is analogous to the structure of Cup1 in which two lobes bury the heptacopper–thiolate cluster (Peterson *et al.*, 1996) (Fig. 17.4).

Only six thiolates are needed to form a tetracopper cluster based on the known Cu_4S_6 synthetic cage clusters. The tetracopper clusters in Ace1 and Amt1 may not contain all bridging sulfurs as is observed in the Cu_4S_6 cage clusters. Our working model is that the Cu–thiolate cage cluster in Amt1 and Ace1 is stabilized by four terminal and four bridging thiolates (Fig. 17.3). Bridging thiolates are expected to be the predominant stabilizing force in the clusters.

Cu(I) activation of Amt1 or Ace1 appears to consist of conversion of this 70-residue CuRD from an apo-conformer or inactive Zn(II) conformer to a structure containing the tetracopper cluster. If the basal state of the regulatory domain is a Zn(II) conformer, then copper activation would occur through a metal exchange reaction shown below. A model for such metal exchange is the facile metal exchange kinetics observed in metallothioneins (Otvos *et al.*, 1989).

$$Zn_4MT + Cu_4Zn_1Ace1 \leftrightarrow Cu_7MT + Zn_4Ace1.$$

The significance of a tetracopper cluster as the structural unit within the activated transcription factors is threefold. First, a polycopper cluster formed by eight cysteinyl residues would organize and stabilize a larger structural unit than a single bound metal ion. A single Cu(I) site would anchor the polypeptide in only three or four places (depending on whether coordination geometry is trigonal or tetrahedral) rather than eight anchor sites as in the candidate copper regulatory domain of Ace1 and Amt1.

A second significant aspect of a tetracopper cluster is provision of metal specificity. Polymetal clusters are also known for Zn(II) and Cd(II) ions, but these clusters are structurally distinct from the polycopper clusters (Dance, 1986).

The third important feature is the observed cooperativity in cluster formation (Dameron *et al.*, 1991; Casas-Finet *et al.*, 1992; Thorvaldsen *et al.*, 1994). The tetracopper center in Amt1 forms in an all-or-nothing manner (Thorvaldsen *et al.*, 1994). Cooperativity in Cu(I) binding also occurs in Ace1 (Casas-Finet *et al.*, 1992).

Cooperativity in cluster formation permits a direct coupling of the intracellular exchangeable copper concentration to transcriptional activation of *CUP1* and, to a lesser extent, *CRS5* and *SOD1*. Cells can respond to small increases in copper ion concentration to activate Ace1 and therefore enhance MT biosynthesis.

17.6 Control over metal-ion binding

The basis for the observed precision in metallation of proteins is being resolved in *S. cerevisiae*. Three low-molecular-weight proteins have been identified that are critical for specific intracellular routing of Cu(I) ions to sites of biosynthesis of three important copper metalloenzymes (Fig. 17.5). Each chaperone participates in a distinct pathway of copper-ion trafficking (Culotta *et al*., 1997). At least one additional chaperone is expected to direct Cu(I) ions to the nucleus for regulation of Mac1 and Ace1 function.

Atx1 is the key cytoplasmic copper chaperone responsible for delivery of Cu(I) ions to the Ccc2 transporter in a late Golgi vesicle (Lin and Culotta, 1995; Yuan *et al*., 1997; Lin *et al*., 1997) (Fig. 17.5). Copper ions transported into the vesicles are incorporated in the ceruloplasmin-like Fet3 molecule (Yuan *et al*., 1995; Lin *et al*., 1997; Klomp *et al*., 1997). Subsequently, the CuFet3 complex is translocated to the plasma membrane for participation in high-affinity Fe ion uptake (Radisky *et al*., 1997). Cells lacking a functional Atx1 are compromised in Fe uptake (Lin *et al*., 1997). *ATX1*, like other genes whose products are involved in high-affinity iron uptake, is regulated by the iron-regulated Atf1 factor (Lin *et al*., 1997). Atx1 was originally isolated as a multi-copy suppressor of oxygen toxicity in yeast lacking a functional Sod1, so it is likely that Atx1 has two distinct functions (Lin and Culotta, 1995).

Atx1 binds a single Cu(I) within a conserved metal-binding motif (MXCXXC) that is found as a repeating motif in copper-transporting ATPases (Bull *et al*., 1993; Pufahl *et al*., 1997; Payne and Gitlin, 1998). A human homolog of Atx1, Hah1, will suppress

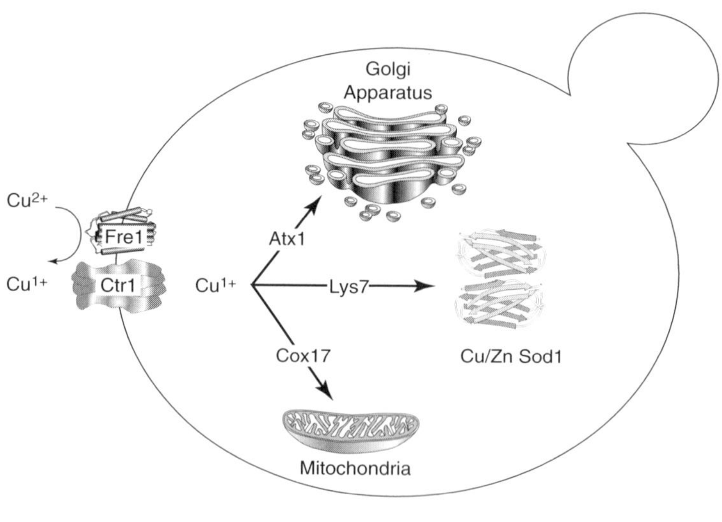

Figure 17.5 Intracellular routing of Cu(I) ions by the three known copper chaperones in yeast. Atx1 shuttles Cu(I) ions to late Golgi vesicles, Lys7 to superoxide dismutase and Cox17 to the mitochondrion.

the phenotypes of yeast lacking a functional Atx1 (Klomp *et al.*, 1997). Hah1 is also a copper-binding protein (Klomp *et al.*, 1997). The two cysteines, but not the methionine, are critical for metal binding and *in vivo* Atx1 function (Hung *et al.*, 1997; Pufahl *et al.*, 1997). Two of those metal-binding motifs are found in the *N*-terminal sequence of Ccc2. A two-hybrid experiment revealed that Atx1 interacts with the metal motifs of Ccc2 in a copper-dependent manner (Pufahl *et al.*, 1997).

Lys7 is a critical factor for the assembly of an active Cu,Zn-superoxide dismutase in yeast (Culotta *et al.*, 1997) (Fig. 17.5). In the absence of Lys7, Sod1 is a stable, inactive apo-protein. An Atx1-like copper-binding motif exists in Lys7. Lys7 is highly specific for copper routing to Sod1; Lys7 does not contribute to copper targeting to the nucleus, Golgi vesicles or the mitochondria (Culotta *et al.*, 1997). The human gene *CCS* is a Lys7 ortholog (Culotta *et al.*, 1997). Apo-Sod1 can be readily reconstituted *in vitro*, yet metallation *in vivo* requires a copper chaperone. A specific copper chaperone for Sod1 may ensure adequate copper levels are presented to Sod1 during biosynthesis. Alternatively, the existence of copper chaperones may be part of the process to limit free copper ion levels in cells, thereby minimizing the risk of copper-catalyzed Fenton chemistry.

Cox17 is the apparent copper chaperone for delivery of Cu(I) ions to the mitochondria (Glerum *et al.*, 1996a) (Fig. 17.5). Cytochrome oxidase contains two hemes, two distinct copper sites, one Zn(II) and one Mg(II) ion in addition to phospholipid constituents (Tsukihara *et al.*, 1995). The complexity of the enzyme appears to require accessory proteins for assembly of the complex. Two genes in *S. cerevisiae* have been shown to be important for incorporation of the copper ions into the enzyme (Glerum *et al.*, 1996a,b). Three copper ions are inserted into cytochrome oxidase, two into subunit II forming the binuclear Cu_A site and a single copper ion into subunit I forming the binuclear Cu-heme a_3 reaction center (Tsukihara *et al.*, 1995). Since both subunits I and II are mitochondrially synthesized, copper ion insertion must occur within the mitochondrion.

Cells harboring a mutant *cox17* were found to be respiratory deficient but able to express both mitochondrial- and nuclear-encoded cytochrome oxidase subunits (Glerum *et al.*, 1996a). The respiratory deficiency of the *cox17* mutant was suppressed by exogenous Cu(II), suggesting that Cox17 was important in copper delivery to the mitochondrion (Glerum *et al.*, 1996a). The point mutation in the *cox17* mutant resulted in a Cys → Tyr substitution. Exogenous Cu(II) did not rescue a subset of other cytochrome oxidase mutants. The apparent defect in copper delivery to cytochrome oxidase was not a general defect in copper balance in the cell. The *cox17* mutant had normal Sod1 activity and Fet3 function, both proteins requiring copper for function (Glerum *et al.*, 1996a). A human ortholog for Cox17 is known (Amaravadi *et al.*, 1997).

Additional factors appear to be important for copper ion insertion into cytochrome oxidase. The respiratory deficient phenotype of the *cox17* point mutant was suppressed by high-copy *SCO1* and *SCO2* (Glerum *et al.*, 1996b). However, high-copy *SCO1* exhibited only limited suppression in a *cox17* null strain. Sco1 was previously characterized as a cytochrome oxidase assembly factor localized within the inner mitochondrial membrane (Buchwald *et al.*, 1991). Like *cox17* mutants, *sco1* mutants exhibit a specific defect in cytochrome oxidase (Glerum *et al.*, 1996b). *sco1* mutants (null and point mutants) cannot be suppressed by high exogenous copper or high copy *cox17*.

Cox17 is an 8057 Da polypeptide. Its localization in the cytosol and intermitochondrial membrane space is consistent with the role of Cox17 in the delivery of copper ions to the mitochondrion for assembly of cytochrome oxidase (Beers *et al.*, 1997). The presence of seven cysteinyl residues and the CCXC sequence motif are consistent with Cox17 being a copper-binding protein.

Cox17 binds copper as a binuclear cuprous-thiolate cluster (Srinivasan *et al.*, 1998). X-ray absorption spectroscopy, together with UV-visible absorption and emission spectroscopy indicated the presence of bound cuprous ions, trigonally coordinated by thiolate ligands. Analysis of the EXAFS of CuCox17 revealed a short Cu–Cu contact at 2.7 Å that is diagnostic of a binuclear cluster (Srinivasan *et al.*, 1998). The cuprous–thiolate cluster in Cox17 is substantially more labile than structurally related clusters in metallothioneins.

The existence of human orthologs of Atx1, Lys7 and Cox17 and the demonstration that the human gene products suppress yeast mutants is a vivid demonstration of how yeast studies have provided insights into control of mammalian copper ion binding. It is anticipated that metal-ion chaperones will also exist for routing of other physiologically important metal ions that are also potentially deleterious to cells. These include Fe(II), Mn(II) and perhaps Zn(II).

References

Agarwal, K., Sharma, A. and Talukder, G. 1989. Effects of copper on mammalian cell components. *Chemico-Biological Interactions* **69**, 1–16.

Amaravadi, R., Glerum, D.M. and Tzagoloff, A. 1997. Isolation of a cDNA encoding the human homolog of COX17, a yeast gene essential for mitochondrial copper recruitment. *Human Genetics* **99**, 329–33.

Askwith, C., Eide, D., Van Ho, A., Bernard, P.S., Li, L., Davis-Kaplan, S., Sipe, D.M. and Kaplan, J. 1994. The FET3 gene of *S. cerevisiae* encodes a multicopper oxidase required for ferrous iron uptake. *Cell* **76**, 403–10.

Babcock, M., De Silva, D., Oaks, R., Davis-Kaplan, S., Jiralerspong, S., Monteermini, L., Pandolfo, M. and Kaplan, J. 1997. Regulation of mitochondrial iron accumulation by Yfh1p, a putative homolog of frataxin. *Science* **276**, 1709–12.

Beeler, T.J., Fu, D., Rivera, J., Monaghan, E., Gable, K. and Dunn, T.M. 1997. SUR1 (CSG1/BCL21), a gene necessary for growth of *Saccharomyces cerevisiae* in the presence of high Ca^{2+} concentrations at 37 °C, is required for mannosylation of inositolphosphorylceramide. *Molecular and General Genetics* **255**, 570–9.

Beers, J., Glerum, D.M. and Tzagoloff, A. 1997. Purification, characterization, and localization of yeast Cox17p, a mitochondrial copper shuttle. *Journal of Biological Chemistry* **272**, 33191–6.

Botstein, D., Chervitz, S.A. and Cherry, J.M. 1997. Yeast as a model organism. *Science* **277**, 1259–60.

Brenes-Pomales, A., Lindegren, G. and Lindegren, C.C. 1955. *Nature* **176**, 841–2.

Buchman, C., Skroch, P., Welch, J., Fogel, S. and Karin, M. 1989. The CUP2 gene product, regulator of yeast metallothionein expression, is a copper-activated DNA-binding protein. *Molecular and Cellular Biology* **9**, 4091–5.

Buchman, C., Skroch, P., Dixon, W., Tullius, T.D. and Karin, M. 1990. A single amino acid change in CUP2 alters its mode of DNA binding. *Molecular and Cellular Biology* **10**, 4778–87.

Buchwald, P., Krummeck, G. and Rodel, G. 1991. Immunological identification of yeast SCO1 protein as a component of the inner mitochondrial membrane. *Molecular and General Genetics* **229**, 413–20.

Bull, P.C., Thomas, G.R., Rommens, J.M., Forbes, J.R. and Cox, D.W. 1993. The Wilson disease gene is a putative copper transporting P-type ATPase similar to the Menkes gene. *Nature Genetics* **5**, 327–337.

Bustin, M. and Reeves, R. 1996. High-mobility-group chromosomal proteins: architectural components that facilitate chromatin function. *Progress in Nucleic Acid Research and Molecular Biology* **54**, 35–95.

Butt, T.R., E.J., Gorman, J.A., Clark, P., Hamer, D., Rosenberg, M. and Crooke, S.T. 1984. Copper metallothionein of yeast, structure of the gene, and regulation of expression. *Proceedings of the National Academy of Sciences (U.S.A.)* **81**, 3332–6.

Casas-Finet, J.R., Hu, S., Hamer, D. and Karpel, R.L. 1992. Characterization of the copper- and silver-thiolate clusters in N-terminal fragments of the yeast ACE1 transcription factor capable of binding to its specific DNA recognition sequence. *Biochemistry* **31**, 6617–26.

Culotta, V.C., Howard, W.R. and Liu, X.F. 1994. CRS5 encodes a metallothionein-like protein in *Saccharomyces cerevisiae. Journal of Biological Chemistry* **269**, 1–8.

Culotta, V.C., Joh, H.D., Lin, S.J., Slekar, K.H. and Strain, J. 1995. A physiological role for *Saccharomyces cerevisiae* copper/zinc superoxide dismutase in copper buffering. *Journal of Biological Chemistry* **270**, 29991–7.

Culotta, V.C., Klomp, L.W.J., Strain, J., Casareno, R.L.B., Krems, B. and Gitlin, J.D. 1997. The copper chaperone for superoxide dismutase. *Journal of Biological Chemistry* **272**, 23469–72.

Dahlman-Wright, K., Baumann, H., McEwan, I.J., Almlof, T., Wright, A.P.H., Gustafsson, J.-A. and Hard, T. 1995. Structural characterization of a minimal functional transactivation domain from the human glucocorticoid receptor. *Proceedings of the National Academy of Sciences (U.S.A.)* **92**, 1699–703.

Dameron, C.T., Winge, D.R., George, G.N., Sansone, M., Hu, S. and Hamer, D. 1991. A copper–thiolate polynuclear cluster in the ACE1 transcription factor. *Proceedings of the National Academy of Sciences (U.S.A.)* **88**, 6127–31.

Dance, I.G. 1986. Structural chemistry of metal thiolate complexes. *Polyhedron* **5**, 1037–104.

Dancis, A., Haile, D., Yuan, D.S. and Klausner, R.D. 1994. The *Saccharomyces cerevisiae* copper transport protein (Ctr1p). Biochemical characterization, regulation by copper, and physiologic role in copper uptake. *Journal of Biological Chemistry* **269**, 25660–7.

Dancis, A., Roman, D.G., Anderson, G.J., Hinnebush, A.G. and Klausner, R.D. 1992. Ferric reductase of *Saccharomyces cerevisiae*: molecular characterization, role in iron uptake, and transcriptional control by iron. *Proceedings of the National Academy of Sciences (U.S.A.)* **89**, 3869–73.

De Silva, D.M., Askwith, C.C., Eide, D. and Kaplan, J. 1995. The FET3 gene product required for high affinity iron transport in yeast is a cell surface ferroxidase. *Journal of Biological Chemistry* **270**, 1098–101.

Dixon, W.J., Inouye, C., Karin, M. and Tullius, T.D. 1996. CUP2 binds in a bipartite manner to upstream activation sequence c in the promoter of the yeast copper metallothionein gene. *Journal of Biological and Inorganic Chemistry* **1**, 451–9.

Dobi, A., Dameron, C.T., Hu, S., Hamer, D. and Winge, D.R. 1995. Distinct regions of Cu(I)·ACE1 contact, two spatially resolved DNA major groove sites. *Journal of Biological Chemistry* **270**, 10171–8.

Ecker, D.J., Butt, T.R., Sternberg, E.J., Neeper, M.P., Debouck, C., Gorman, J.A. and Crooke, S.T. 1986. Yeast metallothionein function in metal ion detoxification. *Journal of Biological Chemistry* **261**, 16895–900.

Farr, S.B. and Kogoma, T. 1991. Oxidative stress responses in *Escherichia coli* and *Salmonella typhimurium. Microbiological Reviews* **55**, 561–85.

Farrell, R.A., Thorvaldsen, J.L. and Winge, D.R. 1996. Identification of the Zn(II) site in the copper-responsive yeast transcription factor, AMT1: a conserved Zn module. *Biochemistry* **35**, 1571–80.

Finegold, A.A., Shatwell, K.P., Segal, A.W., Klausner, R.D. and Dancis, A. 1996. Intramembrane bis-heme motif for transmembrane electron transport conserved in a yeast iron reductase and the human NADPH oxidase. *Journal of Biological Chemistry* **271**, 31021–4.

Fogel, S. and Welch, J.W. 1982. Tandem gene amplification mediates copper resistance in yeast. *Proceedings of the National Academy of Sciences (U.S.A.)* **79**, 5342–6.

Fogel, S., Welch, J.W. and Maloney, D.H. 1988. The molecular genetics of copper resistance in *Saccharomyces cerevisiae* – a paradigm for non-conventional yeasts. *Journal of Basic Microbiology* **28**, 147–60.

Fridovich, I. 1978. The biology of oxygen radicals. *Science* **201**, 875–80.

Furst, P., Hu, S., Hackett, R. and Hamer, D. 1988. Copper activates metallothionein gene transcription by altering the conformation of a specific DNA binding protein. *Cell* **55**, 705–17.

Gadd, G.M. and White, C. 1989. *Metal–Microbe Interactions*, IRL Press, Oxford.

Geierstanger, B.H., Volkman, B.F., Kremer, W. and Wemmer, D.E. 1994. Short peptide fragments derived from HMG-I/Y proteins bind specifically to the minor groove of DNA. *Biochemistry* **33**, 5347–55.

Georgatsou, E. and Alexandraki, D. 1994. Two distinctly regulated genes are required for ferric reduction, the first step of iron uptake in *Saccharomyces cerevisiae. Molecular and Cellular Biology* **14**, 3065–73.

Georgatsou, E., Mavrogiananis, L.A., Fragiadakis, G.S. and Alexandraki, D. 1997. The yeast Fre1p/Fre2p cupric reductases facilitate copper uptake and are regulated by the copper-modulated Mac1p activator. *Journal of Biological Chemistry* **272**, 13786–92.

Glerum, D.M., Shtanko, A. and Tzagoloff, A. 1996a. Characterization of COX17, a yeast gene involved in copper metabolism and assembly of cytochrome oxidase. *Journal of Biological Chemistry* **271**, 14504–9.

Glerum, D.M., Shtanko, A. and Tzagoloff, A. 1996b. SCO1 and SCO2 act as high copy suppressors of a mitochondrial copper recruitment defect in *Saccharomyces cerevisiae. Journal of Biological Chemistry* **271**, 20531–5.

Graden, J.A. and Winge, D.R. 1997. Copper-mediated repression of the activation domain in the yeast Mac1p transcription factor. *Proceedings of the National Academy of Sciences (U.S.A.)* **94**, 5550–5.

Graden, J.A., Posewitz, M.C., Simon, J.R., George, G.N., Pickering, I.J. and Winge, D.R. 1996. Presence of a copper(I)-thiolate regulatory domain in the copper-activated transcription factor Amt1. *Biochemistry* **35**, 14583–9.

Gralla, E.B., Thiele, D.J., Silar, P. and Valentine, J.S. 1991. ACE1, a copper-dependent transcription factor, activates expression of the yeast copper, zinc superoxide dismutase gene. *Proceedings of the National Academy of Sciences (U.S.A.)* **88**, 8558–62.

Hahn, S. 1993. Structure(?) and function of acidic transcription activators. *Cell* **72**, 481–3.

Hamer, D.H., Thiele, D.J. and Lemontt, J.E. 1985. Function and autoregulation of yeast copperthionein. *Science* **228**, 685–90.

Hassett, R. and Kosman, D.J. 1995. Evidence for Cu(II) reduction as a component of copper uptake by *Saccharomyces cerevisiae. Journal of Biological Chemistry* **270**, 128–34.

Henle, E.S. and Linn, S. 1997. Formation, prevention, and repair of DNA damage by iron/hydrogen peroxide. *Journal of Biological Chemistry* **272**, 19095–8.

Hu, S., Furst, P. and Hamer, D. 1990. The DNA and Cu binding functions of ACE1 are interdigitated within a single domain. *New Biologist* **2**, 544–55.

Hung, I.H., Casareno, R.L.B., Labesse, G., Mathews, F.S. and Gitlin, J.D. 1997. HAH1 is a copper-binding protein with distinct amino acid residues mediating copper homeostasis and antioxidant defense. *Journal of Biological Chemistry* **273**, 1749–54.

Imlay, J.A., Chin, S.M. and Linn, S. 1988. Toxic DNA damage by hydrogen peroxide through the Fenton reaction *in vivo* and *in vitro*. *Science* **240**, 640–2.

Jensen, L.T. and Winge, D.R. 1998. Identification of a copper-induced intramolecular interaction in the transcription factor Mac1 from *Saccharomyces cerevisiae*. *EMBO Journal* **17**, 5400–8.

Jensen, L.T., Howard, W.R., Strain, J.J., Winge, D.R. and Culotta, VC. 1996. Enhanced effectiveness of copper ion buffering by CUP1 metallothionein compared with CRS5 metallothionein in *Saccharomyces cerevisiae*. *Journal of Biological Chemistry* **271**, 18514–19.

Jensen, L.T., Posewitz, M.C., Srinivasan, C. and Winge, D.R. 1998. Mapping of the DNA binding domain of the copper-responsive transcription factor Mac1 from *Saccharomyces cerevisiae*. *Journal of Biological Chemistry* **273**, 23805–11.

Jungmann, J., Reins, H.A., Lee, J., Romeo, A., Hassett, R., Kosman, D. and Jentsch, S. 1993. MAC1, a nuclear regulatory protein related to Cu-dependent transcription factors is involved in Cu/Fe utilization and stress resistance in yeast. *EMBO Journal* **12**, 5051–6.

Kampfenkel, K., Kushnir, S., Babiychuk, E., Inze, D. and Van Montagu, M. 1995. Molecular characterization of a putative *Arabidopsis thaliana* copper transporter and its yeast homologue. *Journal of Biological Chemistry* **270**, 28479–86.

Karin, M., Najarian, R., Haslinger, A., Valenzuela, P., Welch, J., and Fogel, S. 1984. Primary structure and transcription of an amplified genetic locus: the CUP1 locus of yeast. *Proceedings of the National Academy of Sciences (U.S.A.)* **81**, 337–41.

Klomp, L.W.J., Lin, S.-J., Yuan, D.S., Klausner, R.D., Culotta, V.C. and Gitlin, J.D. 1997. Identification and functional expression of HAH1, a novel human gene involved in copper homeostasis. *Journal of Biological Chemistry* **272**, 9221–6.

Knight, S.A.B., Labbe, S., Kwon, L. F., Kosman, D.J. and Thiele, D.J. 1996. A widespread transposable element masks expression of a yeast copper transport gene. *Genes and Development* **10**, 1917–29.

Koch, K.A. and Thiele, D.J. 1996. Autoactivation by a *Candida glabrata* copper metalloregulatory transcription factor requires critical minor groove interactions. *Molecular and Cellular Biology* **16**, 724–34.

Labbe, S., Zhu, Z. and Thiele, D.J. 1997. Copper-specific transcriptional repression of yeast genes encoding critical components in the copper transport pathway. *Journal of Biological Chemistry* **272**, 15951–8.

Lefstin, J.A., Thomas, J.R. and Yamamoto, K.R. 1994. Influence of a steroid receptor DNA-binding domain on transcriptional regulatory functions. *Genes and Development* **8**, 2842–56.

Lesuisse, E., Casteras-Simon, M. and Labbe, P. 1996. Evidence for the *Saccharomyces cerevisiae* ferrireductase system being a multicomponent electron transport chain. *Journal of Biological Chemistry* **271**, 13578–83.

Leuther, K.K., Salmeron, J.M. and Johnston, S.A. 1993. Genetic evidence that an activation domain of GAL4 does not require acidity and may form a beta sheet. *Cell* **72**, 575–85.

Lin, C.M. and Kosman, D.J. 1990. Copper uptake in wild type and copper metallothionein-deficient *Saccharomyces cerevisiae*. Kinetics and mechanism. *Journal of Biological Chemistry* **265**, 9194–200.

Lin, S. and Culotta, V.C. 1995. The ATX1 gene of *Saccharomyces cerevisiae* encodes a small metal homeostasis factor that protects cells against reactive oxygen toxicity. *Proceedings of the National Academy of Sciences (U.S.A.)* **92**, 3784–8.

Lin, S.-J., Pufahl, R.A., Dancis, A., O'Halloran, T.V.O. and Culotta, V.C. 1997. A role for the *Saccharomyces cerevisiae* ATX1 gene in copper trafficking and iron transport. *Journal of Biological Chemistry* **272**, 9215–20.

Marbach, K., Fernandez-Larrea, J. and Stahl, U. 1994. Reversion of a long-living, undifferentiated mutant of *Podospora anserina* by copper. *Current Genetics* **26**, 184–6.

Martins, L., Jensen, L.T., Simons, J.R., Keller, G. and Winge, D.R. 1998. Metalloregulation of FRE1 and FRE2 homologs in *Saccharomyces cerevisiae*. *Journal of Biological Chemistry* **273**, 23716–21.

Masters, B.A., Kelly, E.J., Quaife, C.J., Brinster, R.L. and Palmiter, R.D. 1994. Targeted disruption of metallothionein I and II genes increases sensitivity to cadmium. *Proceedings of the National Academy of Sciences* **91**, 584–8.

Mehra, R.K., Garey, J.R. and Winge, D.R. 1990. Selective and tandem amplification of a member of the metallothionein gene family in *Candida glabrata. Journal of Biological Chemistry* **265**, 6369–75.

Mehra, R.K., Thorvaldsen, J.L., Macreadie, I.G. and Winge, D.R. 1992. Disruption analysis of metallothionein-encoding genes in *Candida glabrata. Gene* **114**, 75–80.

Michalska, A.E. and Choo, K.H.A. 1993. Targeting and germ-line transmission of a null mutation at the metallothionein I and II loci in mouse. *Proceedings of the National Academy of Sciences (U.S.A)* **90**, 8088–92.

Mitchell, P.J. and Tjian, R. 1989. Transcriptional regulation in mammalian cells by sequence-specific DNA binding proteins. *Science* **245**, 371–8.

Munder, T. and Furst, P. 1992. The *Saccharomyces cerevisiae* CDC25 gene product binds specifically to catalytically inactive ras proteins *in vivo. Molecular and Cellular Biology* **12**, 2091–9.

Narula, S.S., Mehra, R.K., Winge, D.R. and Armitage, I.M. 1991. Establishment of the metal-to-cysteine connectivities in silver-substituted yeast metallothionein. *Journal of the American Chemical Society* **113**, 9354–8.

Ooi, C.E., Rabinovich, E., Dancis, A., Bonifacino, J.S. and Klausner, R.D. 1996. Copper-dependent degradation of the *Saccharomyces cerevisiae* plasma membrane copper transporter Ctr1p in the apparent absence of endocytosis. *EMBO Journal* **15**, 3515–23.

Osiewacz, H.D. and Nuber, U. 1996. GRISEA, a putative copper-activated transcription factor from *Podospora anserina* involved in differentiation and senescence. *Molecular and General Genetics* **252**, 115–24.

Otvos, J.D., Petering, D.H. and Shaw, C.F. 1989. Structure–reactivity relationships of metallothionein, a unique metal-binding protein. *Comments in Inorganic Chemistry* **9**, 1–35.

Payne, A.S. and Gitlin, J.D. 1998. Functional expression of the Menkes disease protein reveals common biochemical mechanisms among the copper-transporting P-type ATPases. *Journal of Biological Chemistry* **273**, 3765–70.

Peterson, C.W., Narula, S.S. and Armitage, I.M. 1996. 3D solution structure of copper and silver-substituted yeast metallothioneins. *FEBS Letters* **379**, 85–93.

Pickering, I.J., George, G.N., Dameron, C.T., Kurz, B., Winge, D.R. and Dance, I.G. 1993. X-ray absorption spectroscopy of cuprous–thiolate clusters in proteins and model systems. *Journal of the American Chemical Society* **115**, 9498–505.

Posewitz, M.C., Simon, J.R., Farrell, R.A. and Winge, D.R. 1996. Role of the conserved histidines in the Zn module of the copper-activated transcription factors in yeast. *Journal of Biological and Inorganic Chemistry* **1**, 560–6.

Ptashne, M. 1988. How eukaryotic transcriptional activators work. *Nature* **335**, 683–9.

Pufahl, R.A., Singer, C.P., Peariso, K.L., Lin, S.-J., Schmidt, P., Fahrni, C., Culotta, V.C., Penner-Hahn, J.E. and O'Halloran, T.V.O. 1997. Metal ion chaperone function of the soluble Cu(I) receptor Atx1. *Science* **278**, 853–6.

Radisky, D.C., Snyder, W.B., Emr, S.D. and Kaplan, J. 1997. Characterization of VPS41, a gene required for vacuolar trafficking and high-affinity iron transport in yeast. *Proceedings of the National Academy of Sciences (U.S.A.)* **94**, 5662–6.

Reeves, R. and Nissen, M.S. 1990. The A.T-DNA-binding domain of mammalian high mobility group I chromosomal proteins. A novel peptide motif for recognizing DNA structure. *Journal of Biological Chemistry* **265**, 8573–82.

Regier, J.L., Shen, F. and Triezenberg, S.J. 1993. Pattern of aromatic and hydrophobic amino acids critical for one of two subdomains of the VP16 transcriptional activator. *Proceedings of the National Academy of Sciences (U.S.A.)* **90**, 883–7.

Sadowski, I., Triezenberg, S. and Ptashne, M. 1988. GAL4-VP16 is an unusually potent transcriptional activator. *Nature* **335**, 563–4.

Shatwell, K.P., Dancis, A., Cross, A.R., Klausner, R.D. and Segal, A.W. 1996. The FRE1 ferric reductase of *Saccharomyces cerevisiae* is a cytochrome *b* similar to that of NADPH oxidase. *Journal of Biological Chemistry* **271**, 14240–4.

Sievers, C., Deters, D., Hartmann, H. and Weser, U. 1996. Stable thiyl radicals in dried yeast Cu(I)6-thionein. *Journal of Inorganic Biochemistry* **62**, 199–205.

Silverman, N., Agapite, J. and Guarente, L. 1994. Yeast ADA2 protein binds to the VP16 protein activation domain and activates transcription. *Proceedings of the National Academy of Sciences (U.S.A.)* **91**, 11665–8.

Slekar, K.H., Kosman, D.J. and Culotta, V.C. 1996. The yeast copper/zinc superoxide dismutase and the pentose phosphate pathway play overlapping roles in oxidative stress protection. *Journal of Biological Chemistry* **271**, 28831–6.

Srinivasan, C., Posewitz, M.C., George, G.N. and Winge, D.R. 1998. Characterization of the copper chaperone Cox17 of *Saccharomyces cerevisiae. Biochemistry* **37**, 7572–7.

Tamai, K.T., Gralla, E.B., Ellerby, L.M., Valentine, J.S. and Thiele, D.J. 1993. Yeast and mammalian metallothioneins functionally substitute for yeast copper-zinc superoxide dismutase. *Proceedings of the National Academy of Sciences (U.S.A.)* **90**, 8013–17.

Thiele, D.J. 1988. ACE1 regulates expression of the *Saccharomyces cerevisiae* metallothionein gene. *Molecular and Cellular Biology* **8**, 2745–52.

Thorvaldsen, J.L., Sewell, A.K., Tanner, A.M., Peltier, J.M., Pickering, I.J., George, G.N. and Winge, D.R. 1994. Mixed Cu^{+} and Zn^{2+} coordination in the DNA-binding domain of the AMT1 transcription factor from *Candida glabrata. Biochemistry* **33**, 9566–77.

Tsukihara, T., Aoyama, H., Yamashita, E., Tomizaki, T., Yamaguchi, H., Shinzawa-Itoh, K., Hakashima, R., Yaono, R. and Yoshikawa, S. 1995. Structures of metal sites of oxidized bovine heart cytochrome c oxidase at 2.8 Å. *Science* **269**, 1069–74.

Turner, J.S., Morby, A.P., Whitton, B.A., Gupta, A. and Robinson, N.J. 1993. Construction of Zn^{2+}/Cd^{2+} hypersensitive cyanobacterial mutants lacking a functional metallothionein locus. *Journal of Biological Chemistry* **268**, 4494–8.

Turner, R.B., Smith, D.L., Zawrotny, M.E., Summers, M.F., Posewitz, M.C. and Winge, D.R. 1998. Solution structure of a zinc domain conserved in yeast copper-regulated transcription factors. *Nature Structural Biology* **5**, 551–5.

Velculescu, V.E., Zhang, L., Zhou, W., Vogelstein, J., Basrai, M.A., Bassett, D.E., Hieter, P., Volgelstein, B. and Kinzler, K.W. 1997. Characterization of the yeast transcriptome. *Cell* **88**, 243–51.

Wang, D., Hu, Y., Zheng, F., Zhou, K. and Kohlhaw, G. 1997. Evidence that intramolecular interactions are involved in masking the activation domain of transcriptional activator Leu3p. *Journal of Biological Chemistry* **272**, 19383–92.

Winge, D.R., Graden, J.A., Posewitz, M.C., Martins, L.J., Jensen, L.T. and Simon, J.R. 1997. Sensors that mediate copper-specific activation and repression of gene expression. *Journal of Biological Chemistry* **2**, 2–10.

Yamaguchi-Iwai, Y., Dancis, A. and Klausner, R.D. 1995. AFT1: a mediator of iron regulated transcriptional control in *Saccharomyces cerevisiae. EMBO Journal* **14**, 1231–9.

Yamaguchi-Iwai, Y., Stearman, R., Dancis, A. and Klausner D. 1996. Iron-regulated DNA binding by the AFT1 protein controls the iron regulon in yeast. *EMBO Journal* **15**, 3377–84.

Yamaguchi-Iwai, Y., Serpe, M., Haile, D., Yang, W., Kosman, D.J., Klausner, R.D. and Dancis, A. 1997. Homeostatic regulation of copper uptake in yeast via direct binding of MAC1 protein to upstream regulatory sequences of FRE1 and CTR1. *Journal of Biological Chemistry* **272**, 17711–18.

Yu, W., Santhanagopalan, V., Sewell, A.K., Jensen, L.T. and Winge, D.R. 1994. Dominance of metallothionein in metal ion buffering in yeast capable of synthesis of (gamma EC)nG isopeptides. *Journal of Biological Chemistry* **269**, 212010–15.

Yuan, D.S., Stearman, R., Dancis, A., Dunn, T., Beeler, T. and Klausner, R.D. 1995. The Menkes/Wilson disease gene homologue in yeast provides copper to a ceruloplasmin-like oxidase required for iron uptake. *Proceedings of the National Academy of Sciences (U.S.A.)* **92**, 2632–6.

Yuan, D.S., Dancis, A. and Klausner, R.D. 1997. Restriction of copper export in *Saccharomyces cerevisiae* to a late Golgi or post-Golgi compartment in the secretory pathway. *Journal of Biological Chemistry* **272**, 25787–93.

Zawel, L. and Reinberg, D. 1995. Common themes in assembly and function of eukaryotic transcription complexes. *Annual Review of Biochemistry* **64**, 533–61.

Zhao, H. and Eide, D.J. 1997. Zap1p, a metalloregulatory protein involved in zinc-responsive transcriptional regulation in *Saccharomyces cerevisiae. Molecular and Cellular Biology* **17**, 5044–52.

Zhou, B. and Gitschier, J. 1997. hCTR1: a human gene for copper uptake identified by complementation in yeast. *Proceedings of the National Academy of Sciences (U.S.A.)* **94**, 7481–6.

Zhou, P. and Thiele, D.J. 1991. Isolation of a metal-activated transcription factor gene from *Candida glabrata* by complementation in *Saccharomyces cerevisiae. Proceedings of the National Academy of Sciences (U.S.A.)* **88**, 6112–16.

Zhou, P. and Thiele, D.J. 1993. Rapid transcriptional autoregulation of a yeast metalloregulatory transcription factor is essential for high-level copper detoxification. *Genes and Development* **7**, 1824–35.

Zu, Z., Labbe, S., Pena, M.M.O. and Thiele, D.J. 1998. Copper differentially regulates the activity and degradation of yeast Mac1 transcription factor. *Journal of Biological Chemistry* **273**, 1277–80.

18

TOXIC AND ESSENTIAL METALS IN THE CELLULAR RESPONSE TO SIGNALS

James Koropatnick and Rudolfs K. Zalups

18.1 Introduction

There is no environmental niche in which heavy metals do not exist and persist as the most long-lived product of both natural and man-made sources (reviewed by Goyer, 1991). Although many are toxic to cells and organs in varying degrees and through a variety of mechanisms (reviewed in several chapters of this volume), there is still an absolute requirement of some, but not all, metal ions by living systems. In fact, 17 out of the 30 elements essential for life are metals (Cotton and Wilkinson, 1980). Many of these required metals can be damaging under certain circumstances. For example, copper is an essential cofactor for many oxidative enzymes, including catalase, peroxidase, cytochrome oxides, and others – but it is also a dangerous cellular toxicant (Horn, 1984). Iron is required for hemoglobin, but is, like copper, a redox-active metal capable of mediating the production of damaging reactive oxygen species – a capacity that, although exploited by the cell to generate reactive chemical centers for physiologically beneficial purposes, can also lead to detrimental effects (reviewed by Kasprzak and Buzard, Chapter 16 this volume). The capacity of metal ions of this type to generate reactive species that both nurture and destroy has resulted in the development of complex cellular mechanisms to transport, chaperone, localize and detoxify metal ions, to regulate intracellular availability of metals to cellular components requiring them for function, but minimize exposure to other cellular molecules and structures that are sensitive to their harmful effects. Although these processes to regulate metal ion availability have been assumed to function to prevent toxicity, it has also been suggested that they act to regulate the function of proteins that require metals to carry out physiological processes. This may be especially true of metal ions that are essential and almost entirely nontoxic – properties characteristic of zinc. This metal ion is required by a large number of metalloenzymes and structural proteins, and is an essential component of many transcription factors and cytoplasmic receptors that transduce external stimuli and regulate the transcription of specific genes by interaction with gene regulatory sequences (reviewed by O'Halloran, 1989). The potential of metal ions to induce the expression of genes relevant to these signaling pathways, or to interact with and modify the activity of intermediate proteins participating in signal transduction and gene transcription, has important consequences for cell behavior in the presence of metals. In particular, it raises the possibility that metal ions may modify specific cellular functions without overtly toxic effects or inhibitory effects on isolated cellular components not essential for cell viability. In this scenario, low levels of metal ions, previously thought to be inconsequential on the basis of lack of toxicity, may have profound effects on cell function and human health. Evidence for such nontoxic consequences of exposure to metal ions will be reviewed in this chapter.

Many biologically important molecules require metals as cofactors, including proteins that mediate signal transduction and gene transcription (zinc-finger proteins, leucine zippers, helix–turn–helix (HLH) motif proteins and others, that are components of hormone receptors and other transcription factors, and zinc-requiring proteins that transmit signals from the plasma membrane or cytoplasm to the nucleus (reviewed by Pabo and Sauer, 1992)). It is an intriguing, but unresolved, possibility that metals might modulate the level of gene transcription *in vivo* by their regulated addition or removal from signal transduction proteins/transcription factors through homeostatic, metal-ion regulatory processes (Zeng *et al.*, 1991a,b; Vallee and Falchuk, 1993; Maret and Vallee, 1998), and evidence for this is reviewed below. In addition, metals are cofactors in metalloenzymes that are important in messenger RNA processing and enzyme and structural protein function, but which are not directly involved in signal transduction or gene transcription. The potential for regulated availability of essential metal ions (or interference with that regulation by exposure to toxic, nonessential metal ions) must be considered a mechanism for metals to control or influence gene expression. Overall, metal ions may influence or induce expression of eucaryotic genes in at least two ways:

1. as an extracellular signal to induce gene expression, mediated by metal-ion-responsive transcription factor(s);
2. as a cofactor *whose availability is regulated* to mediate the function of signaling proteins that transduce information from the cell exterior to the nucleus, and enzymes and structural proteins requiring metal ions for activity or function.

The transport of metals is reviewed in Chapters 5–10 of this volume, and will not be addressed here. In addition, metal-regulated and metal-influenced gene expression in procaryotes and plants is omitted, and the reader is referred to Chapter 5 in this volume, and an excellent review by Ralston and O'Halloran (1990), for details of this important area. The effects of metals on the function of proteins acting in pathways that transduce signals and mediate gene transcription will be addressed. Metallothionein, as a gene product that is both induced by metals and which may act to regulate essential metal ion concentration and availability, is of central importance. Special attention will be paid to its regulation and potential function in mammals.

18.2 Metal influences on gene expression

Many metal ions have the capacity to stimulate the expression of genes in higher eucaryotes (reviewed by Koropatnick and Leibbrandt, 1995), including metallothionein genes (described in detail below). In addition, nonmetallothionein genes in algae, fungi, plants, rodents and humans are also responsive to metals. These include (among others) copper-containing plastocyanin protein and the heme-protein cytochrome *c* (cyt *c6*). Plastocyanin and cyt *c6* act interchangeably as photosynthetic electron carriers: high copper levels result in increased plastocyanin due to increased stability of the copper–plastocyanin complex, and low copper levels result in decreased plastocyanin and increased cyt *c6* expression as a result of transcriptional derepression of the cyt *c6* gene, probably mediated by a copper-sensing regulatory factor (Merchant and Bogorad, 1987; Hill *et al.*, 1991; Merchant *et al.*, 1991). In yeast, Cu,Zn-superoxide dismutase (Cu,ZnSOD) is induced by copper (Gregory *et al.*, 1974), either by copper incorporation into pre-existing Cu,ZnSOD to activate the protein or by copper-mediated induction of SOD gene transcription. Ace1 (a metal-responsive transcription factor in *Saccharomyces cerevisiae* that controls Cup1 (the yeast counterpart of mammalian metallothionein)) is activated by binding to copper and binds to the *Saccharomyces cerevisiae* Cu,ZnSOD gene promoter (Gralla *et al.*, 1991; Chapter 17, this volume) to induce elevated Cu,ZnSOD mRNA (Greco *et al.*, 1990). In this case, the same metal signal and metal-responsive transcription factor that controls yeast Cup1 protein expression also induces *trans*-activation of Cu,ZnSOD. This coregulation can be explained in two, nonexclusive ways:

1 It is possible that copper:Ace1 association acts as a signal to indicate that intracellular copper levels are sufficient for the formation of active Cu,ZnSOD holoenzyme. High copper levels stimulate the formation of toxic free-radicals through redox interactions with superoxide anion ($O_2^{\cdot -}$) (Halliwell and Gutteridge, 1984), and Cu,ZnSOD is required to detoxify those radicals and prevent cell damage. In this case, metals may play a role in protection against oxidative stress resulting, in part, from the production and accumulation of species such as superoxide anion, hydrogen peroxide (H_2O_2), and hydroxyl radical ($HO_{\cdot}$). Metallothioneins may function in that protection, and Sato and Bremner (1993), Kasprzak and Buzard (Chapter 16, this volume), and Waalkes and Pérez-Ollé (Chapter 14, this volume) review the role of metallothionein in oxidative stress.

2 Yeast metallothionein and Cu,ZnSOD may complement one another, or have separate functions aimed at mediating response to oxidative stress. Both yeast and monkey metallothioneins functionally substitute for yeast Cu,ZnSOD – *S. cerevisiae* lacking the Cu,ZnSOD gene is sensitive to oxidative stress induced by dioxygen and paraquat toxicity (paraquat generates superoxide radicals) (Tamai *et al.*, 1993). Copper-induced metallothionein protected the yeast from oxidative stress, and monkey metallothionein genes expressed under the control of the Cup1 promoter in cells lacking Cu,ZnSOD restored growth on lactate. Therefore, mammalian MTs protect cells from oxygen toxicity, implying that antioxidant function is a feature of higher eucaryotic MT proteins. In a human cervical carcinoma cell line (HeLa), oxidant stress induced by hydrogen peroxide treatment increased the DNA-binding activity of AP-1 (Angel *et al.*, 1987), which actives metallothionein gene expression in response to phorbol esters, DNA-damaging agents and other oxidative stress-inducing conditions (Devary *et al.*, 1991; Alam and Smith, 1992; Gralla and Kosman, 1992). Metallothionein could conceivably play several roles under these conditions, including:

   (a) 'scavenging' free radicals to prevent damage;
   (b) mediating movement of copper (which associates with MT with high affinity) to subcellular locations to minimize toxic effects;

(c) providing copper or zinc to Cu,ZnSOD or other metal-binding antioxidant proteins; or
(d) mediating the zinc status of metal-requiring transcription factors important in regulating the expression of stress-response genes.

With respect to the antioxidant role of metallothionein, it is important to note that low levels of reactive oxygen species act as second messengers to activate nuclear factor κB (NF-κB), an event that can be blocked by antioxidants (Schenk *et al.*, 1994). DNA binding by NF-κB *in vitro* requires reduction of cysteinyl residues, and is stimulated *in vivo* by events that alter cellular redox status (Remacle *et al.*, 1995). Thus, oxygen-radical scavenging can act to alter the activity of signaling proteins, as well as to prevent direct toxic damage to cells. These possibilities, with special emphasis on the implications of possibilities (c) and (d), are discussed in more detail below.

Transcription of the gene for the human 70 kDa heat shock protein (HSP70) is induced by several mechanisms, including heat, amino-acid analogs, and arsenic and other metals. Transcription is mediated by a heat shock element (HSE) in the promoter – metals may interact with a heat shock factor (HSF) to allow productive interaction of HSF and HSE to induce HSP70 transcription (Williams and Morimoto, 1990). Evidence exists for these kinds of events in HSF activity (Larson *et al.*, 1988; Zimarino *et al.*, 1990), and suggests that metal interactions with regulatory protein may mediate that regulation.

In decreasing order of effectiveness, lead, copper, nickel, zinc and magnesium induce the acute-phase protein genes for serum α_1-acid glycoprotein (AGP1) and C-reactive protein (CRP) in mice (Yiangou *et al.*, 1991). The AGP1 promoter can confer $HgCl_2$ induction on a chloramphenicol acetyltransferase reporter gene in HEPG2 cells, possibly through four regions of homology to promoter regions responsible for metallothionein gene expression in response to heavy-metal induction (metal response elements (MREs), discussed below).

The heme oxygenase (HO) gene (which controls a rate-limiting step in heme catabolism) is induced by $CdCl_2$ in rat and mouse (Alam *et al.*, 1989) through the activity of two regions in the HO gene promoter – an enhancer region that contains antioxidant response elements (AREs, also called electrophile response elements), and an upstream stimulatory factor (USF) binding site in the proximal promoter region (Alam, 1994; Maeshima *et al.*, 1996). AREs can mediate induction of genes (for example, those encoding the rat glutathione-*S*-transferase Ya subunit and quinone reductase) in response to redox-active compounds and hydrogen peroxide (Favreau and Pickett, 1993; Jaiswal, 1994). Both AREs and USF binding sites are present in metallothionein gene promoters (Dalton *et al.*, 1994) and are responsive to cadmium induction (Andrews *et al.*, 1999).

Cellular proto-oncogenes *myc* and *jun* are induced by cadmium in rat cells (Jin *et al.*, 1991), and overexpression of c-*ras* in mouse fibroblast NIH 3T3 cells results in increased MT expression (Isonishi *et al.*, 1991). In addition, c-*fos* mRNA levels are enhanced in response to cadmium or zinc (without *de novo* protein synthesis) in Swiss 3T3 cells (Epner and Herschman, 1991). Cadmium is, in fact, a known carcinogen (Friberg *et al.*, 1992), but the relationship between metal induction of oncogene and MT expression, and malignancy, is not known.

Certain metals can interact with specific signal transducers to suppress responsiveness to extracellular signals. As such, these metal species do not act as direct inducers of gene expression, but as modulators of responsiveness to other agents. In this regard, protein kinase C (PKC) has been reported to be a molecular target of lead (Heiman and Tonner, 1995), and exposure of cultured hepatoma cells to lead suppressed glucocorticoid receptor-mediated expression of a reporter gene (Tonner and Heiman, 1997), a phenomenon that was partially reversed by the addition of calcium. An effect on PKC function was implicated by lead-induced changes in PKC translocation to and from cytosolic to particulate fractions. Conversely, at relatively low concentrations (10^{-8}–10^{-11} M), lead has the capacity to activate PKC directly in a cell-free preparation of rat brain – free Ca^{2+} levels in the range of 10^{-8}–10^{-5} were required to directly activate PKC (Long *et al.*, 1994). Taken together, this suggests that PKC enzymes (involved in glucocorticoid and other signal transduction pathways) may be molecular sites of action of lead, and aberrant modulation of PKC activity in hormonal target cells may contribute to overall toxicity resulting from lead exposure, or altered hormonal responses (including altered growth and differentiation) in cells pre-exposed to low lead concentrations. Interestingly, PKC has been reported to participate in metal-induced metallothionein expression (Yu *et al.*, 1997), since treatment of Chinese hamster ovary (CHO) cells is inhibited by PKC inhibitors (H7 or chelerythrine), but not a PKA-specific inhibitor (HA1004). The point at which inhibition occurs (transport of zinc into cells, localization in cellular compartments, induction or activation of metal-dependent metallothionein transcription factors, etc.) has not been elucidated, and the potential role of metallothionein in mediating PKC activity remains to be explored.

Zinc is not the only metal that can affect transcription factor activity in mammalian cells. Intracellular copper levels and the redox activity of copper may participate in the physiological control of function of the human copper-binding tumor suppressor gene, *p53* (Verhaegh *et al.*, 1997). Metallothioneins are the major single binding molecules for both intracellular zinc and copper, and the potential for a role for metallothionein in regulating copper association with p53, and with it the physiological activity of p53, remains open.

18.3 Zinc

Zinc is an essential cofactor in over 300 enzymes, including proteins involved in signal transduction, transcription and replication (Coleman, 1992; Vallee and Falchuk, 1993). The availability of zinc for these proteins constitutes a potential mechanism for control of activity, and homeostatic regulation of zinc within cells is an appropriate starting point to review the effects of metals on signal transduction and gene expression.

Plasma concentrations of zinc in humans are normally quite low (12–16 μM), but this is only a small portion of total zinc in all tissues (Bettger and O'Dell, 1993). Extracellular zinc is predominantly bound to albumin, α_2-macroglobulin and transferrin, while intracellular zinc is bound to a wide variety of macromolecules and to low-molecular-weight metallothionein (the major intracellular zinc-binding protein). According to some reports, only free, unbound zinc appears to be biologically active (Phillips, 1976; Borth and Luger, 1989; Gless *et al.*, 1992; Vallee and Falchuk). On the other hand, zinc may regulate the function of one or more of its binding proteins. For example, α_2-macroglobulin interaction with cytokines and proteases can be altered by interaction with zinc *in vitro* (Borth and Luger, 1989; James, 1990). In view of altered protein function in the presence and absence of zinc, the potential regulation of zinc availability within cells as a homeostatic mechanism to alter the activity of zinc-associated proteins has become a major focus of discussion. Zinc availability can be regulated in the first instance at the level of transport into, and out of, cells and intracellular compartments.

Serum zinc is bound primarily to transferrin, and the transferrin receptor (CD71) was once considered a candidate for zinc uptake (Phillips, 1976; Cunningham-Rundles *et al.*, 1980; Mathe *et al.*, 1985). However, recent evidence describing a family of zinc-transport proteins in *Saccharomyces cerevisiae* (Zrt1, Zrt2, Zip1, Zip2, Zip3, Zip4) and *Arabidopsis thaliana* (Irt1) (Eng *et al.*, 1998) has been matched by description of mammalian zinc transporters relevant to prevention of zinc toxicity and, potentially, zinc homeostasis. A plasma-membrane zinc transporter (ZnT-1), originally discovered in rat kidney and in mouse (Palmiter and Findley, 1995), has been proposed to act to transport zinc out of cells (analogous to the human ATP7B and ATP7A copper-transport proteins in humans (DiDonato and Sarker, Chapter 6, this volume)). The absence or decreased activity of the protein in mammalian cells in culture leads to zinc overaccumulation and increased sensitivity to the toxic effects of zinc exposure. A second mammalian zinc transporter (ZnT-2) has been localized to the vesicular structures other than the plasma membrane, and resembles ZnT-1 topologically: both have six membrane-spanning domains, a histidine-rich intracellular loop and a long C-terminal tail, but the overall amino-acid identity is only 26% (Palmiter *et al.*, 1996). Prolonged exposure of cells expressing high levels of ZnT-2 to zinc causes accumulation of zinc into an endosomal lysozomal compartment, and likely protects cells from toxicity due to high intracellular zinc content. A third zinc transporter (ZnT-3) has been described in regions of mouse and monkey brain that are rich in histochemically detectable zinc, and is most abundant in mossy fibers projecting from dentate granule cells to hilar and CA3 pyramidal neurons (Wenzel *et al.*, 1997). These fibers contain clear, round synaptic vesicles that are both zinc and ZnT-3 enriched. This suggests that ZnT-3 transports zinc into synaptic vesicles: a process critical for the ability to release zinc upon excitation.

Zinc transport into and out of cells is clearly a process regulated by transporter proteins. However, once inside the cell, zinc (unlike copper and iron) is not a strongly redox-active metal (i.e. it does not readily participate in oxidation–reduction reactions leading to the formation of free radicals), but can contribute to the structure and activity of zinc-requiring enzymes and DNA-binding proteins (Vallee, 1995). The regulation of zinc by associated proteins within cells is a second level of potential control of zinc availability, and the role of metallothioneins as potential regulators has been investigated.

18.4 Metallothionein interactions with zinc-dependent proteins

Metallothioneins (MTs) comprise a family of small (typically less than 10 000 Da), cysteine-rich, metal-binding proteins found in all eucaryotes. Humans possess 16 MT genes linked on chromosome 16 (including MT-1_A, MT-1_B, MT-1_F, MT-1_G, MT-2_A,

MT-3, and MT-4) (West *et al.*, 1990). Rodents have only four functional genes (MT-1, MT-2, MT-3 and MT-4) on chromosome 8 (Hamer, 1985). Gene sequences are similar among isoforms and even between species, allowing cross-hybridization of specific probes among species (Hamer, 1985; Koropatnick *et al.*, 1988). MTs are classically induced by a variety of transition metals. However, MTs are also expressed without metal induction in many different tissues and cell types. Intriguingly, some isoforms are preferentially expressed in specific tissues. MT-3 mRNA and protein, for example, is expressed primarily in brain (Uchida *et al.*, 1991) and, at least with respect to mRNA, at much lower levels in testis, prostate, epididymis, tongue, ovary, uterus, stomach, heart and seminal vesicles (Moffatt and Seguin, 1998). MT-4 mRNA and protein expression is restricted to cornified, stratified, squamous epithelium (which provides a protective surface on skin, tongue, the upper part of the alimentary tract and the vagina). MTs are also expressed during retinoic acid-induced differentiation of teratocarcinoma cells *in vitro* (Hamer, 1985), during growth and proliferation in a human prostate stem-cell line (Koropatnick *et al.*, 1995), in rat kidney undergoing compensatory hypertrophy (Zalups *et al.*, 1995), in proliferating human colonic cancer cells (Nagel and Vallee, 1995), in cells on the proliferative edge of epithelial tumors *in situ* (Kontozoglou *et al.*, 1989), the proliferating cell compartment of human breast and colon tumors (Meskel *et al.*, 1993) and in human monocytes undergoing a respiratory burst (Leibbrandt and Koropatnick, 1994; Leibbrandt *et al.*, 1994). In rodents and humans, MT expression in visceral yolk sac and liver is developmentally regulated (Ouellette, 1982; Koropatnick and Duerksen, 1987; Slotkin and Cherian, 1989). Subpopulations of mouse ova and preimplantation cells express MT at a high level (Andrews *et al.*, 1991). The intracellular location of MT is dependent upon the stage of tissue development: MT protein is located in the nuclei of parenchymal cells in fetal and newborn liver, and is primarily cytoplasmic in adult liver (Panemangalore *et al.*, 1983; Nartey *et al.*, 1987; Cherian, 1994). MT isoforms change during differentiation in mice. As inner-layer keratinocytes in stratified epithelium differentiate into the outer stratum spinosum and switch keratin isoform production, they also switch from MT-1 to MT-4 production (Quaife *et al.*, 1994). Cultured embryocarcinoma cells, induced to differentiate by retinoic acid, express MT-1 mRNA and MT protein at increasing levels during differentiation to visceral endoderm. Thus, the expression of different MT isoforms, in differing amounts and in different cellular compartments, is associated with proliferation and hypertrophy, and differentiation and cell activation.

The multiple circumstances under which MTs are expressed has fueled speculation about their function(s). A role in metabolism and detoxication of metals is strongly suggested by the ability of MT to bind to, and be induced by, heavy-metal ions. Other potential functions include control of intracellular redox potential, regulation of thiol metabolism, detoxication of activated oxygen and other reactive nucleophilic species and, of special interest, essential metal ion (primarily zinc) homeostasis. A role for MT in supplying essential zinc to metalloproteins (including transcription factors, hormone receptors, metalloproteinases, superoxide dismutase and catalase, among others), or in sequestering zinc to inactivate metal-dependent molecules has been suggested (Zeng *et al.*, 1991a, b; Maret, 1994, 1995; Vallee, 1995; Maret *et al.*, 1997; Jiang *et al.*, 1998a). The ability of metals to induce signaling events directly related to gene transcription and cellular activation (see below), coupled with the observation that nontoxic metal pretreatment of human monocyte/macrophages leads to altered activation potential (described below), suggests that certain metals can have a modulatory action on cellular reception of signals. That modulation has the potential to affect the expression of genes important in a broad range of physiological events, including those regulating cell growth, differentiation, specialized functions and death. It appears, therefore, that MT is a homeostatic molecule, and alteration in MT levels in cells could affect the activity of zinc-requiring proteins that influence gene expression (Vallee *et al.*, 1991; Czupryn *et al.*, 1992).

18.5 Metallothionein and zinc availability to cellular proteins

18.5.1 Enzymes

Relatively few reports of the participation of MT in transfer of zinc to and from enzymes exist, in spite of the clear requirement for zinc of many of those proteins. However, some critical experiments have been performed that are highly suggestive. Udom and Brady (1980) reported *in vitro* reactivation of zinc-requiring apo-enzymes in the presence of zinc-containing rat MT. In addition, metallothionein has been reported to be able to exchange zinc within each of the two domains (α and β) of MT itself, suggesting that metal-ion transfer between different MT molecules is possible (Otvos *et al.*, 1987). Zinc in solution is able to exchange with zinc in both MT domains (Otvos and Armitage, 1980), and MT can transfer zinc to glutathione (Brouwer and Brouwer-Hoexum, 1991). Furthermore, association between MT and glutathione (which has been reported to vary in relation to the ratio of reduced to oxidized glutathione) can modulate zinc transfer from metallothionein to zinc-depleted sorbitol dehydrogenase (Jiang *et al.*, 1998b). A recent study by Jacob *et al.* (1998) points out that a direct process mediating this process is difficult to envisage, since the affinity for zinc of most enzymes studied is at least 1000 times lower than the affinity of MT for zinc ions. Under these circumstances, direct transfer of zinc between metallothionein and metalloenzymes would be energetically unfavorable, and the authors cite evidence that the transfer of zinc between zinc-binding enzymes and MT sequences requires the participation (at least *in vitro*) of other agents. Oxidative stress can shift the balance of reduced glutathione (GSH) to oxidized glutathione (GSSG) in favor of GSSG (Maret, 1995). GSSG appears to modulate the release of zinc from MT (thus facilitating metal-ion transfer to proteins with less affinity for the ion than MT). They propose that undefined, low-affinity biological chelators might associate with released zinc and transfer the metal ion directly to apo-enzymes, although the number and identities of such agents are presently unknown.

In addition to its interactions with glutathione and glutathione disulfide, MT has been reported to form a 1 : 1 complex with ATP or GTP that enhances the transfer of zinc from MT to zinc-depleted sorbitol dehydrogenase (Jiang *et al.*, 1998b). The formation of the MT : ATP complex appears to be enhanced by glutathione, and decreased by glutathione disulfide. This suggests that the redox state of the cell can trigger association events that allow MT to transfer zinc to and from zinc-requiring proteins whose affinity for the metal ion differs significantly from the affinity of MT for zinc. Therefore, MT may, under physiological conditions, act as a source of zinc for macromolecules that require zinc for activity or, conceivably, sequester zinc from those molecules. In either case, MT might act as a regulator of the activity of zinc-requiring proteins *in vivo*. Confirmation of this possibility awaits the results of experiments in living cells to corroborate the hypothesis raised by these *in vitro* observations.

18.5.2 Transcription factors

Many of the proteins involved in DNA replication, repair, and transcription contain zinc atoms that mediate their function (reviewed by O'Halloran, 1993; Vallee and Falchuk, 1993). In addition, proteins interacting in a series of transduction steps to relay signals from the external cellular environment to the cytoplasm and then to the nucleus (signal transduction proteins), including, for example, protein kinase C isoforms and NF-κB and its inhibitor proteins (IκB), all require zinc for activity (Coleman, 1992).

Transcription factor IIIA (TFIIIA, a *Xenopus laevis* oocyte transcription factor) activates transcription by RNA polymerase III, and was the first to be identified as a zinc protein (Hanas *et al.*, 1983; Shang *et al.*, 1989), and zinc transfer from TFIIIA zinc fingers to metallothionein has been demonstrated *in vitro* (Zeng *et al.*, 1991b). The bacteriophage T4 g32p protein, which stabilizes the conformation of single-stranded DNA to allow effective action of DNA replication and repair enzymes (Alberts and Frey, 1970), is also a zinc-containing protein (Giedroc *et al.*, 1986). Removal of the zinc diminishes, but does not abolish, binding to DNA, and increases susceptibility to tryptic digestion (Giedroc *et al.*, 1987). Glucocorticoid and estrogen receptors (GR and ER) are also zinc proteins, as are other members of the larger multigene family (thyroid hormone receptor, retinoic acid receptor, vitamin D_3 receptor and others) of which GR and ER are members (Evans, 1988). Reversible zinc exchange between metallothionein and the estrogen-receptor zinc finger has been observed *in vitro* (Cano-Gauci and Sarkar, 1996). With respect to the glucocorticoid receptor, removal of zinc by chelating agents produces an apo-protein that does not bind appropriately to DNA glucocorticoid response elements, but binding is fully restored by reconstitution with zinc or, interestingly, cadmium. Cobalt restores binding to some degree, but other metals lack the capacity to mediate GR function (Freedman *et al.*, 1988). Binding of steroid by GR (a step necessary for translocation of active GR to the nucleus to induce GR-dependent gene expression) is inhibited, in rat fibroblast cytosols, by metals that cross-link or otherwise modify vicinal thiols, including arsenite (AsO_2^-), cadmium and selenite (SeO_2^-), but not by zinc (Simons *et al.*, 1990). On the other hand, low, nontoxic concentrations of zinc (10–100 μM) have been reported to inhibit steroid-receptor binding between dexamethasone and GR in mouse liver cytosols (Telford and Fraker, 1997). Many other double-stranded DNA-binding proteins, including Zif268 (Pavletich and Pabo, 1991), Sp1 (Kadonaga *et al.*, 1987), ADR1 (Klevit *et al.*, 1990), Xfin (Lee *et al.*, 1989), Gal4 (Pan and Coleman, 1991), GATA-1 (Omichinski *et al.*, 1993) and CRP (Michelsen *et al.*, 1993), have been identified as requiring zinc for activity. A critical question exists with respect to regulation of activity for these proteins: is modulation of zinc association a physiological mechanism to regulate activity? For this hypothesis to be plausibly entertained, the availability of free zinc within cells must necessarily be limited: a readily available pool of zinc ions for all proteins requiring them for activity would frustrate biological strategies to control accessibility of zinc to zinc-sensing or zinc-requiring proteins. Approximately 50% of total cellular zinc resides in the cytoplasm, 30–40% in the nucleus and 10–20% in the cell membrane or cell wall.

Virtually all of the zinc in these compartments is associated with macromolecules (proteins/enzymes and chromatin) (Vallee and Falchuk, 1993), suggesting that, in order for zinc to be made available for newly synthesized enzymes and other proteins, there must exist mechanisms to transfer zinc from one protein to another.

A zinc-requiring protein that has received recent attention is the metal transcription factor-1 (MTF-1) that interacts with the metal-regulatory elements (MREs) that reside in the proximal 5′ regulatory region of members of the metallothionein gene family. Metallothionein genes respond to multiple regulatory signals interacting with specific sequences in promoter and enhancer regions, including AP2 proteins interacting with basal level elements, SP1 interacting with a GC-box, AP1 interacting with a TPA response element, glucocorticoid and steroid hormones interacting with a hormone response element, and members of the signal transducers and activators of transcription (STAT) transcription factor family (STAT1 and STAT3) interacting with a proximal binding site (TTCTCGTAA) in the promoter to mediate cytokine induction of MT expression (Lee *et al.*, 1999). Generalized 'stress' due to heat, cold, physical trauma, etc. will induce MT expression in whole animals, possibly mediated by glucocorticoid receptor activity (Ghoshal *et al.*, 1998) or cytokines. Through these transcription regulation pathways, and possibly others, metallothionein genes respond to nonmetal-ion signals and indirect effects arising from metal-ion exposure in whole animals.

In direct response to soluble metal ions, both MT-1 and MT-2 gene transcription is rapidly induced: zinc and cadmium are particularly effective (Andrews, 1990). Metal response elements (MREs), present in multiple copies in MT gene promoters, mediate MT gene responsiveness to zinc and cadmium (Stuart *et al.*, 1984; Samson and Gedamu, 1995; Foster and Gedamu, 1991) and to oxidative stress (Dalton *et al.*, 1994, 1996). The transcription factor binding to MREs is a zinc-finger protein called the MRE-binding transcription factor (MTF-1; Fig. 18.1), which specifically binds MREs to induce MT expression (Radtke *et al.*, 1993; Brugnera *et al.*, 1994). MTF-1 contains six zinc fingers of the Cys_2His_2 type in the DNA-binding domain, and at least three separate domains responsible for transcriptional activation. The capacity of MTF-1 to bind to DNA is reversibly and directly activated by zinc in a temperature-dependent fashion (Dalton *et al.*, 1997) and the zinc fingers have been reported to be physically and functionally nonequivalent. Three to four of the fingers participate in MTF-1 folding and in binding to DNA metal-regulatory elements, while one or more appears to act in a metalloregulatory role as a 'zinc sensor' (Chen *et al.*, 1998). Treatment of cells with

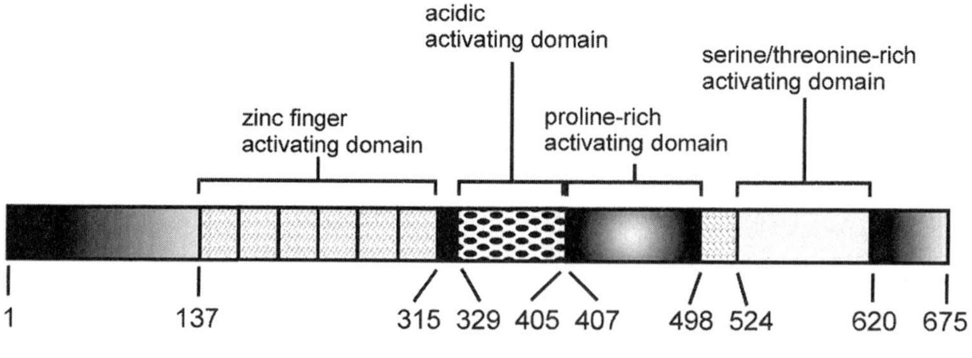

Figure 18.1 Metal transcription factor-1 (MTF-1).

zinc, or with oxidative stress-inducing agents, rapidly increased the amount of MTF-1 binding to DNA *in vitro* (Dalton *et al.*, 1996). Interestingly, cadmium added *in vitro* has been reported to have little or no effect on the amount of DNA binding activity in Hepa cell lysates (Andrews *et al.*, 1999), and the authors hypothesize that non-MTF-1 transcription factors, including upstream stimulatory factor (USF, a member of the bHLH-bZip protein superfamily, which bind to DNA as homo- or heterodimers (Sirito *et al.*, 1992)) are involved in cadmium-responsiveness of metallothionein genes. USFs can interact with other transcription factors (including Maf, which binds to antioxidant response elements (AREs) in DNA regulatory regions) to influence gene expression (Kataoka *et al.*, 1994; Meier *et al.*, 1994; Inamdar *et al.*, 1996). However, MTF-1 null cells (generated by insertional mutation to 'knock out' MTF-1 genes) fail to respond to zinc, cadmium, copper, nickel and lead induction of MT expression (Heuchel *et al.*, 1994). Thus, it is clear that MTF-1 is an obligatory partner in direct MT gene induction by zinc and other metal ions.

An alternative model has been suggested by Palmiter (1994). In this scenario, an inhibitory molecule associated with MTF-1 could be released upon association of zinc ions with the inhibitor, although demonstration of the existence of such an inhibitor has not been reported. Direct activation of MTF-1 by zinc remains the simplest and most obvious possibility. In view of the capacity of MTF-1 to respond to zinc, and only zinc, the precise mechanism by which cadmium, copper, mercury, lead, nickel and other metal ions, and reactive oxygen intermediates, induce MTF-1 activity is not resolved. One possibility, however, is that the activity of MTF-1 (and possibly other zinc-binding transcription factors and proteins) depends upon a metalloregulatory protein or proteins that do not, themselves, have DNA-binding or *trans*-activation capacity (Koropatnick and Leibbrandt, 1995).

Such a metalloregulatory protein (MP) would respond to many different metal ions by removing or adding zinc, and only zinc, from MTF-1 or an intermediate regulator of MTF-1 (Fig. 18.2). Metallothioneins are good candidates for such a role, especially since they exhibit a hierarchy of noncovalent metal-binding affinities: stability constants for copper–MT (10^{19}–10^{17}) and cadmium–MT (10^{17}–10^{15}) are at least ten- to 100-fold greater than that for zinc–MT (10^{4}–10^{11}) (Hamer, 1985). MTs associated with zinc (which is the predominant ligand of MTs in cells uninduced with other heavy metals) release zinc by displacement by added copper or cadmium (Suzuki *et al.*, 1990). Released zinc would then be available to mediate MTF-1 activity, especially if the newly released zinc were present in the nucleus. MT has, in fact, been localized to the nucleus during mammalian development (Panemangalore *et al.*, 1983; Andrews *et al.*, 1987; Nartey *et al.*, 1987), under conditions of toxic metal overload (Deng *et al.*, 1998) and in certain tumors (Zhang and Takenaka, 1998). Such a metal-responsive zinc-carrier model would be part of a multiple factor signal transduction system, and might be particularly important under conditions where zinc availability is limited.

In that regard, mice have been generated with both MT-1 and MT-2 inactivated by targeted insertion of *neo* gene sequences (Michalska and Choo, 1993; Masters *et al.*, 1994), but these animals appear to be normal in almost all respects except for increased sensitivity to cadmium. However, they are also sensitive to both zinc toxicity and deficiency: a severely zinc-deficient diet (1.5 μg/g) retarded kidney development, as seen by retention of the nephrogenous zone and incomplete tubule development, and exposure to high levels of injected zinc (a total of 3700 μmole Zn/kg body weight) led

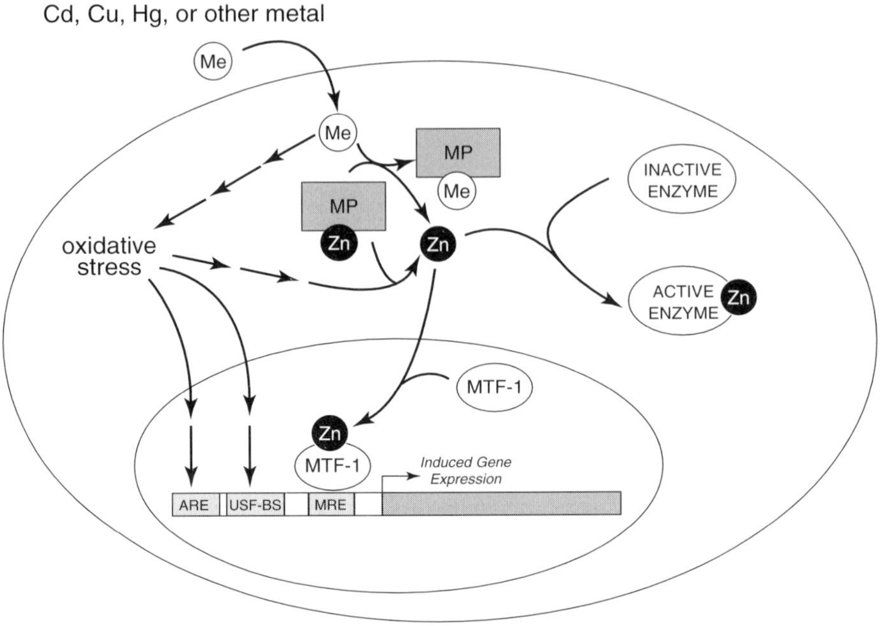

Figure 18.2 Pathways of metal induction of metal-sensitive gene expression. A schematic drawing of a cell delimited by a plasma membrane, with the nucleus containing a metal-responsive gene, is shown. MP = metalloregulatory protein, postulated to mediate intracellular zinc availability in response to entry of metal ions. Metallothionein may, under certain circumstances, act as an MP to control zinc availability to enzymes and transcription factors. ARE, antioxidant response element; USF-BS, upstream stimulatory factor binding site. ARE and USF-BS are *cis*-acting DNA elements mediating enhanced gene expression in response to oxidative stress. Oxidative stress can enhance dissociation of zinc from MT (a potential MP).

to pancreatic acinar cell degeneration in spite of accumulating less zinc in pancreata than control mice (Kelly *et al.*, 1996). Thus, it appears that MT under these conditions may keep zinc available to zinc-requiring proteins under conditions when it is limiting, as well as decreasing the toxic effects of zinc under conditions where massive overdosing occurs.

The capacity of MT to enhance zinc-requiring transcription factor activity in cultured cells has been observed directly. In baby hamster kidney cells transfected with a β-galactosidase reporter gene fused to an MTF-1-responsive metallothionein promoter, exposure to zinc induced β-galactosidase expression. When cells harboring the reporter construct were transfected with an additional vector directing expression of mouse MT-1, constitutive MT overexpression enhanced the capacity of zinc to induce metal-responsive gene expression (Palmiter, 1995). Thus, MT has the ability to mediate zinc-responsive gene expression, particularly under low zinc conditions, possibly by regulating the availability of zinc to MTF-1.

Regulation of zinc availability would have the potential to affect not only transcription factor activity, but DNA : protein structures exerting control over gene expression. Zinc has been reported to be important in chromatin folding in human sperm (Casswall *et al.*, 1987; Bianchi *et al.*, 1992; Kjellberg *et al.*, 1992) and zinc deficiency changes chromatin conformation to a more highly condensed form, along with alteration in the proportions of histone H1 subforms (Castro, 1987). Zinc has also been suggested to play a role in facilitating H1 histone removal from chromatin to expose specific genes for transcription (Chesters, 1992).

18.6 Metals and cellular signal reception

Changes in MT expression due to transfection of an MT expression vector into cultured mammalian cells clearly has an effect on cellular responsiveness to metal signals mediated by a metal-responsive transcription factor (MTF-1)(see above). However, proteins that act as metal 'sensors' (for example, iron response element-binding protein (IRE-BP)/c-aconitase as a regulator of transferrin receptor, ferritin and δ-aminolevulinate synthase (ALAS) mRNA activity (Klausner *et al.*, 1993) are a subset of the larger family of proteins that require metals as cofactors for activity. It has been suggested that the vast majority of metal-binding proteins involved in gene expression may not be involved in monitoring metal ion concentration (O'Halloran, 1993). According to this hypothesis, metal ions required for activity may be freely available within cells, either as unbound zinc, or zinc very loosely associated with intracellular proteins in such a way as to be freely available in appropriate cellular compartments. Access to zinc, or lack of it, would not exert a measure of control on protein activity. On the other hand, metal-ion availability, mediated by essential metal-ion transporters and intracellular chaperones, might be a well-controlled post-translational protein modification that acts as an activity switch ('on' or 'off') to control and modulate gene transcription, mRNA translation, protein transport and activity.

There is evidence to suggest that genetic modulation of essential metal-binding proteins has an effect on signal reception and transduction, and gene expression, in response to nonmetal-induced signals mediated through metal ion-requiring transcription factors. For example, human monocytes respond to extracellular signals exerted by exposure to phorbol 12-myristate 13-acetate (PMA) binding to protein kinase C and gene activation through mitogen-activated protein (MAP)-kinase and NF-κB (Nishizuka, 1992), or by bacterial lipopolysaccharide (LPS)/LPS-binding protein (LBP) association with soluble CD14 (sCD14) to prime cells for activation by cytokines, including tumor necrosis factor-α (TNF-α) and interleukin-1β (IL-1β) (Beutler and Cerami, 1989). Zinc is necessary for the function of signals involved in monocyte/macrophage activation by LPS and other antigen-independent, receptor-mediated signals. In response to these agents, activation proceeds through a G-protein-linked pathway to activate phospholipases C, A2 and D. Phosphorylation of several protein kinases (including tyrosine kinase and MAP kinase) also occurs, and contributes to a cascade of events contributing to cytokine production and arachidonate release. Phospholipase C activation stimulates release of diacylglycerol to activate PKC, and inositol 1,4,5-trisphosphate (IP_3) to increase intracellular calcium levels. PKC activation stimulates release of NF-κB from its cytoplasmic inhibitor (IκBα). Released NF-κB is translocated to the nucleus, where it acts as a transcription factor to participate in activation-associated events, including the induction of genes associated with the respiratory burst and IL-1α, IL-1β and TNF-α (Hancock *et al.*, 1995; Meier *et al.*, 1995). Zinc is required for the function of PKC (Szallasi *et al.*, 1996) and virtually all of the transcription factors activated by these varied signal transduction pathways, including NF-κB (Otsuka *et al.*, 1994). If zinc availability to these transcription factors (especially those induced by metals, or whose activity is mediated by association with exogenous metals) is regulated by intermediate zinc regulatory proteins, then cellular exposure to metals at low, nontoxic levels may have profound implications for the capacity of cells to respond to signals.

18.7 Metal-induced alterations in cellular responsiveness to signals

Metals have the capacity to damage cells and cellular components (reviewed in Goyer and Cherian, 1995; Chang, 1996) but exposure to certain metals can either enhance or inhibit specific cellular functions without inducing toxicity. Leukocyte function has been investigated in this regard, as metals are one group of chemicals that are necessary for the development and maintenance of healthy leukocytes. Copper, iron, selenium and zinc are all required for important processes in leukocyte development and function, including growth, differentiation and activation. Zinc, for example, can regulate the capacity of mouse thymocytes to respond to stimulation by IL-1 (Winchurch, 1988). Leukocyte function is compromised in individuals suffering from zinc deficiency secondary to AIDS (Fabris *et al.*, 1988), acrodermatitis enteropathica (Moynahan, 1981) and marasmus (Castillo-Duran *et al.*, 1987). Zinc-deficient animals and humans have reduced levels of markers of macrophage activation, including increased phagocytic capacity (Wirth *et al.*, 1989), chemotaxis (Vruwink *et al.*, 1991) and chemokinesis (Briggs *et al.*, 1982). However, apparently contradictory effects of zinc supplementation have been reported. For example, superoxide anion production following activation of guinea-pig peritoneal macrophages was depressed following *in vitro* zinc supplementation, but increased when animals were treated with zinc *in vivo* (Nakamura *et al.*, 1987). Both stimulation (Briggs *et al.*, 1982) and depression (Chvapil *et al.*, 1977) of phagocytosis has been reported following zinc treatment. In addition, while zinc supplementation enhanced humoral and cellular immunity in humans (Hertel and Wibbertman, 1990), it has been reported to inhibit leukocyte mobility (Kiremidjian-Schumacher *et al.*, 1981), release of histamine from rat peritoneal mast cells and blood basophils (Kazimierczak and Maslinski, 1974), and LPS-induced monocyte activation *in vitro* (Leibbrandt and Koropatnick, 1994).

In addition to modulating basal functions and activities, metal ions (albeit at relatively high concentrations) can induce events associated with inflammation. Nickel and cobalt salts induce the expression of adhesion molecules (for example, E-selectin and inter-cellular adhesion molecule-1) and IL-6 and IL-8 in human vascular endothelial cells (HUVECS) (Wagner *et al.*, 1998), through activation of NF-κB and AP-1. In human bronchial epithelial cells, high concentrations (500 μM) of arsenic, vanadium and zinc markedly activate MAP kinase, while chromium and copper activated MAP kinase to a minor degree, and iron and nickel had no effect (Samet *et al.*, 1998), indicating that metals can induce expression of inflammatory mediators.

Cadmium, lead and inorganic mercury exert mainly detrimental effects on leukocytes, including direct *in vitro* cytotoxicity as well as inhibition of antibody production and cytokine production, and proliferation, in isolated primary peripheral blood mono-nuclear cells (Lawrence and Slavik, 1992; Lawrence and McCabe, 1995). Some of the inhibitory potential may be due to interference by one metal with the homeostatic activity of another. For example, both copper (Scuderi, 1990) and cadmium (Bendtzen, 1982) inhibit zinc-induced IL-1 production in human monocytes.

The signal-modulating activity of metals generally occurs at concentrations much lower than that required for overt cytotoxicity (Lawrence and McCabe, 1995). Both enhancement and depression of specific activities has been observed. Although lead and mercury are both toxic to leukocytes, certain levels can enhance some functions. Lead (32 μM) can increase IgG and IgM levels in human peripheral blood mononuclear cells (PBMCs) *in vitro*, and low concentrations of mercury (1 μM) increase PBMC IgM production (Lawrence and McCabe, 1995). Cadmium can both suppress (Chowdhury and Chandra, 1989) and enhance (Koller and Roan, 1977) natural killer (NK) cell activity, and suppress (Thomas *et al.*, 1985) and enhance (Koller and Roan, 1977) macrophage activity. Mercury treatment can trigger genetically restricted glomerulonephritis in both BN rats (Pelletier *et al.*, 1990) and H-2^S haplotype mice (Reuter *et al.*, 1989; Hultman *et al.*, 1993; Monestier *et al.*, 1994; Kubicka-Muranyi *et al.*, 1993), probably due to production of autoreactive T lymphocytes that, in turn, induce polyclonal B-cell activation. Mercury is suspected to have the same effect in humans, since low levels of mercury correlate with the appearance of idiopathic crescentic glomerulonephritis (ICG) (Tubbs *et al.*, 1982; Oliviera, 1987). On the other hand, mercury has been reported to inhibit activation of T and B cells at exposure levels as low as 5 ng/ml, indicating a suppressive effect (Shenker *et al.*, 1993). Overall, a wide variety of metal ions can both suppress and enhance different aspects of immune cell function. This may be a consequence of either direct effects of these metal ions or their reaction products on cellular effector proteins (including enzymes, structural proteins and transcription factors). On the other hand, these metals may induce and interact with proteins that regulate the ability of essential metals (primarily zinc) to mediate the function of proteins involved in gene expression. Direct genetic manipulation of the expression of genes normally induced by metals provides the best evidence of the involvement of these metal regulatory proteins in signal transduction.

18.8 Altered metallothionein in signal transduction

In addition to genetic ablation of MT genes in embryonic stem cells and mice (Michalska and Choo, 1993; Masters *et al.*, 1994), antisense RNA or oligonucleotides complementary to functional cellular DNA can be used to downregulate specific gene products (Izant and Weintraub, 1984, 1985; Stout and Caskey, 1987; Khokha *et al.*, 1989). Antisense MT RNA transcribed from a transfected eucaryotic expression vector has been used to transiently downregulate MT gene expression in a human monocyte cell line (THP-1) (Leibbrandt *et al.*, 1994). The cells exhibited decreased MT-1 and MT-2 protein, decreased cadmium resistance, increased adhesion to a reconstituted basement membrane, increased invasiveness through basement membrane, and increased superoxide radical production. The responsiveness of these cells to bacterial lipopolysaccharide (LPS) was also altered. Control THP-1 cells were activated by LPS to undergo a 'respiratory burst' with increased superoxide radical production, increased adhesion to a basement membrane, differentiation into the macrophage line and increased MT mRNA and protein (Leibbrandt and Koropatnick, 1994). THP-1 cells expressing antisense MT RNA and downregulated MT protein showed no increase in any of these parameters in response to LPS, yet remained completely viable (Leibbrandt *et al.*, 1994).

Thus, antisense MT RNA abolished the ability of THP-1 cells to undergo the respiratory burst in response to LPS without inhibiting viability or growth. Disregulation of metallothionein leads to altered ability to respond to activation signals in human monocyte/macrophages.

On the other hand, metal pretreatment of THP-1 cells both induced MT and inhibited activation by LPS (Leibbrandt and Koropatnick, 1994; Leibbrandt *et al.*, 1994; Koropatnick and Zalups, 1997). Treatment of THP-1 cells with zinc (20–40 μM), cadmium (1 μM) or inorganic mercury (2–10 μM) had no effect on cell viability as measured by a metabolic activity assay or proliferation potential, but significantly induced MT mRNA and protein, consistent with previous reports (Hamer, 1985). However, metal pretreatment dramatically reduced the ability of cells to undergo activation as measured by increased production of reactive oxygen intermediates and IL-1β and MT-2 mRNA, and increased adhesion to plastic tissue-culture plates (indicative of differentiation into the macrophage cell line). Zinc chloride (20 μM) pretreatment of primary, freshly isolated human peripheral blood monocytes had a similar inhibitory effect on the capacity of monocytes to undergo LPS-induced activation (Leibbrandt and Koropatnick, 1994; Dale and Koropatnick, 1999) and did not stimulate production of reactive oxygen intermediates, nor was there any synergistic effect of metal ion plus LPS to induce toxicity. Given the apparent importance of MT in mediating activation (revealed by antisense-mediated suppression of MT levels in monocytes), this was unexpected. It is possible that metal treatment affected the activity of receptors mediating LPS signal transduction without decreasing cell viability. This would affect activation potential without reference to MT. However, primary mouse monocytes and THP-1 cells treated with nontoxic doses of mercury have diminished activation potential in response to phorbol ester (an agent that does not act through a receptor-mediated mechanism) (Koropatnick and Zalups, 1997). Alternatively, metal treatment could affect the expression of other genes in addition to those encoding MT, and those non-MT factors might act independently, or in conjunction with MT, to inhibit activation potential.

18.9 Summary

Heavy metals are ubiquitous and long-lived environmental species. Some are essential and nontoxic, some essential but toxic under certain conditions, and others purely toxic and without apparent cellular function. Although they may have toxic effects that diminish cell viability, they can also induce the expression of genes through a variety of transcription factors that are either directly responsive to metals, or to chemical factors generated in the presence of metals. The requirement of hundreds of different transcription factors and other proteins for metal ions (primarily zinc and copper) creates the potential for occupation of metal-binding sites by inappropriate metal ions, leading to alteration of the quantity or quality of the function carried out by that protein. In addition, metal-induced proteins (notably metallothionein) may modulate the function of metal-dependent and metal-responsive transcription factors, enzymes and structural proteins by regulating the availability of essential metal ions (primarily zinc), especially under conditions of restricted zinc availability. Clearly, the viability and apparent normality of MT 'knockout' mice indicate that not all metal-requiring proteins can be regulated by MT under normal laboratory conditions, but there is tantalizing evidence (both *in vitro* and *in vivo*) that suggests that MT plays a role when cells are under stress. Confirmation of this intriguing and exciting hypothesis awaits the results of experiments where metallothionein expression is genetically manipulated in the absence of metal induction, and the capacity of cells to carry out specific signal transduction and gene expression events is observed.

References

Alam, J. 1994. Multiple elements within the 5′ distal enhancer of the mouse heme oxygenase-1 gene mediate induction by heavy metals. *Journal of Biological Chemistry* **269**, 25049–56.

Alam, J. and Smith, A. 1992. Heme-hemopexin-mediated induction of metallothionein gene expression. *Journal of Biological Chemistry* **267**, 16379–84.

Alam, J., Shibahara, S. and Smith, A. 1989. Transcriptional activation of the heme oxygenase gene by heme and cadmium in mouse hepatoma cells. *Journal of Biological Chemistry* **264**, 6371–5.

Alberts, B.M. and Frey, L. 1970. T4 bacteriophage gene 32: a structural protein in the replication and recombination of DNA. *Nature* **227**, 1313–18.

Andrews, G.K. 1990. Regulation of metallothionein gene expression. *Progress in Food and Nutrition Science* **14**, 193–258.

Andrews, G.K., Gallant, K.R. and Cherian, M.G. 1987. Regulation of the ontogeny of rat liver metallothionein mRNA by zinc. *European Journal of Biochemistry* **166**, 527–31.

Andrews, G.K., Huet-Hudson, Y.M., Paria, B.C., McMaster, M.T., De, S.K. and Dey, S.K. 1991. Metallothionein gene expression and metal regulation during preimplantation mouse embryo development (MT mRNA during early development). *Developmental Biology* **145**, 13–27.

Andrews, G.K., Bittel, D., Dalton, T., Hu, N., Chu, W., Daggett, M., Li, Q. and Johnson, J. 1999. New insights into the mechanisms of cadmium regulation of mouse metallothionein-1 gene expression, in *Metallothionein-IV*, (ed. C.D. Klaassen), Birkhauser Verlag, Basel, pp. 227–32.

Angel, P., Imagawa, M., Chiu, R., Stein, B., Imbra, R.J., Rahmsdorf, H.J., Jonat, C., Herrlich, P. and Karin, M. 1987. Phorbol ester-inducible genes contain a common *cis* element recognized by a TPA-modulated *trans*-acting factor. *Cell* **49**, 729–39.

Bendtzen, K. 1982. Induction of antigen-specific lymphocyte unresponsiveness *in vitro:* possible role of divalent cations and defective function of human T-cell-activating factor (TAF). *Cellular Immunology* **66**,152–63.

Bettger, W.J. and O'Dell, B.L. 1993. Physiological roles of zinc in the plasma membrane of mammalian cells. *Journal of Nutritional Biochemistry* **4**, 194–207.

Beutler, B. and Cerami, A. 1989. The biology of cachectin/TNF – a primary mediator of the host response. *Annual Review of Immunology* **7**, 625–55.

Bianchi, F., Rousseaux Prevost, R., Sautiere, P. and Rousseaux , J. 1992. P2 protamines from human sperm are zinc-finger proteins with one CYS2/H1S2 motif. *Biochemical and Biophysical Research Communications* **182**, 540–7.

Borth, W. and Luger, T.A. 1989. Identification of α_2-macroglobulin as a cytokine binding plasma protein. *Journal of Biological Chemistry* **264**, 5818–25.

Briggs, W.A., Pedersen, M.M., Mahajan, S.K., Sillix, D.H., Prasad, A.S. and McDonald, F.D. 1982. Lymphocyte and granulocyte function in zinc-treated and zinc-deficient hemodialysis patients. *Kidney International* **21**, 827–32.

Brouwer, M. and Brouwer-Hoexum, T. 1991. Interaction of copper–metallothionein from the American lobster, *Homarus americanus*, with glutathione. *Archives of Biochemistry and Biophysics* **290**, 207–13.

Brugnera, E., Georgiev, O., Radtke, F., Heuchel, R., Baker, E., Sutherland, G.R. and Schaffner, W. 1994. Cloning, chromosomal mapping and characterization of the human metal-regulatory transcription factor MTF-1. *Nucleic Acids Research* **22**, 3167–73.

Cano-Gauci, D.F. and Sarkar, B. 1996. Reversible zinc exchange between metallothionein and the estrogen receptor zinc finger. *FEBS Letters* **386**, 1–4.

Casswall, T.H., Bjorndahl, L. and Kvist, U. 1987. Cadmium interacts with the zinc-dependent stability of the human sperm chromatin. *Journal of Trace Elements and Electrolytes in Health and Disease* **1**, 85–7.

Castillo-Duran, C., Heresi, G., Fisberg, M. and Uauy, R. 1987. Controlled trial of zinc supplementation during recovery from malnutrition: effects on growth and immune function. *American Journal of Clinical Nutrition* **45**, 602–8.

Castro, C.E. 1987. Nutrient effects on DNA and chromatin structure. *Annual Review of Nutrition* **7**, 407–21.

Chang, L.W. (ed.) 1996. *Toxicology of Metals*, CRC Press, Lewis Publishers, Boca Raton, USA.

Chen, X., Agarwal, A. and Giedroc, D.P. 1998. Structural and functional heterogeneity among the zinc fingers of human MRE-binding transcription factor-1. *Biochemistry* **37**, 11152–61.

Cherian, M.G. 1994. The significance of the nuclear and cytoplasmic localization of metallothionein in human liver and tumor cells. *Environmental Health Perspectives* **102**, 131–5.

Chesters, J.K. 1992. Trace element–gene interactions. *Nutrition Reviews* **50**, 217–23.

Chowdhury, B.A. and Chandra, R.K. 1989. Effect of zinc administration on cadmium-induced suppression of natural killer cell activity in mice. *Immunology Letters* **22**, 287–91.

Chvapil, M., Stankova, L., Zukoski C. IV and Zukoski C. III 1977. Inhibition of some functions of polymorphonuclear leukocytes by *in vitro* zinc. *Journal of Laboratory and Clinical Medicine* **89**, 135–46.

Coleman, J.E. 1992. Zinc proteins: enzymes, storage proteins, transcription factors, and replication proteins. *Annual Review of Biochemistry* **61**, 897–946.

Cotton, F.A. and Wilkinson, G. 1980. *Advanced Inorganic Chemistry. A Comprehensive Text*, John Wiley and Sons, New York.

Cunningham-Rundles, S., Bockman, R.S., Lin, A., Giardina, P.V., Hilgartner, M.W., Caldwell-Brown, D. and Carter, D.M. 1980. Physiological and pharmacological effects of zinc on immune response. *Annals of the New York Academy of Sciences* **567**, 113–22.

Czupryn, M., Brown, W.E. and Vallee, B.L. 1992. Zinc rapidly induces a metal response element-binding factor. *Proceedings of the National Academy of Sciences (USA)* **89**, 10395–9.

Dale, S. and Koropatnick, J. 1999. Altered primary monocyte activation potential in mice with genetically ablated MT-1 and MT-2 genes, manuscript submitted.

Dalton, T., Palmiter, R.D. and Andrews, G.K. 1994. Transcriptional induction of the mouse metallothionein-I gene in hydrogen peroxide-treated Hepa cells involves a composite major late transcription factor/antioxidant response element and metal response promoter elements. *Nucleic Acids Research* **22**, 5016–23.

Dalton, T.P., Li, Q., Bittel, D., Liang, L. and Andrews, G.K. 1996. Oxidative stress activates metal-responsive transcription factor-1 binding activity. Occupancy *in vivo* of metal response elements in the metallothionein-I gene promoter. *Journal of Biological Chemistry* **271**, 26233–41.

Dalton, T.P., Bittel, D. and Andrews, G.K. 1997. Reversible activation of mouse metal response element-binding transcription factor 1 DNA binding involves zinc interaction with the zinc finger domain. *Molecular and Cellular Biology* **17**, 2781–9.

Deng, D.X., Ono, S., Koropatnick, J. Cherian, M.G. 1998. Metallothionein and apoptosis in the toxic milk mutant mouse. *Laboratory Investigation* **78**, 175–83.

Devary, Y., Gottlieb, R.A., Lau, L.F. and Karin, M. 1991. Rapid and preferential activation of the c-*jun* gene during the mammalian UV response. *Molecular and Cellular Biology* **11**, 2804–11.

Eng, B.H., Guerinot, M.L., Eide, D. and Saier, M.H. Jr 1998. Sequence analyses and phylogenetic characterization of the ZIP family of metal ion transport proteins. *Journal of Membrane Biology* **166**, 1–7.

Epner, D.E. and Herschman, H.R. 1991. Heavy metals induce expression of the TPA-inducible sequence (TIS) genes. *Journal of Cellular Physiology* **148**, 68–74.

Evans, R.M. 1988. The steroid and thyroid hormone receptor superfamily. *Science* **240**, 889–95.

Fabris, N., Mocchegiani, E., Galli, M., Irato, L., Lazzarin, A. and Moroni, M. 1988. AIDS, zinc deficiency, and thymic hormone failure. *Journal of the American Medical Association* **259**, 839–40.

Favreau, L.V. and Pickett, C.B. 1993. Transcriptional regulation of the rat NAD(P)H:quinone reductase gene. Characterization of a DNA–protein interaction at the antioxidant responsive element and induction by 12-O-tetradecanoylphorbol 13-acetate. *Journal of Biological Chemistry* **268**, 19875–81.

Foster, R. and Gedamu, L. 1991. Functional analyses of promoter elements responsible for the differential expression of the human metallothionein (MT)-IG and MT-IF genes. *Journal of Biological Chemistry* **266**, 9866–75.

Freedman, L.P., Luisi, B.F., Korszun, Z.R., Basavappa, R., Sigler, P.B. and Yamamoto, K.R. 1988. The function and structure of the metal coordination sites within the glucocorticoid receptor DNA binding domain. *Nature* **334**, 543–6.

Friberg, L., Elinder, C.G. and Kjellstrom, T. 1992. *Cadmium*. Environmental Health Criteria 134, World Health Organization, Geneva, p. 199.

Ghoshal, K., Wang, Y., Sheridan, J.F. and Jacob, S.T. 1998. Metallothionein induction in response to restraint stress. Transcriptional control, adaptation to stress, and role of glucocorticoid. *Journal of Biological Chemistry* **273**, 27904–10

Giedroc, D.P., Keating, K.M., Williams, K.R., Konigsberg, W.H. and Coleman, J.E. 1986. Gene 32 protein, the single-stranded DNA binding protein from bacteriophage T4, is a zinc metalloprotein. *Proceedings of the National Academy of Sciences (USA)* **83**, 8452–6.

Giedroc, D.P., Keating, K.M., Williams, K.R. and Coleman, J.E. 1987. The function of zinc in gene 32 protein from T4. *Biochemistry* **26**, 5251–9.

Gless, U., Schmitt, Y., Ziegler, S. and Druse-Jarres, J.D. 1992. Chromatographic separation of serum proteins and estimation of their zinc and copper content. *Journal of Trace Elements and Electrolytes in Health and Disease* **6**, 245–50.

Goyer, R.A. 1991. Toxic effects of metals, in *Casarett and Doull's Toxicology*, 4th edn, (eds M.O. Amdur, J. Doull and C.D. Klaassen), Pergamon Press, New York, pp. 652–61.

Goyer, R.A. and Cherian, M.G. (eds) 1995. *Toxicology of Metals: Biochemical Aspects*, Springer-Verlag, Berlin.

Gralla, E.B. and Kosman, D.J. 1992. Molecular genetics of superoxide dismutases in yeasts and related fungi. *Advances in Genetics* **30**, 251–319.

Gralla, E.B., Thiele, D.J., Silar, P. and Valentine, J.S. 1991. ACE1, a copper-dependent transcription factor, activates expression of the yeast copper, zinc superoxide dismutase gene. *Proceedings of the National Academy of Sciences (USA)* **88**, 8558–62.

Greco, M.A., Hrab, D.I., Magner, W., and Kosman, D.J. 1990. Cu,Zn superoxide dismutase and copper deprivation and toxicity in *Saccharomyces cerevisiae*. *Journal of Bacteriology* **172**, 317–25.

Gregory, E.M., Goscin, S.A. and Fridovich, I. 1974. Superoxide dismutase and oxygen toxicity in a eucaryote. *Journal of Bacteriology* **117**, 456–60.

Halliwell, B. and Gutteridge, J.M. 1984. Oxygen toxicity, oxygen radicals, transition metals and disease. *Biochemical Journal* **219**, 1–14.

Hamer, D.H. 1985. Metallothionein. *Annual Review of Biochemistry* **55**, 913–51.

Hanas, J.S., Hazuda, D.J., Bogenhagen, D.F., Wu, F.Y. and Wu, C.W. 1983. *Xenopus* transcription factor A requires zinc for binding to the 5S RNA gene. *Journal of Biological Chemistry* **258**, 14120–5.

Hancock, W.W., Grey, S.T., Hau, L., Akalin, E., Orthner, C., Sayegh, M.H. and Salem, H.H. 1995. Binding of activated protein C to a specific receptor on human mononuclear phagocytes inhibits intracellular calcium signaling and monocyte-dependent proliferative response. *Transplantation* **60**, 1525–32.

Heiman, A.S. and Tonner, L.E. 1995. The acute effect of lead acetate on glucocorticoid regulation of tyrosine aminotransferase in hepatoma cells. *Toxicology* **100**, 57–68.

Hertel, R.F. and Wibbertman, A. 1990. Adverse effects of zinc on the immune system, in *Immunotoxicity of Metals and Immunotoxicology*, (eds A.D. Dayan *et al.*), Plenum Press, New York, pp. 223–32.

Heuchel, R., Radtke, F., Georgiev, O., Stark, G., Aguet, M. and Schaffner, W. 1994. The transcription factor MTF-1 is essential for basal and heavy metal-induced metallothionein gene expression. *EMBO Journal* **13**, 2870–5.

Hill, K.L., Li, H.H., Singer, J. and Merchant, S. 1991. Isolation and structural characterization of the *Chlamydomonas reinhardtii* gene for cytochrome c6. Analysis of the kinetics and metal specificity of its copper-responsive expression. *Journal of Biological Chemistry* **266**, 15060–7.

Horn, N. 1984. Copper metabolism in Menkes' disease, in *Metabolism of Trace Metals in Man*, Vol. II, (eds O.M. Rennert and W.-Y. Chan), CRC Press, Boca Raton, pp. 26–52.

Hultman, P., Bell, L.J., Enestrom, S. and Pollard, K.M. 1993. Genetic susceptibility to silver-induced anti-fibrillarin autoantibodies in mice. *Clinical Immunology and Immunopathology* **68**, 9–20.

Inamdar, N.M., Ahn, Y.I. and Alam, J. 1996. The heme-responsive element of the mouse heme oxygenase-1 gene is an extended AP-1 binding site that resembles the recognition sequences for MAF and NF-E2 transcription factors. *Biochemical and Biophysical Research Communications* **221**, 570–6.

Isonishi, S., Horn, D.K., Thiebaut, F.B., Mann, S.C., Andrews, P.A., Basu, A., Lazo, J.S., Eastman, A. and Howell, S.B. 1991. Expression of the c-Ha-*ras* oncogene in mouse NIH 3T3 cells induces resistance to cisplatin. *Cancer Research* **51**, 5903–9.

Izant, J.G. and Weintraub, H. 1984. Inhibition of thymidine kinase gene expression by anti-sense RNA: a molecular approach to genetic analysis. *Cell* **36**, 1007–15.

Izant, J.G. and Weintraub, H. 1985. Constitutive and conditional suppression of exogenous and endogenous genes by anti-sense RNA. *Science* **229**, 345–52.

Jacob, C., Maret, W. and Vallee, B.L. 1998. Control of zinc transfer between thionein, metallothionein, and zinc proteins. *Proceedings of the National Academy of Sciences (USA)* **95**, 3489–94.

Jaiswal, A.K. 1994. Antioxidant response element. *Biochemical Pharmacology* **48**, 439–44.

James, K. 1990. Interaction between cytokines and α_2-macroglobulin. *Immunology Today* **11**, 163–6.

Jiang, L.J., Maret, W. and Vallee, B.L. 1998a. The ATP–metallothionein complex. *Proceedings of the National Academy of Sciences (USA)* **95**, 9146–9.

Jiang, L.J., Maret, W. and Vallee, B.L. 1998b. The glutathione redox couple modulates zinc transfer from metallothionein to zinc-depleted sorbitol dehydrogenase. *Proceedings of the National Academy of Sciences (USA)* **95**, 3483–8.

Jin, P., Sejersen, T. and Ringertz, N.R. 1991. Recombinant platelet-derived growth factor-BB stimulates growth and inhibits differentiation of rat L6 myoblasts. *Journal of Biological Chemistry* **266**, 1245–9.

Kadonaga, J.T., Carner, K.R., Masiarz, F.R. and Tjian, R. 1987. Isolation of cDNA encoding transcription factor Sp1 and functional analysis of the DNA binding domain. *Cell* **51**, 1079–90.

Kataoka, K., Noda, M. and Nishizawa, M. 1994. Maf nuclear oncoprotein recognizes sequences related to an AP-1 site and forms heterodimers with both Fos and Jun. *Molecular and Cellular Biology* **14**, 700–12.

Kazimierczak, W. and Maslinski, C. 1974. Histamine release from mast cells by compound 48/80. The membrane action of zinc. *Agents Actions* **4**, 320–3.

Kelly, E.J., Quaife, C.J., Froelick, G.J. and Palmiter, R.D. 1996. Metallothionein I and II protect against zinc deficiency and zinc toxicity in mice. *Journal of Nutrition* **126**, 1782–90.

Khokha, R., Waterhouse, P., Yagel, S., Lala, P.K., Overall, C.M., Norton, G. and Denhardt. D.T. 1989. Antisense RNA-induced reduction in murine TIMP levels confers oncogenicity on Swiss 3T3 cells. *Science* **243**, 947–50.

Kiremidjian-Schumacher, L., Stotzky, G., Dickstein, R.A. and Schwartz, J. 1981. Influence of cadmium, lead, and zinc on the ability of guinea pig macrophages to interact with macrophage migration inhibitory factor. *Environmental Research* **24**,106–16.

Kjellberg, S., Bjorndahl, L. and Kvist, U. 1992. Sperm chromatin stability and zinc binding properties in semen from men in barren unions. *International Journal of Andrology* **15**, 103–13.

Klausner, R.D., Rouault, T.A. and Harford, J.B. 1993. Regulating the fate of mRNA: the control of cellular iron metabolism. *Cell* **72**, 19–28.

Klevit, R.E., Herriott, J.R. and Horvath, S.J. 1990. Solution structure of a zinc finger domain of yeast ADR1. *Proteins* 7, 215–26.

Koller, L.D. and Roan, J.G. 1977. Effects of lead and cadmium on mouse peritoneal macrophages. *Journal of the Reticuloendothelial Society* **21**, 7–12.

Kontozoglou, T.E., Banerjee, D. and Cherian, M.G. 1989. Immunohistochemical localization of metallothionein in human testicular embryonal carcinoma cells. *Virchows Archivs A: Pathology and Anatomy* **415**, 545–9.

Koropatnick, J. and Duerksen, J.D. 1987. Nuclease sensitivity of alfafetoprotein, metallothionein-1, and immunoglobulin gene sequences in mouse during development. *Developmental Biology* **122**, 1–10.

Koropatnick, J. and Leibbrandt, M.E.I. 1995. Effects of metals on gene expression, in *Handbook of Experimental Pharmacology, Toxicology of Metals – Biochemical Aspects*, (eds R.A. Goyer, and M.G. Cherian), Springer Verlag, Berlin, pp. 93–113.

Koropatnick, J. and Zalups, R.K. 1997. Effect of non-toxic mercury, zinc or cadmium pretreatment on the capacity of human monocytes to undergo lipopolysaccharide-induced activation. *British Journal of Pharmacology* **120**, 797–806.

Koropatnick, J., Pearson-Sharpe, J. and Harris, J.F. 1988. Extensive loss of human DNA accompanies loss of antibody production in heteromyeloma hybridoma cells. *Molecular Biology and Medicine* **5**, 69–83.

Koropatnick, J., Kloth, D.M., Kadhim, S., Chin, J.L. and Cherian, M.G. 1995. Metallothionein expression and resistance to cisplatin in a human germ cell tumor cell line. *Journal of Pharmacology and Experimental Therapeutics* **275**, 1681–7.

Kubicka-Muranyi, M., Behmer, O., Uhrberg, M., Klonowski, H., Bister, J. and Gleichmann, E. 1993. Murine systemic autoimmune disease induced by mercuric chloride: Hg-specific helper T-cells react to antigen stored in macrophages. *International Journal of Immunopharmacology* **15**, 151–61.

Larson, J.S., Schuetz, T.J. and Kingston, R.E. 1988. Activation *in vitro* of sequence-specific DNA binding by a human regulatory factor. *Nature* **335**, 372–5.

Lawrence, D.A. and McCabe, M.J. Jr 1995. Immune modulation by toxic metals, in *Metal Toxicology*, (eds R.A. Goyer, C.D. Klaassen and M.P. Waalkes), Academic Press, New York, pp. 305–37.

Lawrence, D.A. and Slavik, J. 1992. In vitro effects of metals on various reactivities of human peripheral blood and splenic mononuclear cells. *The Toxicologist* **12**, 29.

Lee, D.K., Carrasco, J., Hidalgo, J. and Andrews, G.K. 1999. Identification of a signal transducer and activator of transcription (STAT) binding site in the mouse metallothionein-I promoter involved in interleukin-6-induced gene expression. *Biochemical Journal* **337**, 59–65.

Lee, M.S., Gippert, G.P., Soman, K.V., Case, D.A. and Wright, P.E. 1989. Three-dimensional solution structure of a single zinc finger DNA-binding domain. *Science* **245**, 635–7.

Leibbrandt, M.E.I. and Koropatnick, J. 1994. Activation of human monocytes with lipopolysaccharide induces metallothionein expression and is diminished by zinc. *Toxicology and Applied Pharmacology* **124**, 72–81.

Leibbrandt, M.E.I., Khokha, R. and Koropatnick, J. 1994. Antisense down-regulation of metallothionein in a human monocytic cell line alters adherence, invasion, and the respiratory burst. *Cell Growth and Differentiation* **5**, 17–25.

Long, G.J., Rosen, J.F. and Schanne, F.A. 1994. Lead activation of protein kinase C from rat brain. Determination of free calcium, lead, and zinc by 19F NMR. *Journal of Biological Chemistry* **269**, 834–7.

Maeshima, H., Sato, M., Ishikawa, K., Katagata, Y. and Yoshida, T. 1996. Participation of altered upstream stimulatory factor in the induction of rat heme oxygenase-1 by cadmium. *Nucleic Acids Research* **24**, 2959–65.

Maret, W. 1994. Oxidative metal release from metallothionein via zinc-thiol/disulfide interchange. *Proceedings of the National Academy of Sciences (USA)* **91**, 237–41.

Maret, W. 1995. Metallothionein/disulfide interactions, oxidative stress, and the mobilization of cellular zinc. *Neurochemistry International* **27**, 111–17.

Maret, W. and Vallee, B.L. 1998. Thiolate ligands in metallothionein confer redox activity on zinc clusters. *Proceedings of the National Academy of Sciences (USA)* **95**, 3478–82.

Maret, W., Larsen, K.S. and Vallee, B.L. 1997. Coordination dynamics of biological zinc 'clusters' in metallothioneins and in the DNA-binding domain of the transcription factor Gal4. *Proceedings of the National Academy of Sciences (USA)* **94**, 2233–7.

Masters, B.A., Kelly, E.J., Quaife, C.J., Brinster, R.L. and Palmiter, R.D. 1994. Targeted disruption of metallothionein I and II genes increases sensitivity to cadmium. *Proceedings of the National Academy of Sciences (USA)* **91**, 584–8.

Mathe, G., Blazsek, I., Gil-Delgado, M.A., Canon, C., Misset, J.L., Gaget, H. and Reizenstein, P. 1985. The effect of zinc on normal and neoplastic T-lymphocyte proliferation. *Medical Oncology and Tumor Pharmacotherapy* **2**, 203–10.

Meier, J.L., Luo, X., Sawadogo, M. and Straus, S.E. 1994. The cellular transcription factor USF cooperates with varicella-zoster virus immediate-early protein 62 to symmetrically activate a bidirectional viral promoter. *Molecular and Cellular Biology* **14**, 6896–906.

Meier, R.W., Niklaus, G., Dewald, B., Fey, M.F. and Tobler, A. 1995. Inhibition of the arachidonic acid pathway prevents induction of IL-8 mRNA by phorbol ester and changes the release of IL-8 from HL 60 cells: differential inhibition of induced expression of IL-8, TNFα, IL-1α and IL-1β. *Journal of Cellular Physiology* **165**, 62–70.

Merchant, S. and Bogorad, L. 1987. The Cu(II)-repressible plastidic cytochrome c. Cloning and sequence of a complementary DNA for the pre-apoprotein. *Journal of Biological Chemistry* **262**, 9062–7.

Merchant, S., Hill, K. and Howe, G. 1991. Dynamic interplay between two copper-titrating components in the transcriptional regulation of cyt c6. *EMBO Journal* **10**, 1383–9 [published erratum appears in *EMBO J* (1991) **10**, 23201].

Meskel, H.H., Cherian, M.G., Martinez, V.J., Veinot, L.A. and Frei, J.V. 1993. Metallothionein as an epithelial proliferative compartment marker for DNA flow cytometry. *Modern Pathology* **6**, 756–60.

Michalska, A.E. and Choo, K.H.A. 1993. Targeting and germ-line transmission of a null mutation at the metallothionein I and II loci in mouse. *Proceedings of the National Academy of Sciences (USA)* **90**, 8088–92.

Michelsen, J.W., Schmeichel, K.L., Beckerle, M.C. and Winge, D.R. 1993. The LIM motif defines a specific zinc-binding protein domain. *Proceedings of the National Academy of Sciences (USA)* **90**, 4404–8.

Moffatt, P. and Seguin, C. 1998. Expression of the gene encoding metallothionein-3 in organs of the reproductive system. *DNA Cell Biology* **17**, 501–10.

Monestier, M., Losman, M.J., Novick, K.E. and Aris, J.P. 1994. Molecular analysis of mercury-induced antinucleolar antibodies in H-2S mice. *Journal of Immunology* **152**, 667–75.
Moynahan, E.J. 1981. Acrodermatitis enteropathica and the immunological role of zinc. *Immunodermatology* **30**, 437–47.
Nagel, W.W. and Vallee, B.L. 1995. Cell cycle regulation of metallothionein in human colonic cancer cells. *Proceedings of the National Academy of Sciences (USA)* **92**, 579–83.
Nakamura, T., Shiraishi, N. and Aono, K. 1987, Effects of *in vitro* and *in vivo* supplementation with zinc on superoxide anion production in leukocytes. *Physiological Chemistry and Physical Medicine Nuclear Magnetic Resonance* **19**, 147–51.
Nartey, N.O., Banerjee, D. and Cherian, M.G. 1987. Immunohistochemical localization of metallothionein in cell nucleus and cytoplasm of fetal human liver and kidney and its changes during development. *Pathology* **19**, 233–8.
Nishizuka, Y. 1992. Intracellular signaling by hydrolysis of phospholipids and activation of protein kinase C. *Science* **258**, 607–14.
O'Halloran, T.V. 1989. Metalloregulatory proteins: metal-responses molecular switches governing gene expression, in *Metal Ions in Biological Systems, Vol. 25, Interrelations Among Metal Ions, Enzymes and Gene Expression* (ed. H. Sigel and A. Sigel), Marcel Dekker, New York, p. 105.
O'Halloran, T.V. 1993. Transition metals in control of gene expression. *Science* **261**, 715–25.
Oliviera, D.B.G. 1987. Membranous nephropathy caused by mercury-containing skin lightening cream. *Postgraduate Medical Journal* **63**, 303–4.
Omichinski, J.G., Trainor, C., Evans, T., Gronenborn, A.M., Clore, G.M. and Felsenfeld, G.A. 1993. A small single-'finger' peptide from the erythroid transcription factor GATA-1 binds specifically to DNA as a zinc or iron complex. *Proceedings of the National Academy of Sciences (USA)* **90**, 1676–80.
Otsuka, F., Iwamatsu, A., Suzuki, K., Ohsawa, M., Hamer, D.H. and Koizumi, S.F. 1994. Purification and characterization of a protein that binds to metal responsive elements of the human metallothionein IIA gene. *Journal of Biological Chemistry* **269**, 23700–7.
Otvos, J.D. and Armitage, I.M. 1980. Structure of the metal clusters in rabbit liver metallothionein. *Proceedings of the National Academy of Sciences (USA)* **77**, 7094–8.
Otvos, J.D., Engeseth, H.R., Nettesheim, D.G. and Hilt, C.R. 1987. Interprotein metal exchange reactions of metallothionein. *EXS* **52**, 171–8.
Ouellette, A.J. 1982. Metallothionein mRNA expression in fetal mouse organs. *Developmental Biology* **92**, 240–6.
Pabo, C.O. and Sauer, R.T. 1992. Transcription factors: structural families and principles of DNA recognition. *Annual Review of Biochemistry* **61**, 1053–95.
Palmiter, R.D. 1994. Regulation of metallothionein genes by heavy metals appears to be mediated by a zinc-sensitive inhibitor that interacts with a constitutively active transcription factor, MTF-1. *Proceedings of the National Academy of Sciences (USA)* **91**, 1219–23.
Palmiter, R.D. 1995. Constitutive expression of metallothionein-III (MT-III), but not MT-I, inhibits growth when cells become zinc deficient. *Toxicology and Applied Pharmacology* **135**, 139–46.
Palmiter, R.D. and Findley, S.D. 1995. Cloning and functional characterization of a mammalian zinc transporter that confers resistance to zinc. *EMBO Journal* **14**, 639–49.
Palmiter, R.D., Cole, T.B. and Findley, S.D. 1996. ZnT-2, a mammalian protein that confers resistance to zinc by facilitating vesicular sequestration. *EMBO Journal* **15**, 1784–91.
Pan, T. and Coleman, J.E. 1991. Sequential assignments of the ^{1}H NMR resonances of Zn(II)2 and ^{113}Cd(II)2 derivatives of the DNA-binding domain of the GAL4 transcription factor reveal a novel structural motif for specific DNA recognition. *Biochemistry* **30**, 4212–22.
Panemangalore, M., Banerjee, D., Onosaka, S. and Cherian, M.G. 1983. Changes in the intracellular accumulation and distribution of metallothionein in rat liver and kidney during postnatal development. *Developmental Biology* **97**, 95–102.

Pavletich, N.P. and Pabo, C.O. 1991. Zinc finger-DNA recognition: crystal structure of a Zif268–DNA complex at 2.1 Å. *Science* **252**, 809–17.

Pelletier, L., Rossert, J., Pasquier, R., Vial, M.C. and Druet, P. 1990. Role of CD8+ T cells in mercury induced autoimmunity or immunosuppression in the rat. *Scandinavian Journal of Immunology* **31**, 65–74.

Phillips, J.L. 1976. Specific binding of zinc transferrin to human lymphocytes. *Biochemical and Biophysical Research Communications* **72**, 634–9.

Quaife, C.J., Findley, S.D., Erickson, J.C., Froelick, G.J., Kelly, E.J., Zambrowicz, B.P. and Palmiter, R.D. 1994. Induction of a new metallothionein isoform (MT-4) occurs during differentiation of stratified squamous epithelia. *Biochemistry* **33**, 7250–9.

Radtke, F., Heuchel, R., Georgiev, O., Hergersberg, M., Gariglio, M., Dembic, Z. and Schaffner, W. 1993. Cloned transcription factor MTF-1 activates the mouse metallothionein I promoter. *EMBO Journal* **12**, 1355–62.

Ralston, D.M. and O'Halloran, T.V. 1990. Metalloregulatory proteins and molecular mechanisms of heavy metal signal transduction. *Advances in Inorganic Biochemistry* **8**, 1–31.

Remacle, J., Raes, M., Toussaint, O., Renard, P. and Rao, G. 1995. Low levels of reactive oxygen species as modulators of cell function. *Mutation Research* **316**, 103–22.

Reuter, R., Tessars, G., Vohr, H.W., Gleichmann, E. and Luhrmann, R. 1989. Mercuric chloride induces autoantibodies against U3 small nuclear ribonucleoprotein in susceptible mice. *Proceedings of the National Academy of Sciences (USA)* **86**, 237–41.

Samet, J.M., Graves, L.M., Quay, J., Dailey, L.A., Devlin, R.B., Ghio, A.J., Wu, W., Bromberg, P.A. and Reed, W. 1998. Activation of MAPKs in human bronchial epithelial cells exposed to metals. *American Journal of Physiology* **275**, L551–L558.

Samson, S.L. and Gedamu, L. 1995. Metal-responsive elements of the rainbow trout metallothionein-B gene function for basal and metal-induced activity. *Journal of Biological Chemistry* **270**, 6864–71.

Sato, M. and Bremner, I. 1993. Oxygen free radicals and metallothionein. *Free Radicals in Biology and Medicine* **14**, 325–37.

Schenk, H., Klein, M., Erdbrugger, W., Droge, W. and Schulze-Osthoff, K. 1994. Distinct effects of thioredoxin and antioxidants on the activation of transcription factors NF-kappa B and AP-1. *Proceedings of the National Academy of Sciences (USA)* **91**,1672–6.

Scuderi, P. 1990. Differential effects of copper and zinc on human peripheral blood monocyte cytokine secretion. *Cellular Immunology* **126**, 391–405.

Shang, Z., Liao, Y.D., Wu, F.Y. and Wu, C.W. 1989. Zinc release from *Xenopus* transcription factor IIIA induced by chemical modifications. *Biochemistry* **28**, 9790–5.

Shenker, B.J., Mayro, J.S., Rooney, C., Vitale, L. and Shapiro, I.M. 1993. Immunotoxic effects of mercuric compounds on human lymphocytes and monocytes. IV. Alterations in cellular glutathione content. *Immunopharmacology and Immunotoxicology* **15**, 273–90.

Simons, S.S. Jr, Chakraborti, P.K. and Cavanaugh, A.H. 1990. Arsenite and cadmium(II) as probes of glucocorticoid receptor structure and function. *Journal of Biological Chemistry* **265**, 1938–45.

Sirito, M., Walker, S., Lin, Q., Kozlowski, M.T., Klein, W.H. and Sawadogo, M. 1992. Members of the USF family of helix–loop–helix proteins bind DNA as homo- as well as heterodimers. *Gene Expression* **2**, 231–40.

Slotkin, S.H. and Cherian, M.G. 1989. Elevated metallothionein expression in human fetal liver. *Paediatric Research* **24**, 326–9.

Stout, J.G. and Caskey, C.T. 1987. Antisense RNA inhibition of endogenous genes. *Methods in Enzymology* **151**, 519–30.

Stuart, G.W., Searle, P.F., Chen, H.Y., Brinster, R.L. and Palmiter, R.D. 1984. A 12-base-pair DNA motif that is repeated several times in metallothionein gene promoters confers metal regulation to a heterologous gene. *Proceedings of the National Academy of Sciences (USA)* **81**, 7318–22.

Suzuki, C.A., Ohta, H., Albores, A., Koropatnick, J. and Cherian, M.G. 1990. Induction of metallothionein synthesis by zinc in cadmium pretreated rats. *Toxicology* **63**, 273–84.

Szallasi, Z., Bogi, K., Gohari, S., Biro, T., Acs, P. and Blumberg, P.M. 1996. Non-equivalent roles for the first and second zinc fingers of protein kinase C[d]. Effect of their mutation on phorbol ester-induced translocation in NIH 3T3 cells. *Journal of Biological Chemistry* **271**, 18299–301.

Tamai, K., Gralla, E.B., Ellerby, L.M., Valentine, J.S. and Thiele, D.J. 1993. Yeast and mammalian metallothioneins functionally substitute for yeast copper–zinc superoxide dismutase. *Proceedings of the National Academy of Sciences (USA)* **90**, 8013–17.

Telford, W.G. and Fraker, P.J. 1997. Zinc reversibly inhibits steroid binding to murine glucocorticoid receptor. *Biochemical and Biophysical Research Communications* **238**, 86–9.

Thomas, P.T., Ratajczak, H.V., Aranyi, C., Gibbons, R. and Fenters, J.D. (1985) Evaluation of host resistance and immune function in cadmium-exposed mice. *Toxicology and Applied Pharmacology* **80**, 446–56.

Tonner, L.E. and Heiman, A.S. 1997. Lead may affect glucocorticoid signal transduction in cultured hepatoma cells through inhibition of protein kinase C. *Toxicology* **119**, 155–66.

Tubbs, R.R., Gordon, D.O., Gephardt, N., McMahon, J.T., Pohl, M.C., Vidt, D.G., Barenburg, S.A. and Valenzuels, R. 1982. Membranous glomerulonephritis associated with industrial mercury exposure. Study of pathogenetic mechanisms. *American Journal of Clinical Pathology* **77**, 409–13.

Uchida, Y., Takio, K., Titani, K., Ihara, Y. and Tomonaga, M. 1991. The growth inhibitory factor that is deficient in the Alzheimer's disease brain is a 68 amino acid metallothionein-like protein. *Neuron* **7**, 337–47.

Udom, A.O. and Brady, F.O. 1980. Reactivation *in vitro* of zinc-requiring apo-enzymes by rat liver zinc–thionein. *Biochemical Journal* **187**, 329–35.

Vallee, B.L. 1995. The function of metallothionein. *Neurochemistry International* **27**, 23–33.

Vallee, B.L. and Falchuk, F.H. 1993. The biochemical basis of zinc physiology. *Physiological Reviews* **73**, 79–118.

Vallee, B.L., Coleman, J.E. and Auld, D.S. 1991. Zinc fingers, zinc clusters, and zinc twists in DNA-binding protein domains. *Proceedings of the National Academy of Sciences (USA)* **88**, 999–1003.

Verhaegh, G.W., Richard, M.J. and Hainaut, P. 1997. Regulation of p53 by metal ions and by antioxidants: dithiocarbamate down-regulates p53 DNA-binding activity by increasing the intracellular level of copper. *Molecular and Cellular Biology* **17**, 5699–706.

Vruwink, K.G., Fletcher, M.P., Keen, C.L., Golub, M.S., Hendrickx, A.G. and Gershwin, M.E. 1991. Moderate zinc deficiency in rhesus monkeys. An intrinsic defect of neutrophil chemotaxis corrected by zinc repletion. *Journal of Immunology* **146**, 244–9.

Wagner, M., Klein, C.L., van Kooten, T.G. and Kirkpatrick, C.J. 1998. Mechanisms of cell activation by heavy metal ions. *Journal of Biomedical Materials Research* **42**, 443–52.

Wenzel, H.J., Cole, T.B., Born, D.E., Schwartzkroin, P.A. and Palmiter, R.D. 1997. Ultrastructural localization of zinc transporter-3 (ZnT-3) to synaptic vesicle membranes within mossy fiber boutons in the hippocampus of mouse and monkey. *Proceedings of the National Academy of Sciences (USA)* **94**, 12676–81.

West, A.K., Stallings, R., Hildebrand, C.E., Chiu, R., Karin, M. and Richards, R. 1990. Human metallothionein genes: structure of the functional locus at 16q13. *Genomics* **8**, 513–18.

Williams, G.T. and Morimoto, R.I. 1990. Maximal stress-induced transcription from the human HSP70 promoter requires interactions with the basal promoter elements independent of rotational alignment. *Molecular and Cellular Biology* **10**, 3125–36.

Winchurch, R.A. 1988. Activation of thymocyte responses to interleukin-1 by zinc. *Clinical Immunology and Immunopathology* **47**, 174–80.

Wirth, J.J., Fraker, P.J. and Kierszenbaum, F. 1989. Binding of the specific ligand to Fc receptors on *Trypanosoma cruzi* increases the infective capacity of the parasite. *Immunology* **68**, 114–19.

Yiangou, M., Ge, X., Carter, K.D. and Papaconstantinou, J. 1991. Induction of several acute-phase protein genes by heavy metals: a new class of metal-responsive genes. *Biochemistry* **30**, 3798–806.

Yu, C.W., Chen, J.H. and Lin, L.Y. 1997. Metal-induced metallothionein gene expression can be inactivated by protein kinase C inhibitor. *FEBS Letters* **420**, 69–73.

Zalups, R.K., Fraser, J. and Koropatnick, J. 1995. Enhanced transcription of metallothionein genes in rat kidney: effect of uninephrectomy and compensatory renal growth. *American Journal of Physiology* **268**, F643–F650.

Zeng, J., Heuchel, R., Schaffner, W. and Kagi, J.H. 1991a. Thionein (apometallothionein) can modulate DNA binding and transcription activation by zinc finger containing factor Sp1. *FEBS Letters* **279**, 310–12.

Zeng, J., Vallee, B.L. and Kagi, J.H. 1991b. Zinc transfer from transcription factor IIIA fingers to thionein clusters. *Proceedings of the National Academy of Sciences (USA)* **88**, 9984–8.

Zhang, X.H. and Takenaka, I. 1998. Incidence of apoptosis and metallothionein expression in renal cell carcinoma. *British Journal of Urology* **81**, 9–13.

Zimarino, V., Wilson, S. and Wu, C. 1990. Antibody-mediated activation of *Drosophila* heat shock factor *in vitro*. *Science* **249**, 546–9.

INDEX

Page numbers appearing in **bold** refer to figures and page numbers appearing in *italics* refer to tables.